L. S. PONTRYAGIN
SELECTED WORKS

Volume 1

Selected Research Papers

Classics of Soviet Mathematics

L. S. PONTRYAGIN SELECTED WORKS

Edited by R. V. Gamkrelidze

Volume 1: **Selected Research Papers**
Volume 2: **Topological Groups**
Volume 3: **Algebraic and Differential Topology**
Volume 4: **The Mathematical Theory of Optimal Processes**

ISSN 0743–9199
This book is part of a series. The publishers will accept continuation orders which may be cancelled at any time and which provide for automatic billing and shipping of each title in the series upon publication. Please write for details.

L. S. PONTRYAGIN SELECTED WORKS

Volume 1

Selected Research Papers

Edited by

R. V. Gamkrelidze

V. A. Steklov Institute of Mathematics
USSR Academy of Sciences
Moscow, USSR

Gordon and Breach Science Publishers
New York · London · Paris · Montreux · Tokyo

Gordon and Breach Science Publishers

P.O. Box 786
Cooper Station
New York, NY 10276
United States of America

P.O. Box 197
London WC2E 9PX
England

58, rue Lhomond
75005 Paris
France

P.O. Box 161
1820 Montreux 2
Switzerland

14-9 Okubo 3-chome
Shinjuku-ku, Tokyo 160
Japan

Papers marked by an asterisk in the table of contents were translated from the Russian by P. S. V. Naidu; all other papers are in the language of original publication.

Library of Congress Cataloging-in-Publication Data

Pontryagin. L. S. (Lev Semenovich), 1908–
  Selected research papers.

  (L. S. Pontryagin selected works ; v. 1) (Classics
of Soviet mathematics, ISSN 0743–9199)
  English, French, and German.
  1. Topology, 2. Control theory. 3. Game theory.
I. Gamkrelidze, R. V. II. Title. III. Series:
Pontryagin, L. S. (Lev Semenovich), 1908–
Selections. Polyglot. 1985 ; v. 1) IV. Series:
Classics of Soviet mathematics.
QA3.P76 1985, vol. 1 [QA611.15] 510 s 84–11706
[514]

ISBN 2–88124–105–0 (Switzerland : v. 1)

Volume 1: ISBN 2–88124–105–0; 4-volume set: ISBN 2–88124–134–4. Printed in Great Britain by Bell and Bain Ltd., Glasgow.

Lev Semenovich Pontryagin

Contents

* Translated from the Russian by P. S. V. Naidu.

Preface

On 3 September 1983 Lev Semenovich Pontryagin was seventy-five. To mark this important event in the life of this outstanding contemporary mathematician we are beginning the publication of his scientific works in four volumes, according to a decision taken by the Mathematics Division of the USSR Academy of Sciences. The first volume contains the most important mathematical papers of L. S. Pontryagin and also includes a bibliography of his basic scientific works, the second is his well-known monograph *Topological Groups*, the third comprises two monographs, *Foundations of Algebraic Topology* and *Smooth Manifolds and Their Applications in Homotopy Theory*, and the fourth is a revised edition of *The Mathematical Theory of Optimal Processes* by L. S. Pontryagin, V. G. Boltyanskii, R. V. Gamkrelidze, and E. F. Mishchenko.

The scientific activity of Lev Semenovich Pontryagin has left a deep imprint on many crucial areas of modern mathematics, both pure and applied. His work has had a definitive influence on the development of topology and topological algebra, and because of him optimal control theory is one of the topical trends in present-day applied mathematics. In a brief review we can neither delve deeply into his important works nor describe the profound, multifaceted impact of these works on the advancement of the respective fields. This is, therefore, only a broad outline that may be of help in studying his works.

While still a second-year student at Moscow State University, Pontryagin began his scientific activity under the guidance of P. S. Aleksandrov. In this early period, his interests concentrated mainly on two central topics of algebraic (combinatorial) topology, namely, topological duality theorems and dimension theory, which Pontryagin looked upon as a local variant of duality theory.

The discovery of "Pontryagin duality," the culmination of his work in topological duality theorems, and the construction of the general theory of characters of a locally compact commutative group are Pontryagin's two most notable contributions and are undoubtedly among the finest achievements in modern mathematics.

We begin with a survey of his main works in duality theory and topological algebra. To assess the full value of the advances made by Pontryagin in this area, it is apt to recall here that at the time when Pontryagin had just begun his activity, homology groups were hardly used in topology; instead, Betti numbers with respect to different moduli and torsion coefficients were used, and the Alexander duality theorem was formulated as an equality of Betti numbers (modulo 2) of dimensions $n - r - 1$ and r of a polyhedron

$K \subset R^n$ and its complement $R^n \backslash K$,

$$p^r(R^n \backslash K) = p^{n-r-1}(K).$$

In his first published paper,[1]* Pontryagin improved this theorem by extending the duality between the Betti numbers of a polyhedron and its complement in R^n to the duality between the r- and $(n - r - 1)$-dimensional homology groups (modulo 2) of the polyhedra $(R^n \backslash K)$ and K. The full statement of this theorem follows. In $R^n \backslash K$ and K, two bases

$$z_1^r, \ldots z_s^r \quad \text{and} \quad \zeta_1^{n-r-1}, \ldots \zeta_s^{n-r-1}$$

of homology (mod 2) of dimensions r and $(n - r - 1)$, respectively, can be chosen, such that the square matrix of linking coefficients (mod 2)

$$\| (z_i^r, \zeta_j^{n-r-1}) \| (i, j = 1, \ldots s)$$

is the identity matrix.

Thus, the duality between the homology groups (mod 2) established here by means of the linking coefficients led to a group isomorphism.

The next paper[2] deals with the same problem in modulus 2, but the polyhedron K is now imbedded in an arbitrary closed n-dimensional manifold M^n. Its solution demanded, probably for the first time in the history of topology, a study of homological properties of continuous mappings. That is, Pontryagin was led to the study of kernels and images of homomorphisms of homology groups (mod 2) for the inclusions $K \subset M^n$ and $M^n \backslash K \subset M^n$, and the duality theorem was formulated in terms of the ranks of the corresponding kernels. Later, the study of the homological properties of mappings acquired immense significance in topology and greatly influenced the creation of homological algebra.

This paper also contained a statement, known subsequently as the "Pontryagin cycle removal theorem," that asserted: If an r-dimensional cycle Z^r in M^r intersects every $(n - r)$-dimensional cycle in K with a zero intersection index, then the cycle Z^r can be "homologically removed" from K, i.e., there exists in $M^n \backslash K$ an r-dimensional cycle that is homologous to Z^r in M^n. This theorem found successful applications in the topological theory of variational problems; Pontryagin himself used it in estimating the category of a manifold.

From the foregoing it is clear how far one of the central problems of algebraic topology of the late twenties had been advanced in two short papers of a 19-year old sophomore.

* Reference numbers refer to the bibliography of Pontryagin's publications, pp. 609–618.

The next work concerning duality theorems, his master's thesis,[6] was stimulated by the course in algebra given by E. Noether. It gives a profound analysis of the algebraic nature of topological duality theorems. Duality for an arbitrary modulus $m > 0$ obtained a final solution in the form of an isomorphism of the corresponding groups, in view of the fact, now well understood, that a finite cyclic group is the Pontryagin dual of itself (a concept which Pontryagin had not yet arrived at that time).

A particular corollary of the results of the paper is that, for any $m > 0$, the homology groups (mod m), $H_r^{(m)}(R^n \setminus K)$ and $H_{n-r-1}^{(m)}(K)$ of dimensions r and $n - r - 1$, respectively, are isomorphic, and, consequently, all homology groups (mod m) of the complement $R^n \setminus K$ are invariant, i.e., they depend only on the homology groups of the polyhedron K, but do not depend on the inclusion of K in R^n.

Duality theorems for full homology groups with integral coefficients cannot be formulated in terms of isomorphisms and, therefore, could not be fitted into the framework of the paper. For instance, the full r-dimensional integral homology group $H_r(R^n \setminus K)$ is neither isomorphic to the group $H_{n-r-1}(K)$ nor even determined by it. There exist only isomorphisms (also noted in the paper) separately between the r- and $(n - r - 1)$-dimensional weak homology groups and between the r- and $(n - r - 2)$-dimensional torsion groups of the sets K and $R^n \setminus K$, obviously implying the invariance of the full integral homology groups of the complement $(R^n \setminus K)$.

If, instead of a finite polyhedron K, an arbitrary compact set F is considered in R^n, then the corresponding integral and weak homology groups are, in general, no longer finitely generated, and a special investigation is needed to establish the invariance of the homology groups of the complement $R^n \setminus F$. Pontryagin also studied the duality for an arbitrary compact set $F \subset R^n$ and established the invariance of the groups $H_r^{(m)}(R^n \setminus F)$, $m > 0$, as well as the invariance of weak homology groups of $R^n \setminus F$, thereby significantly advancing the problem.

But the central question of the independence of the full group of integral homology $H_r(R^n \setminus F)$ of the inclusion of the compact set $F \subset R^n$ still remained unsolved. Its solution demanded the introduction of a new homological invariant of the set F, namely, a homology group related not to a discrete but to a compact coefficient group. This permitted him, while rejecting the narrow concept of duality as an isomorphism, to define "Pontryagin duality." In 1931–32, he made this decisive step and completely solved all problems relating to duality and also the longstanding problem of the proper definition of homology groups of compact metric spaces.

In constructing the homology group $H_r(F)$ of the set F, the coefficients are not taken from a discrete group of residues (modulo m) or from the group of integers, but are taken from a compact topological group of rotations of a circle. The group $H_r(F)$ is, in itself, a compact commutative topological group. The group $H_r(F)$ and the $(n - r - 1)$-dimensional integral homology group $H_{n-r-1}(R^n \backslash F)$ proved to be Pontryagin duals, i.e., each is the character group of the other (for a detailed exposition of the theory of characters, see reference 110 or the second volume of the *Selected Works*).

Generally, let Γ, G be a dual group pair, i.e., each is the character group of the other, and let Γ be compact and G discrete. Take Γ as the coefficient group for constructing the homology group $H_r^\Gamma(F)$. Then its dual (i.e., its character group) is the homology group of the complement $H_{n-r-1}^G(R^n \backslash F)$, which is constructed, using G, the dual of Γ, as the coefficient group. Duality is realized through linking coefficients.

The general duality theorem for a closed set $F \subset R^n$ was first reported as a short communication in the Proceedings of the International Mathematics Congress held in Zurich in 1932, while its full exposition is given in reference 18.

This paper actually marks the end of Pontryagin's research into topological duality theorems. These theorems, being a powerful tool for studying general homological problems in topology, resolved the crucial question in algebraic topology of the thirties. Particularly after Pontryagin's duality theorems, homology groups have gained a firm foothold in topology as the basic homological invariants in place of the Betti numbers and torsion coefficients, which had fully served the purpose of homology groups until the main circle of topological problems led to finitely generated groups.

Topological duality theorems for a (finite) polyhedron in an arbitrary closed n-dimensional manifold are given in their final formulation in reference 54.

A logical continuation of the duality theorems is the general theory of characters of locally compact commutative groups created by Pontryagin. The main result of this theory is the assertion that every compact commutative group is the character group of some discrete group. Its proof rests on the construction of the invariant measure introduced by Haar in 1933, which had played a key role in the development of topological algebra.

The general theory of characters had enabled Pontryagin to elucidate the structure of compact and locally compact groups, the results obtained for compact and locally compact commutative groups being final. A positive answer to Hilbert's fifth problem for a compact and locally compact commutative group follows directly from these results. (For a detailed exposition of the structure of compact and locally compact commutative groups, refer to the third edition of *Topological Groups*, Volume 2 of the *Selected Works*.) However, the significance of the theory of characters of locally compact topological groups does not end here. Its creation has indeed laid the foundation of topological algebra as an independent discipline, which has been primarily responsible for the development of general harmonic analysis on topological groups. Pontryagin's works in duality theory and character theory had a deep impact on algebraic-topological reasoning in the thirties and, in particular, made a great contribution to "functorial thinking" in mathematics.

His first publications on the general theory of characters of commutative topological groups, on the structure of compact groups, and on locally compact commutative groups are references 16, 17, and 19, respectively.

His remarkable theorem (see reference 10) that asserts that the field of real numbers, the field of complex numbers, and the division ring of quaternions are the only locally compact connected division rings should also be classified under topological algebra.

The methods developed here were later fully utilized by Pontryagin in elucidating the structure of locally compact commutative groups with the help of the theory of characters, as we have already pointed out.

The outcome of his studies in topological algebra was the famous monograph *Topological Groups*, first published in 1938, which has had several editions both in the USSR and in many other countries, in most of the major European languages. It became a classic that influenced many generations of mathematicians and that has not lost its value even today, forty-five years since its first publication, a rare event in mathematics. Its third English edition forms the second volume of the *Selected Works* of L. S. Pontryagin.

The early works of Pontryagin also deal with dimension theory. He constructed examples of compact metric spaces that have different dimensions in different moduli. He later used these examples (see reference 4) to construct the famous "dimensionally deficient" continuums, which disproved the longstanding hypothesis that the dimension of compact sets is additive under topological multiplication. He found two two-dimensional compact sets whose product is of dimension three, instead of four. His theorem that any n-dimensional compact set is homeomorphically mapped into R^{2n+1} (see reference 7) also fits into the category of dimension theory.

The homological dimension theory due to P. S. Aleksandrov owes much to Pontryagin's work in dimension theory. For Pontryagin himself, his studies in dimension theory had a far-reaching consequence – under their influence he began, in the mid-thirties, a systematic investigation of homotopic problems in topology.

His studies in homotopic topology likewise reached their climax (at the beginning of the forties) in the discovery of methods that basically paved the way for a new field in modern mathematics, differential topology. Here we have in mind his discovery of characteristic classes and his contributions to the theory of fiber bundles.

Prior to taking up the "homotopic period," mention should be made of his outstanding topological paper written in 1935,[21] a full exposition of which is given in reference 21. It gives the solution to the Cartan problem of calculating the homology groups of compact group manifolds for the four main series of compact Lie groups. Historically, in this paper, the homological invariants were first found for a large and extremely important class of manifolds defined, not by triangulation, but by analytical (in this case, by algebraic) relations. To solve this problem, Pontryagin used, instead of Cartan's method based on the algebra of exterior invariant forms on a group (R. Brauer applied this method later), Morse's method of defining a smooth function on a manifold with isolated critical points and constructing trajectories orthogonal to level surfaces of the function. He refined this method further – the critical points were no longer "isolated," but formed "critical manifolds."

The methods developed in this paper were fruitfully used by H. Hopf and others to advance further the topology of group manifolds and homogeneous spaces, and later by Pontryagin himself to solve certain auxiliary problems in homotopy theory, and, in particular, to calculate the homology groups of Grassmann manifolds.

A direct consequence of this work is an elegant result obtained by Pontryagin many years later.[39] The point is that, for all compact simple Lie groups, the Betti numbers are equal to the corresponding Betti numbers of the direct products of spheres of different dimensions. The question therefore naturally arose: is a compact simple Lie group homeomorphic to the product of spheres of appropriate dimensions? Through the use of homotopic techniques, he found the answer to be negative. The special unitary group of third-order matrices has the same Betti numbers as the product of a 3-dimensional sphere and a 5-dimensional sphere, but the group itself is not homeomorphic to the product of the spheres: this was established through the use of the classification of the mappings of S^4 into S^3.

We shall now outline the homotopic works of L. S. Pontryagin. The topical problem in homotopic topology in the early stages of its development centered around the homotopic classification of the mappings of a sphere into a sphere of lesser dimension. Pontryagin encountered this problem while making fruitless attempts at giving a local characterization of the dimension of a compact set in R^n in terms of the homological characteristics of its complement.

In the beginning, he tried to solve the homotopic classification problem of the mappings of the sphere S^{n+k} into S^n using homological methods. But, shortly after learning about Hopf's work on the classes of mappings of S^3 onto S^2, he came to fully appreciate the situation; that was the beginning of a fifteen-year period during which Pontryagin was completely engaged in homotopic topology.

First, he demonstrated that the Hopf invariant is unique and, consequently, that Hopf's construction gives all the classes of the mappings of S^3 into S^2; thus, he obtained the full classification of the mappings of S^3 into S^2. Soon after, in 1936, he discovered an amazing result: the number of classes of mappings of S^{n+1} into S^n, for $n \geqslant 3$, is two (see reference 28). A mistake was made, however, in classifying the mappings of S^{n+2} into S^n, which led to an erroneous result. It was noticed and corrected by Pontryagin in 1950 (see reference 63). For these mappings, too, the number of classes was found to be two.

The initial proofs of these theorems were incredibly cumbersome. Only later, after the discovery of the method of framed manifolds (see below), could they be greatly simplified.

Then followed the solution to a series of problems in the homotopic classification of mappings of polyhedra into spheres and vice versa. Of these papers we mention here only two, reference 40 and 43. These papers introduced such basic concepts in homotopy theory as "obstructions" and "difference cochains" and a new cohomological operation – the Pontryagin square, the predecessor of Steenrod's cohomological operations.

But the major problem, the classification of the mappings of S^{n+k} into S^n for $k \geqslant 3$, still defied solution. This is exactly the problem that led Pontryagin to discover the so-called "framed manifold method," to define new invariants of smooth manifolds – characteristic classes known as "Pontryagin classes," and to create the theory of fiber bundles, i.e., to create a new and very important field in modern mathematics, differential topology.

Among the pioneers in this field, besides L. S. Pontryagin, we should name H. Hopf, E. Stiefel, H. Whitney, and C. S. Chern.

The framed manifold technique was designed to study the homotopic properties of mappings with the help of the information available about the differential-topological structure of a manifold. It was only fruitful in classifying the mappings of S^{n+k} into S^n for $k \leqslant 3$ (as had already been noted at the beginning of the fifties by Pontryagin for $k = 1, 2$, and by Rokhlin for $k = 3$), because, for $k > 3$, information was needed about smooth manifolds of dimensions > 3, which could not be obtained by the methods available in the early fifties. However, the framed manifold technique is equally effective for the opposite purpose, studying smooth manifolds when we have homotopic information at our disposal, which can be more successfully derived with the help of Leray's algebraic (spectral sequence) method. This reversal of the method, known as bordism theory, is due to R. Thom. Most of the far-reaching results in the modern theory of smooth manifolds have been obtained precisely through a combination of the Pontryagin–Thom differential-topological method and Leray's algebraic method.

Today, characteristic classes constitute the central topic not only in differential topology, but also in modern differential geometry as a whole; fiber bundle theory has long since become a common research tool in topology, geometry, and analysis.

The theory of characteristic classes and the closely related theory of singularities of vector fields are presented in three large papers.[56,57,61] The results of these papers were reported in earlier preliminary works.[45,48–50] Reference 49 also reports briefly on the theory of classifying spaces, which subsequently played an important role in the development of fiber bundle theory.

The framed manifold method and a full classification of the mappings of S^{n+k} into S^n for $k = 0$, 1, and 2 are presented in reference 69 (see also Volume 3 of the *Selected Works*), which was the original exposition in the literature of the fundamentals of differential topology.

The "topological period" in the activity of L. S. Pontryagin ends with reference 69; from the early fifties on, he switched over exclusively to applied fields. Up until this time he had turned his attention to applied and nontopological topics only occasionally, but with great success.

We begin the survey of his earlier nontopological works with the famous paper written in collaboration with A. A. Andronov,[29] in which the concept of the structural stability of a dynamical system in a plane was first introduced, using the term "rough system," and the roughness condition was formulated.

In a broad context there are two motives behind the idea of roughness: physical and mathematical. The physical motive arose in connection with Andronov's investigations into auto-oscillations and consists of the following: if a dynamic system describing a physical phenomenon is known only approximately, then the qualitative portrait of the system's phase plane can reflect the phenomenon only if this portrait does not change under small perturbations of the dynamic system. The mathematical motive is related to the idea of "typicality," or "general position," which is not at all specific to differential equations and which is widely used in different fields of mathematics, including some topological works of L. S. Pontryagin. For the "general position" case, the phase portrait should be expected to be simpler than in exceptional cases; thus, the "general position" case deserves the utmost attention.

In this paper, smooth flow (of class C^1) in a domain $O \subset R^2$ bounded by a smooth closed curve everywhere transversal to the trajectories is called rough, if, for any flow sufficiently C^1-close to the initial flow, there exists a homeomorphism of the domain O onto itself, C^0 close to the identity, that sends the trajectories of one flow into the trajectories of another, preserving the direction of motion along these trajectories.

After giving this definition, the authors show that the rough systems on a plane are typical (they form an everywhere dense open set) and that their qualitative portrait is quite simple. Here the three ideas, "simplicity", "roughness", and "typicality", merge together (the corresponding classes of the systems coincide). This merger is specific to the small dimension of the phase space and fails for higher dimensions. But these three ideas are themselves of great interest for higher-dimensional systems also, and the questions of the behavior of trajectories for the corresponding class of systems and of the mutual relations between these classes have dominated the study of dynamic systems through the past twenty or twenty-five years, and go back, in the final analysis, to reference 29.

Still earlier, reference 29 had influenced the development of the two-dimensional qualitative theory of differential equations. First, it outlines the role of "singular" (orbitally unstable) trajectories, subdividing the phase plane into "cells" filled with trajectories of identical behavior. Second, the solution of the problem concerning rough systems on a plane paved the way for studies of "typical" bifurcations of a parameter-dependent dynamic system in the two-dimensional case.

Of his early works on dynamic systems, mention should be made of one more paper,[13] which gives simple conditions, conveniently applied, for the birth of a cycle from a closed trajectory of a plane nonlinear Hamiltonian system under small autonomous (nonconservative) perturbations.

Among the early nontopological works of Pontryagin, reference 47 also deserves special mention, and had a considerable impact on the development of functional analysis on spaces with an indefinite metric. It was written during World War II at Kazan in connection with a purely applied problem of stability in ballistics. Its main result is that any Hermitian operator in a Hilbert space with an indefinite metric of index k has a k-dimensional invariant subspace on which all eigenvalues of the operator have nonnegative imaginary parts, and the main (indefinite) form of the space is nonnegative.

One more work completed during wartime at Kazan concerns stability theory. It formulates the conditions that must be fulfilled for a quasipolynomial to have roots with negative real parts (see reference 42). These conditions were later extended to functions of the type f/g having no poles, where f is a quasi-polynomial and g a polynomial (see reference 66).

We shall now take up the period that dates approximately from the beginning of the fifties, when Pontryagin was basically devoting himself to problems in applied mathematics.

Here, too, he displays with great strength his exceptional talent to perceive amidst the primal chaos in each new problem the main path, which leads to the goal via the shortest route. He forges ahead on this pathway, overcoming technical difficulties that seem, at times, to be insurmountable.

To study new topics, Pontryagin founded a special seminar in oscillation and control theory in 1952 at the Steklov Mathematics Institute. He believed that, to gain success in any applied field of mathematics, one should not confine oneself to the existing mathematical models, but start the study with technical problems, not only to gain a deeper insight into the existing models, but also to formulate new mathematical problems that have a pure mathematical interest as well as a technical interest.

Soon, as a result of this seminar, two basic advances emerged: the theory of relaxation (discontinuous) oscillations and the optimal control theory, which later Pontryagin began to elaborate on with great success jointly with his younger collaborators V. G. Boltyanskii, R. V. Gamkrelidze, and E. F. Mishchenko.

Relaxation oscillations are encountered in physical, and, in particular, in radio engineering systems described by differential equations with a small parameter ε attached to higher derivatives. Mathematically, relaxation oscillations can be defined as the periodic solutions of differential equations (or a system of differential equations) with a small parameter attached to higher derivatives that contain "slow motion" sections traversed by a phase point in a finite time, as well as "junction points" where the "fast motion" sections start and which are traversed in infinitely small time as $\varepsilon \to 0$. A classical example of these oscillations is the Van der Pohl equation. The study of the asymptotic behavior of these oscillations in relation to ε is a very difficult mathematical problem and was only partially solved in some simplest cases. Pontryagin's studies have made much headway with this problem for general systems and are of fundamental value.

Of great help to Pontryagin in these investigations was his phenomenal ability to do long mental calculations and to memorize complicated expressions.

Pontryagin's works on relaxation oscillations are listed in that part of the bibliography which comprises papers published in 1955–1963.

In the mid-fifties, he discovered the famous "Pontryagin maximum principle," which, though universal, is easily formulated and is an effective tool in solving a broad range of optimization problems from purely applied questions in diverse engineering fields to complicated theoretical questions. The maximum principle includes the first-order theory of the classical calculus of variations, which had proved futile in tackling many new technical problems, the analysis of which has led to the discovery of the maximum principle.

The maximum principle is simple to formulate and we state it for the important time-optimal case.

A process is called controlled if it can be described by an n-dimensional vector differential equation

$$\dot{x} = f(x, u),$$

where $x \in R^n$ is the phase point and u is an r-dimensional vector control parameter that takes values from some given subset $U \subset R^r$, which is, as a rule, a closed domain. The problem then is to choose a control $u(t) \in U$, as a function of time t, such that the corresponding trajectory $x(t)$ of the equation

$$\dot{x} = f(x, u(t))$$

is shifted from a given point x_0 to some other given point x_1 in minimum time. This control and its corresponding trajectory are called optimal. Let us introduce the following scalar function

$$H(x, \psi, u) = \psi f(x, u),$$

where $\psi f(x, u)$ is the scalar product of an n-dimensional vector ψ and f, and write the canonical system of equations

$$\dot{x} = f = \frac{\partial H}{\partial \psi}, \qquad \dot{\psi} = -\psi \frac{\partial f}{\partial x} = -\frac{\partial H}{\partial x}.$$

The Pontryagin maximum principle asserts that, for a control $u(t)$, $t_0 \leqslant t \leqslant t_1$, and the corresponding trajectory $x(t)$ to be optimal, it is necessary that there exist a nonzero variable vector $\psi(t)$ such that $u(t)$, $x(t)$, and $\psi(t)$ satisfy the above canonical system of equations and "Pontryagin's maximum condition":

$$H(x(t), \psi(t), u(t)) = \max_{u \in U} H(x(t), \psi(t), u), \; \forall t \in [t_0, t_1].$$

The discovery of the maximum principle proved a startling event that soon gave birth to a new advance, the optimal control theory, which, at present, is a vital and flourishing area in applied mathematics – and the stream of papers brought forth by this theory is truly immense.

Among the works of Pontryagin that have greatly influenced the development of optimal control theory, we may mention his Plenary Address to the International Congress of Mathematicians held in Edinburgh in 1958[89] and his monograph "The Mathematical Theory of Optimal Processes" written jointly with V. G. Boltyanskii, R. V. Gamkrelidze, and E. F. Mishchenko (see Volume 4 of the *Selected Works*).

A natural development of optimal control theory proposed by Pontryagin himself is differential game theory, which he is presently pursuing. A review of this theory is outlined in his Plenary Address to the International Congress of Mathematicians held at Nice in 1970.[126] A full exposition of his theory of linear differential games is given in references 127 and 143, which are also included in this volume.

Since 1934, L. S. Pontryagin has been working at the Steklov Mathematics Institute of the USSR Academy of Sciences; he was made a full-time member of the Institute and given the position of Head of the Topology Division in 1939. From 1961 to the present, he has held the position of Head of the Division of the Theory of Ordinary Differential Equations and Control Theory. At the same time, he has always attached great importance to the teaching of mathematics and has devoted much time to giving lectures at Moscow State University. Being an excellent teacher, he always prepared his lectures with utmost care, even designing notation to the minutest detail. Four of his books, *Topological Groups* (Volume 2 of the *Selected Works*), *Combinatorial Topology*, *Algebraic and Differential Topology* (Volume 3 of the *Selected Works*), and *Ordinary Differential Equations* (English edition[100]), which have been translated into many languages, were based on his lecture courses at Moscow State University; they have greatly influenced the education of many generations of mathematicians all over the world.

R. V. GAMKRELIDZE

Zum Alexanderschen Dualitätssatz*

Gesellschaft der Wissenschaft Göttingen Nachrichten
Mathematik-Physik (1927) Heft 4: 315–22

Vorgelegt von R. Courant in der Sitzung vom 25. November 1927.

1

Herr Alexander hat folgenden Dualitätssatz bewiesen:[1] Wenn K^λ ein im R^n liegender Komplex ist, so ist[2]

$$p^r(K^\lambda) = p^{n-r-1}(R^n - K^\lambda). \tag{1}$$

In der vorliegenden Arbeit will ich zeigen, daß eine gewisse Dualität nicht nur zwischen den Bettischen Zahlen, sondern auch zwischen den Homologiebasen von K^λ und $R^n - K^\lambda$ besteht. Um diese Dualität formulieren zu können, muß ich zuerst an den (im westentlichen von Brouwer herrührenden)[3] Begriff der *Verschlingungszahl* zweier geschlossenen Komplexe (im R^n) erinnern.

---

* Die Resultate dieser Arbeit bilden die Beantwortung einer von Herrn P. Alexandroff in seinem Topologischen Seminar (Wintersemester 1926–27, Universität Moskau) gestellten Frage und wurden von mir (mit Beweis) in jenem Seminar und von Herrn Alexandroff in seiner Göttinger Vorlesung (Sommer 1927) vorgetragen.

2

Es seien K^p und K^q zwei im R^n liegende Komplexe, deren Dimensionszahlen p und q der Gleichung $p + q = n$ genügen. Wir setzen überdies noch voraus, daß jeder dieser Komplexe zum Rande des anderen fremd ist. Es seien ferner $\tilde{K}^p$ und $\tilde{K}^q$ Polyderkomplexe, die die gegebenen Komplexe genügend gut approximieren; insbesondere setzen wir voraus, daß der Rand eines jeden dieser Komplexe zum anderen fremd ist. Außerdem nehmen wir an, daß die Komplexe $\tilde{K}^p$ und $\tilde{K}^q$ sich in einer allgemeinen Lage befinden (d.h. daß das System sämtlicher in diesen beiden Komplexen vorkommenden Eckpunkte linear unabhängig ist). Man beweist dann leicht,[4] daß die Parität der (auf Grund letzterer Voraussetzung endlichen) Anzahl der Schnittpunkte von $\tilde{K}^p$ und $\tilde{K}^q$ von der speziellen Wahl dieser approximierenden Komplexe unabhängig ist und somit eine topologische Invariante der (in ihrer gegenseitigen Lage gegebenen) Komplexe K^p und K^q darstellt. Diese Invariante ist die *Kroneckersche Charakteristik* $\chi(K^p, K^q)$ der Komplexe K^p und K^q modulo 2.

3

Es seien jetzt im R^n zwei zueinander fremde geschlossene Komplexe Γ^r und Γ^s gegeben, unter der Bedingung, daß $r + s = n - 1$ ist. Wenn K^{r+1} und K^{s+1} irgendwelche durch Γ^r bzw. durch Γ^s begrenzte Komplexe sind, so ist, wie man leicht zeigt,

$$\chi(K^{r+1}, \Gamma^s) = \chi(K^{s+1}, \Gamma^r);$$

diese Zahl $\chi(K^{r+1}, \Gamma^s) = \chi(K^{s+1}, \Gamma^r)$ hängt nur von den Komplexen Γ^r und Γ^s und ihrer gegenseitigen Lage im R^n ab und ist die *Verschlingungszahl*[5] $\mathfrak{v}(\Gamma^r, \Gamma^s)$ (mod. 2) von Γ^r und Γ^s. Aus der obigen Definition folgt ohne weiteres, daß $\mathfrak{v}(\Gamma^r, \Gamma^s)$ immer Null ist, wenn einer der beiden Komplexe Γ^r und Γ^s im Komplementärraume zum anderen homolog Null ist.[6]

4

Das Hauptresultat dieser Arbeit besteht nun im folgenden.

Hauptsatz. Es sei K^λ irgendein im R^n liegender λ-dimensionaler Komplex. Wenn

$$\Gamma_1^r, \Gamma_2^r, \ldots, \Gamma_p^r \tag{2}$$

eine r-dimensionale Basis von K^λ und

$$\gamma_1^{n-r-1}, \gamma_2^{n-r-1}, \ldots, \gamma_p^{n-r-1} \tag{3}$$

eine $n-r-1$-dimensionale Basis von $R^n - K^\lambda$ ist, so ist jede lineare. Kombination von Elementen der einen Basis verschlungen mit mindestens einem Elemente der anderen.

Beweis des Hauptsatzes. Wir beweisen zuerst folgenden.

Satz I. Wenn ein γ^{n-r-1} im $R^n - K^\lambda$ nicht homolog Null ist, so ist γ^{n-r-1} mit mindestens einem Elemente der r-dimensionalen Basis von K^λ verschlungen.

Der Beweis stützt sich auf den (zum ersten Mal bei Lebesgue vorkommenden)[7] Begriff eines zu einem beliebigen Elemente[7a] a_k^h von K^λ *dualen* (geschlossenen) Komplexes γ_k^{n-h-1}. Die erste allgemeine und strenge Konstruktion dieser dualen Komplexe ist von Alexander[8] angegeben. Folgende Eigenschaften der Alexanderschen Konstruktion lassen sich (durch Induktion) leicht beweisen:

1. γ_k^{n-h-1} ist zu allen i-dimensionalen Elementen $(0 \leqq i \leqq h)$ von K^λ fremd;

2. Es existiert ein $Q_k^{n-h} \to \gamma_k^{n-h-1}$, so daß Q_k^{n-h} zu allen i-dimensionalen $(0 \leqq i \leqq h)$ Elementen von K fremd ist, *mit der einzigen Ausnahme* des Elementes a_k^h selbst: es ist vielmehr

$$\chi(a_k^h, Q_k^{n-h}) = 1.$$

5

Wir beweisen jetzt den Satz I zuerst für $r = \lambda$. Es seien

$$\Gamma_i^\lambda = \kappa_i^j a_j^\lambda \quad (i = 1, 2, \ldots, s) \tag{4}$$

sämtliche (als Linearformen dargestellten) geschlossene Komplexe, die man aus den λ-dimensionalen Elementen von K^λ bilden kann. Der Rang der Koeffizientenmatrix $\|\kappa_i^j\|$ ist dann die Zahl $p^\lambda(K) = p$. Es existiert also eine (aus den χ_i^j gebildete) quadratische Matrix von demselben Range p. Es sei

$$\begin{array}{c|cccc} & a_1^\lambda & a_2^\lambda & \ldots & a_p^\lambda \\ \hline \Gamma_1^\lambda & \kappa_1^1, & \kappa_1^2, & \ldots, & \kappa_1^p \\ \Gamma_2^\lambda & \kappa_2^1, & \kappa_2^2, & \ldots, & \kappa_2^p \\ \vdots & \cdot & \cdot & \cdots & \cdot \\ \Gamma_p^\lambda & \kappa_p^1, & \kappa_p^2, & \ldots, & \kappa_p^p \end{array} \tag{5}$$

diese Matrix.

Est ist

$$\kappa_i^j = \mathfrak{b}(\Gamma_i^\lambda, \gamma_j^{n-\lambda-1}). \tag{6}$$

(Letztere Formel folgt aus der Tatsache, daß $Q_j^{n-\lambda}$ zu allen

$$a_{j'}^\lambda, \quad (j' \neq j)$$

fremd ist, daß infolgedessen

$$\chi(Q_j^{n-\lambda}, \Gamma_i^\lambda) = \chi(Q_j^{n-\lambda}, \kappa_i^h a_h^\lambda) = \kappa_i^h \chi(Q_j^{n-\lambda}, a_h^\lambda) = \kappa_i^j$$

ist, daß endlich auf Grund von $Q_j^{n-\lambda} \to \gamma^{n-\lambda-1}$

$$\chi(Q_j^{n-\lambda}, \Gamma_i^\lambda) = \mathfrak{b}(\gamma_j^{n-\lambda-1}, \Gamma_i^\lambda)$$

ist.)

Da der Rang von (5) gleich p ist, so ist (wie die nicht sämtlich verschwindenden Koeffizienten μ_j, auch gewählt sein mögen) wenigstens für ein i

$$\mu_j \kappa_i^j \neq 0;$$

letztere Relation heißt aber nichts anderes, als daß jede lineare Kombination der $\gamma_j^{n-\lambda-1}$ mit wenigstens einem Γ_i^λ (und also mit wenigstens einem Elemente einer beliebigen λ-dimensionalen Basis von K^λ) verschlungen ist. Da die Anzahl der $\gamma_j^{n-\lambda-1}$ gleich

$$p = p^{n-\lambda-1}(R^n - K^\lambda)$$

ist, und (wie wir soeben bewiesen haben) die $\gamma_j^{n-\lambda-1}$ in $R^n - K^\lambda$ linear unabhängig sind, so bilden die letztgenannten Komplexe eine $n-\lambda-1$-dimensionale Basis von $R^n - K^\lambda$; jeder $n-\lambda-1$-dimensionale Komplex $\gamma^{n-\lambda-1} \subset R^n - K^\lambda$ ist also (insofern er nicht homolog Null ist) einer linearen Kombination der $\gamma_j^{n-\lambda-1}$ homolog und also mit mindestens einem Elemente der λ-dimensionalen Basis von K^λ verschlungen.

Hiermit ist der Satz I für den Fall $r = \lambda$ bewiesen.

6

Es sei jetzt $r < \lambda$ und γ^{n-r-1} nicht homolog Null in $R^n - K^\lambda$. Auf Grund einer Bemerkung von Alexander[9] kann durch das Tilgen eines a_i^k aus einem K^k ein $\gamma^h \subset R^n - K^k$ (der in $R^n - K^k$ nicht homolog Null ist) nur dann [in $R^n - (K^k - a_i^k)$] homolog Null werden, wenn die Dimensionszahl h gleich $n-k-1$ ist. Durch wiederholte Anwendung dieser Bemerkung beweist man, daß unser γ^{n-r-1} auch in $R^n - K^r$ (wobei K^r den aus allen a_i^r gebildeten Teil-komplex von K^n bedeutet) nicht homolog Null, und also mit wenigstens einem Element der (beliebig gewählten) r-dimensionalen Basis von K^r verschlungen ist. Nun kann man eine r-dimensionale Basis von K^r dadurch erhalten, daß man zu den Elementen

$$\Gamma_1^r, \Gamma_2^r, \ldots, \Gamma_p^r \tag{7}$$

einer beliebigen r-dimensionalen Basis von K^λ evtl. gewisse neue Komplexe

$$\Gamma_{p+1}^r, \Gamma_{p+2}^r, \ldots, \Gamma_{p+q}^r \tag{8}$$

hinzunimmt.

Unser Komplex γ^{n-r-1} ist mindestens mit einem der Komplexe (7) oder (8) verschlungen. Wenn γ^{n-r-1} mit einem $\Gamma_i^r, i \leqq p$ verschlungen ist, ist unser Satz bewiesen. Es sei γ^{n-r-1} mit einem bestimmten Komplexe Γ_{p+h}^r und mit keinem Γ_i^r, $i \leqq p$ verschlungen. Zwei Fälle sind dann a priori möglich:

1. Γ_{p+h}^r ist homolog Null in K^λ;
2. $\Gamma_{p+h}^r \sim \mu^i \Gamma_i^r, i \leqq p.$ (in K^λ).

Der Fall 1. kann aber nicht auftreten, weil in diesem Falle jeder mit Γ_{p+h}^r verschlungene Komplex (also insbesondere unser γ^{n-r-1}) mit (dem durch Γ_{p+h}^r begrenzten Teilkomplex von K^λ, also) mit dem Komplexe K^λ gemeinsame Punkte haben müßte.

Im Fall 2. dagegen ist γ^{n-r-1} mit $\mu^i \Gamma_i^r$, also mit mindestens einem $\Gamma_i^r, i \leqq p$ verschlungen, w.z.b.w.

7

Es seien jetzt:

$$\Gamma_1^r, \Gamma_2^r, \ldots, \Gamma_p^r \tag{2}$$

die r-dimensionale Basis von K^λ und

$$\gamma_1^{n-r-1}, \gamma_2^{n-r-1}, \ldots, \gamma_p^{n-r-1} \tag{3}$$

die $n-r-1$-dimensionale Basis von $R^n - K^\lambda$.

Eine beliebige $\mu^i \gamma_j^{n-r-1}$ (als ein Komplex, der in $R^n - K^\lambda$ gewiß nicht homolog Null ist) ist dann mit mindestens einem F_i^r verschlungen. Infolgedessen ist der Rang der Matrix $\|\mathfrak{b}(\Gamma_i^r, \gamma_j^{n-r-1})\|$ gleich p, d.h. jede lineare Kombination der Γ_i^r ist mit mindestens einem γ_j^{n-r-1} verschlungen. Der Hauptsatz ist hiermit vollständig bewiesen.

8

Folgende (für manche Zwecke nützliche) Verallgemeinerung unsres Hauptsatzes liegt auf der Hand:

Satz II. Es sei

$$\Gamma_1^r, \Gamma_2^r, \ldots, \Gamma_s^r \quad (9) \qquad \gamma_1^{n-r-1}, \gamma_2^{n-r-1}, \ldots, \gamma_s^{n-r-1} \quad (9)$$

ein System von

$$\text{in } K^\lambda \qquad \text{in } R^n - K^\lambda$$

liegenden und daselbst unabhängigen geschlossenen Komplexen.

Man kann dann unter den Elementen der $n-r-1$ bzw. der r-dimensionalen Basis

$$\text{von } R^n - K^\lambda \qquad \text{von } K^\lambda$$

ein System von Komplexen

$$\gamma_1^{n-r-1}, \gamma_2^{n-r-1}, \ldots, \gamma_s^{n-r-1} \,(10) \qquad \Gamma_1^r, \Gamma_2^r, \ldots, \Gamma_s^r \quad (10)$$

finden von der Eigenschaft, daß jede lineare Kombination der Komplexe (9) mit mindestens einem Komplexe (10), und jede lineare Kombination der Komplexe (10) mit mindestens einem Komplexe (9) verschlungen ist.

Wir beweisen nur die linke Seite des Satzes (der Beweis der rechten Seite läuft wörtlich parallel).

Da die Γ_i^r unabhängig sind, kann man sie in eine Basis

$$\Gamma_1^r, \Gamma_2^r, \ldots, \Gamma_s^r; \quad \Gamma_{s+1}^r, \ldots, \Gamma_p^r$$

von K^λ aufnehmen. Man konstruiere jetzt eine $n-r-1$-dimensionale Basis

$$\gamma_1^{n-r-1}, \gamma_2^{n-r-1}, \ldots, \gamma_p^{n-r-1} \tag{3}$$

von $R^n - K^\lambda$ und betrachte die Matrix $\|\mathfrak{b}(\Gamma_i^r, \gamma_j^{n-r-1})\|$. Da der Rang dieser Matrix gleich p ist, ist s der Rang der Matrix

$$\|\mathfrak{b}(\Gamma_i^r, \gamma_j^{n-r-1})\|, \quad (i = 1, 2, \ldots, s;\ j = 1, 2, \ldots, p).$$

Es gibt also eine quadratische Teilmatrix

$$\|\mathfrak{b}(\Gamma_i^r, \gamma_{j_h}^{n-r-1})\|, \quad (h = 1, 2, \ldots, s;\ i = 1, 2, \ldots, s)$$

vom selben Range s; sodann ist jede lineare Kombination der Γ_i^r ($i = 1, 2, \ldots, s$) mit mindestens einem $\gamma_{j_h}^{n-r-1}$ ($h = 1, 2, \ldots, s$) verschlungen und umgekehrt. Der Satz II ist hiermit bewiesen.

Aus dem Hauptsatze folgt unmittelbar auch der.

Satz III. Wenn zwei irreduzible geschlossene Komplexe Γ^r und Γ^{n-r-1} im R^n gegeben sind, so sind nur die beiden folgenden Fälle möglich:

entweder ist jeder dieser Komplexe im Komplementärraum zum andern homolog Null;

oder sind die beiden Komplexe verschlungen.

9

Als letzte Anwendung des Hauptergebnisses dieser Arbeit beweise ich folgenden.

Santz IV. Es sei F eine abgeschlossene Teilmenge des R^n und Γ^r ein geschlossener Komplex, der in $R^n - F$ nicht homolog Null ist. Wenn man F außerhalb von Γ^r mittels einer stetigen Deformation in eine Menge Φ überführt, so ist Γ^r auch in $R^n - \Phi$ nicht homolog Null.

Zum Beweise des Satzes IV brauchen wir folgenden Hilfssatz:

Es seien F und Γ zwei zueinander fremde abgeschlossene Teilmengen des R^n. Es sei außerdem eine stetige Deformation δ von F in eine Menge Φ gegeben, bei der kein zu F gehörender Punkt die Menge Γ berührt. Dann läßt sich die Deformation δ zu einer (auf F mit δ übereinstimmenden) Deformation Δ des ganzen R^n in sich so erweitern, daß bei der Deformation Δ kein Punkt von R^n in einen Punkt von Γ abgebildet wird und die Punkte letzterer Menge fest bleiben.

Um den Hilfssatz zu beweisen, erweitere man zuerst die Deformation δ irgendwie auf den ganzen Raum R^n.[10] Man wähle sodann ein so kleines $\epsilon > 0$, daß $S(F, \epsilon)$[11] während der ganzen (soeben durch Erweiterung von δ definierten) Deformation $\tilde{\Delta}$ zu Γ fremd bleibt. Man nehme jetzt irgendeinen der Bedingung

$$F \subset Q^n \subset S(F, \epsilon)$$

genügenden Polyederkomplex Q^n (dessen Simplexe einer bestimmten Simplizialzerlegung ζ des ganzen Raumes R^n entnommen sind) und unterteile die Zerlegung ζ so fein, daß der Raum R^n nach dieser Unterteilung in einer in bezug auf Q^n "*bequemen*"[12] Simplizialzerlegung vorliegt. Dies erlaubt[12] eine gewisse in $S(F, \epsilon)$ enthaltene Umgebung $U(Q^n)$ von Q^n in ein Vektorfeld Θ zu zerlegen. Man definiere jetzt die Deformation Δ wie folgt:

1. Auf Q^n ist Δ mit $\tilde{\Delta}$ identisch. Wir bezeichnen durch $\Lambda_x = \overrightarrow{xx_1}$ die Bahn des Punktes $x \subset Q^n$ während dieser Deformation.

2. Es sei $V(x)$ ein Vektor $\overline{xy}$ ($x \subset Q^n$) des Feldes Θ und z die Mitte der Strecke xy. Man deformiere die Kurve $\overrightarrow{yx} + \Lambda_x$ stetig in sich dadurch, daß man $\overrightarrow{yz}$ proportional in $\overrightarrow{yx}$ überführt, während die Strecke $\overrightarrow{zx}$ in die Lage $\overrightarrow{xx_1}$ hineingezogen wird.

3. Alle übrigen Punkte des Raumes läßt die Deformation Δ fest. Die auf diese Weise gewonnene Deformation Δ genügt allen Forderungen unsres Hilfssatzes.

10

Der Beweis des Satzes IV bietet jetzt keine Schwierigkeiten dar. Es sei Γ^r in $R^n - F$ nicht homolog Null; es sei ferner eine stetige Deformation δ gegeben, die außerhalb von Γ^r die Menge F in eine abgeschlossene Menge Φ überführt. Vermöge unsres Hilfssatzes kann man δ zu einer im ganzen R^n definierten Deformation Δ erweitern, die auf F mit δ übereinstimmt, alle Punkte des Komplexes Γ^r festhält und keinen Punkt von $R^n - \Gamma^r$ in einen Punkt von Γ^r abbildet. Durch die Deformation Δ wird jede Menge $M \subset R^n$ auf eine Menge $\Delta(M)$ eindeutig und stetig abgebildet. Man kann auch umgekehrt zu jeder Menge M die Menge $\Delta^{-1}(M)$ aller Originalpunkte von M bei der soeben definierten Abbildung betrachten. Wenn M abgeschlossen ist, so sind die beiden Mengen $\Delta(M)$ und $\Delta^{-1}(M)$ ebenfalls abgeschlossen.

Wir nehmen jetzt an, daß $\Gamma^r \sim 0$ (in $R^n - \Phi$) ist; dann gibt es einen durch Γ^r berandeten $K^{r+1} \subset R^n - \Phi$. Es sei $C \subset R^n - F$ die Menge $\Delta^{-1}(K^{r+1})$ und Q^n eine zu F fremde Polyederumgebung von C. Da bei der Abbildung Δ kein Punkt von $R^n - \Gamma^r$ in einen Punkt von Γ^r übergeht, so ist

$$\Gamma^r = \Delta^{-1}(\Gamma^r) \subset C \subset Q^n \subset R^n - F;$$

Γ^r kann also nicht im Komplexe Q^n homolog Null sein. Auf Grund des Satzes II existiert sodann ein mit Γ^r verschlungener Komplex $\gamma^{n-r-1} \subset R^n - Q^n$. Bei der Deformation Δ geht γ^{n-r-1} in einen (eventuell singulären) Komplex $\tilde{\gamma}^{n-r-1}$ über, der mit $\Gamma^r = \Delta(\Gamma^r)$ verschlungen ist. Daraus folgt aber, daß $\tilde{\gamma}^{n-r-1} . K^{r+1} \neq 0$ und mithin auch $\gamma^{n-r-1} . C \neq 0$, also um so mehr $\gamma^{n-r-1} . Q^n \neq 0$ ist, was einen den Satz IV beweisenden Widerspruch bedeutet.

Notes

1. Alexander, *Trans. Amer. Math. Soc.*, **23** (1922), SS. 333–349.

2. Siehe wegen Erklärung von Bezeichnungen die nachstehende Arbeit von Alexandroff.

3. Brouwer "On looping coefficients" (*Kon. Akademie Amsterdam, Proceedings*, 1912, SS. 113–122. Vgl. auch die deselbst zitierte Comptes-Rendus Note von Lebesgue (C. R. 27 mars 1911), in der der Begriff der Verschlingung "modulo 2" ohne den zugehörigen Invarianzbeweis eingeführt wurde.

4. Vgl. Brouwer a.a.O.[3] Die erste Behandlung der Kroneckerschen Charakteristik allgemeiner Komplexe befindet sich bei Lefschetz, *Trans. Amer. Math. Soc.*, **28**, SS. 1–49 (insbesondere §§ 3–6).

5. S. die unter (3) zitierten Arbeiten von Brouwer und Lebesgue.

6. Die umkehrung dieser Behauptung ist nur im Falle der *irreduziblen* geschlossenen Komplexe richtig: siehe Satz III der vorliegenden Arbeit. Ein geschlossener Komplex heißt irreduzibel, wenn er keinen (echten) geschlossenen Teilkomplex von derselben Dimension enthält.

7. a.a.O.[3]

7a. Die h-dimensionalen Elemente eines Komplexes bezeichnen wir durch a^h (mit verschiedenen unteren Indexen).

8. a.a.O.,[1] SS. 345 und 348.

9. a.a.O.,[1] S. 347, Zeilen 8–11 (von oben).

10. Auf Grund des bekannten Erweiterungssatzes für stetige Funktionen ist dies, wie eine leichte Überlegung zeigt, stets möglich. Siehe über diesen Erweiterungssatz Brouwer, *Math. Ann.*, **71**, S. 309 und 79, S. 209 und Urysohn, *Math. Ann.*, **94**, S. 293 (in letzterer Arbeit wird der Satz unter den allgemeinsten bis jetzt bekannten Voraussetzungen bewiesen; es befinden sich dort auch ausführliche Literaturangaben).

11. Durch $S(F, \epsilon)$ wird die Menge aller Punkte bezeichnet, deren Entfernung von der Menge F kleiner als ϵ ist.

12. Im Sinne von Alexandroff, "Über den allgemeinen Dimensionsbegriff usw.", §2 u. 3 (erscheint demnächst in den *Math. Ann.*).

Zum Alexanderschen Dualitätssatz
Zweite Mitteilung*

Gesellschaft der Wissenschaft Göttingen Nachrichten
Mathematik-Physik (1927) Heft 4: 446–56

Vorgelegt von R. Courant in der Sitzung vom 9. März 1928.

1

Der Alexandersche Dualitätssatz[1] besagt, daß die Bettischen Zahlen des zu einem im n-dimensionalen sphärischen Raume R^n liegenden Komplexe C komplementären Raumes $R^n - C$ durch die Bettischen Zahlen von C vollständig bestimmt sind und zwar durch die Formel

---

* Die Kenntnis meiner ersten Mitteilung über denselben Gegenstand (*Gött. Nachr.*, S. 315–322) wird im folgenden worausgesetzt; auch sämtliche Bezeichnungen werden im folgenden im Sinne der ersten Mitteilung gebraucht. Insbesondere sind alle kombinatorische Begriffe (wie Begrenzung, Homologien, Geschlossenheit eines Komplexes, Bettische Zahlen, Kroneckersche Charakteristiken usw.) im Sinne der Definitionen "modulo 2" verstanden. Mit C werden *allgemeine* (d. h. den Polyederkomplexen homöomorphe) Komplexe, durch K und Q dagegen nur Polyederkomplexe selbst bezeichnet. Geschlossene Komplexe werden "*Zyklen*" genannt und durch Γ, γ, Δ, δ bezeichnet. Überall, wo das sinngemäß ist, wird vorausgesetzt, daß Polyederkomplexe sich in "allgemeiner Lage" zueinander befinden (d. h. daß die entsprechenden Unabhängigkeitsbedingungen für Elemente dieser Polyederkomplexe tragende Ebenen verschiedener Dimensionszahl erfüllt sind). Um von Polyederkonstruktionen bequem reden zu können, wird der *sphärische* Raum R^m als der Euklidische Raum, ergänzt durch Hinzufügung eines uneigentlichen (unendlich fernen) Punktes gedacht. Alle Einbettungen sind singularitätenfrei.

$$p^{n-r-1}(R^n - C) = p^r(C). \tag{1}$$

In meiner ersten Mitteilung über den Alexanderschen Satz habe ich gezeigt, daß die durch die Formel (1) ausgedrückte Dualität zwischen den Bettischen Zahlen, sich zu einer Dualität zwischen den Homologiebasen selbst von C und $R^n - C$ erweitern läßt.

Es entsteht nun die ganz natürliche Frage: was wird aus dem Alexanderschen Dualitätssatze und dessen Verallgemeinerung, wenn man sich nicht auf Teilkomplexe des sphärischen Raumes beschränken will, sondern in beliebigen n-dimensionalen Mannigfaltigkeiten M^n gelegene Komplexe zu untersuchen beabsichtigt?

Es sei zunächst bemerkt, daß es prinzipiell unmöglich ist, die $p^k(M^n - C)$ als Funktionen der $p^i(C)$ allein im allgemeinen zu bestimmen, weil bereits auf dem Torus zwei homöomorphe Teilkomplexe Gebiete mit verschiedenen Bettischen Zahlen bestimmen können: es genügt ja zwei Kreise zu nehmen, von denen der eine den Torus zerlegt, während der andere es nicht tut. Es handelt sich also nicht nur um eine Verallgemeinerung der Alexanderschen Formel, sondern vielmehr um eine Verallgemeinerung der ganzen Fragestellung, die zu dieser Formel führt. Letztere Verallgemeinerung kann z. B. auf folgende Weise geschehen.[2]

2

Man betrachte zuerst den sphärischen Raum R^n; jeder geschlossene Teilkomplex von R^n ist homolog Null. Wenn man nun aus R^n einen Komplex C entfernt, so wird die letztere Eigenschaft *zerstört*: es treten im $R^n - C$ Zyklen auf, die daselbst nicht homolog Null sind; die Alexandersche Formel gibt uns gerade die Möglichkeit, die "Größe" dieser Störung zu messen, indem sie besagt, daß die Anzahl der $n-r-1$-dimensionalen Zyklen die (in R^n homolog Null waren und) in $R^n - C$ nicht homolog Null sind gleich der r-ten Bettischen Zahl von C, d.h. der Anzahl der Zyklen, die in R^n, aber nicht in C homolog Null sind. Dieser Gesichtspunkt läßt sich aber unmittelbar auch auf den Fall, wo C nicht notwendig in R^n, sondern in einer beliebigen geschlossenen M^n liegt, übertragen. Um diese Übertragung ordentlich machen zu können, führen wir folgende naheliegende Definition ein:

Es sei $C \subset M^n$. Ein System von in C (bzw. in $M^n - C$) gelegenen und daselbst im Sinne der Homologien unabhängigen r-dimensionalen Zyklen heißt eine Null-basis von C (bzw. von $M^n - C$), wenn folgende beiden Bedingungen erfüllt sind:

1. Jeder Zykel unseres Systems ist in M^n homolog Null;
2. Jeder in C (bzw. in $M^n - C$) liegende r-dimensionale Zykel, der in M^n homolog Null ist, ist in C (bzw. in $M^n - C$) einer linearen Kombination der Zyklen unseres Systems homolog.

Wenn wir nun durch q mit dem entsprechenden Dimensions-index die Anzahl der Elemente einer Nullbasis bezeichnen, so gilt folgender.

Saltz Ia. Es ist:
$$q^{n-r-1}(M^n - C) = q^r(C).$$ [3]

Man kann diesen Satz wesentlich verschärfen.

In det Tat, für zwei in M^n liegende zueinander fremde Zyklen γ^k und δ^{n-k-1} *von denen jeder in M^n homolog Null ist*, läßt sich der Begriff der *Verschlingungszahl* $\mathfrak{b}_{M^n}(\gamma^k, \delta^{n-k-1})$ wörtlich so definieren, wie für im sphärischen (bzw. in Euklidischen) Raume liegende Komplexe;[4] dabei gelten auch die zugehörigen Invariantensätze. Dies erlaubt folgendes Resultat auszusprechen:

Satz I. Es sei eine beliebige r-dimensionale Nullbasis von C und eine beliebige $n-r-1$-dimensionale Nullbasis von $M^n - C$ gegeben. Dann ist jede nicht identisch verschwindende lineare Kombination der Elemente der einen Basis mit mindestens einem Elemente der anderen Basis in M^n verschlungen.

Wenn jetzt überhaupt ein System S von r-dimensionalen Zyklen
$$\gamma_1^r, \gamma_2^r, \ldots, \gamma_s^r \qquad \text{(S)}$$
in C und ein System S' von $n-r-1$-dimensionalen Zyklen
$$\gamma_1^{n-r-1}, \gamma_2^{n-r-1}, \ldots, \gamma_s^{n-r-1} \qquad \text{(S')}$$
von $M^n - C$ gegeben ist, *so daß alle diese Komplexe in M^n homolog Null sind*, so gehört zu S und S' die *Verschlingungsmatrix* $\|\mathfrak{b}_{M^n}(\gamma_i^r, \gamma_j^{n-r-1})\|$, Wir nennen die beiden Systeme S und S' miteinander in M^n *verschlungen*, wenn der Rang dieser Matrix der höchste, d.h. s ist. Insbesondere sagen wir, daß die beiden Systeme *eindeutig* verschlungen sind, wenn die Verschlingungsmatrix die Einheitsmatrix ist. Nun aber kann man bekanntlich jede modulo 2 betrachtete quadratische Matrix höchsten Ranges durch unimodulare Transformationen der Spalten bzw. der Zeilen in die Gestalt der Einheitsmatrix bringen. Mit anderen Worten, wenn man zwei verschlungene Systeme S und S' hat, so kann man stets die Elemente des

einen Systems so durch lineare Kombinationen der Elemente desselben Systems ersetzen, daß die Verschlingung eindeutig ist, d. h., daß jedes Element des einen Systems mit einem und nur einem Elemente des anderen verschlungen ist. Wenn wir diese Definition auf unsre Nullbasen anwenden, können wir zunächst dem Satz I folgende Form geben:

Jede r-dimensionale Nullbasis von C ist mit jeder $n-r-1$-dimensionalen Basis von $M^n - C$ in M^n verschlungen.

Daraus (und aus der soeben gemachten Matrizenbetrachtung) folgt

Satz Ie. Zu jeder r-dimensionalen Nullbasis von C (bzw. von $M^n - C$) gibt es eine mit derselben in M^n eindeutig verschlungene $n-r-1$-dimensionale Nullbasis von $M^n - C$ (bzw. von C).

Es sei endlich bemerkt, daß ein Spezialfall des Satzes Ie nämlich der Fall $M^n = R^n$ mit Hilfe unsrer Matrizenbetrachtung unmittelbar aus dem Hauptzatz der ersten Mitteilung folgt; dieser Spezialfall darf und wird beim Beweise des allgemeinen Satzes I benutzt werden.

3

Der Begriff einer Nullbasis erlaubt aber auch eine zweite Fragestellung, die als Verallgemeinerung des Alexanderschen Dualitätsproblems bezeichnet werden kann, zu erledigen: ich meine die direkte Bestimmung der Zahl $p^{n-r-1}(M^n - C)$; letztere Zahl läßt sich nämlich durch die Zahlen q und die Bettischen Zahlen von C und M^n ausdrücken. Es gilt in der Tat der

Satz II. Wenn $C \subset M^n$ ist, so ist:

$$p^{n-r-1}(M^n - C) = q^r(C) + p^{n-r-1}(M^n) - [p^{r+1}(C) - q^{r+1}(C)];$$

oder wenn man die Zahl

$$p^k(C) - q^k(C)$$

(„die Anzahl der in M^n unabhängigen k-dimensionalen Zyklen von C“) durch $u^k(C)$ bezeichnet:

$$p^{n-r-1}(M^n - C) = q^r(C) + [p^{n-r-1}(M^n) - u^{r+1}(C)].$$

Wir wollen jetzt zuerst den Satz I und dann der Satz II beweisen. Wir brauchen dazu folgenden

Hilfssatz. Es sei M^n eine im R^m $(m > n)$ liegende geschlossene Polyedermannigfaltigkeit, und γ^r irgendein Zyklus in M^n der daselbst ~ 0 and zu einem festen $C \subset M^n$ fremd ist; es sei ferner U eine beliebige Umgebung von γ^r (rel. M^n). Es existiert dann in R^m ein zu C fremder Polyederzyklus γ^{r+m-n} (der sich in bezug auf M^n in einer *allgemeinen Lage* befindet, und) dessen Durchschnitt $\bar{\gamma}^r$ mit M^n ein den Bedingungen

$$\bar{\gamma}^r \subset U; \bar{\gamma}^r \sim \gamma^r \text{ in } U$$

genügender Zyklus ist.

Beweis des Hilfssatzes. Wir führen ein für allemal folgende Abkürzung ein: eine k-dimensionale Basis von C soll durch $B^k(C)$ bezeichnet werden.

Es sei also

$$\gamma_0^r, \gamma_1^r, \ldots, \gamma_s^r \tag{2}$$

eine $B^r(\gamma^r)$, von der wir annehmen, daß sie den Komplex γ^r als ein Element enthält, und zwar, daß $\gamma^r = \gamma_0^r$ ist.

Es sei ferner

$$\delta_0^{m-r-1}, \delta_1^{m\,;r\,;1}, \ldots, \delta_s^{m-r-1} \tag{3}$$

eine mit (2) eindeutig verschlungene $B^{m-r-1}(R^m - \gamma^r)$; dabei setzen wir voraus, daß alle δ_i^{m-r-1} sich in bezug auf M^n in einer allgemeinen Lage befinden, so daß die $M^n \cdot \delta_i^{m-r-1}$ $n-r-1$-dimensionale Zyklen

$$\delta_0^{n-r-1}, \delta_1^{n-r-1}, \ldots, \delta_s^{n-r-1}$$

sind von denen manche verschwinden können, aber jedenfalls, wie man leicht zeigt, in M^n sämtlich homolog Null sind.[5] Man kann deshalb von den Zahlen $\mathfrak{b}_M(\gamma^r, \delta_i^{n-r-1})$ sprechen und zwar ist

$$\mathfrak{b}_M(\gamma^r, \delta_0^{n-r-1}) = 1 \text{ und (für } i \neq 0\text{): } \mathfrak{b}_M(\gamma^r, \delta_i^{n-r-1}) = 0.^6 \tag{5}$$

Nun ist δ_0^{n-r-1} in $M^n - \gamma^r$ von den δ_i^{n-r-1} $(i \neq 0)$ unabhängig[7] (weil aus

$$\delta_0^{n-r-1} \sim \lambda^i \delta_i^{n-r-1} \text{ in } M^n - \gamma^r$$

folgen würde, daß

$$\mathfrak{b}_M(\gamma^r, \delta_0^{n-r-1}) = \lambda^i \mathfrak{b}_M(\gamma^r, \delta_i^{n-r-1})$$

ist, was auf Grund von (5) unmöglich ist). Die δ_i^{n-r-1} $(i \neq 0)$ können dagegen mit Homologierelationen in $M^n - \gamma^r$ verbunden sein.

Es sei dementsprechend

$$\delta_0^{n-r-1}, \delta_1^{n-r-1}, \ldots, \delta_h^{n-r-1} \quad (h \leqslant s) \tag{6}$$

ein Teilsystem von (4), welches eine Basis aller in $M^n - \gamma^r$ linear unabhängigen δ_i^{n-r-1} darstellt. Es ist somit (in $M^n - \gamma^r$)

$$\delta_{h+i}^{n-r-1} \sim \lambda_i^j \delta_j^{n-r-1} \quad (0 < i \leqslant s-h; 0 < j \leqslant h). \tag{6'}$$

4

Man wähle jetzt eine polyedrale Umgebung U von γ^r (in M^n) beliebig, jedoch so klein, daß C und sämtliche δ_i^{m-r-1} außerhalb von U liegen und so, daß alle Homologien (6′) auch in $M^n - U$ richtig bleiben. Man kann dann eine $B^r(U)$

$$\left.\begin{array}{ll} & \Gamma_0^r, \Gamma_1^r, \ldots, \Gamma_\sigma^r \qquad (7) \\ \text{und eine } B^{m-r-1}(R^m - U) & \\ & \Delta_0^{m-r-1}, \Delta_1^{m-r-1}, \ldots, \Delta_\sigma^{m-r-1} \qquad (8) \end{array}\right\} \sigma \geqslant s$$

so wählen, daß (7) mit (8) eindeutig verschlungen und

$$\Gamma_i^r = \gamma_i^r \quad (\text{für } 0 \leqslant i \leqslant s) \tag{9}$$

ist. Da unsere Basen *eindeutig* verschlungen sind, so folgt aus dem Satze I der erstern Mitteilung, daß

$$\Delta_i^{m-r-1} \sim 0 \quad (\text{für alle } i > s) \text{ in } R^m - \gamma^r \tag{10}$$

ist. Es existiert somit eine in U enthaltene ebenfalls polyedrale Umgebung U' von γ^r von der Beschaffenheit, daß ein in U' enthaltener Zyklus unmöglich mit einem Δ_{s+i}^{m-r-1} $(0 < i \leqslant \sigma - s)$, verschlungen sein kann; daraus folgt aber leicht, daß jeder Zyklus $\Gamma^r \subset U'$ einer linearen Kombination der γ_i^r $(0 \leqslant i \leqslant s)$, in U homolog ist.

Nun sind aber die δ_i^{n-r-1} $(i \leqslant h)$ in $M^n - \gamma^r$, also umsomehr in $M^n - U'$ linear unabhängig und können deshalb in eine $B^{n-r-1}(M^n - U')$ aufgenommen werden. Es sei

$$\Gamma_0^{n-r-1}, \Gamma_1^{n-r-1}, \ldots, \Gamma_t^{n-r-1} \quad (\Gamma_i^{n-r-1} = \delta_i^{n-r-1} \text{ für } i \leqslant h) \tag{11}$$

diese Basis. Man betrachte jetzt das Gebiet $R^m - (M^n - U')$ und eine mit (11) in R^m eindeutig verschlungene, in bezug auf M^n sich in allgemeiner Lage befindende $m - (n - r - 1) - 1 = m - n + r$-dimensionale Basis

$$\Gamma_0^{m-n+r}, \Gamma_1^{m-n+r}, \ldots, \Gamma_t^{m-n+r} \tag{12}$$

dieses Gebietes. Ich behaupte, daß das (mit Γ_0^{m-r-1} verschlungene) Element Γ_0^{m-n+r} dieser Basis allen Forderungen unseres Hilfssatzes genügt. In der Tat ist $\Gamma_0^{m-n+r} \cdot C = 0$, $\Gamma_0^{m-n+r} \cdot M^n$ dagegen nicht leer,[8] also (da Γ_0^{m-n+r} sich bezüglich M^n in allgemeiner Lage befindet) ist $\Gamma_0^{m-n+r} \cdot M^n$ ein in U' enthaltener Zyklus $\bar{\gamma}^r$, der, wie man leicht zeigt, in M^n homolog Null ist.

Ferner ist

$$\mathfrak{b}_{R^m}(\bar{\gamma}^r, \delta_i^{m-r-1}) = \mathfrak{b}_{R^m}(\Gamma_0^{m-n+r}, \delta_i^{n-r-1}) \tag{13}$$

(für $i \leqslant s$): es sei in der Tat $K_i^{m-r} \subset R^m$ irgendein durch δ_i^{m-r-1} begrenzter Komplex und K_i^{n-r} – der Durchschnitt von K_i^{m-r} und M^n; dann ist die linke Seite von (13) gleich der Anzahl (mod 2) der Schnittpunkte von K_i^{m-r} mit $\bar{\gamma}^r$, während die rechte Seite derselben Relation die (ebenfalls modulo 2 reduzierte) Anzahl der Schnittpunkte von K_i^{n-r} mit Γ_0^{m-n+r} angibt. Beide Zahlen sind aber mit der Anzahl (mod 2) der Schnittpunkte von K_i^{n-r} mit $\bar{\gamma}^r$ identisch.

Aus (13) folgt

$$\bar{\gamma}^r \sim \gamma^r \quad (\text{in } U); \tag{14}$$

In der Tat, da $\bar{\gamma}^r \subset U'$ is, so ist (in U) $\bar{\gamma}^r \sim \lambda^j \gamma_j^r$, $(0 \leqslant j \leqslant s)$, also ist

$$\mathfrak{b}_{R^m}(\Gamma_0^{m-n+r}, \delta_i^{n-r-1}) = \lambda^j \mathfrak{b}_{R^m}(\gamma_j^r, \delta_i^{m-r-1}) = \lambda^i \tag{14'}$$

da aber $\mathfrak{b}_{R^m}(\Gamma_0^{m-n+r}, \delta_i^{n-r-1})$ dann und nur dann von Null verschieden, also gleich 1 ist, wenn $i = 0$ ist (wegen der Formal (6'), die auch in $M^n - U$ richting ist, und weil für $0 \leqslant i \leqslant h$ $\delta_i^{n-r-1} = \Gamma_i^{m-r-1}$ ist), so ist auch $\lambda^0 = 1$, $\lambda^i = 0$ (für $i \neq 0$), womit die Homologie (14) und der Ganze Hilfssatz bewiesen sind.

5

Wir gehen jetzt zum Beweise des Satzes I über.

Es sei

$$C \subset M^n$$

und

$$\tilde{\gamma}_1^r, \tilde{\gamma}_2^r, \ldots, \tilde{\gamma}_{pr(C)}^r \quad \text{bzw.} \quad \tilde{\Gamma}_1^r, \tilde{\Gamma}_2^r, \ldots, \tilde{\Gamma}_{pr(M^n)}^r$$

die $B^r(C)$ bzw. $B^r(M^n)$. Dann ist für jedes $h \leqslant p^r(C)$: $\tilde{\gamma}_h^r \sim c_h^i \tilde{\Gamma}_i^r$ (in M^n). Der Rang der Matrix $\|c_h^i\|$ sei $u^r = u$; er ist gleich der Anzahl der in M^n unabhängigen r-dimensionalen Zyklen von C. Nun kann man durch eine geeignete lineare Substitution diese Matrix in die „Normalform" bringen, in der alle c_h^i für $h = i$, $i \leqslant u$ gleich 1 und alle übrigen Elemente gleich Null sind. Mann kann mit anderen Worten annehmen, daß unsere Basen die folgende Gestalt haben:

$$B^r(C) = \{\gamma_1^r, \gamma_2^r, \ldots, \gamma_u^r, \gamma_{u+1}^r, \ldots, \gamma_{u+q}^r\}, \quad u^r + q^r = p^r(C), \tag{15}$$

$$B^r(M^n) = \{\Gamma_1^r, \Gamma_2^r, \ldots, \Gamma_u^r, \Gamma_{u+1}^r, \ldots, \Gamma_{u+m}^r\}, \quad u^r + m^r = p^r(M^n), \tag{16}$$

$$\begin{cases} \gamma_i^r \sim \Gamma_i^r \ (\text{in } M^n) & \text{für } i \leqslant u \\ \gamma_{u+i}^r \sim 0 \ \ (\text{in } M^n) & \text{für } i \leqslant q. \end{cases} \tag{17}$$

Zu bemerken ist noch, daß kein Γ_{u+i}^r einem Teilkomplex von C in M^n homolog ist.

Man bette jetzt M^n in einem genügend hoch dimensionalen R^m als

Polyederkomplex ein und bestimme eine mit $B^r(C)$ eindeutig verschlungene $B^{m-r-1}(R^m - C)$

$$\delta_1^{m-r-1}, \ldots, \delta_u^{m-r-1}, \ldots, \delta_{u+q}^{m-r-1}. \tag{18}$$

Analog wird auch eine mit (16) eindeutig verschlungene $B^{m-r-1}(R^m - M^n)$

$$\Delta_1^{m-r-1}, \ldots, \Delta_u^{m-r-1}, \ldots, \Delta_{u+m}^{m-r-1} \tag{19}$$

konstruiert.

Man überzeugt sich leicht, daß $\mathfrak{b}_{R^m}(\Delta_i^{m-r-1}, \gamma_j^r)$, $(i \leqslant u, j$ beliebig) für $j = i$ gleich 1, für $j \neq i$ gleich 0 ist. Das System

$$\Delta_1^{m-r-1}, \ldots, \Delta_u^{m-r-1}, \delta_{u+1}^{m-r-1}, \ldots, \delta_{u+q}^{m-r-1} \tag{20}$$

ist also eine mit (15) eindeutig verschlungene $B^{m-r-1}(R^m - C)$; mit anderen Worten, wir dürfen voraussetzen, daß

$$\delta_i^{m-r-1} = \Delta_i^{m-r-1} \quad \text{für } i \leqslant u \tag{21}$$

ist.

Da wir selbstverständlich annehmen, daß alle Komplexe (18), (19), (20) sich in bezug auf M^n in allgemeiner Lage befinden, und da, wie leicht ersichtlich (vgl. Fußnote 8), $\delta_{u+i}^{m-r-1} \cdot M^n$ jedenfalls nicht leer ist, können wir annehmen, daß $\delta_{u+i}^{m-r-1} \cdot M^n$ ein Zyklus δ_{u+i}^{n-r-1} ist, der, wie man leicht zeigt, in M^n homolog Null ist. Somit ist die Verschlingungsmatrix $\| \mathfrak{b}_{M^n}(\gamma_{u+i}^r, \delta_{u+j}^{n-r-1}) \|$, $1 \leqslant i \leqslant q$, $1 \leqslant j \leqslant q$ definiert und es ist leicht zu zeigen, daß sie die Einheitsmatrix ist.

Da die γ_{u+i}^r, $1 \leqslant i \leqslant q$ offenbar eine Nullbasis von C inbezug auf M^n darstellen, wird Satz I bewiesen sein, wenn wir zeigen, daß die δ_{u+i}^{n-r-1}, $1 \leqslant i \leqslant q$ eine Nullbasis von $M^n - C$ bilden. Da $\| \mathfrak{b}_{M^n}(\gamma_{u+i}^r, \delta_{u+j}^{n-r-1}) \|$ die Einheitsmatrix ist, so sind die δ_{u+j}^{n-r-1} in $M^n - C$ gewiß unabhängig; es genügt also zu zeigen, daß aus $\delta^{n-r-1} \sim 0$ (in M^n), $\delta^{n-r-1} \subset M^n - C$, eine Homologie

$$\delta^{n-r-1} \sim \lambda^i \delta_{u+i}^{n-r-1} \ (\text{in } M^n - C)$$

folgt.

Es sei U eine zu C fremde Umgebung von δ^{n-r-1} in M^n. Vermöge unseres Hilfssatzes gibt es einen $\delta^{m-r-1} \subset R^m - C$, so daß

$$U \supset \bar{\delta}^{m-r-1} \cdot M^n = \bar{\delta}^{n-r-1} \sim \delta^{n-r-1} \ (\text{in } U).$$

Nun ist aber

$$\bar{\delta}^{m-r-1} \sim \lambda^i \delta_i^{m-r-1} \ (\text{in } R^m - C), \quad 1 \leqslant i \leqslant p^r(C),$$

d. h. es gibt einen mit $\bar{\delta}^{m-r-1} + \lambda^i \delta_i^{m-r-1}$ begrenzten $K^{m-r} \subset R^m - C$. Dann ist aber

$$K^{n-r} = K^{m-r} \cdot M^n \to M^n \cdot \bar{\delta}^{m-r-1} + \lambda^i (M^n \cdot \delta_i^{m-r-1}) \text{ in } M^n - C;$$

da aber für $i \leqslant u$

$$\delta_i^{m-r-1} = \Delta_i^{m-r-1} \subset R^m - M^n \text{ und } M^n \cdot \delta_{u+i}^{m-r-1} = \delta_{u+i}^{n-r-1}$$

ist, so ist

$$K^{n-r} \to \bar{\delta}^{n-r-1} + \lambda^{u+i}\delta_{u+i}^{n-r-1},\ 1 \leqslant i \leqslant q^r, \text{ in } M^n - C,$$

woraus $\bar{\delta}^{n-r-1} \sim \lambda^{u+i}\delta_{u+i}^{n-r-1}$, (in $M^n - C$) und also $\delta^{n-r-1} \sim \lambda^{u+i}\delta_{u+i}^{n-r-1}$ (in $M^n - C$) folgt. Der Satz I (und somit auch Ia und Ie) ist hiermit bewiesen.

6

Beweis des Satzes II. Wir halten an den Bezeichnungen des vorigen Paragraphen fest und betrachten zunächst die Δ_i^{m-r-1}; es sei K_i^{m-r} irgendein durch Δ_i^{m-r-1} in R^m begrenzter Komplex, $(i = 1, 2, \ldots, p^r(M^n))$.

Es sei $i > u$; dann kann Δ_i^{m-r-1} mit keinem γ_{u+j}^r verschlungen sein (weil Δ_i^{m-r-1} zu M^n fremd und γ_{u+j}^r in $M^n \sim 0$ ist); Δ_i^{m-r-1} kann aber auch mit keinem γ_j^r, $j \leqslant u$ verschlungen sein, weill dann Δ_i^{m-r-1} auch mit Γ_j^r verschlungen wäre (was der Eindeutigkeit der Verschlingung von (19) und (16) widersprechen würde). Da somit Δ_i^{m-r-1} mit keinem Element der $B^r(C)$ verschlungen ist, so ist $\Delta_i^{m-r-1} \sim 0$ (in $R^m - C$), m. a. W. *mann kann die* K_{u+i}^{m-r}, $i = 1, 2, \ldots, m^r$ *außerhalb von* C *wählen.*

Wir setzen nun

$$K_i^{m-r} \cdot M^n = \Gamma_i^{n-r} \quad (i = 1, 2, \ldots, p^r(M^n)),$$

(wobei

$$\Gamma_{u+i}^{n-r} \subset M^n - C, \quad (i = 1, 2, \ldots, m^r) \tag{22}$$

ist). Da die Kroneckerschen Charakteristiken $\chi_{K^n}(\Gamma_i^{n-r}, \Gamma_j^r)$ und $\chi_{R^m}(K_i^{m-r}, \Gamma_j^r)$ gleich sind, so ist $\|\chi(\Gamma_i^{n-r}, \Gamma_j^r)\|$ die Einheitsmatrix. Daraus folgt leicht, daß *die* Γ_i^{n-r} *in* M^n *unabhängig sind.*

(Es sei in der Tat

$$c^i \Gamma_i^{n-r} \sim 0 \quad (\text{in } M^n).$$

Dann ist *für jedes* j

$$0 = \chi(c^i \Gamma_i^{n-r}, \Gamma_j^r) = c^i \chi(\Gamma_i^{n-r}, \Gamma_j^r) = c^j).$$

Die Anzahl der Γ_i^{n-r} ist aber $p^r(M^n) = p^{n-r}(M^n)$ (für $0 < r < n$), sie bilden also eine $B^{n-r}(M^n)$.

Es sei jetzt ein Zykel $\Gamma^{n-r} \subset M^n - C$ beliebig gegeben; es ist:

$$\Gamma^{n-r} \sim c^h \Gamma_h^{n-r} = c^i \Gamma_i^{n-r} + c^{u+i}\Gamma_{u+i}^{n-r} \quad (\text{in } M^n), \tag{23}$$

$$1 \leqslant h \leqslant p^{n-r}(M^n), \quad 1 \leqslant i \leqslant u^r, \quad 1 \leqslant i \leqslant m^r.$$

Wir betrachten ein $i \leqslant u$; da $\|\chi(\Gamma_i^{n-r}, \Gamma_j^r)\|$ die Einheitsmatrix ist, so ist

$\chi(\Gamma^{n-r}, \Gamma_i^r) = c^i$; da aber andrerseits $\Gamma_i^r \sim \gamma_i^r$ (in M^n) ist, so ist $\chi(\Gamma^{n-r}, \Gamma_i^r) = \chi(\Gamma^{n-r}, \gamma_i^r)$, also $c_i = \chi(\Gamma^{n-r}, \gamma_i^r) = 0$ (weil $\Gamma^{n-r} \cdot C = 0$ ist). So sind in (23) *alle* c_i, $i \leqslant u$, *gleich Null.* Wir sind also zu folgenden Resultate gelangt:

Es gibt m^r in M^n unabhängige, in $M^n - C$ enthaltene Zyklen Γ_{u+i}^{n-r}, $i = 1, 2, \ldots, m^r$ von der Beschaffenheit, daß jeder $\Gamma^{n-r} \subset M^n - C$ einer linearen Kombination dieser Zyklen in M^n homolog ist.

Wir setzen jetzt $n - r = k$; es seien

$$\Gamma_1^k, \Gamma_2^k, \ldots, \Gamma_{m^{n-k}}^k, \quad \Gamma_i^k \subset M^n - C \tag{24}$$

die soeben erwähnten Zyklen (wenn also ein $\Gamma^k \subset M^n - C$ gegeben ist, so ist:

$$\Gamma^k \sim c^i \Gamma_i^k, \quad 1 \leqslant i \leqslant m^{n-k} \text{ in } M^n).$$

Es sei ferner:

$$\gamma_1^k, \gamma_2^k, \ldots, \gamma_q^k, \quad q = q^k(M^n - C) = q_{M^n}^{n-k-1}(C) \tag{25}$$

eine k-dimensionale Nullbasis von $M^n - C$.

Ich behaupte, daß *die* (24) *zusammen mit den* (25) *eine* $B^k(M^n - C)$ *bilden.* Da die (24) in M^n und die (25) in $M^n - C$ unabhängig, andrerseitzs aber die (25) in M^n homolog Null sind, so sind die (24) und (25) zusammen in $M^n - C$ unabhängig. Es bleibt uns also nur übrig zu zeigen, daß jeder $\Gamma^k \subset M^n - C$ einer Linearkombination der (24) und (25) *in* $M^n - C$ homolog ist.

Nun ist zunächst

$$\Gamma^k \sim c^i \Gamma^k \quad (\text{in } M^n),$$

also

$$\Gamma^k + c^i \Gamma_i^k \sim 0 \quad (\text{in } M^n);$$

daraus folgt aber, daß

$$\Gamma^k + c^i \Gamma_i^k \sim \tilde{c}^i \gamma_i^k \quad (\text{in } M^n - C)$$

und somit

$$\Gamma^k \sim c^i \Gamma_i^k + \tilde{c}^i \lambda_i^k \quad (\text{in } M^n - C),$$

womit bewiesen ist, daß (24) und (25) eine $B^k(M^n - C)$ bilden. Daraus folgt aber, daß

$$p^k(M^n - C) = q_{M^n}^{n-k-1}(C) + m^{n-k}$$

ist, d. h. (da $m^{n-k} = p^k(M^n) - u^{n-k}(C)$ ist):

$$p^k(M^n - C) = q^{n-k-1}(C) + (p^k(M^n) - u^{n-k}(C)),[9]$$

was bei $k = n - r - 1$ in

$$p^{n-r-1}(M^n - C) = q^r(C) + (p^{n-r-1}(M^n) - u^{r+1}(C))$$

übergeht, womit der Satz II bewiesen ist.

Korollar. In jeder Mannigfaltigkeit, deren sämtliche Bettischen Zahlen verschwinden, gilt der Alexandersche Dualitätssatz.

Notes

1. Literaturhinweise – in der ersten Mitteilung.

2. Eine andere Verallgemeinerung der Alexanderischen Formel hat neuerdings Lefschetz (*Proceed. Nat. Ac. Sciences U.S.A.*, **13** (Aug. 1927), S. 614) gegeben. In der Lefschetzschen Formel treten jedoch Glieder auf, die sowohl von C als auch von $M^n - C$ abhängen; sie scheint mir deshalb nicht geeignet zu sein, als endgültige Lösung des Einbettungsproblem betrachtet zu werden.

3. Eigentlich sollte man schreiben $q^r_{M^n}(C)$ usw., weil diese Zahl keine Invariante des Komplexes C, sondern eine von der Lage von C abhängende Zahl ist; da aber Mißverständnisse ausgeschlossen zu sein scheinen, werden wir den sich auf M^n beziehenden Index unterdrücken.

4. Der Begriff „der Verschlingungszahl $\mathfrak{v}_{M^n}(\gamma^k, \delta^{n-k-1})$ zweier Zyklen γ^k und δ^{n-k-1} in der Mannifaltigkeit M^n" beruht in unmittelbar klarer Weise auf dem Begriffe der Kroneckerschen Charakteristik (C^{n-k}, γ^k) zweier Komplexe in M^n; letzterer Begriff ist von Lefschetz (*Trans. Amer. Math. Soc.*, **28** (1926), S. 1–49) eingehend untersucht worden. In unserem Falle der Charakteristik „mod 2" würden sich die Lefschetzschen Betrachtungen übrigens wesentlich einfacher darstellen lassen.

5. Man zeigt im allgemeinen: wenn Γ^k sich zu M^n in R^m in allgemeiner Lage befindet und also M^n in einem Γ^{k+n-m} schneidet, so ist

$$\Gamma^{k+n-m} \sim 0 \text{ (in } M^n).$$

In der Tat: Γ^k ist ~ 0 in R^m, also existiert ein durch Γ^k begrenzter $K^{k+1} \subset R^m$, von dem man annehmen kann, daß er sich auch in allgemeiner Lage zu M^n befindet; dann ist aber $K^{k+1} \cdot M^n = K^{k+n+1-m} \to \Gamma^{k+n-m}$, woraus folgt, daß $\Gamma^{k+n-m} \sim 0$ (in M^n), w.z.b.w. Diese Schlußweise wird im folgenden öfters benutzt, ohne explizite erwähnt zu werden.

6. Es sei in der Tat K^{r+1} ein mit γ^r begrenzter Teilkomplex M^n; dann ist sowohl $\mathfrak{v}_M(\gamma^r, \delta_i^{n-r-1})$ als $\mathfrak{v}_{R^m}(\gamma^r, \delta_i^{m-r-1})$ der Anzahl (mod 2) der Schnittpunkte von K^{r+1} mit δ_i^{m-r-1} gleich; also ist $\mathfrak{v}_M(\gamma^r, \delta_i^{n-r-1}) = \mathfrak{v}_{R^m}(\gamma^r, \delta_i^{m-r-1})$, womit unsere Behauptung bewiesen ist.

7. Also insbesondere ist δ_0^{n-r-1} nicht Null.

8. $\Gamma_0^{m-n+r} \cdot M^n$ ist nicht leer, weil $\Gamma_0^{n-r-1} \sim 0$ (in M^n) und Γ_0^{m-n+r} mit Γ_0^{n-r-1} verschlungen ist, folglich mit einem beliebigen durch Γ_0^{n-r-1} begrenzten Teilkomplex von M^n gemeinsame Punkte hat.

9. Diese Formel ist bewiesen für $k > 0$, aber es ist leicht einzusehen, daß sie auch für $k = 0$ richtig ist.

Topologie. Sur une Hypothèse Fondamentale de la Théorie de la Dimension*

Compte Rendus Académie des Sciences Paris (1930)
190(19): 1105–7

1

Soient X et Y deux espaces quelconques, où les voisinages se trouvent définis pour chaque point M.[1] On appelle *produit topologique* de ces espaces l'espace $Z = X \times Y$ formé de la façon suivante: les points z de Z sont les couples (x, y) de points de X et de Y; pour obtenir un voisinage d'ailleurs arbitraire, d'un point $z_0 = (x_0, y_0)$ de Z, on prend tous les points $z = (x, y)$ tels que x et y appartiennent respectivement aux voisinages arbitrairement donnés de x_0 et de y_0.

2

Depuis les origines de la théorie de la dimension, on se proposait de démontrer la loi de l'*addition des dimensions*, c'est-à-dire la formule: $\dim (F' \times F'') = \dim F' + \dim F''$, quels que soient les ensembles fermés F' et F'' (situés dans un espace euclidien à un nombre quelconque de dimensions). On montre facilement que cette loi se trouve vérifiée si l'on définit *dimension* comme la dimension *module m* au sens de M. Alexandroff et cela quel que soit m.[2] Il en résulte que la formule d'addition des dimensions reste exacte (pour la dimension au sens de M. Brouwer), quand il

* Note de M. L. Pontrjagin, Séance du 23 avril, 1930.

s'agit d'ensembles situés dans l'espace tridimensionnel. Je me propose de construire dans cette Note deux ensembles fermés F' et F'' situés dans l'espace à quatre dimensions et tels que $\dim F' = \dim F'' = 2$, tandis que $\dim (F' \times F'') = 3$. *Le problème général d'addition des dimensions se résout donc par la négative, dans la théorie de dimension de Brouwer–Urysohn–Menger.*

3

Considérons un cylindre de révolution A comme une surface ayant pour frontière complète les deux circonférences-bases C_1 et C_2' Divisons C_1 en k arcs égaux et identifions les points correspondants de ces arcs (c'est-à-dire les points dont la coordonnée angulaire prend des valeurs différant de multiples de $2\pi/k$); cette opération (réalisable sans singularités dans l'espace à quatre dimensions) transforme A en une surface A_k que nous appellerons *feuillet de Möbius* mod k; la circonférence C_2 sera dite la frontière (géométrique) du feuillet et désignée par A'_k.

Soit C une surface située dans E et formée d'un nombre fini de triangles (C est donc ce qu'on appelle complexe à deux dimensions). Considérons un quelconque des triangles constituant C, soit T, et traçons à l'intérieur de T un petit triangle τ; enlevons de T l'intérieur de τ et remplaçons-le par un feuillet de Möbius mod k de façon que la frontière du feuillet soit identifiée avec celle du triangle τ et qu'aucune singularité ne surgisse; cette opération effectuée dans chaque triangle de C, cette dernière surface se transformera en une surface que nous désignerons par $f^h(C)$ et qu'on peut évidemment supposer polyédrique. On peut d'ailleurs s'arranger de sorte que tous les feuillets de Möbius soient situés dans des voisinages donnés d'avance de triangles correspondants. Désignons maintenant par C_0^h la surface formée par un seul triangle; C_v^k étant supposé construit, posons $C_{v+1}^k = f^k(C)$, la surface C_v^k étant préalablement décomposée en des triangles suffisamment petits. On effectue toute la construction d'une telle manière que la suite $C_0, C_1^k, \ldots, C_v^k, \ldots$ soit convergente; la limite de cette suite sera désignée par F^k. On démontre sans peine *que, pour l'ensemble F, la dimension* mod k au sens de M. Alexandroff, de même que la dimension brouwerienne, sont égales à 2; $\Delta^k(F^k) = \dim F^k = 2$, tandis que, pour tout $h \neq k$, $\Delta^k(F^k) = 1$. Il en résulte que *la définition de dimension due à M. Alexandroff donne pour des modules différents des invariants topologiques différents; tous ces invariants diffèrent d'ailleurs de la dimension au sens de Brouwer–Urysohn–Menger.*[3]

4

Soient maintenant p et q deux nombres premiers entre eux. Démontrons que dim $(F^p \times F^q) = 3$.

On démontre d'abord en s'appuyant sur un resultat fondamental de M. Hopf[4] le lemme suivant: Soit donnée une représentation continue φ de $B = A_p \times A_q$ sur l'ensemble $\varphi(B)$ situé dans E^4; il existe une autre représentation continue φ' coïncidant avec φ sur $(A'_p \times A_q) + (A'_q \times A_p)$ et telle qu'un point p de E^4 donné d'avance soit étranger à $\varphi'(B)$. [C'est ce lemme qui tombe en défaut quand p et q ne sont pas premiers entre eux, auquel cas on aurait dim $(F^p \times F^q) = 4$.]

Cela posé, considérons $F = F^p \times F^q$ comme sous-ensemble de l'espace E^2; soient H et H' deux hyperplans à 4 dimensions de E^2, n'ayant qu'un seul point p en commun. Soit $B = (A_p \times A_q)$ un des éléments constituant $Q_v = C_v^p \times C_v^q$. Projetons Q_v sur H par des hyperplans parallèles à H'; Q_v étant constitué des éléments B de la forme $A_p \times A_q$, une représentation continue φ se trouve ainsi définie pour chacun de ces éléments; en appliquant le lemme on peut remplacer chacune de ces représentations par une autre φ'' ne touchant pas au point p. Il en résulters une déformation continue de Q_v tout entier en un complexe étranger à H'; celle-ci pourra étre prolongée ensuite à un certain voisinage de Q_v. En appliquant ce procédé aux hyperplans formant les faces à 4 dimensions d'un cubillage donné de E, on obtiendra une déformation de F en un ensemble ne touchant pas les faces à 4 dimensions dudit cubillage; par un choix convenable de ce cubillage (et de l'indice v), cette déformation peut étre rendue aussi petite que l'on voudra; on en tire facilement que dim $F = 3$.

(Extrait des *Comptes rendus des séances de l'Académie des Sciences*, **190**, p. 1105, séance du 12 mai 1930.)

Notes

1. Voir par exemple le livre de M. Fréchet sur les *Espaces abstraits* (Collection Borel).

2. Voir la Note ci-dessus de M. Alexandroff.

3. Il est intéressant de remarquer que si l'on pose, dans la construction précédente, $C_{2v} = f^p(C_{2v-1})$, $C_{2v+1} = f^q(C_{2v})$, p et q étant premiers entre eux, l'ensemble-limite F serait de dimension 1 au sens de M. Brouwer (donc aussi au sens de M. Alexandroff, pour n'importe quel module k).

4. *Math. Annalen*, **100**, 1928, p. 590, théor' IX, *a*.

Über den Algebraischen Inhalt Topologischer Dualitätssätze

Mathematische Annalen (1931) 105(2): 165–205

Zusammenfassung

U und V seien zwei Abelsche Gruppen mit endlich-vielen Erzeugenden, M eine (endliche oder unendliche) zyklische Gruppe. Die Gruppen U und V bilden ein *Gruppenpaar* in bezug auf M, wenn je zwei Elementen u und v von U bzw. V ein Element m von M, das *Produkt* der beiden Elemente u und v, zugeordnet ist und dabei die Distributivgesetze in bezug auf die Addition (als Gruppenverknüpfung in U und V), d. h. die Relationen $(u_1 + u_2) \cdot v = u_1 \cdot v + u_2 \cdot v$ und $u \cdot (v_1 + v_2) = u \cdot v_1 + u \cdot v_2$ gelten. Das Gruppenpaar U, V heißt *primitiv*, wenn zu jedem von Null verschiedenen Element der einen Gruppe ein solches Element der zweiten gefunden werden kann, daß das Produkt der beiden Elemente von Null verschieden ist.

Sodann gilt folgender *algebraischer Dualitätssatz*: zwei Gruppen, die ein primitives Gruppenpaar bilden, sind isomorph.

Wenn U und V die r- und $(n - r)$-dimensionale Bettische Gruppe einer n-dimensionalen geschlossenen Mannifaltigkeit sind und man als $u \cdot v$ die Schnittzahl (bzw. die Schnittzahl nach einem festen Modul μ) der zugehörigen Zyklen versteht, so bilden U und V ein primitives Gruppenpaar (wobei M entweder die Gruppe aller ganzen Zahlen oder der Restklassen mod μ ist), sind also isomorph (Dualitätssatz von Poincaré).

Wenn man für U und V die r- bzw. $(n - r - 1)$-dimensionale Bettische Gruppe eines Komplexes $K \in R^n$ bzw. seiner Komplementärmenge $R^n - K$ wählt und anstatt der Schnittzahlen die Verschlingungszahlen ins Auge faßt, bilden die Gruppen U und V wiederum ein primitives Gruppenpaar, sind also ebenfalls isomorph (Dualitätssatz von Alexander).

In analoger Weise erhält man alle bis jetzt bekannten topologischen Dualitätssätze (die alls Verallgemeinerungen der obigen beiden anzusprechen sind).

Einleitung

Poincaré hat 1895 in seiner berühmten Arbeit „Analysis Situs"[1] das heute seinen Namen tragende Dualitätsgesetz entdeckt, nämlich die Tatsache, daß bei jedem r die r-te und $n-r$-te Bettische Zahl einer orientierbaren n-dimensionalen Mannigfaltigkeit einander gleich sind. Ungefähr um dieselbe Zeit hat Jordan zum erstenmal seinen Kurvensatz ausgesprochen. Damals aber ahnte kein Mensch, daß diese beiden so total verschiedenen Sätze in einen und denselben Ideenkreis gehören und daß insbesondere der zweite zu ganz weiten und äußerst bedeutungsvollen Verallgemeinerungen führen wird. Der Weg dieser Verallgemeinerungen is in aller Kürze der folgende.

1912 hat Brouwer[2] den Satz von der Invarianz der geschlossenen Kurve bewiesen, der den Jordanschen Satz als einen sehr speziellen Fall enthält und allgemein besagt, daß die Anzahl der Gebiete, die eine abgeschlossene Menge in der Ebene bestimmt, nur von den topologischen Eigenschaften dieser Menge selbst abhängt. Auf diese Weise wurde zum erstenmal die prinzipielle Möglichkeit gegeben, den Begriff der geschlossenen Kurve von der Ebene loszulösen und invariant zu definieren. Damit wurde schon der Weg der Übertragung der Invarianten der sogenannten kombinatorischen Topologie all allgemeinste abgeschlossene Mengen angedeutet, ein Weg, der in den letzten fünf Jahren zu einer Fülle neuer Erkenntnisse geführt hat, die man in erster Linie Alexandroff, Lefschetz und Vietoris verdankt.[3] *Alle diese Ergebnisse lassen sich in das allgemeine Dualitätsgesetz für abgeschlossene Mengen einordnen, welchem das dritte Kapitel der vorliegenden Arbeit gewidmet ist.*

Aber es handelt sich bei diesen Untersuchungen nicht nur um Übertragung von für elementare Gebilde bewiesenen Sätzen auf Gebilde allgemeinster Art, sondern auch um eine Verallgemeinerung in bezug auf die dimensionellen Verhältnisse: das, was der Jordansche Satz in bezug auf die Dimensionen 2 (die Ebene) und 1 (die Kurve) behauptet, wird mutatis mutandis für n und r formuliert und bewiesen. Den ersten Schritt in dieser Richtung verdankt man Lebesgue, der 1911 als erster erkannt hat,[4] daß die Eigenschaft einer n-dimensionalen Mannigfaltigkeit (also für $n=1$ einer Jordanschen Kurve), den $n+1$-dimensionalen Raum zu zerlegen, ein Spezialfall der Eigenschaft einer r-dimensionalen Mannigfaltigkeit im n-dimensionalen Raum, eine $n-r-1$-dimensionale Verschlingung zuzulassen, ist. Auf diese Weise hat Lebesgue einen Teil des n-dimensionalen Jordanschen Satzes bewiesen; den Beweis der übrigbleibenden Teile sowie eine vollständige und invariant begründete Theorie der Verschlingung hat um dieseble Zeit Brouwer erbracht.[5]

Einen wesentlich neuen und die weitesten Perspektiven eröffnenden Schritt verdankt man Alexander,[6] der in einer ungewöhnlich einfachen und eleganten Weise bewiesen hat, daß die $(n-r-1)$-te Bettische Zahl des Komplementärraumes zu einem beliebigen Komplexe (im R^n)[7] gleich der r-ten Bettischen Zahl des Komplexes selbst ist (*Alexanderscher Dualitätssatz*). Das war eine gewaltige Verallgemeinerung aller bis damals bekannten Sätze aus dem Ideenkreise des Jordanschen Satzes, insofern sie sich auf topologische Bilder von Polyedern (und nicht auf allgemeinere abgeschlossene Mengen) bezogen. Die Übertragung des Alexanderschen Dualitätssätzes auf beliebige abgeschlossene Mengen wurde dann von Alexandroff[8] 1927 und ungefähr um dieselbe Zeit von Lefschetz und Frankl geliefert.[9] Lefschetz hat dabei Resultate erhalten, die sich auf den Fall der abgeschlossenen Teilmengen beliebiger Mannigfaltigkeiten beziehen; das wesentliche Hilfsmittel, das er gebraucht, ist eine weitere Ausbildung der Verschlingungstheorie, d. h. letzten Endes der Theorie der sogenannten Kroneckerschen Schnittzahlen, die er in einer solchen Allgemeinheit entwickelt, wie für die neueren Probleme der Topologie nur irgend erwünscht werden könnte.

Andererseits hat Veblen[10] bereits 1923 die Theorie der Schnittzahlen zum Beweise und zur Verallgemeinerung des Poincaréschen Dualitätssatzes angewandt: er zeigt nämlich, daß man die r-te und $n-r$-te Bettische Basis einer n-dimensionalen geschlossenen Mannigfaltigkeit stets so wählen kann, daß die Matrix der Schnittzahlen der Elemente der beiden Basen die Einheitsmatrix ist, eine Tatsache, die den Satz von Poincaré offenbar enthält und wesentlich verallgemeinert. Wenn man den so formulierten Poincaré–Veblenschen Dualitätssatz mit einer Verallgemeinerung des Alexanderschen Dualitätssatzes vergleicht, welche behauptet, daß man die r-dimensionale Bettische Basis eines Komplexes im R^n und die $n-r-1$-dimensionale Bettische Basis des Komplementärraumes R^n-K stets so wählen kann, daß die Matrix der Verschlingungszahlen der Elemente der beiden Basen die Einheitsmatrix ist,[11] springt eine gewisse Analogie zwischen diesen beiden Sätzen ohne weiteres in die Augen.

In der vorliegenden Arbeit wird diese Analogie vollständig geklärt, indem die beiden Dualitätssätze – der Alexandersche sowie der Poincaré–Veblensche – auf die Anwendung eines und desselben rein algebraischen Prinzips auf die Bettischen Gruppen der entsprechenden Dimensionen zurückgeführt werden. Dieses algebraische Prinzip besteht darin, daß man für zwei Abelsche Gruppen U und V (die man sich als additive Gruppen denkt, d. h. die Gruppenoperation als Addition deutet) eine neue Operation, die Multiplikation eines beliebigen Elementes u von U mit einem beliebigen Element v von V einführt, wobei das Produkt $u \cdot v$ stets ein Element einer dritten Gruppe, des Moduls M ist; M ist dabei eine endliche oder unendliche zyklische Gruppe.[12]

Die Einführung des soeben geschilderten Multiplikation verwandelt das System der beiden Gruppen U und V in ein *Gruppenpaar* nach dem Modul M; dabei heißt ein Gruppenpaar primitiv, wenn es zu jedem von Null verschiedenen Element u bzw. v der einen Gruppe ein Element w der anderen Gruppe gibt, derart, daß $u \cdot w$ bzw. $w \cdot v$ von Null verschieden ist.

Der Hauptsatz von den primitiven Gruppenpaaren besteht darin, daß die beiden Gruppen eines solchen Paares untereinander isomorph sind. Nun ist ja in den letzteren Jahren ziemlich allgemein erkannt worden, daß nicht die Bettischen Zahlen, sondern darüber hinaus die Bettischen Gruppen den Hauptgegenstand algebraisch topologischer Untersuchungen bilden[13] und daß man dabei auch die sogenannten Bettischen Gruppen modulo μ zu berücksichtigen hat. Um terminologischen Verwirrungen aus dem Wege zu gehen, stelle ich hier kurz diese Grundbegriffe zusammen. Die r-dimensionalen orientierten Teilkomplexe eines gegebenes Komplexes sind Linearformen mit ganzzahlige Koeffizienten in den r-dimensionalen Elementen des Komplexes; sie bilden (in bezug auf die Addition) eine Abelsche Gruppe mit endlichvielen Erzeugenden, welche L^r heißen möge. Wenn man die Koeffizienten in den obigen Linearformen modulo einer ganzen Zahl $\mu > 1$ reduziert, ensteht die Gruppe L^r_μ, die Gruppe aller Teilkomplexe mod μ. Für jeden Teilkomplex ist sein Rand als algebraische Summe der Ränder seiner Elemente definiert,[14] ebenso ist für einen Teilkomplex mod μ der Rand mod μ definiert. Teilkomplexe, die den Rand Null haben, heißen *Zyklen*; dergleichen für Teilkomplexe mod μ. Die Zyklen bzw. die Zyklen mod μ bilden eine Untergruppe Z^r bzw. Z^r_μ von L^r bzw L^r_μ. Die Gruppe Z^r bzw. Z^r_μ enthält eine Untergruppe $\hat{H}^r$ bzw. $\hat{H}^r_\mu$ derjenigen Zyklen, die als Ränder (bzw. als Ränder mod μ) von (r + 1-dimensionalen) Teilkomplexen (bzw. Teilkomplexen mod μ) auftreten. Die Gruppe $\hat{H}^r$ bzw. $\hat{H}^r_\mu$ soll schlechtweg die Gruppe der berandenden r-dimensionalen Zyklen (bzw. der berandenden Zyklen mod μ) heißen. Die Faktorgruppe $Z^r | \hat{H}^r$ heißt die *volle r-dimensionale Bettische Gruppe* des gegebenen Komplexes, während die Gruppe $Z^r_\mu | \hat{H}^r_\mu$ die *Bettische Gruppe mod* μ heißt. Die volle Bettische Gruppe is direkte Summe zweier Untergruppen: der *Torsionsgruppe*, die von allen Elementen endlicher Ordnung erzeught wird, und der *reduzierten* Bettischen Gruppe, die von den Elementen unendlicher Ordnung der vollen Bettische Gruppe erzeugt wird. Einfachheitshalber bezeichnen wir die reduzierte Bettische Gruppe als *die Bettische Gruppe modulo Null*, so daß jetzt die Zahlen 0, 2, 3, . . . als Werte von μ auftreten.[15]

Nach diesen Vorbereitungen können wir die von uns erzielte Verallgemeinerung des Poincaréschen Dualitätssatzes einfach so aussprechen: *die r-te und n − r-te Bettische Gruppe mod* μ *bilden in bezug auf die zyklische Gruppe von der Ordnung* μ *as Modul*[12] *ein primitives Gruppenpaar. Als Produkt zweier Elemente ist die Schnitzahl der in Frage kommenden Zyklen zu betrachten, wobei im Falle* $\mu \neq 0$ *diese Schnittzahl modulo* μ *zu reduzieren ist.* In einer ganz analogen Weise erhält man auch den Alexanderschen Dualitätssatz in folgender Form: wenn K ein im R^n liegender Komplex ist, bilden die r-te Bettische Gruppe von K und die $n - r - 1$-te Bettische Gruppe von $R^n - K$ ein primitives Gruppenpaar, wenn man als Produkt die Verschlingungszahl der in Frage kommenden Zyklen betrachtet. Dabei gelten in bezug auf die verschiedenen Moduln dieselben Verabredungen wie im Falle des Poincaréschen Dualitätssatzes.[16]

Den soeben ausgesprochenen Satz nenne ich *den Alexanderschen Satz im engeren Sinne*: er bezieht sich auf Komplexe, die im R^n liegen. Wir untersuchen aber auch den viel allgemeineren Fall eines Komplexes, welcher in eine beliebige M^n eingebettet ist. Auch hier ergibt sich eine vollständige Lösung des Problems; den entsprechenden Satz nenne ich den *Alexanderschen Dualitätssatz im weiteren Sinne*; er ist als eine Verallgemeinerung der von mir für den Fall „modulo 2“ schon früher bewiesenen Formel zu betrachten.[17] Nebenbei sei bemerkt, daß in der ganzen vorliegenden Arbeit der Mannigfaltigkeitsbegriff im viel allgemeineren Sinne verstanden wird, als bis jetzt üblich: es werden nämlich durchweg die sogenannten h-Mannigfaltigkeiten betrachtet, deren Definition ungefähr gleichzeitig von verschiedenen Autoren, darunter Alexander, van Kampen, Vietoris und dem Verfasser, als eine allein auf dem Homologiebegriff beruhende und deshalb leicht as invariant erkennbare Verallgemeinerung des klassischen Mannigfaltigkeitsbegriffes gefunden und im Sektion 1 des zweiten Kapitels wiedergegeben ist.

Nachdem des sozusagen klassische Fall der in Mannigfaltigkeiten eingebetteten Komplexe erledigt ist, wende ich mich zu dem Fall einer beliebigen abgeschlossenen Menge. Mann könnte auch hier sofort den allgemeinsten Fall einer abgeschlossenen Menge in einer beliebigen Mannigfaltigkeit behandeln („Fall F in M^n“). Da aber all prinzipiellen Schwierigkeiten algebraischer Natur bereits im Falle „K in M^n“ und alle mengentheoretischen Schwierigkeiten im Fall „F in R^n“ auftreten, habe ich mich, um technische Komplikationen zu vermeiden, auf den letzteren Fall beschränkt. Der Fall der abgeschlossenen Mengen wird den algebraischen Methoden dieser Arbeit dadurch zugänglich gemacht, daß man sich konsequent der Darstellung einer abgeschlossenen Menge mittels der Alexandroffschen Projektionsspektren bedient.[18] Man hat auf diese Weise für jede

Dimension r anstatt einer einzigen Abelschen Gruppe eine Folge solcher Gruppen, von denen jede endlichviele Erzeugenden besitzt; diese Gruppen sind die r-ten Bettischen Gruppen der im Projektionsspektrum vorkommenden approximierende Komplexe; die Gruppen sind miteinander durch homomorphe Abbildungen verbunden, die den simplizialen Abbildungen im Projektionsspektrum entsprechen. Auf diese Weise entstehen die sogenannten „inversen Homomorphismenfolgen", die für die Zusammenhangseigenschaften der abgeschlossenen Menge maßgebend sind. Diese Homomorphismenfolgen sind zwar mit Hilfe eines willkürlich gewählten Projektionsspektrums definiert, es erweist sich aber, daß Projektionsspektra, die homöomorphe Mengen definieren, in einem gewissen Sinne *äquivalente* Homomorphismenfolgen besitzen, so daö man berechtigt ist, den Inbegriff aller untereinander äquivalenten Homomorphismenfolgen als eine neue topologische Invariante die *r-dimensionale Zyklosis* der Menge einzuführen. Es ergibt sich ferner, daß die r-dimensionale Zyklosis, die ja selbst keine Gruppe ist, in eindeutiger Weise eine Gruppe definiert: *die zu der Zyklosis duale Gruppe.* Und *diese Gruppe ist mit der $n-r-1$-dimensionalen Bettischen Gruppe des Komplementärraums R^n-F* isomorph. Die ganze Untersuchung läßt sich wieder nach einem beliebigen Modul μ führen, wobei, wie immer, man im Fall $\mu = 0$ unter der Bettischen Gruppe modulo Null des R^n-F die reduzierte Bettische Gruppe zu verstehen hat. Die wichtigste Folge aus dieser Theorie ist zweifellos der in ihr enthaltene Beweis der Tatsache, daß *die reduzierte Bettische Gruppe des Komplementärraumes zu einer abgeschlossenen Menge eine topologische Invariante dieser Menge ist.* Man könnte übrigens mit Hilfe derselben Methoden auch die Invarianz der Torsionsgruppe von R^n-F beweisen; *dagegen bleibt die Frage über die Invarianz der vollen Bettischen Gruppe von R^n-F* gegenüber topologischen Transformationen von F unentschieden: im allgemeinen hat ja eine Bettische Gruppe von R^n-F keine endlichviele Erzeugenden, und sie braucht auch nicht als direkte Summe ihrer reduzierten Gruppe und der Torsionsgruppe darstellbar zu sein.

Die Invarianz der Bettischen Gruppen modulo 2 des R^n-F wurde schon von Alexandroff bewiesen.[19] Der Beweis gilt wörtlich auch für eine beliebige Primzahl als Modul. Für den Fall modulo Null ist ein Invarianzbeweis in den Sätzen von Lefschetz[20] enthalten, *der aber nur dann gilt, wenn die Gruppen endlichviele Erzeugenden haben*: in den erwähnten Fällen folgt nämlich die Invarianz der Gruppen aus der Invarianz ihrer Ränge (es werden ja im Falle mod 0 nur die reduzierten, also in diesem Fall die freien Gruppen betrachtet). *Dagegen folgt im allgemeinen Falle der unendlichvielen Erzeugenden die Isomorphie der Gruppen keineswegs aus der Gleichheit der Ränge*, selbst im Falle, wo es keine Elemente endlicher Ordnung gibt: schon die (additive) Gruppe aller rationalen Zahlen, ebenso wie die Gruppe aller dyadischen Brüche, liefern uns Beispiele von zwei nicht-isomorphen Gruppen, deren Rang Eins ist, die aber kein Element endlicher Ordnung enthalten. Es wird überdies im Anhang III bewiesen, daß *jede* aus abzählbar vielen Elementen bestehende Abelsche Gruppe ohne Elemente endlicher Ordnung als Bettische Gruppe eines $R^n - F$ (sogar für $n = 3$) auftreten kann. Die Invarianz dieser Gruppen könnte mit Methoden, die nur die Bettischen Zahlen, also die Ränge berücksichtigt, prinzipiell nicht bewiesen werden, so daß unser Satz durchaus keine selbstverständliche Erweiterung der bekannten Invarianzsätze für Bettische Zahlen ist, sondern grundsätzlich tiefer liegt. Um so mehr dürfte es von Interesse sein, auch die Invarianz der vollen Bettischen Gruppen der $R^n - F$ zu beweisen.

Ich möchte noch zum Schluß erwähnen, daß diese Arbeit im hohen Maße durch eine Vorlesung von Herrn Alexandroff über kombinatorische Topologie und eine Vorlesung von Fräulein Emmy Noether über abstrakte Algebra angeregt worden ist (beide Vorlesungen wurden im Winter 1928/29 an der Universität Moskau gehalten). Ich verdanke auch Herrn Alexandroff mehrere Ratschläge bei der endgültigen Redaktion der vorliegenden Abhandlung.

Inhaltsübersicht

Kapitel I
Die algebraischen Grundlagen

1. U und V seien zwei Abelsche Gruppen mit endlich-vielen Erzeugenden und M eine zyklische Gruppe von der Ordnung μ; falls $\mu = 0$ ist, so soll das heißen, das M die unendliche zyklische Gruppe, d. h. die freie Gruppe mit einer Erzeugenden ist. Wir denken uns die Gruppen U, V, M additiv geschrieben; da in dieser Untersuchung M durch eine beliebige isomorphe Gruppe ersetzt werden kann, treffen wir ein für allemal die Verabredung, M sie im Falle $\mu = 0$ die (additive) Gruppe aller ganzen Zahlen, während für $\mu > 0$ die Gruppe M durch das System der kleinsten nichtnegativen Reste modulo μ repräsentiert werden soll. In diesem Sinne ist ein Element von M stets eine ganze Zahl.

2. Definition I. Zwei Gruppen U und V bilden ein *Gruppenpaar* in bezug auf (*den Modul*) M, wenn jedem geordneten Elementenpaar x, y – wobei x ein Element von U und y ein Element von V ist – ein Element k von M, das *Produkt* der beiden Elemente x und y:

$$k = x \cdot y$$

zugeordnet ist, wobei stets

$$(x + x') \cdot y = x \cdot y + x' \cdot y \quad \text{(1. Distributivgesetz)}$$

und

$$x \cdot (y + y') = x \cdot y + x \cdot y' \quad \text{(2. Distributivgesetz)}$$

gilt (daraus folgt insbesondere, daß

$$x \cdot 0 = 0 = 0 \cdot y \tag{1}$$

bei jeder Wahl von x bzw. y ist).

3. Definition II. Es sei A eine beliebige Untergruppe von U; die Gesamtheit aller Elemente y von V von der Eigenschaft, daß für jedes x aus A

$$x \cdot y = 0$$

ist, heißt der Annullator von A in V und wird mit (V, A) bezeichnet.

In analoger Weise definiert man den Annullator (U, B) für eine beliebige Untergruppe B von V.

Es folgt zunächst aus dem 1. bzw. 2. Distributivgesetz, daß bei jeder Wahl von x und y

$$x \cdot 0 = 0 = 0 \cdot y,$$

$$(-x) \cdot y = -x \cdot y \quad \text{bzw.} \quad x \cdot (-y) = -xy$$

ist, so daß gleichzeitig mit $x \cdot y = 0$ auch $(-x) \cdot y$ und $x \cdot (-y)$ Null ist; m. a. W.: ein Annullator enthält mit x auch $-x$, mit y auch $-y$, und es gehört zu ihm auch das Nullelement; es gilt somit der Satz:

I. Der Annulator ist eine Untergruppe von U bzw. von V.

Definition III. Ein Gruppenpaar U, V heißt primitiv, wenn der Annullator jeder der beiden Gruppen in der anderen nur aus dem Nullelement besteht:

$$(U, V) = 0, \quad (V, U) = 0.$$

Wir sagen auch öfters, daß U und V *zueinander primitiv* (in bezug auf M) *sind.* Es gilt sodann folgender Satz:

II. Falls U und V ein primitives Gruppenpaar bilden, so lassen sich diese Gruppen derart in direkte Summen von zyklischen Untergruppen

$$U = A_1 + A_2 + \ldots + A_n, \tag{2}$$

$$V = B_1 + B_2 + \ldots + B_n \tag{3}$$

zerlegen, daß für die Erzeugenden $a_1, a_2, \ldots, a_n$ und $b_1, b_2, \ldots, b_n$ der Gruppen A_i und B_i die Relationen

$$a_i \cdot b_j = 0 \quad (\text{für } i \neq j), \tag{4}$$

$$a_i \cdot b_i = k_i > 0$$

mit $k_{i+1} \equiv 0 \pmod{k_i}$ gelten; dabei sind die k_i Teiler von μ, und μ/k_i ist die Ordnung von A_i und B_i.

Bevor wir zum Beweis des Satzes II übergehen, bemerken wir, daß ihm zufolge die Gruppen U und V sich als direkte Summen von einer und derselben Anzahl zyklischer Gruppen bzw. gleicher Ordnung darstellen lassen, so daß man folgenden Zusatz formulieren kann:

Zueinander primitive Gruppen sind isomorph.

4. *Beweis des Satzes II.* Man bezeichne mit a_1 bzw. b_1 solche Elemente von U bzw. V (solche „Werte" von x und y), daß die Zahl $x \cdot y$ einen möglichst kleinen positiven Wert k_1 enthält.

Zunächst ist bei jeder Wahl von y bzw. von x k_1 ein Teiler von $a_1 \cdot y$ und von $x \cdot b_1$; wenn nämlich z. B.

$$a_1 \cdot y = qk_1 + r = q(a_1 \cdot b_1) + r$$

mit $r > 0$ wäre, wäre

$$a_1 \cdot (y - qb_1) = r, \quad r < k_1,$$

also wäre b_1 falsch gewählt.

Man bezeichne mit A_1 bzw. B_1 die von a_1 bzw. von b_1 erzeugte zyklische Gruppe und betrachte ein beliebiges $x \subset U$. Aus dem soeben bewiesenen folgt die Existenz eines solchen q, daß $x \cdot b_1 = qk_1 = q(a_1 \cdot b_1)$, also $(x - qa_1)b_1 = 0$ ist; somit ist $x - qa_1$ ein Element x'' von (U, B_1), und es ist

$$x = x' + x'', \tag{5}$$

mit $x' \subset A_1$, $x'' \subset (U, B_1)$. Ebenso läßt sich jedes Element y von V in der Form $y = y' + y''$ mit $y' \subset B_1$, $y'' \subset (V, A_1)$ darstellen. Es sei a ein gemeinsames Element von A_1 und $(U, B_1) = U_1$; wir wählen irgendein $y \subset V$, $y = y' + y''$. Sodann ist $a \cdot y = a \cdot y' + a \cdot y''$. Nun ist aber $a \cdot y' = 0$, denn a ist in (U, B_1) und y' in B_1 enthalten. Andererseits ist wegen $a \subset A_1$, $y'' \subset (V, A_1)$ auch $a \cdot y'' = 0$; somit ist bei jedem $y \subset V$ $a \cdot y = 0$, woraus vermöge der Primitivität des Gruppenpaares U, V die Identität $a = 0$ folgt. Hiermit ist bewiesen, daß U direkte Summe von A_1 und U_1 ist. In derselben Weise beweist man, daß V direkte Summe von B_1 und $(V, A_1) = V_1$ ist.

Wir beweisen jetzt, daß die Ordnung von A_1 bzw. von B_1 gleich μ/k_1 ist. Für $\mu = 0$ ist dies klar: wenn in der Tat die Ordnung von A_1 gleich $s \neq 0$ wäre, wäre

$$sk_1 = s(a_1 \cdot b_1) = sa_1 \cdot b_1 = 0,$$

was unmöglich ist, denn k_1 ist ja von Null verschieden.

Man betrachte jetzt den Fall $\mu \neq 0$. Es sei s die kleinste positive Zahl von der Eigenschaft, daß $sk_1 \equiv 0 \pmod{\mu}$, d. h. $sa_1 \cdot b_1 = 0$ ist; da dann das Produkt von sa_1 mit einem beliebigen Element von B_1 Null ist; ist $sa_1 \subset (U, B_1)$, folglich ist, da A_1 und (U, B_1) nur die Null im Durchschnitt haben, $sa_1 = 0$. Somit ist die Ordnung von A die kleinste Zahl s von der Eigenschaft, daß μ in sk_1 aufgeht; da aber $\mu \equiv 0 \pmod{k_1}$ ist,[22] ist diese Zahl gleich μ/k_1.

Die Gruppen U_1 und V_1 bilden, wie leicht ersichtlich, wieder ein primitives Gruppenpaar, und das obige Verfahren ergibt die Zerlegungen.

$$U_1 = A_2 + (U_1, B_2), \quad V_1 = B_2 + (V_1, A_2).$$

Auf diese Weise fortfahrend erhalten wir die direkten Summenzerlegungen

$$U = A_1 + A_2 + \ldots + A_n + U_n,$$
$$V = B_1 + B_2 + \ldots + B_n + V_n$$

mit

$$U_{i+1} = (U_i, B_{i+1}), \quad V_{i+1} = (V_i, A_{i+1}); \quad U_i = A_{i+1} + U_{i+1}, \quad V_i = B_{i+1} + V_{i+1},$$

wobei bei jedem i U_i und V_i zueinander primitiv sind. Das Verfahren bricht nach endlich-vielen Schritten ab (denn U und V hatten ja endlich-viele Erzeugende), d. h. für ein bestimmtes n ist etwa V_n die Nullgruppe; da aber U_n und V_n ein primitives Gruppenpaar bilden, muß auch U_n die Nullgruppe sein. Somit bricht das Verfahren in den U_i und in den V_i *gleichzeitig* ab, und man erhält schließlich $U = A_1 + A_2 + \ldots + A_n$, $V = B_1 + B_2 + \ldots + B_n$.

Bleibt nur noch übrig zu zeigen, daß $k_{i+1} \equiv 0 \pmod{k_i}$ ist; nun würde aber aus $k_{i+1} = a_{i+1} \cdot b_{i+1} = d \cdot k_i + r$, mit $0 < r < k_i$, folgen, daß

$$(-da_i + a_{i+1}) \cdot (b_i + b_{i+1}) = -d(a_i \cdot b_i) + a_{i+1} \cdot b_{i+1} = -dk_i + k_{i+1} = r,$$

was der Definition von a_i und b_i widerspricht.

Der Satz II ist somit in allen seinen Teilen bewiesen.

5. Definition IV. Die Summenzerlegungen (2), (3) bilden – falls sie die Bedingungen des Satzes II erfüllen – eine *charakteristische Darstellung des Gruppenpaares U, V.* Die dabei auftretenden Konstanten k_i heißen die *invarianten Faktoren des Gruppenpaares.*

Der Terminus „invariante Faktoren" wird durch folgende Bemerkung gerechtfertigt. Es seien $x_1, x_2, \ldots, x_n$ und $y_1, y_2, \ldots, y_n$ zwei linear unabhängige Erzeugendensysteme der Gruppen U und V. Der Satz II besagt, daß man von den x_i und y_i zu neuen Erzeugenden a_i und b_i übergehen kann, so daß dabei $a_i \cdot b_j = 0$ (bei $i \neq j$), $a_i \cdot b_i = k_i$ und $k_{i+1} \equiv 0 \pmod{k_i}$ ist. Da aber der Übergang von einem Erzeugendensystem zu einem anderen durch unimodulare Substitution erfolgt, bedeutet das nichts anderes, als den folgenden.

Zusatz II zum Satz II. Die Zahlen k_i sind die Elementarteiler der Matrix $(x_i \cdot y_j)$, wobei die x_i bzw. y_i ein beliebiges linear unabhängiges Erzeugendensystem von U bzw. V bilden.

Die invarianten Faktoren des Gruppenpaares sind also durch dieses Gruppenpaar eindeutig bestimmt.

6. Man betrachte jetzt: ein Gruppenpaar U, V, eine Untergruppe A von U und eine Untergruppe B von V. Wenn $A \subset (U, V)$ und $B \subset (V, U)$ ist, und die Elemente x und x' von U, sowie die Elemente y, y' von V zu derselben Restklasse nach A bzw. nach B gehören, so ist (wenn $\alpha = x' - x \subset A$ und $\beta = y' - y \subset B$ gesetzt wird)

$$x \cdot \beta = \alpha \cdot y = \alpha \cdot \beta = 0,$$

folglich

$$x' \cdot y' = (x + \alpha) \cdot (y + \beta) = xy.$$

Im Falle $A \subset (U, V)$, $B \subset (V, U)$ induziert somit das Multiplikationsgesetz des Gruppenpaares U, V ein Multiplikationsgesetz (nach demselben Modul) für die Faktorgruppen $U|A$ und $V|B$. Wenn insbesondere $A = (U, V)$ und $B = (V, U)$ ist, so ist das Gruppenpaar $U|A$, $V|B$ primitiv.

7. Wir betrachten jetzt ausführlicher den Fall $\mu = 0$ und führen zunächst folgende Definition ein.

Definition V. Im Falle $\mu = 0$ heißt ein primitives Gruppenpaar orthogonal, wenn seine invarianten Faktoren sämtlich gleich 1 sind.

Eine einfache Rechnung zeigt, daß, wenn in soeben formulierten Sinne U, V zueinander orthogonal sind und man ein linear unabhängiges Erzeugendensystem, etwa $x_1, x_2, \ldots, x_n$ der einen Gruppe, etwa U, hat, man ein Erzeugendensystem $y_1, y_2, \ldots, y_n$ der anderen Gruppen so finden kann, daß $x_i \cdot y_j = \delta_{ij}$ ist, wobei – wie üblich – $\delta_{ij} = 0$ für $i \neq j$ und $\delta_{ii} = 1$ ist.

(Wenn a_i und b_i die Erzeugenden der charackteristischen Darstellung sind und $x_i = \lambda_i^j a_j$ die x_i durch die a_i ausdrückt, so hat man für die $y_i = \mu_i^h b_h$ die Bestimmungsgleichungen

$$\sum_{(j)} \lambda_i^j \mu_k^j = \delta_{ik},$$

die (wegen $\|\lambda_i^j\| = 1$) eindeutig und ganzzahlig lösbar sind.)

Man könnte natürlich den Begriff der Orthogonalität in genau derselben Weise auch für den Fall $\mu > 0$ einführen; es wird sich aber zeigen, daß dies unzweckmäßig ist, weil in diesem Falle schon die gewöhnlichen primitiven Gruppenpaare dasselbe leisten wie die orthogonal Gruppenpaare im Falle $\mu = 0$. In der Mehrzahl der folgenden Sätze werden also primitive Gruppenpaare mit $\mu > 0$ parallel mit den orthogonalen Gruppenpaaren mit $\mu = 0$ auftreten, was dadurch zum Ausdruck gebracht wird, daß wir „primitiv“ und in Klammern „orthogonal“ schreiben, wobei ein für allemal die Verabredung getroffen wird, daß sich das erstere Adjektiv auf den Fall $\mu > 0$, das letztere auf den Fall $\mu = 0$ bezieht.

Definition VI. Wenn für ein Gruppenpaar U, V im Falle $\mu = 0$ die Faktorgruppen $U|(U, V)$ und $V|(V, U)$ nicht nur zueinander primitiv, sondern auch orthogonal sind, so heißt U, V ein konjugiertes Gruppenpaar.

Im Falle $\mu > 0$ leisten alle Gruppenpaaren bereits dasselbe wie im Falle $\mu = 0$ die konjugierten Gruppenpaare. Dementsprechend wird im folgenden von „Eigenschaften (konjugierter) Gruppenpaare" gesprochen in dem Sinne, daß die betreffende Eigenschaft im Falle $\mu > 0$ für alle Gruppenpaare, im Falle $\mu = 0$ dagegen im allgemeinen nur für konjugierte Gruppenpaare stattfindet.

Wir bemerken schließlich noch, daß im Falle $\mu = 0$ unter einer Untergruppe A von U stets eine Untergruppe *mit Division*[23] verstanden wird.

8. Es sei U, V ein (konjugiertes) Gruppenpaar und $z(v)$ eine homomorphe Abbildung von V in die Gruppe M, bei der sämtliche Elemente von (V, U) auf das Nullelement von M abgebildet werden. Dann läßt sich dieser Homomorphismus durch ein Element x_0 von U erzeugen in dem Sinne, daß für sämtliche $v \subset V$

$$z(v) = x_0 \cdot v$$

ist. Falls dabei das Gruppenpaar U, V primitiv (orthogonal) ist, läßt sich das Element x_0 nur auf eine Weise bestimmen.

Man nehme zuerst an, U und V seien zueinander primitiv (orthogonal).

Es seien $a_1, a_2, \ldots, a_n$ bzw. $b_1, b_2, \ldots, b_n$ die Erzeugenden der charakteristischen Darstellung von U und V, $k_1, k_2, \ldots, k_n$ ihre invarianten Faktoren. Man setze $h_i = z(b_i)$ und beweise zunächst, daß k_i in h_i aufgeht. Für $\mu = 0$ ist dies klar, denn k_i ist ja in diesem Falle $= 1$. Wenn $\mu > 0$ ist, so hat man – da μ/k_i die Ordnung von b_i ist –

$$0 = z\left(\frac{\mu}{k_i} b_i\right) = \frac{\mu}{k_i} z(b_i) = \frac{\mu h_i}{k_i} \pmod{\mu}$$

woraus folgt, daß μ in $\mu h_i/k_i$ aufgeht und somit h_i/k_i eine ganze Zahl ist.

Man setze jetzt

$$x_0 = \sum_{i=1}^{n} \frac{h_i}{k_i} a_i;$$

wenn $v = \Sigma_{j=1}^{n} \mu_j b_j$ ein beliebiges Element von V ist, so ist

$$x_0 \cdot v = \sum_{i,j} \frac{h_i}{k_i} a_i \cdot \mu_j b_j = \sum_{i,j} \frac{h_i}{k_i} \mu_j (a_i \cdot b_j) = \sum_i \frac{h_i}{k_i} \mu_i (a_i \cdot b_i)$$

$$= \sum_i h_i \mu_i = z\left(\sum_i \mu_i b_i\right) = z(v),$$

womit unsere Behauptung bewiesen ist.

Wenn es zwei Elemente x und x' gäbe, die der obigen Bedingung genügen, so wäre für jedes v $(x-x')v=0$, was nur dann mit der Primitivität des Gruppenpaares U, V verträglich ist, wenn $x=x'$ ist.

Es sei jetzt U, V ein (konjugiertes) – jedoch nicht notwending primitives – Gruppenpaar. Aus den Bedingungen unseres Satzes folgt unmittelbar, daß für alle v, die zu derselben Restklasse nach (V, U) gehören, $z(v)$ denselben Wert hat, so daß die Abbildung z eine homomorphe Abbildung der Faktorgruppe $V_1 = V|(V, U)$ definiert, welche unseren Bedingungen genügt; V_1 ist aber zu $U_1 = U|(U, V)$ primitiv (orthogonal), folglich existiert ein Element ξ_0 von U_1 derart, daß für jedes η aus V_1 $\xi_0 \cdot \eta = z(\eta)$ ist. Aus der Multiplikationsdefinition für Restklassen (Sektion 6) folgt sodann, daß für alle zur Restklasse ξ_0 gehörenden Elemente x_0 von U und für jedes Element y von V

$$x_0 \cdot y \;=\; \xi_0 \cdot \eta \;=\; z(\eta) \;=\; z(y)$$

ist, wobei η die Restklasse bedeutet, zu der y gehört. Unser Satz ist hiermit bewiesen.

9. Hilfssatz. Wenn U, V ein primitives (orthogonales) Gruppenpaar und A eine Untergruppe von U ist, so ist A, V ein (konjugiertes) Gruppenpaar.

Für $\mu \neq 0$ ist der Hilfssatz trivial. Im Fall $\mu = 0$ sei $u_1, u_2, \ldots, u_n$ ein linear unabhängiges Erzeugendensystem von U, welches so beschaffen ist, daß etwa $u_1, \ldots, u_r$ die Untergruppe A erzeugen. Nach Sektion 7 läßt sich dann das Erzeugendensystem $v_1, v_2, \ldots, v_n$ so bestimmen, daß $u_i \cdot v_j = \delta_{ij}$ ist. Da bei beliebigen $i \leqslant r$ und $h > r$ $u_i \cdot v_h = 0$ ist, gehören die $v_{r+1}, \ldots, v_n$ sämtlich zu (V, A); wenn andererseits $v = c^i v_i$ irgendein Element von V ist derart, daß für ein $h \leqslant r$ c^h von Null verschieden ist, ist $u_h \cdot v$ von Null verschieden. Somit gehören zu (V, A) alle und nur die von $v_{r+1}, \ldots, v_n$ erzeugten Elemente; m. a. W. $v_{r+1}, \ldots, v_n$ bilden ein unabhängiges Erzeugendensystem von (V, A).

Da offenbar (A, V) nur aus dem Nullelement besteht, haben wir zu zeigen, daß A, $V|(V, A)$ ein orthogonales Gruppenpaar ist. Bei der homomorphen Abbildung von V auf $V|(V, A)$ gehen die von $v_{r+1}, \ldots, v_r$ erzeugten Elemente nach dem soeben bewiesenen in Null, die $v_1, \ldots, v_r$ dagegen in lauter von Null und voneinander verschiedenen Elemente $\beta_1, \ldots, \beta_r$ über, die ein Erzeugendensystem von $V|(V, A)$ bilden. Da ferner (für $i, j \leqslant r$) $u_i \cdot \beta_j = u_i \cdot v_j = \delta_i^j$ gesetzt werden soll, sind die Gruppen A und $V|(V, A)$ zueinander orthogonal, under der Hilfssatz ist bewiesen.

10. Satz III. Wenn U, V ein primitives (orthogonales) Gruppenpaar, A eine Untergruppe von U und $B = (V, A)$ ist, so ist $A = (U, B)$.

Man bezeichne mit A' die Gruppe (U, B); zunächst folgt aus der Definition von B, daß bei jeder Wahl der Elemente $x \subset A$, $y \subset B$ $x \cdot y = 0$ ist, so daß jedenfalls die Relation $A \subset A'$ gilt. Um die umgekehrte Inklusion zu beweisen, betrachte man – in der Annahme, daß die nich zutrifft – irgendein Element z von $A' - A$. Für dieses Element z und ein beliebiges v aus V ist das Produkt $z \cdot v$ eindeutig bestimmt, wobei für sämtliche y aus B $z \cdot y = 0$ ausfällt. Somit ist eine homomorphe Abbildung $z(v) = z \cdot v$ definiert, auf die der Satz des Sektiones 8 angewandt werden kann, wobei jetzt A die Rolle von U übernimmt (was erlaubt ist, denn die Gruppen A und V sind ja nach dem Hilfssatz zueinander konjugiert). Somit existiert ein Element x_0 von A derart, daß für sämtliche $y \subset V$ $x_0 \cdot y = z \cdot y$, also $(x_0 - z) \cdot y = 0$ ist; da die Gruppen U und V zueinander primitiv sind, folgt aus der letzteren Gleichung, daß $z = x_0 \subset A$ ist, entgegen der Definition von z. Durch diesen Widerspruch ist Satz III bewiesen.

11. Als Verallgemeinerung des Hilfssatzes des Sektiones 9 beweisen wir noch den folgenden

Satz IV. U und V mögen ein (konjugiertes) Gruppenpaar bilden; wenn A und B Untergruppen von U und V sind, welche die Annulatoren $A' = (U, B)$ und $B' = (V, A)$ bzw. enthalten, so bilden (auf Grund des für U und V erklärten Multiplikationsgesetzes) auch A und B ein (konjugiertes) Gruppenpaar.

Beweis. Im Falle $\mu \neq 0$ ist die Behauptung trivial. Es sei also $\mu = 0$. Wir beweisen zunächst unseren Satz in der Annahme, daß U und V nich nur konjugiert, sondern orthogonal sind.

Es seien

$$a_1, a_2, \ldots, a_k, a_{k+1}, \ldots, a_{k+r}, a_{k+r+1}, \ldots, a_n, \tag{1}$$

$$b_1, b_2, \ldots, b_k, b_{k+1}, \ldots, b_{k+r}, b_{k+r+1}, \ldots, b_n, \tag{2}$$

zwei Erzeugendensysteme von U bzw. V mit $a_i \cdot b_j = \delta_{ij}$. Wir numerieren die Elemente der Systeme (1) und (2) so, daß $a_1, \ldots, a_k$ ein Erzeugendensystem von A' und $a_1, \ldots, a_{k+r}$ ein Erzeugendensystem von A ist.

Da definitionsgemäß $A' = (U, B)$ ist, so ist vermöge des Satzes III $B = (V, A')$. Ein Element b von V gehört somit dann und nur dann zu B, wenn für alle a_i mit $i \leqslant k$ $a_i \cdot b = 0$ ist; dieser Bedingung genügen aber nur die Linearkombinationen der b_j mit $j > k$; folglich bilden die letztgenannten b_j ein Erzeugendensystem von B. In analoger Weise gehört b zu $B' = (V, A)$, wenn für all a_i mit $i \leqslant k + r$ $a_i \cdot b = 0$ ist, woraus wiederum folgt, daß die $b_{k+r+1}, \ldots, b_n$ ein Erzeugendensystem von B' bilden.

Wir betrachten jetzt die beiden Faktorgruppen $A|A'$ und $B|B'$ und die zugehörigen Homomorphismen. Es seien bei diesen Homomorphismen α_i das Bild von a_i, β_j das Bild von b_j; dabei ist $\alpha_1 = \ldots = \alpha_k = 0$, während die $\alpha_{k+1}, \ldots, \alpha_{k+r}$ voneinander und von Null verschieden sind und ein unabhängiges Erzeugendensystem von $A|A'$ bilden. Ebenso ist $\beta_{k+r+1} = \ldots = \beta_n = 0$, und die $\beta_{k+1}, \ldots, \beta_{k+r}$ bilden ein unabhängiges Erzeugendensystem von $B|B'$. Ferner ist $\alpha_i \cdot \beta_j = a_i \cdot b_j = \delta_i^j$, so daß $A|A'$ zu $B|B'$ orthogonal ist. Im Spezialfall des orthogonalen Gruppenpaares U, V ist der Satz hiermit bewiesen.

Es sei jetzt U, V ein konjugiertes Gruppenpaar, von dessen Orthogonalität nichts vorausgesetzt wird. Wir bezeichnen mit A'' bzw. B'' die beiden Annulatoren (U, V) bzw. (V, U). Man hat dann die Inklusionen

$$A'' \subset A' \subset A, \qquad B'' \subset B' \subset B,$$

wobei alle diese Gruppen Untergruppen mit Division sind.

Wir bilden jetzt die Faktorgruppen $\bar{U} = U|A''$ und $\bar{V} = V|B''$ und betrachten die zugehörigen Homomorphismen f und g:

$$\bar{U} = f(U), \quad \bar{V} = g(V).$$

Es sei dabei

$$\bar{A} = f(A), \ \bar{A}' = f(A') \quad \text{und} \quad \bar{B} = g(B), \ \bar{B}' = g(B');$$

diese Gruppen sind wieder Untergruppen mit Division.

Wegen des ersten Isomorphiesatzes[24] ist $\bar{A}|\bar{A}'$ mit $A|A'$ sowie $\bar{B}|\bar{B}'$ mit $B|B'$ isomorph. Ferner sind $\bar{U}$ und $\bar{V}$ zueinander orthogonal, und es gelten die Relationen

$$\bar{A}' = (\bar{U}, \bar{B}) \subset \bar{A}; \quad \bar{B}' = (\bar{V}, \bar{A}) \subset \bar{B}.$$

Aus der Orthogonalität von $\bar{U}$, $\bar{V}$ folgt sodann auf Grund des soeben Bewiesenen die Orthogonalität von $\bar{A}|\bar{A}'$ und $\bar{B}|\bar{B}'$, woraus sich – vermöge der Isomorphie zwischen $\bar{A}|\bar{A}'$ und $A|A'$ und zwischen $\bar{B}|\bar{B}'$ und $B|B'$ – die Orthogonalität von $A|A'$ und $B|B'$ ergibt. Dadurch ist der Satz IV vollständig bewiesen.

Kapitel II.
Die verallgemeinerten Dualitätssätze von Poincaré–Veblen und von Alexander.

I. Geometrische Hilfsbetrachtungen[25]

1. Es sei K^n ein simplizialer Komplex; a^r sei ein Simplex von K, $a_1^n, a_2^n, \ldots, a_s^n$ seien die Simplexe von K^n, die a^r auf ihrem Rande tragen; diese

Simplexe bilden den *Stern* um a^r; die Gesamtheit der dem Simplex a^r gegenüberliegenden Seiten der a_i^n bildet den *Umgebungskomplex* $Z^{n-r-1}(a^r)$ von a^r in K^n.

Ein zusammenhängender Komplex M^n heißt eine *Mannigfaltigkeit*, wenn sämtliche Umgebungskomplexe in ihm Sphären der entsprechenden Dimension homöomorph sind. Da die topologische Invarianz der soeben formulierten Definition (d. h. ihre Unabhängigkeit von der speziellen Simplizialzerlegung, in der M^n vorliegt) bis jetzt unbewiesen bleibt, werden wir uns im folgenden des allgemeineren Begriffes der sogenannten h-Mannigfaltigkeiten[26] bedienen.

Ein zusammenhängender Komplex M^n heißt eine *h-Mannigfaltigkeit,* wenn an jedes $n-1$-dimensionale Element von M^n genau zwei n-dimensionale Simplexe anschließen, während der Umgebungskomplex eines jeden a^k $(k < n-1)$ folgende Eigenschaft hat: $Z^{n-k-1}(a^k)$ ist ein zusammenhängender Komplex, in dem der r-dimensionale Zyklus $(0 < r < n-k-1)$ berandet und im wesentlichen ein einziger $n-k-1$-dimensionaler Zyklus existiert, welcher dortselbst nicht homolog Null ist. Die Invarianz dieses Begriffes läßt sich leicht nachweisen.[26]

Eine h-Mannigfaltigkeit (die übrigens stets ein Pseudomannigfaltigkeit im Brouwerschen Sinne ist) heißt *orientierbar*, wenn sich ihre n-dimensionalen Elemente so orientier lassen, daß die algebraische Summe ihrer orientierten Ränder gleich Null ist. Im folgenden werden wir nur solche Orientierungen zulassen.

2. Es sei M^n eine n-dimensionale orientierbare h-Mannigfaltigkeit. Man betrachte eine baryzentrische Unterteilung von M^n und orientiere die Elemente derselben folgendermaßen: es seien $a^n = \epsilon\,(a_0, a_1, \ldots, a_n)$ ein positiv-orientiertes n-dimensionales Simplex von M^n, $a^r = \eta\,(a_0, a_1, \ldots, a_r)$ eine ebenfalls positiv orientierte Seite von a^r; es sei endlich β_i der Schwerpunkt von $(a_0, a_1, \ldots, a_i)$ (dabei ist i beliebig, also unabhängig von r). Die positive Orientierung des durch die Eckpunkte $\beta_r, \beta_{r+1}, \ldots, \beta_n$ bestimmten baryzentrischen Simplexes ist dann definitionsgemäß $\epsilon \cdot \eta \cdot (\beta_r, \beta_{r+1}, \ldots, \beta_n)$. Wenn man auf alle möglichen Weisen die Reihenfolge der Eckpunkte $a_{r+1}, a_{r+2}, \ldots, a_n$ variiert, erhält man $(n-r)!$ verschiedene baryzentrische Simplexe; sie liegen in a^n und heißen zu a^r *dual*; wenn man diese Konstruktion in allen an a^r anschließenden Simplexes a^n durchführt, erhält man die Gesamtheit aller zu a^r dualen baryzentrischen Simplexe; die algebraische Summe dieser (nach obiger Vorschrift orientierten) Simplexe bildet den zu a^r *dualen baryzentrischen Stern.* Wir bezeichnen ihn mit $b^{n-r}(a^r)$.

Man sieht leicht ein, daß der Rand eines r-dimensionalen baryzentrischen Sternes aus baryzentrischen Sternen von der Dimension $r-1$ aufgebaut ist. Dabei gilt folgende Vorzeichenregel: Wenn

$$a^k \to \epsilon a^{k-1} + \ldots$$

und b^{n-k} bzw. b^{n-k+1} die zu a^k bzw. a^{k-1} dualen baryzentrischen Sterne sind, so ist

$$b^{n-k+1} \to (-1)^k \cdot \epsilon \cdot b^{n-k} + \ldots .$$

3. Wir betrachten jetzt zweierlei Arten von „Bausteinen", aus denen wir Teilkomplexe von M^n konstruieren werden: die Bausteine erster Art (die zu Teilkomplexen erster Art führen) sind die Elemente der gegebenen Simplizialzerlegung von M^n, d. h. die Simplexe verschiedener Dimension; es seien dies für die Dimension r etwa $a_1^r, a_2^r, \ldots, a_{\alpha_r}^r$. Ein Teilkomplex erster Art ist dementsprechend als eine Linearform von der Gestalt $\lambda^i a_i^r$ aufzufassen. Die Bausteine zweiter Art sind die baryzentrischen Sterne. Da die r-dimensionalen baryzentrischen Sterne eineindeutig den $n-r$-dimensionalen Simplexen von M^n entsprechen, lassen sie sich so numerieren:

$$b_1^r, b_2^r, \ldots, b_{\alpha_{n-r}}^r,$$

dabei ist $b_i^r = b^r(a_i^{n-r})$. Ein r-dimensionaler Teilkomplex zweiter Art ist sodann definitionsgemäß ein Linearform von der Gestalt $\lambda^i b_i^r$.

Es seien jetzt zwei Teilkomplexe $A = A^r = \lambda^i a_i^r$ und $B = B^{n-r} = \mu^i b_i^{n-r}$ bzw. erster und zweiter Art gegeben. Die Zahl

$$\chi(A, B) = \sum_{i=1}^{\alpha_r} \lambda^i \mu^i$$

heißt die *Kroneckersche Charakteristik* oder die *Schnittzahl* der beiden Komplexe A und B.

4. Man betrachte jetzt zwei *stetige Komplexe* A^r und B^{n-r}, die in M^n (mit eventuellen Singularitäten)[27] eingebettet seien; es wird vorausgesetzt, daß keiner der beiden Komplexe A^r und B^{n-r} den Rand des anderen trifft, daß also die Minimalentfernung des einen Komplexes vom Rande des anderen eine positive Zahl σ ist. Man kann sodann annehmen, daß die Simplexe von M^n sämtlich kleiner als etwa $\frac{1}{100}\sigma$ sind. Die Komplexe A^r und B^{n-r} lassen sich durch Teilkomplexe erster bzw. zweiter Art von M^n beliebig gut approximieren.

Man beweist ohne Mühe die folgenden Tatsachen:

1. Wenn A' ein Teilkomplex erster, B' ein Teilkomplex zweiter Art ist und A' bzw. B' die Komplexe A^r und B^{n-r} hinreichend gut approximieren,

so hat die Zahl $\chi(A', B')$ einen von der speziellen Wahl der approximierenden Komplexe A' und B' unabhängigen Wert; dieser heißt die *Kroneckersche Charakteristik* (die *Schnittzahl*) der Komplexe A^r und B^{n-r}, $\chi(A^r, B^{n-r})$.

2. Die Kroneckersche Charakteristik zweier Komplexe A^r und B^{n-r} hängt nicht von der Wahl der Simplizialzerlegung von M^n ab; sie stellt also eine relative Invariante von A und B in bezug auf M^n dar.

3. $\chi(A^r, B^{n-r}) = (-1)^{r(n-r)}\chi(B^{n-r}, A^r)$.

Bemerkung. Wir bezeichnen mit $\dot{K}^{r-1}$ den (orientierten) Rand des (orientierten) Komplexes K^r.

Sodann gilt

Satz I. Wenn $\dot{A}^{r-1}$ zu $\dot{B}^{n-r}$ fremd ist, so ist

$$\chi(A^r, \dot{B}^{n-r}) = (-1)^r\chi(\dot{A}^{r-1}, B^{n-r+1}). \tag{1}$$

Es genügt, diese Behauptung für den Fall zu beweisen, wo A^r ein Simplex, B^{n-r+1} ein baryzentrischer Stern ist. Falls dabei A^r und B^{n-r+1} zueinander fremd sind, ist die Behauptung trivial. Man nehme an, A^r und B^{n-r+1} haben einen nichtleeren Durchschnitt. In diesem Fall ist der baryzentrische Stern B^{n-r+1} zu einer Seite A^{r-1} von A^r dual; wenn dabei etwa

$$A^r \to \epsilon\cdot A^{r-1} + \ldots, \quad B^{n-r} = b^{n-r}(A^r)$$

ist, so ist

$$\chi(\dot{A}^{r-1}, B^{n-r+1}) = \epsilon, \quad B^{n-r+1} \to (-1)^r\cdot\epsilon\cdot B^{n-r} + \ldots,$$

$$\chi(A^r, \dot{B}^{n-r}) = (-1)^r\epsilon,$$

woraus die Behauptung folgt.

Satz II. Wenn A^r und B^{n-r} zwei Zyklen sind, von denen mindestens einer in M^n berandet, so ist

$$\chi(A^r, B^{n-r}) = 0.$$

Beweis. Es sei in der Tat z. B. $B' \to B^{n-r}$; dann ist

$$\chi(A^r, B^{n-r}) = \chi(A^r, \dot{B}') = \pm\chi(\dot{A}^r, B') = 0,$$

w.z.b.w.

4. Es seien A^r und B^s zwei zueinander fremde Zyklen in M mit $r + s = n - 1$; diese Zyklen mögen ferner in M^n beranden, und zwar sei

$$A' \to A^r, \quad B' \to B^s.$$

Die Zahl $\chi(A^r, B')$ heißt die *Verschlingungszahl* von A^r mit B^s und wird mit $\mathfrak{v}(A^r, B^s)$ bezeichnet. *Sie hängt nicht von der Wahl von B' ab*, weil, wenn B'' ein zweiter durch B^s berandeter Komplex ist, $B' - B''$ ein Zyklus, und A^r ein berandender Zyklus ist. Mithin folgt aus Satz II

$$\chi(A^r, B' - B'') = 0, \quad \text{d. h.} \quad \chi(A^r, B') = \chi(A^r, B''),$$

w.z.b.w.

Satz III. Es seien wieder A^r und B^s zwei zueinander fremde berandende Zyklen in M^n mit $r + s = n - 1$; dann ist

$$\mathfrak{v}(A^r, B^s) = (-1)^{rs+n}\mathfrak{v}(B^s, A^r).$$

Beweis. Es sei wie immer $A' \to A^r$, $B' \to B^s$. Man hat vermöge des schon Bewiesenen

$$\mathfrak{v}(A^r, B^s) = \chi(\dot{A}', B') = (-1)^{r+1}\chi(A', \dot{B}') = (-1)^{r+1}(-1)^{(r+1)s}\chi(\dot{B}', A')$$
$$= (-1)^{rs+n}\chi(B^s, A') = (-1)^{rs+n}\mathfrak{v}(B^s, A^r),$$

w.z.b.w.

Wenn die Zyklen A^r und B^s, $r + s = n - 1$, zueinander fremd und in M^n homolog Null sind, so daß z. B. cA^r und dB^s in M^n beranden (wobei c und d passend gewählte ganzzahlige Koeffizienten sind), kann man die Verschlingungszahl $\mathfrak{v}(A^s, B^s)$ als $\mathfrak{v}(cA^r, dB^s)/cd$ definieren; im allgemeinen erhält man auf diese Weise gebrochene Verschlingungszahlen; in unserer weiteren Darstellung werden jedoch die Verhältnisse so liegen, daß wir stets mit ganzzahligen Verschlingungszahlen auskommen werden.

5. Wir schließen diese Hilfsbetrachtungen mit folgenden naheliegenden Sätzen:

Satz IV. Es seien: M^n eine h-Mannigfaltigkeit, K ein aus Simplexen von M^n aufgebauter Komplex, L der aus sämtlichen zu K fremden baryzentrischen Sternen (= Bausteinen zweiter Art) bestehende Komplex, Γ irgendein in $M^n - K$ liegender Zyklus. Unter diesen Begingungen ist Γ einem Teilzyklus Δ von L (der also ein Teilkomplex zweiter Art von M^n ist) in $M^n - K$ homolog. Falls Γ ein Teilzyklus von L ist, welcher in $M^n - K$ berandet, ist Γ Rand eines (aus Bausteinen zweiter Art zusammengesetzten) Teilkomplex von L.

Beweis. Zuerst beweisen wir folgenden Hilfssatz:

Jede abgeschlossene Menge $F \subset M^n - K$ läßt sich mittels einer stetigen Deformation innerhalb von $M^n - K$ in eine in L gelegene Menge F' überführen, und zwar so, daß während des ganzen Deformationsprozesses alle zu L gehörenden Punkte von F fest bleiben.

Es sei $\mathfrak{K}$ das System aller baryzentrischen Sterne von M^n, die mit K einen nicht leeren Durchschnitt haben, also zu den Simplexen von K dual sind. Wir bezeichnen mit s die Höchstdimension der Sterne des Systems $\mathfrak{K}$, die in ihrem Innern Punkte von F enthalten; es sei S einer unter jenen Sternen. Da S zu einem Simplex dual ist, welcher keinen Punkt von F enthält, befördert man durch Zentralprojektion aus dem Schwerpunkte dieses Simplexes (des „Mittelpunktes des Sternes") sämtlich zu S gehörenden Punkte von F auf den Rand des Sternes. Diese „Ausfegung" ist eine Deformation von F, welche alle außerhalb und auf dem Rande von S gelegenen Punkte dieser Menge festläßt und die ganze Zeit F und K fremd zueinander hält. Eine wiederholte Anwendung dieses Ausfegeverfahrens verwandelt F in eine abgeschlossene Menge F', die keinen Punkt im Innern eines zu $\mathfrak{K}$ gehörenden Sternes besitzt. Es sei x ein beliebiger Punkt von F'; dieser ist innerer Punkt eines baryzentrischen Sternes, der – da er nach dem soeben Bewiesenen zu K fremd ist – in L enthalten sin muß, womit der Hilfssatz bewiesen ist.

Es sei jetzt $\bar{L}$ der (geometrisch mit L zusammenfallende) Komplex erster Art, den man erhält, wenn man die Bausteine zweiter Art, aus denen L zusammengesetzt ist, wieder in Simplexe zerschlägt. Aus dem Hilfssatz folgt, daß jeder zu $M^n - K$ gehörende Zyklus Γ^r in einen in $\bar{L}$ liegenden, also weiter in einen aus Simplexen von $\bar{L}$ aufgebauten Zyklus γ^r durch eine Homotopie verwandelt werden kann; man betrachte jetzt die Simplexe von γ^r als Teilsimplexe der Bausteine zweiter Art, die den Komplex L bilden. Wenn ein solcher Baustein S^r in γ^r nur zum Teil enthalten wäre, würde man ihn dadurch wegschaffen können, daß man den in ihm liegenden Teil von γ^r auf den Rand von S^r befördert (was immer mühelos geht). Nachdem man dies endlich oft wiederholt hat, verwandelt sich γ^r in einen aus Bausteinen zweiter Art aufgebauten Zyklus, also in einen Teilzyklus von L. Genau dasselbe Verfahren läßt sich auch auf den am Schluß des Satzes IV erwähnten Homologieträger anwenden und ergibt sodann den Beweis des beiden Behauptungen dieses Satzes.

II. Formulierung und Beweis der beiden Dualitätssätze

1. M^n sei eine orientierbare und orientierte h-Mannigfaltigkeit; $a_1^r, a_2^r, \ldots,$ $a_{\alpha_r}^r$ seien die r-dimensionalen Elemente der gegebenen Simplizialzerlegung von M^n, $\mathfrak{b}_1^{n-r}, \mathfrak{b}_2^{n-r}, \ldots, \mathfrak{b}_{\alpha_r}^{n-r}$ die zu ihnen dualen baryzentrischen Sterne; wir setzen $\chi(a_i^r, \mathfrak{b}_j^{n-r}) = \delta_{ij}$. K sei ein aus Simplexen von M^n aufgebauter Komplex, wobei die in K auftretenden r-dimensionalen Simplexe $a_1^r, a_2^r, \ldots, a_{h^r}^r$ sein mögen. Sodann haben unter den $\mathfrak{b}_k^{n-r}$ die mit $k \leqslant h^r$ und nur diese einen nichtleeren Durchschnitt mit K.

Der ganzen folgenden Untersuchung liege eine feste ganze Zahl μ, die gleich Null oder größer als 1 ist, als Modul zugrunde. Die sich auf K beziehenden Gruppen L_μ^r, Z_μ^r, $\hat{H}_\mu^r$, H_μ^r seien einfach mit L^r, Z^r, $\hat{H}^r$, H^r bezeichnet. Die Bettischen Gruppen mod μ von K seien einfach mit B^r bezeichnet, wobei die in der Einleitung gemachten Verabredungen (insbesondere die der Fußnote[15]) durchweg ihre Geltung behalten.

Wir bezeichnen ferner mit $\mathfrak{L}^r$ die von den Elementen $\mathfrak{b}_k^r$, $k = 1, 2, \ldots, h^{n-r}$ erzeugte Abelsche Gruppe mit der einzigen Relation $\mu \mathfrak{b}_k^r = 0$. Der Rand mod μ eines jeden Elementes von $\mathfrak{L}^r$ (der ja ein Komplex zweiter Art ist) kann als eine Linearform in den $\mathfrak{b}_k^{r-1}$ geschrieben werden; indem wir in dieser Linearform nur die Glieder mit $k \leqslant h^{n-r+1}$ behalten, bekommen wir einen Komplex zweiter Art, den *reduzierten Rand* des gewähltes Elementes von $\mathfrak{L}^r$. Die Komplexe, die als reduzierte Ränder der in $\mathfrak{L}^r$ vorkommenden Komplexe auftreten, bilden eine Untergruppe von $\mathfrak{L}^{r-1}$, die wir mit $\hat{\mathfrak{H}}^{r-1}$ bezeichnen; analog zu den früheren Bezeichnungen setzen wir im Falle $\mu \neq 0$ $\mathfrak{H}^r = \hat{\mathfrak{H}}^r$, während im Falle $\mu = 0$ $\mathfrak{H}^r$ als die kleinste Untergruppe mit Division von $\mathfrak{L}^r$ über $\hat{\mathfrak{H}}^r$ definiert wird. Es liegt ferner eine homomorphe Abbildung von $\mathfrak{L}^r$ auf $\hat{\mathfrak{H}}^{r-1}$ vor; der *Kern*[28] diese Homomorphismus soll mit $\mathfrak{Z}^r$ bezeichnet werden; man sieht leicht ein, daß $\mathfrak{H}^r$ eine Untergruppe von $\mathfrak{Z}^r$ ist.

2. Es soll jetzt ein Multiplikationsgesetz für die beiden Gruppen L^r und $\mathfrak{L}^{n-r}$ festgelegt werden. Dies geschieht einfach dadurch, daß als Produkt der Elemente $a \subset L^r$ und $\mathfrak{b} \subset \mathfrak{L}^{n-r}$ die Schnittzahl $\chi(a, \mathfrak{b})$ mod μ gesetzt wird. Vermöge diese Multiplikationsgesetzes bilden die Gruppen L^r und $\mathfrak{L}^{n-r}$ ein primitives (orthogonales) Gruppenpaar, denn die beiden Gruppen besitzen ja je ein Erzeugendensystem a_i^r bzw. $\mathfrak{b}_j^{n-r}$ mit $a_i^r \cdot \mathfrak{b}_j^{n-r} = \delta_i^j$.

3. Wir beweisen jetzt die Relationen

$$(L^r, \mathfrak{H}^{n-r}) = Z^r; \quad (\mathfrak{L}^{n-r}, H^r) = \mathfrak{Z}^{n-r}, \tag{1}$$

aus denen dann wegen Kap. I, Sektion 10, Satz III die Relationen

$$(\mathfrak{L}^{n-r}, Z^r) = \mathfrak{H}^{n-r}; \quad (L^r, \mathfrak{Z}^{n-r}) = H^r \tag{2}$$

folgen.

A. Es sei $a \subset Z^r$, $\mathfrak{b} \subset \mathfrak{H}^{n-r}$; wir beweisen, daß $\chi(a, \mathfrak{b}) = 0 \pmod{\mu}$ ist.

Wenn $\mu \neq 0$ ist, so existiert ein $\mathfrak{c} \subset \mathfrak{L}^{n-r+1}$, dessen reduzierten Rand $\mathfrak{b}$ ist. Folglich ist

$$\mathfrak{c} \to \mathfrak{b} + \mathfrak{b}', \quad \mathfrak{b}' \text{ zu } K \text{ fremd.}$$

$\mathfrak{b} + \mathfrak{b}'$ ist homolog Null, woraus $\chi(a, \mathfrak{b} + \mathfrak{b}') = 0$ folgt; da aber $\mathfrak{b}'$ außerhalb von K liegt, ist $\chi(a, \mathfrak{b}') = 0$, also $\chi(a, \mathfrak{b}) = 0$, wie wir es haben wollten.

Wenn $\mu = 0$ ist, existiert ein von Null verschiedenes k und ein $\mathfrak{c} \subset \mathfrak{L}^{n-r+1}$ derart, daß

$$\mathfrak{c} \to k\mathfrak{b} + \mathfrak{b}', \quad \mathfrak{b}' \text{ zu } K \text{ fremd;}$$

$kb + b'$ ist wieder homolog Null, also ist $\chi(a, k\mathfrak{b} + \mathfrak{b}') = 0$, was vermöge $\chi(a, \mathfrak{b}') = 0$

$$0 = \chi(a, k\mathfrak{b}) = k\chi(a, \mathfrak{b})$$

und schließlich $\chi(a, \mathfrak{b}) = 0$ ergibt.

B. Es sei a ein nicht zu Z^r gehörendes Element von L^r; wir nehmen uns vor, ein solches Element von $\mathfrak{H}^{n-r}$ zu finden, welches mit a eine von Null verschiedene Schnittzahl besitzt. Da a nicht zu Z^r gehört, also kein Zyklus ist, ist

$$a \to b \subset H^{r-1}, \quad b \neq 0;$$

L^{r-1} und $\mathfrak{L}^{n-r+1}$ bilden aber ein primitives (orthogonales) Gruppenpaar; es existiert folglich ein solches $\mathfrak{c} \subset \mathfrak{L}^{n-r+1}$, daß $\chi(b, \mathfrak{c}) \neq 0$ ist. Wenn wir mit $\dot{\mathfrak{c}}$ den Rand, mit $\hat{\mathfrak{c}}$ den reduzierten Rand von $\mathfrak{c}$ bezeichnen und bemerken, daß wegen des Satzes I dieses Kapitels $\chi(b, \mathfrak{c}) = \pm\, \chi(a, \dot{\mathfrak{c}})$, während andererseits offenbar $\chi(a, \dot{\mathfrak{c}}) = \chi(a, \hat{\mathfrak{c}})$ ist, so sehen wir, daß $\chi(a, \hat{\mathfrak{c}}) \neq 0$ ist. Das Element $\hat{\mathfrak{c}}$ genügt also unseren Forderungen, und die erste der beiden Formeln (1) ist somit bewiesen; die zweite beweist sich auf genau dieselbe Weise.

Wir schließen an diese Ausführungen noch folgende Bemerkung an: da L^r und $\mathfrak{L}^{n-r}$ ein primitives (orthogonales) Gruppenpaar bilden und

$$(L^r, \mathfrak{Z}^{n-r}) \subset Z^r, \quad (\mathfrak{L}^{n-r}, Z^r) \subset \mathfrak{Z}^{n-r}$$

ist, so folgt aus Kap. I, Sektion 6, daß auch die Faktorgruppen $Z^r | H^r$ und $\mathfrak{Z}^{n-r} | \mathfrak{H}^{n-r}$ ebenfalls zueinander primitiv (orthogonal) sind.

4. Wir erhalten jetzt das Theorem von Poincaré–Veblen, wenn wir $K = M^n$ setzen. In diesem Fall ist die Gruppe $Z^r | H^r$ die r-dimensionale und $\mathfrak{Z}^{n-r} | \mathfrak{H}^{n-r}$ die $n-r$-dimensionale Bettische Gruppe von M^n. Wir erhalten somit

Das Verallgemeinerte Poincaré–Veblensche Dualitätsgesetz. Die reduzierten Bettischen Gruppen der Dimensionen r und $n-r$ einer n-dimensionalen

h-Mannigfaltigkeit sind zueinander primitiv (orthogonal), sobald man als Multiplikationsgesetz die Bildung der Schnittzahl betrachtet; insbesondere sind die beiden Gruppen untereinander isomorph.

5. Wir gehen jetzt zum Alexanderschen Dualitätssatz über. Wir beweisen zunächst den Alexanderschen Satz im engeren Sinne (der dennoch eine Verallgemeinerung des von Alexander selbst bewiesenen Resultates bildet), indem wir voraussetzen, daß in der h-Mannigfaltigkeit M^n sämtliche Zyklen beranden. Solche h-Mannigfaltigkeiten nennen wir (verallgemeinerte) *Poincarésche Räume*. Es sei sodann K ein (aus den Elementen der gegebenen Simplizialzerlegung von M^n) aufgebauter Komplex, von dem wir nur voraussetzen, daß er nicht mit M^n zusammenfällt.

Bei der folgenden Untersuchung der Barandungsrelationen in $M^n - K$ können wir uns auf die Betrachtung der aus baryzentrischen Sternen $\mathfrak{b}$ aufgebauten Komplexen beschränken. Wir tun das, ohne es jedesmal zu erwähnen.

Es sei $\mathfrak{a}$ ein s-dimensionaler Zyklus in $M^n - K$. Da $\mathfrak{a}$ in M^n berandet, existiert ein $\mathfrak{c} \to \mathfrak{a}$; $\mathfrak{c}$ ist eine Linearform in den $\mathfrak{b}_i^{s+1}$ und kann in der Gestalt $\mathfrak{c} = \mathfrak{b} + \mathfrak{d}$ dargestellt werden, wobei $\mathfrak{b}$ ein Element von $\mathfrak{Z}^{s+1}$ und $\mathfrak{d}$ zu K fremd ist. Ferner ist

$$\mathfrak{d} = \mathfrak{c} - \mathfrak{b} \to \dot{\mathfrak{c}} - \dot{\mathfrak{b}} = \mathfrak{a} - \dot{\mathfrak{b}},$$

also $\mathfrak{a} \sim \dot{\mathfrak{b}}$ in $M^n - K$.

Es sei $\mathfrak{a}'$ irgendein anderer Zyklus in $M^n - K$, der dortselbst zu $\mathfrak{a}$ homolog ist. Wenn $\mu \neq 0$ ist, so gibt es einen $\mathfrak{e} \subset M^n - K$,

$$\mathfrak{e} \to \mathfrak{a}' - \mathfrak{a}.$$

Andererseits berandet a' in M^n, was zu

$$\mathfrak{c}' \to \mathfrak{a}', \quad \mathfrak{c}' = \mathfrak{b}' + \mathfrak{d}'$$

führt. Ferner ist

$$\mathfrak{c} - \mathfrak{c}' + \mathfrak{e} \to \mathfrak{a} - \mathfrak{a}' + \mathfrak{a}' - \mathfrak{a} = 0,$$

so daß $\mathfrak{c} - \mathfrak{c}' + \mathfrak{e}$ ein Zyklus ist (der in M^n berandet). Folglich gibt es einen $\mathfrak{f}$ mit

$$\mathfrak{f} \to \mathfrak{c} - \mathfrak{c}' + \mathfrak{e}, \qquad \mathfrak{f} = \mathfrak{f}' + \mathfrak{f}'',$$

wobei $\mathfrak{f}' \in \mathfrak{L}^{s+2}$, $\mathfrak{f}'' \subset M^n - K$. Der reduzierte Rand von $\mathfrak{f}'$ ist, wie leicht ersichtlich, $\mathfrak{b} - \mathfrak{b}'$. Mit anderen Worten: *Aus der Homologie $\mathfrak{a} \sim \mathfrak{a}'$ in $M^n - K$ folgt, daß die entsprechenden Komplexe $\mathfrak{b}$ und $\mathfrak{b}'$ (deren Ränder $\mathfrak{a}$ und $\mathfrak{a}'$ bzw. homolog in $M^n - K$ sind) Elemente von $\mathfrak{Z}^{s+1}$ sind, die zur selben Restklasse bezüglich $\mathfrak{H}^{s+1}$ gehören.* Dasselbe Ergebnis findet auch im Falle $\mu = 0$ statt: dann ist nämlich (für ein bestimmtes $k \neq 0$)

$$\mathfrak{e} \rightarrow k(\mathfrak{a}' - \mathfrak{a}) \quad (\text{in}\, M^n - K),$$

$$\mathfrak{f} \rightarrow k(\mathfrak{c} - \mathfrak{c}') + \mathfrak{e}, \qquad \mathfrak{f} = \mathfrak{f}' + \mathfrak{f}'', \qquad \mathfrak{f}' \in \mathfrak{L}^{s+2}, \qquad \mathfrak{f}'' \subset M^n - K;$$

der reduzierte Rand von $\mathfrak{f}'$ ist diesmal $k(\mathfrak{b} - \mathfrak{b}')$, und die obige Behauptung behält ihre Gültigkeit.

Wir fassen nochmals zusammen: Jeder s-dimensionale Zyklus $\mathfrak{a} \subset M^n - K$ ist dortselbst dem Rande eines Elementes $\mathfrak{b}$ von $\mathfrak{Z}^{s+1}$ homolog, wobei, wenn $\mathfrak{a}$ und $\mathfrak{a}'$ in $M^n - K$ homolog sind, $\mathfrak{b}$ und $\mathfrak{b}'$ zur selben Restklasse in bezug auf $\mathfrak{H}^{s+1}$ gehören; hieraus folgt aber ein Isomorphismus zwischen der s-dimensionalen Bettischen Gruppe von $M^n - K$, $\mathfrak{B}^s(M^n - K)$ und der Gruppe $\mathfrak{Z}^{s+1} | \mathfrak{H}^{s+1}$. Nun ist aber die r-dimensionale Bettische Gruppe von K, d. h. die Gruppe $Z^r | H^r$ zu $\mathfrak{Z}^{n-r} | \mathfrak{H}^{n-r}$ primitiv (orthogonal), und die Multiplikation ist dabei durch die Schnittbildung gegeben. Folglich bilden auch $Z^r | H^r$ und $\mathfrak{B}^{n-r-1}(M^n - K)$ ein primitives (orthogonales) Gruppenpaar, wobei die Schnittzahl des entsprechenden $\mathfrak{b}$ mit dem r-dimensionalen Zyklus aus K nichts anderes als die Verschlingungszahl dieses Zyklus mit dem entsprechen Zyklus $\mathfrak{a}$ aus $M^n - K$ definiert. Wir erhalten somit folgenden verallgemeinerten

Dualitätssatz von Alexander im engeren Sinne. Es sei M^n ein verallgemeinerter Poincaréscher Raum von der Dimension n; wenn K einen Komplex in M^n bedeutet, so ist die r-dimensionale Bettische Gruppe von K zu der $n - r - 1$-dimensionalen Bettischen Gruppe von $M^n - K$ primitiv (orthogonal), sobald man als Multiplikation die Bildung der Verschlingungszahl der entsprechenden Zyklen betrachtet; insbesondere sind die beiden Gruppen untereinander isomorph.

6. Es sei jetzt M^n eine beliebige h-Mannigfaltigkeit. Wir betrachten irgendeinen aus Simplexen von M^n aufgebauten Komplex K, von dem nur vorausgesetzt wird, daß er nicht mit M^n zusammenfällt. B^r, $\mathfrak{B}^r$, W^r seien bzw. die r-dimensionalen Bettischen Gruppen von K, $M^n - K$ und M^n. Wegen $K \subset M^n$ wird B^r auf eine Untergruppe $\hat{V}^r$ und $\mathfrak{B}^r$ auf die Untergruppe $\hat{\mathfrak{B}}^r$ von W^r homomorph abgebildet; die Kerne dieser Homomorphismen seien A^r und $\mathfrak{A}^r$. Bei $\mu \neq 0$ setzen wir $V^r = \hat{V}^r$ und $\mathfrak{B}^r = \hat{\mathfrak{B}}^r$, während bei $\mu = 0$ durch V^r bzw. $\mathfrak{B}^r$ die kleinste Untergruppe mit Division von W^r über $\hat{V}^r$ bzw. $\hat{\mathfrak{B}}^r$ bezeichnet wird. Offenbar besteht A^r aus allen denjenigen Elementen – d. h. Restklassen von Z^r mod H^r –, die nur Zyklen, welche in M^n homolog Null sind, enthalten. Dementsprechend bezeichnen wir mit $\overset{*}{A}{}^r$ die Gruppe aller derjenigen Elemente von B^r, die Restklassen sind, welche Zyklen enthalten, die in M^n beranden. Offenbar ist $\overset{*}{A}{}^r$ eine Untergruppe von A^r, welche im Fall $\mu \neq 0$ mit A^r zusammenfällt;

ferner ist im Falle $\mu = 0$ A^r die kleinste in B^r enthaltene Obergruppe mit Division für die Gruppe $\overset{*}{A}{}^r$. Analog wird auch die Gruppe $\overset{*}{\mathfrak{A}}{}^r$ definiert. Der verallgemeinerte Alexandersche Dualitätssatz läßt sich sodann in der Gestalt der folgenden beiden Behauptungen aussprechen:

Dualitätssatz von Alexander im weiteren Sinne.

Vorbemerkung. Wegen des Poincaré–Veblenschen Satzes ist W^r, W^{n-r} ein Gruppenpaar.

Erste Behauptung:

$$(W^r, \mathfrak{B}^{n-r}) = V^r; \quad (W^{n-r}, V^r) = \mathfrak{B}^{n-r}. \tag{3}$$

Zweite Behauptung:

A^r und $\overset{*}{\mathfrak{A}}{}^{n-r-1}$ bzw. $\overset{*}{A}{}^r$ und $\mathfrak{A}^{n-r-1}$ bilden ein primitives (orthogonales) Gruppenpaar (wobei als Produkte die Verschlingungszahlen der entsprechenden Zyklen auftreten).

Beweis der ersten Behauptung. Vermöge des Satzes III des Kap. I (Sektion 10) genügt es, die eine der beiden Formeln (3) zu beweisen, z. B. die zweite. Zu diesem Zweck bemerken wir zunächst, daß jeder außerhalb von K gelegene $n-r$-dimensionale Zyklus (sowie jeder ihm in M^n homologe Zyklus) selbstverständlich mit jedem r-dimensionalen Zyklus aus K die Schnittzahl Null hat. Daraus folgt unmittelbar, daß $\mathfrak{B}^{n-r} \subset (W^{n-r}, V^r)$ ist. Um die umgekehrte Inklusion nachzuweisen, betrachten wir irgendeinen $n-r$-dimensionalen Zyklus $\mathfrak{a}$ von M^n und zeigen, daß, falls $\mathfrak{a}$ mit sämtlichen r-dimensionalen Zyklen von K die Schnittzahl Null hat, ein Zyklus $\mathfrak{a}'$ in $M^n - K$ existiert, welcher im Falle $\mu = 0$ mit $\mathfrak{a}$, im Falle $\mu = 0$ mit $k\mathfrak{a}$, $k \neq 0$, in M^n homolog ist. Man kann sich dabei auf den Fall beschränken, daß $\mathfrak{a}$ eine Linearform in den zu den r-dimensionalen Simplexen von M^n dualen baryzentrischen Sternen ist. Sodann ist

$$\mathfrak{a} = \mathfrak{b} + \mathfrak{b}', \quad \mathfrak{b} \in \mathfrak{B}^{n-r}, \quad \mathfrak{b}' \subset M^n - K.$$

Da die Schnittzahl von $\mathfrak{a}$ mit sämtlichen Zyklen aus K Null ist, ist $\mathfrak{b} \subset \mathfrak{H}^{n-r}$, folglich existiert ein Element $\mathfrak{c}$ von $\mathfrak{L}^{n-r+1}$ und eine höchstens im Fall $\mu = 0$ von 1 verschiedene positive ganze Zahl k derart, daß $k\mathfrak{b}$ den reduzierten Rand von $\mathfrak{c}$ bildet. Dann ist aber $\mathfrak{a}' = k\mathfrak{a} - \dot{\mathfrak{c}}$ ein in $M^n - K$ gelegener Zyklus und (da offenbar $\dot{\mathfrak{c}} \sim 0$ in M^n ist) ist

$$\mathfrak{a}' \sim k\mathfrak{a} \quad (\text{in } M^n),$$

w.z.b.w.

Beweis der zweiten Behauptung. Wir beweisen in erster Linie, daß A^r und $\mathfrak{A}^{*n-r-1}$ zueinander primitiv (orthogonal) sind. Zunächst ist klar, daß jeder s-dimensionale Zyklus aus $M^n - K$, welcher in M^n berandet, in $M^n - K$ dem Rande eines Elementes von $\mathfrak{Z}^{s+1}$ homolog ist. Es genügt deshalb solche Zyklen zu betrachten, die Ränder der Elemente von $\mathfrak{Z}^{n-r}$ sind. Wir bezeichnen ferner mit $\bar{H}^r$ bzw. $\dot{\mathfrak{H}}^s$ diejenige Untergruppe von Z^r bzw. von $\mathfrak{Z}^s$, welche aus allen Elementen besteht, die in M^n, bzw. deren Ränder in $M^n - K$ homolog Null sind. Es soll jetzt die Formel

$$(\mathfrak{Z}^{n-r}, \bar{H}^r) = \dot{\mathfrak{H}}^{n-r} \tag{4}$$

bewiesen werden. Die Inklusion $\dot{\mathfrak{H}}^{n-r} \subset (\mathfrak{Z}^{n-r}, \bar{H}^r)$ ist wieder trivial; um die umgekehrte Inklusion zu beweisen, hat man zu zeigen, daß, wenn für ein bestimmtes $\mathfrak{a} \subset \mathfrak{Z}^{n-r}$ für sämtliche γ aus $\bar{H}^r$ $\chi(\mathfrak{a}, \gamma) = 0 \pmod{\mu}$ ist, $\dot{\mathfrak{a}}$ in $M^n - K$ notwending homolog Null ist. Wir definieren zu diesem Zweck für jedes Element $y \subset B^r$ das Produkt $y \cdot \mathfrak{a}$ als die Schnittzahl $(\bmod\ \mu)$ eines beliebigen Zyklus der Restklasse y mit $\mathfrak{a}$. Aus dieser Definition folgt, daß für die Gruppe (Restklasse) A^r, als Element y_0 von B^r betrachtet, $y_0 \cdot \mathfrak{a} = 0$ ist, woraus sich eine natürliche Definition der Produkte $x \cdot \mathfrak{a}$ ergibt, wenn x ein beliebiges Element der Faktorgruppe $B^r | A^r = \hat{V}^r$ bedeutet. Im Falle $\mu \neq 0$ ist dadurch automatisch ein Multiplikationsgesetz $x \cdot \mathfrak{a}$ für die Elemente x von V^r definiert (denn in diesem Fall ist ja V^r mit $\hat{V}^r$ identisch); wenn dagegen $\mu = 0$ ist, definiert man das analoge Multiplikationsgesetz folgendermaßen: x sei ein beliebiges Element von V^r; es existiert dann ein $z \subset \hat{V}^r$ mit $h \cdot x = z$, und wir setzen $x \cdot \mathfrak{a} = (z \cdot \mathfrak{a})/h$. Dabei können Brüche auftreten, da aber V endlichviele Erzeugenden besitzt, sind alle Nenner beschränkt, so daß für ein passend gewähltes k bei jeder Wahl von $x \subset V^r$ $x \cdot k\mathfrak{a}$ eine ganze Zahl ist. Der Einheitlichkeit der Formel halber führen wir auch im Fall $\mu \neq 0$ den Koeffizienten k ein, nur soll er dann definitionsgemäß gleich 1 sein. Da W^r und W^{n-r} zueinander primitiv (orthogonal) sind, bilden V^r und W^{n-r} (kraft des Hilfssatzes des Sektiones 9 des Kap. I) ein konjugiertes Gruppenpaar. Da überdies $(W^r, W^{n-r}) = 0$ ist, sind die Voraussetzungen des Sektiones 8 des Kap. I reichlich erfüllt, und es gibt ein solches Element $b \subset W^{n-r}$, daß bei jedem $x \subset V^r$ $x \cdot k\mathfrak{a} = x \cdot b$ ist, dies bedeutet aber, daß, wenn β ein beliebiger Zyklus aus der Restklasse b und ξ ein beliebiger Zyklus einer beliebigen Restklasse $x \subset V^r$, also einfach ein beliebiger r-dimensionaler Zyklus von K ist, $\chi(\xi, k\mathfrak{a} - \beta) = 0$ ausfällt. Wir zerlegen jetzt den Zyklus β wie üblich in die Summe zweier Komplexe $\mathfrak{a}'$ und $\mathfrak{a}''$, wobei $\mathfrak{a}'$ ein Element von $\mathfrak{Z}^{n-r}$ und $\mathfrak{a}''$ zu K fremd ist. Aus dem soeben Bewiesenen folgt, daß $k\mathfrak{a} - \mathfrak{a}'$ ein Element von $(\mathfrak{Z}^{n-r}, Z^r)$ ist; also ist (bei einem nur im Fall $\mu = 0$ von 1 eventuell verschiedenen k') $k'(k\mathfrak{a} - \mathfrak{a}')$ der reduzierte Rand eines gewissen $\mathfrak{c}$ aus $\mathfrak{L}^{n-r+1}$ ist. Wir haben ferner die Berandungsrelation

$$k'k\mathfrak{a}-\dot{\mathfrak{c}}-k'\beta \rightarrow k'k\dot{\mathfrak{a}}, \tag{5}$$

andererseits ist aber nach der Definition von $\mathfrak{a}'$ und $\mathfrak{c}$ die linke Seite von (5) ein in $M^n - K$ liegender Komplex, so daß aus (5) die Homologie

$$k'k\dot{\mathfrak{a}} \sim 0 \quad \text{in } M^n - K$$

folgt, wie wir es beweisen wollten.

Nachdem die Formel (4) bewiesen ist, bleibt nur noch ein leichter Schritt, und die zweite Behauptung des Alexanderschen Dualitätssatzes wird erledigt. Zunächst ist nach dem Sektion 2 diese Abschnittes L^r zu $\mathfrak{L}^{n-r}$ primitiv (orthogonal); da (wegen Sektion 3, Form. (2))

$$(L^r, \mathfrak{Z}^{n-r}) = H^r \subset Z^r, \quad (\mathfrak{L}^{n-r}, Z^r) = \mathfrak{H}^{n-r} \subset \mathfrak{Z}^{n-r}$$

ist, bilden nach Satz IV des Kap. I Z^r und $\mathfrak{Z}^{n-r}$ ein konjugiertes Gruppenpaar; da ferner $(Z^r, \mathfrak{Z}^{n-r}) = H^r \subset \bar{H}^r$ ist, sind wegen des Satzes IV, Kap. I auch die Gruppen $\bar{H}^r$ und $\mathfrak{Z}^{n-r}$ konjugiert, so daß die Faktorgruppen nach den Annullatoren $(\bar{H}^r, \mathfrak{Z}^{n-r}) = H^r$ und $(\mathfrak{Z}^{n-r}, \bar{H}^r) = \dot{\mathfrak{H}}^{n-r}$ primitiv (orthogonal) sind. M. a. W.: die Gruppen $\bar{H}^r | H^r$ und $\mathfrak{Z}^{n-r} | \dot{\mathfrak{H}}^{n-r}$ sind zueinander primitiv (orthogonal. Die erste dieser Gruppen ist aber definitionsgemäß die Gruppe A^r; ein Element $\mathfrak{z}$ der zweiten Gruppe ist eine Restklasse von $\mathfrak{Z}^{n-r}$ mod $\dot{\mathfrak{H}}^{n-r}$; jedem Element $\mathfrak{z}$ dieser Restklasse wird sein Rand $\mathfrak{a}^{n-r-1}$ zugeordnet, und alle diese Ränder sind untereinander homolog in $M^n - K$ (die Restklassen sind ja modulo $\dot{\mathfrak{H}}^{n-r}$ genommen!); mithin wird der ganzen Restklasse $\mathfrak{z}$ ein Element von $\mathfrak{B}^{n-r-1}$, und sogar ein Element von $\overset{*}{\mathfrak{A}}{}^{n-r-1}$ zugeordnet (die $\mathfrak{a}^{n-r-1}$ beranden ja nach ihrer Konstruktion in M^n, weil sie als Ränder definiert waren). Wenn umgekehrt ein Element von $\overset{*}{\mathfrak{A}}{}^{n-r-1}$ gegeben ist, so ist es eine Restklasse modulo $\mathfrak{H}^{n-r-1}$ der Gruppe aller Zyklen von $M^n - K$, und in dieser Restklasse ist ein Zyklus enthalten, der in M^n berandet; dieser ist dem Rand eines Elementes von $\mathfrak{Z}^{n-r}$ homolog; folglich ist in der Restklasse der Rand eines Elementes aus $\mathfrak{Z}^{n-r}$ enthalten; auf die Weise werden Zyklen, die einem und demselben Element von $\overset{*}{\mathfrak{A}}{}^{n-r-1}$ entspringen, Elemente von $\mathfrak{Z}^{n-r}$ zugeordnet, die zur selben Restklasse modulo $\dot{\mathfrak{H}}^{n-r}$ gehören, folglich ein und dasselbe Element von $\mathfrak{Z}^{n-r} | \dot{\mathfrak{H}}^{n-r}$. Somit wird ein Isomorphismus zwischen $\overset{*}{\mathfrak{A}}{}^{n-r-1}$ und $\mathfrak{Z}^{n-r} | \dot{\mathfrak{H}}^{n-r}$ hergestellt, welcher erlaubt, die für $A^r = \bar{H}^r | H^r$ und $\mathfrak{Z}^{n-r} | \dot{\mathfrak{H}}^{n-r}$ erklärte Multiplikation unmittelbar auf A^r und $\overset{*}{\mathfrak{A}}{}^{n-r-1}$ zu übertragen; dieses Multiplikationsgesetz ergibt als Produkt der Elemente $a \subset A^r$ und $\mathfrak{a} \subset \overset{*}{\mathfrak{A}}{}^{n-r-1}$ die Verschlingungszahl eines Zyklus aus der Restklasse a mit einem solchen aus der Restklasse $\mathfrak{a}$.

7. Somit ist bewiesen, daß A^r und $\overset{*}{A}{}^{n-r-1}$ ein primitives (orthogonales) Gruppenpaar bilden. Man kann dies als eine sich auf die beiden Komplexe K und L beziehende Aussage auffassen, denn jede in $M^n - K$ stattfindende Berandung läßt sich ja nach L wegziehen. Nun aber treten in unserem Beweise die Elemente (erster Art) von K und die Elemente (zweiter Art) von L in durchaus symmetrischer Weise auf, so daß man ihre Rollen tauschen könnte. Dann aber würden die Überlegungen der letzteren Paragraphen uns zum Beweis der Primitivität (Orthogonalität) des Gruppenpaares $\overset{*}{A}{}^r$ und $\mathfrak{A}^{n-r-1}$ führen, w.z.b.w.

8. Wir wollen zum Schluß noch zeigen, wie man aus dem soeben Bewiesenen die Dualitätsformeln ableitet, die für den Fall mod 2 noch früher aufgestellt worden sind.[29] Dabei ist unter dem Modul μ Null oder eine beliebige Primzahl zu verstehen.

Wenn man im allgemeinen mit $\mathfrak{r}(G)$ den Rang der Abelschen Gruppe G bezeichnet, so hat man zunächst:

$$\mathfrak{r}(\overset{*}{\mathfrak{A}}{}^{n-r-1}) = \mathfrak{r}(\mathfrak{A}^{n-r-1}) = \mathfrak{r}(\overset{*}{A}{}^r) = \mathfrak{r}(A^r) \tag{6}$$

und

$$\mathfrak{r}(\hat{\mathfrak{B}}^{n-r}) = \mathfrak{r}(\mathfrak{B}^{n-r}), \quad \mathfrak{r}(\hat{V}^r) = \mathfrak{r}(V^r). \tag{6'}$$

Da ferner $(W^r, \mathfrak{B}^{n-r}) = V^r$ ist, während $(\mathfrak{B}^{n-r}, W^r)$ aus dem Nullelement besteht, ist $W^r | V^r$ mit $\mathfrak{B}^{n-r}$ isomorph, was unter Beachtung der allgemeinen Relation $\mathfrak{r}(W^r) = \mathfrak{r}(W^r | V^r) + \mathfrak{r}(V^r)$

$$\mathfrak{r}(W^r) = \mathfrak{r}(V^r) + \mathfrak{r}(\mathfrak{B}^{n-r}) \tag{7}$$

ergibt. Wir haben also

$$\mathfrak{r}(\mathfrak{B}^{n-r-1}) = \mathfrak{r}(\mathfrak{A}^{n-r-1}) + \mathfrak{r}(\mathfrak{B}^{n-r-1} | \mathfrak{A}^{n-r-1}) = \mathfrak{r}(\mathfrak{A}^{n-r-1}) + \mathfrak{r}(\hat{\mathfrak{B}}^{n-r-1}). \tag{8}$$

Aus (6′) und (7) folgt ferner

$$\mathfrak{r}(\hat{\mathfrak{B}}^{n-r-1}) = \mathfrak{r}(W^{r+1}) - \mathfrak{r}(V^{r+1}); \tag{9}$$

wenn man dies und (6) in (8) einsetzt und beachtet, daß

$$\mathfrak{r}(B^r) = \mathfrak{r}(A^r) + \mathfrak{r}(B^r | A^r) = \mathfrak{r}(A^r) + \mathfrak{r}(V^r)$$

ist, erhält man schließlich

$$\mathfrak{r}(\mathfrak{B}^{n-r-1}) = \mathfrak{r}(B^r) + \mathfrak{r}(W^{r+1}) - \mathfrak{r}(V^r) - \mathfrak{r}(V^{r+1}),$$

w.z.b.w.

Kapitel III
Das allgemeine Dualitätsgesetz für abgeschlossene Mengen

I. Direkte und inverse Folgen von Homomorphismen

1. Es sei

$$U_1, U_2, \ldots, U_m, \ldots \tag{1}$$

eine unendliche Folge von Gruppen, von denen jede Gruppe U_m in[30] ihren Nachfolger U_{m+1} mittels des Homomorphismus φ_m abgebildet wird; die Folge

$$\varphi_1, \varphi_2, \ldots, \varphi_m, \ldots \tag{2}$$

heißt sodann eine *direkte Homomorphismenfolge.* Sie bestimmt folgendermaßen eine neue Gruppe U – den *Limes der Folge* (1) *in bezug auf die Folge* (2) oder kurz – die Limesgruppe. Zunächst nenne man eine *Fundamentalfolge* jede Folge von der Gestalt

$$x_k, x_{k+1}, \ldots, x_m, \ldots \tag{3}$$

wobei x_m ein Element von U_m und dabei stets $x_{m+1} = \varphi_m(x_m)$ ist. Zwei Fundamentalfolgen (3) und

$$y_h, y_{h+1}, \ldots, y_m, \ldots, \tag{4}$$

heißen *konfinal*, wenn es ein x derart gibt, daß für $m > x$ $x_m = y_m$ ist. Offenbar zerfällt die Gesamtheit der Fundamentalfolgen in Klassen untereinander konfinaler Fundamentalfolgen. Diese Klassen werden zu Elementen der Gruppe U gemacht. Die Gruppenoperation in U wird sodann wie folgt erklärt. Es seien α und β zwei Klassen; man wähle in jeder je eine Fundamentalfolge, etwa

$$a = (x_k, x_{k+1}, \ldots, x_m, \ldots)$$

und

$$b = (y_h, y_{h+1}, \ldots, y_m, \ldots)$$

mit etwa $h \geqslant k$; die durch die Fundamentalfolge

$$c = (x_h \cdot y_h, x_{h+1} \cdot y_{h+1}, \ldots, x_m \cdot y_m, \ldots)$$

bestimmte Klasse γ heißt dann das Produkt (im Sinne der Gruppenoperation in U) der Elemente α und β. Offenbar ist γ durch α und β eindeutig bestimmt (d. h. γ hängt von der Wahl der Folgen a und b in den Klassen α und β nicht ab). Wenn e_m das Einheitselement von U_m ist, so ist das durch $(e_1, e_2, \ldots, e_m, \ldots)$ definierte Element von U das Einheitselement in bezug auf die soeben definierte Gruppenoperation; man erhält ferner zu jedem Element α von U das inverse α^{-1}, wenn man in den Folgen der Klasse α alle Elemente durch ihre Inverse ersetzt. Da überdies unsere Operation dem assoziativen Gesetz genügt, sind alle Gruppenpostulate erfüllt, und U ist eine Gruppe. Bevor wir weiter gehen, führen wir folgende Bezeichnung ein: φ_s^r soll die Abbildung $\varphi_{s-1}(\ldots(\varphi_{r-1}(\varphi_r)))$ von U_r in U_s bedeuten (dabei ist natürlich $s > r$).

Wenn

$$U_{m_1}, U_{m_2}, \ldots, U_{m_q}, \ldots \tag{5}$$

eine Teilfolge der Folge (1) ist, so entspricht ihr die Homomorphismenfolge

$$\varphi_1', \varphi_2', \ldots, \varphi_q', \ldots \quad \text{mit } \varphi_q' = \varphi_{m_{q+1}}^{m_q}, \tag{6}$$

und die Gruppe U', die als Limes von (5) in bezug auf (6) auftritt, ist, wie leicht ersichtlich, mit dem Limes von (1) in bezug auf (2) isomorph. Wir sagen in diesem Falle von der direkten Homomorphismenfolge (2), sie *umfasse* die Folge (6). Zwei direkte Homomorphismenfolgen heißen ferner äquivalent, wenn man in ihnen zwei solche Teilfolgen finden kann, welche von einer dritten Folge umfaßt werden. Diesen Äquivalenzbegriff genügt den Gleichheitsaxiomen (der Reflektivität, der Symmetrie und der Transitivität),[31] so daß man von Klassen untereinander äquivalenter Homomorphismenfolgen sprechen kann. Da ferner zwei Folgen, von denen die eine die andere umfaßt, im Limes isomorphe Gruppen bestimmen, ergibt sich folgender Satz:

I. *Äquivalente Homomorphismenfolgen haben isomorphe Limesgruppen*

Eine direkte Homomorphismenfolge (1), (2) wird von jetzt an stets mit $F(U_m, \varphi)$ bezeichnet.

2. Wir betrachten wieder eine Folge

$$U_1, U_2, \ldots, U_m, \ldots, \tag{1}$$

nehmen aber jetzt an, daß U_{m+1} in U_m mittels des Homomorphismus π_m abgebildet ist; die Folge

$$\pi_1, \pi_2, \ldots, \pi_m, \ldots \tag{7}$$

(dabei wird die Abbildung $\pi_r(\ldots\pi_{s-2}(\pi_{s-1}))$ von U_s in U_r $(s > r)$ mit π_r^s bezeichnet) heißt sodann eine *inverse Homomorphismenfolge.* Genau wie vorher lassen sich auch für inverse Homomorphismenfolgen die Begriffe der umfassenden bzw. äquivalenten Folgen einführen. Es gibt dabei kein

Analogon zum Satze I, da ja eine inverse Homomorphismenfolge keine Limesgruppe besitzt.

Eine inverse Homomorphismenfolge (1), (7) wird von jetzt an stets mit $\bar{\mathcal{F}}(U_m, \pi)$ bezeichnet. Eine Homomorphismenfolge in (1), von der man nicht weiß, ob sie direkt oder invers ist, soll schlechtweg mit $\mathcal{F}(U_m)$ bezeichnet werden.

3. Von jetzt an beschränken wir uns auf kommutative Gruppen und bedienen uns dabei wir früher der additiven Schreibweise.

Hilfssatz. Es seien U, A bzw. V, B zwei primitive (orthogonale) Gruppenpaare in bezug auf den Modul M. Es sei ferner ein Homomorphismus φ von U in V gegeben. Es gibt dann eine und nur eine homomorphe Abbildung ψ von B in A, welche folgender Bedingung genügt: wenn u bzw. b irgendein Element von U bzw. B ist, so ist

$$u \cdot \psi(b) = \varphi(u) \cdot b. \tag{8}$$

Beweis. Wenn u die ganze Gruppe U durchläuft, nimmt $\varphi(u) \cdot b$ gewisse Werte aus M an, wobei stets $\varphi(u)b + \varphi(u') \cdot b = \varphi(u + u') \cdot b$ ist.

Man betrachte jetzt die homomorphe Abbildung

$$z(u) = \varphi(u) \cdot b$$

von U in M. Aus Kap. I, Sektion 9 folgt sodann daß es ein einziges Element a von A gibt, welches für sämtliche u

$$u \cdot a = z(u) = \varphi(u) \cdot b$$

leistet. Dieses a bezeichnen wir mit $\psi(b)$. Aus

$$\begin{aligned} u \cdot (\psi(b) + \psi(b')) &= u \cdot \psi(b) + u \cdot \psi(b') = \varphi(u) \cdot b + \varphi(u) \cdot b' \\ &= \varphi(u) \cdot (b + b') = u \cdot \psi(b + b') \end{aligned}$$

folgt

$$u \cdot ((\psi(b) + \psi(b')) - \psi(b + b')) = 0,$$

also (wegen der Primitivität des Gruppenpaares U, A)

$$\psi(b) + \psi(b') = \psi(b + b');$$

ψ ist also ein Homomorphismus, w.z.b.w.

4. Definition. Es seien $\mathcal{F}(U_m, \varphi)$ und $\bar{\mathcal{F}}(V_m, \pi)$ gegeben. Diese beiden Homomorphismenfolgen heißen zueinander *orthogonal* (in bezug auf den Modul M), wenn folgende Bedingungen erfüllt sind:

1. U_m und V_m bilden ein primitives (orthogonales) Gruppenpaar (in bezug auf M);

2. wenn u und v beliebige Elemente von U_m bzw. V_{m+1} sind, so ist

$$\varphi_m(u) \cdot v = u \cdot \pi_m(v).$$

Aus dem Hilfssatz des Sektiones 3 folgt unmittelbar der

Satz II. Es sei $\digamma(U_m)$ gegeben, wobei die U_m so beschaffen sind, daß zu jeder Gruppe U_m eine zu ihr primitive (orthogonale) Gruppe V_m existiert (die dann bis auf Isomorphismen eindeutig bestimmt ist). In den V_m läßt sich eine Homomorphismenfolge $\bar{\digamma}(V_m)$ eindeutig (ebenfalls bis auf Isomorphismen) so definieren, daß $\digamma$ und $\bar{\digamma}$ zueinander orthogonal (in bezug auf den Modul M) sind.

Bemerkung. Wenn zwei Homomorphismenfolgen zueinander orthogonal sind, so gilt dasselbe von je zwei Teilfolgen, die aus zueinander entsprechenden (d. h. mit derselben Nummer m versehenen) Gliedern der beiden Folgen bestehen.

Satz III. Es seien $\digamma(U_m)$ und $\digamma(V_m)$ zueinander orthogonal ebenso wie $\digamma'(U'_m)$ und $\digamma'(V'_m)$; wenn überdies $\digamma(U_m)$ mit $\digamma'(U'_m)$ äquivalent ist, so ist auch $\digamma(V_m)$ mit $\digamma'(V'_m)$ äquivalent.

Beweis. Es gibt in $\digamma(U_m)$ und $\digamma'(U'_m)$ je eine Teilfolge $\digamma_0$ und $\digamma'_0$, welche Teilfolgen eines und desselben $\digamma_1$ sind; von der Homomorphismenfolge $\digamma_1$ dürfen wir dabei voraussetzen, daß sie nur aus den Elementen von $\digamma_0$ und $\digamma'_0$ besteht (indem wir all übrigen eventuell vorhandenen Elementen einfach streichen); wir wählen ferner aus $\digamma(V_m)$ und $\digamma'(V'_m)$ die den Teilfolgen $\digamma_0$ und $\digamma'_0$ entsprechenden Teilfolgen $\bar{\digamma}_0$ und $\bar{\digamma}'_0$. Da man zu jedem Element von $\digamma_1$ ein primitives (orthogonales) Element konstruieren kann, existiert kraft des Satzes II eine zu $\digamma_1$ orthogonale Homomorphismenfolge $\bar{\digamma}_1$; $\bar{\digamma}_0$ und $\bar{\digamma}'_0$ können als Teilfolgen von $\bar{\digamma}_1$ betrachtet werden, $\digamma(V_m)$ und $\digamma'(V'_m)$ sind also äquivalent.

5. Wir heben besonders hervor: eine inverse Homomorphismenfolge $\bar{\digamma}(U_m, \pi)$ definiert – unter der Voraussetzung, daß es zu jedem U_m eine (also im wesentlichen nur eine) primitive (orthogonale) Gruppe V_m gibt – eindeutig eine zu $\bar{\digamma}(U_m, \pi)$ orthogonale direkte Homomorphismenfolge $\digamma(V_m, \varphi)$, und diese definiert eindeutig die Limesgruppe V. *Diese durch die Homomorphismenfolge $\bar{\digamma}(U_m, \pi)$ eindeutig* ***bestimmte Gruppen heißt***

die zu $\overline{F}(U_m, \pi)$ duale Gruppe. Äquivalente Homomorphismenfolgen besitzen isomorphe duale Gruppen.

II. Formulierung und Beweis des allgemeinen Dualitätssatzes

6. Es sei F eine im R^n gelegene kompakte abgeschlossene Menge. Man betrachte irgendein Projektionsspektrum[32]

$$A = (A_1, A_2, \ldots, A_m, \ldots) \tag{1}$$

von F und die zugehörigen simplizialen Abbildungen π_m von A_{m+1} auf A_m.

$B_m = B^r(A_m)$ sie die r-te Bettische Gruppe von A_m. Vermöge der simplizialen Abbildung π_m entsteht ein Homomorphismus von B_{m+1} in B_m (den wir ebenfalls mit π_m bezeichnen werden) und folglich eine inverse Homomorphismenfolge $\overline{F}(B_m, \pi)$. Man hat zunächst den

Hilfssatz. Es seien zwei verschiedene Projektionsspektra (1) und

$$A' = (A'_1, A'_2, \ldots, A'_m, \ldots) \tag{2}$$

der Menge F gegeben. Die diesen Spektra entsprechenden Homomorphismenfolgen $\overline{F}(B_m, \pi)$ und $\overline{F}(B'_m, \pi)$ sind äquivalent.

Nehmen wir für einen Augenblick an, der Hilfssatz wäre schon bewiesen.

Den Inbegriff aller untereinander äquivalenten Homomorphismenfolgen $F(B_m, \pi)$, die durch die Projektionsspektra von F bestimmt werden, ist offenbar eine topologische Invariante der Menge F. Wir nennen sie *die r-dimensionale Zyklosis der Menge F.* Die Zyklosis bestimmt eindeutig eine Gruppe, nämlich die (bis auf Isomorphie) einzige zu allen Homomorphismenfolgen der Zyklosis duale Gruppe, die wir kurz *die zur r-dimensionalen Zyklosis duale Gruppe* nennen. Diese Gruppe hat offenbar ebenfalls eine invariante topologische Bedeutung für die Menge F. Nun besteht der Hauptzweck dieses Kapitels im Beweise des folgenden Satzes.

Allgemeiner Dualitätssatz. Wenn F eine abgeschlossene kompakte Menge des R^n ist, so ist die zur r-dimensionalen Zyklosis von F duale Gruppe mit der $n-r-1$-ten Bettischen Gruppe des Komplementärraumes $R^n - F$ isomorph.

Hieraus ergibt sich ohne weiteres der

Invarianzsatz. Die Bettischen Gruppen[33] des Komplementärraumes zu einer abgeschlossenen Menge F des R^n sind topologische Invarianten der Menge F.

Beweis. Wir führen den Beweis des Hilfssatzes und des Dualitätssatzes gleichzeitig.

Es sei

$$Q_1, Q_2, \ldots, Q_i, \ldots \tag{3}$$

eine abnehmende Folge von Polyederumgebungen der Menge F, welche sich auf diese Menge zusammenziehen.

G_i sei die zu Q_i komplementäre offene Menge. Diese Mengen wachsen mit i, und ihre Vereinigungsmenge ist mit $G = R^n - F$ identisch. Wir bezeichnen mit β_i die r-te Bettische Gruppe von Q_i, mit $\bar{\beta}_i$ die $n - r - 1$-te Bettische Gruppe von G_i. Aus $Q_i \supset Q_{i+1}$ folgt eine homomorphe Abbildung $\tilde{\omega}_i$ von β_{i+1} in β_i, aus $G_{i+1} \supset G_i$ eine homomorphe Abbildung φ_i von $\bar{\beta}_i$ in $\bar{\beta}_{i+1}$. Die so entstandenen Homomorphismenfolgen $\bar{\mathcal{F}}(\beta_i, \tilde{\omega})$ und $\mathcal{F}(\bar{\beta}_i, \varphi)$ sind zueinander orthogonal, denn β_i ist ja zu $\bar{\beta}_i$ primitiv (orthogonal). Die durch die Homomorphismenfolge $\mathcal{F}(\bar{\beta}_i, \varphi)$ bestimmte Limesgruppe $\bar{\beta}$ ist, wie leicht ersichtlich, zur $n - r - 1$-ten Bettischen Gruppe von G isomorph. Wir brauchen also – um alles zu beweisen – nur noch folgendes zu zeigen:

Bei jeder Wahl des Projektionsspektrums (1) *der Menge F* ist die Homomorphismenfolge $\bar{\mathcal{F}}(B_i, \pi)$ mit $\bar{\mathcal{F}}(\beta_i, \tilde{\omega})$ äquivalent.

8. Wir wenden uns jetzt dem Beweise der letzteren Behauptung zu.

Zunächst ist es vorteilhaft, durch einen leichten Kunstgriff zu erreichen, daß die Komplexe A_i geometrisch, und zwar ohne Singularitäten realisiert seien. Zu diesem Zweck betrachten wir das topologische Produkt $Z = R^n \times E$ von R^n mit einem hinreichend hoch dimensionalen Simplex E. Der Schwerpunkt von E sei ξ; wir nehmen an, R^n sei mit $R^n \times \xi$ identisch, und betrachten eine Folge konzentrischer Simplexe $E_1, E_2, \ldots, E_h, \ldots$ um ξ, welche sich auf diesen Punkt zusammensiehen. Die

$$Q_1 \times E_1, Q_2 \times E_2, \ldots, Q_h \times E_h, \ldots$$

bilden sodann eine Folge von sich auf F zusammenziehender Polyederumgebungen von F in Z. Da $Q_i \times E_i$ innerhalb von sich selbst in Q_i stetig übergeführt werden kann (so daß dabei Q_i punktweise festbleibt), sind die Bettischen Gruppen von $Q_i \times E_i$ mit denen von Q_i isomorph, die entsprechenden Homomorphismen sind dieselben, und man kann ruhig Q_i durch $Q_i \times E_i$ – also eine Umgebung von F in R^n durch eine Umgebung von F in Z – ersetzen. In den $Q_i \times E_i$ lassen sich aber – wenn nur die Dimension von E hinreichend groß ist – sämtliche A_m (mit eventueller Ausnahme von endlich vielen) singularitätenfrei realisieren. Von jetzt an bezeichnen wir die $Q_i \times E_i$ einfach mit Q_i.

Es sei i beliebig, q so groß, daß A_q in Q_i liegt; daraus folgt die Existenz einer homomorphen Abbildung f_1 von B_q in β_i; wir schreiben dies so:

$$f_1(B_q) \subset \beta_i.$$

Es ist nun bekannt, daß, wenn ein Komplex K in einer hinreichend engen Umgebung von F liegt und aus genügend kleinen Simplexen aufgebaut ist, man diesen Komplex mittels einer kleinen Verschiebung seiner Eckpunkte simplizial in A_q abbilden kann.[34] Wenn man j hinreichend groß und die Simplexe von Q_j hinreichend klein nimmt, kann Q_j für einen solchen K gewählt werden. Die erwähnte simpliziale Abbildung von Q_j in A_q bezeichne man mit g; wir wollen sie folgendermaßen konstruieren. Zunächst sei s so groß, daß A_s in Q_j liegt. Man wähle jetzt eine solche Simplizialzerlegung $\mathfrak{z}$ von Q_j, daß eine gewisse Unterteilung A_s^* von A_s als Teilkomplex der Zerlegung $\mathfrak{z}$ betrachtet werden kann. Der simplizialen Abbildung g von Q_j in A_q soll die Simplizialzerlegung $\mathfrak{z}$ zugrunde gelegt werden. Es sei also a irgendein Eckpunkt von $\mathfrak{z}$. Wir untersuchen in erster Linie den Fall, wo a gleichzeitig ein Eckpunkt von A_s^* ist. In diesem Fall ist a ein innerer Punkt eines bestimmten (eventuell auch nulldimensionalen) Simplex T von A_s; als Bild von a soll ein Eckpunkt a' von A_q gewählt werden, in den vermöge der Projektion von A_s auf A_q irgendein Eckpunkt von T abgebildet wird; die übrigen (also nicht zu A_s^* gehörenden) Eckpunkte von $\mathfrak{z}$ mögen in je einen der zu ihnen am nächsten liegenden Eckpunkte von A_q abgebildet werden. Wie leicht ersichtlich, stimmt die so definierte Abbildung g auf den Komplex A_s mit der Projektion von A_s auf A_q algebraisch überein.[35] Endlich kann von g noch vorausgesetzt werden, daß sie mittels einer stetigen Deformation innerhalb von Q_i realisierbar sei.

Man erhält also folgenden Homomorphismen:

$$f_2(B_s) \subset \beta_j$$

(denn A_s ist ja in Q_j enthalten);

$$f_3(\beta_j) \subset B_q$$

(vermöge der Abbildung g von Q_j in A_q);

$$\tilde{\omega}(\beta_j) \subset \beta_i$$

(denn Q_j ist in Q_i enthalten),

$$\pi(B_s) \subset B_q$$

(Projektion). Dabei ist

$$f_1 f_3(\beta_j) = \tilde{\omega}(\beta_j), \tag{4}$$

$$f_3 f_2(B_s) = \pi(B_s), \tag{5}$$

so daß die Gruppen $\beta_i, B_q, \beta_j, B_s$ wie folgt ineinander abgebildet werden:

$$\beta_i \leftarrow B_q \leftarrow \beta_j \leftarrow B_s.$$

Wenn man dieses Verfahren mit $i = 1$ beginnt und immer weiter fortsetzt, erhält man ohne Mühe eine Homomorphismenfolge, welche wegen (4) und (5) gewisse Teilfolgen der beiden Folgen $\bar{F}(B_m, \pi)$ und $\bar{F}(\beta_m, \tilde{\omega})$ umfaßt. Somit sind die beiden letztgenannten Homomorphismenfolgen äquivalent, w.z.b.w.

Anhang I. Der Dualitätssatz für stetige Komplexe

Es sei K ein (singularitätenfreier) stetiger Komplex im R^n. Man betrachte einen zu K homöomorphen Polyederkomplex Q in einem hinreichend hoch dimensionalen R^m. Da K und Q homöomorphe abgeschlossene Mengen sind, sind die zu den r-dimensionalen Zyklosen derselben dualen Gruppen, also auch die $n-r-1$-te Bettische Gruppe von $R^n - K$ und die $m-r-1$-te Bettische Gruppe von $R^m - Q$ zueinander isomorph. Nun ist aber letztere Gruppe zur r-dimensionalen Bettischen Gruppe von Q (also auch zu der von K) isomorph.

Mit anderen Worten:

Wenn $K \subset R^n$ ein stetiger Komplex ist, so ist die $n-r-1$-te Bettische Gruppe von $R^n - K$ zu der r-dimensionalen Bettischen Gruppe von K isomorph.

Um aus dieser Isomorphie die (auf Verschlingung) beruhende Primitivität (Orthogonalität)[66] der beiden Gruppen abzuleiten, betrachte man eine Polyederumgebung V von K, die den folgenden beiden Bedingungen genügt:

(a) jeder Zyklus von K, welcher in V homolog Null ist, ist es auch in K selbst;

(b) zu jedem Zyklus $z \subset R^n - K$ gibt es einen Zyklus $z' \subset R^n - V$ von der Eigenschaft, daß $z \sim z'$ in $R^n - K$ ist.

Es sei ferner $U \subset V$ eine solche Polyederumgebung von K, daß es zu jedem Zyklus $z \subset U$ einen Zyklus $z' \subset K$ gibt derart, daß $z \sim z'$ in V ist.

Wir bezeichnen mit B bzw. B' die r-dimensionale Bettische Gruppe von K bzw. U, mit β bzw. β' die $n-r-1$-dimensionale Bettische Gruppe von $R^n - K$ bzw. $R^n - U$. Wegen der Bedingung (a) folgt aus $K \subset U$ eine *iso*morphe Abbildung von B auf eine Untergruppe von B', die wir ebenfalls mit B bezeichnen. Wir beweisen zunächst, daß im Falle $\mu = 0$ *diese Untergruppe eine Untergruppe mit Division von B' ist.* Es sei in der Tat $x \subset B$, $y \subset B'$, $x = ky$, $k \neq 0$; wir zeigen, daß auch $y \subset B$ ist. Zu dem Zweck bezeichnen wir mit $\bar{x}$ bzw. $\bar{y}$ Zyklen in K bzw. U, die zu den Homologieklassen x bzw. y gehören. Sodann gilt

$$\bar{x} \sim k\bar{y} \quad (\text{in } U).$$

Aus der Definition von U folgt, daß es einen Zyklus $\bar{y}'$ in K gibt, der $\sim \bar{y}$ in V ist, so daß

$$\bar{x} \sim k\bar{y}' \quad (\text{in } V),$$

also (wegen a))

$$\bar{x} \sim k\bar{y}' \quad (\text{in } K)$$

ist. Wenn wir mit y' das durch den Zyklus $\bar{y}'$ erzeugte Element von B bezeichnen, so ist $x = ky'$; da aber andererseits $x = ky$ und die Gruppe B' frei ist, ist $y = y'$, also $y \subset B$.

Die Gruppen B' und β' sind zueinander primitiv (orthogonal), während B und β isomorph sind. Aus der Inklusion $R^n - U \subset R^n - K$ und der Bedingung (b) ergibt sich eine homomorphe Abbildung von β' auf die ganze Gruppe β. Den Kern dieses Homomorphismus bezeichnen wir etwa mit A und beweisen, daß $A = (\beta', B)$ ist. Zunächst ist klar, daß $A \subset (\beta', B)$ ist, denn ein Zyklus von $R^n - K$, der mit einem Zyklus von K verschlungen ist, kann in $R^n - K$ nicht homolog Null sein. A kann aber auch nicht eine echte Untergruppe von (β', B) sein, denn sonst hätte die Gruppe $\beta' | A \simeq \beta$ (je nachdem $\mu \neq 0$ oder $= 0$ ist) eine größere Ordnung bzw. einen größeren Rang als die Gruppe $\beta' | (\beta', B) \simeq B$, was der Isomorphie zwischen β und B widersprechen würde. Somit ist $A = (\beta', B)$, und, da $\beta \simeq \beta' | (\beta', B)$ ist, bilden B und β ein primitives (orthogonales) Gruppenpaar, w.z.b.w.

Anhang II
Einordnung des Lefschetzschen Dualitätssatzes für abgeschlossene Mengen in die Theorie des Kapitels III[37]

Definition I. Es sei $\bar{F}(V_n, \pi)$ eine inverse Homomorphismenfolge. Eine Folge

$$y_k, y_{k+1}, \ldots, y_n, \ldots \tag{1}$$

wobei y_n ein Element von V_n ist, heißt eine *Kette*, wenn für jedes $n \geqslant k$

$$\pi_n(y_{n+1}) = t_n y_n$$

und dabei t_n eine positive ganze Zahl ist.

Definition II. Ein System von Ketten heißt *linear unabhängig*, wenn die entsprechenden (d. h. die gleiche Nummer habenden) Elemente dieser Ketten, as Elemente der Gruppen, zu denen sie gehören, betrachtet, in diesen Gruppen linear unabhängig sind. Wenn man in einer gegebenen Homomorphismenfolge beliebig viele untereinander unabhängige Ketten finden kann, sagen wir, daß diese Homomorphismenfolge einen *unendlichen Rang* hat, sonst aber heißt die Höchstzahl der in der Homomorphismenfolge vorkommenden unabhängigen Ketten der *Rang* derselben. Als Rang einer direkten Homomorphismenfolge soll der Rang ihrer Limesgruppe definiert werden.

Die Einordnung des Lefschetzschen Satzes in unsere Theorie ist somit durch folgenden Satz gegeben:

Zwei (modulo Null) orthogonale Homomorphismenfolgen haben denselben Rang.

Zunächst bemerken wir, daß *äquivalente Homomorphismenfolgen denselben Rang haben.* Für direkte Homomorphismenfolgen folgt dies daraus, daß äquivalente Folgen isomorphen Limesgruppen haben; was inverse Homomorphismenfolgen betrifft, so folgt für diese unsere Behauptung aus der analogen Behauptung für zwei Folgen, von denen die erste eine Teilfolge der zweiten ist; für diese Spezialfall läßt sich aber die Behauptung sofort begründen.

Wir nennen jetzt eine inverse Homomorphismenfolge *voll*, falls bei jedem n die einzige Untergruppe mit Division von V_n, welche $\pi_n(V_{n+1})$ enthält, die Gruppe V_n selbst ist.

Wenn $\bar{F}(V_n, \pi)$ eine volle Homomorphismenfolge und a_n irgendein freies Element von V_n ist, so gibt es ein Element a_{n+1} von V_{n+1}, so daß $\pi_n(a_{n+1}) = ta_n$ bei passend gewähltem ganzzahligem $t \neq 0$ ist. Das heißt aber, daß man zu jedem freien Element a_n einer beliebigen Gruppe V_n aus $\bar{F}(V_n, \pi)$ mindestens eine Kette finden kann, die mit a_n beginnt. Wenn man ferner ein System von linear-unabhängigen Elementen $a_n^1, a_n^2, \ldots, a_n^k$ von V_n hat, so sind die in ihnen beginnenden Ketten offenbar auch linear unabhängig. Daraus folgt, daß der Rang einer vollen Homomorphismenfolge gleich der endlichen oder unendlichen oberen Grenze der Ränge der einzelnen Gruppen V_n ist.

Jetzt beweisen wir die folgenden beiden Hilfssätze.

Hilfssatz I. Zu jeder inversen Homomorphismenfolge $\bar{F}(V_n, \pi)$ gibt es eine mit ihr äquivalente volle Homomorphismenfolge.

Es sei in der Tat V_{nk} die kleinste Untergruppe mit Division von V_n, welche $\pi_n^{n+k} V_{n+k}$ enthält. Wenn k wächst, kann V_{nk} nur abnehmen; unter den V_{nk} (n fest!) existiert infolgedessen (V_n hat ja endlichviele Erzeugende) eine kleinste Untergruppe, und diese soll V'_n heißen. Man setze jetzt $V_{s_1} = V_1$, also auch $V'_{s_1} = V'_1$ und nehme an, V_{s_i} (also auch V'_{s_i}) wäre bereits gefunden. V'_{s_i} ist definitionsgemäß die kleinste den Gruppen $V_{s_i k}$, ist also eine bestimmte Gruppe $V_{s_i s}$. Wir setzen $V_{s_{i+1}} = V_{s_i + s}$. Für jedes $h > s_{i+1}$ ist V'_{s_i} die kleinste Untergruppe mit Division von V_{s_i}, welche $\pi^h_{s_i} V_h$ oder auch nur $\pi^h_{s_i} V'_h$ enthält. Daraus folgt, daß

$$\begin{cases} V'_{s_1}, V'_{s_2}, \ldots, V'_{s_n}, \ldots \\ \pi^{s_2}_{s_1}, \pi^{s_2}_{s_2}, \ldots, \pi^{s_{n+1}}_{s_n}, \ldots \end{cases} \tag{2}$$

eine volle Homomorphismenfolge ist. Man kann aber auch in

$$V_{s_1}, V'_{s_1}, V_{s_2}, V'_{s_2}, \ldots, V_{s_n}, V'_{s_n}, \ldots \tag{3}$$

eine Homomorphismenfolge erklären, indem man V_{s_n} in $V'_{s_{n-1}}$ mittels der Abbildungen $\pi^{s_n}_{s_{n-1}}$ und V'_{s_n} in V_{s_n} mittels der identischen Abbildung abbildet. Die Folge (3) umfaßt (2) und die Teilfolge $V_{s_1}, V_{s_2}, \ldots, V_{s_n}, \ldots$ der ursprünglich gegebenen Folge $\bar{F}(V_n, \pi)$, also ist (2) mit $\bar{F}(V_n, \pi)$ äquivalent, womit der Hilfssatz I bewiesen ist.

Hilfssatz II. Wenn $\bar{F}(V_n, \pi)$ eine volle und $F(U_n, \varphi)$ eine zu $\bar{F}(V_n, \pi)$ orthogonale Homomorphismenfolge ist, so sind die Homomorphismen φ lauter Isomorphismen.

Es sei in der Tat a ein von Null verschiedenes Element von U_n, das vermöge φ_n auf das Nullement von U_{n+1} abgebildet ist. Infolge der Orthogonalität des Gruppenpaares U_n, V_n gibt es ein Element b von V_n derart, daß $ab \neq 0$ ist; da $\bar{F}(V_n, \pi)$ voll ist, existiert ein $b' \subset V_{n+1}$, so daß $\pi_n(b') = tb$ (mit $t > 0$) ist. Nun hat man aber $\varphi_n(a)b' = a\pi_n(b') = tab \neq 0$, folglich auch $\varphi_n(a) \neq 0$; durch diesen Widerspruch ist Hilfssatz II bewiesen.

Wenn man jetzt zwei orthogonale Homomorphismenfolgen F und $\bar{F}$ hat, so ersetze man sie zunächst durch zwei ebenfalls orthogonale Folgen F_1 und $\bar{F}_1$, die zu F bzw. $\bar{F}$ äquivalent sind, von denen die erste eine volle Homomorphismenfolge, die zweite also eine direkte Isomorphismenfolge ist. Der Rang einer Isomorphismenfolge ist offenbar gleich der oberen Grenze der Ränge der betreffenden Gruppen; da das analoge auch von vollen Homomorphismenfolgen gilt und die einander entsprechenden Gruppen der beiden Folgen orthogonal, also isomorph sind, haben die beiden Folgen denselben Rang, w.z.b.w.

Anhang III
Beispiel einer Kurve im R^3, deren Komplementärraum eine beliebige abzählbare Abelsche Gruppe ohne Elemente endlicher Ordnung als erste Bettische Gruppe hat

U sei eine beliebige Abelsche Gruppe, welche aus abzählbarvielen freien Elementen $a_1, a_2, \ldots, a_n, \ldots$ besteht; U_n sei die von $a_1, a_2, \ldots, a_n$ erzeugte Untergruppe von U. Da U_n Untergruppe von U_{n+1} ist, liegt ein Homomorphismus φ_n von U_n in U_{n+1} und somit eine direkte Homomorphismenfolge $\mathfrak{F}(U_n, \varphi)$ for. Die entsprechende Limesgruppe ist, wie leicht ersichtlich, zu U isomorph.

Da U_n eine freie Gruppe mit endlichvielen Erzeugenden ist, so existiert eine zu U_n orthogonale Gruppe V_n und folglich eine zu $\mathfrak{F}(U_n, \varphi)$ orthogonale (inverse) Homomorphismenfolge $\overline{\mathfrak{F}}(V_n, \pi)$. Wenn $x_n^1, x_n^2, \ldots, x_n^{\alpha_n}$ ein System von freien Erzeugenden der Gruppe V_n ist, so ist (bei passend gewählten ganzzahligen ${}_nc_j^i$)

$$\pi_n(x_{n+1}^i) = {}_nc_j^i x_n^j. \tag{1}$$

Es sei jetzt K_n der Streckenkomplex, den man erhält, wenn man die Eckpunkte eines α_n-Ecks alle untereinander identifiziert. Wir bezeichnen mit $x_n^1, x_n^2, \ldots, x_n^{\alpha_n}$ das in der einfachsten Weise gewählte System von eindimensionalen Zyklen von K_n, die eine eindimensionale Basis dieses Komplexes bilden. Es läßt sich nun leicht eine stetige Abbildung f_n von K_{n+1} in K_n angeben, bei der

$$f_n(x_{n+1}^i) \sim {}_nc_j^i x_n^j$$

ist. Man bette jetzt K_1 als einen singularitätenfreien polygonalen Streckenkomplex in den R^3 ein und wähle eine solche Polyederumgebung Q_1 von K_1, daß jeder Zyklus von Q_1 dortselbst einem Zyklus von K_1 homolog ist und dadurch eine Isomorphie zwischen den Bettischen Gruppen von K_1 und Q_1 erreicht wird. Das Bild $f_1(K_2)$ von K_2 liegt auf K_1, und man kann durch eine beliebige kleine Abänderung von f_1 diese Abbildung in eine solche überführen, welche K_2 auf einen in Q_1 singularitätenfrei gelegenen, zu K_2 homöomorphen polygonalen Komplex $\overline{K}_2$ abbildet. Man wähle jetzt eine in Q_1 enthaltene Polyederumgebung Q_2 von $\overline{K}_2$ derart, daß jeder in Q_2 gelegene Zyklus dortselbst einem Zyklus von $\overline{K}_2$ homolog ist und die Bettischen Gruppen von Q_2 und K_2 isomorph sind, und setze diese Verfahren fort. Es entsteht dabei eine Folge von ineinander geschachtelten zusammenhängenden Polyederbereichen $Q_1, Q_2, \ldots, Q_n, \ldots,$ die so gewählt werden konnten, daß ihr Durschnitt eine Kurve F ist. Dabei wird die Bettische Gruppe von Q_{n+1} in die von Q_n gemäß den Formeln (1) homomorph abgebildet. Die zu der entstehenden inversen Homomorphismenfolge duale Gruppe (die mit der zur Zyklosis von F dualen Gruppe übereinstimmt) ist offenbar die Gruppe U, welche somit zu der ersten Bettischen Gruppe von $R^3 - F$ isomorph ist.

Es möge zum Schluß noch bemerkt werden, daß schon die Menge der aus abzählbarvielen freien Elementen bestehenden, im Sinne der Isomorphie voneinander paarweise verschiedenen Abelschen Gruppen vom Range 1 die Mächtigkeit des Kontinuums hat. Man sieht hieraus, in wie hohem Maße der Satz von der Invarianz der Bettischen Gruppen inhaltsreicher ist als der Invariansatz für die Bettischen Zahlen.

Eingegangen am 24. 12. 1930

Notes

1. *Journ. Ec. Pol.*, 1895.

2. *Math. Annalen*, **72** (1912), S. 422–425.

3. Alexandroff, Gestalt und Lage abgeschlossener Mengen, *Ann. of Math.*, (2) **30** (1928), S. 101–187. Dort befinden sich auch Hinweise auf die einschlägigen Arbeiten von Lefschetz, Vietoris u. a.

4. *Comptes Rendus Acad. Sciences Paris*, **154**, séance du 27 mars 1911.

5. Brouwer, Beweis der Jordanschen Satzes für n-Dimensionen, *Math. Annalen*, **71** (1911), S. 314–319 und On looping coefficients, *Proc. Akad. Amsterdam*, **15** (1912), S. 113–122.

6. A Proof and Generalisation of the Jordan–Brouwer Theorem, *Trans. Amer. Math. Soc.*, **23** (1922), S. 333–349.

7. Mit R^n wird in dieser Arbeit durchweg der durch den unendlich fernen Punkt ergänzte n-dimensionale Euklidische Raum bezeichnet.

8. *Gött. Nachr., Math.-Phys. Kl.*, 25. Nov. 1927.

9. Lefschetz, *Ann. of Math.*, (2) **29** (1928), S. 232; Frankl, *Wien. Ber., Dez.*, 1927, S. 689.

10. *Trans. Amer. Math. Soc.*, **25** (1923), S. 540.

11. Pontrjagin, *Gött. Nachr., Math.-Phys. Kl.* 25. Nov. 1927 und Frankl a. a. O.[9]

12. Die unendliche zyklische Gruppe (also die Gruppe aller ganzen Zahlen) wird gelegentlich auch als die zyklische Gruppe von der Ordnung Null bezeichnet. Diese Redeweise wird sich im Laufe dieser Arbeit wiederholt als sehr bequem erweisen.

13. Vgl. hierzu z B. H. Hopf, Eine verallgemeinerung der Euler–Poincaréschen Formel, *Gött. Nachr., Math.-Phys. Kl.*, 1928.

14. Siehe die Literaturangaben unter.[25]

15. Offenbar können dabei die Gruppen L^n, Z^n, $\hat{H}^n$ als die Gruppen L^r_μ, Z^r_μ, $\hat{H}^r_\mu$ mit $\mu = 0$ betrachtet werden (Teilkomplexe, Zyklen, Berandungen modulo Null). Es Empfiehlt sich ferner im Falle $\mu = 0$ noch die Gruppe $H^r = H^r_0$ aller derjenigen Zyklen Γ^r einzuführen, für die es positive ganze Zahlen $k \neq 0$ derart gibt, daß $k\ \Gamma^r$ berandet (also in $\hat{H}^r = \hat{H}^r_0$ enthalten ist). Wenn man ganz allgemein eine Untergruppe

U einer Abelschen Gruppe G *Untergruppe mit Division* nennt, falls aus der Inklusion $kx \subset U$ (x ein Element von G, k eine positive ganze Zahl) die Inklusion $x \subset U$ folgt, kann man H^r_0 als die kleinste Untergruppe mit Division über $\hat{H}^r_0$ definieren. Man sieht leicht ein, daß die reduzierte Bettische Gruppe (also die Bettische Gruppe modulo Null) nichts anderes als die Faktorgruppe $Z^r_0 | H^r_0$ ist. Wir setzen jetzt definitionsgemäß für $\mu \neq 0$ $H^r_\mu = \hat{H}^r_\mu$ und führen im Falle eines *beliebigen* $\mu = 0, 2, 3, \ldots$ für sämtliche Elemente von H^r_μ die Bezeichnung *homolog Null* (in Zeichen ~ 0) ein. Im Falle $\mu \neq 0$ ist also ein Zyklus dann und nur dann homolog Null, wenn er (mod μ) berandet, während wir im Falle $\mu = 0$ sagen werden, daß Γ^r homolog Null ist, wenn es eine von Null verschiedene ganze Zahl k gibt, so daß $k\ \Gamma^r$ berandet. *Bei jedem μ kann also die r-te Bettische Gruppe modulo μ (den Fall $\mu = 0$ nicht ausgeschlossen) als die Faktorgruppe $Z^r_\mu | H^r_\mu$ definiert werden.*

16. Im Falle $\mu = 0$ wird sogar ein schärferes Ergebnis bewiesen, nämlich die sogenannte Orthogonalität der beiden Gruppen, ein stärkere Eigenschaft als die Primitivität, auf die es uns aber im Augenblick nicht weiter ankommt. Es sei hier noch bemerkt, daß man in der Formulierung der beiden Dualitätssätze notwendig auf die hier gegebene Definition der Bettischen Gruppen modulo 0 als der reduzierten Gruppen angewiesen ist: es läßt sich nämlich mit denselben Methoden zeigen, daß die r-te Torsionsgruppe von M^n nicht zu der $n-r$-ten, sondern zu der $n-r-1$-ten Torsionsgruppe isomorph ist; ebenso ist die r-te Torsionsgruppe von K zu der $n-r-2$-ten Torsionsgruppe von $R^n - K$ isomorph (wenn K ein in R^n liegender Komplex ist). Infolgedessen ist im allgemeinen weder die $n-r$-te volle Bettische Gruppe einer M^n zu der r-ten vollen Bettischen Gruppe derselben Mannigfaltigkeit, noch die $n-r-1$-te volle Bettische Gruppe von $R^n - K$ zu der r-ten Bettische Gruppe von K isomorph.

17. *Gött. Nachr.*, 1927, S. 323.

18. Alexandroff, a.a.O.,[3] S. 107.

19. a.a.O.[8]

20. Lefschetz, a.a.O.[9]

21. Die Aufführungen des Kap. II beziehen sich auf Polyederkomplexe; der Fall stetiger Komplexe (d. h. topologischer Bilder von Polyederkomplexen) wird im Anhang I behandelt.

22. Wenn $\mu \not\equiv 0 \bmod k_1$, also etwa $\mu = qk_1 - r$ (mit positivem $r < k_1$) wäre, hätte man $qk_1 = q(a_1 \cdot b_1) = \mu + r$, d. h. $qa_1 \cdot b_1 = r < k_1$, die Elemente a_1, b_1 wären also falsch gewählt.

23. Untergruppen mit Division sind in der Fußnote[15] definiert.

24. Vgl. etwa E. Noether, *Math. Zeitschr.* **30**, S. 648.

25. Wegen der Grundbegriffe der Topologie der Komplexe und Mannigfaltigkeiten vgl. etwa E. R. van Kampen, Die kombinatorische Topologie und die Dualitätssätze, Dissertation, Leiden 1929, sowie Alexander, Combinatorial Analysis Situs, *Trans. Amer. Math. Soc.*, **28** (1926), S. 301–329 und *Ann. of Math.*, (2) **31** (1930), S. 292–320 und Lefschetz, Intersections and transformations of complexes and manifolds, *Trans. Amer. Math. Soc.*, **28** (1926), S. 1–49. Weitere Literatur bei van der Waerden, Kombinatorische Topologie, Jahresber. d. D. M. V. **39** (1930), S. 121. Man beachte im folgenden die terminologischen Festsetzungen der Einleitungen, insbesondere die der Fußnote.[15]

26. Vgl. Alexander, *Ann. of Math.*, (2) **31**, S. 307 und Vietoris, *Monatsh. f. Math. u. Phys.*, **35** (1927), S. 165, sowie van Kampen a.a.O., S. 13.

27. Ein in M^n liegendes eindeutiges (jedoch nicht notwendig eineindeutiges)

stetiges Bild eines Polyederkomplexes heißt ein in M^n (mit eventuellen Singularitäten) eingebetteter Komplex. Falls die Abbildung eineindeutig ist, spricht man von singularitätenfreier Einbettung.

28. Wenn eine Gruppe A auf eine Gruppe B homomorph abgebildet ist, so heißt die Untergruppe von A, die aus allen Elementen besteht, welche auf das Einheits-(Null-)element von B abgebildet werden, der *Kern* der homomorphen Abbildung.

29. Pontrjagin, Zum Alexanderschen Dualitätssatz, zweite Mitteilung, *Gött. Nachr.*, 1927, S446.

30. Wenn jedem Element a einer Menge A ein Element b einer Menge B zugeordnet ist, und dabei jedes b mindestens einem a entspricht, spricht man von einer Abbildung von A *auf* B. Eine Abbildung von A auf eine echte oder unechte Teilmenge von B heißt (nach Herrn van der Waerden) eine Abbildung von A *in* B.

31. Daß unsere Äquivalenzbegriff transitiv ist, ergibt sich aus den folgenden beiden Bemerkungen, von denen die erst selbstverständlich und die zweite mühelos nachzuweisen ist:

1. wenn $I \equiv II$ und $I' \supset I, II' \supset II$ ist, so ist $I' \equiv II'$,
2. wenn $I \equiv II$ und $II'' \subset II$ ist, so ist $I \equiv II''$.

Zunächst folgt aus $I \equiv II$, $II \equiv III$ die Existenz der Teilfolgen I_1 und II_1, II'' und III'', sowie der Folgen IV und V mit

$$I_1 + II_1 \subset IV, \quad II'' + III'' \subset V.$$

Vermöge der Bemerkungen 1 und 2 ergibt sich sodann der Reihe nach

$$I = II'', \quad I = V, \quad I \equiv III'', \quad I \equiv III,$$

w.z.b.w.

32. Alexandroff, a.a.O.,[3] S. 107.

33. Vgl. Fußnote 15 und Anhang III.

34. Alexandroff, a.a.O.,[3] S. 117 (Cor. I).

35. Beweis durch Induktionsschluß nach der Dimension der Elemente der Komplexe; vgl. z. B. Alexander, *Trans. Amer. Math. Soc.*, **28**.

36. Vgl. die Formulierung des Alexanderschen Dualitätssatzes im engeren Sinne (Kap. II, II, Sektion 5).

37. Wir betrachten nur den Fall, in dem M^n ein verallgemeinerter Poincaréscher Raum ist; der Fall einer allgemeinen M^n läßt sich in analoger Weise erledigen.

Beweis des Mengerschen Einbettungssatzes*

Mathematische Annalen (1931) 105(5): 734–45

1

Der Mengersche Einbettungssatz lautet: Jeder n-dimensionale kompakte metrisierbare Raum ist einer Teilmenge des R^{2n+1} homöomorph.

Seinen Beweis für diesen Satz hat Menger für $n = 1$ durchgeführt, für ein allgemeines n dagegen nur skizziert.[1] Selbst im Falle $n = 1$ ist der Beweis sehr unständlich und bedarf mühevoller mengentheoretischer und elementargeometrischer Überlegungen, die den einfachen und – wie wir sehen werden – im Grunde genommen kombinatorischen Kern der Sache nur verschleiern.

Im folgenden soll ein einfacher und – bis auf einen dem Wesen der Sache nach notwendigen Grenzübergang – rein kombinatorischer Beweis des Einbettungssatzes gegeben werden. Es wird sich dabei zeigen, daß dieser Satz als eine unmittelbare Anwendung des Alexandroffschen allgemeinen Approximationsverfahrens (für kompakte Räume)[2] zu betrachten ist und

---

* In Gemeinschaft mit G. Tolstowa. Der Inhalt dieser Arbeit wurde im Winter 1929/30 im Topologischen Seminar von Prof. Alexandroff (Universität Moskau) vorgetragen.

Nach Fertigstellung der Arbeit erfuhren die Verfasser, daß inzwischen Herr Nöbeling in den *Math. Annalen*, **104** (1930) einen Beweis des Einbettungssatzes gegeben hat. Da aber der Beweis von Herrn Nöbeling auf ganz anderen (nämlich vorwiegend mengentheoretischen) Methoden beruht, dürfte der voliegende Beweis dennoch sein Interesse behalten.

sich mit Hilfe dieses Verfahrens leicht auf die entsprechende Behauptung für n-dimensionale Komplexe zurückführen läßt. Letztere Behauptung aber (d. h. die Tatsache, daß man einen n-dimensionalen Komplex in den $2n+1$-dimensionalen Euklidischen Raum topologisch einbetten kann) folgt aus den Elementen der mehrdimensionalen analytischen Geometrie und ist längst bekannt.

Als Zusatz zum Einbettungssatz ergibt sich bie unserer Beweismethode von selbst folgendes neue Resultat: *Jede stetige Abbildung eines n-dimensionalen kompakten metrischen Raumes in einen mindestens $2n+1$-dimensionalen Euklidischen Raum läßt sich durch eine beliebig kleine stetige Abänderung in eine Homöomorphie verwandeln.*

2

Es werden benutzt: aus der kombinatorischen Topologie – nur die Begriffe des Komplexes, der baryzentrischen Unterteilung und der simplizialen Abbildung; aus der Alexandroffschen Approximationstheorie – nur der Begriff des Projektionsspektrums sowie die Tatsache, daß ein n-dimensionaler kompakter Raum durch ein n-dimensionales Projektionsspektrum approximiert werden kann.[3]

Hilfssatz I. Es sei ein Projektionsspektrum

$$K_1, K_2, \ldots, K_m, \ldots, K_m = \pi(K_{m+1}) \tag{1}$$

gegeben. Man nehme an, daß auf Grund irgendeiner Eigenschaft gewisse Simplexpaare (t_m, t'_m) aus K_m *ausgezeichnet* werden, und zwar soll die Eigenschaft, die die Auszeichnung bedingt, die einzige Forderung erfüllen, daß die Projektion der Simplexe eines ausgezeichneten Paares (t_m, t'_m) wieder ein ausgezeichnetes Paar bilden. Dann gilt die Behauptung: wenn es für beliebig große n ausgezeichnete Paare (t_n, t'_n) gibt, so gibt es zwei Projektionsfolgen

$$t_1, t_2, \ldots, t_n, \ldots$$
$$t'_1, t'_2, \ldots, t'_n, \ldots$$

deren Elemente t_n, t'_n für jedes n ein ausgezeichnetes Paar bilden.

Beweis. Aus unseren Voraussetzungen folgt unmittelbar, daß es für jedes n in K_n ausgezeichnete Paare gibt, und zwar solche, in die sich ausgezeichnete Paare aus K_h ($h > n$, sonst beliebig) projizieren. Somit projizieren sich unendlichviele ausgezeichnete Paare in (ausgezeichnete) Paare von K_n;

da es aber in K_n überhaupt nur endlichviele Simplexpaare gibt, muß mindestens ein (ausgezeichnetes) Simplexpaar (t_n, t'_n) in K_n existieren, welches als Projektion von unendlichvielen ausgezeichneten Paaren aufgefaßt werden kann. Man wähle ein solches Paar (t_1, t'_1) insbesondere für $n = 1$.

Man nehme an, die ausgezeichneten Paare

$$(t_1, t'_1), (t_2, t'_2), \ldots, (t_m, t'_m) \tag{2}$$

seien bereits konstruiert, und zwar so, daß für $1 \leqq i < m$

$$t_i = \pi(t_{i+1}), \quad t'_i = \pi(t'_{i+1})$$

und jedes der Paare (2) Projektion von unendlichvielen ausgezeichneten Paaren ist. Man betrachte nun alle ausgezeichneten Paare (t_n, t'_n), $n \geqq m+1$, die sich auf (t_m, t'_m) projizieren und nenne sei für einen Augenblick P-Paare. Da es unendlichviele P-Paare, aber nur endlichviele Simplexpaare in K_{m+1} gibt, existiert in K_{m+1} mindestens ein (ausgezeichnetes) Paar (t_{m+1}, t'_{m+1}), in welches unendlichviele P-Paare (also unendlichviele ausgezeichnete Paare) projiziert werden. Die Induktion geht also weiter.

Auf diese Weise fortfahrend, erhält man die ausgezeichneten Paare

$$(t_1, t'_1), (t_2, t'_2), \ldots, (t_n, t'_n), \ldots,$$

wobei für jedes n das Simplexpaar (t_n, t'_n) die Projektion von $(t_{n+1}; t'_{n+1})$ ist. Die Folgen $(t_1, t_2, \ldots, t_n, \ldots)$ und $(t'_1, t'_2, \ldots, t'_n, \ldots)$ sind also Projektionsfolgen, w.z.b.w.

3

Man betrachte wieder das Projektionsspektrum (1) und ein festes Simplex t_h von K_h; ein Simplex t_n von K_n, $n > h$, heißt *Nachfolger* von t_h, wenn $\pi(t_n) = t_h$ ist. Sodann besteht folgender.

Hilfssatz II. Das Projektionsspektrum (1) und ein Simplex t_1 von K_1 seien gegeben. Für alle hinreichend großen n und jedes Simplex t_n aus K_n, welches mit einem Nachfolger von t_1 benachbart its, ist $\pi_1^n(t_n)$ im Stern[4] um t_1 enthalten.

Wir nehmen an, der Hilfssatz sei falsch, und nennen ein Simplexpaar (t_n, t'_n), $n > 1$, ausgezeichnet, wenn t_n und t'_n benachbart sind und

$t_1 = \pi_1^n(t_n)$ ist, während $\pi_1^n(t_n')$ nicht zum Stern um t_1 gehört; ein Paar (t_1, t_1') soll ausgezeichnet heißen, wenn es Projektion eines ausgezeichneten Paares ist. Sodann sind alle Bedingungen des Hilfssatzes I erfüllt, und es gibt zwei Projektionsfolgen

$$t_1, t_2, \ldots, t_n, \ldots \tag{3}$$

und

$$t_1', t_2', \ldots, t_n', \ldots, \tag{3'}$$

so daß für jedes n die Simplexe t_n, t_n' ein ausgezeichnetes Paar bilden, und folglich t_1' nicht zum Stern um t_1 gehört. Da aber zwei Simplexe eines ausgezeichneten Paares benachbart sind, sind auch die Folgen (3) und (3′) benachbart, so daß es eine diese beiden Folgen umfassende Projektionsfolge

$$T_1, T_2, \ldots, T_n, \ldots$$

gibt.[5] Somit sind die beiden Simplexe t_1 und t_1' Seiten desselben Simplexes T_1, folglich gehört t_1' zu dem Stern um t_1, was unserer Annahme widerspricht. Der Hilfssatz II ist damit bewiesen.

4

Es sei T^n ein Simplex mit den Eckpunkten $a_0, a_1, \ldots, a_n$ (wir schreiben kurz $T^n = (a_0, a_1, \ldots, a_n)$).[6] Unter der *Elementarzerlegung* von T^n in bezug auf seine Seite $t^r = (a_0, a_1, \ldots, a_r)$ verstehen wir die Zerlegung von T^n in die $r+1$ Simplexe

$$\begin{gathered}(b, a_1, \ldots, a_r, a_{r+1}, \ldots, a_n)\\ (a_0, b, \ldots, a_r, a_{r+1}, \ldots, a_n)\\ \cdots\cdots\cdots\\ (a_0, a_1, \ldots, b, a_{r+1}, \ldots, a_n),\end{gathered}$$

wobei b der Schwerpunkt von t^r ist.

Dabei wird ausdrücklich auch das Simplex T^n selbst als seine einzige n-dimensionale („uneigentliche“) Seite betrachtet.

Man nehme eine Elementarzerlegung von T^n in bezug auf sich selbst vor; es entstehen Simplexe von der Form $(b, a_0, \ldots, a_{i_1-1}, a_{i_1+1}, \ldots, a_n)$. Wenn man jedes dieser Simplexe in bezug auf seine a-Seite (d. h. die Seite $(a_0, \ldots, a_{i_1-1}, a_{i_1+1}, \ldots, a_n)$) elementar-zerlegt, entstehen Simplexe $(b, b_{i_1}, a_0, \ldots, a_{i_1-1}, a_{i_1+1}, \ldots, a_{i_2-1}, a_{i_2+1}, \ldots, a_n)$. Wenn man jedes dieser Simplexe in bezug auf seine a-Seite zerlegt, entstehen Simplexe, deren erste drei Eckpunkte Schwerpunkte[7] $b, b_{i_1}, b_{i_1 i_2}$ sind und deren übrige Eckpunkte zu den ursprünglichen Eckpunkten a_i gehören. Durch

Fortsetzung dieses Verfahrens gelangt man schließlich zu den Simplexen von der Form $(b, b_{i_1}, b_{i_1 i_2}, \ldots, b_{i_1 i_2 \ldots i_n})$, d. h. den Simplexen der baryzentrischen Unterteilung von T^n.

5

Es sei jetzt K ein Komplex, t ein Simplex von K. Wenn man mit jedem Simplex von K, welches t als eigentliche oder uneigentliche Seite enthält, eine Elementarzerlegung in bezug auf t vornimmt und die übrigen Simplexe von K intakt läßt, entsteht eine *Elementarzerlegung von* K in bezug auf t; aus der Überlegung der vorigen Nummer ergibt sich sodann der eigentlich selbstverständliche

Hilfssatz III. Es seien

$$t_1, t_2, \ldots, t_s$$

sämtliche Simplexe aller Dimensionen des Komplexes K, in einer solchen Reihenfolge geschrieben, daß stets $\dim t_i \geqq \dim t_{i+1}$ ist; wenn man $K = K_0$ setzt und mit K_{i+1} den Komplex bezeichnet, welcher aus K_i durch Elementarzerlegung in bezug auf t_{i+1} entsteht, so ist K_s mit der baryzentrischen Unterteilung von K identisch.

6

Jetzt gehen wir zum Beweis des wichtigsten Hilfssatzes über.

Hilfssatz IV. Man betrachte ein Projektionsspektrum

$$K_1, K_2, \ldots, K_m, \ldots, \qquad K_m = \pi(K_{m+1}) \tag{1}$$

und ein Simplex t von K_1. Es sei K_1' der Komplex, der durch die Elementarzerlegung von K_1 in bezug auf t entsteht. Man bezeichne mit g eine (festzudenkende) simpliziale Abbildung von K_1' auf K_1, welche den Schwerpunkt b von t auf einen bestimmten (aber beliebigen) Eckpunkt a von t abbildet und alle übrigen Eckpunkte von K_1' (die ja zugleich Eckpunkte von K_1 sind) festläßt. Es existiert sodann ein K_m des Spektrums (1) und eine simpliziale Abbildung f von K_m auf K_1' von der Eigenschaft, daß die Abbildung gf von K_m auf K_1 mit der Projektion π_1^m identisch ist.

Beweis. Man wähle m so groß, daß für jedes Simplex t_m von K_m, welches mit einem Nachfolger von t benachbart ist, $\pi_1^m(t_m)$ zum Stern um t gehört. Nach Hilfssatz II ist ein solches m stets vorhanden. Es sei $a^{(m)}$ ein beliebiger Eckpunkt von K_m. Zwei Fälle sind möglich:

1. $\pi_1^m(a^{(m)}) \neq a$; dann setzen wir $f(a^{(m)}) = \pi_1^m(a^{(m)})$;

2. $\pi_1^m(a^{(m)}) = a$; dann unterscheiden wir die weiteren beiden Fälle:

2'. $a^{(m)}$ gehört zu keinem Nachfolger von t; dann setzen wir

$$f(a^{(m)}) = a \quad (\text{also } f(a^{(m)}) = \pi_1^m(a^{(m)});$$

2''. $a^{(m)}$ gehört zu einem Nachfolger von t; dann sei

$$f(a^{(m)}) = b.$$

Für jeden Eckpunkt $a^{(m)}$ von K_m ist offenbar

$$gf(a^{(m)}) = \pi_1^m(a^{(m)}),$$

es bleibt also nur zu zeigen, daß die obige Eckpunktzuordnung tatsächlich eine simpliziale Abbildung von K_m auf K_1' bestimmt.

Jedes Simplex t_m von K_m gehört zu einer der drei folgenden Kategorien:

I. t_m enthält keinen Nachfolger des Eckpunktes a. In diesem Falle gehört $\pi_1^m(t_m)$ nicht nur zu K_1, sondern auch zu K_1', und es ist $f(t_m) = \pi_1^m(t_m)$.

II. t_m enthält mindestens einen Nachfolger des Eckpunktes a, aber keiner dieser Nachfolger gehört zu einem Nachfolger von t. In diesem Falle wird keine Seite von t_m auf t abgebildet, folglich gehört $\pi_1^m(t_m)$ nicht zu den Simplexen von K_1, die t als Seite enthalten, ist also unter den Simplexen von K_1' vorhanden; es ist sodann nach der Vorschrift 1 und 2':

$$f(t_m) = \pi_1^m(t_m).$$

III. t_m enthält einen Nachfolger von a, und zwar einen solchen, der zu einem Nachfolger von t gehört. Somit ist t_m mit einem Nachfolger von t benachbart, und laut unserer Voraussetzung über die Zahl m gehört $\pi_1^m(t_m)$ zum Stern um t. Unter den Eckpunkten von $\pi_1^m(t_m)$ ist ferner bestimmt der Eckpunkt a enthalten. Wir unterscheiden nun zwie weitere Fälle:

III'. t_m enthält keinen Nachfolger von t (d. h. $\pi_1^m(t_m)$ enthält t nicht). Dann gehört $\pi_1^m(t_m) = (a, a_1, \ldots, a_r)$ zum Rande des Sternes um t, folglich ist das Simplex $(b, a, a_1, \ldots, a_r)$ gewiß in K_1' enthalten. Nun sind aber die Bilder der Eckpunkte von t_m bei der Abbildung f unter den Punkten $b, a, a_1, \ldots, a_r$ enthalten, so daß $f(t_m)$ eine echte oder unechte Seite des Simplexes $(b, a, a_1, \ldots, a_r)$ und folglich ein Simplex von K_1' ist.

III''. t_m enthält einen Nachfolger von t (d. h. $\pi_1^m(t_m)$ enthält t). Dann wird jeder Eckpunkt von t_m, der ein Nachfolger von a ist, bei der Abbildung f auf b abgebildet; wenn $\pi_1^m(t_m) = (a, a_1, \ldots, a_s, a_{s+1}, \ldots, a_r)$ und

$t = (a, a_1, \ldots, a_s)$ ist, so ist $f(t_m) = (b, a_1, \ldots, a_s, a_{s+1}, \ldots, a_r)$, und das ist ein Simplex von K'_1.

Die Eckpunkte von t_m werden also in allen Fällen mittels f auf das Eckpunktsystem eines Simplexes von K'_1 abgebildet; f ist also tatsächlich eine simpliziale Abbildung, under der Hilfssatz IV ist bewiesen.

7

Wir beweisen nun den letzten Hilfssatz, aus dem dann der Einbettungssatz ganz leicht folgen wird.

Hilfssatz V. Zu jedem Projektionsspektrum

$$K_1, K_2, \ldots, K_m, \ldots \quad (K_m = \pi(K_{m+1})) \tag{1}$$

läßt sich ein äquivalentes (d. h. denselben Raum approximierendes) Projektionsspektrum

$$Q_1, Q'_1, Q_2, Q'_2, \ldots, Q_h, Q'_h, \ldots \tag{4}$$

unter folgenden Bedingungen konstruieren. Die Q_h bilden eine Teilfolge der K_m, $Q_h = K_{m_h}$, während für jedes h Q'_h ein Teilkomplex der baryzentrischen Unterteilung von Q_h ist. Die Projektionen werden als Abbildungen g_h von Q'_h auf Q_h bzw. Abbildungen f_h von Q_{h+1} auf Q'_h definiert, die folgenden Bedingungen genügen:

1. Wenn b ein Eckpunkt von Q'_h, also der Schwerpunkt eines Simplexes t_h von Q_h ist, so ist $g_h(b)$ ein Eckpunkt von t_h. (Daraus folgt insbesondere, daß ein Eckpunkt von Q'_h, der gleichzeitig Eckpunkt von Q_h ist, bei der Abbildung g_h in sich selbst übergeht.)

2. Wenn a_{h+1} ein beliebiger Eckpunkt von $Q_{h+1} = K_{m_{h+1}}$ ist, so gilt

$$g_h f_h(a_{h+1}) = \pi_{m_h}^{m_{h+1}}(a_{h+1}).$$

Beweis. Es sei $K_1^{(0)}, K_1^{(1)}, \ldots, K_1^{(s-1)}, K_1^{(s)}, K_1^{(0)} = K_1, K_1^{(s)} = K'_1$ die endliche Folge von sukzessiven Elementarzerlegungen, die sich dem Hilfssatz III gemäß zwischen $K_1 = Q_1$ und seiner baryzentrischen Unterteilung $K'_1 = Q'_1$ einschalten läßt. Durch sukzessive Anwendung des Hilfssatzes IV erhält man ein Projektionsspektrum

$$Q_1, Q'_1, K_{m_1}, K_{m_1+1}, \ldots,$$

welsches auf Grund des Hilfssatzes IV den Bedingungen 1 und 2 des Hilfssatzes V für Q_1, Q'_1 und $K_{m_1} = Q_2$ genügt und im übringen mit (1) zusammenfällt, Man nehme an, das Projektionsspektrum

$$Q_1, Q'_1, \ldots, Q_h, Q'_h, K_{m_h}, K_{m_h+1}, \ldots$$

wäre bereits konstruiert, so daß für $Q_1, Q'_1, \ldots, Q_h, Q'_h, K_{m_h} = Q_{h+1}$ die Bedingungen 1 und 2 erfüllt sind und die Projektionen für K_m $(m > m_h + 1)$ mit denen des Spektrums (1) identisch sind. Man schaltet zwischen $K_{m_h} = Q_{h+1}$ und seiner baryzentrischen Unterteilung Q'_{h+1} wiederum eine Kette von Elementarzerlegungen ein und erhält durch Anwendung des Hilfssatzes IV ein Spektrum

$$Q_1, Q'_1, \ldots, Q_{h+1}, Q'_{h+1}, K_{m_{h+1}}, K_{m_{h+1}+1}, \ldots,$$

wobei wieder alle Bedingungen erfüllt sind. Die durch Fortsetzen des Verfahrens entstehende unendliche Folge

$$Q_1, Q'_1, \ldots, Q_h, Q'_h, \ldots$$

mit den zugehörigen Abbildungen g_h und f_h genügt sodann den beiden Bedingungen 1 und 2. Aus der Bedingung 2 folgt, daß diese Folge ein dem Spektrum (1) äquivalentes Projektionsspektrum bildet, w.z.b.w.

8

Beweis des Einbettungssatzes. Es sei F ein n-dimensionaler kompakter metrisierbarer Raum und

$$K_1, K_2, \ldots, K_m, \ldots \quad (K_m = \pi(K_{m+1}), \dim K_m = n) \qquad (1)$$

ein Projektionsspektrum von F. Auf Grund des Hilfssatzes V besitzt F ein Projektionsspektrum

$$Q_1, Q'_1, Q_2, Q'_2, \ldots, Q_h, Q'_h, \ldots, \qquad (4)$$

welches den Bedingungen dieses Hilfssatzes genügt.

Man betrachte jetzt einen polyedralen Komplex C_1 des R^{2n+1}, welcher ein eineindeutiges simpliziales Bild von Q_1 ist; man darf dabei voraussetzen, daß die Simplexe von C_1 sämtlich einen Durchmesser <1 haben. Die baryzentrische Unterteilung von C_1 bezeichnen wir mit C'_1; die Simplex von C'_1 sind kleiner als $n/(n+1) = q$.[8] Indem man C_1 bzw. C'_1 mit Q_1 bzw. Q'_1 als identisch betrachtet, kann man von der Abbildung f_1 von Q_2 auf C'_1 sowie von der Abbildung g_1 von C'_1 auf C_1 sprechen.

Jedem Simplex T_1 von C_1 ordne man eine feste Umgebung $U(T_1)$ zu, deren Durchmesser kleiner als 1 ist. Diese Umgebungen sollen so gewählt sein, daß zwei disjunkte Simplexe Umgebungen bekommen, die voneinander einen positiven Abstand haben.

Das Bild $f_1(t_2)$ eines beliebigen Simplexes t_2 von Q_2 ist ein Simplex t_1' von C_1', also ein baryzentrisches Teilsimplex eines gewissen Simplexes T_1 von C_1. Das Simplex $g_1 f_1(t_2)$ ist dabei eine echte order unechte Seite von T_1, und zwar ist diese Seite jedenfalls unecht (d. h. sie fällt mit T_1 zusammen), wenn die Dimensionszahlen von t_2 und πt_2 gleich sind.

Man ändere jetzt die Abbildung f_1 von Q_2 auf C_1' unter Wahrung ihres simplizialen Charakters durch eine kleine Verrückung der Bildpunkte der Eckpunkte von Q_2 zu einer *topologischen* Abbildung f_1' von Q_2 auf einen Komplex C_2 ab, und zwar sollen die Verrückungen der Bildeckpunkte so gering sein, daß das Bildsimplex $f_1'(t_2) = T_2$ in $U(T_1)$ liegt und noch immer kleiner als q ist. Der Projektionsabbildung $\pi = g_1 f_1$ von Q_2 auf Q_1 bzw. C_1 entspricht (vermöge der Homöomorphie f_1' zwischen Q_2 und C_2) eine Abbildung von C_2 auf C_1, die ebenfalls Projektion genannt und mit π bezeichnet werden soll.

Dabei gilt die Relation

$$T_2 \subset U(T_1), \tag{5_1}$$

und $\pi(T_2)$ ist eine echte oder unechte Seite von T_1, die notwendig mit T_1 zusammenfällt, wenn die Dimensionszahlen von T_2 und von $\pi(T_2)$ gleich sind.

Schleißlich wählen wir eine $U(T_2) \subset U(T_1)$ so, daß dabei disjunkten Simplexen positiv-entfernte Umgebungen zugeordnet werden und daß der Durchmesser von $U(T_2)$ stets $< q$ ist.

Man nehme jetzt an, daß dei Komplexe $Q_1, Q_2, \ldots, Q_m$ durch eineindeutige simpliziale Abbildungen als Polyederkomplexe $C_1, C_2, \ldots, C_m$ in den R^{2n+1} bereits eingebettet sind; daß den Projektionen von Q_{i+1} auf Q_i $(1 \leqq i \leqq m-1)$ Projektionen von C_{i+1} auf C_i entsprechen und für jedes Simplex T_i von C_i $(1 \leqq i \leqq m)$ eine feste Umgebung $U(T_i)$ gewählt ist, die kleiner als q^{i-1} und so beschaffen ist, daß disjunkte Simplex Umgebungen bekommen, die voneinander einer positive Entfernung haben. Ferner soll $U(T_{i+1})$ in einer $U(T_i)$ liegen, wobei $T_i \supset \pi(T_{i+1})$ und insbesondere $T_i = \pi(T_{i+1})$ ist, wenn die Dimensionen von T_{i+1} und $\pi(T_{i+1})$ zusammenfallen.

Indem wir zwischen Q_m und C_m nicht unterscheiden, betrachten wir die Abbildung f_m von Q_{m+1} auf die baryzentrische Unterteilung C_m' von C_m. Da die Simplexe von C_m kleiner als q^{m-1} waren, sind diese von C_m' kleiner als q^m; wenn t_{m+1} ein beliebiges Simplex von Q_{m+1} ist, so ist $f_m(t_{m+1})$ ein baryzentrisches Teilsimplex eines bestimmten Simplexes T_m von C_m, zu dem auch $g_m f_m(t_{m+1}) = \pi(t_{m+1})$ als echte oder unechte Seite gehört; wenn dabei t_{m+1} die gleiche Dimensionszahl wie $\pi(t_{m+1})$ hat, ist $\pi(t_{m+1}) = T_m$. Man ändere jetzt die Abbildung f_m unter Wahrung ihres simplizialen Charakters durch eine kleine Verrückung der Bildpunkte der

Eckpunkte von Q_{m+1} zu einer *topologischen* Abbildung f'_m von Q_{m+1} auf einen Komplex C_{m+1} ab; dabei sollen die Verrückungen der Bildeckpunkte so gering sein, daß $f'_m(t_{m+1}) = T_{m+1}$ in $U(T_m)$ liegt und noch immer kleiner als q^m ist. Die Projektion $\pi = g_m f_m$ von Q_{m+1} auf Q_m bzw. C_m geht in die Projektion von C_{m+1} auf C_m über, und, wenn man mit T_{m+1} ein willkürliches Simplex von C_{m+1} bezeichnet, gibt es in C_m ein solches T_m. daß

$$T_{m+1} \subset U(T_m), \qquad T_m \supset \pi(T_{m+1}), \tag{5_m}$$

wobei, wenn die Dimensionszahlen von T_{m+1} und $\pi(T_{m+1})$ gleich sind, die Simplexe T_m und $\pi(T_{m+1})$ zusammenfallen. Schließlich wählen wir für jedes T_{m+1} von C_{m+1} eine $U(T_{m+1}) \subset U(T_m)$; diese Umbegungen sollen $<q^m$ sein und die Eigenschaft haben, daß disjunkten Simplexen positiv entfernte Umgebungen entsprechen.

Die Induktion geht also weiter; wir merken uns noch, daß aus (5_m) die Ungleichung

$$\rho(T_{m+1}, \pi(T_{m+1})) < q^{m-1} \tag{6}$$

folgt.

9

Jedem Punkt x von F entspricht eine Kette des Projektions spektrums

$$Q_1, Q_2, \ldots, Q_m, \ldots,$$

also eine Kette

$$(T_1^x, T_2^x, \ldots, T_m^x, \ldots) \tag{7}$$

des Projektionsspektrum

$$C_1, C_2, \ldots, C_m, \ldots. \tag{8}$$

Da die Dimensionszahlen der Simplexe von (7) mit wachsendem m jedenfalls nicht abnehmen können und andererseits immer höchstens gleich n bleiben, muß von einem bestimmten m_x an

$$\dim T_m^x = \dim T_{m+1}^x = \ldots$$

und folglich für alle $m \geqq m_x$

$$U(T_{m+1}^x) \subset U(T_m^x)$$

sein. Da $\delta U(T_m^x)$ kleiner als q^{m-1} ist, konvergieren die Simplexe T_m^x gegen den einzigen Durchschnittspunkt der $U(T_i^x)$, $i \geqq m_x$; diesen bezeichnen wir mit $\varphi(x)$. Dabei gilt nach (6) folgende Abschätzung

$$\rho(\varphi(x), T_m^x) \leqq \sum_{i=m}^{\infty} [\rho(T_i^x, T_{i+1}^x) + \delta(T_{i+1}^x)]$$

$$< \sum_{i=m}^{\infty} (q^{i-1} + q^{i-1}) < 2\frac{q^{m-1}}{1-q}. \tag{9}$$

10

Wir beweisen jetzt der Reihe nach folgendes.

1. *Die Abbildung φ ist stetig.* Man wähle in der Tat einen beliebigen Punkt x_0 von F und eine positive Zahl ϵ. Ihm entspricht eine Kette des Projektionsspektrums (8):

$$x_0 \equiv (T_1^{(0)}, T_2^{(0)}, \ldots, T_m^{(0)}, \ldots).$$

Wir wählen m so groß, daß $\dim T_m^{(0)} = \dim T_{m+1}^{(0)} = \ldots$ und

$$\frac{q^{m-1}}{1-q} < \epsilon$$

ist. Es sei x ein beliebiger Punkt der m-ten Umgebung[9] von x_0, d. h. es sei

$$x \equiv (T_1^x, T_2^x, \ldots, T_m^x, \ldots),$$

wobei für $1 \leqq i \leqq m$ das Simplex T_i^x eine echte oder unechte Seite von $T_i^{(0)}$ ist. Dann ist zunächst

$$\varphi(x_0) \subset U(T_m^{(0)}), \quad \delta(U(T_m^{(0)})) < q^{m-1} < \epsilon,$$

also

$$\rho(\varphi(x_0), T_m^{(0)}) < \epsilon. \tag{10}$$

Wegen (9) ist (unter Berücksichtigung der Relation $T_m^x \subset T_m^{(0)}$)

$$\rho(\varphi(x), T_m^{(0)}) \leqq \rho(\varphi(x), T_m^x) < 2\frac{q^{m-1}}{1-q} < 2\epsilon, \tag{11}$$

also nach (10) und (11)

$$\rho(\varphi(x), \varphi(x_0)) \leqq \rho(\varphi(x), T_m^{(0)}) + \delta(T_m^{(0)}) + \rho(T_m^{(0)}, \varphi(x_0)) < 4\epsilon,$$

womit die Stetigkeit von φ bewiesen ist.

Die Bildmenge Φ von F ist als stetiges Bild eines kompakten metrischen Raumes in sich kompakt. Φ is also eine beschränkte abgeschlossene Menge des R^{2n+1}.

2. *Die Abbildung φ ist eineindeutig.* Denn sind x und x' zwei verschiedene Punkte von F, so sind die betreffenden Ketten nicht benachbart, d.h. von einem gewissen m an sind die Simplexe T_m^x und $T_m^{x'}$ zueinander fremd. Wir nehmen dieses m sogleich so groß, daß dim $T_m^x = \dim T_{m+1}^x = \ldots$ und dim $T_m^{x'} = \dim T_{m+1}^{x'} = \ldots$ ist. Da die zugehörigen Umgebungen $U(T_m^x)$ und $U(T_m^{x'})$ voneinander einen positiven Abstrand haben und die Punkte x und x' in $U(T_m^x)$ bzw. $U(T_m^{x'})$ liegen, sind sie gewiß verschieden.

Eine eineindeutige und in einer Richtung stetige Abbildung eines kompakten Raumes ist bekanntlich eine Homöomorphie.[10] Somit sind F und Φ homöomorph, und der Einbettungssatz ist bewiesen.

Korollar. Für jeden kompakten metrischen Raum F der Dimension n kann man stets im $(2n+1)$-dimensionalen Euklidischen Raum eine Teilfolge $C_1, C_2, \ldots, C_m, \ldots$ von n-dimensionalen polyedralen Komplexen finden, deren topologischer Limes existiert und mit F homöomorph ist, dabie kann man die Folge $C_1, C_2, \ldots, C_m, \ldots$ so konstruieren, daß, wenn ein endlicher Abschnitt von ihr $(C_1, C_2, \ldots, C_m)$ schon konstruiert ist, der Komplex C_{m+1} in eine beliebig kleine Umgebung von C_m verlegt wird.

Diese Bemerkung folgt unmittelbar aus dem obigen Beweis des Satzes.

11

Zusatz. Eine stetige Abbildung f eines n-dimensionalen kompakten metrischen Raumes F in einen Euklidischen Raum R von einer Dimension $\geqq 2n+1$ kann durch eine beliebig kleine stetige Abänderung in eine Homöomorphie verwandelt werden.

Man betrachte wieder die Projektionsspektra (1) bzw. (4) von F. Jedes Q_m kann als Nerv einer ϵ_m-Überdeckung P_m von F, bestehend aus den Mengen

$$F_m^{(1)}, F_m^{(2)}, \ldots, F_m^{(s_m)},$$

aufgefaßt werden. Diesen Überdeckungen entsprechen Überdeckungen

$$f(F_m^{(1)}), f(F_m^{(2)}), \ldots, f(F_m^{(s_m)})$$

der Bildmenge $f(F)$, und man kann von vorherein die ϵ_m so klein wählen, daß schon die $f(F_1^{(i)})$ sämtlich kleiner als das beliebig vorgeschriebene $\alpha > 0$ sind. Den Eckpunkten von Q_m sind eineindeutig die Mengen $F_m^{(i)}$, folglich auch die Mengen $f(F_m^{(i)})$ zugeordnet. Man wähle nun die den Eckpunkten von Q_1 entsprechenden Eckpunkte c_1^i von C_1 (Nr. 8) so nahe zu den betreffenden Mengen $f(F_1^i)$, daß die Entfernung von c_1^i bis $f(F_1^i)$ kleiner als α und jedes Simplex T_1 von C_1 kleiner als 2α ist; auch die Umgebunen $U(T_1)$ wähle man kleiner als 2α. Sodann führe man die Konstruktion des Projektionsspektrums

$$C_1, C_2, \ldots, C_m, \ldots \tag{8}$$

wörtlich wie in Nr. 8 aus, mit dem einzigen Unterschied, daß jetzt die Umgebungen $U(T_m)$ der Simplexe von C_m sinngemäß kleiner als $2\alpha q^{m-1}$ sein sollen, was zur Folge hat, daß auf der rechten Seite der Ungleichung (6), sowie im letzten Glied der Formel (9) der Faktor 2α hinzukommt.

12

Genau wie früher unterscheiden wir zwischen Q_m und C_m nicht mehr.

Das Projektionsspektrum (8) definiert nach Nr. 9 und 10 eine topologische Abbildung φ von F auf eine abgeschlossene beschränkte Menge Φ des Raumes R. Wir behaupten, *daß für jeden Punkt x von F die Entfernung zwischen $f(x)$ und $\varphi(x)$ kleiner als $9\alpha/(1-q)$ ist*; mit dieser Behauptung ist auch der Zusatz bewiesen: um die stetige Überführung von f in φ zu erhalten, braucht man ja nur jeden Punkt $f(x)$ bis zum Punkte $\varphi(x)$ während derselben Zeiteinheit gleichförmig und geradlining laufen zu lassen.

Unsere Behauptung beweist man aber so. Ist

$$x \equiv (T_1^x, T_2^x, \ldots, T_m^x, \ldots)$$

ein Punkt von F, c_1 ein Eckpunkt von T_1^x, F_1 die diesem Eckpunkt entsprechende Menge der Überdeckung P_1, so liegt x in F_1, $f(x)$ in $f(F_1)$. Unter Berücksichtigung der am Schluß der voringen Nummer gemachten Bemerkung folgt aus (9)

$$\rho(\varphi(x), T_1^x) < \frac{4\alpha}{1-q}, \quad \text{also} \quad \rho(\varphi(x), c_1) < \frac{4\alpha}{1-q} + 2\alpha$$

$$\begin{aligned}\rho(\varphi(x), f(x)) &< \rho(\varphi(x), c_1) + \rho(c_1, f(F_1)) + \delta f(F_1)\\ &< \left(\frac{4\alpha}{1-q} + 2\alpha\right) + \alpha + 2\alpha < \frac{9\alpha}{1-q},\end{aligned}$$

w.z.b.w.

NOTES

1. Menger, *Dimensionstheorie* (Leipzig, Teubner, 1928), S. 287–303.

2. Alexandroff, Untersuchungen über Gestalt und Lage abgeschlossener Mengen (*Annals of Mathematics*, (2) **30** (1928), S. 101–187); insbesonder Einleitung, § 10 und Kap. II.

3. Alexandroff a.a.O., S. 107 und S. 130–133.

4. Unter dem Stern des Komplexes K um das Simplex t dieses Komplexes verstehen wir hier die Vereinigungsmenge aller (abgeschlossenen) Simplexe von K, die t als ihre Seite haben. Wenn t dem Simplex T als eine seiner Seiten angehört und t' eine andere Seite von T ist, so heißt t' ein inneres oder ein Randelement des Sternes, je nachdem t eine Seite von t' ist oder nicht.

5. Alexandroff a.a.O., S. 107.

6. Orientierung des Simplexes kommt hier nicht in Frage, die Reihenfolge der Eckpunkte ist also gleichgültig.

7. Wenn t eine Seite von T ist, und $a_{i_1}, a_{i_2}, \ldots, a_{i_k}$ diejenigen Eckpunkte von T sind, die *nicht* zu t gehören, so bezeichnen wir den Schwerpunkt von t mit $b_{i_1 i_2 \ldots i_k}$.

8. Wenn der Durchmesser eines n-dimensionalen Simplexes α ist, so ist der Durchmesser eines Simplexes seiner baryzentrischen Unterteilung $\leqq [n/(n+1)]\alpha$.

9. Alexandroff, a.a.O., S. 108.

10. Seihe z. B. Hausdorff, *Mengenlehre* (zweite Auflage), S. 196.

Eingegangen am 5. 2. 1931.

Über Stetige Algebraische Körper*

Annals of Mathematics (1932) 33(1): 163–74

1

Der Zweck dieser Arbeit ist, einer topologische Charakterisierung der Körper der reellen und der komplexen Zahlen zu geben (Satz II). Wenn wir auf die Kommutativitätsbedingung, die in dieser Charakterisierung eine wesentliche Rolle Spielt, verzichten, erhalten wir noch den Quaternionenkörper (Satz I).

Die Fragestellung, die zu dieser Arbeit guführt hat, sowie zahlreiche Anregungen bei der Beweisführung und bei der Redaktion der vorliegenden Abhandlung verdanke ich Herrn A. Kolmogoroff.

2

Definition I. Eine Menge K von Elementen a, b, $c, \ldots$ bildet einen *stetigen Körper*, wenn folgende Bedingungen erfüllt sind:

(1) K ist ein topologischer Raum,[1]

(2) K ist ein algebraischer, im allgemeinen nicht kommutativer Körper,

(3) die Funktionen $f_1(a,b) = a + b$, $f_2(a,b) = ab$, $f_3(a) = -a$, $f_4(a) = a^{-1}$ sind stetig (überall, wo sie definiert sind, d. h. f_1, f_2 und f_3 im ganzen Raum und f_4 überall mit Ausnahme des Nullpunktes).

* Received May 17, 1931.

Im folgenden wird unter einem *Körper* stets ein stetiger Körper verstanden.

Definition II. Zwei stetige Körper heißen isomorph, wenn sie aufeinander eineindeutig abgebildet werden können, und zwar so, daß dabei sowohl die algebraischen als auch die topologischen Relationen in beiden Körpern einander entsprechen (d. h. daß die Abbildung gleichzeitig einen Isomorphismus der Körper und einen Homöomorphismus der Räume darstellt).

Satz I. Jeder stetige lokalbikompakte,[2] zusammenhängende Körper K ist dem Körper der reellen, dem Körper der komplexen Zahlen oder dem Quaternionenkörper isomorph.

Satz II. Jeder stetige lokalbikompakte, zusammenhängende kommutative Körper K ist entweder dem Körper der reellen oder dem Körper der komplexen Zahlen isomorph.

In der Formulierung der Sätze I und II kann die Bikompaktheit durch die Bedingung ersetzt werden, daß K ein Hausdorffscher lokalkompakter topologischer Raum mit erstem Abzählbarkeitsaxiom ist:

Satz Ia. Jeder stetige lokalkompakte, zusammenhängende Körper K, in dem das erste Abzählbarkeitsaxiom und das hausdorffsche Trennbarkeitsaxiom erfüllt sind, ist mit dem Körper der reellen Zahlen, dem Körper der komplexen Zahlen oder dem Quarternionenkörper isomorph.

Satz IIa. Unter den Bedingungen des Satzes Ia mit Hinzufügung der Kommutativität von K ist K entweder dem Körper der reellen Zahlen oder dem Körper der komplexen Zahlen isomorph.

Wir werden beweisen, daß die Bedingungen von Satz Ia aus den Bedingungen von Satz I folgen, worauf wir nur noch den Satz Ia beweisen müssen: II und IIa folgen ja unmittelbar aus I und Ia infolge der Nichtkommutativität des Quaternionenkörpers.

3

In jedem stetigen Körper K gilt folgender Satz, der keines Beweises bedarf:

Hilfssatz I. Wenn $a \neq 0$ ist, so ist die Abbildung

$$x' = ax + b$$

von K auf sich eine topologische (d. h. eineindeutige und beiderseits stetige) Abbildung.

Dieser Satz zeigt, daß K topologisch *homogen* ist, d. h. daß man mittels einer topologischen Abbildung des ganzen Körpers auf sich selbst jeden Punkt a in jeden Punkt b überführen kann: es genügt in der Tat, die Abildung $x' = x - a + b$ heranzuziehen.

4

In diesem Paragraphen setzen wir voraus, daß der Körper K lokalbikompakt und zusammenhängend ist.

Hilfssatz II. Es existiert eine abzählbare Menge $a_1, a_2, \ldots, a_r, \ldots$ (1) von Elementen von K, die 0 zum Häufungspunkt hat.

Beweis. Daraus, daß K zusammenhängend ist, folgt, daß alle seine Punkte Häufungspunkte sind (K enthält ja mindestens zwei Elemente: Null und Eins). Sei G eine offene Menge, deren abgeschlossene Hülle $\bar{G}$ kompakt ist. Da alle Elemente von K Häufungspunkte sind, enthält $\bar{G}$ unendlich viele Elemente. Sei $b_1, b_2, \ldots, b_r, \ldots$ (2) eine abzählbar unendliche Menge von Elementen aus $\bar{G}$ und b ein Häufungspunkt dieser Menge. Man sieht leicht, daß die Menge $b_1 - b, b_2 - b, \ldots, b_r - b, \ldots$ Null zum Häufungspunkt hat.

Hilfssatz III. Sei F eine bikompakte Menge[3] und

$$a_1, a_2, \ldots, a_r, \ldots \tag{1}$$

eine Folge von Elementen, die Null zum Häufungspunkt hat und G eine Umgebung der Null, so existiert ein n, so daß $F_n = Fa_n \subset G$.[3a]

Beweis. Für jeden Punkt a gibt es stets eine Umgebung $G(a)$ von a und eine Umgebung $G'(a)$ von Null, so daß, wenn $x \subset G(a)$ und $y \subset G'(a)$ ist, xy in G liegt. Wenn nun a die ganze Menge F durchläuft, so überdecken die Gebiete $G(a)$ die ganze Menge F; infolge der Bikompaktheit von F kann man aus diesem Überdeckungssystem ein endliches Teilsystem $G(a^{(1)}), G(a^{(2)}), \ldots, G(a^{(k)})$ herausgreifen, welches ebenfalls F überdeckt; bezeichnen wir nun mit G' den Durchschnitt $G'(a^{(1)}), G'(a^{(2)}), \ldots, G'(a^{(k)})$. Man sieht leicht, daß das Produkt eines beliebigen Elementes aus F mit einem beliebigen Element aus G' in G liegt. Da aber G' eine Umgebung der Null ist, so gibt es unter den Elementen der Folge (1) sicher ein Element $a_n \subset G'$; daraus folgt aber $F_n \subset G$, w. z. b. w.

Hilfssatz IV. K ist ein regulärer hausdorffscher Raum mit erstem Abzählbarkeitsaxiom.

Beweis. Es sei U eine Umgebung des Nullpunktes mit bikompakter abgeschlossener Hülle $\bar{U}$ und b_n eine Folge, die Null zum Häufungspunkt hat. Bei beliebiger Wahl der Umgebung V des Nullpunktes ist bei entsprechend egwähltem n $\bar{U}\cdot b_n \subset V$, woraus folgt, daß die offenen Mengen $U\cdot b_n$ ein der Regularitätsbedingung genügendes abzählbares Umgebungssystem des Nullpunktes bilden.

Wegen der Homogenität von K (Hilfssatz I) existiert ein solches Umgebungssystem auch für jeden Punkt x. Daraus folgt bekanntlich auch das Hausdorffsche Trennungsaxiom (denn, wenn x und y zwei Punkte sind, $V(x)$ zu y fremd, $\bar{U}(x)$ in $V(x)$ enthalten ist, so bilden $U(x)$ und das Komplement $K - \bar{U}(x)$ zwei disjunkte Umgebungen von x und y).

Auf diese Weise sehen wir, daß in K die Voraussetzungen des Satzes Ia erfüllt sind. Wir gehen also zum Beweis dieses Satzes über.

5

Wir setzen voraus, daß K lokalkompakt und zusammenhängend ist und dem ersten Abzählbarkeitsaxiom und dem Hausdorffschen Trennungsaxiom genügt. Ist x Häufungspunkt einer Menge E, so können wir aus E eine Folge herausgreifen, die gegen x *konvergiert.*[4] Dies ermöglicht uns, die ganze Untersuchung mit Hilfe des Begriffes der Konvergenz einer Folge zu führen.

Hilfssatz V. Wenn die folge b_n gegen Null konvergiert, so ist die Folge b_n^{-1} divergent, d. h. sie besitzt keinen einzigen Häufungspunkt.

Wenn nämlich c ein Häufungspunkt der Folge b_n^{-1} wäre, so hätte die Folge $x_n = b_n \cdot b_n^{-1} = 1$ den Punkt $0 \cdot c = 0$ zum Häufungspunkt, was offenbar unmöglich ist.

Hilfssatz VI. Zu jedem Punkt a gibt es eine gegen diesen Punkt konvergierende Folge von a verschiedenen Punkten.

Beweis. Dies folgt daraus, daß a kein isolierter Punkt ist, und aus dem ersten Abzählbarkeitsaxiom.

Hilfssatz VII. K ist nicht kompakt.

Es gibt nämlich nach Hilfssatz VI eine Folge b_n mit $b_n \neq 0$, $\lim b_n = 0$. Die Folge b_n^{-1} ist dann nach Hilfssatz V divergent.

Hilfssatz VIII. Wenn F kompakt,[5] V eine Umgebung des Nullpunktes und $\lim b_n = 0$ ist, so ist für hinreichend großes $nF \cdot b_n \subset V$.

Man nehme das Gegenteil an. Dann gäbe es unendlich viele Produkte $x_n = f_n b_{k_n}$ $(f_n \subset F)$, die zu V fremd sind. Nun haben aber die f_n einen Häufungspunkt f, während die b_{k_n} gegen 0 konvergieren. Die x_n haben daher den Häufungspunkt $f \cdot 0 = 0$, was unmöglich ist, da sie außerhalb V liegen.

Hilfssatz IX. Es seien U und V zwei Umgebungen des Nullpunktes mit compakten abgeschlossenen Hüllen. Wenn a_n eine divergente Folge ist, so hat für hinreichend großes n die Begrenzung U' von U mit $\bar{V} \cdot a_n$ notwendig gemeinsame Punkte.

Beweis. Wir betrachten den Durchschnitt $D(\bar{U}, \bar{V}a_n)$. Diese Menge ist kompakt und enthält den Nullpunkt. Da K selbst nicht kompakt ist, bildet sie eine echte Teilmenge von K. Da ferner K zusammenhängend ist, ist der Rand von $D(\bar{U}, \bar{V}a_n)$, d. h. die Menge $D(U', \bar{V}a_n) + D(\bar{U}, V'a_n)$, gewiß nicht leer. Wenn für unendlich viele n der erste Summand leer wäre, so wäre für dieselben n die Menge $D(\bar{U}, V'a_n)$ von 0 verschieden. Es würden m. a. W. die Folgen $u_k \subset \bar{U}$, $v_k \subset V'$ und eine Teilfolge a_{n_k} von a_n existieren, so daß $u_k = v_k a_{n_k}$ ist. Offenbar kann man überdies voraussetzen, daß u_k und v_k konvergente Punktfolgen mit

$$\lim u_k = u \subset \bar{U} \quad \text{und} \quad \lim v_k = v \subset V'$$

sind. Dabei ist v wegen der letzten Relation gewiß von 0 verschieden (V' ist ja die Grenze von V). Folglich ist $\lim a_{n_k} = \lim v_k^{-1} u_k = v^{-1} u$, was der Divergenz der Folge a_n widerspricht.

Hilfssats X. Wenn die Folge a_n divergiert, so konvergiert die Folge a_n^{-1} gegen 0.

Es genügt zu beweisen, daß, falls a_n divergiert, die Folge a_n^{-1} Null zum

Häufungspunkt hat: daraus folgt nämlich, daß auch jede Teilfolge von a_n^{-1} den Nullpunkt zum Häufungspunkt hat, woraus sich vermöge des ersten Abzählbarkeitsaxioms ergibt, daß die ganze Folge a_n gegen Null konvergiert.

Es seien U und V zwei Umgebungen des Nullpunktes mit kompakter abgeschlossener Hülle (vgl. den vorigen Hilfssatz). Für alle hinreichend großen n ist $D(U', \bar{V}a_n)$ nicht leer, so daß man die Punkte v_n in $\bar{V}$ so wählen kann, daß für alle Punkte $u_n = v_n a_n$ zu U' gehören. Man kann Teilfolgen u_{n_k}, v_{n_k} auswählen, so daß u_{n_k} gegen $u \subset U'$ und v_{n_k} gegen $v \subset \bar{V}$ konvergieren, wobei $u \neq 0$ ist. Somit müßte $a_{n_k}^{-1} = u_{n_k}^{-1} v_{n_k}$ gegen $u^{-1}v$ konvergieren; da aber a_{n_k} divergent ist, muß $u^{-1}v$ der Nullpunkt sein, w. z. b. w.

Hilfssatz XI. Wenn a_n divergiert, F kompakt und V eine Umgebung des Nullpunkts ist, so ist für hinreichend großes $nF \subset Va_n$.

In der Tat: a_n^{-1} konvergiert wegen Hilfssatz X gegen Null und daher ist für alle hinreichend großen $nFa_n^{-1} \subset V$, waraus der Satz folgt.

Hilfssatz XII. Wenn F kompakt und V offen ist, p_n gegen p konvergiert und Fp in V enthalten ist, so gilt für hinreichend großes n: $Fp_n \subset V$.

Sonst gäbe es eine Folge $f_k \subset F$, so daß für unendlich viele n, etwa n_k die Punkte $x_k = f_k p_{n_k}$ außerhalb V liegen würden. Da die p_{n_k} gegen p konvergieren und die f_{n_k} einen Häufungspunkt $f \subset F$ besitzen, so müssen die x_k den Punkt $fp \subset V$ als Häufungspunkt haben, was der Definition der x_k widerspricht.

Hilfssatz XIII. Wenn F beliebig, U abgeschlossen und $\lim p_n = p$ ist, so folgt aus $Fp_n \subset U$, daß auch $Fp \subset U$ ist.

Andernfals gäbe es ein f aus F, so daß $fp \not\subset U$; dann lägen aber auch fast alle fp_n außerhalb U, was unmöglich ist.

Hilfssatz XIV. Es sei a irgendein Punkt von K. Wenn die Folge

$$a, a^2, a^3, \ldots, a^n, \ldots$$

(die "Potenzfolge von a") den Nullpunkt als Häufungspunkt besitzt, so convergiert sie gegen diesen Punkt, und dasselbe gilt auch von der Potenzfolge eines jeden Punktes b, der zu einer passend gewählten Umgebung von a gehört.

Beweis. Für jede Umgebung G des Nullpunktes, deren abgeschlossene Hülle $\bar{G}$ kompakt ist, existiert vermöge Hilfssatz VIII ein solches m, daß

$$\bar{G}a^m \subset G$$

ist. Wegen Hilfssatz XII gilt die analoge Inklusion

$$\bar{G}b^m \subset G \tag{1}$$

für jeden Punkt b einer gewissen Umgebung $U(a)$ von a.

Wir wollen zeigen, daß unter der Bedingung (1) die Folge b^n gegen den Nullpunkt konvergiert. Damit wird der Hilfssatz XIV offenbar bewiesen sein.

Wiederholte Anwendung der Formel (1) gibt zunächst für jedes ganzzahlige q

$$\bar{G}b^{qm} \subset \bar{G}b^{(q-1)m} \subset \ldots \subset \bar{G}b^m \subset G. \tag{2}$$

Wenn $c \subset \bar{G}$, $c \neq 0$ ist, so gilt somit

$$cb^{qm} \subset G \subset \bar{G} \quad \text{und daher} \quad b^{qm} \subset c^{-1}\bar{G}.$$

Alle Punkte von der Form b^{qm} liegen somit in der kompakten Menge $c^{-1}\bar{G}$, so daß die Folge b^{qm} mindestens einen Häufungspunkt besitzt.

Wir nehmen jetzt an, die Folge besäße einen von Null verscheidenen Häufungspunkt. Dann gäbe es eine gegen diesen Punkt konvergierende Teilfolge b^{mq_k}, wobei man offenbar voraussetzen kann, daß $q_k - q_{k-1} = v_k$ mit k unendlich groß wird. Offenbar konvergiert dann die Folge $b^{v_k m}$ gegen den Einheitspunkt 1 von K. Andererseits haben wir nach (2)

$$\bar{G}b^{v_k m} \subset \bar{G}b^m \subset G.$$

Anwendung des Hilfssatzes XIII ergibt dann (indem man für das dortige U die abgeschlossene Menge $\bar{G}b^m$ wählt)

$$\bar{G} = \bar{G}\cdot 1 \subset \bar{G}b^m \subset G,$$

was unmöglich ist, denn K ist ja zusmmenhängend.

Dasselbe gilt natürlich auch für jede Folge der Gestalt b_{mq}^{mq+k} mit konstantem k. Da aber die ganze Folge b^n aus m Folgen der obigen Gestalt besteht (die man erhält, indem man k alle Werte von 0 bis $m-1$ gibt), konvergiert auch diese Folge gegen 0, w. z. b. w.

Hilfssatz XV. Wenn die Folge a^n eine divergente Teilfolge enthält, so ist auch die ganze Folge a^n divergent. Dasselbe gilt auch von der Potenzfolge eines jeden Elementes b aus einer gewissen Umgebung $U(a)$.

Beweis. Die Folge a^{-n} hat nach Hilfssatz X den Nullpunkt unter ihren Häufungspunkten; folglich konvergiert nicht nur a^{-n}, sondern auch jede Folge b^n mit b aus einer gewissen Umgebung von a^{-1} gegen Null. Vermöge Hilfssatz V folgt hieraus, daß a^n bzw. b^{-n} divergiert.

Hilfssatz XVI. Der Körper K zerfällt in drei paarweise zueinander fremde Mengen λ, μ und ρ, wobei ein Punkt zu λ, bzw. zu μ gehört, wenn seine Potenzfolge gegen Null konvergiert bzw. divergent ist. Die Menge ρ besteht aust allen Punkten, deren Potenzfolgen keine divergenten Teilfolgen enthalten und trotzdem den Nullpunkt nicht zum Häufungspunkt haben.

Beweis. Folgt aus Hilfssätzen XIV und XV.

Hilfzzatz XVII. $\lambda + \rho$ ist kompakt.

Beweis. Man nehme an, die Punktfolge x_n aus $\lambda + \rho$ sei divergent. Dann konvergiert x_n^{-1} wegen Hilfssatz X gegen Null und, da λ eine Umgebung des Nullpunktes ist, sind alle x_n^{-1} bis auf endlich viele in λ enthalten. Es sei $x_n^{-1} \subset \lambda$. Vermöge Hilfssatz V ist dann x_n^m divergent, so daß x_n zu μ gehört, was unserer Voraussetzung widerspricht.

Hilfssatz XVIII. Wenn $a \subset \lambda + \rho$ ein mit jedem Element aus K vertauschbares Element ist, so gilt: $\bar{\lambda} a \subset \bar{\lambda}$.

Beweis. Wenn $x \subset \lambda$ ist, so konvergiert die Folge $x, x^2, \ldots, x^n, \ldots$ gegen Null; da aber die Folge $a, a^2, \ldots, a^n, \ldots$ Keine divergente Teilfolge enthält, so konvergiert auch die Folge $ax, a^2x^2, \ldots, a^nx^n, \ldots$ gegen Null. Da aber diese Folge wegen $ax = xa$ mit der Folge $xa, (xa)^2, \ldots, (xa)^n, \ldots$ zusammenfällt , ist auch $ax \subset \lambda$. Es ergibt sich also $\lambda a \subset \lambda$, wenn wir von beiden Seiten dieser Inklusion die abgeschlossene Hülle bilden, erhalten wir: $\bar{\lambda} a \subset \bar{\lambda}$.

Hilfssatz XIX. Sei q ein beliebiges, von Null verschiedenes Element von K und F eine kompakte Teilmenge von K. Dann existiert ein endliches System von Elementen $a_1, a_2, \ldots, a_m$ aus F, so daß es für jedes Element x aus F ein a_i gibt, so daß $x - a_i \subset \lambda q$ ist.

Beweis. Im gegenteiligen Falle würde eine unendliche Folge

$$a_1, a_2, \ldots, a_n, \ldots \tag{1}$$

aus F existieren, so daß für $i \neq j$ stets $a_i - a_j \not\subset \lambda q$ ist. Dies ist aber unmöglich: die Flge (1) hat nämlich einen Häufungspunkt a und wenn die Teilfolge a_{n_i} gegen a konvergiert, so konvergiert die Folge $a_{n_i} - a_{n_{i+1}}$ gegen Null, und daher ist für genügend großes i

$$a_{n_i} - a_{n_{i+1}} \subset \lambda q.$$

Hilfssatz XX. Wenn F eine kompakte Teilmenge von K und p ein beliebiges, von Null verschiedenes Element von K ist. dann gibt es für jede ganze Zahl k eine ganze Zahl N, so daß, wenn

$$x_1, x_2, \ldots, x_N \tag{1}$$

beliebige N Elemente aus F sind, es unter ihnen k Elemente, etwas x_{n_1}, $x_{n_2}, \ldots, x_{n_k}$ gibt, so daß $x_{n_i} - x_{n_j} \subset \lambda p$ gilt für beliebiges i und j $(i, j \leqq k)$.

Beweis. Wegen Hilfssatz VIII bilden die λq bei beliebigem q ein vollständiges Umgebungssystem des Nullpunktes. Daher kann man q so bestimmen, daß aus $x_1 \subset \lambda q$, $x_2 \subset \lambda q$ folgt: $x_1 + x_2 \subset \lambda p$. Für das auf diese Weise bestimmte q bestimmen wir ein System von Elementen a_1, $a_2, \ldots, a_m$, das den Bedingungen des Satzes XIX genügt, und setzen $N = km$. Dann entspricht jedem Element x_i des Systems (1) ein Element a_r, so daß $x_i - a_r \subset \lambda q$ ist. Da aber $N = km$ ist, so gibt es ein solches Element a_r, dem auf diese Weise mindestens k Elemente des Systems (1) entsprechen, etwa $x_{n_1}, x_{n_2}, \ldots, x_{n_k}$. Auf diese Weise haben wir für $(i, j) \leqq k$: $x_{n_i} - a_r \subset \lambda q$, $a_r - x_{n_j} \subset \lambda q$ und daher $x_{n_i} - x_{n_j} = (x_{n_i} - a_r) + (a_r - x_{n_j}) \subset \lambda p$.

Hilfssatz XXI. Bezeichnen wir mit λ_n die Menge aller Elemente der Form $x_1 + x_2 + \ldots + x_n$, wobei $x_i \subset \lambda$ ist, dann ist:

1. λ_n eine offene Menge;
2. ihre abgeschlossene Hülle $\bar{\lambda}_n$ besteht aus allen Elementen der Form $x_1 + x_2 + \ldots + x_n$, wobei $x_i \subset \bar{\lambda}$ ist; $\bar{\lambda}_n$ ist daher kompakt;
3. $\lambda'_n = \bar{\lambda}_n - \lambda_n$ ist nicht leer.

Beweis. 1. und 2. erkennt man unmittelbar, 3. folgt daraus, daß K zusammenhängend und nicht kompakt ist.

Hilfssatz XXII. Wenn $x \subset \lambda_r$ und $y \subset \lambda_s$ ist, so ist $x + y \subset \lambda_{r+s}$; wenn $x \subset \bar{\lambda}_r$ und $y \subset \bar{\lambda}_s$ ist, so ist $x + y \subset \bar{\lambda}_{r+s}$.

Hilfssatz XXIII. Wenn $x \subset \lambda_r$ und $y \subset \bar{\lambda}_s$ ist, so ist $x + y \subset \lambda_{r+s}$.

Beweis. Da $y \subset \bar{\lambda}_s$ ist, so gibt es für jede Umgebung G des Nullpunktes ein

$z \subset G$, so daß $y - z \subset \lambda_s$ ist. Da aber λ_r eine offene Menge ist, so ist für genügend kleines $G: x + z \subset \lambda_r$, also

$$x + y = (x + z) + (y - z), \quad x + z \subset \lambda_r, \quad y - z \subset \lambda_s$$

und mithin nach Hilfssatz XXII: $x + y \subset \lambda_{r+s}$.

Hilfssatz XXIV. Für jedes ganzzahlige $k > 1$ gibt es ein $\omega_k \subset \bar{\lambda}$, so daß $k\omega_k \not\subset \bar{\lambda}_{k-1}$.

Beweis. Sei $p \neq 0$ so bestimmt, daß aus $y_i \subset \lambda p$, $i = 1, 2, \ldots, k-1$ folgt: $y_1 + y_2 + \ldots + y_{k-1} \subset \lambda$. (Ein solches p existiert, da die λp ein vollständiges Umgebungssystem des Nullpunkts bilden.) Die ganze Zahl N sei entsprechend Hilfssatz XX bestimmt für p, k und $F = \bar{\lambda}$. Sei weiter $z \subset \lambda'_N$. (Infolge Hilfssatz XXI existiert ein solches z.) Dann ist

$$z = x_1 + x_2 + \ldots + x_N, \quad x_n \subset \bar{\lambda}.$$

Auf Grund von Hilfssatz XX können wir annehmen, daß für die ersten k Elemente dieser Summe $x_i - x_j \subset \lambda p$ gilt $(i, j \leqq k)$. Wir erhalten:

$$z = (x_1 - x_k) + (x_2 - x_k) + \ldots + (x_{k-1} - x_k) + kx_k + \\ + x_{k+1} + x_{k+2} + \ldots + x_N.$$

Wegen $x_i - x_k \subset \lambda p$ $(i < k)$ ergibt sich:

$$x' = (x_1 - x_k) + (x_2 - x_k) + \ldots + (x_{k-1} - x_k) \subset \lambda.$$

Aus der Inklusion $x_n \subset \bar{\lambda}$ folgt weiter:

$$x'' = x_{k+1} + x_{k+2} + \ldots + x_N \subset \bar{\lambda}_{N-k}.$$

Folglich haben wir nach Hilfssatz XXIII:

$$x' + x'' \subset \lambda_{N-k+1}.$$

Hieraus und aus $kx_k \subset \bar{\lambda}_{k-1}$ ergäbe sich dann nach Hilfssatz XXIII: $z \subset \lambda_N$, was der Voraussetzung widerspricht. Wir erhalten also $kx_k \not\subset \bar{\lambda}_{k-1}$ und $\omega_k = x_k$.

Hilfssatz XXV. Der Körper K hat die Charakteristik Null.

Beweis. Nehmen wir an, der Körper K habe eine Charakteristik $p \neq 0$. Dann gilt für beliebiges $x \subset K: px = 0$. Dies ist aber unmöglich, denn nach Hilfssatz XXIV gilt $p\omega_p \not\subset \bar{\lambda}_{p-1}$; $\bar{\lambda}_{p-1}$ enthält aber für $p > 1$ den Nullpunkt.

Hilfssatz XXVI. Wenn e die Einheit des Körper K ist und m und n zwei ganze positive Zahlen sind ($m \geqq n$), dann ist $\bar{\lambda}_m \cdot (e/n) \supset \bar{\lambda}$.

Beweis. Jedes Element der Menge $\bar{\lambda} \cdot ne$ können wir in der Form $x + x + \ldots + x + 0 + 0 + \ldots + 0$ (xn-mal, 0 ($m - n$)-mal genommen), wobie $x \subset \bar{\lambda}$. Folglich ist nach Hilfssatz XXI $\bar{\lambda} \cdot ne \subset \bar{\lambda}_m$. Indem wir diese Inklusion mit e/n multiplizieren, erhalten wir: $\bar{\lambda} \subset \bar{\lambda}_m \cdot (e/n)$.

Hilfssatz XXVII. $(m/n)e$ liegt in λ, ρ oder μ, je nachdem ob $m < n$, $m = n$ oder $m > n$ ist.

Beweis. Wir betrachten zunächst den Fall $m > n$. Wäre $(m/n)e \subset \lambda + \rho$, so hätten wir nach Hilfssatz XVIII $\bar{\lambda} \cdot (m/n)e \subset \bar{\lambda}$, da $(m/n)e$ mit allen Elementen von K vertauschbar ist. Da $\omega_m \subset \bar{\lambda}$ ist, ergibt sich $(m/n) \cdot \omega_m \subset \bar{\lambda}$ und folglich $m\omega_m \subset \bar{\lambda}_n$; nach Hilfssatz XXIV wissen wir aber, daß $m\omega_m$ nicht in $\bar{\lambda}_{m-1}$ liegt. Daraus ergäbe sich aber $n > m - 1$, also $n \geqq m$ (anderenfalls hätten wir nämlich $m\omega_m \subset \bar{\lambda}_n \subset \bar{\lambda}_{m-1}$), was der Voraussetzung widerspricht. Also ist $(m/n)e \subset \mu$.

Ist $m < n$, so ist $(n/m)e \subset \mu$ und folglich nach Hilfssatz X $(m/n)e \subset \lambda$.

Ist endlich $m = n$, so ist $(m/n)e = e \subset \rho$.

Hilfssatz XXVIII. Die Folge

$$e, 2e, \ldots, ne, \ldots \tag{1}$$

divergiert und die Folge $e, e/2, \ldots, e/n, \ldots$ konvergiert gegen Null.

Beweis. Anderenfalls gäbe es eine konvergente Teilfolge der Folge (1) $n_1e, n_2e, \ldots, n_ke, \ldots$ wobei die ganzen Zahlen n_k eine wachsende Folge bilden. Die Folge $(n_2 - n_1)e$, $(n_3 - n_2)e, \ldots, (n_k - n_{k-1})e, \ldots$ würde dann gegen Null konvergieren, was unmöglich ist, da alle Elemente $(n_k - n_{k-1})e$ nach Hilfssatz XXVII außerhalb λ liegen und λ eine Umgebung des Nullpunktes ist.

Aus der Divergenz der Folge $e, 2e, \ldots, ne, \ldots$ folgt nach Hilfssatz X, daß die Folge $e, e/2, \ldots, e/n, \ldots$ gegen Null konvergiert.

Hilfssatz XXIX. K enthält einen Körper R, der isomorph ist mit dem gewöhnlichen stetigen Körpen der rationalen Zahlen.

Beweis. Nach Hilfssatz XXV enthält K einen Körper R, der im rein algebraischen Sinn isomorph ist mit dem Körper der rationalen Zahlen; dabei schreibt sich iedes Element von R in der Form re, wobei r eine beliebige rationale Zahl ist. Nach Hilfssatz XXVIII bilden die offenen Mengen $\lambda \cdot (e/n)$ ein vollständiges Umgebungssystem des Nullpunktes im Körper K. Daraus folgt aber, daß die Durchschnitte der $\lambda \cdot (e/n)$ mit R ein vollständiges Umgebungssystem des Nullpunktes in R bilden. Diese Durchschnitte bestehen aber nach Hilfssatz XXVII aus denjenigen Elementen re, für die $|r| < 1/n$ ist. R ist also tatsächlich isomorph mit dem stetigen Körper der rationalen Zahlen.

Hilfssatz XXX. Im Körper K gilt die Cauchysche Konvergenzbedingung, d. h. für die Konvergenz der Folge

$$a_1, a_2, \ldots, a_n, \ldots$$

ist notwendig und hinreichend, daß es zu einer jeden Umgebung U des Nullpunktes eines ganze Zahl N gibt, so daß aus $r > N$, $s > N$ folgt $a_r - a_s \subset U$.

Beweis. Die Notwendigkeit der Bedingung erkennt man unmittelbar; beweisen wir, daß sie hinreichend ist. Wenn die Folge (1) dem Cauchyschen Konvergenzkriterium genügt, so gibt es ein N, so daß $a_{N+r} - a_N \subset \lambda$ ist; da $\bar{\lambda}$ kompakt ist, hat daher die Folge $a_{N+1} - a_N, a_{N+2} - a_N, \ldots, a_{N+k} - a_N, \ldots$ mindestens einen Häufungspunkt a und enthält keine divergente Teilfolge; infolgedessen hat die Folge (1) einen Häufungspunkt $a_N + a$; andererseits hat sie aber, wie man leicht erkennt, auch *höchstens* einen Häufungspunkt und enthält keine divergente Teilfolge. Sie ist also konvergent.

Hilfssatz XXXI. Der Körper K enthält einen Körper D, der isomorph ist mit dem gewöhnlichen stetigen Körper der reellen Zahlen; dabei ist jedes Element von D mit jedem Element von K vertauschbar.

Beweis. Auf Grund der Hilfssätze XXIX und XXX überzeugt man sich leicht, daß die abgeschlossene Hülle $\bar{R} = D$ isomorph ist mit dem stetigen Körper der reellen Zahlen. Da jedes Element von R mit jedem Element von K vertauschbar ist, so gilt das gleiche auch für die abgeschlossene Hülle $D = \bar{R}$.

Hilfssatz XXXII. Sei A_n die Menge der Elemente aus K der Gestalt

$$x = \alpha_1 x_1 + \alpha_2 x_2 + \ldots + \alpha_n x_n,$$

wobei $x_1, x_2, \ldots, x_n$ gegebene Elemente von K sind und die variablen Koeffizienten α_i in D liegen. Wenn dann A_n nicht mit K zusammenfällt, so existiert ein Element $z \subset \rho$, das nicht auf folgende Weise dargestellt werden kann:

$$z = x + z', \quad x \subset A_n, \quad z' \subset \lambda.$$

Beweis. Die Menge A_n ist abgeschlossen in K. Wenn daher $a \not\subset A_n$ ist, so gibt es unter den Umgebungen von a der Gestalt $a + \lambda\alpha$, $(\alpha \subset D)$, welche ein vollständiges Umgebungssystem von a bilden, eine solche, die zu A_n fremd ist. Setzen wir dann $a' = a\alpha^{-1}$, so ergibt sich, daß $a' + \lambda = (a + \lambda\alpha)\alpha^{-1}$ zu $A_n = A_n\alpha^{-1}$ fremd ist. Sei nun B_n die Menge der Elemente der Gestalt $x + z$, $x \subset A_n$, $z \subset \lambda$. Offenkundig ist B_n eine offene Menge und a' gehört nicht zu B_n. Da K zusammenhängend ist, existiert also mindestens ein Randpunkt von B_n. Sei b ein solcher Punkt und möge die Folge $x_k + z_k$ $(x_k \subset A_n,\ z_k \subset \lambda)$ gegen b konvergieren. Die z_k haben mindestens einen Häufungspunkt z und die x_k daher den Häufungspunkt $x = b - z$. Wegen $x \subset A_n$, $b = x + z$ kann z nicht zu λ gehören, als Häufungspunkt der Punkte $z_k \subset \lambda$ gehört es aber zu $\bar{\lambda}$; folglich ist $z \subset \rho$. Wenn sich nun z in folgender Form darstellen ließe: $z = x' + z'$, $x' \subset A_n$, $z' \subset \lambda$, so ergäbe sich: $b = x + x' + z' = x'' + z'$, $x'' \subset A_n$. Dann müßte aber b zu B_n gehören, was unmöglich ist. Folglich genügt z allen Bedingungen des Hilfssatzes.

Hilfssatz XXXIII. Es existiert eine linear unabhängige endliche Basis

$$x_1, x_2, \ldots, x_m \tag{1}$$

des Körpers K bezüglich D, d. h. jedes Element von K kann auf eine und nur eine Weise in der Form

$$\sum_{i=1}^{m} \alpha_i x_i \quad (\alpha_i \subset D) \tag{2}$$

dargestellt werden.

Beweis, Sie $x_1 = e$. Nehmen wir an, wir hätten bereits in K ein System

$$x_1, x_2, \ldots, x_n \tag{3}$$

von bezüglich D linear unabhängigen Elementen derart, daß $x_i \subset \rho$ und $x_i - x_j \not\subset \lambda$ für $i \neq j$. Entweder fällt die Menge aller Elemente der Form

$$\sum_{i=1}^{n} \alpha_i x_i \quad (\alpha_i \subset D) \tag{4}$$

bereits mit K zusammen, oder man kann das System nach Hilfssatz XXXII zu einem System $x_1, x_2, \ldots, x_n, x_{n+1}$ erweitern, das denselben

Bedingungen genügt, Auf Grund der Kompaktheit von ρ und wegen Hilfssatz XX erkennt man, daß diese Erweiterung des Systems (3) nicht endlos fortgesetzt werden kann, und folglich gelangen wir schließlich zu einem System (1), welches eine linear unabhängige Basis von K bezüglich D bilded.

Der Hilfssatz XXXIII bedeutet, daß K isomorph ist mit einer Divisions algebra im Körper der reellen Zahlen im Sinne von Dickson. In Diesem Falle muß K mit einem der drei in Satz I angegebenen Körper isomorph sein, was zuerst Frobenius bewiesen hat.[6] Ein Beweis dieses Satzes, der nicht de komplizierten Hilfsapparat von Frobenius benutzt, findet sich in dem Buche "Algebren und ihre Zahlentheorien" von Dickson (deutsche Ausgabe S. 46 Satz V).

NOTES

1. Eine Menge R von irgendwelchen Elementen soll ein topologischer Raum heißen, (vgl. Alexandroff, *Math. Ann.* **96,** S. 555), wenn für jede Teilmenge M von R die abgeschlossene Hülle $\bar{M}$ definiert ist, und zwar so, daß folgende Bedingungen erfüllt sind:

I. Eine aus einem einzigen Punkt bestehende Menge ist mit ihrer abgeschlossenen Hülle identisch.

II. $\overline{A+B} = \bar{A} + \bar{B}$.

III. $\overline{(\bar{A})} = \bar{A}$.

Mengen, die mit ihrer abgeschlossenen Hülle zusammenfallen, heißen abgeschlossen, ihre Komplementärmengen offen. Jede offene Menge, die einen gegebenen Punkt enthält, heißt eine *Umgebung* dieses Punktes.

2. Wegen der Kompaktheit bzw. Kompaktheit im kleinen, sowie der weiteren hier gebrauchten topologischen Begriffe (wie Tennungsaxiome, insb. Regularität, Abzählbarkeitsaxiome u. dgl.), siehe Alexandroff, *Math. Ann.* **96**, S. 555, sowie Alexandroff und Urysohn, *Math. Ann.* **92**.

3. d. h., daß F als Raum betrachtete bikompakt ist.

3a. Wenn X eine Teilmenge von K und y ein Punkt von K ist, so bedeutet Xy die Menge aller Punkte xy, wobei x zu X gehört.

4. Eine Folge $x_1, x_2, \ldots, x_n, \ldots$ konvergiert gegen x, wenn jede Umgebung von x alle bis auf endlich viele dieser Punkte enthält.

5. "Kompakt" bedeutet *in sich* kompakt (d. h. kompakt und abgeschlossen).

6. Journal für reine und angewandte Mathematik, Bd. 84.

On Dynamical Systems Close to Hamiltonian Systems

Zhurnal Eksperimental'noi i Teoreticheskoi Fiziki (1934)
4.(9): 883–85

A theorem is proved to show the existence of limit cycles for systems close to Hamiltonian systems.

Lemma 1. Let $f(u,\mu)$ be a function with a continuous derivative $\partial^2 f(u,\mu)/\partial u\partial\mu$ for $|u-u_0|<\epsilon$, $|\mu|<\epsilon$, such that

$$f(u,0) = u; \quad \frac{\partial f(u_0,0)}{\partial\mu} = 0; \quad \frac{\partial^2 f(u_0,0)}{\partial u\partial\mu} \neq 0.$$

Then, for small μ, there exists a uniquely defined function $u(\mu)$ which is continuous and is equal to u_0 when $\mu=0$, satisfying the equation $f[u(\mu),\mu] = u(\mu)$; $u(\mu)$ is analytic, if $f(u,\mu)$ is analytic.

Proof. It is readily seen that the function $v(u,\mu) = [f(u,\mu)-u]/\mu$ is continuous and has a continuous derivative with respect to u. Since

$$v(u_0,0) = \frac{\partial f(u_0,0)}{\partial\mu} = 0 \quad \text{and} \quad \frac{\partial v(u_0,0)}{\partial u} = \frac{\partial^2 f(u_0,0)}{\partial u\partial\mu} \neq 0,$$

then from the general existence theorem, it follows that there exists a function $u(\mu)$ satisfying the equation $v[u(\mu),\mu] = 0$.

Lemma 2. Let

$$\frac{d\rho}{d\varphi} = r(\rho,\varphi,\mu) \tag{1}$$

be a differential equation in polar coordinates; $r(\rho, \varphi, 0) = 0$. The right-hand side is a periodic function of φ that has several continuous partial derivatives for $|\rho - \rho_0| < \epsilon$, $|\mu| < \epsilon$.

If

$$R(\rho, \mu) = \int_0^{2\pi} r(\rho, \varphi, \mu)\, d\varphi, \quad \frac{\partial R(\rho_0, 0)}{\partial \mu} = 0, \quad \frac{\partial^2 R(\rho_0, 0)}{\partial \rho \partial \mu} = q \neq 0,$$

then there exists one limit cycle, which depends continuously on μ with a characteristic number having the same sign as μq, and which tends to $\rho = \rho_0$ as $\mu \to 0$.

Proof. Let $w(u, \varphi, \mu)$ be a solution of the differential equation (1) with the initial value u, $w(u, 0, \mu) = u$. We readily find that $f(u, \mu) = w(u, 2\pi, \mu)$ satisfies the conditions of Lemma 1, because

$$\frac{\partial w(\rho_0, 2\pi, 0)}{\partial \mu} = \frac{\partial R(\rho_0, 0)}{\partial \mu} = 0, \quad \frac{\partial^2 w(\rho_0, 2\pi, 0)}{\partial u \partial \mu} = \frac{\partial^2 R(\rho_0, 0)}{\partial \rho \partial \mu} \neq 0.$$

Thus, Lemma 2 has been proved.

Theorem. Let

$$\frac{dx}{dt} = -\frac{\partial H(x, y)}{\partial y} + p(x, y, \mu); \quad \frac{dy}{dt} = \frac{\partial H(x, y)}{\partial x} + q(x, y, \mu);$$

$$p(x, y, 0) = 0; \quad q(x, y, 0) = 0 \tag{2}$$

be a system of equations; and

$$p_1(x, y) = \frac{\partial p(x, y, 0)}{\partial \mu}; \quad q_1(x, y) = \frac{\partial q(x, y, 0)}{\partial \mu}.$$

Let C_0 be a closed curve, the points of which satisfy the equation $H(x, y) = h_0$, and along which $\partial H/\partial x$ and $\partial H/\partial y$ do not vanish simultaneously. Then near C_0 there exists a closed curve C_h whose points satisfy the equation $H(x, y) = h$ when $|h - h_0|$ is sufficiently small. Let $\psi(h)$ denote the double integral

$$\iint \left(\frac{\partial p_1}{\partial x} + \frac{\partial q_1}{\partial y} \right) dx\, dy,$$

taken over the interior of the curve C_h. If $d\psi(h_0)/dh = e \neq 0$ and $\psi(h_0) = 0$, then there exists one and only one limit cycle for the system (2) that depends continuously on μ, having a characteristic number* of the same sign as μe, and tending to the curve C_0 as $\mu \to 0$.

Proof. On the curve C_0 let us introduce a cyclic parameter φ such that $x = \xi(\varphi)$, $y = \eta(\varphi)$ is the parametric equation of the curve C_0, where the right-hand sides have a period equal to 2π. Choose the direction of φ in such a manner that it increases with increasing t. Through the point $\xi(\varphi)$, $\eta(\varphi)$ draw a small segment of a curve orthogonal to the curve C_h, and let the equation of this segment be:

$$n(x, y, \varphi) = 0.$$

Using the equations $H(x,y) = h$, $n(x,y,\varphi) = 0$, we can introduce new coordinates h and φ near C_0, where φ is a periodic coordinate, and where h only takes values which are close to h_0.

Near C_0 the system (2) in new coordinates assumes the form

$$\frac{dh}{dt} = r(h,\varphi,\mu); \quad \frac{d\varphi}{dt} = s(h,\varphi,\mu), \quad r(h,\varphi,0) = 0.$$

Now let us calculate

$$r(h,\varphi,\mu) = \frac{\partial H}{\partial x}\cdot\frac{dx}{dt} + \frac{\partial H}{\partial y}\cdot\frac{dy}{dt} = \frac{\partial H}{\partial x}p(x,y,\mu) + \frac{\partial H}{\partial y}q(x,y,\mu), \quad (3)$$

where x and y should be replaced by their expressions h and φ by means of the coordinate transformation

$$\frac{dh}{d\varphi} = \frac{r(h,\varphi,\mu)}{s(h,\varphi,\mu)}.$$

The derivative of the right-hand side of equation (3) with respect to μ when $\mu = 0$ is

$$\frac{dr(h,\varphi,0)/d\mu}{s(h,\varphi,0)} = \frac{(\partial H/\partial x)p_1 + (\partial H/\partial y)q_1}{s(h,\varphi,0)} = g(h,\varphi).$$

It is readily seen that when $h = h_0$

$$\frac{\partial H}{\partial x} = \frac{d\eta}{d\varphi}s(h_0,\varphi,0); \quad \frac{\partial H}{\partial y} = -\frac{d\xi}{d\varphi}s(h_0,\varphi,0);$$

consequently,

$$g(h_0,\varphi) = \frac{d\eta}{d\varphi}p_1 - \frac{d\xi}{d\varphi}q_1; \quad \int_0^{2\pi} g(h_0,\varphi)\,d\varphi = \psi(h_0),$$

* I.e., winding number (Ed.).

and since h_0 is not special, we have

$$\int_0^{2\pi} g(h, \varphi)\, d\varphi = \psi(h).$$

Thus, the theorem has been proved with the help of Lemma 2.

Laboratory of Oscillations,
Scientific Research Institute of Physics,
Moscow State University,
Moscow

Received 15 July 1934

The Theory of Topological Commutative Groups*

Annals of Mathematics (1934) 35(2): 361–88

Introduction

The purpose of the present paper is to make an exhaustive investigation into the structure of continuous, locally compact, commutative groups, satisfying the second axiom of countability.[1]

1. Contents. First of all I discuss compact groups. The principal method here is the determination of the correspondence between continuous and discrete groups: To every discrete commutative group $\mathfrak{G}$ corresponds a compact commutative group X, which is called the group of characters of the group $\mathfrak{G}$. The first chapter of this paper is devoted to the study of the connection between the groups $\mathfrak{G}$ and X. It is also written so as to be applicable to combinatorial topology. In the second chapter I show that every compact, commutative group is the group of characters of some discrete group, using results of a paper by F. Peter and H. Weyl,[2] and also results of Haar.[3] The one-to-one correspondence thus established between compact commutative groups and discrete commutative groups enables us to reduce the study of the former to that of the latter. This fact is illustrated by the first and second appendices to this paper.

* Received November 22, 1933; Revised March 6, 1934. The results of this paper were reported in condensed form at the International Mathematical Congress in Zurich, 1932.

In the first appendix I establish a complete equivalence between the question of decomposition of compact, commutative groups into a direct sum, and the corresponding question concerning discrete commutative groups. I also give here some counterexamples. The second appendix gives the connection between the simplest topological properties of the group X and the algebraic properties of the group $\mathfrak{G}$.

In Chapter III I consider locally compact groups (which need not be compact) and confine myself to the case of connected groups. I show, first of all, that every such group Ω admits a subgroup Δ which has no limiting elements in Ω, and which is such that the factor group Ω/Δ is compact. This result, which is proved in a purely topological way, enables us to reduce the study of noncompact groups to the consideration of compact groups, which have already been discussed. In the same way I show that the group Ω is decomposible into the sum of a certain compact group and a system of groups isomorphic to the additive group of real numbers.

In the fourth and last section I investigate the structure of locally connected, locally compact, connected groups. It turns out that every such group is the direct sum of a finite or denumerable number of groups K and a finite number of groups isomorphic to the additive group of real numbers. The group K is the so-called continuous cyclic group, that is an additive group of real numbers determined up to an additive integer.

2. Notation. A continuous cyclic group, namely the additive group of real numbers determined up to an additive integer, is of fundamental importance in this paper, since with its aid we obtain the notion of the group of characters. We designate this group by K.

All the groups considered in this paper are commutative and are written additively except in Chapter II. All the continuous groups discussed are locally compact and satisfy the second axiom of countability. All the discrete groups contain at most a denumerable number of elements.

Continuous groups are designated by capital Greek letters and their elements by small Greek letters. Discrete groups are designated by capital German letters, and their elements by small German letters.

I. Groups of Characters

Here I discuss the connection between discrete and continuous groups and show that to each discrete group there corresponds a certain continuous group and conversely; the correspondence being reciprocal. Moreover, this pairing of the groups enables me to establish a correspondence between their sub-groups.

All continuous groups considered here are compact.

1. Definition 1. Let K be the continuous group of rotations of a circle, which we will consider as an additive group of real numbers defined up to an additive integer, and let $\mathfrak{G}$ be a discrete group.

Let us denote by X the set of all homomorphic representations of the group $\mathfrak{G}$ into the group K. The set X forms a continuous group in the following natural way. A sequence

$$\alpha_1, \alpha_2, \ldots, \alpha_n, \ldots$$

of elements of X is said to converge to the element α, if for each $\mathfrak{r} \subset \mathfrak{G}$,

$$\lim_{n\to\infty} \alpha_n(\mathfrak{r}) = \alpha(\mathfrak{r}).$$ [4]

We call the sum of two elements α' and α'' an element α such that

$$\alpha'(\mathfrak{r}) + \alpha''(\mathfrak{r}) = \alpha(\mathfrak{r}),$$

for each $\mathfrak{r} \subset \mathfrak{G}$. It is easy to see that the operations thus defined satisfy all the axioms of a continuous group. The group X so obtained we shall call the *group of characters* of the group $\mathfrak{G}$.

Remark 1. If $\mathfrak{G}$ is a finite group the definition of a group of characters given here coincides with the usual one. In fact, usually the group of characters of $\mathfrak{G}$ is defined as the group of homomorphisms of $\mathfrak{G}$ on the multiplicative group of complex numbers; but as all the elements of $\mathfrak{G}$, by virtue of the finiteness of $\mathfrak{G}$, have a finite order, they must be mapped onto complex numbers of absolute value 1. But the multiplicative group of complex numbers of absolute value 1 is isomorphic to the group K: In fact, let z be a complex number $|z| = 1$. Let us bring it into correspondence with a real number, equal to $\log z/2\pi i$. This number is defined up to an additive integer and constitutes an element of the group K; it is easy to see that the correspondence obtained in this way is an isomorphism.

Theorem 1. The group of characters of a discrete group is always compact and satisfies the second axiom of countability.

Proof. Let

$$\alpha_1, \alpha_2, \ldots, \alpha_n, \ldots$$

be an arbitrary sequence of elements of the group X and

$$\mathfrak{x}_1, \mathfrak{x}_2, \dots, \mathfrak{x}_n, \dots$$

be the ordered sequence of all elements of the group $\mathfrak{G}$. Consider the sequence

$$\alpha_1(\mathfrak{x}_1), \alpha_2(\mathfrak{x}_1), \dots, \alpha_n(\mathfrak{x}_1), \dots$$

of elements of the group K. Since K is compact there exists a sub-sequence

$$\alpha_{i_1}(\mathfrak{x}_1), \alpha_{i_2}(\mathfrak{x}_1), \dots, \alpha_{i_n}(\mathfrak{x}_1), \dots$$

converging to a limit element in K. Let us now consider the sequence

$$\alpha_{i_1}(\mathfrak{x}_2), \alpha_{i_2}(\mathfrak{x}_2), \dots, \alpha_{i_n}(\mathfrak{x}_2), \dots$$

from it we can again choose a convergent sub-sequence

$$\alpha_{i_{j_1}}(\mathfrak{x}_2), \alpha_{i_{j_2}}(\mathfrak{x}_2), \dots, \alpha_{i_{j_n}}(\mathfrak{x}_2), \dots .$$

Continuing in this way and applying, finally the diagonal process, that is, constructing the sequence

$$\alpha_1, \alpha_{i_2}, \alpha_{i_{j_2}}, \dots$$

we arrive at a converging sequence of elements of X, thereby proving the compactness of X.

To prove the second axiom of countability let us give the topological structure of X by means of neighborhoods. This can be done in the following way: Let $\alpha \subset$ X and let us define a certain neighborhood U of the element α as follows: let I_s be a certain interval on K containing $\alpha(\mathfrak{x}_s)$ ($s = 1, 2, \dots, m$), such that the ends of the interval are rational numbers. Then, by definition, $\beta \subset U$, if

$$\beta(\mathfrak{x}_s) \subset I_s.$$

It is easy to see that the definition of neighborhoods given here leads to the same topological structure of X as given in Definition 1, but the axiom of countability is here evidently fulfilled, since only a countable system of neighborhoods of X is given.

2. **Lemma 1.** Let $\mathfrak{G}$ be a group, $\mathfrak{H}$ a sub-group of $\mathfrak{G}$, $\mathfrak{x}$ a certain element of $\mathfrak{G}$, $\not\subset \mathfrak{H}$, and $\alpha \neq 0$ a certain homomorphism of $\mathfrak{H}$ in K (as always, K is the same as in Definition 1). The homomorphism α can be extended to a homomorphism β of the whole group $\mathfrak{G}$ in K, $\beta(\mathfrak{x}) \neq 0$.

Proof. Let

$$\mathfrak{x}_1 = \mathfrak{x}, \mathfrak{x}_2, \mathfrak{x}_3, \dots, \mathfrak{x}_n, \dots$$

be the totality of all elements of the group $\mathfrak{G}$. Let us construct the sequence

$$\mathfrak{H}_0 = \mathfrak{H}, \mathfrak{H}_1, \mathfrak{H}_2, \dots, \mathfrak{H}_n,$$

of sub-groups of $\mathfrak{G}$, where $\mathfrak{H}_{n+1}$ is the minimal sub-group containing $\mathfrak{G}_n$ and $\mathfrak{r}_{n+1}$. We may then construct further the sequence of homomorphisms

$$\alpha_0 = \alpha, \alpha_1, \alpha_2, \ldots, \alpha_n, \ldots$$

where α_{n+1} is a homomorphism of the group $\mathfrak{H}_{n+1}$ in K, which is an extension of the homomorphism of $\mathfrak{H}_n$. Suppose that α_n is already constructed, and let us construct α_{n+1}. If $\mathfrak{r}_{n+1} \subset \mathfrak{H}_n$ then we set $\alpha_{n+1} = \alpha_n$, since in this case $\mathfrak{H}_{n+1} = \mathfrak{H}_n$. If $\mathfrak{r}_{n+1} \not\subset \mathfrak{H}_n$ then two cases arise: (1) $\mathfrak{r}_{n+1}$ is linearly independent of the elements of $\mathfrak{H}_n$, that is, there do not exist non-trivial relations of the form $\mathfrak{b} + k\mathfrak{r}_{n+1} = 0$, where $\mathfrak{b} \subset \mathfrak{H}_n$, and k is an integer. (2) $\mathfrak{r}_{n+1}$ depends linearly on the elements of the group $\mathfrak{H}_n$. In the first case each element of $\mathfrak{H}_{n+1}$ is uniquely expressed in the form $\mathfrak{b} + k\mathfrak{r}_{n+1}$. Then we put

$$\alpha_{n+1}(\mathfrak{b} + k\mathfrak{r}_{n+1}) = \alpha_n(\mathfrak{b}) + k\gamma$$

where γ is an arbitrary element of K (when $n = 0$ we choose $\gamma \neq 0$). In the second case there are non-trivial relations of the form $\mathfrak{b} + k\mathfrak{r}_{n+1} = 0$. We will choose that relation for which k has the least positive value l, and let it be $\mathfrak{b} + l\mathfrak{r}_{n+1} = 0$, where $l > 1$, for otherwise $\mathfrak{r}_{n+1} \subset \mathfrak{H}_n$. In this case each element of $\mathfrak{H}_{n+1}$ is uniquely expressed in the form $\mathfrak{b} + k\mathfrak{r}_{n+1}, 0 \leqq k < l$. Let

$$\gamma = \alpha_n(\mathfrak{b})/l$$

(in the group K division is always possible, though perhaps not unique. In fact each element of K is a real number, defined up to an additive integer; therefore, if we divide by an element of unity, it can always be supposed that the quotient is not equal to 0, even when the dividend is 0). (When $n = 0$ we choose $\gamma \neq 0$.) Let us put

$$\alpha_{n+1}(\mathfrak{b} + k\mathfrak{r}_{n+1}) = \alpha_n(\mathfrak{b}) + k\gamma.$$

In this way α_{n+1} is constructed. The sequence of homomorphisms

$$\alpha_1, \alpha_2, \ldots, \alpha_n, \ldots$$

defines a homomorphism β of the group $\mathfrak{G}$ in K which does not transform $\mathfrak{r}_1$ into 0.

Remark 2. Lemma 1 is applicable also in the case when $\mathfrak{G}$ has more than a countable number of elements, except that, instead of the usual induction, it is necessary to apply transfinite induction in the proof.

Definition 2. Let $\mathfrak{G}$ be a group and X be the group of its characters, $\mathfrak{H}$ a sub-group of $\mathfrak{G}$, and Φ a sub-group of X (as was stated, we shall consider only compact continuous groups, consequently Φ is compact and closed in X). The totality of all elements of X mapping $\mathfrak{H}$ on 0, we will designate by $(X, \mathfrak{H})$, the totality of all elements of $\mathfrak{G}$ that are mapped on 0 by all homomorphisms of Φ we will designate by $(\mathfrak{G}, \Phi)$. It is easy to see, that $(X, \mathfrak{H})$ and $(\mathfrak{G}, \Phi)$ are groups, and that the first of them is closed in X.

Theorem 2. Let $\mathfrak{G}$ be a group, X the group of its characters and $\mathfrak{H}$ be a sub-group of the group $\mathfrak{G}$. If $\Phi = (X, \mathfrak{H})$, then $\mathfrak{H} = (\mathfrak{H}, \Phi)$.

Proof. Let $\bar{\mathfrak{H}} = (\mathfrak{G}, \Phi)$. It is obvious that $\mathfrak{H} \subset \bar{\mathfrak{H}}$. Suppose that there exists an element $\mathfrak{x} \subset \bar{\mathfrak{H}}$, such that $\mathfrak{x} \not\subset \mathfrak{H}$. Let α be a homomorphism of group $\mathfrak{H}$ in K mapping all the elements of $\mathfrak{H}$ into 0. According to Lemma 1, there exists a homomorphism β of group $\mathfrak{G}$ in K, which is an extension of α and such that $\beta(\mathfrak{x}) \neq 0$. Since β is an extension of α it transforms $\mathfrak{H}$ into 0, that is, $\beta \subset \Phi$. But, by Definition 2 $\beta(\mathfrak{x}) = 0$, since $\mathfrak{x} \subset \bar{\mathfrak{H}} = (\mathfrak{G}, \Phi)$. Hence, we arrived at a contradiction.

Theorem 3. Let $\mathfrak{G}$ be a group, $\mathfrak{H}$ a sub-group of $\mathfrak{G}$, and X the group of characters of $\mathfrak{G}$. Let us put $\Phi = (X, \mathfrak{H})$. Then the group of characters of $\mathfrak{H}$ is isomorphic to the factor group X/Φ, and the group of characters of the factor group $\mathfrak{G}/\mathfrak{H}$ is isomorphic to the group Φ.

Proof. Each homomorphism of the group $\mathfrak{G}$ in K gives a certain homomorphism of the group $\mathfrak{H}$ in K, while, according to Lemma 1, every homomorphism of the group $\mathfrak{H}$ may be obtained in this way. On the other hand, two homomorphisms α and β of the group $\mathfrak{G}$ coincide in $\mathfrak{H}$ if and only if their difference $\alpha - \beta$ maps $\mathfrak{H}$ on 0, that is $\alpha - \beta \subset \Phi$. In this way we have a homomorphism of the group X on the group of characters of the group $\mathfrak{H}$, while the totality of all ements of X that are mapped on 0 by this homomorphism, coincides with Φ. Hence, the first part of the theorem is proved.

Each homomorphism of Φ, that is, one by which $\mathfrak{H}$ is mapped on 0, gives a certain homomorphism of the group $\mathfrak{G}/\mathfrak{H}$ in K, but every homomorphism of the group $\mathfrak{G}/\mathfrak{H}$ may be obtained in this way, consequently the second part of the theorem is proved.

3. **Lemma 2.** Each sub-group Ψ of the group K either coincides with K or is a finite cyclic group with the numerical generator $1/r$, where r is the order of the group Ψ. (As usual, we consider only compact groups; that is, Ψ is a closed sub-group of K.)

Proof. If Ψ contains an infinity of elements, then by virtue of the compactness of K there are two arbitrarily near elements α and β in Ψ. The element $\gamma = \beta - \alpha \subset \Psi$ is arbitrarily small and its integer multiples fill K arbitrarily densely. Thus Ψ is everywhere dense in K and, as Ψ is closed, $\Psi = \mathrm{K}$.

If Ψ is of a finite order r, $r\alpha = 0$ for all $\alpha \subset \Psi$. Consequently, each element of Ψ has the form q/r, where q is an integer, but as there are only r such elements in K, for $q = 0, 1, \ldots, r-1$, and they form a group with the generator $1/r$, the lemma is proved.

Lemma 3. If the group $\mathfrak{G}$ is finite and of order r, the group of its characters is also of order r.

Remark 3. In reality the group of characters of the finite group $\mathfrak{G}$ is isomorphic to $\mathfrak{G}$. This is a well-known fact, but we will make no use of it.

Proof. Express the group $\mathfrak{G}$ in the form of a direct sum of cyclic groups $\mathfrak{G}_1, \mathfrak{G}_2, \ldots, \mathfrak{G}_n$, where $\mathfrak{G}_i$ has a generator $\mathfrak{g}_i$ of order r_i. In the mapping on K, the generator of $\mathfrak{G}_i$ is transformed into an element whose order is a divisor of r_i, but as the number of such elements in K, according to Lemma 2 is exactly r_i, the number of possible representations of the generator of $\mathfrak{G}_i$ is equal to r_i, and consequently, the number of possible mappings of $\mathfrak{G}$ in K is equal to the product $r_1 r_2 \ldots r_n$ which is equal to r.

Lemma 4. Let $\mathfrak{G}$ be a group, X the group of its characters, and Φ a subgroup of X, such that $(\mathfrak{G}, \Phi) = (0)$, then $\Phi = \mathrm{X}$.

To prove this lemma we must prove first three preliminary lemmas.

Lemma 5. The assertion of Lemma 4 holds when $\mathfrak{G}$ is finite.

Proof. Each element $\mathfrak{x}$ of the group $\mathfrak{G}$ gives a homomorphism of the group Φ in K. In fact, when α runs over Φ, $\alpha(\mathfrak{x})$ is a homomorphism of Φ in K. Since $(\mathfrak{G}, \Phi)$ equals (0), all these homomorphisms are distinct, and if r designates the orders of $\mathfrak{G}$, their number is r. If the group Φ does not coincide with X, its order $s < r$, so that according to Lemma 3, the group Φ admits only s different homomorphisms in K, of which there exist r, so that $s = r$, a contradiction.

Lemma 6. Lemma 4 holds when $\mathfrak{G}$ admits a finite system of linearly independent generators.

Proof. Let $\mathfrak{r}_1, \mathfrak{r}_2, \ldots, \mathfrak{r}_n$ be a system of linearly independent generators of $\mathfrak{G}$. Let us show that for every k, $0 \leqq k \leqq n$, no matter what elements δ_i, $(i = 1, 2, \ldots, k)$ we choose in K there exists an element $\gamma \subset \Phi$ such that $\gamma(\mathfrak{r}_i) = \delta_i$. When $k = n$ this assertion is equivalent to the assertion of the lemma, for, in this case, it follows that Φ contains all the homomorphisms of $\mathfrak{G}$ in K.

When $k = 0$ our statement is trivial; let us suppose that it holds for a given k and let us prove it for $k + 1$.

Consider the totality Ψ of elements of K expressible as $\alpha(\mathfrak{r}_{k+1})$, $\alpha \subset \Phi$, with the condition that $\alpha(\mathfrak{r}_i) = 0$, $(i = 1, 2, \ldots, k)$.

It is obvious that Ψ is a group. By Lemma 2, Ψ either coincides with K or all of its elements are expressible in the form q/r, where q is an integer and r is the order of Ψ.

If $\Psi = \mathrm{K}$, then the supposition is proved. In fact, let δ_j, $(j = 1, 2, \ldots, k + 1)$ be an arbitrary system of elements of K. By the supposition there exists an element $\beta \subset \Phi$, such that $\beta(\mathfrak{r}_i) = \delta_i$, $(i = 1, 2, \ldots, k)$. Suppose, that $\beta(\mathfrak{r}_{k+1}) = \delta'_{k+1}$. Since $\Psi = \mathrm{K}$ there exists an element $\alpha \subset \Phi$, such that $\alpha(\mathfrak{r}_i) = 0$, $\alpha(\mathfrak{r}_{k+1}) = \delta_{k+1} - \delta'_{k+1}$. For $\gamma = \alpha + \beta$ we have $\gamma(\mathfrak{r}_j) = \delta_j$, $(j = 1, 2, \ldots, k + 1)$.

Let us now suppose that all the elements of Ψ are expressible in the form q/r, with fixed r. From this it follows, that if $\alpha(\mathfrak{r}_i) = \beta(\mathfrak{r}_i)$, $\alpha \subset \Phi$, $\beta \subset \Phi$, $(i = 1, 2, \ldots, k)$, then

$$\beta(\mathfrak{r}_{k+1}) - \alpha(\mathfrak{r}_{k+1}) = q/r.$$

Let ϵ_i, $(i = 1, 2, \ldots, k)$, be a system of linearly independent elements of K. Let $\gamma(\mathfrak{r}_i) = \epsilon_i$, $\gamma \subset \Phi$, $(i = 1, 2, \ldots, k)$, which is possible according to the induction. Put $\gamma(\mathfrak{r}_{k+1}) = \epsilon_{k+1}$. Let us show that ϵ_j, $(j = 1, 2, \ldots, k + 1)$ are linearly dependent. If it were otherwise, then, by the Kronecker theorem, no matter what the system of elements δ_j of K, there could be found an infinite sequence of integers n_m, $(m = 1, 2, \ldots)$, such that

$$\lim_{m \to \infty} n_m \epsilon_j = \delta_j.$$

If we denote by λ one of the limit elements of the sequence $n_m \gamma$, we will have $\lambda(\mathfrak{r}_j) = \delta_j$ where $\lambda \subset \Phi$, since Φ is closed, that is, the induction is proved for $k + 1$, and Ψ coincides with K. In this way we have the relation

$$\sum_{j=1}^{k+1} r_j \epsilon_j = 0, \quad r_j \text{ integers.}$$

Let us show that from this relation there follows

$$\sum_{j=1}^{k+1} rr_j\alpha(\mathfrak{r}_j) = 0$$

for all $\alpha \subset \Phi$. In fact, since ϵ_i are linearly independent, by Kronecker's theorem there exists an infinite sequence $\bar{n}_m$ of integers, such that $\lim_{m\to\infty}\bar{n}_m\epsilon_i = \alpha(\mathfrak{r}_i)$. The sequence $\bar{n}_m$ can be chosen in such a way that $\bar{n}_m\gamma$ has a limit. Let it be equal to β. It is obvious, then, that

$$\sum_{j=1}^{k+1} r_j\beta(\mathfrak{r}_j) = 0.$$

Since $\alpha(\mathfrak{r}_i) = \beta(\mathfrak{r}_i)$, then, according to what was previously proved

$$\beta(\mathfrak{r}_{k+1}) - \alpha(\mathfrak{r}_{k+1}) = q/r;$$

thus, we have for α:

$$\sum_{j=1}^{k+1} rr_j\alpha(\mathfrak{r}_j) = 0.$$

Consequently;

$$\alpha\left(\sum_{j=1}^{k+1} rr_j\mathfrak{r}_j\right) = 0;$$

but, since α is an arbitrary element of Φ therefore

$$(\mathfrak{G}, \Phi) \supset \sum_{j=1}^{k+1} rr_j\mathfrak{r}_j = \mathfrak{r}.$$

But then, according to the condition of the lemma, $\mathfrak{r} = 0$, which is impossible since the generators $\mathfrak{r}_1, \mathfrak{r}_2, \ldots, \mathfrak{r}_n$ are linearly independent.

Lemma 7. Lemma 4 holds for any $\mathfrak{G}$ with a finite system of generators.

Proof. Let $\mathfrak{H}$ be the group composed of all elements of finite order of the group $\mathfrak{G}$. Let us put $\Psi = (\mathrm{X}, \mathfrak{H})$, and let us show that $\Psi \subset \Phi$. To this effect let us denote by Δ the intersection of Φ and Ψ and let us show that $\Delta = \Psi$. Let r be the order of $\mathfrak{H}$; let us show that if $\alpha \subset \mathrm{X}$, then $r\alpha \subset \Psi$. In fact, for any $\mathfrak{r} \subset \mathfrak{H}$ we have $r\alpha(\mathfrak{r}) = \alpha(r\mathfrak{r}) = \alpha(0) = 0$. Let us show, further, that $(\mathfrak{G}, \Delta) = \mathfrak{H}$. In fact, if we suppose the contrary, there exists a free element $\mathfrak{h}$ of $\mathfrak{G}$ such that for any $\beta \subset \Delta$, $\beta(\mathfrak{h}) = 0$. From this supposition it follows that for any $\alpha \subset \Phi$, $\alpha(r\mathfrak{h}) = r\alpha(\mathfrak{h}) = 0$. $r\alpha \subset \Delta$, because it is contained, as was previously proved, in Ψ and is a member of Φ and so belongs to their intersection. Thus, $(\mathfrak{G}, \Phi) \supset r\mathfrak{h} \neq 0$, which is impossible. Thus, $(\mathfrak{G}, \Delta) = \mathfrak{H}$. By Theorem 3, Ψ is the group of characters of the group $\mathfrak{G}/\mathfrak{H}$. From $(\mathfrak{G}, \Delta) = \mathfrak{H}$ follows $(\mathfrak{G}/\mathfrak{H}, \Delta) = (0)$. Since $\mathfrak{G}/\mathfrak{H}$ is a free group with a finite number of generators it follows, by Lemma 6, that $\Delta = \Psi$, that is, $\Psi \subset \Phi$. Let us set $X/\Psi = \bar{X}$, $\Phi/\Psi = \bar{\Phi}$. Then let us prove that $\bar{\Phi} = \bar{X}$. By Theorem 3, $\bar{X}$ is the group of characters of the group $\mathfrak{H}$. From $(\mathfrak{G}, \Phi) = 0$ it follows that $(\mathfrak{H}, \bar{\Phi}) = 0$, but, since $\mathfrak{H}$ is a finite group, we may apply Lemma 5, from which it follows that $\bar{\Phi} = \bar{X}$.

From $\Psi \subset \Phi$ and from $\Phi/\Psi = X/\Psi$ follows $\Phi = X$.

Proof of Lemma 4. Let $\mathfrak{H}_1, \mathfrak{H}_2, \ldots, \mathfrak{H}_n, \ldots$ be an expanding sequence of sub-groups of $\mathfrak{G}$, where each $\mathfrak{H}_n$ has a finite number of generators while the sum of all the $\mathfrak{H}_n$ coincides with $\mathfrak{G}$; such a sequence always exists since $\mathfrak{G}$ has not more than a countable number of elements. Let us put $\Psi_n = (X, \mathfrak{H}_n)$. It is obvious that $\Psi_{n+1} \subset \Psi_n$. From the fact that the sum of all $\mathfrak{H}_n = \mathfrak{G}$ it follows that the intersection of all the Ψ_n contains only zero.

Let Φ_n be the group composed of all elements of the form $\alpha + \beta$, where $\alpha \subset \Psi_n$ and $\beta \subset \Phi$. By Theorem 3, X/Ψ_n is the group of characters of the group $\mathfrak{H}_n$. From $(\mathfrak{G}, \Phi) = (0)$ it follows that $(\mathfrak{H}_n, \Phi_n/\Psi_n) = (0)$, but since $\mathfrak{H}_n$ has a finite system of generators it follows, from what has been said, in accordance with Lemma 7, that $\Phi_n/\Psi_n = X/\Psi_n$, that is $\Phi_n = X$. Thus, the intersection of all Φ_n coincides with X, but the same intersection must coincide with Φ, since the intersection of all Ψ_n contains only zero, that is $\Phi = X$.

Theorem 4. If $\mathfrak{G}$ is a group, X the group of its characters, Φ a sub-group of X, $\mathfrak{H} = (\mathfrak{G}, \Phi)$, then $(X, \mathfrak{H}) = \Phi$.

Proof. Let $\bar{\Phi} = (X, \mathfrak{H})$. It is evident that $\bar{\Phi} \supset \Phi$. By Theorem 3, $\bar{\Phi}$ is the group of characters of the group $\mathfrak{G}/\mathfrak{H}$. From $(\mathfrak{G}, \Phi) = \mathfrak{H}$ it follows that $(\mathfrak{G}/\mathfrak{H}, \Phi) = (0)$; from this by Lemma 7 we have $\Phi = \bar{\Phi}$.

4. **Definition 1′.**[5] Let X be a continuous compact group. The set $\mathfrak{G}$ of its homomorphic mappings in K, forms, in a natural way, just as in Definition 1, a group which we will call the group of characters of group X. We introduce only algebraic operations in $\mathfrak{G}$ and do not consider topological operators; but the mapping of X in K must be continuous.

Definition 3. Let X and $\mathfrak{G}$ be two groups for which there is defined a law of multiplication, that is, to each pair of elements $\alpha \subset X$, $\mathfrak{x} \subset \mathfrak{G}$ corresponds as their product the element $\gamma \subset K$, $\alpha\mathfrak{x} = \gamma$. Furthermore,

$$(\alpha' - \alpha'')\mathfrak{x} = \alpha'\mathfrak{x} - \alpha''\mathfrak{x}, \quad \alpha(\mathfrak{x}' - \mathfrak{x}'') = \alpha\mathfrak{x}' - \alpha\mathfrak{x}'',$$

and if $\lim \alpha_n = \alpha$ then $\lim \alpha_n \mathfrak{r} = \alpha\mathfrak{r}$. Under these conditions we shall say, that the groups X and $\mathfrak{G}$ form a couple. Further we shall introduce notations, analogous to those given in Definition 2; namely: If $\mathfrak{H}$ is a sub-group of $\mathfrak{G}$ and Φ a sub-group of X we shall designate by $(X, \mathfrak{H})$ the set of all elements of X, the products of which with each element of $\mathfrak{H}$ are equal to zero, in the same way $(\mathfrak{G}, \Phi)$ shall designate the set of all elements of $\mathfrak{G}$ the products of which with any element of Φ are equal to zero. In case X and $\mathfrak{G}$ form a couple, such that $(X, \mathfrak{G}) = (0)$, and $(\mathfrak{G}, X) = (0)$ we shall call X and $\mathfrak{G}$ orthogonal.

Remark 4. It is easy to see, that if X is the group of characters of the group $\mathfrak{G}$, then X and $\mathfrak{G}$ form an orthogonal couple; it is sufficient to put $\alpha\mathfrak{r} = \alpha(\mathfrak{r})$, $\alpha \subset X$, $\mathfrak{r} \subset \mathfrak{G}$ and we obtain the law of multiplication, satisfying all the conditions of Definition 3.

Theorem 5. If X and $\mathfrak{G}$ are orthogonal, each of these groups is the group of characters of the other; in fact, if $\alpha \subset X$, $\mathfrak{r} \subset \mathfrak{G}$, the homomorphism $\alpha(\mathfrak{r})$ is defined by the equation $\alpha(\mathfrak{r}) = \alpha\mathfrak{r}$ and the homomorphism $\mathfrak{r}(\alpha)$ by the condition $\mathfrak{r}(\alpha) = \alpha\mathfrak{r}$.

Proof. Each element $\alpha \subset X$ gives a homomorphism of $\mathfrak{G}$ in K. Moreover to distinct elements of X correspond distinct homomorphisms, since $(X, \mathfrak{G}) = 0$. Thus X is a sub-group of the group $\bar{X}$ of characters of the group $\mathfrak{G}$. On account of the orthogonality of X and $\mathfrak{G}$ it follows that $(\mathfrak{G}, X) = (0)$, and from this by Lemma 4 follows that $X = \bar{X}$.

Similarly let us prove that $\mathfrak{G}$ is a group of characters for X; here, however, the proof becomes somewhat more complicated, since, *a priori*, the group of characters for X could turn out to be uncountable. Therefore, let us carry out the proof somewhat differently. Let $\mathfrak{g}$ be a certain homomorphism of X in K. Let us show that it is realized with the help of one of the elements of group $\mathfrak{G}$. To this effect let us construct a group $\bar{\mathfrak{G}}$, each element of which is expressible in the form $\mathfrak{r} + k\mathfrak{g}$, where $\mathfrak{r} \subset \mathfrak{G}$, and k is an integer; let us define the law of multiplication of an arbitrary element $\alpha \subset X$ by an arbitrary element $\mathfrak{r} + k\mathfrak{g} \subset \bar{\mathfrak{G}}$ by the condition $\alpha(\mathfrak{r} + k\mathfrak{g}) = \alpha\mathfrak{r} + k\mathfrak{g}(\alpha)$ and let us, moreover, consider that in the group $\bar{\mathfrak{G}}$ there exist relations such that two of its elements coincide if and only if, they give the same homomorphisms of X in K. By virtue of these conditions the groups X and $\bar{\mathfrak{C}}$ are orthogonal, and by the first part of the theorem already proved X is the group of characters for $\bar{\mathfrak{G}}$; but, since $(X, \mathfrak{G}) = (0), \mathfrak{G} = \bar{\mathfrak{G}}$; that is, the homomorphism $\mathfrak{g}$ is already contained among the homomorphisms of the group given by the elements of the group $\mathfrak{G}$.

Lemma 8. Let X and $\mathfrak{G}$ be orthogonal groups, $\mathfrak{r}$ an element of $\mathfrak{G}$ of order g ($g = 0$ or $g > 1$). If $\gamma \subset$ K and has an order which is a divisor of g, then there exists an $\alpha \subset$ X, such that $\alpha\mathfrak{r} = \gamma$.

Proof. Let $\mathfrak{X}$ be the cyclic group with generator $\mathfrak{r}$. Evidently there exists a homomorphism β of the group $\mathfrak{X}$ in K such that $\beta(\mathfrak{r}) = \gamma$. By Lemma 1 the homomorphism β can be extended into a homomorphism $\bar{\alpha}$ of the entire group $\mathfrak{G}$ in K. By Theorem 5 there exists in X such an element α that $\bar{\alpha}(\mathfrak{h}) = \alpha(\mathfrak{h})$, for each $\mathfrak{h} \subset \mathfrak{G}$, that is, $\alpha\mathfrak{r} = \gamma$.

Remark 5. If the group $\mathfrak{G}$ has a system of n linearly independent generators $\mathfrak{r}_1, \mathfrak{r}_2, \ldots, \mathfrak{r}_n$, then the group X of its characters has a particularly simple structure. In fact it decomposes into a direct sum of n continuous cyclic groups. Topologically, the group X is in this case the product of n circles, that is, an n-dimensional torus and we shall therefore call it a toroidal group. For $n = 1$, this is almost obvious. In fact, let $\alpha \subset$ X, then $\alpha(\mathfrak{r}_1) = \gamma \subset$ K but for a given γ, this determines α uniquely since $\mathfrak{r}_1$ is a generator of $\mathfrak{G}$. In this way X and K are isomorphic. For a general n, each of the generators $\mathfrak{r}_1, \mathfrak{r}_2, \ldots, \mathfrak{r}_n$ can be mapped on K entirely arbitrarily and we obtain n continuous cyclic summands.

II. Structure of a Compact Group

In this chapter I make use of the results of F. Peter and H. Weyl,[2] which have become applicable to any compact group satisfying the second axiom of countability, thanks to the results of Haar.[3] I show that every compact commutative group satisfying the second axiom of countability is the group of characters of some discrete group.[6] Thus all the results of the preceding chapter become applicable to arbitrary compact, commutative groups.

All the continuous groups considered here are compact. The group is written multiplicatively.

5. In order to obtain the above mentioned result I formulate those results of Peter and Weyl, which I shall use.

These are as follows:

If Ω is a continuous compact group satisfying the second axiom of countability, then there exists on it a finite or denumerable system

$$f_{r,k_r}(\xi), \quad (r = 1, 2, \ldots; \; k_r = 1, 2, \ldots, n_r)$$

of real numerical continuous functions depending on the argument ξ, which is an element of the group Ω. Moreover the following conditions are satisfied:

(1) For any element α of the group Ω

$$f_{r,i}(\xi \cdot \alpha) \equiv \sum_{j=1}^{n_r} m_{r,i}^j(\alpha) f_{r,j}(\xi), \quad (i = 1, 2, \ldots, n_r)$$

where $m_{r,i}^j(\alpha)$ are real numbers forming an orthogonal matrix $M_r(\alpha)$ which depends continuously on α, and the number of whose rows and columns is equal to n_r.

(2) If ϵ is the unit of the group Ω, then for every element α such that $\epsilon \neq \alpha \subset \Omega$ there exists a function $f_{r,i}(\xi)$ of our system such that $f_{r,i}(\xi \cdot \alpha) \not\equiv f_{r,i}(\xi)$, that is, there exists an r such that the matrix $M_r(\alpha)$ is not a unit matrix. It is easy to see that if α and β are two elements of Ω, then $M_r(\alpha \cdot \beta) = M_r(\beta) \cdot M_r(\alpha)$, where on the right we have matrix multiplication. Hence if the group Ω is commutative, then all the matrices $M_r(\alpha)$ are commutative among themselves for a fixed r.

In order to apply the above results to commutative groups, I will have to consider orthogonal matrices in greater detail. In what follows it is not fruitful to confine ourselves to matrices with real elements, but it is better to make use of orthogonal matrices with complex elements. As is well known, a matrix M with complex elements is said to be orthogonal if the following condition is fulfilled. Let M' be the matrix obtained from M by transposing rows and columns, and let $\bar{M}'$ be the complex conjugate of M', then $M^{-1} = \bar{M}'$. It is easy to see that if M and N are two orthogonal matrices with complex elements, then their product MN is also orthogonal. It is further known that if M is an orthogonal matrix, then there exists an orthogonal matrix T such that TMT^{-1} is a diagonal matrix, all of whose diagonal elements have absolute value unity. Making use of this fact we can prove the following lemmas:

Lemma 9. Let α run over the elements of any set, and let $M(\alpha)$ be a system of orthogonal n-rowed matrices, any two of which are commutative with each other, then there exists an n-rowed orthogonal matrix S, such that for any α, $SM(\alpha)S^{-1}$ is a diagonal matrix all of whose diagonal elements have absolute value unity.

Proof. Let $M(\alpha_1)$ be a non-diagonal matrix of our system and let T_1 be an orthogonal matrix such that $T_1 M(\alpha_1) T_1^{-1}$ is diagonal. Let us designate its diagonal elements by e_i, $i = 1, 2, \ldots, n$, where the e_i are arranged in such a way that all equal ones are consecutive. Let us further denote by $M_1(\alpha)$ the matrix $T_1 M(\alpha) T_1^{-1}$. Its elements we designate by $m_i^j(\alpha)$. Since the matrices $M(\alpha_1)$ and $M(\alpha)$ are commutative, it follows that the matrices $M_1(\alpha_1)$ and $M_1(\alpha)$ have the same property. In this way we have $e_i m_i^j(\alpha) = m_i^j(\alpha) e_j$, that is $m_i^j(\alpha) = 0$ whenever $e_i \neq e_j$. It follows that the matrix $M_1(\alpha)$ decomposes into a system of square matrices along the diagonal of $M_1(\alpha)$, whose number s is not less than the number of distinct elements e_i. The number of distinct e_i, however exceeds unity, since if all e_i were equal, the matrix $M(\alpha_1)$ would be diagonal, which is contrary to hypothesis.

Let $M_1(\alpha_2)$ be a non-diagonal matrix. Transforming separately every square matrix of which it is composed, we obtain the general transforming matrix T_2, which is such that $T_2 M_1(\alpha_2) T_2^{-1}$ is a diagonal matrix, while $T_2 M_1(\alpha_1) T_2^{-1} = M_1(\alpha)$. Further, let us denote by $M_2(\alpha)$ the matrix $T_2 M_1(\alpha) T_2^{-1}$. Writing out the conditions that $M_2(\alpha)$ and $M_2(\alpha_2)$ commute, we see as before that the matrix $M_2(\alpha)$ decomposes into separate square matrices, lying along its diagonal, whose number must now exceed s. Continuing this process we will reduce all our matrices to a diagonal form in a finite number of steps.

Lemma 10. Let Ω be a continuous compact group; then there exists a finite or denumerable system of continuous complex valued functions $\mathfrak{g}_n(\xi)$, $n = 1, 2, \ldots$, where the argument ξ is an element of the group Ω, having the following properties:

(1) For any element α of Ω, $\mathfrak{g}_n(\xi\alpha) = e_n(\alpha)\mathfrak{g}_n(\xi)$, where $e_n(\alpha)$ is a complex number of absolute value unity, and $e_n(\alpha^{-1}) = (e_n(\alpha))^{-1}$.

(2) For any element α of Ω with $\alpha = \epsilon$, there exists a number m such that $\mathfrak{g}_m(\xi\alpha) \neq \mathfrak{g}_m(\xi)$, that is, $e_m(\alpha) \neq 1$.

Proof. Let us examine the system of functions $f_{r,i}(\xi)$ on Ω for a fixed r (which were mentioned in the beginning of this section) and the corresponding system of matrices $M_r(\alpha)$. Since the group Ω is commutative, the matrices $M_r(\alpha)$ also commute. Hence we can apply Lemma 9 to the system of matrices $M_r(\alpha)$ and there exists a transforming matrix S which reduces the matrices $M_r(\alpha)$ to diagonal form. We shall denote the elements of S by S_i^j. It is easy to see that the function

$$\sum_{j=1}^{n_r} S_i^j f_{r,j}(\xi), \quad (i = 1, 2, \ldots, n_r)$$

satisfies condition (1) of our lemma.

Applying the above operation to all the systems of functions $f_{r,i}(\xi)$ for every r, we obtain a totality of functions satisfying both conditions of the lemma.

First Fundamental Theorem. Let Ω be a continuous compact commutative group satisfying the second axiom of countability, and $\mathfrak{G}$ the discrete group of its characters, then the group Ω is isomorphic to the group of characters of $\mathfrak{G}$. (See Definitions 1 and 1′.)

Proof. Let $\mathfrak{g}_n(\xi)$, $(n = 1, 2, \ldots)$ be a system of functions defined on Ω and satisfying the conditions of Lemma 10. Let us show, first of all, that no function $\mathfrak{g}_n(\xi)$ can be equal to zero without being identically zero. Let us suppose that $\mathfrak{g}_n(\alpha) = 0$, then

$$\mathfrak{g}_n(\xi) = \mathfrak{g}_n(\alpha\alpha^{-1}\xi) = e_n(\alpha^{-1}\xi)\mathfrak{g}_n(\alpha) = 0.$$

We shall suppose that from the system $\mathfrak{g}_n(\xi)$ all identically zero functions have been removed. This does not violate the conditions of Lemma 10. Since none of the functions $\mathfrak{g}_n(\xi)$ now admit the value zero, they can be normalized in such a way that $\mathfrak{g}_n(\epsilon) = 1$, where ϵ is the unit of the group Ω. Let us suppose that this normalisation has been carried out. This, likewise will not affect the conditions of Lemma 10. We will have

$$\mathfrak{g}_n(\xi) = \mathfrak{g}_n(\epsilon\xi) = e_n(\xi)\mathfrak{g}_n(\epsilon) = e_n(\xi).$$

We now build a multiplicative group $\mathfrak{G}$ of functions defined on Ω, taking as generators the functions $\mathfrak{g}_n(\xi)$. Every element $\mathfrak{g}(\xi)$ of the group $\mathfrak{G}$ can be expressed in the form $(\mathfrak{g}_1(\xi))^{m_1} \cdot (\mathfrak{g}_2(\xi))^{m_2} \ldots (\mathfrak{g}_n(\xi))^{m_n}$ where n is sufficiently large and m_i is an integer. Since

$$\mathfrak{g}_n(\xi\eta^{-1}) = e_n(\eta^{-1})\mathfrak{g}_n(\xi) = {}_n(\xi)(\mathfrak{g}_n(\eta))^{-1},$$

we have

$$\mathfrak{g}(\xi\eta^{-1}) = \mathfrak{g}(\xi)(\mathfrak{g}(\eta))^{-1}.$$

Let us further formulate the law of multiplication between the elements of the groups Ω and $\mathfrak{G}$. We shall define the product of $\alpha \subset \Omega$ by $\mathfrak{g}(\xi) \subset \mathfrak{G}$ as the element of the group K equal to

$$\frac{\log \mathfrak{g}(\alpha)}{2\pi i}.$$

From the conditions

$$\frac{\log \mathfrak{g}(\xi\eta^{-1})}{2\pi i} = \frac{\log \mathfrak{g}(\xi)}{2\pi i} - \frac{\log \mathfrak{g}(\eta)}{2\pi i}$$

and

$$\frac{\log [\mathfrak{g}'(\xi)(\mathfrak{g}''(\eta))^{-1}]}{2\pi i} = \frac{\log \mathfrak{g}'(\xi)}{2\pi i} - \frac{\log \mathfrak{g}''(\eta)}{2\pi i},$$

and also because of the continuity of the function $\mathfrak{g}_n(\xi)$, the groups Ω and $\mathfrak{G}$ form a couple (see Definition 3). Moreover the groups Ω and $\mathfrak{G}$ are orthogonal. In fact from condition 2 of Lemma 10, there is an $\alpha \neq \epsilon$ such that there is a $\mathfrak{g}(\xi)$ for which $\mathfrak{g}(\alpha) \neq 1$. Therefore we have $(\Omega, \mathfrak{G}) = (0)$. Moreover if $\mathfrak{g}(\xi)$ is not the unit of the group $\mathfrak{G}$, that is if $\mathfrak{g}(\xi)$ is not identically equal to unity, then there exists an element α, such that $\mathfrak{g}(\alpha) \neq 1$, and hence $(\mathfrak{G}, \Omega) = (0)$. It follows from Theorem 5 that each of the groups Ω and $\mathfrak{G}$ is the group of characters of the other.

III. Structure of a Locally Compact Connected Group

The purpose of this chapter is to prove the second Fundamental Theorem, that is to reduce the investigation of a locally compact, connected, commutative group, satisfying the second axiom of countability, to that of a compact group, already considered in Chapter II.

The groups are written additively, as in Chapter I.

6. We pass to the consideration of locally compact connected, but not necessarily compact groups by means of the following lemma:

Lemma 11. Let Ω be a locally compact, connected group. Then there exists a subgroup A of the group Ω, having not more than a denumerable number of elements, and having no limiting elements in Ω such that the factor group Ω/A is compact.

In order to prove Lemma 11, we first establish:

Lemma 12. Let Ω be a non-compact group, satisfying the conditions of Lemma 11. Then there exists an element α in Ω which is such that the cyclic sub-group of Ω generated by α is free and has no limiting elements in Ω.[7]

Proof. (a) Let U be some neighborhood of zero of the group Ω, whose closure $\bar{U}$ is compact. Let us denote by U_n the set of all the elements of Ω, which can be written in the form: $\alpha_1 + \alpha_2 + \ldots + \alpha_n$ where $\alpha_i \subset U$, $(i = 1, 2, \ldots, n)$. Since U is a domain it follows that U_n is likewise a domain. It is easy to see that the closure $\bar{U}_n$ of the domain U_n is composed of all the elements of the form: $\beta_1 + \beta_2 + \ldots + \beta_n$ where $\beta_i \subset \bar{U}$, $(i = 1, 2, \ldots, n)$. Since $\bar{U}$ is compact, it follows easily that $\bar{U}_n$ is also compact. Since Ω is non-compact and connected, while $\bar{U}_n$ is compact, it follows that the boundary $U_n' = \bar{U}_n - U_n$ of the domain U_n is not empty.

(b) Let $\alpha \subset \bar{U}_r, \beta \subset \bar{U}_s$ and $\gamma = \alpha + \beta \subset U'_{r+s}$. We will show that $\alpha \subset U'_r$, $\beta \subset U'_s$. Let us suppose on the contrary that α, for instance, belongs to U_r. Since β is a limiting element of U_s, an arbitrarily small element δ can be found, such that $\beta - \delta \subset U_s$. Since $U_r \supset \alpha$ is a domain, δ can be chosen sufficiently small so that $\alpha + \delta \subset U_r$. In this way we have $\gamma = (\alpha + \delta) + (\beta - \delta)$ where $\alpha + \delta \subset U_r$ and $\beta - \delta \subset U_s$. That is, from the construction of the domains U_n given in (a), $\gamma \subset U_{r+s}$ which contradicts the assumption that $\gamma \subset U'_{r+s}$.

(c) Let $\alpha_n \subset U'_n$. The set of elements α_n, $(n = 1, 2, \ldots)$ has no limiting points. Let us assume on the contrary that we have found two elements α_i and α_j, $i<j$ such that $\alpha_j - \alpha_i \subset U$. Then we would have $\alpha_j = \alpha_i + (\alpha_j - \alpha_i)$ that is, by (b) $\alpha_j \subset U_{i+1}$, which is impossible since $\alpha_j \subset U'_j$, and U'_j does not intersect U_{i+1}.

(d) Let us build up a system of elements ϵ_n $n = 1, 2, \ldots$, belonging to $\bar{U}$, and such that $\epsilon_1 + \epsilon_2 + \ldots + \epsilon_n \subset U'_n$. Let us choose an element $\gamma_n \subset U'_n$. Such an element exists since, by (a), U'_n is not empty and can be represented in the form $\epsilon_{1,n} + \epsilon_{2,n} + \ldots + \epsilon_{n,n} = \gamma_n$ where $\epsilon_{i,n} \subset \bar{U}$, $(i = 1, 2, \ldots, n)$. Since $\bar{U}$ is compact, we can use the diagonal process to choose a sequence of numbers n_k, $(k = 1, 2, \ldots)$, such that $\lim_{k\to\infty} \epsilon_{i,nk}$ exists. Let us denote this limit by ϵ_i, $(i = 1, 2, \ldots)$. Since $\epsilon_{i,n} \subset \bar{U}$, $(i = 1, 2, \ldots, n)$, then $\epsilon_{1,n} + \epsilon_{2,n} + \ldots + \epsilon_{j,n} \subset \bar{U}_j$ and $\epsilon_{j+1,n} + \epsilon_{j+2,n} + \ldots + \epsilon_{n,n} \subset \bar{U}_{n-j}$, $j<n$ but $(\epsilon_{1,n} + \epsilon_{2,n} + \ldots + \epsilon_{j,n}) + (\epsilon_{j+1,n} + \epsilon_{j+2,n} + \ldots + \epsilon_{n,n}) \subset U'_n$. In this way using (b) we have

$$\epsilon_{1,n} + \epsilon_{2,n} + \ldots + \epsilon_{j,n} \subset U'_j.$$

$$\lim_{k\to\infty} (\epsilon_{1,n_k} + \epsilon_{2,n_k} + \ldots + \epsilon_{j,n_k}) + \epsilon_1 + \epsilon_2 + \ldots + \epsilon_j.$$

The left-hand side of the last equation belongs to U'_j since U'_j is closed. In this way the construction is completed.

(e) Let us select a system of m elements from the system constructed in (d) and let us denote these elements by $\epsilon_{n_1}, \epsilon_{n_2}, \ldots, \epsilon_{n_m}$, where we suppose that $n_i \neq n_j$ with $i \neq j$. Then $\epsilon_{n_1} + \epsilon_{n_2} + \ldots + \epsilon_{n_m} \subset U'_m$. In fact there exists an n sufficiently large so that $n > n_j$; $j = 1, 2, \ldots, m$. Let us split the sum $\epsilon_1 + \epsilon_2 + \ldots + \epsilon_n$ into two parts. The first part shall consist of $\epsilon_{n_1} + \epsilon_{n_2} + \ldots + \epsilon_{n_m}$ and the second part of all the remaining summands. The first part belongs to $\bar{U}_m$, and the second part to $\bar{U}_{n-m}$. Hence, from (b) $\epsilon_{n_1} + \epsilon_{n_2} + \ldots + \epsilon_{n_m} \subset U'_m$.

Let α be a limiting element of the set ϵ_n. Then there exists a system $\epsilon_{n_1}, \epsilon_{n_2}, \ldots, \epsilon_{n_m}$, with an arbitrary m, in an arbitrarily small neighborhood of the element α. From this we conclude that for an arbitrary m, the element $m\alpha \subset U'_m$. Hence all the elements $m\alpha$ are distinct for distinct

m's and have no limiting elements in Ω by (c). Hence Lemma 12 is proved.

Proof of Lemma 11. Let U be an arbitrary neighborhood of zero of the group Ω, whose closure $\bar{U}$ is compact. Let us construct an increasing, transfinite sequence of subgroups A_i of the group Ω, each of which has only zero belonging to $\bar{U}$. Let A_1 be the group which contains only zero. Let us suppose further that all the groups A_i for $i<j$ have already been constructed. If j has no predecessor, we take for A_j the sum of all A_i with $i<j$. The group thus obtained has only zero in common with $\bar{U}$. Now let $j=k+1$. Since A_k has only zero in common with $\bar{U}$, it follows A_k has no limiting points in Ω, and, hence, A_k is closed. We can now consider the factor group Ω/A_k which we denote by Ω', and the set $\bar{U}'$ which is obtained from $\bar{U}$ in forming the factor group. It is obvious that $\bar{U}'$ is compact. If Ω' is a compact group then we will stop the construction of the sequences A_i at the number k. If Ω' is not compact, we will apply to it Lemma 12. Let α be an element of the group Ω' constructed in Lemma 12. Since $\bar{U}'$ is compact, there exists only a finite number of elements of the form $m\alpha$, where m is an integer, which belongs to $\bar{U}'$. For in the contrary case the set of elements of the form $m\alpha$ would have a limiting element in $\bar{U}'$. Hence there exists a sufficiently large number n, such that all the elements of the form $m\alpha$ with $|m|>n$ lie outside $\bar{U}'$. We denote the subgroup with the generator $n\alpha$ by A'. The complete image of the subgroup A' in the group Ω we denote by A_{k+1}. It is easy to see that A_{k+1} does not coincide with A_k, and has only zero in common with $\bar{U}$.

Since Ω satisfies the second axiom of countability and is locally compact, it can be covered with a countable system of domains, the closure of each of which is compact. Consequently every non-denumerable subset of the group Ω, must have a limiting point in Ω. But A_i has no limiting points in Ω, hence A_i is denumerable. That is, our construction can be stopped at a countable step, and we obtain in the end, a group A which satisfies the conditions of Lemma 11.

7. **Definition 4.** We will call the group of vectors in the n-dimensional affine space, an n-dimensional vector group. This group decomposes into a direct sum of subgroups, isomorphic to the additive group of real numbers.

Lemma 13. Let T be a connected group, having a subgroup A, which has no limiting points in T, and which is such that the factor group $T/A = T'$

is a toroidal group (see Remark 5), then T is the direct sum of a certain vector group M and a toroidal group Λ.

Proof. Let r be the dimensionality of T′. Let us denote by N, an r-dimensional vector group, and let B be that subgroup of N all of whose elements have integral coordinates in some coordinate system. It is easy to see that T′ is isomorphic to the factor group N/B. Hence using Schreier's[8] result, N covers universally the groups T′ and T, and hence is isomorphic to the factor group N/B′, where B′ is a subgroup of B. By an integer transformation of the coordinates of N we can obtain a system of generators $\beta_1, \beta_2, \ldots, \beta_k$, $k \leqq r$ of B′, such that β_i has all its coordinates equal to zero, except the i-th one, which is equal to some integer m_i. From this it is easy to conclude that T is isomorphic to a direct sum of a k-dimensional toroidal group and an $(r-k)$-dimensional vector group.

Lemma 14. Let Ω be a locally compact, connected group, and U an arbitrary neighborhood of zero in Ω. Then there exists a compact subgroup Λ situated in U, and such that the factor group Ω/Λ decomposes into a direct sum of a vector group and a toroidal group.

Proof. Let A be the subgroup of the group Ω, constructed in Lemma 11. Since A has no limiting element in Ω, there exists a sufficiently small neighborhood of zero V in Ω, such that if α and β are two elements of V with $\alpha-\beta \subset$ A, then $\alpha=\beta$. Let us also suppose that $V \subset U$.

Since Ω is connected, the compact factor group $\Omega' = \Omega/\mathrm{A}$ is also connected. From the First Fundamental Theorem it follows that the group Ω′ is the group of characters of a certain discrete group $\mathfrak{G}$. Since Ω′ is connected, $\mathfrak{G}$ contains no elements of finite order. (See Appendix 2, Theorem 1c.) Let $\mathfrak{H}_1, \mathfrak{H}_2, \ldots, \mathfrak{H}_n$ be an increasing sequence of subgroups of the group $\mathfrak{G}$, such that each of the subgroups has a finite number of generators, and the sum of the subgroups coincides with $\mathfrak{G}$. Let $\Phi_n = (\Omega', \mathfrak{H}_n)$. Obviously Φ_n forms a decreasing sequence of subgroups, whose intersection contains only zero. Let V' be the image of V in forming the factor group Ω/A. From the construction, V is reflected on V' in a one to one way and V' forms a neighborhood of zero in Ω′. Since the Φ_n intersect in the zero of Ω′, there exists a sufficiently large number m, such that $\Phi_m \subset V'$. The factor group Ω'/Φ_m is the group of characters of the group $\mathfrak{H}_m$ (see Theorem 3). Since the group $\mathfrak{H}_m$ has no elements of finite order and has a finite number of generators, it follows that $\mathfrak{H}_m$ has a finite system of linearly-independent generators, that is, the group $\Omega'/\Phi_m = \mathrm{T}'$ is toroidal (see Remark 5).

Let Γ be that portion of the image of the group Φ_m in the group Ω, which lies entirely in V. Since the mapping of V on V' is single valued, the group Γ is isomorphic to Φ_m. Since $\Gamma \subset V$, the intersection of Γ and A is a zero of the group Ω. Let $\Omega'' = \Omega/\Gamma$, and let A$'$ be the subgroup of the group Ω' into which A is mapped. Is is easy to see that A$'$ has no limiting elements in Ω'', and that Ω''/A' is isomorphic to T$'$. Since T$'$ is toroidal, it follows from Lemma 13 that Ω'' is a direct sum of a vector group and a toroidal group.

Lemma 15. Let Ω be a locally compact connected group. Then it contains a maximal compact subgroup Δ (that is, Δ is such that every other compact subgroup of Ω is a subgroup of Δ), and the factor group Ω/Δ is a vector group.

Proof. Let U be a neighborhood of zero in Ω, whose closure is compact, and let Γ be the subgroup constructed in Lemma 14. By the same lemma the factor-group Ω/Γ decomposes into the direct sum of the toroidal group Λ, and the vector group M. Let Δ be the complete image of the group Λ in the group Ω. Since Γ and Λ are compact, Δ is also compact and the factor-group Ω/Δ is isomorphic to M and is a vector group. Let us suppose that there exists a subgroup Δ', which is compact, but which does not belong to Δ. Let $\alpha \subset \Delta'$ and $\alpha \not\subset \Delta$. Let α' be the element into which α goes in forming the factor-group Ω/Δ. Since $\alpha \not\subset \Delta$, $\alpha \neq 0$, this means that the sequence of multiples of the element α' has no limiting elements in Ω/Δ. But this is impossible since all the multiples of α belong to Δ' and their set must have a limiting element.

The Second Fundamental Theorem. A locally compact connected group Ω satisfying the second axiom of countability decomposes into a direct sum of a compact subgroup Δ, and a vector subgroup N (see Definition 4), where the subgroup Δ is determined uniquely, and the dimensionality r of the group N is an invariant of the group Ω.

Proof. Let Δ be the maximal compact subgroup of the group Ω, constructed in Lemma 15. Let us choose a decreasing sequence $U_1, U_2, \ldots, U_n, \ldots$ of neighborhoods of zero of the group Ω, whose closures are compact, and whose intersection contains only zero. Let us construct the sequence $\Omega_0 = \Omega, \Omega_1, \Omega_2, \ldots, \Omega_n, \ldots$ of closed, connected subgroups of the group Ω, such that the intersection Γ_n of the group Ω_n and the group Δ is contained in U_n, and such that every element of the group Ω can be written in the form $\alpha + \beta$, where $\alpha \subset \Delta$, and $\beta \subset \Omega_n$.

Let us suppose that Ω_n is already constructed. It is easy to see that Γ_n is the maximal, compact subgroup of the group Ω_n. In fact every compact subgroup of Ω_n belong to Δ by Lemma 15, and hence to the intersection of Δ and Ω_n, which is Γ_n. The factor-group Ω_n/Γ_n is isomorphic to Ω/Δ, and hence the dimensionality of Ω_n/Γ_n is r. By Lemma 14, there exists a subgroup $\Gamma'_{n+1} \subset U_{n+1}$ of the group Ω_n which is such that Ω_n/Γ'_{n+1} decomposes into a direct sum of a toroidal subgroup Λ and a vector subgroup M. Since Λ is the maximal compact subgroup of Ω_n/Γ'_{n+1}, the subgroup Γ_n is the complete image of Λ in Ω_n. Let us denote by Ω'_{n+1} the complete image of the group M in Ω_n. Since the intersection of Λ and M in the group Ω_n/Γ'_{n+1} is zero, the intersection of Ω'_{n+1} and Γ_n is Γ'_{n+1}. Hence the intersection of Ω'_{n+1} and Δ is Γ'_{n+1}. Let Ω_{n+1} be the zero component of the group Ω'_{n+1}; such a group is, as is well known, a closed subgroup of the group Ω'_{n+1}. We denote by Γ_{n+1} the intersection of Ω_{n+1} and Γ'_{n+1}, then the intersection of Ω_{n+1} and Δ is Γ_{n+1}. We have then Ω/Δ is isomorphic to Ω_n/Γ_n which is isomorphic to $\Omega'_{n+1}/\Gamma'_{n+1}$ which in turn is isomorphic to $\Omega_{n+1}/\Gamma_{n+1}$. Hence the latter factor-group is an r-dimensional vector space. In forming the factor, group Ω/Δ, the subgroup Ω_{n+1} goes over into a subgroup isomorphic to $\Omega_{n+1}/\Gamma_{n+1}$ but since Ω/Δ and $\Omega_{n+1}/\Gamma_{n+1}$ are vector groups of the same dimensionality, $\Omega_{n+1}/\Gamma_{n+1}$ cannot be a proper subgroup of the group Ω/Δ. Hence, in forming the factor-group Ω/Δ, the subgroup Ω_{n+1} goes over into the whole group Ω/Δ. In this way every element of Ω can be represented in the form $\alpha + \beta$, where $\alpha \subset \Delta$ and $\beta \subset \Omega_{n+1}$.

This last property is also true for the intersection N of all the groups Ω_n, but the intersection of N and Δ contains only zero. Hence Ω is the direct sum of the subgroups Δ and N, with N isomorphic to Ω/N.

IV. Structure of a Locally Connected Group

The purpose of the present chapter is to investigate the structure of a connected, locally connected, locally compact group, satisfying the second axiom of countability. We solve the problem in the Third Fundamental Theorem, which shows that a group of the above type reduces to a direct sum of a finite or infinite number of continuous cyclic groups and a finite number of groups isomorphic to the additive group of real numbers. In the case of compact groups, the latter summands do not exist and we get only one compact, connected and locally connected group for every finite or infinite dimensionality.

A space is said to be locally connected if any two of its points which are sufficiently close together can be connected by a short path. Hence a locally Euclidean space is a locally connected finite-dimensional space, so that the Third Fundamental Theorem solves the question of the structure of a locally Euclidean group, as a special case.

8. **Lemma 16.** Let $\mathfrak{G}$ be a discrete group which has no elements of finite order, and which is such that any increasing sequence of subgroups $\mathfrak{H}_1, \mathfrak{H}_2, \ldots, \mathfrak{H}_n, \ldots$ of the same finite rank stops. Then there exists a finite or infinite system of linearly independent generators of the group $\mathfrak{G}$.

Proof. Since every increasing sequence of subgroups of $\mathfrak{G}$ of a fixed finite rank stops, every subgroup of finite rank can be included in the maximal subgroup of the same rank, having a finite number of generators.

Let us construct an increasing sequence $\mathfrak{G}_1, \mathfrak{G}_2, \ldots, \mathfrak{G}_n, \ldots$ of subgroups of $\mathfrak{G}$, such that $\mathfrak{G}_n$ is the maximal subgroup of rank n having a finite number of generators (this means that $\mathfrak{G}_n$ cannot be included in any subgroup of the same rank). Moreover the sum of all the groups $\mathfrak{G}_n$ coincides with $\mathfrak{G}$.

Let $\mathfrak{x}_1, \mathfrak{x}_2, \ldots, \mathfrak{x}_m, \ldots$ be a numbered sequence of all the elements of the group $\mathfrak{G}$. Let $\mathfrak{G}_1$ be the maximal subgroup of $\mathfrak{G}$ of rank 1, containing $\mathfrak{x}_1$. Let us suppose that $\mathfrak{G}_n$ is already constructed. Let m be the least number such that $x_m \not\subset \mathfrak{G}_n$, and let the group $\mathfrak{G}_{n+1}$ be the maximal subgroup of rank $n+1$ containing $\mathfrak{G}_n$ and x_m. Let us construct a sequence of elements $\mathfrak{g}_1, \mathfrak{g}_2, \ldots, \mathfrak{g}_n, \ldots$ of the group $\mathfrak{G}$, which is such that the system $\mathfrak{g}_1, \mathfrak{g}_2, \ldots, \mathfrak{g}_n$ forms a linearly independent basis of the group $\mathfrak{G}_n$. Since $\mathfrak{G}$ has no elements of finite order, the subgroup $\mathfrak{G}_1$, having a finite number of generators, is a free cyclic group. Let us denote its generator by $\mathfrak{g}_1$. Let us show that the factor-group $\mathfrak{G}_{n+1}/\mathfrak{G}_n$ is cyclic. Since $\mathfrak{G}_{n+1}/\mathfrak{G}_n$ has a finite number of generators and is of rank 1, it will suffice to show that $\mathfrak{G}_{n+1}/\mathfrak{G}_n$ has no elements of finite order. Let us suppose that there exists a subgroup $\mathfrak{H}$ of finite order, of the group $\mathfrak{G}_{n+1}/\mathfrak{G}_n$. Then the image of the group $\mathfrak{H}$ in $\mathfrak{G}_{n+1}$ is a group of rank n containing $\mathfrak{G}_n$, which is impossible since $\mathfrak{G}_n$ is a maximal subgroup. Hence $\mathfrak{G}_{n+1}/\mathfrak{G}_n$ has a generator $\mathfrak{g}'_{n+1}$. Let us designate one of the images of $\mathfrak{g}'_{n+1}$ in $\mathfrak{G}_{n+1}$ by $\mathfrak{g}_{n+1}$. The system $\mathfrak{g}_1, \mathfrak{g}_2, \ldots, \mathfrak{g}_n, \ldots$ thus obtained is a linearly independent system of generators of the group.

Lemma 17. Let $\mathfrak{G}$ be a group of finite rank r, having no elements of finite order. Let $\mathfrak{H}_1, \mathfrak{H}_2, \ldots, \mathfrak{H}_n, \ldots$ be an increasing infinite sequence of subgroups of rank r of the group $\mathfrak{G}$, which are such that each $\mathfrak{H}_n$ has a finite number of generators and the sum of all $\mathfrak{H}_n$ coincides with $\mathfrak{G}$. Then the group of characters X of the group $\mathfrak{G}$ cannot be locally connected.

Proof. Let $\Phi_n = (X, \mathfrak{H}_n)$ and $T_n = X/\Phi_n$. T_n is isomorphic to the group of characters of $\mathfrak{H}_n$, and since $\mathfrak{H}_n$ has a finite number of generators, and no elements of finite order, it must have a finite system of linearly independent generators, that is T_n is a toroidal group (see Remark 5). Let U be a neighborhood of zero in T_1, homomorphic to a sphere. Such a neighborhood always exists since T_1 is a manifold. Let V be a sufficiently small neighborhood of zero in X, such that its image on reflection in X/Φ_1 lies completely in U. Since the sequence $\mathfrak{H}_1, \mathfrak{H}_2, \ldots, \mathfrak{H}_n, \ldots$ is infinite, the group $\mathfrak{G}/\mathfrak{H}_1$ is infinite and hence the group of its characters Φ_1 (see Theorem 3) is also infinite. But Φ_1 is compact and hence has no isolated elements. In this way there exists an element α in Φ_1, which is arbitrarily close to zero. Let us suppose that X is locally connected and let α be sufficiently close to zero so that α and zero can be connected by a continuous path $l \subset V$. Since the sum of all $\mathfrak{H}_n$ coincides with $\mathfrak{G}$, the intersection of all Φ_n coincides with zero and hence there exists a sufficiently large number m such that $\alpha \not\subset \Phi_m$. Let Φ_1', α' and l' be the images of Φ_1, α and l in T_m with a homomorphism $X/\Phi_m, \alpha' \neq 0$. Since $X/\Phi_1 = T_1$, $T_m/\Phi_1' = T_1$. Let us denote by l'' the image of l' in T_1 with the homomorphism T_m/Φ_1'.

Since $l \subset V$, $l'' \subset U$ and hence the closed path l'' in T_1 is homotopic to zero since U is homomorphic to a sphere. But this is impossible, because, using Schreier's[8] results, T_m covers T_1, and the open path l' in T_m cannot go over into a closed path l'', homotopic to zero in T_1.

Theorem 6. The connected, locally connected, compact group X satisfying the second axiom of countability decomposes into a direct sum of a finite or denumerable number of continuous cyclic groups. (See Appendix 1 for the notion of a direct sum.)

Proof. From the First Fundamental Theorem, X is the group of characters of a certain group $\mathfrak{G}$. Since X is connected, $\mathfrak{G}$ does not contain any elements of finite order (see Appendix 2, Theorem 1c). If the group $\mathfrak{G}$ satisfies the conditions of Lemma 16, it decomposes into a direct sum of free cyclic groups and hence by Theorem 1a of Appendix 1, X decomposes into a direct sum of continuous cyclic groups.

Let us now suppose that $\mathfrak{G}$ does not satisfy the conditions of Lemma 16, then there exists in $\mathfrak{G}$ an increasing infinite sequence $\mathfrak{H}_1, \mathfrak{H}_2, \ldots, \mathfrak{H}_n, \ldots$ of groups of rank r, each of which has a finite system of generators. Let $\mathfrak{H}$

be the sum of all the groups $\mathfrak{H}_n$. The rank of $\mathfrak{H}$ is r. Let $\Phi = (\mathrm{X}, \mathfrak{H})$. Since X is locally connected, its continuous image X/Φ has the same property. But by Theorem 3, the group X/Φ is the group of characters of the group $\mathfrak{H}$, which contradicts Lemma 17.

Third Fundamental Theorem. A connected, locally connected, locally compact group Ω, satisfying the second axiom of countability decomposes into the direct sum of a finite or denumerable number of continuous cyclic groups and a vector group.

Proof. By the Second Fundamental Theorem, the group Ω decomposes into the direct sum of the compact group X and the vector group N. It is easy to see that X is isomorphic to Ω/N. Hence X is locally connected and connected and decomposes into the direct sum of continuous cyclic groups by Theorem 6.

Corollary. By imposing on the Third Fundamental Theorem the further condition that Ω is of finite dimensionality, or the condition of shrinkage to zero of every sufficiently small closed contour, we obtain only a finite number of cyclic summands.

Proof. The first statement is obvious, the second follows from the fact that the presence of an infinite number of cyclic summands enables us to choose from among them an arbitrarily small subgroup, which is itself a closed contour but cannot shrink to a point, since it is a direct summand.

Appendix I. On the Decomposition of Groups into direct Sums

Here I give the connection between the decomposition of a group $\mathfrak{G}$ into a direct sum, and that of the group X of its characters. This enables me to establish the connection between the decompositions into direct sums of discrete and continuous compact groups. This allows me to give a number of examples not devoid of interest; namely I give examples which contradict the results of J. W. Alexander and L. W. Cohen,[9] St. Pietrkowsky,[10] D. v. Dantzig.[11] We recall first of all the concept of a direct sum.

Definition 1a. Let $\mathfrak{G}$ be a group and $\{\mathfrak{U}_m\}$ a system, of any power, of subgroups of the group $\mathfrak{G}$. We shall say that the group $\mathfrak{G}$ is decomposed into a direct sum of the groups $\mathfrak{U}_m$, if (a) the minimum subgroup, which contains all the groups $\mathfrak{U}_m$, coincides with $\mathfrak{G}$; (b) let $\mathfrak{B}_i$ be a minimum subgroup which contains all the subgroups $\mathfrak{U}_m$ with the exception of one, $\mathfrak{U}_i$. Then the intersection of the groups $\mathfrak{B}_i$ and $\mathfrak{U}_i$ contains only zero.

Remark 1a. The power of the set of groups $\mathfrak{U}_m$, which give the decomposition of the group $\mathfrak{G}$ into a direct sum, is denumerable or finite. In fact, from every group $\mathfrak{U}_m$ we can choose an element different from zero. It is easy to see that all these elements are distinct, and since, by a general assumption of this paper, the group $\mathfrak{G}$ contains not more than a denumerable number of elements, the statement is proved. Therefore we can suppose that $m = 1, 2, \ldots$.

Remark 2a. Every element $\mathfrak{r} \subset \mathfrak{G}$ is decomposed uniquely into a sum $\mathfrak{r} = \Sigma_{m=1}^{n} \mathfrak{r}_m$, where $\mathfrak{r}_m \subset \mathfrak{U}_m$, and n is an arbitrary integer, but sufficiently large depending on $\mathfrak{r}$. The totality of all the sums of this form is a group, which contains all the groups $\mathfrak{U}_m$ and must therefore, by Definition 1a, coincide with $\mathfrak{G}$. Suppose that we have two decompositions of $\mathfrak{r}$ into a sum $\mathfrak{r} = \Sigma_{m=1}^{n} \mathfrak{r}_m = \Sigma_{m=1}^{n} \mathfrak{r}'_m$; from this we have $\mathfrak{r}'_i - \mathfrak{r}_i = \Sigma_{m \neq i} (\mathfrak{r}_m - \mathfrak{r}'_m)$. But the left hand side of this equality belongs to $\mathfrak{U}_i$ and the right hand side to $\mathfrak{B}_i$, hence by Definition 1a, $\mathfrak{r}'_i = \mathfrak{r}_i$ for every i. Hence the uniqueness of the decomposition is established.

Definition 1b. Let X be a compact group, and let Φ_m be a system, of any power, of the subgroups of X. We shall say that X is decomposed into a direct sum of the groups Φ_m if (a) the minimum closed subgroup which contains all the groups Φ_m coincides with X. (b) Let Ψ_i be the minimum closed subgroup which contains all the groups Φ with the exception of one group Φ_i, then the intersection of Ψ_i and Φ_i contains only zero.

Remark 1b. The power of the set of groups Φ_m, which give the decomposition of X into a direct sum, is denumerable or finite. To prove this let us suppose that the index m runs over a well ordered set of values. Let us denote by Ξ_j the minimum closed group which contains all the groups Φ_i for $i \geqq j$. It is obvious that Ξ_j forms a well ordered descending sequence of groups, but from a general assumption of this paper, the second axiom of countability for X, it follows that the sequence Ξ_j stops at a finite or countable term. In this way we can say without loss of generality that m assumes integer values.

Remark 2b. Every element $\alpha \subset X$ can be uniquely represented as an infinite sum $\alpha = \Sigma_{m=1}^{\infty} \alpha_m$ (the infinite sum has a meaning, since X is a continuous group). The totality of all the convergent sums of the above type is easily seen to form a compact subgroup of the group X, which contains all the Φ_m and therefore must, in accordance with Definition 1b, coincide with X. The uniqueness of decomposition is proved in the same way as in Remark 2a.

Theorem 1a. If $\mathfrak{G}$ and X are orthogonal (see Definition 3) and if $\mathfrak{G}$ is decomposed into a direct sum of groups $\mathfrak{U}_m$, then there exists a decomposition of the group X into a direct sum of groups Φ_m, which is such that $(\mathrm{X}, \mathfrak{U}_m) \supset \Phi_n$ for $n \neq m$, and $\mathfrak{U}_m$ and Φ_m are orthogonal.

Theorem 1b. If $\mathfrak{G}$ and X are orthogonal, and if Φ_m gives the decomposition of X into a direct sum, then there exists a decomposition of the group $\mathfrak{G}$ into subgroups $\mathfrak{U}_m$ under the same conditions as in Theorem 1a.

In order to prove both these theorems we first prove two preparatory lemmas.

Lemma A. Let $\mathfrak{G}$ and X be orthogonal and $\mathfrak{H}_m$ an arbitrary system of subgroups of $\mathfrak{G}$, whose intersection we will designate by $\mathfrak{H}$. If $\Xi_m = (\mathrm{X}, \mathfrak{H}_m)$ and if the minimum closed subgroup which contains all Ξ_m be designated by Ξ, then $\Xi = (\mathrm{X}, \mathfrak{H})$.

Proof. It is easily seen that $(\mathrm{X}, \mathfrak{H}) \supset \Xi_m$, moreover since $(\mathrm{X}, \mathfrak{H})$ is closed, it contains Ξ. Let $\mathfrak{x} \subset \mathfrak{G}$, and $\mathfrak{x} \not\subset \mathfrak{H}$, then there exists an i, such that $\mathfrak{x} \not\subset \mathfrak{H}_i$, hence there exists an $\alpha \subset \Xi_i \subset \Xi$ such that $\alpha\mathfrak{x} \neq 0$, so that $(\mathfrak{G}, \Xi) = \mathfrak{H}$.

Lemma B. This lemma is analogous to Lemma A with $\mathfrak{G}$ and X interchanged. Its proof is analogous to the proof of Lemma A.

Proof of Theorem 1a. With the notation of Definitions 1a and 1b, let us suppose that $\Phi_m = (\mathrm{X}, \mathfrak{B}_m)$ and prove that Φ_m give the desired decomposition of the group X. Let $\mathfrak{x} \subset \mathfrak{B}_i$, in accordance with Remark 2a, $\mathfrak{x} = \Sigma_{m=1}^{n} \mathfrak{x}_m$. Since $\mathfrak{x} \subset \mathfrak{B}_i$, $\mathfrak{x}_i = 0$. If $\mathfrak{x}$ belongs simultaneously to all the $\mathfrak{B}_m$, then $\mathfrak{x} = 0$: that is the intersection of all the $\mathfrak{B}_m$ contains only zero. From this it follows, from Lemma A that the product of all the Φ_m coincides with X. Analogously with intersection of all the $\mathfrak{B}_m$ with the exception of $\mathfrak{B}_i$ coincides with $\mathfrak{U}_i$, from which it follows by Lemma A that $\Psi_i = (\mathrm{X}, \mathfrak{U}_i)$. Since the product of $\mathfrak{U}_i$ and $\mathfrak{B}_i$ coincides with $\mathfrak{G}$, it follows from Lemma B and Theorem 3 and 4 that the intersection of Φ_i and Ψ_i contains only zero. In this way, Φ_m give a decomposition of the group X into a direct sum. From $\Psi_i = (\mathrm{X}, \mathfrak{U}_i)$ it is obvious that $(\mathrm{X}, \mathfrak{U}_m) \supset \Phi_n$, $m \neq n$.

We shall show now that $\mathfrak{U}_m$ and Φ_m are orthogonal. Let $0 \neq \mathfrak{r} \subset \mathfrak{U}_i$. Because of the orthogonality of $\mathfrak{G}$ and X there exists an $\alpha \subset X$ with $\alpha\mathfrak{r} \neq 0$. In accordance with Remark 2b we have $\alpha = \Sigma_{m=1}^{\infty} \alpha_m$ but $\alpha_m \mathfrak{r} = 0$ with $m \neq i$, hence $\alpha_i \mathfrak{r} = \alpha\mathfrak{r} \neq 0$, that is $(\mathfrak{U}_i, \Phi_i) = (0)$. In the same way we can prove that $(\Phi_i, \mathfrak{U}_i) = (0)$.

The proof of Theorem 1b is entirely analogous.

Remark 3ab. From Theorems 1a and 1b, there exists a complete duality between the decomposition into a direct sum of a discrete group and that of the group of its characters. Moreover the decomposition of the one group into *a* not further decomposible sum corresponds to the decomposition of the other group into a sum which is likewise not decomposible any further. This fact enables us, starting with examples for discrete groups, to arrive automatically at example for continuous groups. It also shows that the question of decomposition of continuous compact groups is equivalent to the question of decomposition of discrete groups. (See First Fundamental Theorem.)

Example 1. H. Prüfer[12] gave an example of a group $\mathfrak{G}$, all of whose elements are of a finite order and which cannot be decomposed into a direct sum of indecomposible subgroups, although it admits of a decomposition into a sum of further decomposible subgroups. If we construct for $\mathfrak{G}$ its group of characters X we get an example of a compact, zero-dimensional group (see Appendix 2, Theorem 2c), which possesses the same properties of decomposibility. A more detailed study of the group X shows that it satisfies the conditions of the paper of Alexander and Cohen[9] and also the conditions of the paper by St. Pietrkowsky.[10] In this way it gives an example which contradicts the results of these papers, since these papers assert the decomposibility of groups satisfying certain given conditions into non-decomposible subgroups.

The group $\mathfrak{G}$ is given by its generators $\mathfrak{h}, \mathfrak{r}_i$, $(i = 1, 2, \ldots)$ and relations between them $p\mathfrak{h} = 0$, $p^i \mathfrak{r}_i = \mathfrak{h}$, where p is any prime number. It is easy to see that the order of the element $\mathfrak{r}_i$ is p^{i+1}, and that the order of the element $\mathfrak{h}$ is p.

Let us form the group of characters X of the group $\mathfrak{G}$ by means of its generators, β, α_i, $(i = 1, 2, \ldots)$ where β and α_i are determined by the conditions $\beta(\mathfrak{h}) = 1/p$, $\beta(\mathfrak{r}_i) = 1/p^{i+1}$, $\alpha_i(\mathfrak{h}) = 0$, $\alpha_i(\mathfrak{r}_j) = 0$ with $i \neq j$, $\alpha_i(\mathfrak{r}_1) = 1/p^i$. The right hand sides of these equalities are numbers, which are considered as elements of a group K. (See Definition 1.) It is easily seen that every element of the group X can be expressed as an infinite sum $b\beta + \Sigma_{i=1}^{\infty} a_i \alpha_i$ with $0 \leqq b < p$ and $0 \leqq a < p^i$, since the conditions

$p^i\alpha_i = 0$, and $p\beta = \Sigma_{i=1}^{\infty} \alpha_i$ are fulfilled. Also the kth neighborhood of zero in X is determined by the condition $b = 0$, $\alpha_i = 0$, with $i \leqq k$. The group X which is determined by its generators and given relations, and which is the group of characters of $\mathfrak{G}$, obviously satisfies the conditions of Alexander and Cohen. It is not difficult to show that it likewise satisfies the conditions of St. Pietrkowsky.

Example 2. Let us construct a group $\mathfrak{G}$ without any elements of finite order, whose rank is two and let $\mathfrak{G}$ be such that it is not decomposible into a direct sum. The group of characters X of the group $\mathfrak{G}$ will be a compact connected group of dimensionality 2, (see Appendix 2, Theorems 1c and 2c) which is likewise not decomposible into a direct sum (see Theorem 1b), which contradicts the statements of the paper of D. v. Dantzig.[11]

We will give the group $\mathfrak{G}$ by its generators $\mathfrak{h}, \mathfrak{r}_i$, $(i = 0, 1, 2, \ldots)$ and the relation,

$$2^{k_{i+1}}\mathfrak{r}_{i+1} = \mathfrak{r}_i + \mathfrak{h}, \quad (i = 0, 1, 2, \ldots), \tag{*}$$

where k_i are positive integers satisfying only the condition that some k's are arbitrarily large.

It is easy to see that the element $\mathfrak{h}$ of the group $\mathfrak{G}$ is free, since from the relation (*) we cannot eliminate all the $\mathfrak{r}_i$ in such a way as to leave a relation in $\mathfrak{h}$ alone. Let us designate by $\mathfrak{H}$ the subgroup of the group $\mathfrak{G}$ with the generator $\mathfrak{h}$. In forming the factor group $\mathfrak{G}/\mathfrak{H}$ if we denote by $\mathfrak{r}'_i$ the element of $\mathfrak{G}/\mathfrak{H}$ into which $\mathfrak{r}_i$ goes, then $\mathfrak{r}'_i$ is the system of generators of $\mathfrak{G}/\mathfrak{H}$ with the relation $2^{k_{i+1}}\mathfrak{r}'_{i+1} = \mathfrak{r}_i$, so that the group contains no elements of finite order. Since $\mathfrak{H}$ and $\mathfrak{G}/\mathfrak{H}$ do not contain any elements of finite order, the group $\mathfrak{G}$ itself has the same property.

Because of the absence in $\mathfrak{G}$ of elements of finite order, division in $\mathfrak{G}$ is unique, that is the equation $g\mathfrak{g} = \mathfrak{b}$, where $\mathfrak{a} \subset \mathfrak{G}$, $\mathfrak{b} \subset \mathfrak{G}$, and g an integer, has not more than one solution. Therefore in $\mathfrak{G}$ we can use fractional coefficients without fear.

Now we shall show that $\mathfrak{G}$ does not contain any infinitely divisible elements, that is for every element $\mathfrak{b}$, a sufficiently large integer k can be found such that if $q > k$ the equation $q\mathfrak{a} = \mathfrak{b}$ has no solution or, which is the same, there exists no element $(1/q)\mathfrak{b}$. From this the non-decomposibility of $\mathfrak{G}$ can be derived without much effort.

Making use of fractional coefficients we can express all the generators of $\mathfrak{G}$ by means of $\mathfrak{h}$ and $\mathfrak{r}_0 = \mathfrak{r}$. Simple calculations show, that

$$\mathfrak{r}_i = \frac{\mathfrak{r} + \mathfrak{h}\,(1 + 2^{k_1} + 2^{k_1+k_2} + \ldots + 2^{k_1+k_2+\ldots+k_{i+1}})}{2^{k_1+k_2+\ldots+k_i}}.$$

In this way every element of the group $\mathfrak{G}$ can be written in the form $r\mathfrak{x}+s\mathfrak{h}$, where r and s are rational numbers, but of course not every expression of this type represents an element of $\mathfrak{G}$. In particular, if there are two elements $r\mathfrak{x}+s'\mathfrak{h}$ and $r\mathfrak{x}+s''\mathfrak{h}$, then $s'-s''$ is an integer: In fact $(s''-s')\mathfrak{h}$ is an element of $\mathfrak{G}$ since it is the difference of two of its elements. But since $\mathfrak{G}/\mathfrak{H}$ has no elements of finite order, $s''-s'$ must be an integer, for otherwise $\mathfrak{G}/\mathfrak{H}$ would have an element of finite order. Let us suppose now that some element $\mathfrak{b}$ admits of infinite division. From what we just said, $\mathfrak{b}$ cannot be a multiple of $\mathfrak{h}$. Therefore in the transition to the factor group, $\mathfrak{b}$ will go into an element $\mathfrak{b}'$, but if $\mathfrak{b}$ admits an infinite division, so will $\mathfrak{b}'$. However, in $\mathfrak{G}/\mathfrak{H}$ infinite division can only take place by arbitrary high powers of 2. Therefore $\mathfrak{b}$ admits division by arbitrary high powers of 2. But if $\mathfrak{b}$ admits such a division, an arbitrary multiple of $\mathfrak{b}$ will have the same property, and therefore it would be possible to find an element $\bar{\mathfrak{b}}=m\mathfrak{x}+n\mathfrak{h}$ which admits a division by arbitrarily high powers of 2 with m and n integers. In this way $\mathfrak{G}$ contains an element $(m\mathfrak{x}+n\mathfrak{h})/2^{k_1+k_2+\ldots+k_j}=\mathfrak{b}_j$ with an arbitrarily large j. Let us form the difference $m\mathfrak{x}_j-\mathfrak{b}_j$; it is equal to

$$\mathfrak{h}\frac{m(1+2^{k_1}+2^{k_1+k_2}+\ldots+2^{k_1+k_2+\ldots+k_{j+1}})-n}{2^{k_1+k_2+\ldots+k_j}}=\mathfrak{h}\cdot t.$$

From the preceding remarks the number t must be an integer. Let us show that this is impossible. First of all it is clear that with a sufficiently large j, it is positive. Let us show that for some arbitrarily large j, $t<1$. To prove this statement let us estimate t. Replacing the sum which is in the numerator of t by a geometrical progression we find easily that

$$t<\frac{(|m|+|n|)2^{k_1+k_2+\ldots+k_{j+1}+1}}{2^{k_1+k_2+\ldots+k_j}}=\frac{2(|m|+|n|)}{2^{k_j}}$$

but as k_j takes on arbitrarily large values, it is true that for some arbitrarily large j we have $t<1$. In this way $\mathfrak{G}$ does not contain any elements which admit infinite division.

Let us suppose now that $\mathfrak{G}$ is decomposible into a direct sum of two groups $\mathfrak{G}'$ and $\mathfrak{G}''$. The groups $\mathfrak{G}'$ and $\mathfrak{G}''$ have to be of rank 1 and without elements of finite order. Let us study the structure of such a group $\bar{\mathfrak{G}}$ without any elements of finite order and of rank 1. Let $\mathfrak{u}$ and $\mathfrak{v}$ be elements of $\bar{\mathfrak{G}}$. Since the rank of $\bar{\mathfrak{G}}$ is 1 we have $n\mathfrak{v}=m\mathfrak{u}$ that is $\mathfrak{v}=(m/n)\mathfrak{u}$. Putting every element $\mathfrak{v}$ in correspondence with a rational number m/n we obtain an isomorphic mapping of the group $\bar{\mathfrak{G}}$ into a group of rational numbers, that is $\bar{\mathfrak{G}}$ is isomorphic with some subgroup of the group of rational numbers. But a subgroup of the group of rational numbers is either cyclic or else contains elements which allow infinite division. But since $\mathfrak{G}$ has no elements which admit infinite division, it follows that $\mathfrak{G}'$ and $\mathfrak{G}''$ will have to be cyclic, which is obviously impossible.

Appendix 2. Connectivity and Dimensionality of the Group of Characters

Here I endeavor to explain the simplest topological properties of the group of characters X of some discrete group $\mathfrak{G}$ and to establish a connection between these properties and the algebraic properties of the group $\mathfrak{G}$.

Theorem 1c. The group of characters X of the group $\mathfrak{G}$ is connected if and only if the group $\mathfrak{G}$ has no elements of finite order.

Proof. Let us suppose that $\mathfrak{G}$ has an element $\mathfrak{r}$ of finite order. Let us denote the group which has $\mathfrak{r}$ for a generator by $\mathfrak{H}$: this group is finite. Let $\Phi = (X, \mathfrak{H})$. By Theorem 3, the group X/Φ is the group of characters of the group $\mathfrak{H}$. Since $\mathfrak{H}$ is finite, its group of characters X/Φ is also finite, by Lemma 3. Hence X can be continuously mapped by means of the homomorphism X/Φ into a finite system of elements. Hence X is not connected.

Let us suppose X can be divided into two non-intersecting compact parts U and V. Let $\mathfrak{H}_1, \mathfrak{H}_2, \ldots, \mathfrak{H}_n, \ldots$ be an increasing sequence of subgroups of the group $\mathfrak{G}$, whose sum coincides with $\mathfrak{G}$, and each of which has a finite number of generators. Let $\Phi_n = (X, \mathfrak{H}_n)$. The intersection of all Φ_n contains only zero, hence there exists a sufficiently large number m, such that the following condition is satisfied for the group Φ_m. If $\alpha \subset U$ and $\beta \subset V$, then $\beta - \alpha \not\subset \Phi_m$. Because of this condition, U and V go into two non-intersecting compact sets in forming the factor group X/Φ_m. Hence X/Φ_m is not connected. But X/Φ_m is the group of characters of $\mathfrak{H}_m$ with a finite system of generators, by Theorem 3. If $\mathfrak{H}_m$ has no elements of finite order, it would have a finite system of linearly independent generators, and X/Φ_m would be a toroidal, that is a connected group (see Remark 5). Hence $\mathfrak{H}_m$ contains elements of finite order and so also does $\mathfrak{G}$.

Corollary 1c. Let $\mathfrak{G}$ be a discrete group, and let X be a group of its characters. Let $\mathfrak{H}$ be the subgroup of $\mathfrak{G}$ composed of all the elements of finite order in $\mathfrak{G}$, and let $\Phi = (X, \mathfrak{H})$. Then Φ is the component of zero in X.

Proof. By Theorem 3, Φ is the group of characters of $\mathfrak{G}/\mathfrak{H}$, and since $\mathfrak{G}/\mathfrak{H}$ has no elements of finite order, Φ is connected, by Theorem 1c. At the same time, Φ is the maximal connected subgroup of $\mathfrak{G}$. For suppose that Φ is a proper subgroup of Φ', and let $\mathfrak{H}'=(\mathfrak{G},\Phi')$. Then $\mathfrak{H}'$ is a proper subgroup of $\mathfrak{H}$ (see Theorem 4), and hence $\mathfrak{G}/\mathfrak{H}'$ has elements of finite order, which means that its group of characters Φ' is not connected.

Theorem 2c. If $\mathfrak{G}$ is a discrete group, and X the group of its characters, the dimensionality of X equals the rank of $\mathfrak{G}$.

Proof. Let us prove that the dimensionality of X cannot exceed the rank r of $\mathfrak{G}$. If r is infinite, the statement has no meaning. Suppose that r is finite. Let $\mathfrak{H}_1, \mathfrak{H}_2, \ldots, \mathfrak{H}_n, \ldots$ be an increasing sequence of subgroups of $\mathfrak{G}$, having the rank r. Let the sum of all the $\mathfrak{H}_n$ equal $\mathfrak{G}$, and let each $\mathfrak{H}_n$ have a finite number of generators. Let $\Phi_n=(\mathrm{X},\mathfrak{H}_n)$. The intersection of all Φ_n contains only zero, that is, there exist arbitrarily small subgroups among the groups Φ_n. Since $\mathfrak{H}_n$ has a finite number of generators, it decomposes into a direct sum of r free cyclic subgroups and a finite group. Hence the group of its characters X/Φ_n is a direct sum of a toroidal group of rank r and a finite group (see Appendix 1, Theorem 1a and Remark 5), that is, the dimensionality of X/Φ_n is r. But for a sufficiently large n, X/Φ_n can be obtained from X by an arbitrarily small displacement, which, as is well known, cannot lower the dimensionality. Hence the dimensionality of X does not exceed r.

Let us show now that the dimensionality of X is not less than the rank r of $\mathfrak{G}$. To do this, it is sufficient to show that the dimensionality of the component of zero in X is not less than the rank of $\mathfrak{G}$, that is, using the preceding corollary, it is sufficient to discuss only the case where $\mathfrak{G}$ has no elements of finite order. In this case there exists a finite or infinite system $\mathfrak{r}_1, \mathfrak{r}_2, \ldots, \mathfrak{r}_n, \ldots$ of linearly independent elements of $\mathfrak{G}$, which is such that every element of $\mathfrak{G}$ can be expressed uniquely as a finite linear form in $\mathfrak{r}_1, \mathfrak{r}_2, \ldots, \mathfrak{r}_n, \ldots$ with rational coefficients. Let us now build a homomorphic mapping of the group $\mathfrak{G}$ into an additive group of real numbers, putting every element $\mathfrak{r}$ of the group $\mathfrak{G}$ into correspondence with a real number $\alpha(t_1, t_2, \ldots, t_n; \mathfrak{r})$ where $t_1, t_2, \ldots, t_n$ are real numbers upon which the homomorphism depends. This homomorphism is determined by the conditions $\alpha(t_1, t_2, \ldots, t_n; \mathfrak{r}_i)=t_i$ with $i=1,2,\ldots,n$ and $\alpha(t_1, t_2, \ldots, t_n; \mathfrak{r}_j)=0$ with $j>n$. It is easy to see that from these conditions α is uniquely determined. Reducing the number $\alpha(t_1, t_2, \ldots, t_n; \mathfrak{r})$ modulo unity, we obtain the homomorphism of the group $\mathfrak{G}$ into the group K. It is easy to see that if $0<t_i<1$, $i=1,2,\ldots,n$, then to every system of values $t_1, t_2, \ldots, t_n$ corresponds its own special homomorphism. Hence there is a subset in X homomorphic to the interior of an n-dimensional cube, but n is an arbitrary number not exceeding the rank of $\mathfrak{G}$.

Corollary 2c. The group of characters X of the discrete group $\mathfrak{G}$ is zero-dimensional if, and only if, all the elements of $\mathfrak{G}$ are of finite order, since in this case the rank of $\mathfrak{G}$ is zero.

Appendix 3

(Added in proof May 14, 1934.)

The reason why we do not introduce topological relations in definition I' of the character group $\mathfrak{G}$ of a compact group X is that in $\mathfrak{G}$ there cannot exist any convergent sequence. We must prove that if X and $\mathfrak{G}$ form an orthogonal pair $\mathfrak{G}$ cannot contain a sequence $\mathfrak{g}_1, \mathfrak{g}_2, \ldots, \mathfrak{g}_n, \ldots$ converging to zero. We will prove that this is impossible by constructing a charater χ of $\mathfrak{G}$ such that $\chi\mathfrak{g}_n$ does not converge to 0.

Let $\mathfrak{G}_n$ be the subgroup of $\mathfrak{G}$ generated by $\mathfrak{g}_1, \ldots, \mathfrak{g}_n$ and $\mathfrak{G}'$ the subgroup generated by $\mathfrak{g}_1, \mathfrak{g}_2, \ldots, \mathfrak{g}_n, \ldots$. We suppose first that $\mathfrak{G}'$ does not have a finite basis. Then there are arbitrarily big numbers n such that $\mathfrak{g}_n$ is not contained in $\mathfrak{G}_{n-1}$. Suppose that $\chi(\mathfrak{G}_{n-1})$ has already been constructed. If $\mathfrak{g}_n$ is independent of $\mathfrak{G}_{n-1}$ then we put $\chi\mathfrak{g}_n = \frac{1}{2}$. If $\mathfrak{g}_n$ is linearly dependent on $\mathfrak{G}_{n-1}$ we take a relation $k\mathfrak{g}_n = \mathfrak{g}', \mathfrak{g}' \subset \mathfrak{G}_{n-1}$, with lowest possible positive integer k. Then we put $\chi\mathfrak{g}_n = \chi\mathfrak{g}'/k$, taking the value of $\chi\mathfrak{g}'/k$ that is as far away from zero as possible. The character χ now defined over $\mathfrak{G}'$ can be extended over $\mathfrak{G}$. As we have for arbitrarily big values of n either $\chi\mathfrak{g}_n = \frac{1}{2}$ or $\chi\mathfrak{g}_n = \chi\mathfrak{g}'/k$ with $k > 1$ so that $\chi\mathfrak{g}_n$ is at least $\frac{1}{4}$ away from 0, we do not have $\lim \chi\mathfrak{g}_n = 0$.

Now we take the case where $\mathfrak{G}'$ possesses a finite basis $\mathfrak{a}_1, \ldots, \mathfrak{a}_p$, $\mathfrak{b}_1, \ldots, \mathfrak{b}_q$, where $\mathfrak{a}_1, \ldots \mathfrak{a}_p$, are linearly independent and $\mathfrak{b}_1, \ldots \mathfrak{b}_q$ have finite order. Putting

$$\mathfrak{g}_n = r_1^n\mathfrak{a}_1 + \ldots + r_p^n\mathfrak{a}_p + s_1^n\mathfrak{b}_1 + \ldots + \mathfrak{b}_q^n$$

we can suppose that $r_1^n \neq 0$ for arbitrarily high n. Then we make $\chi\mathfrak{a}_i = 0$, $(i = 2, \ldots, p)$, $\chi\mathfrak{b}_i = 0$, $(i = 1, \ldots, q)$ and $\chi\mathfrak{a}_1 = d$. It is easy to see that d can be chosen in such a way that $\lim_{n\to\infty} r_1^n d \neq 0$.

Notes

1. For the fundamental concepts of the theory of continuous, i.e., topological groups see D. van Dantzig, Zur topologischen Algebra **1**, *Math. Ann.*, **107** (1932), pp. 587–626.

2. F. Peter and H. Weyl, Die Vollständigkeit der primitiven Darstellungen einer geschlossenen kontinuierlichen Gruppe, *Math. Ann.*, **97** (1927) pp. 737.

3. A. Haar, Der Massbegriff in der Theorie der kontinuierlichen Gruppen, *These Annals* (2) **34** (1933) pp. 147–169.

4. By $\alpha(\mathfrak{r})$ I understand the function which puts into correspondence any element $\mathfrak{r} \subset \mathfrak{G}$ with the element $\alpha(\mathfrak{r})$ of the group K, which is determined by the homomorphism.

5. This definition was given in my paper (see footnote on page ■■■).

6. To understand my paper it is sufficient to know the paper of Haar (note 3).

7. Pontrjagin, Über den algebraischen Inhalt topologischer Dualitätssätze, *Math. Ann.*, **105** (1931) pp. 165–205.

8. O. Schreier, Die Verwandlschaft stetiger Gruppen im Grossen, *Abh. Math. Seminar der Hamburg Univ.*, **5** (1926–7) 233–244.

9. J. W. Alexander and L. W. Cohen, A classification of the homology groups of compact spaces, *These Annals* (2) **33** (1932) pp. 538–566.

10. St. Pietrkowsky, Theorie der unendlichen Abelschen Gruppen, *Math. Ann.*, **104** (1931) pp. 535–569.

11. D. v. Dantzig, Groupes monoboloques et functions presque périodiques, *C. R. Paris*, 1933, pp. 1074–1076, théorème IV, p. 1076.

12. H. Prüfer, Untersuchungen über die Zerlegbarkeit der abzählbaren primären Abelschen Gruppen, *Math. Zeitschr.*, **17** (1923) pp. 35–61 (§18).

The General Topological Theorem of Duality for Closed Sets*

Annals of Mathematics (1934) 35(4): 904–14

Introduction**

In recent years there has appeared a series of papers devoted to topological theorems of duality. Leaving out the theorems of duality of Poincaré, which, by the way, are closely related to these questions, the general statement of the question is the following: Let M be a manifold and F one of its compact subsets. It is required to study the topological properties of the space $M-F$, starting from those of the set F. In particular, we may suppose that M is a Euclidean space or, what is essentially the same, a sphereical manifold, F is homeomorphic to some complex, and the topological properties studied are the Betti numbers modulo two. In this form the problem was solved by Alexander.[1] Namely, he showed that the r-dimensional Betti number modulo two of the space $M-F$ equals the $(n-r-1)$-dimensional Betti number of the complex F, where n is the dimensionality of the spherical manifold M.

This work of Alexander gave strong impetus to further development of the question, which went along the following lines:

(1) The spherical manifold was replaced by a general manifold.

(2) The assumption that F was homeomorphic to a complex was dropped.

* Received 8 August 8, 1934.

** The results of this paper were reported in condensed form at the International Mathematical Congress in Zürich, 1932.

(3) Not only Betti groups modulo 2, but others as well, began to be studied.

(4) The theory of linking was made use of in these studies; it is an application of the general theory of intersections developed by Lefschetz;[2] the last, together with the theories of homologies, is at present the basic well-developed apparatus of combinatorial topology. It is just because of their applications, it seems to me, that we owe the generalizations of the first three points. I will not dwell any more in detail on the history of this question; it is given in one of my papers.[3]

However, in all the work done so far, there remains unsolved the final question: letting F be an arbitrary compact subset of M, where M is a Euclidean space, the reduction of the full Betti group of the space $M-F$ to the invariants of the space F is not accomplished. It is to the solution of this question that the present paper is devoted. In it I limit myself to the case where the manifold M is a Euclidean space, since, with existing methods, the generalization to the case of an arbitrary manifold does not present any great difficulty.

For the solution of the question I have to introduce a new invariant of a compact metrizable space F in the form of a commutative continuous group connected with a definite dimension r, which I call the r-dimensional Betti group of the space F. In view of the introduction of continuous groups I have devoted considerable attention to their direct study.[4] I must note that the initiative for using continuous groups in combinatorial topology belongs not to me but to Alexander and Cohen[5] who study continuous groups, having in view their application to combinatorial topology; moreover they concentrate their attention on the reduction of such a group to a system of numerical invariants, which seems to me to be impossible in essence, and which leads them to an error (see Appendix 1, T. T. G.).

In conclusion, I must express my gratitude to P. Alexandroff who has many times pointed out to me the desirability of solving the question to whose solution the present work is devoted.

1

Here I give the fundamental topological concepts in the somewhat generalized form in which I will have to use them in what follows.

Let K be an arbitrary finite or infinite complex, the set of whose r-dimensional oriented simplexes we shall denote by T_i^r, $i=1,2,\ldots$, and let

$$T_i^r \to \sum e_{ij}^r T_j^{r-1}$$

be the system of incidence relations for K, where e_{ij}^r is ± 1 or 0.

Further, let $\mathfrak{G}$ be an arbitrary discrete commutative group, written additively, with not more than a countable number of elements, which we shall make the basis for building the invariants of the complex K. The finite linear form $\Sigma_i \mathfrak{x}_i T_i^r$ with coefficients in $\mathfrak{G}$, $\mathfrak{x}_i \subset \mathfrak{G}$, we shall call an r-dimensional complex of K modulo $\mathfrak{G}$. The totality of all r-dimensional complexes of K modulo $\mathfrak{G}$ forms a group $\mathfrak{L}_{\mathfrak{G}}^r$ if we take for our composition the addition of linear forms. By the boundary of $\Sigma_i \mathfrak{x}_i T_i^r$ we shall mean the linear form $\Sigma_{i,j} \mathfrak{x}_i e_{ij}^r T_j^{r-1}$, which is an element of the group $\mathfrak{L}_{\mathfrak{G}}^{r-1}$. In this way, to each element of the group $\mathfrak{L}_{\mathfrak{G}}^r$ there corresponds, as its boundary, some element of the group $\mathfrak{L}_{\mathfrak{G}}^{r-1}$, and because of this correspondence we have a homomorphism of the group $\mathfrak{L}_{\mathfrak{G}}^r$ on a subgroup $\mathfrak{H}_{\mathfrak{G}}^{r-1}$ of the group $\mathfrak{L}_{\mathfrak{G}}^{r-1}$ with the divisor $\mathfrak{Z}_{\mathfrak{G}}^r$. The elements of the group $\mathfrak{Z}_{\mathfrak{G}}^r$ are the cycles modulo $\mathfrak{G}$, and the elements of $\mathfrak{H}_{\mathfrak{G}}^r$ are cycles homologous to zero modulo $\mathfrak{G}$. It is easy to see that $\mathfrak{H}_{\mathfrak{G}}^r \subset \mathfrak{Z}_{\mathfrak{G}}^r$. The factor group $\mathfrak{Z}_{\mathfrak{G}}^r / \mathfrak{H}_{\mathfrak{G}}^r = \mathfrak{B}_{\mathfrak{G}}^r$ we shall call the r-dimensional Betti group of the complex K modulo $\mathfrak{G}$; it is the common invariant of the complex K and the group $\mathfrak{G}$.

In the event $\mathfrak{G}$ is the additive group of integers, then $\mathfrak{B}_{\mathfrak{G}}^r$ is the full Betti group of the complex K. If, further, $\mathfrak{G}$ is the group of residues of the group of whole numbers modulo m, then $\mathfrak{B}_{\mathfrak{G}}^r$ is the Betti group of the complex K modulo m.

Let us now, in the previous consideration, substitute instead of a discrete group $\mathfrak{G}$ a continuous compact commutative group X, with the second countability axiom. Then, as before, every linear form of the type $\Sigma_{i=1}^{\infty} \alpha_i T_i^r$ with coefficients in X, $\alpha_i \subset \mathrm{X}$, will be called an r-dimensional complex modulo X. The totality of all r-dimensional complexes modulo X forms a group Λ_{X}^r, where we can now define continuity in Λ_{X}^r; namely, if U is a certain neighborhood of the zero of X and k is some positive whole number, then we define a neighborhood of the zero of Λ_{X}^r as the totality of all elements of the form $\Sigma_1^{\infty} \alpha_i T_i^r$, where $\alpha_j \subset U$, with $j \leqslant k$. In exactly the same way as above we define the boundary of a complex modulo X, and we obtain a homomorphism of the group Λ_{X}^r on the subgroup $\mathrm{H}_{\mathrm{X}}^{r-1}$ with the divisor $\mathrm{Z}_{\mathrm{X}}^r$. In this way the entire theory of homologies is applicable also for the modulus X. Making use of the diagonal process it is easy to show that Λ_{X}^r is compact and that, since the homomorphism of Λ_{X}^r in $\Lambda_{\mathrm{X}}^{r-1}$ is continuous, $\mathrm{Z}_{\mathrm{X}}^r$ and $\mathrm{H}_{\mathrm{X}}^{r-1}$, and also $\mathrm{B}_{\mathrm{X}}^r = \mathrm{Z}_{\mathrm{X}}^r / \mathrm{H}_{\mathrm{X}}^r$ are compact groups.

2

It is known that the theorem of duality of Alexander does not hold for the full Betti group; that is, if K is a complex, located in Euclidean space R^n, the full r-dimensional Betti group $\mathfrak{B}^r(R^n - K)$ is not isomorphic to the full Betti group $\mathfrak{B}^{n-r-1}(K)$; moreover, the group $\mathfrak{B}^r(R^n - K)$ is in general not defined by the group $\mathfrak{B}^{n-r-1}(K)$. This was precisely the basic obstable to the proof of invariance of the full Betti group of the domain complementary to a certain set F relative to various positions of F in Eucledian space. It turns out, however, that if the Betti groups of the space $R^n - K$ are constructed on the basis (see Section 1) of a certain discrete group $\mathfrak{G}$, that is, if we consider groups $\mathfrak{B}^r_{\mathfrak{G}}$, and if the groups of the complex K are constructed on the basis of the group of characters X of the group $\mathfrak{G}$ (see Definition 1, T. T. G.), that is, we consider groups $\mathrm{B}^{n-r-1}_{\mathrm{X}}$, then these are already in one-to-one correspondence with each other. Thus, the groups $\mathrm{B}^{n-r-1}_{\mathrm{X}}$ and $\mathfrak{B}^r_{\mathfrak{G}}$ form an orthogonal couple, that is, each of them is a group of characters of the other (see Definitions 1′ and 3, T. T. G.). In particular, if $\mathfrak{G}$ is taken as the additive group of integers, then X transforms into the group K (see Remark 5, T. T. G.) and since $\mathfrak{B}^r_{\mathfrak{G}}$ in this case transforms in the full Betti group $\mathfrak{B}^r(R^n - K)$, the invariance of the latter will be proved. Precisely this case is of greatest interest to us. If we are to take, further, for the group $\mathfrak{G}$ the group of residues of integers modulo m, then the group X turns out also to be isomorphic to the group of residues modulo m; in this case the groups $\mathrm{B}^r_{\mathrm{X}}$ and $\mathfrak{B}^r_{\mathfrak{G}}$ are the usual Betti groups modulo m, and as they are both finite, according to Remark 3, T. T. G., we obtain their isomorphisms, that is, the theorem of duality of Alexander modulo m.

For the proof of the facts we have stated it is necessary to generalize certain basic concepts of combinatorial topology to the algebraic complexes not constituting linear forms relative to simplexes with integer coefficients but linear forms with coefficients that are elements of more general discrete and continuous groups. Sections 3 and 4 will be devoted to this, while Sections 5 and 6 will be devoted to the proof of the theorems of duality themselves.

3

Definition 1. Let X and $\mathfrak{G}$ be two groups forming an orthogonal couple (see Definition 3, T. T. G.). Consider two algebraic polyhedral complexes α and $\mathfrak{a}$ of dimensions r and $n-r$ located in Eucledian space R^n in a general position.[2] Here complex α is a linear form with coefficients in X, while complex $\mathfrak{a}$ is a linear form with coefficients in $\mathfrak{G}$:

$$\alpha = \sum_{i=1}^{p} \alpha_i \epsilon_i, \qquad \mathfrak{a} = \sum_{j=1}^{q} \mathfrak{a}_j \epsilon_j',$$

where ϵ_i and ϵ_j' are oriented simplexes, $\alpha_i \subset \mathrm{X}$, $\mathfrak{a}_j \subset \mathfrak{G}$. Let us define $\mathrm{X}(\alpha, \mathfrak{a})$, the index of intersection of complexes α and $\mathfrak{a}$, by putting

$$\mathrm{X}(\alpha, \mathfrak{a}) = \sum_{i=1}^{p} \sum_{j=1}^{q} \alpha_i \mathfrak{a}_j \mathrm{X}(\epsilon_i, \epsilon_j').$$

By virtue of the fact that X and $\mathfrak{G}$ are orthogonal, the product $\alpha_i \mathfrak{a}_j$ is an element of the group K, while the index of intersection $\mathrm{X}(\epsilon_i, \epsilon_j')$ of the two simplexes is understood in the usual sense[2] and is an integer,[8] so that $\mathrm{X}(\alpha, \mathfrak{a})$ is an element of the group K.

It is easy to see that if β and $\mathfrak{b}$ are two complexes, the first with coefficients in X, the second with coefficients in $\mathfrak{G}$, the dimensions of which are r and $n-r+1$, then the usual formula[2] holds:

$$\mathrm{X}(\dot{\beta}, \mathfrak{b}) = \epsilon(r)\mathrm{X}(\beta, \dot{\mathfrak{b}}),$$

where $\epsilon(r) = +1$ or -1 depending upon the value of r, and $\dot{\beta}$ and $\dot{\mathfrak{b}}$ are respectively the boundaries of β and $\mathfrak{b}$.

Definition 2. Let γ and $\mathfrak{g}$ be two non-intersecting cycles in space R^n, of dimensions r and $n-r-1$, the first of these with coefficients in X, the second with coefficients in $\mathfrak{G}$ (X and $\mathfrak{G}$, as always, form an orthogonal couple). Let $\mathfrak{d}$ be an arbitrary complex with boundary $\mathfrak{g}$. Let us define the coefficient of linking $\mathfrak{B}(\gamma, \mathfrak{g})$ of cycles γ and $\mathfrak{g}$ by putting $\mathfrak{B}(\gamma, \mathfrak{g}) = \mathrm{X}(\gamma, \mathfrak{d})$. In this way $\mathfrak{B}(\gamma, \mathfrak{g})$ is an element of the group K. The invariance of $\mathfrak{B}(\gamma, \mathfrak{g})$ is proved in the usual way.

A series of properties of $\mathrm{X}(\alpha, \mathfrak{a})$ and $\mathfrak{B}(\gamma, \mathfrak{g})$ may be proved in the usual way;[6] I shall make use of them in what follows without giving proofs.

4

Definition 3. Let F be a compact metric space. We shall call an r-dimensional simplex on F the total of $r+1$ points on F (the vertices of the simplex). A simplex on F we shall call an ϵ-simplex if the totality of its vertices forms a set of diameter smaller than ϵ. A simplex on F we shall call oriented if, as usual, a certain order of its vertices is given to within an even permutation while an odd permutation gives an orientation of opposite sign. A finite linear form relative to oriented r-dimensional simplexes on F with coefficients in X we shall call a complex of dimension r on F, modulo X. The sum of two r-dimensional complexes on F is defined as the sum of the corresponding linear forms. If the complex consists of one simplex with coefficient α, we will define its boundary as the sum of all faces, taken with suitable orientation and with coefficient α. The boundary of an arbitrary complex on F we shall define as the sum of the boundaries of the terms in the linear form. A complex, the boundary of which is equal to zero, will be called a cycle. We shall call a complex an ϵ-complex if all its simplexes are ϵ-simplexes. A cycle is called ϵ-homologous to zero if it is the boundary of a certain ϵ-complex, $\zeta \underset{\epsilon}{\sim} 0$. Two cycles are ϵ-homologous to each other if their difference is ϵ-homologous to zero.

Definition 4. Let F be a compact metric space. We shall call the sequence $\zeta_1, \zeta_2, \ldots, \zeta_m, \ldots$ of r-dimensional cycles on F, modulo X, an r-dimensional true cycle on F modulo X, if (a) ζ_m is an ϵ_m-cycle, (b) ζ_m and ζ_{m+1} are ϵ_m-homologous to each other, where $\lim \epsilon_m = 0$. The sum of two r-dimensional true cycles $\mathrm{Z} = (\zeta_1, \zeta_2, \ldots, \zeta_m, \ldots)$ and $\mathrm{Z}' = (\zeta'_1, \zeta'_2, \ldots, \zeta'_m, \ldots)$ shall be called the true cycle

$$\mathrm{Z} + \mathrm{Z}' = (\zeta_1 + \zeta'_1, \zeta_2 + \zeta'_2, \ldots \zeta_m + \zeta'_m, \ldots).$$

A true cycle Z shall be called homologous to zero on F, $\mathrm{Z} \sim 0$, if there exist numbers δ_m, $\lim \delta_m = 0$, such that $\zeta_m \underset{\delta_m}{\sim} 0$. Two true cycles are homologous to each other if their difference is homologous to zero.

Definition 5. Let F be a compact subset of n-dimensional Euclidean space. Further, let $\mathrm{Z} = (\zeta_1, \zeta_2, \ldots, \zeta_m, \ldots)$ be an r-dimensional true cycle on F modulo X and $\mathfrak{z}$ an $n-r-1$-dimensional cycle in space $R^n - F$ modulo $\mathfrak{G}$ (X and $\mathfrak{G}$ are orthogonal groups). Let us define the coefficient of linking $\mathfrak{B}(\mathrm{Z}, \mathfrak{z})$ of cycles Z and $\mathfrak{z}$. Each simplex of ζ_m is given only by its vertices. Let us realize it in the form of the polyhedral geometrical simplex, taking it with the same coefficient with which it enters in ζ_m and let us form the sum of all the simplexes entering in ζ_m, in this way obtaining a cycle $\bar{\zeta}_m$ giving a geometrical realization of ζ_m. Let us define $\mathfrak{B}(\mathrm{Z}, \mathfrak{z})$ as $\mathfrak{B}(\bar{\zeta}_m, \mathfrak{z})$ for sufficiently large m; obviously, for sufficiently large m, $\mathfrak{B}(\bar{\zeta}_m, \mathfrak{z})$ is independent of m, since all the cycles of $\bar{\zeta}_m$ with sufficiently large m are homologous to each other outside the cycle $\mathfrak{z}$. Thus, $\mathfrak{B}(Z, \mathfrak{z})$ is an element of the group K.

Definition 6. Let us define an r-dimensional Betti group B^r_X modulo X of a compact metric space F. Each class α of true r-dimensional cycles homologous to each other in pairs modulo X on F is an element of the group B^r_X. If Z is a certain true cycle of class α and Z' is a certain true cycle of class α', we shall call the sum $\alpha + \alpha'$, a class β which contains the true cycle $Z + Z'$. It is easy to see that the sum $\alpha + \alpha'$ defined in this way does not depend upon the arbitrariness of choice of Z and Z', but depends only upon the elements α and α' themselves. In this way the algebraic operations in B^r_X are defined. Let us now define the topological operations in B^r_X in order to make this group continuous; for this it is sufficient to give a full system of neighborhoods of zero in B^r_X. Let us define a neighborhood V of zero in B^r_X as a function of a certain positive whole number k and a certain neighborhood U of zero of the group X. The element α of the group B^r_X we shall regard as belonging to V if: (a) there exists a true cycle Z of class α, $Z = (\zeta_1, \zeta_2, \ldots, \zeta_m, \ldots)$, where ζ_m is an ϵ_m-cycle, and $\zeta_m \underset{\epsilon_m}{\sim} \zeta_{m+1}$, such that $\epsilon_m < 1/k$, for all m; (b) α_i, $i = 1, 2, \ldots l$, being a totality of all elements of the group X which are coefficients of simplexes of ζ_1, any element of X of the form $\Sigma^l_{i=1} a_i\alpha_i$ belongs to U if a_i are arbitrary integers whose absolute values do not exceed unity. As k runs over all positive integral values and U a full system of neighborhoods of zero of the group X, V ranges over a full system of neighborhoods of zero of the group B^r_X. Since there exists a countable full system of neighborhoods of zero in X and the set of integers is countable, the group B^r_X also admits a countable full system of neighborhoods of zero. Let us now define the neighborhood W of an arbitrary element $\alpha \subset B^r_X$ as a function of a certain neighborhood of zero V. We shall consider that $\beta \subset W$ if $\beta - \alpha \subset V$. When V ranges over a full system of neighborhoods of zero, W runs over a full system of neighborhoods of α. Since the full system of neighborhoods of zero in B^r_X, by what was proved, is countable, the full system of neighborhoods of α is also countable; consequently, the first axiom of countability in B^r_X is fulfilled.

Theorem 1. A Betti group $B^r_X = B$ modulo X of a compact metric space F is compact and satisfies the second axiom of countability.

Proof. Let $A_1, A_2, \ldots, A_m, \ldots$ be a sequence of complexes giving a projective spectrum F,[7] and let Π_m be the simplicial reflection of A_{m+1} in A_m arising from the projective spectrum. It is easy to see that there exists a sequence of positive numbers $\epsilon_1, \epsilon_2, \ldots, \epsilon_m, \ldots$ such that: (a) $\lim \epsilon_m = 0$; (b) the simplexes A_m have a diameter less than ϵ_m; (c) if ζ is a cycle of A_{m+1}, then ζ and $\Pi_m(\zeta)$ are ϵ_m-homologous on F.

It is likewise not difficut to show that from every class of homologies α, a true cycle $Z = (\zeta_1, \zeta_2, \ldots, \zeta_m, \ldots)$ can be chosen in such a way that Z_m is a cycle of A_m, and ζ_{m+1} and ζ_m are ϵ_m-homologous to each other on F. Let $\beta_1, \beta_2, \ldots \beta_i, \ldots$ be an arbitrary sequence of elements of B. We shall show that this sequence has a limit element. To do this we choose from β_i a true cycle $Z_i = (\zeta_{1,i}, \zeta_{2,i}, \ldots, \zeta_{m,i} \ldots)$ of the type indicated. Since the group of cycles modulo X of the complex A_m is compact for every m, we can make use of the diagonal process to choose from the sequence $\bar{Z}_i$ the sequence Z_{ij} such that for every m, $\lim_{j\to\infty} \zeta_{m,ij} = \bar{\zeta}_m$, where $\bar{\zeta}_m$ is a cycle of A_m and moreover $\bar{\zeta}_m$ and $\bar{\zeta}_{m+1}$ are ϵ_m-homologous to each other on F. In this way the sequence of cycles $\bar{\zeta}_1, \bar{\zeta}_2, \bar{\zeta}_3, \ldots, \bar{\zeta}_m, \ldots$ forms a true cycle $\bar{Z}$. We shall show that $\lim_{j\to\infty} \beta_{ij} = \bar{\beta}$, where $\bar{\beta}$ is the class of homologies containing $\bar{Z}$. Let V be an arbitrary neighborhood of zero in B, and let U and k be that neighborhood of zero and that integer, which by Definition 6 determine the neighborhood V. Let m' be sufficiently large that for $n \geqslant m'$, $\epsilon_n < 1/k$. Let us designate by s the number of r-dimensional simplexes of $A_{m'}$. An let us further choose a neighborhood G of zero in X such that if $\alpha_i \subset G$ then $\Sigma_{i=1}^{s} a_i \alpha_i \subset U$, where a_i are integers and $|a_i| \leqslant 1$. Let Z^r be the group of r-dimensional cycles of the complex A_m, and let W be a neighborhood of zero in Z^r such that W contains only cycles the coefficients of all of whose simplexes are contained in G. Since $\lim_{j\to\infty} z_{m',ij} = \bar{\zeta}_{m'}$, there exists a p sufficiently large that with $j > p$, $\bar{\zeta}_{m'} - \zeta_{m',ij} \subset W$. If we denote by γ_i, $i = 1, 2, \ldots, l$, the totality of coefficients in the simplexes of the cycle $\bar{\zeta}_{m'} - z_{m',ij}$ where $l \leqslant s$, then from the construction of W we shall have $\gamma_i \subset G$, and from the definition of G, $\Sigma_{i=1}^{l} a_i \gamma_i \subset U$ for arbitrary integers a_i, whose absolute values do not exceed unity.

Since the obviously true cycle $(\bar{\zeta}_{m'} - \zeta_{m,ij}), (\bar{\zeta}_{m'+1} - \zeta_{m'+1,ij}), \ldots,$ belongs to the class $\bar{\beta} - \beta_{ij}$, it follows from definition 6 that for $j > p$, $\bar{\beta} - \beta_{ij} \subset V$. In this way we have $\lim_{j\to\infty} \beta_{ij} = \bar{\beta}$ but the sequence β_{ij} is a subsequence of an arbitrary sequence β_i. Hence the compactness of B is established.

Now we shall show that if B is an arbitrary compact group for which the first axiom of countability holds, then the second axiom of countability holds likewise. We shall show first of all that B contains a denumerable everywhere dense set N. It is easily seen that in every neighborhood V of zero in B there exists a finite set M such that for every $\alpha \subset$ B there is a $\mu \subset M$ such that $\alpha - \mu \subset V$. Taking a complete denumerable system of neighborhoods of zero in B and constructing the corresponding sets M, and then suming them, we shall get for the sum a set N which is everywhere dense on B. The totality of the complete denumerable systems of neighborhoods for all the elements of N now forms a complete denumerable system of neighborhoods for the whole B. In fact, let G be any domain of B and let $\alpha \subset G$. Then there exists a neighborhood of zero V such that if $\beta \subset V$, $\gamma \subset V$, then $\alpha + \beta - \gamma \subset G$. Furthermore we can find a $\delta \subset V$ such that $\alpha - \delta = \nu \subset N$. The totality of all the elements of the form $\nu + \eta$, where $\eta \subset V$, constitutes the neighborhood W of the point ν with $\alpha \subset W \subset G$. In this way for every domain G and element $\alpha \subset G$ there exists a neighborhood W of our denumerable system such that $\alpha \subset W \subset G$, that is, the system is complete.

5. Theorem of Duality for a Complex

Theorem 2. Let the complex K be polyhedrally situated in R^n, let B^r_X be an r-dimensional Betti group of K modulo X and let $\mathfrak{B}^s_{\mathfrak{G}}$ be an s-dimensional Betti group in the space $(R^n - K)$ modulo $\mathfrak{G}$. (X and $\mathfrak{G}$ are orthogonal.) Then B^r_X and $\mathfrak{B}^{n-r-1}_{\mathfrak{G}}$ are orthogonal, with the product of $\alpha \subset B^r_X$ and $\mathfrak{a} \subset \mathfrak{B}^{n-r-1}_{\mathfrak{G}}$ determined as the coefficient of linking (see Definition 2) of some cycle of class α with a cycle of class $\mathfrak{a}$.[8]

Proof. Let us suppose that K is composed of simplexes of some subdivision of R^n. Let T^r_i, $i = 1, 2, \ldots, a_r$, be the totality of r-dimensional oriented simplexes of K. Let us denote further by S^s_j, $j = 1, 2, \ldots, a_{n-s}$, the totality of s-dimensional barycentric stars of R^n which intersect K. It is obvious that the number of r-dimensional simplexes of K is equal to the number of $n-r$-dimensional stars of S^{n-r}_j, which is foreseeen in the nomenclature. Moreover, the notation can be chosen in such a way that the index of intersection $X(T^r_i, S^{n-r}_j) = \delta^j_i$, where $\delta^j_i = 0$ if $i \neq j$ and $\delta^i_i = 1$.

Let us denote by Λ^r the group of all linear forms relative to all r-dimensional simplexes of the complex K with coefficients in X, that is, the group of r-dimensional subcomplexes K modulo X. To every element in Λ^r corresponds, as its boundary, some element of Λ^{r-1} (see Section 1). In this way we have a homomorphic reflection of the group Λ^r on the subgroup H^{r-1} of the group Λ^{r-1} with the divisor Z^r. The elements of Z^r are cycles, and the elements of H^{r-1} are cycles homologous to zero. Analogously we designate by $\mathfrak{L}^s$ the group of all the linear forms of the type $\sum_{j=i}^{a_{n-s}} \mathfrak{r}_j S^s_j$, where $\mathfrak{r}_i \subset \mathfrak{G}$; in this way $\mathfrak{L}^s$ is a group of complexes modulo $\mathfrak{G}$. To every element $\mathfrak{a}$ of $\mathfrak{L}^s$ corresponds, as its boundary, some linear form relative to barycentric stars; we shall choose only those terms of the form which are composed of S^{s-1}_j and we shall call their sum a relative boundary of $\mathfrak{a}$. In this way to every element of $\mathfrak{L}^s$ corresponds, as its relative boundary, some element of $\mathfrak{L}^{s-1}$, and we get a homomorphic reflection of the group $\mathfrak{L}^s$ on the subgroup $\mathfrak{H}^{s-1}$ of the group $\mathfrak{L}^{s-1}$, with the divisor $\mathfrak{B}^s$. It is easy to see that $\mathfrak{H}^s \subset \mathfrak{B}^s$.

We shall now establish the law of multiplication between the elements of Λ^r and $\mathfrak{L}^{n-r}$. We shall suppose that if $\alpha \subset \Lambda^r$ and $\mathfrak{a} \subset \mathfrak{L}^{n-r}$ then $\alpha\mathfrak{a} = X(\alpha, \mathfrak{a})$ (see Definition 1). We shall show that the groups Λ^r and $\mathfrak{L}^{n-r}$, because of this law of multiplication, form an orthogonal pair. That the multiplication is distributive and continuous is obvious. Let us show that $(\Lambda^r, \mathfrak{L}^{n-r}) = (0)$ (see Definition 3, T. T. G.). Let $0 \neq \alpha \subset \Lambda^r$, $\alpha = \Sigma_{i=1}^{a_r} \alpha_i T_i^r$, with at least one of the coefficients α_i different from zero. Let it be α_ϵ. Since X and $\mathfrak{G}$ are orthogonal there exists an $\mathfrak{r} \subset \mathfrak{G}$ such that $\alpha_\epsilon \mathfrak{r} \neq 0$.

Let $\mathfrak{a} = \mathfrak{r} S_\epsilon^{n-r}$; we shall have $\alpha\mathfrak{a} = \alpha_\epsilon \mathfrak{r} \neq 0$. In exactly the same way we can show that $(\mathfrak{L}^{n-r}, \Lambda^r) = (0)$.

We also show that $(\Lambda^r, \mathfrak{H}^{n-r}) = Z^r$ and $(\mathfrak{L}^{n-r}, H^r) = \mathfrak{B}^{n-r}$. Let $\alpha \subset Z^r$ and $\mathfrak{a} \subset \mathfrak{H}^{n-r}$. There exists an element $\mathfrak{c} \subset \mathfrak{L}^{n-r+1}$ whose relative boundary is $\mathfrak{a}$, that is, $\mathfrak{c} \rightarrow \mathfrak{a} + \mathfrak{b}$, where $\mathfrak{b}$ does not intersect K. $\alpha\mathfrak{a} = X(\alpha, \mathfrak{a}) = X(\alpha, \mathfrak{a} + \mathfrak{b}) = 0$, since the index of intersection of twocycles in R^n is equal to zero. In this way $(\Lambda^r, \mathfrak{H}^{n-r}) \supset Z^r$. Now let $\alpha \subset \Lambda^r$, $\alpha \not\subset Z^r$, then $\alpha \rightarrow \beta \neq 0$, $\beta \subset \Lambda^{r-1}$. Since Λ^{r-1} and $\mathfrak{L}^{n-r+1}$ are orthogonal, there exists a $\mathfrak{c} \subset \mathfrak{L}^{n-r+1}$ such that $\beta\mathfrak{c} \neq 0$. Let $\mathfrak{c} \rightarrow \mathfrak{a} + \mathfrak{b}$, $\mathfrak{a} \subset \mathfrak{H}^{n-r}$, $\mathfrak{b}$ does not intersect K. Then $\alpha\mathfrak{a} = X(\alpha, \mathfrak{a} + \mathfrak{b}) = \pm X(\beta, \mathfrak{c}) = \pm \beta\mathfrak{c} \neq 0$. In this way $(\Lambda^r, \mathfrak{H}^{n-r}) = Z^r$. Quite analogously we can prove that $(\mathfrak{L}^{n-r}, H^r) = \mathfrak{B}^{n-r}$.

From what we have just shown together with Theorems 2 and 4 (T. T. G.) it follows that

$$(\mathfrak{L}^{n-r}, Z^r) = \mathfrak{H}^{n-r}, \quad (\Lambda^r, \mathfrak{B}^{n-r}) = H^r.$$

From which, taking into consideration that Z^r and $\mathfrak{B}^{n-r}$ form a pair (see Definition 3, T. T. G.), we conclude that

$$(Z^r, \mathfrak{B}^{n-r}) = H^r, \quad (\mathfrak{B}^{n-r}, Z^r) = \mathfrak{H}^{n-r}.$$

But from this follows the orthogonality of the groups $B_X^r = Z^r/H^r$ and $\mathfrak{B}^{n-r}/\mathfrak{H}^{n-r}$. The first of these is a Betti group of the complex K, the second is isomorphic with the Betti group $\mathfrak{B}^{n-r-1}$ in the space $R^n - K$. This last statement is demonstrated as follows: let $\mathfrak{z} \subset \mathfrak{B}^{n-r}$, $\mathfrak{z} \rightarrow \mathfrak{z}'$; the boundary $\mathfrak{z}'$ of the complex $\mathfrak{z}$ is a cycle in the space $R^n - K$ and determines some element of the group $\mathfrak{B}_{\mathfrak{G}}^{n-r-1}$; at the same time if $\mathfrak{n} \subset \mathfrak{B}^{n-r}$, and $\mathfrak{z} - \mathfrak{n} \subset \mathfrak{H}^{n-r}$, then the boundaries of $\mathfrak{z}$ and $\mathfrak{n}$ will be cycles homologous to each other in R^n. Conversely if $\mathfrak{z}$ and $\mathfrak{n}$ belong to $\mathfrak{B}^{n-r}$ and if their boundaries are homologous to each other in R^n, then $\mathfrak{z} - \mathfrak{n} \subset \mathfrak{H}^{n-r}$. Moreover, for every $(n-r-1)$-dimensional cycle of $R^n - K$, a cycle can be found, homologous to it, which forms the boundary of some element of $\mathfrak{B}^{n-r}$. In this way the isomorphism between the groups $\mathfrak{B}^{n-r}/\mathfrak{H}^{n-r}$ and $\mathfrak{B}_{\mathfrak{G}}^{n-r-1}$ is established. It is obvious that by going from elements of the group $\mathfrak{B}^{n-r}$ to their boundaries, the indices of their intersections with the elements of Z^r go over into the coefficients of linking. Hence the theorem is proved.

6. The General Theorem of Duality for a Compact Set F Located in R^n

Fundamental Theorem. “Let F be in R^n, let B_X^r be an r-dimensional Betti group of F modulo X, and let $\mathfrak{B}_{\mathfrak{G}}^s$ be an s-dimensional Betti group modulo $\mathfrak{G}$ of the space $R^n - F$. (X and $\mathfrak{G}$ are orthogonal.) Then the groups B_X^r and $\mathfrak{B}_{\mathfrak{G}}^{n-r-1}$ are orthogonal, the product of $\alpha \subset B_X^r$ and $\mathfrak{a} \subset \mathfrak{B}_{\mathfrak{G}}^{n-r-1}$ being determined as the coefficient of linking of some true cycle of class α with a cycle of class $\mathfrak{a}$.

Corollary. It follows from this Theorem, in view of Theorem 5, T. T. G., that each of the groups B_X^r and $\mathfrak{B}_{\mathfrak{G}}^{n-r-1}$ is the group of characters of the other, that is, each of them is determined uniquely by the other.

Proof of Fundamental Theorem. The law of multiplication formulated in the theorem is obviously distributive. Let us show now that it satisfies the condition of continuity of multiplication. To do this, it is sufficient to show that if $\mathfrak{a} \subset \mathfrak{B}_{\mathfrak{G}}^{n-r-1}$, then for any neighborhood J of zero of the group K, there is always a neighborhood V of zero in the group B_X^r, such that if $\alpha \subset V$, then $\alpha\mathfrak{a} \subset J$. Let $\mathfrak{z}$ be a cycle of the class $\mathfrak{a}$ and let $\mathfrak{b} \to \mathfrak{z}$. Let $\mathfrak{r}_i$, $i = 1, 2, \ldots, p$, be the totality of all the coefficients in the simplexes of the complex $\mathfrak{b}$. Let us denote by U a sufficiently small neighborhood of zero in X, such that if $\beta_i \subset U$, then $\Sigma_{j=1}^{p} \beta_j \mathfrak{r}_j \subset J$. Further let K be such a small polyhedral neighborhood of F that K does not intersect $\mathfrak{z}$; let ϵ be a sufficiently small number so that every ϵ-complex on F lies in K, and let k be a whole number so large that $1/k < \epsilon$. Let V be the neighborhood of zero in B_X^r which is determined by the neighborhood U and the number k according to Defition 6. We shall show that if $\alpha \subset V$, then $\alpha\mathfrak{a} \subset J$. Let $Z = (\zeta_1, \zeta_2, \ldots, \zeta_m, \ldots)$ be a true cycle of class α such that, by Definition 6, α belongs to the neighborhood V. Since by force of this condition all the cycles ζ_m, and all the complexes which establish homologies between them, belong to K, we have from Definition 5, $\mathfrak{B}(Z \cdot \mathfrak{z}) = \mathfrak{B}(\zeta_1, \mathfrak{z})$. Let α_i, $i = 1, 2, \ldots l$, be the totality of all the coefficients of the simplexes of the cycle ζ_1; then we have from Definitions 1 and 2, $\mathfrak{B}(Z, \mathfrak{z}) = X(\zeta_1, \mathfrak{b}) = \Sigma_{i=1}^{l} \Sigma_{j=1}^{p} \alpha_i \mathfrak{r}_j a_{ij}$, where a_{ij} are integers whose absolute values do not exceed unity. By Definition 6, $\Sigma_{i=1}^{l} a_{ij}\alpha_i = \beta_j \subset U$. In this way from the construction of U, $\Sigma_{i=1}^{l} \Sigma_{j=1}^{p} \alpha_i \ {}_j a_{ij} = \Sigma_{j=1}^{p} \beta_j \ {}_j \subset J$. From the multiplication law B_X^r and $\mathfrak{B}_{\mathfrak{G}}^{n-r-1}$ form a pair (see Definition 3, T. T. G.).

We shall now prove that B_X^r and $\mathfrak{B}_{\mathfrak{G}}^{n-r-1}$ are orthogonal. Let $0 \neq \alpha \subset B_X^r$, we shall show that there is an element $\mathfrak{a} \subset \mathfrak{B}_{\mathfrak{G}}^{n-r-1}$ such that $\alpha\mathfrak{a} \neq 0$. Let $(\zeta_1, \zeta_2, \ldots, \zeta_m, \ldots) = Z$ be a true cycle belonging to the class α. Since $\alpha \neq 0$, there exists a polyhedral neighborhood K of the set F such that the cycles ζ_m are not homologous to zero in it. We can suppose without loss of generality that all the cycles ζ_m and all the complexes, which establish homologies between them, are in K. Since $\zeta_1 \subset K$, and is not homologous to zero in K, it follows from Theorem 2 that there is a cycle $\mathfrak{z} \subset R^n - K$ such that $V(\zeta_1, \mathfrak{z}) \neq 0$. Let $\mathfrak{a} \subset \mathfrak{B}_{\mathfrak{G}}^{n-r-1}$ be a class to which $\mathfrak{z}$ belongs; then $\alpha\mathfrak{a} \neq 0$. Hence $(B_X^r, \mathfrak{B}_{\mathfrak{G}}^{n-r-1}) = (0)$. We shall now show that if $0 \neq \mathfrak{a} \subset \mathfrak{B}_{\mathfrak{G}}^{n-r-1}$, there exists an $\alpha \subset B_X^r$ such that $\alpha\mathfrak{a} \neq 0$. Let $\mathfrak{z}$ be a cycle of class $\mathfrak{a}$, and let $K_1, K_2, \ldots, K_m, \ldots$ be a decreasing sequence of complexes containing F whose intersection coincides with F. We also suppose that $\mathfrak{z} \subset R^n - K_1$. Let g be the order of a, and let γ be some element of K of order g. From Theorem 2 and Lemma 8, T. T. G., there exists in K_m a cycle ζ_m such that $V(\zeta_m, \mathfrak{z}) = \gamma$. Let ζ_m^i for $j \geqslant m$ be a cycle composed of simplexes K_m homologous in K_m with the cycle ζ_j. Since the group of cycles of the complex K_m is compact, we can choose from the sequence ζ_i^j $(j = 1, 2, \ldots)$ a converging subsequence $\zeta_1^{j_k}$ $(k = 1, 2, \ldots)$. In the same way from the sequence $\zeta_2^{j_k}$ a convergent subsequence can be chosen. Continuing this process further and using the diagonal process we shall arrive at a sequence of numbers $i_1, i_2, \ldots, i_q, \ldots$ such that the sequence $\zeta_m^{i_q}$, with m fixed, converges for every m to some cycle $\bar{\zeta}_m$. It is obvious that $\bar{\zeta}_{m+1} \sim \bar{\zeta}_m$ in K_m, and that $V(\bar{\zeta}_m, \mathfrak{z}) = \gamma$. Pushing the vertices of the cycles $\bar{\zeta}_m$ on to F, we obtain a true cycle Z belonging to some class $\alpha \subset B_X^r$ with $\alpha\mathfrak{a} = \gamma$. Hence $(\mathfrak{B}_{\mathfrak{G}}^{n-r-1}, B_X^r) = (0)$ and the theorem is completely proved.

Notes

1. Alexander, A proof and generalization of the Jordan–Brouwer Theorem, *Trans. Amer. Math. Soc.* **23** (1922), p. 333–349.

2. Lefschetz, Intersections and transformations of complexes and manifolds, *Trans. Amer. Math. Soc.* **28** (1926), p. 1–39.

3. Pontrjagin, Über den algebraischen Inhalt topologischer Dualitätssätze, *Math. Ann.*, 105 (1931), p. 165–205.

4. To understand this paper it is necessary to know the first chapter of my paper, The theory of topological commutative groups, *Ann. of Math.* (2) **35**, 1934, p. 361–370. In the following I will denote this last paper by T. T. G. It must be noted here that in T. T. G. there is a mistake, which was pointed out to me by J. v. Neumann, namely, on page 381 at the end of Definition 1b, it is written "the intersection of ψ_i and φ_i contains only zero" where it should be written "the intersection of all ψ_i contains only zero".

5. Alexander, J. W., and Cohen, L. W., A classification of the homology groups of compact spaces, *Ann. of Math.* (2) **33**, 1932, p. 538–566.

6. For the fundamental concepts of combinatorial topology, see the book of Lefschetz, Topology.

7. Alexandroff, Gestalt und Lage abgeschlossener Mengen, *Ann. of Math.* (2) **30**, 1928, p. 107. But instead of a projective spectrum, it is possible here to use any approximating sequence of complexes.

8. The proof of this theorem is taken from my paper Über den algebraischen Inhalt . . . (see note 3), p. 186, II Formulierung und . . . .

Théorie des Groupes Continus. Sur les Groupes Topologiques Compacts et le Cinquième Problème de M. Hilbert* †

Compte Rendus Académie des Sciences Paris (1934)
198(3): 238–40

1

M. J. von Neumann a démontré récemment[1] que tout groupe continu clos (c'est-à-dire groupe topologique compact, connexe et localement euclidien) est un groupe de Lie; il a donc résolu par l'affirmative le problème de M. Hilbert (pour les groupes compacts). Je me propose de donner ici un résultat bien plus général que celui de M. von Neumann et cela par des moyens qui me semblent plus simples. Je prouverai notamment le théorème suivant:

Théorème. Tout groupe topologique compact, localement connexe et de dimension finie, est un groupe de Lie, d'ailleurs isomorphe à un sous-groupe du groupe de rotations de l'espace euclidien à un nombre de dimension suffisamment grand.

2

Ce théorème est contenu dans le théorème suivant:

---

* Hilbert, *Problémes mathématiques. Conférence faite au Congrès international des Mathématiciens à Paris en 1900.*

† Note de M. L. Pontrjagin, présentée par M. Hadamard, Séance du 3 janvier 1934.

Théorème Fondamental. Soit W un groupe topologique compact, connexe et de dimension finie r. Il existe dans W un sous-groupe invariant (diviseur normal) U, abélien et de dimension nulle, tel que le groupe quotient W/U ("Faktorgruppe") est un groupe de Lie connexe. Si le groupe U est fini, alors W est lui-méme un groupe de Lie isomorphe à un groupe de rotations (d'un espace euclidien E^n). Si le groupe U est infini, il est homéomorphe à l'ensemble parfait discontinu de Cantor; dans ce cas l'élément unité du groupe W possède un voisinage homéomorphe au produit topologique de l'ensemble de Cantor par un cube r-dimensionnel, de sorte que le groupe W ne peut pas étre localement connexe.

On peut d'ailleurs préciser ce résultat. Soit G le groupe fondamental (groupe de connexion d'aprés la terminologie de M. Cartan) de la variété $V = W/U$. D'après les résultats de O. Schreier,[2] le groupe G est abélien. Or il existe une suite non croissante $G > G_1 > G_2 > \ldots > G_i > \ldots$ de sous-groupes de G, tels que *l'indice de* G_{i+1} dans G_i est un nombre fini k_i.

Si tous les k_i, à partir d'une certaine valeur de i, sont égaux à 1, on a le premier cas envisagé dan l'énoncé du théorème fondamental, tandis que le second cas de ce théorème correspond au cas où il y a, parmi les k_i, une infinité d'entiers différents de l'unité.

3

La démonstration du théorème fondamental repose sur la notion d'une *suite convergente du groupe.*[3]

Soit $W_1, W_2, \ldots, W_i, \ldots$ une suite de groupes topologiques compacts. Supposons définie une suite $f_1, f_2, \ldots, f_i, \ldots$, où f_i est une représentation homomorphe (= mériédriquement isomorphe) de W_{i+1}, W_i. Les suites $\{w_1, w_2, \ldots, w_i, \ldots\}$, où $w_i = f(w_{i+1})$ forment un groupe topologique W: il suffit de poser $\{w_i\} \cdot \{w'_1\} = \{w_i \cdot w'_1\}$ et de considérer comme voisinage de $\{w_i\}$ l'ensemble de toutes les suites $\{w'_1\}$ dont les n premiers membres $w'_1, \ldots, w'_r$ (n est un entier quelconque) appartiennent respectivement aux voisinages arbitrairement donnés $w_1, \ldots, w_n$. On s'aperçoit facilement que *la limite d'une suite de groupes compacts est, elle aussi, un groupe compact.*

Cela posé, on tire tout d'abord des résultats de MM. Weyl et Peter[4] [rendus applicables aux groupes compacts arbitraires par les recherches de M. Haar[5]] le suivant:

Théorème Auxilliaire. Tout groupe topologique compact W est limite d'une suite convergente $\{W_i\}$ de groupes de Lie compacts. Si W a la dimension finie r, tous les groupes W_i ont eux aussi (pour i assez grand) la dimension r at les homomorphismes f_i correspondent aux sous-groupes invariants finis.

Cela posé, voici la marche à suivre pour démontrer le théorème fondamental. La suite $\{W_i\}$ de groupes de Lie à r dimensions convergente vers W étant donnée, désignons par U_1 le sous-groupe de W_1 ne contenant que l'élément unité. Le sous-groupe U_i de W_i étant déjà défini, désignons par U_{i+1} le sous-groupe de W_{i+1} constitué par tous les éléments de W_{i+1} dont les images (données par f_i) appartiennent à U_i. Les groupes U_i (abéliens d'après les résultats de M. Schreier) convergent vers un groupe U, et l'on a $W/U = W_1$. Nous désignons par G_i le groupe fondamental de W_i. Le groupe G_{i+1} est un sous-groupe du groupe G_i; tous ces groupes sont abéliens. L'indice de G_{i+1} dans G_i est par définition k_i; ce nombre est égal à l'ordre du sous-groupe invariant de W_{i+1} constitué de tous les éléments de ce groupe qui sont transformés par f_i en l'élément unité de W_i' Il en résulte que tous les groupes W_i sont, pour i assez grand, isomorphes au groupe W, si les k_i correspondants sont égaux à 1.

La démonstration du fait que U est homéomorphe à l'ensemble de Cantor, si $k_i \neq 1$ pour une infinité de valeurs de i, ne présente aucune difficulté sérieuse.

Une exposition complète de la théorie ci-dessus paraîtra dans un autre Recueil.

Notes

1. *Annals of Mathematics*, 2e série, **34**, 1933, p. 170.
2. *Abh. Math. Seminar Hambourg*, **4**, 1926, p. 15–32, et **5**, 1927, p. 233–244.
3. Cette notion fondamentale n'est qu'une application à la théorie des groupes de la notion de *spectre projectif* due à M. Alexandroff (*Annals of Math.,* 3e série, **30**, 1928, p. 101–187), les W_i forment en effet un spectre projectif de W.
4. *Math. Annalen*, 97, 1927, p. 737–755.
5. *Annals of Math.*, 2e série, **34**, 1933, p. 147–169.

Théorie des Groupes. Sur les Nombres de Betti des Groupes de Lie*

Compte Rendus Académie des Sciences Paris (1935)
200(15): 1277–80

Dans sa Conférence à Moscou au printemps 1934, M. E. Cartan a attiré l'attention sur le problème de la recherche des nombres de Betti dans les groupes clos de Lie. Il est facile de démontrer que la question se réduit à la recherche des nombres de Betti dans les groupes clos simples de Lie. La Note présente est consacrée à la résolution de ce problème.

Ma méthode, absolument élémentaire, permet non seulement de définir les nombres de Betti, mais encore de construire les cycles qui forment les bases d'homologie. Je ne me suis occupé justqu'à présent que des quatre grandes classes A_r, B_r, C_r, D_r, où r désigne le rang du groupe correspondant.[1] Voici le résultat obtenu:

Théorème Fondamental. Soit M une multiplicité quelconque à n dimensions dont le $i^{\text{ième}}$ nombre de Betti soit p_i ($i = 0, 1, \ldots, n$). Appelons polynome de Poincaré de la multiplicité M le polynome $p_0 + p_1 t + \ldots + p_n t^n$. Pour les quatre grandes classes de groupes simples le polynome de Poincaré est représenté par $(1 + t^{h_1})(1 + t^{h_2}) \ldots (1 + t^{h_r})$.

Les valeurs des nombres h_s ($s = 1, 2, \ldots, r$) sont diffèrentes pour les diffèrents groupes: pour le groupe A_r on a $h = 2s + 1$; pour B_r on a $h_s = 4s - 1$; pour C_r on a $h_s = 4s - 1$; pour D_r on a $h_s = 4s - 1$ si $s \neq r$ et $h_r = 2r - 1$.

* Note de M. L. Pontrjagin, présentée par M. Élie Cartan.

Ce résultat est une conséquence immédiate d'un théorème dont l'énoncé utilise la définition suivante:

Definition. Soit G un groupe de Lie et a et b deux cycles dans G de dimensions p et q. Appelons produit $c = ab$ des cycles a et b le cycle c obtenu par le procédé suivant: soient x et y des éléments de G; si x parcourt le cycle a et y parcourt le cycle b, xy parcourra le cycle c. La dimension de c est égale à $p + q$.

Théorème. Soit G un groupe simple d'une des quatre grandes classes. Il existe dans G une suite de cycles z_s $(s = 1, 2, \ldots, r)$, r désignant le rang de G; la dimension de z, est égale à h_s. Désignons par e l'élêment-unité du groupe G et considérons e comme le cycle o-dimensionnel. Formons le produit formel $(e + z_1)(e + z_2) \ldots (e + z_r)$. Si l'on développe ce produit, chacun de ses membres sera un cycle dans G (voir la définition). L'ensemble de tous les cycles obtenus de cette manière constitue la base complète des homologies dans G.

Ici je ne m'occuperai que du groupe A_r; la démonstration est analogue pour les autres classes.

A_r est formé de toutes les matrices unitaires et unimodulaires $\|a_{ij}\|$ $(i, j = 0, 1, \ldots, r)$. Les nombres complexes a_{ij} satisfont aux relations $\Sigma_{t=0}^{r} a_{ij}\bar{a}_{ik} = \delta_{jk}$, $\|\delta_{jk}\|$ désignant la matrice-unité. Le déterminant $|a_{ij}| = 1$. La dimension de A_r est égale à $r^2 + 2r$.

Désignons par A'_{r-1} l'ensemble de tous les éléments de A_r satisfaisant à la condition $a_{00} = +1$. A'_{r-1} forme un sous-groupe de A_r qui sera isomorphe à A_{r-1}. Soient f_k $(k = 0, 1, \ldots, r)$ les composantes d'un vecteur complexe unitaire, c'est-à-dire des nombres complexes satisfaisant aux conditions

$$\sum_{k=0}^{r} f_k \bar{f}_k = 1,$$

et soit α un nombre complexe de module égal à 1. Posons

$$f_{ij} = \delta_{ij} + (\alpha - 1) f_i f_j.$$

La matrice $\|f_{ij}\|$, unitaire avec le déterminant égal à α, dépend des paramètres α et f_k. Il est à remarquer que si l'on multiplie tous les f_k par le même nombre β de module égal à 1, les f_{ij} ne changent pas. Changeons le signe de la première ligne de la matrice $\|f_{ij}\|$ et multiplions sa dernière ligne par $-\alpha^{-1}$. La matrice $\|f_{ij}\|$ obtenue de cette manière sera unitaire unimodulaire. L'ensemble de toutes les matrices $\|f_{ij}\|$ constitue dans A_r

un cycle orienté z_r de dimension $2r+1$, $2r+1$ étant la différence des dimensions de A_r et de A'_{r-1}. Un calcul immédiat montre que les cycles A'_{r-1} et z_r se coupent en un seul point et que l'indice de leur intersection est égal à ± 1. Soient a et b deux cycles de A'_{r-1}, dont l'indice de l'intersection dans A'_{r-1} est égal à ν. On voit facilement que l'indice de l'intersection de az_r et de b dans A_r sera alors égal à $\pm \nu$. Ainsi, dans le cas où c est un cycle de A'_{r-1} non homologue à o dans A'_{r-1}, les cycles c et cz_r ne seront pas homologues à o dans A_r. Soient a_l $(l = 1, 2, \ldots, m)$ la base complète des homologies dans A'_{r-1} ; de ce fait démontré suit que les cycles a_l et $a_l z_r$ constituent un système indépendant dans A_r. Démontrons que ce système donne la base des homologies dans A_r.

Définissons sur A_r la fonction réelle $f(x)$, où x désigne la matrice a_{ij} [de A_r, comme étant égale à la partie réelle de a_{00}]. Un calcul immédiat montre que la différentielle totale est nulle alors, et alors seulement, que $a_{00} = \pm 1$. L'ensemble A''_{r-1} des matrices pour lesquelles $a_{00} = -1$ est un cogroupe (Nebengruppe) pour A'_{r-1} : Puisque A_r est connexe, A''_{r-1} peut être transformé en A'_{r-1} par une déformation continue, et par conséquent tout cycle de A''_{r-1} est homologue dans A_r à un cycle correspondant de A'_{r-1}. Construisons des trajectoires orthogonales dans A_r pour la fonction $f(x)$.[2] Alors, à l'aide de déformations continues le long de ces trajectoires, on peut réduire à A''_{r-1} chaque ensemble de A_r ne se coupant pas avec A'_{r-1}. De cette manière, tout cycle de $A_r - A'_{r-1}$ sera homologue à un certain cycle de A''_{r-1} et par suite à un certain cycle de A'_{r-1}. Soit y un cycle quelconque de A_r. Il est facile de construire une combinaison linéaire y' des cycles du système $a_l z_r$ telle que le cycle $y - y'$ ait pour indice d'intersection o avec tout cycle de A'_{r-1} ; par conséquent $y - y'$ est homologue à un certain cycle de $A_r - A'_{r-1}$,[3] et, d'après ce qui fut démontré plus haut, à un certain cycle de A'_{r-1}. Ainsi les systèmes a_l et $a_l z_r$ $(l = 1, 2, \ldots, n)$, constituent une base complète de A_r.

Le théorème reçoit ainsi une démonstration inductive.

Notes

1. E. Cartan, *Mémorial Sc. Math.*, fasc. 42, 1930.
2. M. Morse, *Trans. Amer. Math. Soc.*, **27**, 1925, p. 345.
3. L. Pontrjagin, *Math. Ann.*, **105**, 1931, p. 190.

Rough Systems*

Doklady Akademii Nauk SSSR (1937) 14(5): 247–350

Consider a dynamical system defined by two first order equations:

$$\frac{dx}{dt} = P(x,y); \quad \frac{dy}{dt} = Q(x,y), \tag{A}$$

where x and y are Cartesian coordinates; $P(x,y)$ and $Q(x,y)$ are analytic functions of x,y.

We shall confine ourselves to systems of the type (A) for which there exists a so-called "contactless cycle", i.e., a simple closed curve g with a continuous tangent such that all trajectories passing through the points on this curve intersect it and none of the trajectories touch it. The plane domain inside the curve g is called *domain G.* Without loss of generality, we may assume that the phase trajectories intersecting the curve g enter the domain G with increasing time t.

Along with the system (A), consider a modified system:

$$\frac{dx}{dt} = P(x,y) + p(x,y); \quad \frac{dy}{dt} = Q(x,y) + q(x,y) \tag{B}$$

where $p(x,y)$ and $q(x,y)$ are also analytic functions of x and y.

Definition I. The system (A) is said to be "rough" (in contrast to systems which are "non-rough") in the domain G, if for any $\eta > 0$, we can find

* In collaboration with A. Andronov. (Communicated by Academician L. I. Mandel'shtam on 2 January 1936).

an $\epsilon > 0$, such that for arbitrary analytic functions $p(x,y)$ and $q(x,y)$, satisfying the conditions

$$|p(x,y)| < \epsilon, \quad |q(x,y)| < \epsilon;$$

$$|p'_x(x,y)| < \epsilon, \quad |p'_y(x,y)| < \epsilon, \quad |q'_x(x,y)| < \epsilon, \quad |q'_y(x,y)| < \epsilon,$$

in the domain G, there exists such an one-to-one and bi-continuous transformation T of G into itself, under which

(1) corresponding points are at a distance less than η;

(2) points on the same trajectory of the system (B) correspond to points on the same trajectory of the system (A), and vice versa.*

Necessary conditions for the system (A) to be rough in the domain G may be formulated in the following three theorems.

Theorem 1. If the system (A) is rough in the domain G, then in the domain G the system (A) can only have equilibrium states for which the real parts of the roots of the characteristic equation are nonzero.

In other words, in the domain G the system (A) cannot have an equilibrium state $x = x_0, y = y_0$, for which

$$\text{(a)} \quad \Delta = \begin{vmatrix} P'_x(x_0,y_0) P'_y(x_0,y_0) \\ Q'_x(x_0,y_0) Q'_y(x_0,y_0) \end{vmatrix} = 0, \quad \text{or}$$

(b) for $\Delta > 0$, $\sigma = -[P'_x(x_0,y_0) + Q'_y(x_0,y_0)] = 0$.

---

* This definition of roughness of a system may be regarded as the definition of stability of a set of trajectories of a dynamical system with respect to sufficiently small changes in the right-hand sides of Eq. (A). This type of stability is of great interest in physics. In particular, the condition of stability for periodic motion with respect to sufficiently small changes in the right-hand sides of the system (A) was postulated and used by Andronov [1] to prove that only limit cycles can describe real auto-oscillating processes in systems of the type (A).

In this connection it should be mentioned that for the motion of a point on trajectories, Bohl [2] gave a different definition of stability of a dynamical system with respect to small changes in the right-hand sides of the equations. Such stability, in the sense of Bohl, imposes rather stringent conditions on the initial system. For instance, an initial system of the type (A) may be stable in the sense of Bohl, only if the domain G contains only one equilibrium state and only if all other motions tend to this equilibrium state when $t \to +\infty$.

Theorem 2. If the system (A) is rough in the domain G, then in the domain G the system (A) can only have periodic motions for which the characteristic exponent is not zero.

In other words, in the domain G the system (A) cannot have a periodic motion $x = \varphi(t), y = \psi(t)[\varphi(t+\tau) = \varphi(t); \psi(t+\tau) = \psi(t)]$ for which

$$h = \frac{1}{\tau} \int_0^\tau [P'_x(\varphi, \psi) + Q'_y(\varphi, \psi)]\, dt = 0.$$

Theorem 3. If the system (A) is rough in the domain G, then in the domain G the system (A) can only have separatrices which do not join one saddle point to another.

In other words, in the domain G the system (A) cannot have separatrices which join one saddle point to another.

A system (A) is said to satisfy the conditions Γ in the domain G, if inside a contactless cycle g the system (A) has

(1) only equilibrium states for which $\Delta \neq 0$, and if $\Delta > 0$, then $\sigma \neq 0$;

(2) only periodic trajectories for which $h \neq 0$, and

(3) only separatrices which do not join one saddle point to another.

By virtue of the results obtained by Bendixson [3], a system (A) satisfying the conditions Γ in the domain G can have in the domain G only trajectories of the following eleven types:

I.	Equilibrium states:	Nodes (foci) $[\Delta > 0, \sigma \neq 0]$.	1
		Saddle points $[\Delta < 0]$.	2
II.	Limit cycles	$[h \neq 0]$	3
III.	Separatrices:	Leaving a node (focus) or entering a node (focus).	4
		Unwinding from a limit cycle or tending to a limit cycle.	5
		Entering into the domain G.	6
IV.	Trajectories whose limiting cycles are either nodes (foci) or limiting cycles inside the domain G:	Leaving a node (focus) and entering a node (focus).	7
		Unwinding from a limit cycle and tending to a limit cycle.	8
		Leaving a node (focus) and tending to a limit cycle (or vice versa).	9

V. Trajectories entering the domain G that are not separatrices:	Tending to a node (focus).	10
	Tending to a limit cycle.	11

Definition II. A trajectory is said to be positively (negatively) stable in the sense of Lyapunov in the domain G [4], if, for any point M_0 $(M_0 \subset G)$, the corresponding motion $M = \varphi(t - t_0, M_0)$ satisfies the following two conditions:

(1) for all $t > t_0$ $[t < t_0], M \subset G$;

(2) for any $\epsilon > 0$, a $\delta(\epsilon, M_0)$ can be found such that for all $t > t_0$ $[t < t_0]$, $\rho[\varphi(t - t_0, M_0), \varphi(t - t_0, M'_0)] < \epsilon$ as soon as $\rho[M_0, M'_0] < \delta$.*

A trajectory is called positively (negatively) unstable in the sense of Lyapunov† in the domain G, if it satisfies the condition 1 but does not satisfy the condition 2.

Definition III. A trajectory is said to be singular, if it is positively or negatively unstable in the sense of Lyapunov.

A trajectory is said to be ordinary, if it is

(1) either positively or negatively stable in the sense of Lyapunov,

(2) or positively stable in the sense of Lyapunov and leaves the domain G when $t \to -\infty$.

Theorem IV. For a system (A) satisfying the conditions Γ in the domain G, trajectories of the type I, II, III are singular and all other trajectories are ordinary.

Theorem V. In a dynamical system (A) for which the conditions Γ are satisfied in the domain G, the totality of singular trajectories subdivides the domain G into a finite number of connected components ("cells") covered by ordinary trajectories.‡ These components fall into two classes: a class of components adjacent to a contactless cycle g and a class of internal components. In each internal component every trajectory is positively or negatively stable in the sense of Lyapunov; the boundary of each such component contains a singular trajectory positively stable in the sense of Lyapunov which is an "attraction element" or a "sink", and a

---

* Here ρ denotes Euclidean distance (Ed.).

† For the sake of brevity, we shall henceforth omit the phrase "in the domain G".

‡ Note that there are only a finite number of possible topological types of cells taken together with their boundaries.

singular trajectory negatively stable in the sense of Lyapunov which is a "repulsion element" or a "source".

In each component adjacent to a contactless cycle, every trajectory is positively stable in the sense of Lyapunov; the boundary of each such component contains a singular trajectory positively stable in the sense of Lyapunov which is a "sink".

Let there be given two dynamical systems (A_1) and (A_2) of the type (A), and accordingly two partitions of the phase plane into the trajectories S_1 and S_2. Assume that the conditions Γ are satisfied in the domains G_1 and G_2.

We say the partitions of the phase plane into trajectories S_1 and S_2 are of identical qualitative structure in the domains G_1 and G_2, if there exists a one-to-one and bicontinuous transformation T of the domain G_1 into the domain G_2, for which points of the domain G_2 lying on the same trajectory in the partition S_2 correspond to points of the domain G_1 on the same trajectory in the partition S_1, and vice versa.

By virtue of the theorem stated below, we can assert that in a system of the type (A), for which the conditions Γ are satisfied in the domain G, the qualitative structure of the partitions of these domains into trajectories is determined by the singular trajectories, the direction of motion on closed singular trajectories and by the type of stability of the limiting singular trajectories.

Theorem VI. The qualitative structures of the partitions S_1 and S_2, for which the conditions Γ are satisfied inside the corresponding contactless cycles, are identical in the domains G_1 and G_2, if and only if there exists a one-to-one and bicontinuous transformation T_1 of the domain G_1 into the domain G_2, under which points of G_2 lying on a singular trajectory of partition S_2 correspond to points of G_2 lying on a singular trajectory of partition S_1, and vice versa, and when

(1) to all sources and sinks correspond sources and sinks (or vice versa sinks and sources);

(2) the direction of rotation (in time) over all closed singular trajectories is conserved (or reversed).

Under the conditions of this theorem, points (1) and (2) can be reformulated as follows:

In each component of the partition S_1 there exists an ordinary trajectory whose points under the transformation T_1 are mapped into the points of an ordinary trajectory in the corresponding component of the partition S_2, and vice versa.

Sufficient conditions for the system (A) to be rough in the domain G are formulated in the following:

Theorem VII. If, in the domain G, the system (A) satisfies the conditions Γ, the system is then rough in the domain G.

Evidently, rough systems do exist.

Theorems IV, V, and VI relating to a system of the type (A) satisfying the conditions Γ in the domain G may be formulated for systems of the type (A) rough in the corresponding domains G.

References

1. A. Andronow, *C. R.* **189**, 559 (1929).
2. P. Bohl, *J. f. reine u. angew. Mathem.* **144**, 284 (1914).
3. I. Bendixson, *Acta Mathem.* **24**, 1 (1901).
4. A. Markoff, *Math. ZS.* **36**, 708 (1933).

Institute of Physics and
Institute of Mathematics,
Moscow State University.

Received 2 January 1937

The Classification of Continuous Mappings of a Complex into a Sphere, Communication I*

Doklady Akademii Nauk SSSR (1938) 19(3): 147–49

Two continuous mappings f and g of a topological space R into an n-dimensional sphere S_n are said to be *equivalent*, if these mappings are such that one can be transformed into the other by a continuous deformation. Thus all continuous mappings of a space R into the sphere S_n fall into equivalence classes. The determination of these classes is a problem of current interest in topology. It was completely solved by Hopf for the case in which R is an n-dimensional complex K_n. The case $R = K_r$ $(r > n)$ still awaits a solution.

In this note and in several notes to follow I shall report certain results on this classification. The present paper gives the complete classification of mappings of a sphere S_{n+1} into a sphere S_n $(n = 2, 3, \ldots)$.

First we state certain general preliminary propositions.

Two continuous mappings f and g of a space R into the sphere S_n are said to coincide on a domain $V \subset S_n$, if $f^{-1}(V) = g^{-1}(V) = U \subset R$, and for every $x \in U, f(x) = g(x)$.

(A) If two continuous mappings f and g of a space R into the sphere S_n coincide on some non-empty domain $V \subset S_n$, then these mappings are equivalent.

---

* Theorems 2′ and 2″ were reported on my behalf by S. Lefschetz at the International Mathematical Congress held in Oslo in 1936. (Communicated by Academician I. M. Vinogradov on 25 November 1938.)

(B) Let V be some spherical domain in S_n, and V' its boundary, F some closed subset of a normal space R and F' its boundary. Furthermore, let f be some continuous mapping of F into $V+V'$ such that $f(F') \subset V'$. Then there exists a continuous mapping g of the space R into S_n coinciding with f on $V \subset S_n$.

(C) Let f be a simplicial mapping of a complex K into the sphere S_n (S_n is assumed to be subdivided into simplices), V some open n-dimensional simplex in S_n and $U = f^{-1}(V)$. Then U can be represented as the topological product of the simplex V and a certain complex P, i.e., each point $z \in U$ can be expressed as (x, y) where $x \in V, y \in P$, and $f(z) = f(x, y) = x$. If K is assumed to be a manifold, then P will also be a manifold.

Propositions (A), (B), and (C) are proved trivially.

Consider the sphere S_3 as a group of quarternions and let H denote an one-parameter connected subgroup of the group S_3. Then, as can be readily seen, the totality of all right cosets of the group S_3 with respect to the subgroup H is the sphere S_2. Let φ denote the resulting mapping of the sphere S_3 onto the sphere S_2.

A mapping f of a complex K into the sphere S_n is said to be *homologous to zero*, if each n-dimensional cycle Z of K modulo arbitrary m is transformed into a zero cycle in S_n: $f(Z) = 0$.

Lemma. Let f be a mapping of the complex K into S_2 which is homologous to zero. Then there exists a continuous mapping g of the complex K into S_3 such that $f = \varphi g$, i.e., $f(x) = \varphi[g(x)]$ for every $x \in K$.

The proof is based on Proposition (C) and a complicated construction involving certain special properties of the mapping φ.

Hurewicz has shown that two mappings g and h of a complex K into S_3 are equivalent, if and only if the mappings φg and φh are equivalent. Hurewicz's theorem and the above lemma reduce the classification of zero-homologous mappings of the complex K into S_2 to the classification of all mappings of the complex K into S_3. Hence we directly obtain the classification of the mappings of S_3 into S_2.

Let f be a *simplicial* mapping of S_3 into S_2. Take any two points a and b interior to the simplices of S_2, then $f^{-1}(a)$ and $f^{-1}(b)$ are cycles in S_3. Let $v(f)$ denote the linking number of these cycles. Hopf has shown that $v(f)$ is an invariant of the class of mappings. It turns out that $v(f)$ is the only invariant of this class.

Theorem 1. Two mappings f and g of the sphere S_3 into S_2 are equivalent, if and only if $v(f) = v(g)$.

Now let us take up the case of $n \geqslant 3$.

Theorem 2′. For $n \geqslant 3$, there exist at most two classes of mappings of the sphere S_{n+1} into S_n.

Proof. Let f be a simplicial mapping of S_{n+1} into S_n. Let V denote a ball with center a in one of the n-dimensional simplices of the sphere S_n, and take $U = f^{-1}(V)$. By virtue of Proposition (C), U can be represented as the topological product of the ball V and a finite system of circles P. By modifying f in a simple manner we can transform P into a circle. Modifying f further, we can transform (a, P) into a smooth circle* in S_{n+1} and (V, y) normal to (a, P), where f maps (V, y) isometrically onto V. Modify the second mapping g of the sphere S_{n+1} into S_n in a similar manner. Since any two circles in S_{n+1} are isotopic, we can assume that (a, P) and (V, y) are coincident for both these mappings. Moreover, as the mappings f and g transform (V, y) isometrically into V, we have $f(x, y) = g[\varphi_y(x), y]$, where φ_y is a rotation of the ball V. Thus, by comparing the two mappings f and g, we obtain a family φ_y of rotations of an n-dimensional ball V, where y traverses a circle. Since there are only two distinct homotopy types of circles in the manifold of all rotations of the ball V, we find that there exist at most two classes of mappings of S_{n+1} into S_n.

Let us give one more general remark.

(D) Every continuous mapping of a sphere S_r into a sphere S_n can be approximated by an analytic transformaton. If two analytic transformations of S_r into S_n are equivalent, then there exists an analytic deformation of one mapping into the other.

The proof is based on the approximation of a function of several variables by polynomials.

Theorem 2″. For $n \geqslant 3$, there exist at least two classes of mappings of the sphere S_{n+1} into the sphere S_n.

Proof. Let f be an analytic mapping of S_{n+1} into S_n. Then there exists a

---

* I.e., a circle with a continuous tangent (Ed.).

point $a \in S_n$, such that $P = f^{-1}(a)$ is a finite system of analytic non-intersecting circles in S_{n+1}. Let β denote the number of these circles. Take n orthogonal vectors at a; with these vectors, we can, by virtue of the mapping f, put into correspondence n linearly independent vectors $u_1, u_2, \ldots, u_n$ at the point $y \in P$ orthogonal to P. Supplement this system of vectors with a vector u_0 tangent to P at y. Thus a system U_y consisting of $(n+1)$ linearly independent vectors has been defined at each point $y \in P$. At each point $z \in S_{n+1}$, take a certain system of $(n+1)$ linearly independent vectors. Let M denote the manifold of all such systems of vectors at an arbitrary point z. We can readily show that the one-dimensional Betti group of the manifold M is a cyclic group of order two. Let Z denote a one-dimensional cycle in M that is not homologous to zero. The set of all systems U_y defines in M a cycle homologous (in M) to the cycle αZ, where α is a residue modulo 2. Put $\gamma = \alpha + \beta$ modulo 2. It turns out that the residue of γ does not change under analytic deformations of the mapping f. If in the course of deformation there is no critical moment at which β suffers a change, our assertion is evident because the set of all U_y is then deformed continuously. Such critical moments of deformation call for a special, though quite elementary, consideration. To prove Theorem 2″, it is now sufficient to find two mappings under which the residue of γ takes different values: 0 or 1. This can be done easily with the help of Proposition (B).

In the next communication we shall give the classification of the mappings of the sphere S_{n+2} into S_n.

Mathematical Institute,
Moscow State University

Received 4 March 1938

The Classification of Continuous Mappings of a Complex into a Sphere, Communication II*

Doklady Akademii Nauk SSSR (1938) 19(5): 361–63

This communication gives the classification of the mappings of a sphere S_{n+2} into a sphere S_n for $n \geqslant 2$, as well as certain general relations for the mappings of a sphere S_r of dimension r into a sphere S_n of dimension n.

Theorem 1. There exist exactly two classes of mappings of a sphere S_4 into a sphere S_2.

Proof. In Communication I it was shown that the number of classes of mappings of a sphere S_4 into a sphere S_2 is equal to the number of classes of mappings of S_4 into S_3. As a consequence of the results obtained in Communication I, the latter number is equal to two.

Theorem 2. For $n \geqslant 3$, there exists only one class of mappings of a sphere S_{n+2} into a sphere S_n.

Proof. Without loss of generality, we can confine ourselves to analytic mappings of S_{n+2} into S_n (see Proposition (D) in Communication I). Let f be an analytic mapping of S_{n+2} into S_n and a be a point in S_n for which $f^{-1}(a)$ is a surface in S_{n+2} without singularities. Furthermore,

* This paper is a continuation of Communication I, *Doklady Akademii Nauk SSSR*, **XIX**, N 1–2 (1938). Theorems 1 and 2 were reported on my behalf by S. Lefschetz at the International Mathematical Congress held in Oslo in 1936. (Communicated by Academician I. M. Vinogradov on 25 March 1938.)

assume that V is a sufficiently small spherical neighbourhood of the point a. Then $U = f^{-1}(V)$ can be represented as the topological product of the ball V and a certain given surface P. Each point $z \in U$ can be represented as $z = (x, y)$, where $x \in V, y \in P$, and $f(z) = f(x, y) = x$. We can so modify the mapping f that it isometrically transforms the ball (V, y) into the ball V and that the ball (V, y) is orthogonal to the surface (a, P) for any $y \in P$.

First we consider only those mappings for which the surface P is homeomorphic to the sphere. If $n \geqslant 4$, and f, g are two mappings of the type under consideration, then since any two differentiable spheres are isotopic in S_{n+2}, for $n \geqslant 4$, we may suppose that (a, P) and (V, y) coincide for these mappings. Thus, $f(x, y) = g[\varphi_y(x), y]$, where φ_y is a certain rotation of the ball V, depending on the point $y \in P$. Since in the manifold of all rotations of the ball V, every image of a two-dimensional sphere is homotopic to zero, the mappings f and g are equivalent (see Proposition (A) in Communication I). The case of $n = 3$ requires a special complicated investigation, but the same results are obtained.

It now remains to reduce the general case to a case where P is a sphere. The surface P, as can be easily verified, is always orientable. Let L be some simple closed curve on P. Some neighbourhood of the curve L can be covered by a one-parameter family $\{L_t\}$ of simple closed curves, $-1 < t < 1, L_0 = L$. In S_{n+2} consider the family of curves $\{(x, L_t)\}, x \in V, 0 < t < 1$. This family is evidently homeomorphic to an $(n+1)$-dimensional ball V'. Putting each curve (x, L_t) of the family into correspondence with an appropriate point in V', we obtain a mapping h' of some set of S_{n+2} into V'. Assuming that the ball V' belongs to some $(n+1)$-dimensional sphere S_{n+1}, we can extend the mapping h' to a mapping of the whole sphere S_{n+2} into the sphere S_{n+1} (see Proposition (B) in Communication I). The mapping h thus obtained for transforming the sphere S_{n+2} into the sphere S_{n+1} may or may not be homotopic to zero (see Communication I). In the former case, the curve L is said to have an index equal to zero, and in the latter case, an index equal to unity. Curves which are homologous to each other on P have the same index, and when two curves are added in the sense of homology, their indices are also added modulo 2. Hence we easily infer that on P there exists a curve not homologous to zero whose index is equal to zero, of course only if the surface P is not a system of spheres. Let L be a fixed curve having the aforementioned properties. Let P' denote a certain differentiable surface in S_{n+2} homeomorphic to a disc which intersects (a, P) only along its boundary (a, L). By virtue of this construction, the mapping f can be so modified that the surface P for the new mapping has a genus which is one less than the genus of the surface P

for the initial mapping. Repeatedly applying this modification, we finally arrive at a surface P which is a finite system of spheres; and a finite system of spheres can be easily reduced to one sphere. Thus, the theorem has been demonstrated.

It is useful to introduce the operation of addition of classes of equivalent transformations of S_r into S_n, and thus transform the totality of all such classes into a group.

Definition. Let S_r' and S_r'' be two oriented spheres, and f', f'' two mappings which transform them into the sphere S_n. The mappings f' and f'' are said to be *equivalent*, if there exists a homeomorphic orientation-conserving mapping h of the sphere S_r'' into the sphere S_r' such that the mappings $f'h$ and f'' of the sphere S_r'' into the sphere S_n are equivalent. By this definition, all mappings of oriented spheres S_r into a sphere S_n fall into different equivalence classes.

We now define a group operation of addition in the system of these classes. Let A and B be two classes of mappings: $f \in A, g \in B$. Assume that f maps the sphere S_r' and g the sphere S_r''. Let V' and V'' be some spherical neighbourhoods in S_r' and S_r'', respectively. Since f and g have been defined only up to equivalence, without loss of generality, we may take $f(V') = g(V'') = a$, where a is a point in S_n. From the oriented manifolds $S_r' - V'$ and $S_r'' - V''$ we can construct an oriented sphere S_r by joining these manifolds along their boundaries in such a way that their orientations are preserved. The mappings f and g together define a map h of the sphere S_r. Let C denote the class of mappings containing the map h. It can be shown that the class C is defined uniquely by the classes A and B; hence we can take $C = A + B$. Thus we have constructed a commutative group from the classes of mappings of all possible oriented spheres of dimension r into a sphere S_n of dimension n. The zero of this group is the class of mappings homotopic to zero. If A is some class of mappings, and $f \in A$ maps the oriented sphere S_r', then the mapping f, when applied to a sphere S_r'' differing from S_r' only in orientation, belongs to a class opposite to A, that is, $-A$.

The group thus constructed is isomorphic to one of the homotopy groups constructed by Hurewicz.

The classification of the mappings of a sphere S_{n+k} into a sphere S_n can now be formulated as the problem of determining the group of mappings of S_{n+k} into S_n. Let P_k^n denote this group.

From the results established above, it follows that P_1^2 is a free cyclic group, P_1^n is a cyclic group of order 2 for $n \geqslant 3$, P_2^2 is a cyclic group of order 2, and P_2^n, for $n \geqslant 3$, contains only the identity.

There is no justification for believing that the group P_k^n is always cyclic. But the following conjecture that arises from the examples considered above is true.

Theorem 3. For $n \geqslant k+2$, the group P_k^n is isomorphic to the group $P_k^{k+2} = P_k$.

The proof is comparatively simple and is based on Propositions (D) and (A) in Communication I.

In connection with Theorem 3, we are faced with the problem of the determination of the group P_k. This problem is presumably closely related to the study of the homotopic properties of the group of orthogonal matrices.

Received 28 March 1938

Homologies in Compact Lie Groups

Matematicheskii Sbornik, Novaya Seriya (1939)
6(3): 389–422

The problem of finding topological properties of manifolds given analytically, geometrically or in some other way is very important for the topology itself as well as for its applications. In so general a formulation this problem seems to be somewhat indeterminate, since it is not quite clear precisely which manifolds are meant. For its solution in each concrete case it is necessary to point out explicitly which type of manifolds is to be considered and which topological properties are asked for. É. Cartan has proposed as such a concrete problem the question on the homological invariants of group spaces and homogeneous spaces closely connected with them. He developed a general method of dealing with this question consisting in the study of certain skew-symmetrical forms and thus transposed a geometrical question into the domain of pure algebra [1]. Cartan's method yielded, however, so far comparatively few results, since the arising algebraical problems are of considerable difficulty.

In the present paper Cartan's problem of finding homological properties of compact group manifolds is solved by a purely topological method independent from that proposed by Cartan. As it is known (cf. Section 6), the problem may be reduced to finding homological invariants of simple compact Lie groups. This question is solved only for the four fundamental series of compact Lie groups (cf. Section 2), whereas for the five singular groups it still remains open.

The method applied by me consists in defining on each group G of a real numerical function $f(x), x \in G$, such that its total differential vanishes only on the maximum G' and the minimum G''; G' is a subgroup of the group G and G'' is the co-set of the group G with respect to the subgroup G'. Further we find in the manifold G the cycle Z, the index of whose intersection with the manifold G' is equal to ± 1; if such a cycle does not exist, we find a complex X, the index of whose intersection with G' is equal to ± 1 and whose boundary lies in G''. Assuming that the topology of the group G' is sufficiently well known and using the orthogonal trajectories to the level surfaces of the function $f(x)$ and the cycle Z, I investigate the topology of the manifold G. The general devices of this method and different topological propositions relating to them form the contents of Section 1. The definitions of functions $f(x)$ on the groups under consideration are given in Section 3, and in Section 4 we construct the cycles Z and the complexes X. A complete study of the homology of the four mentioned series of simple groups is contained in Section 5, and Section 6 contains a new purely topological proof of the well known result of Cartan that the Betti numbers of locally isomorphic compact groups are equal.

The results of the present paper were published in a concise exposition in 1935 [2]. After this publication was made, Brauer published a short paper [3], in which he, knowing my results, found them anew by the application of Cartan's method.

1. Some General Remarks

In the present paragraph we shall establish some general propositions and some general notions, on ground of which we shall further below calculate the homology groups of compact group manifolds [4].

In order to avoid misunderstandings with the terminology, we restate here the following generally adopted definitions:

Let $a_0, a_1, \ldots, a_r$ be a system of $r+1$ points of a certain Euclidean space not lying in any linear space of dimensionality less than r. The minimal convex body T containing the points $a_0, a_1, \ldots, a_r$ is called the polyhedral r-dimensional simplex. The aggregate consisting of the simplex T and of its continuous map f into a certain manifold M is called the r-dimensional simplex of the manifold M, $E = (T, f)$. If the simplex T is orientated, then the simplex E is also called orientated. Two simplexes $E = (T, f)$ and $E' = (T, f')$ of the manifold M are considered to be coinciding, if there exists such a non-degenerated affine superposition of the simplexes T and T' on each other that the functions f and f' coincide. If the simplexes E and E' are orientated, then they are considered to be coinciding, if the described superposition preserves the orientation. A linear form of the type $C = c_1E_1 + c_2E_2 + \ldots + c_nE_n$, where $c_1, c_2, \ldots, c_n$ are integers and $E_1, E_2, \ldots, E_n$ are r-dimensional orientated simplexes from M is called an r-dimensional algebraical complex from M. By the boundary of an orientated r-dimensional simplex E is meant the sum of its $(r-1)$-dimensional orientated faces mapped into M by means of f. Thus to each orientated r-dimensional simplex from M and hence, in virtue of additiveness, to each algebraical r-dimensional complex C from M is correlated an algebraical $(r-1)$-dimensional complex $\dot{C}$ called the boundary of the original complex. An algebraical complex from M is called a cycle, if its boundary is equal to zero. As is known, the boundary of every algebraical complex from M is a cycle. A cycle Z which is the boundary of a certain complex is said to be homologic to zero, in symbols: $Z \sim 0$. Two cycles X and Y, the difference of which is homologic to zero are said to be homologic to each other, in symbols: $X \sim Y$. A cycle W is said to be weakly homologic to zero, if there exists a positive integer q such that $qW \sim 0$; then we write: $W \approx 0$. The cycles U and V are said to be weakly homologic to each other, if their difference is weakly homologic to zero, in symbols: $U \approx V$.

If the manifold M is divided into simplexes forming the complex K, then the simplexes of the complex K may be considered as affine nondegenerated maps of polyhedral simplexes and are thus simplexes from M. We shall call the algebraical complexes of the manifold, M formed from such simplexes the polyhedral algebraical complexes from K.

Let C be an r-dimensional algebraical complex from M. Then there exist such an r-dimensional polyhedral complex C^* from K and such an r-dimensional complex C_* from M that C^* is situated in the neighbourhood of C and C_* – in the neighbourhood of $\dot{C}$ and, moreover, the boundary of the complex C^* is equal to $\dot{C}$–$\dot{C}^*$ and the cycle C–C_*–C^* is the homologic to zero in the neighbourhood of C. The proximity which may be here achieved depends on the smallness of the simplexes of the complex K. The complex C^* is said to be polyhedrally approximating the complex C.

Let K' be the barycentric subdivision of the complex K. Every barycentric star of the complex K is a polyhedral algebraical complex from K'. We shall call a linear form of the type $c_1B_1 + c_2B_2 + \ldots + c_nB_n$, where $c_1, c_2, \ldots, c_n$ are integers and $B_1, B_2, \ldots, B_n$ are r-dimensional barycentric stars of the complex K the r-dimensional star complex from K.

If C is an r-dimensional polyhedral complex from K', then there exist such an r-dimensional star complex C^* from K and such an r-dimensional polyhedral complex C_* from K' that C^* is situated in the neighbourhood of C and C_* – in the neighbourhood of $\dot{C}$ and, moreover, the boundary of the complex C_* is equalt to $\dot{C}$–$\dot{C}^*$ and the cycle C–C^*–C_* is homologic to zero in the neighbourhood of C. The proximity which may be achieved here depends on the smallness of the simplexes of the complex K. The complex C^* is said to be star-approximating the complex C.

If C is a complex from M, then approximating it first by a polyhedral complex from K' and then by a star complex from K we obtain a star approximation of an arbitrary complex C from the manifold M by means of a star complex from K.

Lemma 1. Let M be an orientable compact manifold and N its orientable closed sub-manifold situated in it without self-intersections. Suppose that there exists such a division of the manifold M into simplexes forming a complex K that the sub-manifold N represents a sub-complex L of the complex K. Let further X be a certain R-dimensional algebraical complex from M whose boundary does not meet N and whose simplexes are sufficiently small. If the algebraical intersection of the complex X with the manifold N is homologic to zero in N, then there exists in any given neighbourhood U of the manifold N such an $(r+1)$-dimensional complex Y from M that the complex $X-Y$ already does not meet N.

Proof. Let V be such a neighbourhood of the manifold N, the closure of which lies in U. Denote by C the algebraical complex formed of all simplexes of the complex X lying in V. If the simplexes of the complex X are sufficiently small, then the complex $X-C$ already does not meet N. Denote by C^* the star approximation of the complex C· If the simplexes of the complex K are sufficiently small, then the complex C_*, with the boundary $\dot{C}-\dot{C}^*$, lying in the neighbourhood of $\dot{C}$ does not meet N and the complex D, with the boundary $C-C^*-C_*$, lies in U. The complex $X'=X-\dot{D}$ is representable in the form $X'=X-\dot{D}-C^*+C^*$. It is easily seen that the complex $X-\dot{D}-C^*$ does not meet N. Denote by X_2 the star complex formed of all stars of the complex C^* meeting N; then $X'=X_1+X_2$, X_1 does not meet N and X_2 consists of stars meeting N.

Put $X_2=d_1B_1+d_2B_2+\ldots+d_nB_n$, where $d_1,d_2,\ldots,d_n$ are integers and $B_1,B_2,\ldots,B_n$ – barycentric stars of the complex K, which meet L. Denote by T_j the simplex from K dual to the star B_j, i.e. such that the index of intersection of T_j with B_j in M is equal to $+1$. Since B_j intersects with L, T_j must belong to L. Denote by $\bar{B}_j$ the barycentric star of the complex L dual to the simplex T_j, i.e., such that the index of intersection of T_j with $\bar{B}_j$ in N is equal to $+1$. The cycle $\bar{X}=d_1\bar{B}_1+d_2\bar{B}_2+\ldots+d_n\bar{B}_n$ is homologic to N to the algebraical intersection of X with N and hence is homologic to zero in N. Let $\bar{Y}=c_1\bar{A}_1+c_2\bar{A}_2+\ldots+c_m\bar{A}_m$ be that star complex from L, the boundary of which is equal to $\bar{X}$. Denote by S_i the simplex from L dual to the star $\bar{A}_i$ in L and by A_i the barycentric star of the complex K dual to the simplex S_i. Put $Y' = c_1A_1 + c_2A_2 + \ldots + c_mA_m$. It is easily seen that the complex formed from all stars entering into the boundary $\dot{Y}'$ of the complex Y and meeting L coincides with X_2. Thus the complex $X' - \dot{Y}' = X_1 + X_2 - \dot{Y}'$ does not meet N. Passing from the complex X' to the complex X we obtain that $X - \dot{D} - \dot{Y}' = X' - \dot{Y}' = X - \dot{Y}$ does not meet N, while the complex $Y = D + Y'$ lies completely in U. Lemma 1 is thus proved.

Lemma 2. Let M be a compact orientable differentiable manifold, in which it is possible to introduce a positive nowhere degenerated Riemannian metric, and $f(x)$ – a real numerical differentiable function defined on $M, x \in M$. Suppose that the function f attains its maximum on a differentiable orientable sub-manifold M' of the manifold M and its minimum on a differentiable orientable sub-manifold M'' of the manifold M. Suppose, further, that the total differential of the function f vanishes only on the manifolds M' and M''. If X is an algebraical complex from M, the boundary of which lies in M'' and the algebraical intersection of which with M' is homologic to zero in M', then there exists such a complex X' lying in M'' that $X - X'$ is a cycle homologic to zero in M.

Proof. We shall assume that a positive nowhere degenerated Riemannian metric is already introduced in M. Denote by W a neighbourhood of the manifold M'' in the manifold M so small that through every point $x \in W - M''$ there passes one and only one geodetic normal to the manifold M''. Let now the point x uniformly travel along the geodetic normal from the position x to the foot of the normal on M'' with such a velocity that the whole way is traversed in a unit of time. We have thus defined a continuous deformation φ of the neighbourhood W on M'', at which every point belonging to M'' remains fixed.

Let d be the minimum of the function f. Denote by ϵ a positive number so small that from $f(x) \leqslant d + \epsilon$ follows $x \in W$.

In virtue of Lemma 1 there exists such a complex Y situated in $M - M''$ that $X - \dot{Y} = X_1$ already does not meet M'. Let c be the maximum of the function f. Denote by δ a positive number so small that from $f(x) > c - \delta$ follows that x does not belong to X_1. In virtue of the known Morse's construction [5] there exists a continuous deformation ψ of the set of all x, for which $f(x) \leqslant c - \delta$, into the set of all x, for which $x \leqslant d + \epsilon$, i.e., into W, such that during this deformation all points from M'' remain fixed. The deformation ψ transforms the complex X_1 into a complex X_2 situated in W. Applying to the complex X_2 the deformation φ, we obtain a complex X' satisfying the conditions of Lemma 2. Lemma 2 is thus proved.

We recall now a number of known conceptions from the theory of homologies.

If X is a cycle from the manifold M not weakly homologic to zero, then it is said that it is free, or that it has the order zero. If a cycle Y from the manifold M is weakly homologic to zero, then its order is said to be the minimal natural number q such that $qY \sim 0$.

A system $\Delta = \{Z_1, Z_2, \ldots, Z_k\}$ of r-dimensional cycles of the manifold M is said to be independent in the sense of homologies, if it does not contain cycles homologic to zero and if from the relation

$$c_1 Z_1 + c_2 Z_2 + \ldots + c_k Z_k \sim 0$$

follows

$$c_1 Z_1 \sim 0, \quad c_2 Z_2 \sim 0, \ldots, c_k Z_k \sim 0.$$

A system Δ is called the r-dimensional basis of the manifold M, if for every r-dimensional cycle Z from M we have the relation

$$Z \sim a_1 Z_1 + a_2 Z_2 + \ldots + a_k Z_k,$$

where $a_1, a_2, \ldots, a_k$ are properly chosen integers. A system Δ is called the r-dimensional independent basis of the manifold M, if it is independent and forms a basis.

A similar definition is introduced also for weak homologies.

A system Ω of cycles of different numbers of dimensions of the n-dimensional manifold M is said to be independent, if the aggregate of all r-dimensional cycles of the system Ω is independent for any $r = 0, 1, \ldots, n$. The notion of the independent basis of the manifold M is defined similarly.

If Σ and Ω are two systems of cycles from M, then by $\Sigma \vee \Omega$ we shall denote the aggregate of all cycles entering into the systems Σ and Ω, i.e., the sum of the systems Σ and Ω.

In what follows I shall take for granted the notion of the index of intersection $I_M(X, Y)$ of two algebraical complexes X and Y of dimensionalities r and $n-r$ from an n-dimensional orientable manifold M. If X and Y are cycles, then the index of their intersection is an invariant of weak homologies and upt to the sign does not depend on the order in which the cycles are taken. If X and Y are two cycles of dimensionalities r and $n-r-1$ from an n-dimensional orientable manifold M and X and Y are weakly homologic to zero, i.e. have finite orders p and q, then we define their linkage coefficient $V_M(X, Y)$ in the following way: let C be a complex with the boundary qY; then we put $V_M(X, Y) = (1/q)I_M(X, C)$ and consider only the fractional part of $V_M(X, Y)$. The so defined linkage coefficient is an invariant of homologies and up to the sign does not depend on the order in which the cycles are taken.

The following theorem of Poincaré–Veblen will be repeatedly applied in the sequel:

Poincaré–Veblen's Theorem: Let M be an n-dimensional orientable manifold, $X_1, X_2, \ldots, X_k$ its r-dimensional independent basis of weak homologies and $Y_1, Y_2, \ldots, Y_l$ its $(n-r)$-dimensional independent basis of weak homologies. If we put $a_{ij} = I_M(X_i, Y_j)$, then the matrix $\|a_{ij}\|$ is a square matrix and its determinant is equal to ± 1.

From the above it follows that in order that an r-dimensional cycle X from M should not be weakly homologic to zero, it is necessary and sufficient that there should exist in M and $(n-r)$-dimensional cycle Y such that $I_M(X, Y) \neq 0$. For cycles weakly homologic to zero independently from what has been said we have: in order that a weakly homologic to zero r-dimensional cycle X of an n-dimensional orientable manifold M should not be homologic to zero, it is necessary and sufficient that there should exist in M a weakly homologic to zero $(n-r-1)$-dimensional cycle Y such that $V_M(X, Y) \neq 0$.

Observe that if G is a Lie group, the manifold G is orientable and, consequently, Lefshetz's theory of intersections is applicable to it. In fact, in order to define an orientation in any point of the manifold G it suffices to define it in the point e – the unity of the group G – and then transfer it to any point by multiplication of all elements of the group on the right by that element a, to which the orientation must be transferred.

Definition 1. Let G be a Lie group and $E = (T, f)$ and $E' = (T', f')$ – two orientated simplexes of the manifold G. Here in T and T' are orientated polyhedral simplexes and f and f' their maps in G. Denote the properly orientated topological product of the simplexes T and T' by P. Then P represents an orientated convex body which – if we divide it into simplexes – may be considered as an algebraical complex. Every element $z \in P$ is represented in the form of a pair $z = (x, x')$, where $x \in T$ and $x' \in T'$. We define now a continuous mapping h of the body P into G depending on the mappings f and f'. Put $h(z) = f(x)f'(x')$, where on the right-hand side stands the group product. We shall call the so obtained algebraical complex (P, h) the group product of the simplexes E and E' and denote it by EE'. Let $X = a_1E_1 + a_2E_2 + \ldots + a_mE_m$ and $Y = b_1E_1' + b_2E_2' + \ldots + b_nE_n'$ be two algebraical complexes of dimensionalities r and s from G. Their group product XY we define as

$$XY = \sum_{i=1}^{m} \sum_{j=1}^{m} a_i b_j E_i E_j'.$$

Thus the group product XY of two algebraical complexes X and Y from G of dimensionalities r and s is an algebraical complex XY of dimensionality $r+s$ from G.

It may be directly verified that the boundary $\dot{Z}$ of the group product $Z = XY$ is given by the formula

$$\dot{Z} = \delta \dot{X} Y + \epsilon X \dot{Y}, \quad \delta = \pm 1, \quad \epsilon = \pm 1, \tag{1}$$

where the values of the numbers δ and ϵ depend on the dimensionalities r and s.

Thus the group product of two cycles is a cycle and the group product of two cycles of which one is homologic to zero is a cycle homologic to zero.

Let us now state some important although geometrically obvious relations connecting the group product and the algebraical intersection of complexes from a group.

Let G be a Lie group, G' its subgroup, C an algebraical complex from G' situated in G' in a general position and, finally, W an algebraical complex from G meeting G' only in the one point e – the unity of the group, such that the index of intersection of G' with W is equal to $+1$. Then it is geometrically obvious that the algebraical intersection of the complex CW with the manifold G' is either equal to C or differs from C only in the sign:

$$G' \times CW = \delta C, \quad \delta = \pm 1, \tag{2}$$

where δ depends on the dimensionalities of the manifolds G and G' and also on the dimensionality of the complex C.

If W is a cycle and X and Y are two cycles from G' such that the index of their intersection in G' is defined, then from the relation (2) it directly follows that

$$I_G(X, YW) = \epsilon I_{G'}(X, Y), \quad \epsilon = \pm 1, \tag{3}$$

where ϵ depends on the dimensionalities of the complexes under consideration.

Again, if W is a cycle and X and Y are two cycles from G' such that their linkage coefficient is defined, then from the relation (2) it directly follows that

$$V_G(X, YW) = \epsilon' V_{G'}(X, Y), \quad \epsilon' = \pm 1, \tag{4}$$

where ϵ' depends on the dimensionalities of the complexes under consideration.

The following Lemma 3 plays an important role in the study of homologies in group manifolds:

Lemma 3. Let G be a compact Lie group and G' one of its subgroups. Suppose that there exists in G a cycle W meeting G' only in the one point e – the unity of the group – and such that the index of intersection of the cycles G' and W is equal to ± 1. If Ω is a system of cycles from the manifold G' independent in G', then the system $\Omega(e \vee W)$ is independent in G. If the system Ω is independent in the sense of weak homologies, then the system $\Omega(e \vee W)$ is independent also in the sense of weak homologies. Further, $\Omega(e \vee W) = \Omega \vee \Omega W$, where the system ΩW is composed of all cycles of the form ZW, $Z \in \Omega$ (cf. Definition 1). Moreover, if the cycle $Z \in \Omega$ has in G' the order r, then the cycles Z and ZW in G have the same order r.

Proof. Using the relations (3) and (4) and also Poincaré–Veblen's theorem we shall show that if X is a cycle from G' not homologic to zero in G', then the cycles X and XW are not homologic to zero in G.

Suppose that the cycle X is not even weakly homologic to zero in G'. Then there exists a cycle Y from G' such that $I_{G'}(X, Y) \neq 0$. But from this in virtue of the relation (3) we obtain $I_G(X, YW) \neq 0$, $I_G(XW, Y) \neq 0$, and this means that the cycles X and XW are not weakly homologic to zero in G.

Suppose that the cycle X is weakly homologic to zero in G', but not homologic to zero. Then there exists such a cycle Y that $V_{G'}(X, Y) \neq 0$, but from this in virtue of the relation (4) it follows that $V_G(X, YW) \neq 0$, $V_G(XW, Y) \neq 0$, and this means that the cycles X and XW are not homologic to zero in G.

Let us now show that if X and Y are cycles from G', then the relation $X \sim YW$ in G is possible only under the condition $X \sim Y \sim 0$ in G'.

Let a be an element of the group G not belonging to G' and such that a may be connected with e by a continuous path in G. Then $X \sim Xa$ in G. If $YW - X \sim 0$ in G, then also $YW - Xa \sim 0$ in G, but the intersection of a cycle homologic to zero in G with the manifold G' is homologic to zero in G'. On the other hand, the intersection of the cycle $YW - Xa$ with G is equal to $\pm Y$ [cf. (2)]. Thus $Y \sim 0$ in G' and, consequently, YW is also homologic to zero.

From what has been proved directly follows that the system of cycles $\Omega \vee \Omega W$ is independent.

Lemma 3 is thus proved.

Lemma 4. Let G be a connected compact Lie group, G' one of its subgroups and G'' the co-set of the group G with respect to the subgroup G'. Every cycle Z'' from G'' is homologic in G to a certain cycle Z' from G'.

Proof. For the sake of definiteness let us assume that G'' is a right co-set, i.e., that $G'' = G'a$. Since the group G is connected, there exists a path $x(t)$ in G connecting the point e with the point a^{-1}, $x(0) = e, x(1) = a^{-1}$. When t varies from 0 to 1 the cycle $Z''x(t)$ is deformed from the position Z'' into the position $Z''a^{-1} \subset G'$. Thus the cycle Z'' is homologic to the cycle $Z' = Z''a^{-1}$ and Lemma 4 is proved.

Theorem 1. Let G be a compact connected Lie group and $f(x)$, $x \in G$, a real numerical differentiable function defined on it. Suppose that the function f attains its maximum on the subgroup G' of the group G and its minimum on the co-set G'' of the group G with respect to the subgroup G' and that the total differential of the function f vanishes only on the manifolds G' and G''. Suppose, further, that in G there exists a cycle W meeting G' only in e and such that the index of intersection of the cycles G' and W is equal to ± 1. If Ω is the independent basis of homologies of the manifold G'. Then the system of cycles $\Omega(e \vee W)$ is the independent basis of homologies of the manifold G. If Ω is the independent basis of weak homologies, then $\Omega(e \vee W)$ is also the independent basis of weak homologies.

Proof. In virtue of Lemma 3 $\Omega(e \vee W)$ is an independent system of cycles from G. We shall show that this system forms a basis of the manifold G. Let Z be an arbitrary cycle from G. Denote by Y the algebraical intersection of the manifold G' with the cycle Z. The algebraical intersection of the manifold G' with the cycle YW is, in virtue of the relation (2), equal to δY. Thus the algebraical intersection of the cycle $Z - \delta YW$ with G' is homologic to zero in G' and, consequently, in virtue of Lemma 2 the cycle $Z - \delta YW$ is homologic to a certain cycle from G''. But this last cycle is, in virtue of Lemma 4, homologic to a certain cycle X from G'. Hence $Z \sim X + \delta YW$ and Theorem 1 is proved.

Corollary. If the independent basis of homologies of the manifold G' consists of free cycles, i.e., if the manifold G' has no torsions, then the manifold G too has no torsions.

Proof. If the independent basis Ω of the manifold G' consists only of free cycles, then it is simultaneuously the independent basis of weak homologies. But then the system $\Omega(e \vee W)$, being the independent basis of homologies of the manifold G, is for it simultaneously also the independent basis of weak homologies. Thus the manifold G has no torsions.

In the following paragraphs we shall construct for the concrete simple groups of the four fundamental series the functions and the cycles which give us the possibility to apply to these groups the Theorem 1.

2. The Groups A_n, B_n, C_n, D_n [6]

In the present paragraph we shall define the four fundamental series of compact simple Lie groups in the matrix form and discuss their finite as well as infinitesimal structure.

If G is a Lie group, we shall denote by **G** its infinitesimal group. Similarly we shall denote the matrices and their elements relating to the group **G** by the same letters as for the group G, but underlined. If x is a complex number, we shall denote by $\bar{x}$ its conjugated. If a is a square matrix, we shall denote by Da its determinant.

The Groups B_n and D_n. In order to define the groups B_n and D_n we put

$$B_n = H_{2n+1}, \quad D_n = H_{2n}. \tag{1}$$

The group H_m we define as the set of all orthogonal matrices of the order m with a positive determinant. Thus $h \in H_m$ is a matrix $\|h_{ij}\|$, $(i,j) = 1, \ldots, m$, with real elements satisfying the following conditions:

$$\sum_{k=1}^{m} h_{ik}h_{jk} = \delta_{ij}, \tag{2}$$

$$\sum_{k=1}^{m} h_{ki}h_{kj} = \delta_{ij}, \tag{3}$$

$$\mathrm{D}\,\|h_{ij}\| = +1. \tag{4}$$

The conditions (2) and (3) are, as is known, equivalent to each other. In the infinitesimal form the conditions (2) and (3) may be written as follows:

$$\mathbf{h}_{ij} + \mathbf{h}_{ji} = 0. \tag{5}$$

The condition following from (4) is automatically satisfied in virtue of the condition (5).

The group A_n. The group A_n we define as the set of all unitary unimodular matrices of the order $n + 1$. Thus $a \in A_n$ is a matrix $\|a_{ij}\|$, $(i,j) = 0, \ldots, n$, with complex elements satisfying the conditions:

$$\sum_{k=0}^{n} a_{ik}\bar{a}_{jk} = \delta_{ij}, \tag{6}$$

$$\sum_{k=0}^{n} a_{ki}\bar{a}_{kj} = \delta_{ij}, \tag{7}$$

$$\mathrm{D}\,\|a_{ij}\| = +1. \tag{8}$$

The conditions (6) and (7) are, as is known, equivalent to each other and mean that the transposed conjugated matrix $\bar{a}^*$ is inverse to the matrix a,

$$a^{-1} = \bar{a}^*. \tag{9}$$

In the infinitesimal form these conditions may be written as follows:

$$\mathbf{a}_{ij} + \bar{\mathbf{a}}_{ji} = 0, \tag{10}$$

$$\sum_{k=0}^{n} \mathbf{a}_{kk} = 0; \tag{11}$$

the conditions (6) and (7) are expressed by (10) and the condition (8) – by (11).

The group C_n. The group C_n we define as the set of all unitary unimodular matrices of the order $2n$, which leave invariant the bilinear form

$$\sum_{i,j=1}^{2n} f_{i,j}x_iy_j, \tag{12}$$

where $f_{p,n+p} = -f_{n+p,p} = 1$, $p = 1, \ldots, n$, and all other $f_{i,j}$ are equal to zero.

Thus $c \in C_n$ is a matrix $\|c_{ij}\|$ with complex elements satisfying the conditions:

$$\sum_{k=1}^{2n} c_{ik}\bar{c}_{jk} = \delta_{ij}, \tag{13}$$

$$\sum_{k=1}^{2n} c_{ki}\bar{c}_{kj} = \delta_{ij}, \tag{14}$$

$$\mathrm{D}\,\|c_{ij}\| = +1, \tag{15}$$

$$\sum_{i,j=1}^{2n} f_{i,j}c_{ik}c_{jl} = f_{k,l}. \tag{16}$$

In order to give the condition (16) a more comprehensible form, we put

$$C_{pq}^{\delta\epsilon} = c_{\delta n+p,\,\epsilon n+q}, \quad \delta = 0, 1, \quad \epsilon = 0, 1, \quad (p, q) = 1, \ldots, n. \tag{17}$$

Then the condition (16) may be replaced by the conditions:

$$\sum_{r=1}^{n} c_{rp}^{00}c_{rq}^{10} - c_{rq}^{00}c_{rp}^{10} = 0, \tag{18}$$

$$\sum_{r=1}^{n} c_{rp}^{01}c_{rq}^{11} - c_{rq}^{01}c_{rp}^{11} = 0, \tag{19}$$

$$\sum_{r=1}^{n} c_{rp}^{00}c_{rq}^{11} - c_{rp}^{10}c_{rq}^{01} = \delta_{pq}. \tag{20}$$

Since C_n is a group, it contains together with c also c^{-1}. In virtue of the conditions (13) and (14) $c^{-1} = \bar{c}^*$ [cf. (9)]. Writing out the conditions (18), (19) and (20) for the matrix $\bar{c}^*$ and omitting in them the dash, we obtain the conditions:

$$\sum_{r=1}^{n} c_{pr}^{00}c_{qr}^{01} - c_{qr}^{00}c_{pr}^{01} = 0, \tag{21}$$

$$\sum_{r=1}^{n} c_{pr}^{10}c_{qr}^{11} - c_{qr}^{10}c_{pr}^{11} = 0, \tag{22}$$

$$\sum_{r=1}^{n} c_{pr}^{00}c_{qr}^{11} - c_{pr}^{01}c_{qr}^{10} = \delta_{pq}, \tag{23}$$

whicn, in virtue of the conditions (13) and (14), are equivalent to the conditions (18), (19) and (20).

In the infinitesimal form the conditions (13) and (14) may be written as follows:

$$\mathbf{c}_{ij} + \bar{\mathbf{c}}_{ji} = 0. \tag{24}$$

The conditions (18), (19), (20) coinciding with the conditions (21), (22), (23) obtain the form:

$$\mathbf{c}_{pq}^{10} - \mathbf{c}_{qp}^{10} = 0, \tag{25}$$

$$\mathbf{c}_{pq}^{01} - \mathbf{c}_{qp}^{01} = 0, \tag{26}$$

$$\mathbf{c}_{pq}^{00} + \mathbf{c}_{qp}^{11} = 0. \tag{27}$$

The infinitesimal condition implied by the condition (15) follows from the condition (27).

As is well known, all compact non-commutative simple groups with the exception of five isolately standing groups fall up to a local isomorphism into four infinite series

$$\left.\begin{aligned} A_n, B_n, C_n, n = 1, 2, \ldots, D_n, n = 3, 4, \ldots, \\ A_1 \doteq B_1 = C_1, B_2 = C_2, A_3 = D_3. \end{aligned}\right\} \tag{28}$$

Above we have defined these series in the form of groups of matrices. Observe that the so defined groups A_n and C_n are simply connected, whereas the groups B_n and D_n are doubly connected. We also note that the group $D_2 \doteq H_4$ is not contained in any of the series because it is not simple, but only semi-simple; however, this circumstance will not prevent us to study its topological properties, which is the more opportune as the series of groups H_m will be studied inductively.

Finally, let us point out an important topological property of the constructed groups, which we state without proof:

$$\text{The groups } A_n, B_n, C_n, D_n \text{ are connected.} \tag{29}$$

The dimensionalities of the groups may be easily calculated from the infinitesimal relations. If we denote by RG the dimensionality of the group G, then we find:

$$\left.\begin{aligned} RH_m &= \tfrac{1}{2}m(m-1), \quad RB_n = 2n^2 + n, \quad RD_n = 2n^2 - n, \\ RA_n &= n^2 + 2n, \quad RC_n = 2n^2 + n. \end{aligned}\right\} \tag{30}$$

3. Auxiliary Functions on Groups

In this paragraph we shall construct on the groups considered in Section 2 auxiliary real numerical functions and shall consider the critical points of these functions, i.e., those elements of the group in which the total differential of the function vanishes.

Let G be the group of all square, real or complex, matrices of the order m with a non-vanishing determinant and $f(z)$, $z \in G$, a real numerical function defined on G. If G consists of real matrices, then for the coordinates of the element $z \in G$ we shall take the elements of the matrix $z = \|z_{ij}\|$, i.e., the numbers z_{ij}. If G consists of complex matrices, then for the coordinates of the element $z \in G$ we shall take the elements of the matrix $z = \|z_{ij}\|$ and their conjugated, i.e., the numbers z_{ij} and $\bar{z}_{ij}$. In the first case $f(z)$ is a function of the variables z_{ij} and in the second case – of the variables $z_{ij}, \bar{z}_{ij}$.

Let us calculate the total differential $df(z)$ of the function $f(z)$ at the point $z = a$. To this end we introduce an auxiliary matrix x defining it by the equation $z = ax$. In the case of areal group G we compute the total differential of z_{ij} through the differentials of the variables x_{ij} at $x = e$. In the case of a complex group we compute the total differentials of z_{ij} and $\bar{z}_{ij}$ through the differentials of the variables x_{ij} and $\bar{x}_{ij}$ at $x = e$. It is easily seen that in the case of real groups we have

$$dz_{ij} = \sum_{k=1}^{m} a_{ik} dx_{kj}.$$

In the case of complex groups we have

$$dz_{ij} = \sum_{k=1}^{m} a_{ik} dx_{kj}, \quad d\bar{z}_{ij} = \sum_{k=1}^{m} \bar{a}_{ik} d\bar{x}_{kj}.$$

Observe that $\|dx_{ij}\|$ at $x = e$ is nothing else but the infinitesimal matrix, i.e., the element of the infinitesimal group $\mathbf{G}$ of the Lie group G. Therefore, according to the denotations of Section 2, it is natural to put $dx_{ij} = \mathbf{x}_{ij}$. If we denote the elements of the matrix $a\mathbf{x}$ by $(a\mathbf{x})_{ij}$, then the obtained formulae take the form

$$dz_{ij} = (a\mathbf{x})_{ij};$$

$$dz_{ij} = (a\mathbf{x})_{ij}, \quad d\bar{z}_{ij} = (\overline{a\mathbf{x}})_{ij}.$$

The total differential of the function $f(z)$ at $z = a$ takes in the real and in the complex case respectively the form

$$df(z) = \sum_{i,j=1}^{m} \frac{\partial f(z)}{\partial z_{ij}} (a\mathbf{x})_{ij} \quad \text{at } z = a, \tag{1}$$

$$df(z) = \sum_{i,j=1}^{m} \frac{\partial f(z)}{\partial z_{ij}} (a\bar{x})_{ij} + \frac{\partial f(z)}{\partial \bar{z}_{ij}} (\overline{a\mathbf{x}})_{ij} \quad \text{at } z = a. \tag{2}$$

Suppose now that $f(z)$ is a linear function of the variables z_{ij} in the case of a real group and a linear function of the variables z_{ij}, $\bar{z}_{ij}$ in the case of a complex group. Then the relations (1) and (2) take the form

$$df(z) = f(a\mathbf{x}) \quad \text{at } z = a. \tag{3}$$

If H is a subgroup of the group G and we wish to investigate the behaviour of the function $f(z)$ on the subgroup H, then we must observe that the elements of the infinitesimal matrices $\|\mathbf{x}_{ij}\|$ satisfy certain linear relations. If G is a real group and $H = H_m$ (cf. Section 2), then these relations have the form (5), Section 2. If G is a complex group and $m = n + 1$, $H = A_n$, then these relations have the form (10), (11), Section 2. If G is a complex groups and $m = 2n$, $H = C_n$, then, these relations have the form (24)–(27), Section 2.

We proceed now to the definition of the auxiliary functions on the groups $(B_n, D_n) = H_n, A_n, C_n$.

The Group H_m. On the group H_m (cf. Section 2) we define the function $f(h)$, $h \in H_m$, by putting $f(h) = h_{11}$, $h = \|h_{ij}\|$, $(i,j) = 1, \ldots, m$. The total differential of the function $f(h)$ on the group H_m vanishes for those and only those $h \in H_m$, for which $h_{11} = \pm 1$. We denote the set of those. $h \in H_m$ for which $h_{11} = +1$ by H'_{m-1} and the set of those $h \in H_m$ for which $h_{11} = -1$ by H''_{m-1}. Then the set H'_{m-1} is a subgroup of the group

H_m isomorphic to the group H_{m-1} and on it the function $f(h)$ attains its maximum $+1$. The set H''_{m-1} is a co-set of the group H_m with respect to the subgroup H'_{m-1}, and on it the function $f(h)$ attains its minimum -1.

In virtue of the relation (3) we have

$$df(h) = \sum_{i=1}^{m} h_{1i}\mathbf{h}_{i1}, \tag{4}$$

The infinitesimal matrix $\|h_{ij}\|$ satisfies the condition (5), Section 2. For the elements $\mathbf{h}_{11}, \ldots, \mathbf{h}_{m1}$ entering into (4) this gives only $\mathbf{h}_{11} = 0$, whereas the remaining $\mathbf{h}_{21}, \ldots, \mathbf{h}_{m1}$ are arbitrary. Hence from $df(h) = 0$ we obtain

$$h_{12} = \ldots = h_{1m} = 0. \tag{5}$$

Now in virtue of (2), Section 2 it follows that

$$h_{11} = \pm 1 \tag{6}$$

and then, in virtue of (3), Section 2, we obtain

$$h_{21} = \ldots = h_{m1} = 0. \tag{7}$$

Thus $df(h)$ vanishes only when $h_{11} = \pm 1$, wherefrom, in virtue of the relations (2) and (3), Section 2, it follows that the relations (5) and (7) are satisfied, i.e., in particular, that $df(h) = 0$. The other stated properties of the sets H'_{m-1} and H''_{m-1} are obvious.

The Group A_n. On the group A_n (cf. Section 2) we define the function $f(a)$, $a \in A_n$, by putting $f(a) = \frac{1}{2}(a_{00} + \bar{a}_{00})$, $a = \|a_{ij}\|$, $(i, j) = 0, \ldots, n$. The total differential of the function $f(a)$ vanishes for those and only for those elements $a \in A_n$, for which $a_{00} = \pm 1$. Denote the set of those $a \in A_n$ for which $a_{00} = +1$ by A'_{n-1} and the set of those $a \in A_n$ for which $a_{00} = -1$ by A''_{n-1}. Then the set A'_{n-1} is a subgroup of the group A_n isomorphic to the group A_{n-1} and on it the function $f(a)$ attains its maximum $+1$. The set A''_{n-1} is the co-set of the group A_n with respect to the subgroup A'_{n-1} and on it the function $f(a)$ attains its minimum -1.

In virtue of the formula (3) we have

$$df(a) = \sum_{i=0}^{n} \tfrac{1}{2}(a_{0i}\mathbf{a}_{i0} + \bar{a}_{0i}\bar{\mathbf{a}}_{i0}). \tag{8}$$

The infinitesimal matrix $\|\mathbf{a}_{ij}\|$ satisfies the conditions (10) and (11), Section 2. For the $\mathbf{a}_{00}, \ldots, \mathbf{a}_{n0}$ entering into (8) herefrom follows only that $\mathbf{a}_{00} + \bar{\mathbf{a}}_{00} = 0$. The remaining $\mathbf{a}_{10}, \ldots, \mathbf{a}_{n0}$ are independent. Thus from $df(a) = 0$ follows

$$a_{00} - \bar{a}_{00} = 0, \quad a_{01} = \ldots = a_{0n} = 0. \tag{9}$$

Hence, in virtue of the relations (6), Section 2, we obtain

$$a_{00} = \pm 1, \tag{10}$$

and hence it follows that

$$a_{10} = \ldots = a_{n0} = 0. \tag{11}$$

Thus $df(a)$ vanishes only for those a, for which $a_{00} = \pm 1$. But if $a_{00} = \pm 1$, then in virtue of the relations (6) and (7), Section 2, the equations (9) and (11) hold, i.e., in particular, $df(a) = 0$. Thus $df(a)$ vanishes then and only then when $a_{00} = \pm 1$. The other formulated properties of the sets A'_{n-1} and A''_{n-1} are obvious.

The Group C_n. On the group C_n (cf. Section 2) we define the function $f(c)$, $c \in C_n$, by putting

$$f(c) = \tfrac{1}{2}(c_{11} + \bar{c}_{11}), \quad c = \|c_{ij}\|, \quad (i,j) = 1, \ldots, 2n.$$

The total differential of the function $f(c)$ vanishes for those and only for those $c \in C_n$, for which $c_{11} = \pm 1$. The set of those $c \in C_n$ for which $c_{11} = +1$ we shall denote by C'_{n-1} and the set of those $c \in C_n$ for which $c_{11} = -1$ by C''_{n-1}. The set C'_{n-1} is a subgroup of the group C_n isomorphic to the group C_{n-1} and on it the function $f(c)$ attains its maximum $+1$. The set C''_{n-1} is the co-set of the group C_n with respect to the subgroup C'_{n-1} and on it the function $f(c)$ attains its minimum -1.

In virtue of the formula (3) we have

$$df(c) = \sum_{i=1}^{2n} \tfrac{1}{2}(c_{1i}\mathbf{c}_{i1} + \bar{c}_{1i}\bar{\mathbf{c}}_{i1}). \tag{12}$$

The infinitesimal matrix $\|\mathbf{c}_{ij}\|$ must satisfy the conditions (24)–(27), Section 2. For the $c_{11}, \ldots, c_{2n,1}$ entering into (12) from these conditions follows only $\mathbf{c}_{11} + \bar{\mathbf{c}}_{11} = 0$. Thus from $df(c) = 0$ follows

$$c_{11} - \bar{c}_{11} = 0, \quad c_{12} = \ldots = c_{1,2n} = 0, \tag{13}$$

wherefrom, in virtue of the relation (13), Section 2, we obtain

$$c_{11} = \pm 1. \tag{14}$$

Herefrom, in virtue of the relations (14), Section 2, follows in its turn

$$c_{21} = \ldots = c_{2n,1} = 0. \tag{15}$$

Thus $df(c)$ vanishes only under the condition that $c_{11} = \pm 1$; but if $c_{11} = \pm 1$, then, in virtue of the relations (12) and (14), Section 2, we obtain the relations (13) and (15), i.e., in particular, $df(c) = 0$. Thus $df(c) = 0$ for those and only for those $c \in C_n$, for which $c_{11} = \pm 1$.

In order to investigate the structure of the sets C'_{n-1} and C''_{n-1} observe that from the relations (13), (14) and (15), in virtue of the relations (18), (20) and (21), (23), Section 2, it follows that

$$\left.\begin{array}{c} c_{n+1,1} = \ldots = c_{n+1,n} = 0, \quad c_{n+1,n+2} = \ldots = c_{n+1,2n} = 0, \\ c_{n+1,n+1} = \pm 1, \quad c_{1,n+1} = \ldots = c_{n,n+1} = 0, \\ c_{n+2,n+1} = \ldots = c_{2n,n+1} = 0, \end{array}\right\} \tag{16}$$

where $c_{11} = c_{n+1,n+1}$. The structure of the matrix c for $c_{11} = \pm 1$ thus revealed shows that C'_{n-1} is a subgroup of the group C_n isomorphic to the group C_{n-1}. Similarly we deduce that C''_{n-1} is the co-set of the group C_n with respect to the subgroup C'_{n-1}.

Hereby we conclude the construction of auxiliary functions on all groups under consideration.

4. On Certain Cycles in Groups

Here we shall construct those cycles in the groups of the four fundamental series (cf. Section 2), the role of which was described in Section 1 (cf. Theorem 1).

The natural device for finding the necessary cycle in the group H_m consists in the following. From the unity e of the group H_m we draw all one-parametrical subgroups orthogonal to the subgroup H'_{m-1} (cf. Section 3, the group H_m). Their locus may form a certain cycle, the index of whose intersection with the cycle H'_{m-1} is equal to ± 1. Instead of the one-parametrical subgroups we may draw from e the trajectories orthogonal to the level surfaces of the function $f(h)$ (cf. Section 3, the group H_m). In order to realize this some calculations are necessary, but since we are interested only in their result, we shall confine ourselves here to giving the obtained cycle by means of formulae. In the groups A_n and C_n we could probably apply the same device, but in view of its cumbersomeness I prefer to obtain the necessary cycles on ground of analogies in the formulae.

$\boldsymbol{H_m}$. Let P_{m-1} be the real projective space of the dimensionality $m-1$. We shall define the point $p \in P_{m-1}$ by its homogeneous coordinates

$$p_1, p_2, \ldots, p_m, \tag{1}$$

normed by the condition

$$p_1^2 + p_2^2 + \ldots + p_m^2 = 1. \tag{2}$$

Thus two systems of coordinates of the type (1) define one and the same point then and only then when they differ by the factor ± 1. For an even m the manifold P_{m-1} is, as is well known, orientable and being subdivided into simplexes may be considered as a cycle. In the case of an uneven m denote by P^*_{m-2} the linear subspace of the space P_{m-1} defined by the relation

$$p_1 = 0. \tag{3}$$

Then P_{m-1}, after a proper subdivision into simplexes, may be considered as an orientated complex with the boundary $2P^*_{m-2}$.

Let us correlate to every point $p \in P_{m-1}$ the matrix $\|p_{ij}\|$, where

$$p_{ij} = \delta_{ij} - 2p_i p_j, \quad (i,j) = 1, 2, \ldots, m. \tag{4}$$

It is obvious that this matrix is uniquely defined by the point p. Below it will be shown that it is orthogonal and that its determinant is equal to -1. Let us now change the sign of the first column of the matrix $\|p_{ij}\|$ and denote the elements of the so obtained matrix by $h^m_{ij}(p)$,

$$h^m(p) = \|h^m_{ij}(p)\|. \tag{5}$$

In view of the stated properties of the matrix $\|p_{ij}\|$ the matrix $h^m(p)$ is orthogonal and has the determinant $+1$.

Correlating now to every point $p \in P_{m-1}$ the matrix $h^m(p) \in H_m$ (cf. Section 2), we define a continuous mapping h^m of the manifold P_{m-1} into the manifold H_m. The orientated complex P_{m-1} together with the mapping h^m gives us an algebraical complex X_{m-1} in H_m, which is a cycle for an even m and has a boundary lying in H''_{m-1} (cf. Section 3) for an uneven m. We shall show that the manifolds H'_{m-1} and X_{m-1} meet only in the one point e – the unity of the group H_m and that the index of intersection of the complexes H'_{m-1} and X_{m-1} is equal to ± 1.

We shall begin the proof of the above statements by establishing the orthogonality of the matrix $\|p_{ij}\|$. We have

$$\sum_{k=1}^{m} (\delta_{ik} - 2p_i p_k)(\delta_{jk} - 2p_j p_k) = \delta_{ij} - 2p_j p_i - 2p_i p_j + \\ + 4p_i p_j \sum_{k=1}^{m} p_k p_k = \delta_{ij}.$$

In order to calculate the determinant of the matrix $\|p_{ij}\|$ consider the linear transformation of the projective space P_{m-1} given by the relation $p'_i = \Sigma^m_{k=1} t_{ik} p_k$ with the orthogonal matrix $t = \|t_{ik}\|$. It is obvious that this transformation transforms an arbitrary point p into an arbitrary point p' provided that the matrix t is chosen adequately. It is easily seen that

$$p'_{ij} = \sum_{k,l=1}^{m} t_{ik} t_{jl} p_{kl} \text{ or } \|p'_{ij}\| = t\|p_{ij}\|t^{-1}.$$

Thus the determinants of the matrices $\|p'_{ij}\|$ and $\|p_{ij}\|$ coincide. If for the point p' we take the point p_0 with the coordinates $1, 0, \ldots, 0$, then we see immediately that the determinant of the matrix $\|p'_{ij}\|$ is equal to -1. But in view of what has been said it is also equal to -1 for an arbitrary point p.

In order to find the common points of the manifolds H'_{m-1} and X_{m-1} observe that the manifold H'_{m-1} is defined by the condition $h_{11} = 1$ (cf. Section 3). Thus we must solve the equation $h^m_{11}(p) = 1$. Writing out this equation explicitly, we obtain $-1 + 2p_1p_1 = 1$, i.e., $2p_1p_1 = 2$, or $p_1^2 = 1$. In virtue of the condition (2) the point p satisfying the equation $h^m_{11}(p) = 1$ has the coordinates $\pm 1, 0, \ldots, 0$, i.e., the equation $h^m_{11}(p) = 1$ has for its unique solution the point p_0; moreover, as we see directly, $h^m_{ij}(p_0) = \delta_{ij}$, i.e. $h^m(p_0) = e$.

In order to prove that the manifolds H'_{m-1} and X_{m-1} have in the point e the index of intersection ± 1, consider the linear space tangent to the manifold H_m at the point e. As is well known, it represents the infinitesimal group $\mathbf{H}_m$ of the group H_m (cf. Section 2). In the linear space H_m consider the linear subspace $\mathbf{H}'_{m-1}$ tangent to the manifold H'_{m-1} at e and the linear subspace $\mathbf{X}_{m-1}$ tangent to X_{m-1} at e. It is sufficient to show that the linear space $\mathbf{H}_m$ falls into the direct sum of its subspaces $\mathbf{H}'_{m-1}$ and $\mathbf{X}_{m-1}$.

The linear space $\mathbf{H}_m$ consists of all skew-symmetrical matrices $\|\mathbf{h}_{ij}\|$ [cf. (5), Section 2]. The linear space H'_{m-1} consists, as may be easily seen, of all skew-symmetrical matrices $\|h'_{ij}\|$ satisfying to the additional condition that the first row and the first column of each matrix $\|h'_{ij}\|$ vanish. In order to comprehend the structure of the linear space $\mathbf{X}_m$ observe that the differentials of the coordinates of the point p in P_{m-1} are connected by the condition

$$\sum_{i=1}^{m} p_i dp_t = 0. \tag{6}$$

Denoting as above by p_0 the point with the coordinates $1, 0, \ldots, 0$ and the differentials of the coordinates at the point p_0 by $\mathbf{p}_1, \mathbf{p}_2, \ldots, \mathbf{p}_m$, we obtain for them from (6) the relation $\mathbf{p}_1 = 0$. Denoting by $\mathbf{h}^m_{ij}$ the differential of the element of the matrix $h^m(p)$ at $p = p_0$, we obtain for it the following form: $\mathbf{h}^m_{11} = 0$ and for $i > 1, j > 1$: $\mathbf{h}^m_{1j} = -2p_j$, $\mathbf{h}^m_{i1} = 2\mathbf{p}_i$, $\mathbf{h}^m_{ij} = 0$. From the investigated structure of the linear spaces $\mathbf{H}_m$, $\mathbf{H}'_{m-1}$ and $\mathbf{X}_{m-1}$ it now becomes clear that $\mathbf{H}_m$ is the direct sum of $\mathbf{H}'_{m-1}$ and $\mathbf{X}_{m-1}$. Thus the index of intersection of H'_{m-1} and X_{m-1} in H_m is equal to ± 1.

H_ma. The complex $X_{m-1}X_{m-1}$ (cf. Definition 1) in the manifold H_m is degenerated, i.e., the set $X_{m-1}X_{m-1}$ has a dimensionality less than $2m-2$. This means that the set M composed of all matrices of the form $h^m(p)h^m(p^1)$, where p and p^1 are arbitrary points from P_{m-1}, has a dimensionality less than $2m-2$.

Denote by a the diagonal matrix, in which at the first place in the diagonal stands -1 and at all other places $+1$. The set aMa has the same dimensionality as the set M. Thus we have only to find the dimensionality of the set of all matrices of the form $ah^m(p)h^m(p^1)a$. Put $p_1=-p_1^0$, $p_2=p_2^0, \ldots, p_m=p_m^0$; then the matrix $ah^m(p)$ becomes the matrix $\|p_{ij}^0\|$ and the matrix $h^m(p^1)a$ coincides with the matrix $\|\mathbf{p}_{ij}^1\|$, where $p_{ij}^0=\delta_{ij}-2p_i^0p_j^0, p_{ij}^1=\delta_{ij}-2p_i^1p_j^1$.

Now we have to determine the dimensionality of the set of all matrices of the form $\|\hat{p}_{ij}\|$, where $\hat{p}_{ij}=\Sigma_{k=1}^m p_{ik}^0p_{kj}^1$. We have

$$\hat{p}_{ij} = \delta_{ij} + 2\left(2p_i^0p_j^1 \sum_{k=1}^m p_k^0p_k^1 - p_i^0p_j^0 - p_i^1p_j^1\right).$$

Let R be the Euclidean space of dimensionality m. By p_0 we shall mean the unit vector from R with the coordinates $p_1^0, p_2^0, \ldots, p_m^0$ and by p^1 – the unit vector with the coordinates $p_1^1, p_2^1, \ldots, p_m^1$. If we denote by c the cosine of the angle between the vectors p^0 and p^1, then $c=\Sigma_{k=1}^m p_k^0p_k^1$. Denote, further, by σ_{ij} the area of the projection into the plane (i,j) of the parallelogram defined by the vectors p^0 and p^1; then we have $\sigma_{ij}=p_i^0p_j^1-p_i^1p_j^0$. With the help of these denotations the expression $\hat{p}_{ij}$ may be written as follows:

$$\hat{p}_{ij} = \delta_{ij} + 2(cp_i^0p_j^1 + cp_i^1p_j^0 - p_i^0p_j^0 - p_i^1p_j^1) + 2c(p_i^0p_j^1 - p_i^1p_j^0) =$$

$$= \delta_{ij} + 2\sum_{k=1}^m \sigma_{ik}\sigma_{kj} + 2c\sigma_{ij}.$$

If the vectors p^0 and p^1 are linearly dependent, then $\hat{p}_{ij}=\delta_{ij}$. If the vectors p^0 and p^1 are linearly independent, denote by P the plane defined by them. It is obvious that if the vectors p^0 and p^1 are simultaneously turned in the plane P by an arbitrary angle φ, the expressions c and σ_{ij} do not change and, consequently, the expression $\hat{p}_{ij}$ will also remain unchanged. Thus the matrix $\|\hat{p}_{ij}\|$ depends only on the plane P spanned by the vectors p^0 and p^1 and on the angle between them. Hence it directly follows that the set of all matrices of the form $\|\hat{p}_{ij}\|$ has a dimensionality not exceeding $2m-3$.

Thus the assertion H_ma is proved.

H_mb. In the group H_{m-1} we have the algebraical complex X_{m-2} constructed above (cf. H_m). Let us map the group H_{m-1} by means of the natural isomorphism on the subgroup H'_{m-1} of the group H_m; the complex X_{m-2} will be hereby transformed into a certain complex from H_m, which we shall again denote by X_{m-2}. Denote further by b the diagonal matrix from H_m, in which at the first two places in the diagonal stand -1 and at all other places $+1$. Put $X'_{m-2} = X_{m-2}b$, $X''_{m-1} = X_{m-1}b$. Then in the case of an uneven m the boundary of the complex X_{m-1} is the cycle $2X'_{m-2}$, $\dot{X}_{m-1} = 2X'_{m-2}$, and, consequently, the boundary of the complex X''_{m-1} is the cycle $2X_{m-2}$, $\dot{X}''_{m-1} = 2X_{m-2}$.

The formulated assertion follows directly from the definition of the complex X_{m-1} and its boundary (cf. H_m).

H_mc. In the case of an uneven m we put $U_{2m-3} = X_{m-2}X_{m-1}$ (cf. Definition 1). Then U_{2m-3} is a cycle and the index of its intersection with the subgroup H'_{m-2} is equal to ± 1. Here by H'_{m-2} is denoted the subgroup defined by the two conditions $h_{11} = h_{22} = 1$.

In virtue of the formula (1), Section 1, the boundary of the complex U_{2m-3} is equal to $\pm 2X_{m-2}X_{m-2}b$ (cf. H_mb). Since the cycle $X_{m-2}X_{m-2}$ is degenerated (cf. H_ma), the complex U_{2m-3} may be considered to be a cycle. In order to determine the index of intersection of the cycle U_{2m-3} with the manifold H'_{m-2} we must use the formula (3), Section 1, and the fact that the index of intersection of the cycle X_{m-2} with the cycle H'_{m-2} in H'_{m-1} is equal to ± 1 (cf. H_m).

A_n. Let P_n be the complex projective space of the complex dimensionality n. We define the point $p \in P_n$ by its homogeneous coordinates

$$p_0, p_1, \ldots, p_n \tag{7}$$

normed by the condition

$$p_0\bar{p}_0 + p_1\bar{p}_1 + \ldots + p_n\bar{p}_n = 1. \tag{8}$$

Thus two systems of coordinates of the type (7) define one and the same point then and only then when they differ by a factor with the modulus equal to 1. We denote by K the unit circumferences and a point on it by the complex number α,

$$\alpha\bar{\alpha} = 1. \tag{9}$$

By Q_{2n+1} denote the topological product of the manifolds p_n and K. The dimensionality of the manifold Q_{2n+1} is easily seen to be $2n+1$; it is orientable, since P_n is orientable. The point $q \in Q_{2n+1}$ is thus given by the coordinates (7) subjected to the condition (8) and the number α satisfying the condition (9).

Let us now correlate to every point $q \in Q_{2n+1}$ the matrix $\|q_{ij}\|$,

$$q_{ij} = \delta_{ij} - (1+\alpha)p_i\bar{p}_j, \quad (i,j) = 0, 1, \ldots, n. \tag{10}$$

It is easily seen that the matrix $\|q_{ij}\|$ is defined by the point q uniquely. Below we shall show that the matrix $\|q_{ij}\|$ is unitary and that its determinant is equal to $-\alpha$. Change the sign of the first column of the matrix $\|q_{ij}\|$ and multiply its last row by $\bar{\alpha}$. Denote the elements of the so obtained matrix by $a_{ij}^n(q)$, $a^n(q) = \|a_{ij}^n(q)\|$. From the stated properties of the matrix $\|\dot{q}_{ij}\|$ it follows that the matrix $a^n(q)$ is unitary and unimodular. Correlate now to every point $q \in Q_{2n+1}$ the element $a^n(q)$ of the group A_n (cf. Section 2). Then we obtain a mapping a^n of the orientable manifold Q_{2n+1} into the group A_n, which gives us a cycle Y_{2n+1} of dimensionality $2n+1$ in the group A_n. It turns out that the manifolds A'_{n-1} and Y_{2n+1} meet only in the one point e – the unity of the group A_n and that the index of intersection of the cycles A'_{n-1} and Y_{2n+1} in the manifold A_n is equal to ± 1.

Passing now to the proof of the above assertions, we shall in the first place establish that the matrix $\|q_{ij}\|$ is unitary. We have

$$\sum_{k=0}^{n} (\delta_{ik} - (1+\alpha)p_i\bar{p}_k)(\delta_{jk} - (1+\bar{\alpha})\bar{p}_j p_k) = \delta_{ij} - (1+\bar{\alpha})\bar{p}_j p_i -$$
$$- (1+\alpha)p_i\bar{p}_j + (1+\alpha)(1+\bar{\alpha})p_i\bar{p}_j \sum_{k=0}^{n} \bar{p}_k p_k =$$
$$= \delta_{ij} - \{(1+\bar{\alpha}) + (1+\alpha) - (1+\alpha)(1+\bar{\alpha})\}p_i\bar{p}_j = \delta_{ij}.$$

In order to calculate the determinant of the matrix $\|q_{ij}\|$ we introduce a linear transformation of the complex projective space P_n by putting

$$p'_i = \sum_{k=0}^{n} t_{ik}p_k,$$

where the matrix $t = \|t_{ik}\|$ is unitary. It is easily seen that this transformation transforms an arbitrary point $p \in P_n$ into an arbitrary point $p' \in P_n$ provided that the matrix t is adequately chosen. We extend this transformation to the manifold Q_{2n+1} by assuming that the coordinate α is not transformed at all. Then we have

$$q'_{ij} = \sum_{k,l=0}^{n} t_{ik}\bar{t}_{jl}q_{kl} \quad \text{or} \quad \|q'_{ij}\| = t\|q_{ij}\|t^{-1}.$$

Thus the matrices $\|q'_{ij}\|$ and $\|q_{ij}\|$ corresponding to the points q' and q have equal determinants. Take now for the point q' the point with the coordinates $1, 0, \ldots, 0; \alpha$. For this point we may immediately verify that the determinant of the matrix $\|q'_{ij}\|$ is equal to $-\alpha$. Thus it must have the value $-\alpha$ for every point $q \in Q_{2n+1}$.

Proceeding now to the proof of the fact that the manifolds A'_{n-1} and Y_{2n+1} meet only in the one point e, let us recall that the manifold A'_{n-1} is determined by the condition $a_{00} = 1$ (cf. Section 3). Thus we have to solve the equation

$$a^n_{00}(q) = 1. \tag{11}$$

Writing out this equation explicitly, we obtain $-1 + (1+\alpha)p_0\bar{p}_0 = 1$, i.e.,

$$(1+\alpha)p_0\bar{p}_0 = 2. \tag{12}$$

Since $p_0\bar{p}_0$ is a non-negative number not exceeding 1 [cf. (8)] and $1+\alpha$ does not exceed 2 in modulus [cf. (10)], from (12) it follows that $p_0\bar{p}_0 = 1$, $\alpha = 1$. Thus the only point q_0 satisfying the equation (11) has the coordinates $1, 0, \ldots 0; 1$. Moreover, it is easily seen that $a^n(q_0) = e$.

In order to prove that the index of intersection of the manifolds A'_{n-1} and Y_{2n+1} in A_n is equal to ± 1, consider the linear space tangent to A_n at e; as is well known, it is the infinitesimal group $\mathbf{A}_n$ of the Lie group A_n. In the linear space $\mathbf{A}_n$ consider the linear subspace $\mathbf{A}'_{n-1}$ tangent to the manifold A'_{n-1} at e and the linear space $\mathbf{Y}_{2n+1}$ tangent to Y_{2n+1} at e. It suffices to show that the linear space $\mathbf{A}_n$ falls into the direct sum of its subspaces $\mathbf{A}'_{n-1}$ and $\mathbf{Y}_{2n+1}$.

The linear space $\mathbf{A}_n$ consists of all matrices,

$$\|\mathbf{a}_{ij}\| \tag{13}$$

satsifying the conditions (10) and (11), Section 2. The linear space $\mathbf{A}'_{n-1}$ consists, as may be easily seen, of all matrices of the form (13), satisfying the additional condition that the first row and column of each matrix vanish. In order to investigate the structure of the space $\mathbf{Y}_{2n+1}$ observe that the differentials of the coordinates $p_0, p_1, \ldots, p_n$; α of the point $q \in Q_{2n+1}$ are connected by the relations

$$\sum_{k=0}^{n} (p_k \cdot d\bar{p}_k + dp_k \cdot \bar{p}_k) = 0, \tag{14}$$

$$\alpha \cdot d\bar{\alpha} + d\alpha\bar{\alpha} = 0. \tag{15}$$

Denoting as above by q_0 the point with the coordinates $1, 0, \ldots, 0; 1$ and the differentials of the coordinates at this point by $\mathbf{p}_0, \mathbf{p}_1, \ldots, \mathbf{p}_n$; $\boldsymbol{\alpha}$, we obtain from (14) and (15) the relations

$$\mathbf{p}_0 + \bar{\mathbf{p}}_0 = 0, \quad \boldsymbol{\alpha} + \bar{\boldsymbol{\alpha}} = 0.$$

Denoting by $\mathbf{a}^n_{ij}$ the differential of the element $a^n_{ij}(q)$ of the matrix $a^n(q)$ at $q = q_0$, we obtain

$$\mathbf{a}^n_{00} = \boldsymbol{\alpha}, \quad \mathbf{a}^n_{nn} = \bar{\boldsymbol{\alpha}};$$

for $i \neq 0, j \neq 0$ we have

$$\mathbf{a}_{0j}^{n} = -2\bar{\mathbf{p}}_j, \quad \mathbf{a}_{i0}^{n} = 2\mathbf{p}_i;$$

all other $\mathbf{a}_{kl}^{n}$ vanish. From the structure of the linear spaces $\mathbf{A}_n$, $\mathbf{A}'_{n-1}$, $\mathbf{Y}_{2n+1}$ thus explored it follows that the linear space $\mathbf{A}_n$ falls into the direct sum of its subspaces $\mathbf{A}'_{n-1}$ and $\mathbf{Y}_{2n+1}$. Thus we have established that the index of intersection of the manifolds A'_{n-1} and Y_{2n+1} in the manifold A_n is equal to ± 1.

C_n. Let P_{2n-1} be the complex projective space of the complex dimensionality $2n-1$. As is well known, the manifold P_{2n-1} is orientable. The point $p \in P_{2n-1}$ we shall define by its homogeneous coordinates

$$p_1, p_2, \ldots, p_{2n} \tag{16}$$

normed by the condition

$$p_1\bar{p}_1 + p_2\bar{p}_2 + \ldots + p_{2n}\bar{p}_{2n} = 1. \tag{17}$$

Thus two systems of coordinates of the type (16) determine one and the same point then and only then when they differ by a factor of modulus 1. Denote by f the mapping of the manifold P_{2n-1} into itself correlating to the point p with the coordinates $p_1, \ldots, p_n, p_{n+1}, \ldots, p_{2n}$ the point $f(p)$ with the coordinates

$$\left.\begin{aligned} f_1(p) &= \bar{p}_{n+1}, \ldots, f_n(p) = \bar{p}_{2n}, \\ f_{n+1}(p) &= -\bar{p}_1, \ldots, f_{2n}(p) = -\bar{p}_n. \end{aligned}\right\} \tag{18}$$

It is directly verifiable that $f\{f(p)\} = p$, so that f is a homeomorphic involuntary mapping; moreover, as will be shown below, the mapping f changes the orientation. Let L be the segment $0 \leqslant \varphi \leqslant \tau$ of the numerical φ-axis. Form the topological product R' of the manifold P_{2n-1} and the segment L. Then every point $r' \in R'$ is determined by a pair (p, φ). Every pair of the form $(p, 0)$ we identify with the pair $(f(p), 0)$ and every pair of the form (p, π) – with the pair $(f(p), \pi)$. The so obtained closed manifold R_{4n-1} is orientable in virtue of the fact that the mapping f changes the orientation. The dimensionality of R_{4n-1} is equal to $4n-1$.

Instead of the number φ introduce the number $e^{i\varphi} = \alpha$. Correlate now to every point $r' = (p, \alpha) \in R'$ the matrix $\|r_{ij}\|$, by putting

$$r_{ij} = \delta_{ij} - (1+\alpha)p_i\bar{p}_j - (1+\bar{\alpha})f_i(p)\overline{f_j}(p), \quad (i,j) = 1, 2, \ldots, 2n. \tag{19}$$

It is easily seen that the matrix $\|r_{ij}\|$ may be considered as corresponding to the piint $r \in R_{4n-1}$, since to identified points from R' correspond equal matrices. Below we shall show that the matrix $\|r_{ij}\|$ belongs to the group C_n. Multiply now the first column of the matrix $\|r_{ij}\|$ by $i = \sqrt{-1}$ and the $(n+1)$-st by $-i = -\sqrt{-1}$; the so obtained matrix belongs, as may be be easily seen, again to the group C_n and we shall denote its elements by $c_{ij}^n(r)$, $c^n(r) = \|c_{ij}^n(r)\|$. Thus we obtained a mapping c^n of the manifold R_{4n-1} into the group C_n. This mapping determines a cycle in C_n, which we shall denote by Z_{4n-1}. It turns out that the cycles Z_{4n-1} and C'_{n-1} (cf. Section 3) meet only in the one point e – the unity of the group C_n – and that the index of their intersection is equal to ± 1.

Passing now to the proof of the formulated assertions let us first of all establish that the mapping f changes the orientation. To this end decompose the mapping f into the product of two mappings f'' and f'. The mapping f' we define as transforming the point p with the coordinates $p_1, \ldots, p_n, p_{n+1}, \ldots, p_{2n}$ into the point $f'(p)$ with the coordinates $p_{n+1}, \ldots, p_{2n}, -p_1, \ldots, -p_n$. The mapping f'' we define as transforming the point p with the coordinates $p_1, \ldots, p_n, p_{n+1}, \ldots, p_{2n}$ into the point $f''(p)$ with the coordinates $\bar{p}_1, \ldots, \bar{p}_n, \bar{p}_{n+1}, \ldots, \bar{p}_{2n}$. In order to investigate the mapping f' we introduce the mapping f'_ψ, defining it as transforming the point p into the point $f'_\psi(p)$ with the coordinates

$$\cos\psi p_1 + \sin\psi p_{n+1}, \ldots, \cos\psi p_n + \sin\psi p_{2n},$$

$$-\sin\psi p_1 + \cos\psi p_{n+1}, \ldots, -\sin\psi p_n + \cos\psi p_{2n}.$$

It is easily seen that $f'_0(p) = p, f'_{\pi/2}(p) = f'(p)$. Thus the mapping f' may be obtained from the identity by a continuous deformation and, consequently, it does not change the orientation. In order to investigate the mapping f'' introduce the non-homogeneous coordinates

$$q_1 = \frac{p_1}{p_{2n}}, \quad q_2 = \frac{p_2}{p_{2n}}, \ldots, \quad q_{2n-1} = \frac{p_{2n-1}}{p_{2n}}$$

in the projective space P_{2n-1}. Then the mapping f'' transforms the point p with the coordinates $q_1, q_2, \ldots, q_{2n-1}$ into the point $f''(p)$ with the coordinates $\bar{q}_1, \bar{q}_2, \ldots, \bar{q}_{2n-1}$. Thus the mapping f'' simply changes the sign of the coefficients of i in each coordinate, and since there are $2n-1$ coordinates, i.e., an uneven number, the mapping f'' changes the orientation of the manifold P_{2n-1}. Hence the mapping f also changes the orientation.

Proceeding now to prove that the matrix $\|r_{ij}\|$ belongs to the group C_n, we point out the following relations:

$$\left.\begin{aligned} \sum_{k=1}^{2n} f_{ik}p_k &= \bar{f}_i(p), \quad \sum_{k=1}^{2n} f_{ik}f_k(p) = -\bar{p}_i, \\ \sum_{i,j=1}^{2n} f_{ij}p_i f_j(p) &= -1, \quad \sum_{k=1}^{2n} p_k\bar{f}_k(p) = 0, \end{aligned}\right\} \tag{20}$$

where f_{ij} are the coefficients of the bilinear form defining the group C_n

[cf. (12), Section 2]. The relations (20) follow directly from the condition (17) and the definition of the transformation f [cf. (18)].

We shall show that the matrix $\|r_{ij}\|$ is unitary. We have

$$\sum_{k=1}^{2n} r_{ik}\bar{r}_{jk} =$$

$$= \sum_{k=1}^{2n} \{\delta_{ik} - (1+\alpha)p_i\bar{p}_k - (1+\bar{\alpha})f_i(p)\bar{f}_k(p)\}\{\delta_{jk} - (1+\bar{\alpha})\bar{p}_j p_k -$$

$$- (1+\alpha)\bar{f}_j(p)f_k(p)\} = \delta_{ij} + (1+\alpha)(1+\bar{\alpha})p_i\bar{p}_j \sum_{k=1}^{2n} \bar{p}_k p_k +$$

$$+ (1+\bar{\alpha})(1+\alpha)f_i(p)\bar{f}_j(p) \sum_{k=1}^{2n} \bar{f}_k(p)f_k(p) +$$

$$+ (1+\alpha)^2 p_i\bar{f}_i(p) \sum_{k=1}^{2n} \bar{p}_k f_k(p) + 1 + \bar{\alpha})^2 f_i(p)\bar{p}_j \sum_{k=1}^{2n} \bar{f}_k(p)p_k -$$

$$- (1+\alpha)p_i\bar{p}_j - (1+\bar{\alpha})f_i(p)\bar{f}_j(p) - (1+\bar{\alpha})\bar{p}_j p_i - (1+\alpha)\bar{f}_j(p)f_i(p) =$$

$$= \delta_{ij} + \{(1+\alpha)(1+\bar{\alpha}) - (1+\alpha) - (1+\bar{\alpha})\}\{p_i\bar{p}_j + f_i(p)\bar{f}_j(p)\} = \delta_{ij}.$$

Now let us show that the matrix $\|r_{ij}\|$ satisfies the condition (16), Section 2:

$$\sum_{i,j=1}^{2n} f_{ij} r_{ik} r_{jl} =$$

$$= \sum_{i,j=1}^{2n} f_{ij}\{\delta_{ik} - (1+\alpha)p_i\bar{p}_k - (1+\bar{\alpha})f_i(p)\bar{f}_k(p)\}\{\delta_{jl} - (1+\alpha)p_j\bar{p}_l -$$

$$- (1+\bar{\alpha})f_j(p)\bar{f}_l(p)\} = f_{kl} + (1+\alpha)^2\bar{p}_k\bar{p}_l \sum_{i,j=1}^{2n} f_{ij}p_i p_j +$$

$$+ (1+\bar{\alpha})^2\bar{f}_k(p)\bar{f}_l(p) \sum_{i,j=1}^{2n} f_{ij}f_i(p)f_j(p) +$$

$$+ (1+\alpha)(1+\bar{\alpha})\bar{p}_k\bar{f}_l(p) \sum_{i,j=1}^{2n} f_{ij}p_i f_j(p) +$$

$$+ (1+\bar{\alpha})(1+\alpha)\bar{f}_k(p)\bar{p}_l \sum_{i=1}^{2n} f_{ij}f_i(p)p_j - (1+\alpha)\bar{p}_k \sum_{j=1}^{2n} p_i f_{il} -$$

$$-(1+\bar{\alpha})\bar{f}_k(p)\sum_{i=1}^{2n} f_i(p)f_{il} - (1+\alpha)\bar{p}_l\sum_{j=1}^{2n} p_j f_{kj} - (1+\bar{\alpha})\bar{f}_l(p)\sum_{j=1}^{2n} f_j(p)f_{kj} =$$

$$= f_{kl} - \{(1+\alpha)(1+\bar{\alpha}) - (1+\alpha) - (1+\bar{\alpha})\}\{\bar{p}_k\bar{f}_l(p) - \bar{f}_k(p)\bar{p}_l\} = f_{kl}.$$

But from the facts that the matrix $\|r_{ij}\|$ is unitary and leaves the bilinear form (12), Section 2, invariant it follows, as is well known, that its determinant is equal to ± 1. If we take for α the number -1, then we immediately find that the matrix $\|r_{ij}\|$ has the determinant $+1$. But since the determinant of the matrix $\|r_{ij}\|$ can not pass from the value $+1$ to the value -1 while α varies continuously from -1 to $+1$, the determinant of the matrix $\|r_{ij}\|$ must be always equal to $+1$. Thus the matrix $\|r_{ij}\|$ belongs to the group C_n.

We proceed now to prove that the cycles C'_{n-1} and Z_{4n-1} meet only at the point e. We know that the manifold C'_{n-1} is determined by the condition $c_{11} = 1$ (cf. Section 3.) Thus we have to solve the equation

$$c^n_{11}(r) = 1. \tag{21}$$

Writing out this equation explicitly, we obtain

$$i\{1 - (1+\alpha)p_1\bar{p}_1 - (1+\bar{\alpha})p_{n+1}\bar{p}_{n+1}\} = 1,$$

or

$$(1+\alpha)p_1\bar{p}_1 + (1+\bar{\alpha})p_{n+1}\bar{p}_{n+1} = 1 + i.$$

If we put $p_1\bar{p}_1 = \lambda$, $p_{n+1}\bar{p}_{n+1} = \mu$, we easily see that λ and μ are non-negative real numbers satisfying the condition $\lambda + \mu \leqslant 1$ [cf. (17)]. Denote by K the circumference in the plane of the complex variable z with the centre at $+1$ and the radius equal to 1. Then the point $1+\alpha$ lies on the upper half of the circumference K and the point $1+\bar{\alpha}$ — on its lower half. Thus the point $(1+\alpha)\lambda + (1+\bar{\alpha})\mu = z$ always lies inside of the circumference K or on it. In the case when $z = 1 + i$ we have $\alpha = i$, $\lambda = 1$. We consequently obtain the unique solution of the equation (21) in the form of the pair $(p_0, i) = r_0$, where p_0 has the coordinates $1, 0, \ldots, 0$. It is easily seen that $\|c^n_{ij}(r_0)\| = e$. Thus the cycles C'_{n-1} and Z_{4n-1} meet only at the point e.

Finally, in order to prove that the index of intersection of the cycles C'_{n-1} and Z_{4n-1} in the manifold C_n is equal to ± 1, consider the linear space tangent to C_n at e. As is well known, it represents the infinitesimal group $\mathbf{C}_n$ (cf. Section 2) of the Lie group C_n. In the space $\mathbf{C}_n$ consider the linear subspace $\mathbf{C}'_{n-1}$ tangent to C'_{n-1} at e and the linear subspace $\mathbf{Z}_{4n-1}$ tangent to Z_{4n-1} at e. It suffices to show that the linear space $\mathbf{C}_n$ falls into the direct sum of the subspaces $\mathbf{C}'_{n-1}$ and $\mathbf{Z}_{4n-1}$.

The linear space $\mathbf{C}_n$ consists of all matrices

$$\|\mathbf{c}_{ij}\| \tag{22}$$

satisfying the conditions (24)–(27), Section 2. Into the linear space $\mathbf{C}'_{n-1}$ enters every matrix of the form (22), in which the first and the $(n+1)$-st column as well as the first and the $(n+1)$-st row vanish. For the investigation of the structure of the linear space $\mathbf{Z}_{4n-1}$ observe in the first place that the differentials of the coordinates in the space P_{2n-1} are connected by the condition

$$\sum_{k=1}^{2n} (p_k \cdot d\bar{p}_k + dp_k \cdot \bar{p}_k) = 0 \tag{23}$$

[cf. (17)]. Denoting by p_0 the point with the coordinates $1, 0, \ldots, 0$ and the differentials of the coordinates at this point by $\mathbf{p}_1, \mathbf{p}_2, \ldots, \mathbf{p}_{2n}$, we obtain for them from (23) the relation

$$\mathbf{p}_1 + \bar{\mathbf{p}}_1 = 0. \tag{24}$$

Observe, further, that the differential of the coordinate α satisfies the condition

$$\alpha \cdot d\bar{\alpha} + d\alpha \cdot \bar{\alpha} = 0. \tag{25}$$

Denoting by the differential of the coordinate α at $\alpha = i$, we obtain for it from (25).

$$\boldsymbol{\alpha} - \bar{\boldsymbol{\alpha}} = 0. \tag{26}$$

Denoting by $\mathbf{c}_{ij}^n$ the differentials of the functions $c_{ij}^n(r)$ at $r = r_0 = (p_0, i)$, we obtain: for $1 < j \leqslant n$

$$c_{11}^n = -\boldsymbol{\alpha} i, \quad \mathbf{c}_{1j}^n = -(1+i)\bar{\mathbf{p}}_j, \quad \mathbf{c}_{j1}^n = (1-i)\mathbf{p}_j,$$

$$\mathbf{c}_{n+1,n+1}^n = \boldsymbol{\alpha} i, \quad \mathbf{c}_{n+1,n+j}^n = -(1-i)\mathbf{p}_j, \quad \mathbf{c}_{n+j,n+1}^n = (1+i)\bar{\mathbf{p}}_j,$$

$$\mathbf{c}_{1,n+1}^n = -2\bar{\mathbf{p}}_{n+1}, \quad \mathbf{c}_{1,n+j}^n = -(1+i)\bar{\mathbf{p}}_{n+j}, \quad \mathbf{c}_{j,n+1}^n = -(1+i)\bar{\mathbf{p}}_{n+j},$$

$$\mathbf{c}_{n+1,1}^n = +2\mathbf{p}_{n+1}, \quad \mathbf{c}_{n+1,j}^n = (1-i)\mathbf{p}_{n+j}, \quad \mathbf{c}_{n+j,1}^n = (1-i)\mathbf{p}_{n+j},$$

and all other $\mathbf{c}_{ij}^n$ are equal to zero. By a direct computation the differentials of the functions $c_{ij}^n(r)$ are obtained here only in the case of the first row of the matrix $\|\mathbf{c}_{ij}^n\|$; the vanishing of the indicated elements of the matrix $\|\mathbf{c}_{ij}^n\|$ is also proved by direct computation, but all other calculated elements are obtained with the help of the formulae (24)–(27), Section 2. From that structure of the linear spaces $\mathbf{C}'_{n-1}$ and $\mathbf{Z}_{4n-1}$ thus established it follows that the linear space $\mathbf{C}_n$ in fact falls into the direct sum of its linear subspaces $\mathbf{C}'_{n-1}$ and $\mathbf{Z}_{4n-1}$. Thus the index of intersection of the cycles C'_{n-1} and Z_{4n-1} is equal to ± 1.

5. Homologies in the Groups A_n, B_n, C_n, D_n

In the present paragraph we shall study the homologies in the simple groups of the four fundamental series. In the case of the groups A_n and C_n

we have only to apply automatically the Theorem 1, since it gives us the possibility of an inductive transition from the group A_{n-1} to the group A_n and from the group C_{n-1} to the group C_n. The transition from the group $B_{n-1} = H_{2n-1}$ to the group $D_n = H_{2n}$ is as simple. In all these transitions the conditions of Theorem 1 are satisfied in virtue of the existence of the necessary cycles. But when we pass from the group $D_n = H_{2n}$ to the group $B_n = H_{2n+1}$, we find ourselves in a different position, since here we have not at our disposal the necessary cycle and the transition can not be so simply realized. We have to consider the groups of the series B_n and D_n simultaneously as one series H_m, but the method of transition from H_{m-1} to H_m depends on the residue of the number m to the modulus 2. This complication may be probably traced to the fact that the groups H_m are not simply connected, whereas the groups A_n and C_n are simply connected. If instead of the groups H_m we should consider their universal coverings, then it is possible that with them everything would be as simple as with the groups A_n and C_n. Since we are primarily interested in those topological properties, which are invariant with respect to local isomorphisms, we could consider the universal coverings of the groups H_m, but, firstly, this would lead us to groups of a less known type than the groups of orthogonal matrices and, secondly, these last groups are very important in themselves and a complete study of homologies in them may be very useful.

In objective terms independent from the method the simplicity of the structure of the groups A_n and C_n is expressed by the fact that they have no torsions. The groups of the series H_m, on the contrary, have torsions and we shall in the first place find in them the basis of weak homologies and only then the complete basis.

Theorem 2. In the group A_n or, correspondingly, in the group C_n there is a system of cycles $W_1, W_2, \ldots, W_n$ such that the independent basis of homologies in the group A_n or, correspondingly, in the group C_n is given in the form

$$(e \vee W_1)(e \vee W_2) \ldots (e \vee W_n), \tag{1}$$

where all cycles of the basis are free. The product is taken here in the group sense (cf. Definition 1) and the sign of sum $\vee$ simply separates from each other the different elements of the basis. At developing the expression (1) the unity e disappears in all summands except the first, which is equal to e and is considered here as a nul-dimensional cycle consisting of one positively orientated point. For the group A_n the dimensionality of the cycle W_k is equal to $2k + 1$. For the group C_n the dimensionality of the cycle W_k is equal to $4k - 1$.

Proof. In the group A_n consider the subgroup A'_k defined by the conditions $a_{00} = 1, a_{11} = 1, \ldots, a_{n-k-1,n-k-1} = 1$ (cf. the group A_n, Section 2). The group A'_k is in a natural manner isomorphic to the group A_k. The group A_1 is, as is well known, homeomorphic to the three-dimensional sphere. Thus the basis of homologies of the manifold A_1 may be composed of the unity e, i.e., the nul-dimensional cycle, and the however orientated manifold A_1, which we denote by W_1. For A_1 the basis of homologies has thus the form $e \vee W_1$, i.e., precisely the form indicated in the formulation of the theorem. The proof shall proceed by induction. Suppose that the assertion of the theorem is already proved for the group A_{n-1}. Mapping the group A_{n-1} on the subgroup A'_{n-1} by means of the natural isomorphism, we obtain the basis of the subgroup A'_{n-1} in the form $(e \vee W_1)(e \vee W_2) \ldots (e \vee W_{n-1})$. In Section 3 we have already constructed the function f on the group A_n (cf. the group A_n, Section 3). In Section 4 we have constructed the cycle Y_{2n+1} in the group A_n (cf. A_n, Section 4). Denote the cycle Y_{2n+1} by W_n. We are now in the condition to apply Theorem 1 to the group A_n, for which the Theorem 2 is thus proved.

In the group C_n consider the subgroup C'_k defined by the conditions $c_{11} = 1, c_{22} = 1, \ldots, c_{n-k,n-k} = 1$. The group C'_k is in a natural manner isomorphic to the group C_k. The group C_1 is, as is well known, homeomorphic to the three-dimensional sphere. The function f on the group C_n has been constructed in Section 3 (cf. the group C_n, Section 3). The cycle $W_n = Z_{4n-1}$ has been constructed in Section 4 (cf. C_n, Section 4). Thus the proof of the theorem for the group C_n may proceed on ground of Theorem 1 on the same lines as in the case of the group A_n. The proof of Theorem 2 is thus accomplished.

Theorem 3. In the group B_n or, correspondingly, in the group D_n there is a system of cycles $W_1, W_2, \ldots, W_n$ such that the independent basis of weak homologies in the group B_n or, correspondingly, in the group D_n is given in the form

$$(e \vee W_1)(e \vee W_2) \ldots (e \vee W_n) \tag{2}$$

(for the explanation of these denotations cf. Theorem 2). In the group B_n the dimensionality of the cycle W_i is equal to $4i - 1$. In the group D_n the dimensionality of the cycle W_i is equal to $4i - 1$ for $i \leqslant n - 1$ and to $2n - 1$ for $i = n$.

Proof. Let us recall that $B_n = H_{2n+1}, D_n = H_{2n}$. In the group H_m consider the subgroup H'_k defined by the conditions $h_{11} = h_{22} = \ldots = h_{n-k,n-k} = 1$. Since the group H'_k is isomorphic to the group H_k and there exists a definite natural isomorphism of the group H_k on the subgroup H_k', we shall denote the subgroup H_k' simply by H_k. The function f constructed in Section 3 (cf. the group H_k, Section 3) on the group H_k we shall denote by f_k. We denote the independent basis of homologies of the group H_k by Ω_k and the independent basis of weak homologies of the same group by Σ_k.

If k is an even number, then the algebraical complex X_{k-1} (cf. H_m, Section 4) is a cycle and hence, in virtue of theorem 1, we have

$$\Omega_k = \Omega_{k-1} \vee \Omega_{k-1} X_{k-1}, \tag{3}$$

$$\Sigma_k = \Sigma_{k-1} \vee \Sigma_{k-1} \Sigma_{k-1}. \tag{4}$$

This means, in particular, that for every cycle Z from H_k we have

$$Z \sim X + YX_{k-1} \text{ in } H_k, \tag{5}$$

where the cycles X and Y belong to H_{k-1}.

The group H_3 is, as is well known, homeomorphic to the three-dimensional projective space; hence the basis of weak homologies of H_3 may be composed of the unity e considered as a nul-dimensional cycle and the however orientated manifold H_3 considered as a three-dimensional cycle, which we shall denote by W_1. Thus $\Sigma_3 = e \vee W_1$, i.e., for $H_3 = B_1$ the theorem is true. We shall now proceed inductively from $m-1$ to m, but here we have to distinguish the cases of an even and odd m.

If m is even and the theorem is true for H_{m-1}, then, in virtue of the relation (4), it is also true for H_m.

Therefore we shall henceforth assume that m is an odd number. If U is a certain cycle from H_{m-1}, then there exists such a cycle V from H_{m-2} that

$$U \approx V \text{ in } H_m \quad (m \text{ is odd}). \tag{6}$$

In fact, we have $\dot{X}''_{m-1} = 2X_{m-2}$ (cf. $H_m b$, Section 4), i.e., $X_{m-2} \approx 0$ in H_m; but then every cycle of the form YX_{m-2} is weakly homologic to zero in H_m. Hence in virtue of (5) we have (6).

Let X be an arbitrary cycle from H_{m-2}. If the boundary of the complex XX_{m-1} is homologic to zero in H''_{m-1} (cf. group H_m, Section 3), then

$$2X \sim 0 \text{ in } H_{m-2} \quad (m \text{ is odd}). \tag{7}$$

In vritue of the formula (1), Section 1, the boundary of the complex XX_{m-1} is equal to $\pm X\dot{X}_{m-1} = \pm 2XX_{m-2}b$ (cf. $H_m b$, Section 4). Since X and X_{m-2} belong to H_{m-1}, XX_{m-2} belongs to H_{m-1} and $XX_{m-2}b$ — to H''_{m-1}. If $2XX_{m-2}b \sim 0$ in H''_{m-1}, then $2XX_{m-2} \sim 0$ in H_{m-1}. Herefrom in virtue of (3) follows (7).

Let X be a cycle and C an algebraical complex from H_{m-2}, so that

$\dot{C} = 2X$. Then $XX_{m-1} + \alpha CX_{m-2}b$ ($\alpha = \pm 1$ depending on the dimensionality of the cycle X and the number m) is a cycle such that $2(XX_{m-1} + \alpha CX_{m-2}b) \sim 0$ in H_m. We have namely

$$2(XX_{m-1} + \alpha CX_{m-2}b) = \pm(C\dot{X}_{m-1}) \quad (m \text{ is odd}). \tag{8}$$

The relation (8) follows directly from the formula (1), Section 1 (cf. $H_m b$, Section 4).

In Section 4 (cf. $H_m c$, Section 4) we have constructed the cycle $U_{2m-3} = X_{m-2}X_{m-1}$. In virtue of Lemma 3 the system $\Sigma_{m-2} \vee \Sigma_{m-2}U_{2m-3}$ is independent in the sense of weak homologies. Let us show that it forms the basis of weak homologies in H_m,

$$\Sigma_m = \Sigma_{m-2} \vee \Sigma_{m-2}U_{2m-3} \quad (m \text{ is odd}). \tag{9}$$

Let Z be an arbitrary cycle from H_m and denote its algebraical intersection with H_{m-1} by Z'. In virtue of (5) we have $Z' \sim X + YX_{m-2}$ in H_{m-1}, where X and Y belong to H_{m-2}. The complex $D = Z - \delta(X + YX_{m-2})\,X_{m-1} = Z - \delta XX_{m-1} - \delta YU_{2m-3}$ has an algebraical intersection with H_{m-1} homologic to zero in H_{m-1} [cf. (2), Section 1]. Since the complexes Z and YU_{2m-3} are closed, the boundary of the complex D coincides with the boundary of the complex $-\delta XX_{m-1}$ and consequently belongs to H''_{m-1}. Thus, according to the Lemma 2, $\dot{D} \sim 0$ in H''_{m-1} and in virtue of (7) $2X \sim 0$ in H_{m-2}, i.e. there exists in H_{m-2} a complex C such that $\dot{C} = 2X$. The complex

$$E = D - \delta\alpha CX_{m-2}b = Z - \delta(XX_{m-1} + \alpha CX_{m-2}b) - \delta YU_{2m-3}$$

is a cycle [cf. (8)] and since $CX_{m-2}b$ lies in H''_{m-1}, the algebraical intersection of the cycle E with H_{m-1} is homologic to zero in H_{m-1}. Consequently, in virtue of the Lemma 2, the cycle E is homologic to a certain cycle from H''_{m-1} and the latter is homologic to a certain cycle U from H_{m-1} (cf. the Lemma 4). Thus we have

$$Z \sim \delta(XX_{m-1} + \alpha CX_{m-2}b) + \delta YU_{2m-3} + U \quad (m \text{ is odd}). \tag{10}$$

in virtue of (8) $XX_{m-1} + \alpha CX_{m-2}b \approx 0$ in H_m. In virtue of (6) $U \approx V$, in H_m where V belongs to H_{m-2}. Thus from (10) we obtain $Z \approx \delta YU_{2m-3} + V$ and the relation (9) is proved.

If m is odd and the theorem is true for H_{m-2}, then from (9) follows that it is also true for H_m. The Theorem 3 is thus proved.

The Theorems 2 and 3 enable us to find without any difficulties the Betti numbers of the groups under consideration. In order to formulate the results concerning the Betti numbers in a compact form, we introduce the following:

Definition 2. Let K be an n-dimensional complex and p_i its i-dimensional Betti number. The polynomial $f(t) = p_0 + p_1 t + p_2 t^2 + \ldots + p_n t^n$ in the variable t shall be called the Poincaré polynomial of the complex K.

Note the following known result on the Poincaré polynomial of a topological product:

If K and L are two complexes and $f(t)$ and $g(t)$ their Poincaré polynomials then the Poincaré polynomial $h(t)$ of the topological product KL of the complexes K and L is equal to $f(t)g(t)$,

$$h(t) = f(t)g(t). \tag{11}$$

Theorem 4. For each of the groups A_n, B_n, C_n, D_n (cf. Section 2) the Poincaré polynomial has the form

$$(1 + t^{d_1})(1 + t^{d_2}) \ldots (1 + t^{d_n}).$$

For the group A_n we have $d_i = 2i + 1$; for the group B_n we have $d_i = 4i - 1$; for the group C_n we have $d_i = 4i - 1$; for the group D_n we have $d_i = 4i - 1$ for $i \leqslant n - 1$ and $d_n = 2n - 1$.

Proof. The formulated result follows directly from the Theorems 2 and 3.

Thus for each of the groups A_n, B_n, C_n, D_n the Poincaré polynomial has the same form as for the topological product of spheres of dimensionalities $d_1, d_2, \ldots, d_n$. The hypothesis that these groups are indeed homeomorphic to the products of the spheres, suggested by this fact, turns out, however, to be false. Using the homotopical invariants I have shown that the group A_2 cannot be decomposed at all into the product of two manifolds, whereas its Poincaré polynomial has the form $(1 + t^3)(1 + t^5)$.

The question on the complete independent basis of homologies of the manifold H_m is answered by the following:

Theorem 5. The independent basis of homologies of any group H_m consists of free cycles and of cycles of the second order. Suppose now that m is odd and denote by $S_1, S_2, \ldots, S_p; T_1, T_2, \ldots, T_q$ the cycles of the independent basis of homologies of the manifold H_{m-2}, where S_i is a free cycle and T_j – a cycle of the order two. The independent basis of homologies of the manifold H_{m-1} consists of free cycles of the form

$$S_i, S_i X_{m-2} \tag{12}$$

and cycles of the order two of the form

$$T_j, T_j X_{m-2}. \tag{13}$$

Herein $i = 1, 2, \ldots, p; j = 1, 2, \ldots, q$. The independent basis of homologies of the manifold H_m consists of free cycles of the form

$$S_i, S_i U_{2m-3} \tag{14}$$

and cycles of the order two of the form

$$T_j, S_i X_{m-2}, T_j X_{m-2}, T_j U_{2m-3}, T_j X_{m-1} + \alpha_j C_j X_{m-2} b. \tag{15}$$

Herein $i = 1, 2, \ldots, p$; $j = 1, 2, \ldots, q$. X_{m-2} and U_{2m-3} are cycles of dimensionalities $m - 2$ and $2m - 3$ and X_{m-1} is an algebraical complex of dimensionality $m - 1$. The algebraical complex C_j from H_{m-2} is defined by the condition $\dot{C}_j = 2T_j$, $\alpha_j = \pm 1$ depending on the dimensionality of T_j (cf. the proof of the Theorem 3).

Proof. The truth of the assertion concerning the basis of homologies of the manifold H_{m-1} follows from the formula (3).

In order to prove that the indicated cycles form the basis of homologies of the manifold H_m we use the relation (10). For an arbitrary cycle Z from H_m we have:

$$Z \sim (XX_{m-1} + \alpha CX_{m-2}b) + YU_{2m-3} + U.$$

Here U is a cycle from H_{m-1} and hence $U \sim V + WX_{m-2}$ in H_{m-1}, where V and W belong to H_{m-2}. Thus

$$Z \sim (XX_{m-1} + \alpha CX_{m-2}b) + YU_{2m-3} + V + WX_{m-2}. \tag{16}$$

Herein X, Y, V, W are cycles from H_{m-2} and C is an algebraical complex from H_{m-2} such that $\dot{C} = 2X$, $\alpha = \pm 1$ depending on the dimensionality of the cycle X.

Assume now that all cycles X, Y, V, W are homologic to zero in H_{m-2} and consider what represents the cycle Z^* defined by the formula (16). In this case we have obviously $Z^* \sim (XX_{m-1} + \alpha CX_{m-2}b)$. Since $X \sim 0$ in H_{m-2}, there exists in H_{m-2} a complex D such that $\dot{D} = X$. It is easily seen that the boundary of the complex DM_{m-1} is given by the formula $\pm(XX_{m-1} + \alpha 2DX_{m-2}b)$ [cf. (1), Section 1]. Thus $(XX_{m-1} + \alpha 2DX_{m-2}b)$ is a cycle homologic to zero in H_m. Consequently we have

$$Z^* \sim (XX_{m-1} + \alpha CX_{m-2}b) - (XX_{m-1} + \alpha 2DX_{m-2}b) \sim$$
$$\sim \alpha(C - 2D)X_{m-2} = W^* X_{m-2} \tag{17}$$

(cf. Lemma 4), where W^* is a cycle from H_{m-2}.

Let now Z be an arbitrary cycle from H_m. According to what has been proved it may be expressed by means of the formula (16). Let us now express all cycles X, Y, V, W in homologies by means of the cycles S_i and T_j, namely put $X \sim X'$, $Y \sim Y'$, $V \sim V'$, $W \sim W'$ in H_{m-2}, where the cycles X', Y', V', W' we shall consider to be simply linear forms in the cycles S_i and T_j. Define, moreover, C' as a linear form in the C_j such that $\dot{C}' = 2X'$. Put

$$Z' = (X'X_{m-1} + \alpha C'X_{m-2}b) + Y'U_{2m-3} + V' + W'X_{m-2}.$$

Then, according to what has been proved [cf. (17)], $Z^* = Z - Z'$ is a cycle homologic to the cycle W^*X_{m-2}. Expressing the cycle W^* again with the help of the cycles of the systems S_i and T_j we represent the cycle Z as a linear form in the cycles of the systems (14) and (15). But this proves that the systems (14) and (15) taken together form the basis of homologies of the manifold H_m.

Let us now show that the systems (14) and (15) taken together form the independent basis of homologies of the manifold H_m and that the orders of the cycles of the system (14) are equal to zero and the orders of the cycles of the system (15) are equal to two. To this end put

$$Z = \sum_{i=1}^{p} a_iS_i + b_iS_iU_{2m-3} + c_iS_iX_{m-2} + \sum_{j=1}^{q} d_jT_j + e_jT_jX_{m-2} + \\ + f_jT_jU_{2m-3} + g_j(T_jX_{m-1} + \alpha_jC_jX_{m-2}b).$$

We shall show that from

$$Z \sim 0 \text{ in } H_m \tag{18}$$

follows

$$a_i = b_i = 0, \tag{19}$$

$$c_i, d_j, e_j, f_j, g_j \text{ are even}, \quad i = 1, 2, \ldots, p, \quad j = 1, 2, \ldots, q. \tag{20}$$

It is easily seen that (18) follows from (19) and (20).

Since the independence in the sense of weak homologies of the cycles of the system (14) has been already proved [cf. (9)], from (18) follows (19).

From (18) follows that the algebraical intersection $\hat{Z}$ of the cycle Z with the manifold H_{m-1} is homologic to zero in H_{m-1}. Thus we have

$$\hat{Z} \sim \sum_{j=1}^{q} f_jT_jX_{m-2} + g_jT_j \sim 0 \text{ in } H_{m-1}. \tag{21}$$

Since the system (13) is independent in H_{m-1} and its cycles have the order two, from (21) follows that f_j, g_j are even.

From (18) follows that $ZU_{2m-3} \sim 0$ in H_m. Thus

$$\left.\begin{aligned} ZU_{2m-3} = ZX_{m-2}X_{m-1} \sim \sum_{i=1}^{p} c_iS_iX_{m-2}X_{m-2}X_{m-1} + \\ + \sum_{j=1}^{q} d_jT_jU_{2m-3} + e_jT_jX_{m-2}X_{m-2}X_{m-1} \sim 0 \text{ in } H_m \end{aligned}\right\} \tag{22}$$

(cf. H_mC, Section 4). Since the cycle $X_{m-2}X_{m-2}$ is degenerated (cf. H_ma, Section 4), the terms in (22) containing c_i and e_j are homologic to zero and we obtain

$$\sum_{j=1}^{q} d_j T_j U_{2m-3} \sim 0$$

in H_m. In virtue of the Lemma 3 herefrom follows that d_j is even.

Put now

$$S = \sum_{i=1}^{p} c_i S_i, \quad T = \sum_{j=1}^{q} e_j T_j.$$

We have already shown that $Z \sim SX_{m-2} + TX_{m-2}$. Assume now that not all c_i are even. Since the system S_i forms the independent basis of weak homologies of H_{m-2}, this assumption leads in virtue of Poincaré-Veblen's Theorem to the existence in H_{m-2} of a cycle S', the index of intersection of which with S is odd:

$$I_{H_{m-2}}(S'', S') \text{ is odd.}$$

From this we obtain:

$$I_{H_{m-1}}(S, S'X_{m-2}) = I_{H_{m-1}b}(Sb, S'X_{m-2}b) \text{ is odd} \tag{23}$$

[cf. (3), Section 1]. The cycles Z and $S'X_{m-2}b$ have the order two in H_m and hence their linkage coefficient $V_{H_m}(Z, S'X_{m-2}b)$ is defined. Let us calculate it. The cycle $2SX_{m-2}$ bounds the complex $SX_{m-1}b$ and the cycle $2TX_{m-2}$ bounds a certain complex D situated in H_{m-1}. Thus the cycle $2Z$ bounds the complex $SX_{m-1}b + D$. We have now to calculate the index of intersection of this complex with the cycle $S'X_{m-2}b$ situated in $H_{m-1}b$. To this end let us first find the algebraical intersection of the complex $SX_{m-1}b + D$ with $H_{m-1}b$. We easily see that it is equal to $\pm Sb$. Further we have to find the index of intersection of the obtained complex $\pm Sb$ with the cycle $S'X_{m-2}b$. Then we have

$$V_{H_m}(Z, S'X_{m-2}b) = \tfrac{1}{2} I_{H_{m-1}b}(Sb, S'X_{m-2}b)$$

and thus the obtained linkage coefficient is not an integer [cf. (23)]. Consequently the cycle Z is not homologic to zero in H_m (cf. the Poincaré-Veblen's Theorem). Starting thus from the assumption that not all c_i are even, we came to a contradiction with (18). Consequently all c_i are even.

Put, as above,

$$T = \sum_{j=1}^{q} e_j T_j.$$

We have already proved that $Z \sim TX_{m-2}$. In order to prove that all e_j are even, it is sufficient to prove that $D \sim 0$ in H_{m-2}. Assume the contrary. Then, in virtue of the Poincaré-Veblen's Theorem there exists in H_{m-2} a cycle T' of the second order such that $V_{H_{m-2}}(T, T')$ is not an integer. Hence it follows that

$$V_{H_{m-1}}(TX_{m-2}, T') \text{ is not an integer.} \tag{24}$$

Let D be the complex from H_{m-1} with the boundary $2TX_{m-2}$. From (24) it follows that

$$I_{H_{m-1}}(D, T') \text{ is odd.} \tag{25}$$

Let $\dot{C}' = 2T'$, where C' belongs to H_{m-2}. From (25) it follows that

$$I_{H_m}(D, T'X_{m-1} + \alpha C'X_{m-2}b)$$

is odd. Thus $V_{H_m}(TX_{m-2}, T'X_{m-1} + \alpha C'X_{m-2}b)$ is not an integer and, consequently, the cycle $Z \sim TX_{m-2}$ is not homologic to zero in H_m. The Theorem 5 is thus completely proved.

The Theorem 5 shows that all torsions of the manifold H_m are equal to two and gives the possibility of their recurrent calculation. In order to give the number of torsions of the manifold H_m in a more comprehensible form, we introduce the following denotations. If the number of torsions of dimensionality s of the manifold H_m is equal to q_s, we put

$$\varphi_m(t) = q_0 + q_1 t + q_2 t^2 + \ldots + q_r t^r,$$

where r is the dimensionality of the manifold H_m. We call the polynomial $\varphi_m(t)$ the polynomial of torsions of the manifold H_m and denote the Poincaré polynomial of the manifold H_m, which we have already calculated, by $f_m(t)$. Then we have:

$$\varphi_m(t) = \frac{(1+t)(1+t^2)\ldots(1+t^{m-1}) - f_m(t)}{1+t}. \tag{26}$$

In order to prove the formula (26) observe in the first place that for an odd m we have, in virtue of the Theorem 5, the following recurrent relations:

$$\varphi_{m-1}(t) = \varphi_{m-2}(t)(1 + t^{m-2}), \tag{27}$$

$$\varphi_m(t) = f_{m-2}(t)t^{m-2} + \varphi_{m-2}(t)(1 + t^{m-2} + t^{m-1} + t^{2m-3}). \tag{28}$$

Further, the manifold H_3, homeomorphic to the three-dimensional projective space, has for its polynomial of torsions

$$\varphi_3(t) = t. \tag{29}$$

The relations (27)–(29) uniquely determine the sequence $\varphi_3(t), \varphi_4(t), \ldots, \varphi_k(t), \ldots$ of rational functions of the variable t. It is easily verified that the sequence of functions given by the formula (26) satisfies the conditions (27)–(29). Thus the formula (26) gives indeed the polynomial of torsions of the manifold H_m.

6. Homologies in Local Isomorphic Groups

In this last paragraph we shall prove the well known theorem of Cartan that the Betti numbers of two locally isomorphic compact connected Lie groups are equal by a purely topological method. Moreover, we shall show how, knowing the Betti numbers of simple Lie groups, we may find them for any compact Lie group.

Theorem 6. Let G' and G'' be two locally isomorphic connected compact Lie groups. Then the s-dimensional Betti number of the manifold G' is equal to the s-dimensional Betti number of the manifold G'' for an arbitrary s.

Proof. It is known that for two locally isomorphic connected compact Lie groups there exists a common connected compact covering group. Thus there exists a connected compact Lie group G such that $G' = G/C$ and $G'' = G/C'$, where C and C' are discrete invariant subgroups of the group G. It obviously suffices to show the equality of the Betti numbers for the groups G' and G. Observe that since the group G is connected and compact, C is a central normal divisor containing only a finite number γ of elements.

Denote by f the natural homeomorphic mapping of the group G on the group G'. To every algebraical complex U from G we correlate in a natural manner the algebraical complex $f(U($ from G'. The mapping f^{-1} inverse to f is not single-valued. If T' is an oriented simplex from G' of a sufficiently small diameter, then we naturally correlate to it the set of γ orientated simplexes of the manifold G. Denote their sum by $f^{-1}(T')$. In virtue of additiveness this operation may be also extended to an arbitrary algebraical complex U' from G'. We have thus defined the operation $f^{-1}(U')$. It may be directly verified that

$$f\{f^{-1}(U')\} = \gamma U'. \tag{1}$$

Let us now show that if Z' is a cycle from G', then

$$\text{from } f^{-1}(Z') \approx 0 \text{ in } G \text{ follows } Z' \approx 0 \text{ in } G'. \tag{2}$$

In fact, if there exists an algebraical complex U from G with the boundary $kf^{-1}(Z')$, where k is an integer different from zero, then

$$f(U) = f(\dot{U}) = f\{kf^{-1}(Z')\} = k\gamma Z'.$$

From the relation (2) it follows that the Betti numbers of the manifold G' do not exceed the Betti numbers of the manifold G.

Let us now show that if X and Y are two cycles of the manifold G of complementary dimensionalities, then

$$I_{G'}(f(X), f(Y)) = \gamma I_G(X, Y). \tag{3}$$

Excluding the obvious case when the dimensionality of one of the cycles X, Y is equal to zero, we may suppose that the dimensionalities of both cycles are less than the dimensionality of the manifold G. If this is so, we may assume that the cycles X and Y are situated in such a way that at the mapping f no three points, of which one belongs to the one cycle and the two others to the other, are mapped into one and the same point. Let now a' be a point common to the cycles $X' = f(x)$ and $Y' = f(Y)$. Then there exists a unique original a of the point a' in the cycle X and a unique original b of the point a' in the cycle Y and, moreover, $a = bc$, where $c \in C$. It is, further, obvious that the index of intersection of the cycles X' and Y' at the point a' is equal to the index of intersection of the cycles X and Yc at the point $a = bc$. Assume, on the contrary, that the cycles X and Yc have a common point $a = bc$. Then the cycles X' and Y' have a common point $a' = f(a) = f(b)$ and the index of intersection of the cycles X and Yc at the point $a = bc$ is equal to the index of intersection of the cycles X' and Y' at the point a'. From the above it follows that the index of intersection of the cycles X' and Y' is equal to the index of intersection of the cycle X with the sum of all cycles of the form Yc, where $c \in C$. Observe that every cycle of the form Yc is homologic to the cycle Y, since the group G is connected. Thus the sum of all cycles of the form Yc is homologic to the cycle γY and we have

$$I_{G'}(X', Y') = I_G(X, \gamma Y).$$

The relation (3) is thus proved.

Now let us show that if X is a cycle from G, then

$$\text{from } f(X) \approx 0 \text{ in } G' \text{ follows } X \approx 0 \text{ in } G. \tag{4}$$

In fact, if the cycle X is not weakly homologic to zero, then, in virtue of the Poincaré-Veblen's Theorem, there exists a cycle Y such that $I_G(X, Y) \neq 0$. But then, in virtue of (3), $I_{G'}(f(X), f(Y)) \neq 0$, i.e. the cycle $f(X)$ is not weakly homologic to zero in G'.

From the relation (4) it follows that the Betti numbers of the manifold G do not exceed the Betti numbers of the manifold G' and the Theorem 6 is proved.

The Theorem 6 shows that the Betti numbers of a compact connected Lie group are its local invariants.

It is known that for every compact connected Lie group G' there exists a connected compact covering group G such that already it falls into the direct product of simple Lie groups. The manifold G falls thus into the topological product of manifolds of simple Lie groups. In virtue of the relation (1), Section 5, the Poincaré polynomial of the manifold G is the product of the Poincaré polynomials of the simple Lie groups. The Theorem 6 shows that the Poincaré polynomials of the manifolds G and G' coincide. In the present paper we have calculated the Poincaré polynomials for the simple groups of the four fundamental series A_n, B_n, C_n, D_n. Beside the groups contained in these series there exists a one-parametrical compact connected Lie group homeomorphic to the circumference; its Poincaré polynomial is equal to $1 + t$. The question on the Poincaré polynomials of the five singular simple groups remains as yet open.

Bibliography

1. É. Cartan, Sur les invariants intégraux de certains espaces homogènes clos et les propriétés topologiques de ces espaces, *Annales Soc. pol. Math.*, 8 (1929). Cf. also É. Cartan, La topologie des espaces représentatifs des groupes de Lie, *Conférences internationales de Topologie*, Genève, 1935.

2. L. Pontrjagin, On Betti numbers of compact Lie groups, *C. R. Acad. Sc. U.R.S.S.*, **1** (1935). Cf. also L. Pontrjagin, Sur les nombres de Betti des groupes de Lie, *C.R., Paris*, **200** (1935).

3. R. Brauer, Sur les invariants intégraux des variétés des groupes de Lie simples clos, *C.R., Paris*, **201** (1935).

4. For the fundamental conceptions of combinatorial topology cf. S. Lefschetz., *Topology*.

5. M. Morse, Relations between the critical points of a real function of n independent variables, *Trans. Amer. Math. Soc.*, **27** (1925).

6. On the Lie groups cf. É. Cartan, Sur la structure des groupes des transformations finis et continus, Thès, 2. éd., Paris, Vuibert, 1933

The V. A. Steklov Institute of Mathematics
of the Academy of Sciences of the USSR

Received 28 June 1939

Über die Topologische Struktur der Lieschen Gruppen

Comment Mathematische Helvetica (1940/1941)
13(4): 277–83

Die Poincaréschen Polynome der Kompakten Lieschen Gruppen haben die Gestalt

$$(1 + t^{\alpha_1})\cdot(1 + t^{\alpha_2})\cdot \ldots \cdot(1 + t^{\alpha_n}),$$

also dieselbe Gestalt wie die Poincaréschen Polynome der topologischen Produkte der Sphären der Dimensionen $\alpha_1, \alpha_2, \ldots, \alpha_n$;[1] unter den einfachen Gruppen aus den vier großen Klassen in der Killing-Cartanschen Aufzählung haben die Gruppen A_n und C_n überdies, wie ich gezeigt habe, keine Torsion,[2] und ihre Fundamentalgruppen sind, wie man leicht sieht, trivial (d.h. von der Ordnung 1); für diese Gruppen erhebt sich daher die Frage, ob sie mit den entsprechenden Sphärenprodukten homöomorph seien.[3]

Im folgenden wird gezeigt, daß diese Frage im allgemeinen zu verneinen ist; es wird nämlich bewiesen:

Die Gruppe A_n, $n \geqslant 2$, deren Poincarésches Polynom

$$(1 + t^3)\cdot(1 + t^5)\cdot \ldots \cdot(1 + t^{2n+1})$$

lautet,[1] ist nicht mit einem Produkt $S^3 \times M$ homöomorph, worin S^3 die 3-dimensionale Sphäre, M irgend ein topologischer Raum ist.

Der Beweis beruht darauf, daß die 4-dimensionale Homotopiegruppe[4] der Mannigfaltigkeit A_n, $n \geqslant 2$, trivial (d.h. von der Ordnung 1), die

4-dimensionale Homotopiegruppe eines Produktes $S^3 \times M$ aber nicht trivial ist. Übrigens spliet der Begriff der Homotopiegruppe selbst in dem Beweis eigentlich keine Rolle; es wird einfach gezeight, daß jede Abbildung der 4-dimensionalen Sphäre Σ^4 in die Mannigfaltigkeit A_n, $n \geqslant 2$, homotop 0 in A_n ist, während es eine Abbildung von Σ^4 in $S^3 \times M$ gibt, die nicht homotop 0 in $S^3 \times M$ ist.

1

Die Gruppe A_n ist definiert als die Gruppe aller unitären unimodularen Matrizen der Ordnung $n + 1$,

$$a = (a_{jk}), \quad j, k = 0, 1, \ldots, n.$$

Durch die Bedingung $a_{00} = 1$ greifen wir aus A_n eine Untergruppe A'_{n-1} heraus, und durch die Bedingung $a_{00} = -1$ eine Restklasse (Nebengruppe) A''_{n-1} der Gruppe A_n nach der Untergruppe A'_{n-1}. Die Dimension einer Gruppe A_r ist $(r + 1)^2 - 1$, und daher gilt für die Dimensiondifferenzen:

$$\operatorname{Dim} A_n - \operatorname{Dim} A'_{n-1} = \operatorname{Dim} A_n - \operatorname{Dim} A''_{n-1} = 2n + 1. \qquad (1)$$

Wie ich gezeigt habe,[1] kann jede Menge $F \subset A_n - A'_{n-1}$ durch eine stetige Deformation innerhalb von $A_n - A'_{n-1}$ in eine Teilmenge von A''_{n-1} übergeführt werden; und das Analoge gilt für die Mengen $F \subset A_n - A''_{n-1}$.

f sei eine stetige Abbildung der 4-dimensionalen Sphäre Σ^4 in die Mannigfaltigkeit A_n. Wenn $n \geqslant 2$ ist, so kann man, infolge von (1), durch eine kleine Deformation von f zu einer solchen Abbildung f, übergehen. daß $f'(\Sigma^4) \subset A_n - A''_{n-1}$ ist; nach dem zitierten Deformationssatz kann man f' weiter stetig in eine Abbildung f'' überführen, welche die Bedingung erfüllt: $f''(\Sigma^4) \subset A'_{n-1}$.

Durch wiederholte Anwendung dieser Operation erhalten wir eine Abbildung g von Σ^4 mit $g(\Sigma^4) \subset A_2$, und indem wir dieselbe Operation noch einmal wiederholen, eine Abbildung g', die homotop zu f und so beschaffen ist, daß $g'(\Sigma^4) \subset A''_1$ ist. Unser Ziel – zu zeigen, daß f homotop 0 in A_n ist – ist daher erreicht, sobald bewiesen ist: jede Abbildung von Σ^4 in A''_1 ist homotop 0 in A_2.

Bekanntlich ist A''_1 mit der 3-dimensionalen Sphäre S^3 homöomorph; nun weiß man aber, daß es nur zwei Abbildungsklassen von Σ^4 in S^3 gibt;[5] folglich genügt es, folgendes zu beweisen: *Es gibt eine wesentliche Abbildung von Σ^4 auf A''_1, die homotop 0 in A_2 ist.* Dieser Beweis wird im §2 mit Hilfe algebraischer Rechnungen geführt werden.

Betrachten wir jetzt andererseitz ein topologisches Produkt $S^3 \times M$;

seine Punkte $z = x \times y$ entsprechen den Punktepaaren $x \in S^3$, $y \in M$; wir setzen $\psi(z) = x$. Wir behaupten: ist f eine wesentliche Abbildung von Σ^4 auf S^3, so ist die Abbildung g von Σ^4 in $S^3 \times M$, welche durch $g(u) = f(u) \times b$ erklärt ist, wobei b ein fester Punkt von M ist, nicht homotop 0 in $S^3 \times M$; in der Tat: wenn eine stetige Deformation g_t der Abbildung $g_0 = g$ existiert, die $g(\Sigma^4)$ in einen Punkt von $S^3 \times M$ zusammenzieht, so its $\psi g_t = f_t$ eine Deformation der Abbildung f, welche $f(\Sigma^4)$ in einen Punkt von S^3 zusammenzieht.

Nun is bekannt,[5] daß es eine wesentliche Abbildung f von Σ^4 auf S^3 gibt; nach dem eben Gesagten gibt es daher auch eine Abbildung g von Σ^4 in $S^3 \times M$, welche nicht homotop 0 in $S^3 \times M$ ist.

2

Die Eigenschaften von A_2, die ich brauche, führe ich für A_n an, da diese allgemeineren Eigenschaften nicht komplizierter sind und für weitere Zwecke Verwendung finden können.

R^n sei ein komplexer unitärer Raum der komplexen Dimension n. Ein Vektor $\gamma \in R^n$ sei durch seine Komponenten als einzeilige Matrix $\gamma = (\gamma_1, \gamma_2, \ldots, \gamma_n)$ gegeben. Ist d eine beliebige Matrix, so bezeichnen wir durch d^* die dazu konjugiert-komplexe und transponierte Matrix. Es ist also γ^* eine Matrix aus einer Kolonne; weiter ist $\gamma\gamma^*$ eine Matrix der Ordnung 1, deren Zahlwert gleich dem Quadrat der Länge von γ ist; $\gamma^*\gamma$ ist eine quadratische Matrix der Ordnung n:

$$\gamma^*\gamma = (\bar{\gamma}_j \cdot \gamma_k).$$

Wir fügen zu R^n noch eine reelle Achse mit der Koordinate α hinzu und bezeichnen den so entstandenen Raum mit R_n; ein Punkt von R_n ist durch eine Matrix $x = (\alpha, \gamma)$ gegeben. Durch

$$\alpha^2 + \gamma\gamma^* = 1 \tag{2}$$

wird im R_n eine $2n$-dimensionale Sphäre Σ^{2n} bestimmt. Wenn $x = (\alpha, \gamma)$ auf Σ^{2n} liegt, so ist $\gamma\gamma^* = 1 - \alpha^2$; für $\gamma \neq 0$, also $\alpha \neq \pm 1$, bezeichnen wir durch δ den Einheitsvektor in der Richtung von γ, also $\delta = 1/(\sqrt{1-\alpha^2}) \cdot \gamma$; er wird ebenso wie γ durch eine einzeilige Matrix der Länge n beschrieben.

Mit ϵ werde die Einheitsmatrix der Ordnung $n+1$ bezeichnet: $\epsilon = (\delta_{jk})$, $j, k = 0, 1, \ldots, n$; mit ϵ' die Einheitsmatrix der Ordnung n: $\epsilon' = (\delta_{jk}), j, k = 1, \ldots, n$.

Es sei nun $x = (\alpha, \gamma) \in \Sigma^{2n}$; wir betrachten die Matrix der Ordnung $n + 1$

$$u = \begin{pmatrix} \alpha i & \gamma \\ -\gamma^* & -\alpha i\epsilon' \end{pmatrix}$$

in ihr ist γ ein Block aus einer Zeile, $-\gamma^*$ ein Block aus einer Kolonne, $-\alpha i\epsilon'$ ein quadratischer Block der Ordnung n, αi eine Zahl; u ist also quadratisch von der Ordnung $n + 1$; es gilt

$$u + u^* = 0. \tag{3}$$

Ist t eine reelle Zahl, so ist

$$b(x, t) = e^{tu} \tag{4}$$

eine quadratische Matrix der Ordnung $n + 1$, für die aus (3) folgt, daß sie unitär ist; durch Rechnung findet man:

$$b(x, t) = \begin{pmatrix} \cos t + \alpha i \cdot \sin t & \sin t \cdot \gamma \\ -\sin t \cdot \gamma^* & e^{-i\alpha t} \cdot \epsilon' + (\cos t - \alpha i \cdot \sin t - e^{-i\alpha t}) \cdot \delta^* \delta \end{pmatrix}; \tag{5}$$

hierbei ist $\gamma \neq 0$, $\alpha \neq \pm 1$ vorausgesetzt, da sonst δ nicht erklärt ist; die Formel (5) bleibt aber auch für $\gamma = 0$, $\alpha = \pm 1$ gültig, wenn man unter δ einen beliebigen Vektor versteht. Daß $b(x, t) = e^{tu}$ ist, ist nicht wichtig; es kommt nur auf einige Eigenschaften dieser Matrix an, die sich aus (4) oder aus (5) durch direkte Rechnung ergeben: nämlich, daß sie unitär und daß ihre Determinante

$$\operatorname{Det} b(x, t) = e^{-(n-1)i\alpha t} \tag{6}$$

ist.

Beweis. s' sei eine quadratische unitäre Matrix der Ordnung n; wir setzen

$$s = \begin{pmatrix} 1 & 0 \\ 0 & s' \end{pmatrix};$$

dann ist

$$s^{-1} \cdot b(x, t) \cdot s = b(y, t) \tag{7}$$

mit $y = (\alpha, \gamma s')$, wobei $\gamma s'$ das einzeilige Matrixprodukt von γ und s' ist; da s unitär ist, sind $b(x, t)$ und $b(y, t)$ gleichzeitig unitär und haben gleiche Determinanten. Wählen wir nun, zu dem gegebenen Vektor γ, die Matrix s' so, daß

$$\gamma s' = (\beta, 0, \ldots, 0), \beta \text{ reell},$$

wird, dann hat $b(y, t)$ eine besonders einfache Form, für welche man, unter Berücksichtigung der Beziehung $\alpha^2 + \beta^2 = 1$, unsere Behauptungen leicht bestätigt.

Wir betrachten noch die Matrix

$$c(\alpha, t) = \begin{pmatrix} 1 & 0 \\ 0 & \exp\left(\frac{n-1}{n} i\alpha t\right)\epsilon' \end{pmatrix};$$

sie ist unitär, ihre Determinante ist $e^{(n-1)i\alpha t}$. Das Produkt

$$a(x, t) = b(x, t) \cdot c(\alpha, t)$$

ist daher eine unitäre unimodulare Matrix: $a(x, t) \in A_n$.

Bei festem t entspricht jedem Punkt x von Σ^{2n} ein Punkt $a(x, t)$ von A_n, d.h. $a(x, t)$ vermittelt eine Abbildung von Σ^{2n} in A_n; wird t variiert, so entsteht eine Deformation dieser Abbildung. Man sieht unmittelbar, daß

$$a(x, 0) = \epsilon, \quad a(x', \pi) \in A''_{n-1}$$

ist. $a(x, \pi) = a(x)$ ist also eine stetige Abbildung von Σ^{2n} in A''_{n-1}, die in A_n homotop 0 ist.

Es bleibt zu zeigen, daß für $n = 2$ die Abbildung $a(x)$ von Σ^4 in die 3-dimensionale Sphäre A''_1 wesentlich ist.

Wir setzen

$$a(x) = (a_{jk}(x)), \quad j, k = 0, 1, 2;$$

dann ist

$$a_{00}(x) = -1, \quad a_{j0}(x) = a_{0k}(x) = 0, \quad j, k = 1, 2,$$

$$a_{jk}(x) = \left(\cos\frac{\pi}{2}\alpha - i\sin\frac{\pi}{2}\alpha\right)\delta_{jk} - 2\cos\frac{\pi}{2}\alpha \cdot \bar{\delta}_j\delta_k, \quad j, k = 1, 2.$$

Führen wir in A''_1 Koordinaten x_1, x_2, x_3, x_4 ein, indem wir

$$a_{11}(x) = x_2 - ix_1, \quad a_{12}(x) = -x_3 - ix_4$$

setzen, so ist die mit einer 3-dimensionalen Sphäre homöomorphe Mannigfaltigkeit A''_1 durch $x_1^2 + x_2^2 + x_3^2 + x_4^2 = 1$ gegeben, und für unsere Abbildung $a(x)$ ist

$$\begin{aligned} x_1 &= \sin\frac{\pi\alpha}{2} \\ x_2 &= (1 - 2\bar{\delta}_1\delta_1)\cdot\cos\frac{\pi\alpha}{2} \\ x_3 + ix_4 &= 2\bar{\delta}_1\delta_2\cdot\cos\frac{\pi\alpha}{2} \end{aligned} \tag{8}$$

Auf der Sphäre A''_1 werden durch die Bedingungen $x_1 > 0$, $x_1 < 0$, $x_1 = 0$ eine nördliche Halbkugel, eine südliche Halbkugel und eine Äquatorsphäre S^2 ausgezeichnet; analog auf Σ^4 durch $\alpha > 0$, $\alpha < 0$, $\alpha = 0$ eine nördliche und eine südliche Halbkugel sowie eine Äquatorsphäre Σ^3. Aus (8) sieht man: wenn x auf Σ^4 einen Halbmeridian vom Nordpol zum Südpol durchläuft, so durchläuft $a(x)$ auf A_1'' einen Halbmeridian vom Nord- zum Südpol, und wenn x auf Σ^3 ist, so ist $a(x)$ auf S^2.

In Anbetracht dieser Eigenschaften der Abbildung a und in Anbetracht bekannter allgemeiner Sätze über Sphärenabbildungen[6] genügt es für den Beweis der Wesentlichkeit der Abbildung a von Σ^4 auf A_1'', zu zeigen, daß für die durch a gelieferte Abbildung von Σ^3 auf S^2 die Hopfsche Zahl[7] C ungerade ist.

Nun ist aber diese Abbildung des durch $\alpha = 0$, $\delta_1\bar{\delta}_1 + \delta_2\bar{\delta}_2 = 1$, bestimmten Äquators Σ^3 von Σ^4 auf den durch $x_1 = 0$, $x_2^2 + x_3^2 + x_4^2 = 1$, bestimmten Äquator S^2 von A_1'' nach (8) durch die Formeln

$$x_2 = 1 - 2\bar{\delta}_1\delta_1, \quad x_3 + ix_4 = 2\bar{\delta}_1\delta_2 \tag{9}$$

gegeben; for der durch (9) gegebenen Abbildung aber ist es bekannt und leicht su sehen,[8] daß $C = \pm 1$ ist.

Damit ist bewiesen: Die Abbildung a der Sphäre Σ^4 auf die Mannigfaltigkeit A_1'' ist wesentlich auf A_1'', aber homotop 0 in A_2.

3

Im Vorstehenden ist bewiesen worden, daß die Mannigfaltigkeit A_2 nicht mit dem topologischen Produkt $S^3 \times S^5$ homöomorph ist, obwohl die beiden Mannigfaltigkeiten triviale Fundamentalgruppen und isomorphe Homologiegruppen, und sogar isomorphe Schnittringe haben. Der Beweis beruhte darauf, daß die vierte Homotopiegruppe von A_2 trivial ist, während die vierte Homotopiegruppe von $S^3 \times S^5$ nicht trivial (sondern von der Ordnung 2) ist. Es war also nich nötig, für den Beweis neue topologische Invarianten einzuführen; jedoch wird durch die geometrischen Überlegungen, die zu dem Beweis geführt haben, die Einführung einer neuen topologischen Invariante nahegelegt.

Die Elemente der r-ten Homotopiegruppe $H^r(Q)$ eines zusammenhängende Polyeders Q sind die Klassen der Abbildungen der Sphäre S^r in das Polyeder Q, wobei nur solche Abbildungen betrachter werden, welche einen auf S^r fest ausgezeichneten „Pol" p in einen in Q ausgezeichneten „Nullpunkt" o überführen; die Definition der Addition in $H^r(Q)$ darf wohl als bekannt gelten.[9]

Neben der Gruppe $H^r(Q)$ betrachten wir die Gruppe $H^{r+k}(S^r)$; wir nehmen an, daß der für die Definition von $H^r(Q)$ auf S^r ausgezeichnete Pol mit dem für die Definition von $H^{r+k}(S^r)$ auf S^r ausgezeichneten

Nullpunkt zusammenfällt. Ist nun x ein Element von $H^r(Q)$, y ein Element von $H^{r+k}(S^r)$, so gehören diejenigen Abbildungen fg von S^{r+k} in Q, für welche $f \in x$, $g \in y$ ist, einer Klasse von Abbildungen von S^{r+k} in Q an; diese Klasse repräsentiert ein Element xy der Gruppe $H^{r+k}(Q)$.

Die Struktur der hiermit erklärten Multiplikation der Gruppen $H^{r+k}(S^r)$ und $H^r(Q)$, die übrigens das Gesetz $x(y \pm y') = xy \pm xy'$ erfüllt, ist eine Invariante des Polyeders Q, und es liegt kein Grund dafür vor, anzunehmen, daß sie sich auf bekannte Invarianten zurückführen läßt.

Notes

1. Für die Gruppen A_n, B_n, C_n, D_n der vier großen Klassen einfacher Gruppen mit Bestimmung der Exponenten a_i zuerst von mir, *C. R. Acad. Sc. U.R.S.S.*, **1** (1935), 433–437, und *C. R. Paris*, **200** (1935), 1277–1280, bewiesen; eine ausführliche Darstellung mit Bestimmung der Torsionsgruppen habe ich im *Recueil math. de Moscou*, **6** (1939), 389–422, gegeben. Weitere Beweise: R. Brauer, *C. R. Paris*, **201** (1935), 419–421; C. Ehresmann, *C. R. Paris*, **208** (1939), 321–323 und 1263–1265; ferner – als Spezialfall eines allgemeineren Satzes, aber ohne Bestimmung der a_i: H. Hopf, *Annals of Math.*, **42** (1941), 22–52.
2. Recueil math., wie;[1] sowie Ehresmann, l. c.
3. Auf diese Frage hat E. Cartan am Schluß seines Genfer Vortrages „*La Topologie des Groupes de Lie*" hingewiesen (*L'Enseignement math.*, **35** [1936], 177–200; sowie: *Actualités Scient. et Industr.*, **358** [Paris 1936]; sowie: *Selecta, Jubilé Scientifique* [Paris 1939], 253–258).
4. W. Hurewicz, *Proc. Akad. Amsterdam*, **38** (1935), 112–119, 521–528.
5. L. Pontrjagin, *C. R. Congrès intern. des math.*, Oslo 1936, t. II, 140; sowie: *C. R. Acad. Sc. U.R.S.S.*, **19** (1938), 147–149. H. Freudenthal, *Compositio Math.*, **5** (1937), 299–314.
6. Freudenthal, l. c., Satz III.
7. H. Hopf, *Math. Annalen*, **104** (1931), 637–665.
8. Hopf, l. c., §5.
9. Hurewicz, l. c.;[4] sowie Freudenthal, l. c.,[6] §§1, 2.

Eingegangen den 4. April 1941.

A Classification of Mappings of the Three-Dimensional Complex into the Two-Dimensional Sphere

Matematicheskii Sbornik, Novaya Seriya (1941) 9(2): 331–63

Let K and L be two complexes. The family f_t, where t is a real number ($0 \leqslant t \leqslant 1$), of continuous mappings of the complex K into the complex L is called a continuous deformation of mappings of the complex K into the complex L if the function $f_t(x)$ ($x \in K$) is a continuous function of the pair of arguments x, t. Two continuous mappings g and h of the complex K into the complex L are said to be homotopic or equivalent if there exists a continuous deformation f_t transforming the mapping g into the mapping h, i.e., such that $g = f_0$, $h = f_1$. In virtue of this criterium of equivalency all continuous mappings of the complex K into the complex L fall into classes of equivalent mappings. A classification of mappings from this point of view, i.e., the determination of more or less effective criteria of equivalency, forms one of the fundamental problems of topology.

The present state of topology leaves no hopes for the solution of the formulated problem in the near future. At present only certain particular cases have been investigated and solved.

Hopf[1] gave the classification of mappings of the n-dimensional complex K^n into the n-dimensional sphere S^n. The necessary and sufficient conditions of equivalency are given by him in this case in terms of homologies, which is the best way to solve the problem, since homologies admit of a rather effective computation. Hopf[2] has also shown that there exists an enumerable number of classes of the mappings of the $(4n-1)$-dimensional sphere S^{4n-1} into the $2n$-dimensional sphere S^{2n}; the criterium of non-equivalency in this case has been given by him in terms related to homologies. I[3] have given a classification of mappings of the $(n+k)$-dimensional sphere S^{n+k} into the n-dimensional sphere S^n in the case $k = 1, 2$. Freudenthal,[4] knowing my results, but not knowing my proof, gave for a part of them a new proof and established, moreover, that for $k = 3, 7$ there are at least two classes of mappings. I[5] gave, further, a classification of mappings of an $(n+1)$-dimensional complex into the n-dimensional sphere – these results have been published only in a brief exposition. This is, so far as I know, all that has been done in the question on the classification of mappings.

The first and the most important question in the general problem of classification of mappings is undoubtedly the question on the classification of mappings of the $(n+k)$-dimensional sphere into the n-dimensional sphere. Having solved it, it would be possible to attempt a classification of mappings of an $(n+k)$-dimensional complex K^{n+k} into the n-dimensional sphere S^n, as well as a classification of mappings of S^{n+k} into K. It may be surmised that on the way to the solution of these two questions new invariants of complexes of the type of homologies and intersections will arise. It is possible to approach the classification of mappings of K^{n+k} into S^n and of S^{n+k} into K in a different way, without classifying firstly the mappings of S^{n+k} into S^n, but simply assuming that the classification of mappings of S^{n+k} into S^n is already carried out, or, more exactly, that the group[6] of mappings of S^{n+k} into S^n is known.

In the present paper is given a complete exposition of my earlier published results on mappings of the three-dimensional complex K^3 into the two-dimensional sphere S^2;[5] besides, in this paper is partly touched the question on mappings of the four-dimensional complex K^4 into the two-dimensional sphere S^2. So particular a question as the classification of mappings of K^3 into S^2 represents a certain interest due to the fact that in its solution we obtain certain indications as to how the question should be solved on the classification of mappings of K^{n+k} into S^k. Moreover, we give here for the first time an application of the theory of products (intersections) in complexes[7] to the solution of a purely geometrical question, in the formulation of which homologies, not to speak of products, are not even mentioned.

In Sections 3, 4 and 5 are essentially used the results of my preceding paper "Products in complexes"; I shall refer to this paper in the sequel as to P.C.[7]

In the whole of the present paper we shall consider only continuous mappings and continuous deformations of mappings, and therefore the word continuous will be omitted in the sequel.

1. Mappings of the Three-dimensional Sphere into the Two-dimensional One

In the present paragraph is given a classification of mappings of the three-dimensional sphere into the two-dimensional one. The classification is based on the invariant introduced by Hopf;[2] we shall see that this invariant uniquely determines the class of mappings. The fundamental role in the proof is played by Lemma 3. The results of W. Hurewicz,[6] in particular Lemma 1, do not play here an essential role, but they are important by themselves and form an essential complement to Lemma 3.

The fundamental role for all further constructions of this paragraph is played by the standard mapping ϑ of the three-dimensional sphere S^2 on the two-dimensional sphere S^2. We proceed now in the first place to construct this mapping.

(A) Let us construct the mapping ϑ of the three-dimensional sphere S^3 on the two-dimensional sphere S^2. We shall consider the sphere S^3 as the set of all quaternions equal to one in modulus, i.e., every point $z \in S^3$ we shall write in the form

$$z = a + bi + cj + dk,$$

where i, j, k are quaternion units and a, b, c, d are real numbers connected by the relation

$$a^2 + b^2 + c^2 + d^2 = 1.$$

The set S^3 of quaternions forms a group with respect to multiplication. Denote by H the subgroup composed of all quaternions of the form

$$\cos\alpha + \sin\alpha \cdot i, \tag{1}$$

The aggregate S^3/H of all right co-sets of the group S^3 with respect to the subgroup H forms naturally a certain manifold; it turns out that this manifold is homoemorphic to the two-dimensional sphere S^2. Correlating to every element $z \in S^3$ the co-set $Z \in S^3/H$, to which z belongs, we obtain the mapping ϑ.

Let us show that S^3/H is homeomorphic to the two-dimensional sphere and let us consider the mapping ϑ more detailed.

We introduce in the metrical sphere S^2 polar coordinates. To this end denote by p its north and by q its south pole and choose a certain fixed meridian pmq; the centre of the sphere S^2 we denote by o. For the radius vector of the point $y \in S^2$ we take the angle poy divided by π and for the amplitude the angle mpy between the meridians pm and py. Then the point q will have an indefinite amplitude.

The set of all quaternions from S^2 of the form

$$a + cj + dk, \tag{2}$$

where $a \geqslant 0$, we denote by A. Since every quaternion from S^3 has a modulus equal to one, every element from A may be represented in the form

$$\sigma + \rho(\cos\beta \cdot j + \sin\beta \cdot k), \tag{3}$$

where $0 \leqslant \rho \leqslant 1$, $\sigma = +\sqrt{1-\rho^2}$. Here ρ and β may be interpreted as polar coordinates introduced in A, from which immediately follows that A is homoeomorphic to a circle, the boundary A of which is composed of all elements of the form

$$\cos\beta \cdot j + \sin\beta \cdot k. \tag{4}$$

Further we have

$$(\cos\alpha + \sin\alpha \cdot i)(\sigma + \rho(\cos\beta \cdot j + \sin\beta \cdot k)) =$$

$$= \sigma(\cos\alpha + \sin\alpha \cdot i) + \rho(\cos(\alpha+\beta) \cdot j + \sin(\alpha+\beta) \cdot k). \tag{5}$$

From this we see that every $z \in S^3$ is representable in the form (5), i.e., that

$$z = x \cdot y, \tag{6}$$

where $x \in H$, $y \in A$. From the same relation (5) follows that the decomposition (6) is unique for every z not belonging to $\dot{A}$; if, however, $z \in \dot{A}$, then y becomes an arbitrary element from $\dot{A}$, and α is determined from the relation (5). Correlate now to every point $z \in S^3$, represented in the form (5), the point from S^2 with polar coordinates ρ, β. Then we obtain the mapping ϑ transforming every co-set from S^3 into a point from S^2.

The following propositions (B) and (C), as well as Lemma 1, belong to W. Hurewicz.[6] I give them with full proofs.

(B) Let K be a compact metrical space and f_0 and f_1 two of its mappings into S^3. If

$$\vartheta f_0 = \vartheta f_1 \quad \text{[cf. (A)]}, \tag{7}$$

then the mappings f_0 and f_1 are equivalent.

For the proof we consider the sphere S^3 as the group of quaternions [cf. (A)] and the possibility of multiplication of its points.

From relation (7) follows that for every $x \in K$ the elements $f_0(x)$ and $f_1(x)$ belong to one and the same co-set of the group S^3 with respect to the subgroup H. Thus

$$h(x) = f_0(x) f_1(x)^{-1} \in H. \tag{8}$$

Since the mapping h transforms the whole space K into the circumference $H \subset S^3$, there exists a continuous deformation h_t of mappings of K into S^3 such that $h_0 = h$, $h_1(K) = \{e\}$, where e is the unit of the group S^3. From this and the relation (8) follows that

$$f_0(x) = h_0(x) f_1(x), \quad f_1(x) = h_1(x) f_1(x);$$

hence it is natural to put $f_t(x) = h_t(x) f_1(x)$, and f_t gives a continuous deformation of the mapping f_0 into the mapping f_1.

(C) Let f_0 be a mapping of a compact metrical space K into the sphere S^3 and φ_t a continuous deformation of the mappings of the space K into the sphere S^2 such that $\varphi_0 = \vartheta f_0$. Then there exists such a continuous deformation f_t of mappings of the space K into S^3 that

$$\varphi_t = \vartheta f_t \quad \text{[cf. (A)]}. \tag{9}$$

For the proof we consider again the sphere S^3 as a group of quaternions and interprete the points of the sphere as right co-sets of the group S^3 with respect to the subgroup H [cf. (A)]. We recall that the set A constructed in (A) intersects with every right co-set only in one point, with the only exception of the co-set $\dot{A}$, which coincides with the boundary of the topological circle A.

Let ϵ be a positive number so small that if $u \in S^3$, $v \in S^3$, and the distance between the points $\vartheta(u)$ and $\vartheta(v)$ in the sphere S^2 is less than ϵ, then uv^{-1} does not belong to $\dot{A}$. The existence of such an ϵ is easily established. By n we denote a natural number so great that for $|t' - t''| \leqslant 1/n$ the distance between the points $\varphi_{t'}(x)$ and $\varphi_{t''}(x)$ is less than ϵ, where x is an arbitrary element from K.

We shall carry out the construction of the mapping f_t inductively. Suppose that the mapping f_t, satisfying the condition (9), is already constructed for $0 \leqslant t \leqslant m/n$. Starting from this assumption, let us construct the mapping f_t, satisfying the condition (9), for $m/n \leqslant t \leqslant (m+1)/n$.

Let $m/n \leqslant t \leqslant (m+1)/n$, then $\varphi_t(x)$ is a definite right co-set. Since the mapping $f_{m/n}$ is already constructed, $f_{m/n}(x)$ is a definite element from S^3. Thus $\varphi_t(x) f_{m/n}(x)^{-1}$ is a definite co-set. In virtue of the choice of the numbers ϵ and n this co-set does not coincide with $\dot{A}$, and hence the point of its intersection with A is determined; we shall denote it by $h_t(x)$. Put $f_t(x) = h_t(x) f_{m/n}(x)$. Then $f_t(x)$ enters into the co-set $\varphi_t(x)$ and, consequently, $\varphi_t(x) = \vartheta(f_t(x))$. Thus the mapping f_t is constructed. The continuity of the deformation f_t is obvious.

Lemma 1. Let f and g be two mappings of a compact metrical space K into the sphere S^3. The mappings f and g are then and only then equivalent when the mappings ϑf and ϑg are equivalent [cf. (A)].

Proof. Suppose that the mappings $\varphi_0 = \vartheta f$ and $\varphi_1 = \vartheta g$ are equivalent. Then there exists a continuous deformation φ_t connecting them. Put $f_0 = f$. Thus we can apply the proposition (C), i.e., there exists a continuous deformation f_t of mappings of the space K into the sphere S^3, and $\vartheta f_t = \varphi_t$. In virtue of this last, f_1 and g are equivalent [cf. (B)]. Since, moreover, $f = f_0$ and f_1 are equivalent, f and g are equivalent.

Suppose that the mappings f and g are equivalent. Then there exists a continuous deformation g_t such that $g_0 = f, g_1 = g$. The continuous deformation ϑg_t connects ϑf and ϑg. The lemma is thus proved.

The following Lemma 2 forms the base for the proof of Lemma 3.

Lemma 2. Let K be a certain complex of arbitrary dimensionality and f its simplicial mapping into the n-dimensional complex L. Denote by V a certain open n-dimensional simplex of the complex L and put $U = f^{-1}(V)$. Then U naturally falls into the topological product of the simplex V and a certain complex P, $U = V \cdot P$, i.e., every point $z \in U$ is uniquely and continuously representable in the form of a pair $z = x \cdot y$, where $x \in V, y \in P$, and $f(z) = f(x \cdot y) = x$. Further, it turns out that if K is a manifold, then P is also a manifold.

Proof. Denote by Ω the set of all simplexes from K which are mapped under f on V. Let p be an inner point of the simplex V, and denote the complete original of the point p in a certain r-dimensional simplex $T^r \in \Omega$ by $\psi(T^r)$. It is easily seen that $\psi(T^r)$ is a convex body of dimensionality $r - n$. Further, if $T^{r-1} \in \Omega$ is an $(r-1)$-dimensional face of the simplex T^r, then $\psi(T^{r-1})$ is an $(r-n-1)$-dimensional face of the convex body $\psi(T^r)$. Conversely, every $(r-n-1)$-dimensional face of the convex body $\psi(T^r)$ is obtained as $\psi(T^{r-1})$, where T^{r-1} is an $(r-1)$-dimensional face of the simplex T^r entering into Ω.

Put $P = f^{-1}(p)$; then P is composed of all convex bodies of the form $\psi(T)$, where $T \in \Omega$, and we shall interprete P as the geometrical complex composed of these convex bodies. From the just established relation between T^r, T^{r-1} and $\psi(T^r)$, $\psi(T^{r-1})$ follows that the relations of incidences in P are the same as in Ω, only their dimensionalities are reduced by n. Taking into account that if $T^{r-1} \in \Omega$ and $T^r \in K$, and T^{r-1} is incidentic with T^r, then $T^r \in \Omega$; we conclude that if K is a manifold, then P is also a manifold.

Denote by $a_0, a_1, \ldots, a_n$ the vertices of the simplex V, and let Y_i be the complete original of the point a_i in the simplex $T \in \Omega$ under the mapping f. Then Y_i is a certain face of the simplex T. Let $y_i \in Y_i, i = 0, 1, \ldots, n$, be a system of points from the simplexes $Y_0, Y_1, \ldots, Y_n$. Denote by

$$V(y_0, y_1, \ldots, y_n) \tag{10}$$

the open simplex from T with vertices $y_0, y_1, \ldots, y_n$. From elementary geometrical considerations follows that through every point $z \in T$, satisfying the condition $f(z) \in V$, passes one and only one simplex of the form (10). In particular, if $z = y \in \psi(T)$, then we denote the simplex of the form (10) passing through y by $V(y)$. Since the mapping f transforms the simplex $V(y)$ on the whole simplex V, every simplex of the form (10) may be written in the form $V(y)$, where $y \in \psi(T)$. From this, in particular, follows that $\psi(T)$ is the topological product of simplexes $Y_0, Y_1, \ldots, Y_n$.

Let z be an arbitrary point from U. Put $f(z) = x$ and denote by T such a simplex of the system Ω that $z \in T$. Then there exist a simplex $V(y)$ of the form (10) containing z, such that $y \in P$ and the point y is uniquely determined. Thus to every $z \in U$ uniquely corresponds a pair x, y, where $x \in V, y \in P$. Conversely, to every pair x, y uniquely corresponds a z. Thus U falls into the topological product $V \cdot P$ and $f(z) = f(x \cdot y) = x$. Thus the lemma is proved.

In connection with the proof of Lemma 2 it is convenient to formulate the following remark (C$'$), which is not necessary for the proof of Lemma 3, but will be used later.

(C$'$) Let f be a simplicial mapping of the complex K into the n-dimensional complex L, V a certain open orientated n-dimensional simplex from L and p a point belonging to V. If T^r is an r-dimensional orientated simplex from K, then the complete original $\psi(T^r)$ of the point p under the mapping f, which is a convex body of dimensionality $r - n$ or the void set, may be naturally considered as an oriented element. The orientation of the body $\psi(T^r)$ we define inductively. If $r = n$, then we assign to the point $\psi(T^n)$ the sign coinciding with the sign of the power of the mapping f of the simplex T^n on the simplex V. If now the orientated original $\psi(T^r)$ is defined for every simplex T^r with $r < s$, then the function ψ may be by additivity extended to an arbitrary algebraical complex C, which is a linear form of orientated simplexes from K of dimensionalities less than s. Thus the function ψ is defined also for the boundary $\dot{T}^s$ of the simplex T^s. The orientation of the element $\psi(T^s)$ we define by the condition $\psi(T^s)^{\cdot} = \psi(\dot{T}^s)$. If we now assume that the function ψ is already defined for all simplexes from K and, by addivity, for all algebraical complexes from K, then we have $\psi(C)^{\cdot} = \psi(\dot{C})$.

The following propositions (D), (E) and (F) serve for the proof of Lemma 3. I formulate them without proof with reference to the corresponding literature.

(D) Let G be a commutative group with a finite number of generators taken in the additive notation. Under the integral character χ of the group G we understand its homoeomorphic mapping into the additive group of all integers. Let $u_1, u_2, \ldots, u_n$ be an arbitrary finite system of elements from G and $h_1, h_2, \ldots, h_k$ an arbitrary system of integers. We ask under what conditions there exists an integral character χ of the group G satisfying the conditions

$$\chi(u_i) = h_i, \quad i = 1, 2, \ldots, k.$$

It turns out that the character χ exists then and only then, when the following condition is satisfied:

Whatever be the system of integers $a_1, a_2, \ldots, a_k; m$, where $m \geqslant 2$, such that $a_1u_1 + a_2u_2 + \ldots + a_ku_k = mu$, where $u \in G$, the integer $a_1h_1 + a_2h_2 + \ldots + a_kh_k$ is divisible by m.

This assertion is easily proved. For its proof see note 1.

(E) Let S^1 be an orientated circumference, K a complex of arbitrary dimensionality, G its one-dimensional Betti group and f a mapping of the complex K into S^1. Let, further, $u \in G$ and z be a cycle from the class of homologies u; then $f(z) = \chi_f(u)S^1$, where $\chi_f(u)$ is an integer equal to the power of the mapping of the cycle z, which depends only on u and not on the incidental choice of $z \in u$. It is easy to show that χ_f is an integral character of the group G. It turns out that two mappings f and g of the complex K into S^1 are then and only then equivalent, when the characters χ_f and χ_g corresponding to them coincide, $\chi_f = \chi_g$. Further, for every given character χ of the group G there exists such a mapping h that $\chi = \chi_h$.

For the proof see note 8.

(F) Let K be a certain complex and

$$u_1^r, u_2^r, \ldots, u_{p_r}^r \tag{11}$$

the r-dimensional basis of its weak homologies, i.e., such a system of r-dimensional cycles that every cycle of dimensionality r from K is weakly homologic to a linear form of cycles of the system (11) and, moreover, only to one such form. Similarly, let L be another complex, and $v_1^s, v_2^s, \dots, v_{q_s}^s$ its s-dimensional basis of weak homologies. Then the t-dimensional basis of weak homologies of the topological product $K \cdot L$ is composed of all cycles of the form $u_i^r \cdot v_j^s$, where $r + s = t$.

For the proof cf. note 9.

For us only the case $t = 1$ will be important.

Definition 1. A mapping f of the complex K into the n-dimensional sphere S^n is called homologically unessential, if whatever be the integer $m \geqslant 2$ and whatever be the n-dimensional cycle z, taken from K to the modulus m, its image under the mapping f is equal to zero to the modulus m.

Observe that if the mapping f is homologically unessential, then every mapping equivalent to it will be also homologically unessential.

Lemma 3. Let K be a complex of arbitrary dimensionality and φ its homologically unessential mapping into the two-dimensional sphere S^2. Then there exists such a mapping f of the complex K into the three-dimensional sphere S^3 that $\varphi = \vartheta f$ [cf. (A)].

Proof. We triangulate the sphere S^2 so that not one of the edges of triangulation passes through the north pole p of the sphere S^2 [cf. (A)] and choose an open circle V so small that its boundary $\dot{V}$ does not intersect with the edges of triangulation; then $\bar{V} = V + \dot{V}$ will also not intersect with the edges of triangulation. Let us further approximate the mapping φ by a simplicial mapping ψ and put

$$U = \psi^{-1}(V), \quad \dot{U} = \psi^{-1}(\dot{V}), \quad \bar{U} = \psi^{-1}(\bar{V});$$

then $\dot{U}$ is the boundary of the domain U in K and $\bar{U} = U + \dot{U}$ is its closure.

Since the closed circle $\bar{V}$ lies inside a simplex of triangulation, the set $\bar{U}$ falls, in virtue of Lemma 2, into the topological product $\bar{V} \cdot P$, so that if $x \cdot y = z \in \bar{U}$, where $x \in \bar{V}, y \in P$, then $\psi(z) = x$.

We shall assume that the circumference $\dot{V}$ is orientated in accordance with the angle coordinate introduced in it [cf. (A)]. Choose in every component of the complex P one point and denote these points by $p_1, p_2, \dots, p_k$. The orientated circumferences

$$\dot{V} \cdot p_1, \ \dot{V} \cdot p_2, \dots, \dot{V} \cdot p_k \tag{12}$$

are cycles in the complex $K-U$. Let us show that there exists a mapping λ of the complex $K-U$ into the orientated S^1 mapping every cycle of the system (12) with the power one.

In virtue of the propositions (D) and (E) for the proof of the stated assertions it is sufficient to show that from every relation of the form

$$a_1\dot{V}\cdot p_1 + a_2\dot{V}\cdot p_2 + \ldots + a_k\dot{V}\cdot p_k \sim mW \tag{13}$$

in $K-U$ follows that $a_1 + a_2 + \ldots + a_k$ is divisible by m. Suppose that the relation (13) has place. Then

$$mV - (a_1\dot{V}\cdot p_1 + a_2\dot{V}\cdot p_2 + \ldots + a_k\dot{V}\cdot p_k) = \dot{c},$$

where $\dot{c}$ is the boundary of a certain complex c from $K-U$. The complex $c + a_1 V\cdot p_1 + a_2 V\cdot p_2 + \ldots + a_k V\cdot p_k$ is evidently a cycle to the modulus m from K, and the power of the mapping ψ on it is equal to $a_1 + a_2 + \ldots + a_k$. Since the mapping ψ is homologically unessential, the number $a_1 + a_2 + \ldots + a_k$ is divisible by m, and our assertion on the existence of the mapping λ is proved.

On the circumference $\dot{V}$ there is an angle coordinate, and in the sequel we shall not distinguish between the point from $\dot{V}$ itself and its angle coordinate. On the circumference S^1 we shall also introduce the angle coordinate and shall make no distinction between the point from S^1 and its angle coordinate.

In virtue of this agreement $o \in \dot{V}$ is the point from $\dot{V}$ with the angle coordinate O. The point $O\cdot y$, where $y \in P$, belongs to $\dot{U} = \dot{V}\cdot P$, and hence is determined the function

$$\mu(y) = \lambda(O\cdot y), \tag{14}$$

where $\mu(y) \in S^1$, or the angle coordinate of the point $\mu(y)$. For $x \in \dot{V}$, $y \in P$ put

$$\nu(x\cdot y) = x + \mu(y). \tag{15}$$

Thus we have defined a mapping ν of the complex $\dot{U}$ into the circumference S^1. Let us show that this mapping is equivalent to the mapping λ of the complex $\dot{U} \subset K-U$.

For the proof we use the proposition (E). Since every weakly homologic to zero one-dimensional cycle from $\dot{U}$ is transformed under any mapping of the complex $\dot{U}$ into S^1 into the zero cycle, it is sufficient to show that the mappings λ and ν are algebraically equal on a certain basis of weak homologies from $\dot{U}$. For the construction of the basis of weak homologies of the complex $\dot{U}$ we use the proposition (F). Let $z_1, z_2, \ldots, z_l$ be a one-dimensional basis of weak homologies of the complex P. Then the basis of weak homologies of the complex $\dot{U}$ is

$$0 \cdot z_1, 0 \cdot z_2, \ldots, 0 \cdot z_l, \quad \dot{V} \cdot p_1, \dot{V} \cdot p_2, \ldots, \dot{V} \cdot p_k.$$

On every cycle of the form $0 \cdot z_j$ the mappings λ and ν simply coincide. On every cycle of the form $\dot{V} \cdot p_i$ both of them have the power one. Thus the mappings λ and ν of the complex $\dot{U}$ are equivalent.

Since the mappings λ and ν of the complex $\dot{U}$ are equivalent, there exists a continuous deformation transforming the mapping λ of the complex $\dot{U}$ into the mapping ν. This deformation may be extended into a deformation of the mapping λ of the whole complex $K-U$, transforming the mapping λ into a certain new mapping η, and this latter coincides on the complex $\dot{U}$ with ν.

Thus we have constructed a mapping η of the complex $K-U$ into S_1^1, coinciding with ν on $\dot{U}$ [cf. (15)].

The circumference $\dot{V}$ is given in polar coordinates, which are on the sphere S^2 [cf. (A)], by the equation $\rho = \epsilon$, where ϵ is a constant and ρ the radius vector.

The mapping g of the complex $\bar{U} = \bar{V} \cdot P$ into S^3 we define by putting

$$g(x \cdot y) = (\cos(\mu(y)) + \sin(\mu(y))i)(\sigma + \rho(\cos\beta \cdot j + \sin\beta \cdot k)) \quad \text{[cf. (A)]}, \tag{16}$$

where $x \in V$ and has the coordinates $\epsilon\rho$, β and $y \in P$. For $x \in \dot{V}$ we have

$$g(x \cdot y) = \cos(\mu(y) + \beta) \cdot j + \sin(\mu(y) + \beta) \cdot k. \tag{17}$$

The mapping g of the complex $K-U$ into S^3 we define by putting

$$g(z) = \cos\eta(z) \cdot j + \sin\eta(z) \cdot k. \tag{18}$$

From the relations (15), (17) and (18) directly follows that the so constructed on two parts of the complex K mapping g is coordinated on the intersection $\dot{U}$ of these parts, and hence we have a mapping g of the whole complex K into the sphere S^3 [cf. (A)].

Let us show that the mappings ψ and ϑ_g of the complex K into the sphere S^2 are equivalent.

Let us, in the first place, investigate the structure of the mapping ϑg. If x is a point from $\bar{V}$ with coordinates $\epsilon\rho$, β, and $y \in P$, then $\vartheta(g(x \cdot y))$ is the point from S^2 with coordinates ρ, β. The mapping ψ transforms the same point $x \cdot y$ into the point from S^2 with coordinates $\epsilon\rho$, β. The mapping ϑg is thus obtained from the mapping ψ for the point from $\bar{U}$ by a simple elongation of the radius vector. If the point $z \in K-U$, then $\vartheta(g(z)) = q$ [cf. (A)]. Thus, in order to obtain the mapping ϑg from the mapping ψ for $z \in K-U$, it is necessary to make the point $\psi(z)$ slide from its original position along the radius vector into the point

q. From what has been said we see that the mappings ϑg and ψ are equivalent.

Since the mappings φ and ψ are equivalent, we conclude, by what has been just proved, that the mappings φ and ϑg are equivalent. Hence, in virtue of (C), follows that there exists a mapping f satisfying the condition $\varphi = \vartheta f$. The lemma is thus proved.

In addition to Lemma 3 we make the following obvious remark.

(G) If f is a mapping of the complex K into S^3, then the mapping $\varphi = \vartheta f$ of the complex K into S^2 is homologically unessential.

From Lemmas 1 and 3 and the remark (G) we can deduce now the following important

Theorem 1. Let D be a class of mappings of the complex K into the three-dimensional sphere S^3. If f is a mapping of the class D, then denote by Δ the class of mappings of the complex K into the two-dimensional sphere S^2, which contains the mapping ϑf [cf. (A)]. Then the class Δ is determined by the class D and not by the incidental choice from D of the mapping f; hence we may put $\Delta = \vartheta(D)$. It turns out that so obtained correspondence ϑ is a one-to-one correspondence between all classes of mappings of the complex K into S^3 and all homologically unessential classes of mappings of the complex K into S^2.

Since we have already a classification of mappings of the three-dimensional sphere Σ^3 into the three-dimensional sphere S^3, Theorem 1 gives us a classification of mappings of the sphere Σ^3 into the sphere S^2.

If f is a mapping of the sphere Σ^3 into S^3, denote by $\omega_0(f, \Sigma^3)$ the power of this mapping. As is known, two mappings f and g of the sphere Σ^3 into the sphere S^3 are then and only then equivalent, when

$$\omega_0(f, \Sigma^3) = \omega_0(g, \Sigma^3). \tag{19}$$

In order to be able to give a more concrete classification of the mappings of the sphere Σ^3 into S^2, we recall the following definition due to Hopf (cf. note 2):

Definition 2. Let φ be a simplicial mapping of the oriented sphere Σ^3 into the orientated sphere S^2. Choose in the sphere S^2 two points $a \neq b$ not belonging to the edges of triangulation. Then $\varphi^{-1}(a)$ and $\varphi^{-1}(b)$ are naturally one-dimensional cycles from Σ^3 [cf. (C′)]. Denote the linkage coefficient of these cycles by $\omega_1(\varphi, \Sigma^3)$. Let, further, c be a certain two-dimensional algebraical complex from Σ^3 with the boundary $\varphi^{-1}(a)$; then the power of its mapping under φ on S^2 is equal to $\omega_1(\varphi, \Sigma^3)$.

H. Hopf, to whom this construction belongs, has shown (cf. note 2) that for two equivalent mappings φ and ψ we have $\omega_1(\varphi, \Sigma^3) = \omega_1(\psi, \Sigma^3)$. He has also shown that if f is a mapping of the sphere Σ^3 into the sphere S^3 then

$$\omega_1(\vartheta f, \Sigma^3) = \omega_0(f, \Sigma^3) \quad \text{[cf. (A)]}. \tag{20}$$

From this we deduce on groun of Lemma 3 and the condition of equivalency (19) the following

Theorem 2. Two mappings φ and ψ of the three-dimensional sphere Σ^3 into the two-dimensional sphere S^2 are then and only then equivalent, when

$$\omega_1(\varphi, \Sigma^3) = \omega_1(\psi, \Sigma^3)$$

(cf. Definition 2).

2. Preliminary Notions and Remarks

For a classification of mappings of a three-dimensional complex K^3 into the two-dimensional sphere S^2 (cf. Theorem 3) we have to introduce certain invariants of the mappings of K^3 into S^2 [cf. Section 2, (F)], as well as certain invariants of pairs of mappings of K^3 into S^2 [cf. Section 2, (A′)]. The present paragraph is devoted to the introduction of these invariants necessary for the formulation of Theorem 3 itself, as well as of invariants necessary for its proof.

In the first place let us introduce certain denotations and terms. Let $\{f_t\}$ be a family of continuous mappings of the space F into the space R, where t is an arbitrarily element of the topological space Δ. Denote by $F \cdot \Delta$ the topological product of the spaces F and Δ. Then every element $z \in F \cdot \Delta$ is representable in the form of a pair $z = x \cdot t$, where $x \in F$, $t \in \Delta$. We define the mapping f_Δ of the space $F \cdot \Delta$ into R by putting $f_\Delta(z) = f_\Delta(x \cdot t) = f_t(x)$. If this mapping is continuous, then we shall call the family $\{f_t\}$ also continuous. If, conversely, a certain continuous mapping f_Δ of the product $F \cdot \Delta$ is given, then it generates a continuous family $\{f_t\}$ of mappings of the space F.

In the sequel two cases will be essential for us:

(a) Δ is composed of all real numbers $0 \leqslant t \leqslant 1$,

(b) Δ assumes two values 0 and 1.

If the question requires an algebraical interpretation and F is an algebraical complex, then in both cases we shall consider the product $F \cdot \Delta$ also as an algebraical complex. In the case (a) we orientate $F \cdot \Delta$ so that $F \cdot 0$ should enter into the boundary of $F \cdot \Delta$ with the negative sign. In the case of (b) we orientate $F \cdot \Delta$ so that $F \cdot \Delta = F \cdot 1 - F \cdot 0$.

If two mappings f and g of the space F coincide on a closed subset $E \subset F$ and there exists a continuous deformation of the mapping f into the mapping g not changing the mapping on E, we shall say that the mappings f and g are equivalent with respect to E.

Let f_t and g_t be two continuous deformations of the mappings of the space F, coinciding on a closed subset $E \subset F$, such that f_0 and g_0 coincide, as well as f_1 and g_1. It is easily seen that the mappings f_Δ and g_Δ [cf. (a)] coincide then on the set $E^* = F \cdot 0 \vee F \cdot 1 \vee E \cdot \Delta \subset F \cdot \Delta$. We shall say that the continuous deformation f_t and g_t are equivalent with respect to E, if the mappings f_Δ and g_Δ of the space $F \cdot \Delta$ are equivalent with respect to E^*.

If f is a continuous mapping of the orientated n-dimensional ($n = 2, 3$) sphere Σ^n into the orientated two-dimensional sphere S^2, then by $\omega_{n-2}(f, \Sigma^n)$ we shall denote for $n = 2$ the power and for $n = 3$ Hopf's number of the mapping f (cf. Definition 2).

(A) Let f_0 and f_1 be two mappings of the n-dimensional ($n = 2, 3$) orientated element T^n into the two-dimensional orientated sphere S^2, coinciding on the boundary $\dot{T}^n$ of the element T^n. Let us introduce the index $\omega_{n-2}(f_0, f_1, T^n)$ estimating the difference of the mappings f_0 and f_1. Denote by Δ the aggregate of two numbers 0 and 1. Identity in the space $T^n \cdot \Delta$ in one point every pair of points $x \cdot 0$ and $x \cdot 1$, where $x \in \dot{T}^n$; then we obtain from the complex $T \cdot \Delta$ the orientated sphere Σ^n. Since the mappings f_0 and f_1 coincide on the boundary $\dot{T}^n$, the mapping f_Δ may be interpreted as a mapping of the sphere Σ^n. Put

$$\omega_{n-2}(f_0, f_1, T^n) = \omega_{n-2}(f_\Delta, \Sigma^n).$$

It is easily seen that in order that the mappings f_0 and f_1 should be equivalent with respect to T^n, it is necessary and sufficient that $\omega_{n-2}(f_0, f_1, T^n) = 0$. It is as easily seen that the index introduced above does not vary at a simultaneous deformation of the mappings f_0 and f_1, if they remain coinciding on the boundary $\dot{T}^n$.

(B) Let f_t and g_t be two continuous deformations of the mappings of the $(n-1)$-dimensional ($n = 2, 3$) orientated element E^{n-1} into the orientated two-dimensional sphere S^2, coinciding on the boundary $\dot{E}^{n-1}$ of the element E^{n-1}, such that the mappings f_0 and g_0 coincide, as well as the mappings f_1 and g_1. Introduce the index $\omega_{n-2}(f_t, g_t, E^{n-1})$ estimating the difference of deformations f_t and g_t. Let Δ be the set of all numbers $0 \leqslant t \leqslant 1$. It is easily seen that the mappings f_Δ and g_Δ of the element $T^n = E^{n-1} \cdot \Delta$ coincide on the boundary $\dot{T}^n$ of the element T^n. Put

$$\omega_{n-2}(f_t, g_t, E^{n-1}) = \omega_{n-2}(f_\Delta, g_\Delta, T^n).$$

It is easily seen that the deformations f_t and g_t are equivalent with respect to $\dot{E}^{n-1}$ then and only then, when the index is equal to zero.

(C) Let f_t and g_t be two continuous deformations of mappings of the orientated n-dimensional ($n = 2, 3$) simplex T^n into the orientated two-dimensional sphere S^2. Denote by $\dot{T}^n$ the boundary of T^n and by T' the aggregate of $(n-2)$-dimensional faces of the simplex T^n. Suppose that the mappings f_0 and g_0 coincide on $\dot{T}^n$, as well as the mappings f_1 and g_1. Suppose, further, that the deformations f_t and g_t coincide on T'. Denote by E_i^{n-1}, $i = 0, \ldots, n$, the faces of the simplex T^n, properly orientated. Then we have

$$\omega_{n-2}(f_1, g_1, T^n) = \omega_{n-2}(f_0, g_0, T^n) + \sum_{i=0}^{n} \omega_{n-2}(f_t, g_t, E_i^{n-1}).$$

Let us prove the assertion (C). The set of all numbers $0 \leqslant t \leqslant 1$ denote by Δ. We shall consider the complex $T^n \cdot \Delta$ in two copies; the first we shall denote by $T^n \cdot \Delta$, the second by $[T^n \cdot \Delta]$. Similarly we shall distinguish between all possible subsets and algebraical subcomplexes of the complexes $T^n \cdot \Delta$ and $[T^n \cdot \Delta]$. By P we shall denote the complex consisting of the two components $T^n \cdot \Delta$ and $[T^n \cdot \Delta]$. Define the mapping ψ of the complex P as coinciding with f_Δ on $T^n \cdot \Delta$ and with g_Δ on $[T^n \cdot \Delta]$. The aggregate of all $(n-1)$-dimensional faces of the prism $T^n \cdot \Delta$ we denote by A. If $x \in A$, then it is easily seen that $\psi(x) = \psi([x])$. Let us identify in the complex P every pair of points x, $[x]$, where $x \in A$. Then we obtain a complex Q. The mapping ψ of the complex P may be now interpreted as a mapping of the complex Q.

Observe that the following algebraical complexes from Q are orientated spheres:

$$T^n \cdot 0 - T^n \cdot 1 + \sum_{i=0}^{n} E_i^{n-1} \cdot \Delta = U,$$

$$[T^n \cdot 0] - [T^n \cdot 1] + \sum_{i=0}^{n} [E_i^{n-1} \cdot \Delta] = [U],$$

$$[T^n\cdot 0]-T^n\cdot 0 = V_0,\ [T^n\cdot 1]-T^n\cdot 1 = V_1,$$

$$[E_i^{n-1}\cdot\Delta]-E_i^{n-1}\cdot\Delta = W_i,$$

Hence we have

$$[U]-U = V_0-V_1+\sum_{i=0}^{n} W_i,$$

Observe that the sphere U is the boundary of the prism $-T^n\cdot\Delta$, and the sphere $[U]$ – the boundary of the prims $-[T^n\cdot\Delta]$.

Consider now the case $n=2$. Since U and $[U]$ are homotopic to zero in Q,

$$\omega_0(\psi, V_0-V_1+\sum_{i=0}^{2} W_i) = \omega_0(\psi,[U]-U) = 0.$$

But

$$\omega_0(\psi, V_0) = \omega_0(f_0,g_0,T^2),\ \omega_0(\psi,V_1) = \omega_0(f_1,g_1,T^2),$$

$$\omega_0(\psi, W_i) = \omega_0(f_t,g_t,E_i^1).$$

Thus, for $n=2$ the assertion is proved.

Consider the case $n=3$. It is easily seen that in this case every two-dimensional cycle from Q is homological to zero, and hence the mapping ψ is homologically unessential; therefore, there exists a mapping χ of the complex Q into the three-dimensional sphere S^3 such that $\psi=\vartheta\chi$ (cf. Section 1, Lemma 3). Since $\omega_0(\chi,\Sigma^3)=\omega_1(\psi,\Sigma^3)$ for an arbitrary sphere Σ^3 from Q, we obtain the required result by applying to the mapping χ the same argument, as we applied above to ψ.

(D) Let T^n be an n-dimensional ($n=2,3$) orientated element and f,g,h three of its mappings into the orientated two-dimensional sphere S^2 such that all these mappings coincide on the boundary $\dot{T}^n$ of the element T^n. Then we have

$$\omega_{n-2}(f,g,T^n)+\omega_{n-2}(g,h,T^n) = \omega_{n-2}(f,h,T^n).$$

In the case $n=2$ the proof follows directly by computation of the powers of the mappings. Consider the case $n=3$. Take three copies of the element T^3 and denote them by T^3, $[T^3]$, $\{T^3\}$. Compose the complex P of the three components T^3, $[T^3]$, $\{T^3\}$ and defines the mapping ψ of the complex P as coinciding with f on T^3, with g on $[T^3]$ and with h on $\{T^3\}$. Identify in the complex P in one point every triple of points x, $[x]$, $\{x\}$, where $x\in\dot{T}^3$; the so obtained complex denote by Q. The mapping ψ may be obviously interpreted as a mapping of the complex Q. Since in the complex Q every two-dimensional cycle is homologic to zero, there exists a mapping χ of the complex Q into the three-dimensional sphere S^3 such that $\psi = \vartheta\chi$ (cf. Section 1, Lemma 3). Thus the question is again, as in (C), reduced to consideration of the power of the mapping.

We apply now the established definitions and results to the mappings of the n-dimensional complex K^n ($n = 2, 3$) into the two-dimensional orientated sphere S^2. By K^r we shall denote hereby the aggregate of all simplexes of the complex K^n of dimensionalities less than or equal to r.

(A′) Let f_0 and f_1 be two mappings of the complex K^n ($n = 2, 3$) into S^2, coinciding on K^{n-1}. We introduce the n-dimensional ∇-complex estimating the difference between the mappings f_0 and f_1 and denote it by $\omega_{n-2}(f_0, f_1, K^n)$.

If T^n is an orientated n-dimensional simplex from K^n, then we define the value of the n-dimensional ∇-complex to be introduced as

$$\omega_{n-2}(f_0, f_1, T^n)$$

on T^n.

It is easily seen that mappings f_0 and f_1 of the complex K^n are then and only then equivalent with respect to K^{n-1}, when $\omega_{n-2}(f_0, f_1, K^n) = 0$. It is as easily seen that if the mappings f_0 and f_1 are subjected to one and the same simultaneous deformation, while they remain coinciding on K^{n-1}, then the complex $\omega_{n-2}(f_0, f_1, K^n)$ does not vary.

(B′) Let f_t and g_t be two continuous deformations of the mappings of the complex K^{n-1} into S^2, coinciding on K^{n-2}, such that the mappings f_0 and g_0 coincide, as well as f_1 and g_1. Introduce the $(n-1)$-dimensional ∇-complex $\omega_{n-2}(f_t, g_t, K^{n-1})$ estimating the difference of the deformations f_t and g_t. If T^{n-1} is an $(n-1)$-dimensional orientated simplex from K^{n-1}, then the value on it of the ∇-complex to be introduced we define as

$$\omega_{n-2}(f_t, g_t, T^{n-1}).$$

It is easily seen that continuous deformations f_t and g_t are then and only then equivalent with respect to K^{n-2}, when $\omega_{n-2}(f_t, g_t, K^{n-1}) = 0$.

(C′) Let f_t and g_t be two continuous deformations of the mappings of the complex K^n into S^2, coinciding on K^{n-2}, such that the mappings f_0 and g_0, as well as f_1 and g_1, coincide on K^{n-1}. Then we have

$$\omega_{n-2}(f_1, g_1, K^n) = \omega_{n-2}(f_0, g_0, K^n) + \nabla\omega_{n-2}(f_t, g_t, K^{n-1}),$$

where the sign ∇ denotes the ∇-boundary.

(D′) Let f, g and h be three mappings of the complex K^n into S^2, coinciding on K^{n-1}. Then we have

$$\omega_{n-2}(f, g, K^n) + \omega_{n-2}(g, h, K^n) = \omega_{n-2}(f, h, K^n).$$

(E) In addition to (A) we observe that if f_0 is a certain mapping of the n-dimensional ($n = 2, 3$) element T^n into S^2, then there exists a mapping f_1 of the same element into S^2 such that the number $\omega_{n-2}(f_0, f_1, T^n)$ is defined and has a given value. Hence, in addition to (A′), follows that if f_0 is a mapping of the n-dimensional ($n = 2, 3$) complex K^n into S^2, then there exists a mapping f_1 of the same complex into S^2 such that the ∇-complex $\omega_{n-2}(f_0, f_1, K^n)$ is defined and coincides with the given one. In precisely the same manner we observe, in addition to (B), that if f_t is a deformation of the mappings of the $(n-1)$-dimensional ($n = 2, 3$) element E^{n-1} into S^2, then there exists a continuous deformation g_t of mappings of the same element into S^2 such that the number

$$\omega_{n-2}(f_t, g_t, E^{n-1})$$

is defined and has a given value. In addition to (B′) hence follows that if f_t is a deformation of the mappings of the $(n-1)$-dimensional ($n = 2, 3$) complex K^{n-1} into S^2, then there exists a continuous deformation g_t of the mappings of the same complex into S^2 such that the ∇-complex $\omega_{n-2}(f_t, g_t, K^{n-1})$ is defined and coincides with the given one.

Let us prove (E). Let f_0 be a given mapping of the element T^n and let R^n be an element from T^n not intersecting with the boundary $\dot{T}^n$ of the element T^n. Let us deformate the mapping f_0, not changing it on the boundary, into such a mapping f_0' that $f_0'(R^n)$ contains only one point $p \in S^2$. Identify now in one point q all points of the boundary $\dot{R}^n$ of the element R^n and denote the so obtained sphere from R^n by Σ^n. Let us now determine the mapping f' of the sphere Σ^n into S^2 such that $f'(q) = p$ and $\omega_{n-2}(f', \Sigma^n)$ has a given value. The mapping f' of the sphere Σ^n we shall interprete as a mapping of the element R^n. The mapping f_1 of the element T^n we define as coinciding with f_0' on $T^n - R^n$ and as coinciding with f' on R^n. The continuous deformation g_t is constructed in precisely the same way by starting from the mapping f_Δ (Δ being the set of all numbers $0 \leqslant t \leqslant 1$) of the element $T^n = E^{n-1} \cdot \Delta$.

(F) Let K^2 be a certain two-dimensional complex, K^1 – the complex composed from all nul-dimensional and one-dimensional simplexes of the complex K^2, and f a mapping of K^2 into S^2. Suppose that there exists a point $p \in S^2$ such that $f(K^1)$ does not contain p. Let us now define the ∇-complex $\omega_0(f, K^2)$ characterizing the mapping f. The value of the complex $\omega_0(f, K^2)$ on the simplex T^2 from K^2 we define as the power of the mapping f of the simplex T^2 at the point p. It turns out that two mappings f and g are equivalent then and only then, when

$$\omega_0(f, K^2) \mathrel{\underset{\nabla}{\widetilde{}}} \omega_0(g, K^2).$$

It is evident that if f_0 and f_1 are two mappings of the complex K^2 into S^2, coinciding on K^1, then $\omega_0(f_0, f_1, K^2) = \omega_0(f_1, K^2) - \omega_0(f_0, K^2)$.

The proposition (F) which is a particular case of Whitney's theorem,[10] is given here without proof; it follows also very easily from what has been proved already in the present paragraph.

(G) Let K^3 be a three-dimensional complex and K^2 the complex composed of all simplexes of the complex K^3 of dimensionality not greater than 2. The mapping f of the complex K^2 into S^2 may be then and only then extended to the whole complex K^3, when $\omega_0(f, K^2)$ is a ∇-cycle in K^3 [cf. (F)].

Suppose that the mapping f is already defined on the whole complex K^3, and that T^3 is a three-dimensional orientated simplex from K^3. Then the power of the mapping f of the boundary $\dot{T}^3$ of the simplex T^3 is equal to zero. On the other hand this power is obviously equal to the sum of powers of the mappings of the faces of the simplex T^3. Thus $\nabla \omega_0(f, K^2) = 0$.

Suppose that f is given on K^2 and $\nabla \omega_0(f, K^2) = 0$. Then the power of the mapping f of the boundary $\dot{T}^3$ of a certain simplex T^3 from K^3 is equal to zero, and consequently the mapping f may be extended to T^3, and we obtain an extension of the mapping f to the whole complex K^3.

3. The Mapping of a Three-dimensional Complex into the Two-dimensional Sphere

By K^n we shall, as above, denote an n-dimensional complex and by K^r the aggregate of all simplexes from K^n, whose dimensionality does not exceed r. By S^r we shall denote the r-dimensional orientated sphere.

If f and g are two mappings of the complex K^3 into S^2, then for the solution of the question on their equivalency we have first of all to solve the question on the equivalency of these mappings on K^2. In fact, if it turns out that the mappings f and g are not equivalent already on K^2, then the question on their equivalency on K^3 is by this answered in the negative. The criterium of equivalency of the mappings f and g on K^2 has been already given in Section 2 [cf. Section 2, (F)]. Thus it remains to consider the question on the equivalency of the mappings f and g of the complex K^3 in the case, when these mappings are equivalent on K^2. Under this assumption we can transform the mapping g of the complex K^3 by a continuous deformation into the mapping f and then extend this deformation t_0 the whole complex K^3. We come so to the case, when the mappings f and g of the complex K^3 simply coincide on K^2. The question on the equivalency in this case is completely answered by the following theorem.

Theorem 3. Let f and g be two mappings of the complex K^3 into S^2, coinciding on K^2. Put $\omega_0(f_1, K^2) = \omega_0(g, K^2) = z^3$ [cf. Section 2, (E)]. Put, further, $\omega_1(f, g, K^3) = z^3$ [cf. Section 2, (A′)]. z^2 and z^3 are a two-dimensional and three-dimensional ∇-cycles from K^3 [cf. Section 2, (G)]. The mappings f and g of the complex K^3 are equivalent then and only then, when there exists in K^3 a one-dimensional ∇-cycle x^1 such that

$$z^3 \underset{\nabla}{\approx} 2x^1 \times z^2 \quad \text{(cf. P. C.).} \tag{1}$$

Before we proceed to the proof of Theorem 3, we prove Lemma 4, which in substance solves already the question.

Lemma 4. Let f_t be a continuous deformation of mappings of the complex K^3 into the sphere S^2 such that the mappings f_0 and f_1 coincide on K^2, and the mapping f_t coincides on K^0 with the mapping f_0 for arbitrary t. Put

$$\omega_0(f_0, K^2) = \omega_0(f_1, K^2) = z^2, \tag{2}$$

$$\omega_1(f_0, f_1, K^3) = z^3. \tag{3}$$

Then z^2 and z^3 are a two-dimensional and a three-dimensional ∇-cycles from K^3. Denote, further, by e_t such a continuous deformation of the mappings of the complex K^3 that the mapping e_t coincides with f_0 for arbitrary t and put

$$\varpi(e_t, f_t, K^1) = x^1. \tag{4}$$

Then it turns out that x^1 is a ∇-cycle from K^3 and

$$z^3 \underset{\nabla}{\approx} 2\epsilon x^1 \times z^2 \quad \text{(cf. Section 2),} \tag{5}$$

Proof. Let Δ be the set of all numbers $0 \leqslant t \leqslant 1$; then the mapping f_Δ of the complex $K^3 \cdot \Delta$ into S^2 is defined (cf. Section 2). The complex $K^3 \cdot \Delta$ is not simplicial, but without limiting the generality we may suppose that f_Δ is a simplicial mapping of a certain simplicial subdivision of the complex $K^3 \cdot \Delta$.

By p^0 and p^1 we denote two points of the sphere S^2 lying inside one of the simplexes of the taken triangulation of the sphere S^2. Let T^r be an oriented simplex of the complex K^3. By $P^i(T^r \cdot \Delta)$ we denote the complete orientated original of the point p^i in $T^r \cdot \Delta$ under the mapping f_Δ, $i = 0, 1$ [cf. Section 1, (C′)]. By $P^i_t(T^r)$ denote the complete orientated original of the point p^i in T^r under the mapping f_t, $i = 0, 1$, $t = 0, 1$ [cf. Section 1, (C′)]. In view of the fact that the mappings f_0 and f_1 coincide on K^2, we have

$$P^i_0(T^2) = P^i_1(T^2) = P^i(T^2).$$

In the sequel we shall for shortness use the denotations introduced in my preceding paper [cf. P.C., (A)]. For computation of the relations of bounding we shall use the relation obtained in the present paper [cf. Section 1, (C′)].

We note the following relations of bounding:

$$P^i(T^2 \cdot \Delta)^{\cdot} = P^i(T^2) \cdot 1 - P^i(T^2) \cdot 0 - P^i(\dot{T}^2 \cdot \Delta), \tag{6}$$

$$P^i(\dot{T}^3 \cdot \Delta)^{\cdot} = P^i(\dot{T}^3) \cdot 1 - P^i(\dot{T}^3) \cdot 0 \quad \text{[cf. (6)]}, \tag{7}$$

$$P^i_0(T^3)^{\cdot} = P^i_1(T^3)^{\cdot} = P^i(\dot{T}^3). \tag{8}$$

If C is a certain nul-dimensional complex, then under the index $I(C)$ of this complex we shall understand the algebraical number of points entering into it. In virtue of the very definition of ∇-complexes z^3 and x^1 we have

$$z^2(\mathrm{T}^2) = I(P^i(T^2)), \tag{9}$$

$$x^1(T^1) = I(P^i(T^1 \cdot \delta)). \tag{10}$$

Let now every point $x \cdot t$ ($x \in T^r$, t a number) from the complex $P^i(T^r \cdot \Delta)$ slide along a straight line and uniformly in time into the point $x \cdot 0$ (the straightness and uniformness is understood in the sense of affine geometry which is in the prism $T^r \cdot \Delta$). The complex, situated in $T^r \cdot \Delta$, described by this motion of the whole complex $P^i(T^r \cdot \Delta)$ we denote by $Q^i(T^r \cdot \Delta)$. The complex from $T^r \cdot 0$, into which the complex $P^i(T^r \cdot \Delta)$ passes at the end of the motion, we denote by $Q^i(T^r) \cdot 0$, where the complex $Q^i(T^r)$ belongs to T^r.

Observe that the complex $Q^i(T^r)$ is a projection of the complex $P^i(T^r \cdot \Delta)$; in particular, for $r = 1$ both these nul-dimensional complexes have an equal index, and consequently

$$x^1(T^1) = I(Q^i(T^1)) \quad \text{[cf. (10)]}. \tag{11}$$

Note the following relations of bounding:

$$Q^i(T^2\cdot\Delta)^{\cdot} = P^i(T^2\cdot\Delta) - P^i(T^2)\cdot\Delta - Q^i(T^2)\cdot 0 - Q^i(\dot{T}^2\cdot\Delta), \tag{12}$$

$$Q^i(\dot{T}^3\cdot\Delta)^{\cdot} = P^i(\dot{T}^3\cdot\Delta) - P^i(\dot{T}^3)\cdot\Delta - Q^i(\dot{T}^3)\cdot 0 \quad \text{[cf. (12)]}, \tag{13}$$

$$Q^i(T^2)^{\cdot} = -Q^i(\dot{T}^2) \quad \text{[cf. (6)]}, \tag{14}$$

$$Q^i(\dot{T}^3)^{\cdot} = 0 \quad \text{[cf. (14)]}. \tag{15}$$

From the relations (14) and (11) follows that x^1 is a ∇-cycle in K^3. This fact could have been also established in a more direct way on ground of the proposition (C′), Section 2. Similarly, from the relations (8) and (9) follows that z^2 is a ∇-cycle in K^3.

From the relation (8) follows that the nul-dimensional cycle $P^i(\dot{T}^3)$ situated in $\dot{T}^3$ has the index zero and, consequently, bounds a certain one-dimensional complex $A^i(\dot{T}^3)$ also situated in $\dot{T}^3$,

$$A^i(\dot{T}^3)^{\cdot} = P^i(\dot{T}^3). \tag{16}$$

In virtue of (8) and (16) the one-dimensional complex $P^i_t(T^3) - A^i(\dot{T}^3)$ is for $t = 0, 1$, a cycle in the simplex T^3 and, consequently, bounds in it a certain two-dimensional complex $B^i_t(T^3)$, $t = 0, 1$,

$$B^i_t(T^3)^{\cdot} = P^i_t(T^3) - A^i(\dot{T}^3) \quad (t = 0, 1). \tag{17}$$

In virtue of the relations (7) and (16) the one-dimensional complex

$$P^i(\dot{T}^3\cdot\Delta) + A^i(\dot{T}^3)\cdot 0 - A^i(\dot{T}^3)\cdot 1$$

situated in $\dot{T}^3\cdot\Delta$ is a cycle and, consequently, bounds a certain two-dimensional complex $B^i(\dot{T}^3\cdot\Delta)$ situated in $\dot{T}^3\cdot\Delta$,

$$B^i(\dot{T}^3\cdot\Delta)^{\cdot} = P^i(\dot{T}^3\cdot\Delta) + A^i(\dot{T}^3)\cdot 0 - A^i(\dot{T}^3)\cdot 1. \tag{18}$$

Consider in the complex $K^3\cdot\Delta$ the orientated three-dimensional sphere

$$\Sigma = (T^3\cdot\Delta)^{\cdot} = T^3\cdot 1 - T^3\cdot 0 - \dot{T}^3\cdot\Delta. \tag{19}$$

Since this sphere bounds in $K^3\cdot\Delta$ the element $T^3\cdot\Delta$, the mapping f_Δ is for it unessential, i.e., $\omega_1(f_\Delta, \Sigma) = 0$. Let us calculate this vanishing invariant by means of the introduced algebraical complexes; to this end denote by U^i the complete original of the point p^i in Σ under this mapping f_Δ. It is easily seen that

$$U^i = P^i_1(T^3)\cdot 1 - P^i_0(T^3)\cdot 0 - P^i(\dot{T}^3\cdot\Delta). \tag{20}$$

Thus the linkage coefficient of the cycles U^0 and U^1 in Σ is equal to zero, and we have

$$V_\Sigma(U^0, U^1) = I_\Sigma(B_1^0(T^3)\cdot 1 - B_0^0(T^3)\cdot 0 - B^0(\dot{T}^3\cdot\Delta),\ P_1^1(T^3)\cdot 1 - \\ - P_0^1(T^3)\cdot 0 - P^1(\dot{T}^3\cdot\Delta)) = I_{T^3}(B_1^0(T^3), P_1^1(T^3)) - \\ - I_{T^3}(B_0^0(T^3), P_0^1(T^3)) - I_{\dot{T}^3\cdot\Delta}(B^0(\dot{T}^3\cdot\Delta), P^1(\dot{T}^3\cdot\Delta)) = 0.$$

Hence we obtain

$$w(T^3) = I_{T^3}(B_1^0(T^3), P_1^1(T^3)) - I_{T^3}(B_0^0(T^3), P_0^1(T^3)) = \\ = I_{\dot{T}^3\cdot\Delta}(B^0(\dot{T}^3\cdot\Delta), P^1(\dot{T}^3\cdot\Delta)). \tag{21}$$

Let us now show that

$$w(T^3) = z^3(T^3) \quad \text{[cf. (3)]}. \tag{22}$$

To this end denote by Δ' the aggregate of two numbers 0 and 1. Then is defined the mapping $f_{\Delta'}$ of the complex $T^3\cdot\Delta'$ (cf. Section 2). The complex $T^3\cdot\Delta'$ consists of two elements $T^3\cdot 1$ and $-T^3\cdot 0$, and $f_{\Delta'}(x\cdot 0) = f_{\Delta'}(x\cdot 1)$ for every $x\in\dot{T}^3$. Identify in one point every pair of points $x\cdot 0$ and $x\cdot 1$; then the complex $T^3\cdot\Delta'$ will be transformed into the sphere Σ', and the mapping $f_{\Delta'}$ may be considered as a mapping of the sphere Σ'. In virtue of the very definition of the ∇-cycle z^3 we have $z^3(T^3) = \omega_1(f_{\Delta'}, \Sigma')$. Let us calculate $\omega_1(f_{\Delta'}, \Sigma')$. To this end denote by w^i the original of the point p^i in Σ' under the mapping $f_{\Delta'}$. It is easily seen that

$$w^i = P_1^i(T^3)\cdot 1 - P_0^i(T^3)\cdot 0.$$

Thus we have

$$z^3(T^3) = V_{\Sigma'}(w^0, w^1) = \\ = I_{\Sigma'}(B_1^0(T^3)\cdot 1 - B_0^0(T^3)\cdot 0, P_1^1(T^3)\cdot 1 - P_0^1(T^3)\cdot 0) = \\ = I_{T^3}(B_1^0(T^3), P_1^1(T^3)) - I_{T^3}(B_0^0(T^3), P_0^1(T^3)).$$

So the relation (22) is proved and we obtain

$$z^3(T^3) = I_{\dot{T}^3\cdot\Delta}(B^0(\dot{T}^3\cdot\Delta), P^1(\dot{T}^3\cdot\Delta)). \tag{23}$$

Let us now calculate $z^3(T^3)$ starting from the relation (23). To this end identify in the complex $\dot{T}^3\cdot\Delta$ in one point every pair of points $x\cdot 0$ and $x\cdot 1$ and denote the obtained manifold by M; this manifold is, as may be easily seen, homoeomorphic to the product of the two-dimensional sphere and the circumference. The one-dimensional complex $P^i(\dot{T}^3\cdot\Delta)$ represents in M a cycle [cf. (7)] and bounds in M the two-dimensional complex $B^i(\dot{T}^3 \cdot \Delta)$ [cf. (18)]. Thus the linkage coefficient

$$V_M(P^0(\dot{T}^3 \cdot \Delta), P^1(\dot{T}^3 \cdot \Delta)) \tag{24}$$

is defined, since both involved cycles are homologic to zero in M. The relation (23), on the other hand, shows that $z^3(T^3)$ is nothing else but the linkage coefficient (24),

$$z^3(T^3) = V_M(P^0(\dot{T}^3 \cdot \Delta), P^1(\dot{T}^3 \cdot \Delta)). \tag{25}$$

Let us now calculate this linkage coefficient.

Observe that the complexes $P^i(\dot{T}^3) \cdot \Delta$ and $Q^i(\dot{T}^3) \cdot 0$ are cycles in M [cf. (15)]. In virtue of the relation (13) we have

$$\begin{aligned} &V_M(P^0(\dot{T}^3 \cdot \Delta) - P^0(\dot{T}^3) \cdot \Delta - Q^0_0(\dot{T}^3) \cdot 0, P^1(\dot{T}^3 \cdot \Delta)) = \\ &= I_M(Q^0(\dot{T}^3 \cdot \Delta), P^1(\dot{T}^3 \cdot \Delta)) = \\ &= \sum_{T^2 \in \dot{T}^2} I_{T^2 \in \dot{T}^2}(Q^0(T^2 \cdot \Delta), P^1(T^2 \cdot \Delta)) = \sum_{T^2 \in \dot{T}^2} y^2(T^2), \end{aligned} \tag{26}$$

where y^2 is a function defined on an arbitrary two-dimensional simplex T^2, i.e., a two-dimensional ∇-complex from K^3 depending on the deformation f_t. Put now

$$z^{*3}(T^3) = V_M(P^0(\dot{T}^2) \cdot \Delta + Q^0(\dot{T}^3) \cdot 0, P^1(\dot{T}^3 \cdot \Delta)). \tag{27}$$

In virtue of the relations (25), (26) and (27) we have

$$z^3 - z^{*3} = \nabla y^2,$$

and this means that

$$z^3 \underset{\nabla}{\sim} z^{*3}. \tag{28}$$

Now we shall calculate $z^{*3}(T^3)$ on ground of the relation (27).

In virtue of the relation (27) we have

$$z^{*3}(T^3) = V_M(P^0(\dot{T}^3) \cdot \Delta, P^1(\dot{T}^3 \cdot \Delta)) + V_M(Q^0(\dot{T}^3) \cdot 0, P^1(\dot{T}^3 \cdot \Delta)). \tag{29}$$

Since the cycle $Q^0(\dot{T}^3) \cdot 0$ lies totally in $\dot{T}^3 \cdot 0$, and the cycle $P^1(\dot{T}^3 \cdot \Delta)$ intersects $T^3 \cdot 0$ along the nul-dimensional complex $P^1(\dot{T}^3) \cdot 0$,

$$V_M(Q^0(\dot{T}^3) \cdot 0, P^1(\dot{T}^3 \cdot \Delta)) = V_{\dot{T}^3}(Q^0(\dot{T}^3), P^1(\dot{T}^3)). \tag{30}$$

Further,

$$V_M(P^0(\dot{T}^3) \cdot \Delta, P^1(\dot{T}^3 \cdot \Delta)) = V_M(P^1(\dot{T}^3 \cdot \Delta), P^0(\dot{T}^3) \cdot \Delta). \tag{31}$$

In virtue of the relation (13) we have

$$V_M(P^1(\dot{T}^3\cdot\Delta)-P^1(\dot{T}^3)\cdot\Delta-Q^1(\dot{T}^3)\cdot 0, P^0(\dot{T}^3)\cdot\Delta) = \\ = I_M(Q^1(\dot{T}^3\cdot\Delta), P^0(\dot{T}^3)\cdot\Delta) = 0, \tag{32}$$

since the complex $Q^1(\dot{T}^3\cdot\Delta)$ does not intersect with $P^0(\dot{T}^3)\cdot\Delta$. Thus, from (32) we obtain

$$V_M(P^1(\dot{T}^3\cdot\Delta), P^0(\dot{T}^3)\cdot\Delta) = \\ = V_M(P^1(\dot{T}^3)\cdot\Delta, P^0(\dot{T}^3)\cdot\Delta)+V_M(Q^1(\dot{T}^3)\cdot 0, P^0(\dot{T}^3)\cdot\Delta) = \\ = V_M(P^1(\dot{T}^3)\cdot\Delta, P^0(\dot{T}^3)\cdot\Delta)+V_{\dot{T}^3}(Q^1(\dot{T}^3), P^0(\dot{T}^3)). \tag{33}$$

The cycle $P^1(T^3)\cdot\Delta$ bounds in M the complex $-A^1(T^3)\cdot\Delta$, which does not intersect with $P^0(\dot{T}^3)\cdot\Delta$, and hence

$$V_M(P^1(\dot{T}^3)\cdot\Delta, P^0(\dot{T}^3)\cdot\Delta) = 0. \tag{34}$$

Thus we finally obtain

$$z^{*3}(T^3) = V_{\dot{T}^3}(Q^0(\dot{T}^3), P^1(\dot{T}^3))+V_{\dot{T}^3}(Q^1(\dot{T}^3), P^0(\dot{T}^3)). \tag{35}$$

Taking into account the relations (9), (11) and (14), we obtain from (35)

$$z^{*3} = \epsilon 2x^1\times z^2, \tag{36}$$

where $\epsilon=\pm 1$ has a quite determined value, but is not calculated here (cf. P.C., theorem 2). From (36) and (28) follows (5), i.e., the assertion of Lemma 4. The Lemma 4 is thus proved.

Proof of Theorem 3. Let g_t be such a deformation of mappings of the complex K^3 into the sphere S^2 that the mappings g_0 and g_1 coincide on K^2. Put

$$z^3 = \omega_1(g_0, g_1, K^3) \quad \text{[cf. Section 2, (B')]},$$

$$z^2 = \omega_0(g_0, K^2) = \omega_0(g_1, K^2) \quad \text{[cf. Section 2, (F)]}.$$

Let us show that in K^3 there exists an one-dimensional ∇-cycle u^1 satisfying the condition

$$z^3 \underset{\nabla}{\sim} 2u^1\times z^2. \tag{37}$$

Denote by Δ the set of all numbers $0\leqslant t\leqslant 1$. Then the mapping g_Δ of the complex $K^3\cdot\Delta$ into S^2 is defined (cf. Section 2). Denote by L the subcomplex of the complex $K^3\cdot\Delta$, composed of $K^3\cdot 0$, $K^3\cdot 1$ and all segments of the form $x\cdot\Delta$, where $x\in K^0$. The mapping g_Δ is defined on L, and every segment of the form $x\cdot\Delta$ is mapped in such a way that

$g_\Delta(x \cdot 0) = g_\Delta(x \cdot 1)$. Let us define a continuous deformation χ_t of every segment $x \cdot \Delta$ such that χ_0 coincides with g_Δ and the mapping χ_1 transforms the whole segment $x \cdot \Delta$ into the point $g_\Delta(x \cdot 0) = g_\Delta(x \cdot 1)$, where $\chi_t(x \cdot 0) = \chi_t(x \cdot 1) = g_\Delta(x \cdot 0)$ for every t. On $K^3 \cdot 0$ and $K^3 \cdot 1$ we define the deformation χ_t so that the mapping χ_t should coincide with g_Δ for every t. The so obtained deformation χ_t of the mappings of the complex L is continuous and it may be continuously extended to the whole complex $K^3 \cdot \Delta$. We denote the mapping χ_1 by f_Δ; it is defined on $K^3 \cdot \Delta$. The mapping f_Δ determines a deformation f_t of mappings of the complex K^3, satisfying the condition of Lemma 3. We thus obtain $z^3 \mathrel{\tilde{\nabla}} 2\epsilon x^1 \times z^2$, and the existence of the ∇-cycle u^1 satisfying the condition (37) is proved.

Let f_0 and l be two mappings of the complex K^3 coinciding on K^2. Put

$$z^3 = \omega_1(f_0, l, K^3), \quad z^2 = \omega_0(f_0, K^2) = \omega_0(l, K^2).$$

We shall show that if in K^3 there exists a ∇-cycle u^1 satisfying the condition $z^3 \mathrel{\tilde{\nabla}} 2u^1 \times z^2$, then the mappings f_0 and l are equivalent. This will complete the proof of Theorem 3.

Let $u^1 = \epsilon x^1$ (cf. Lemma 4). Denote, further, by e_t, $0 \leqslant t \leqslant 1$, a mapping of the complex K^3 into S^2, coinciding with f_0. In virtue of the remark (E), Section 2, there exists a continuous deformation f_t of mappings of the complex K^1 into S^2 such that $\omega_0(e_t, f_t, K^1) = x^1$ [cf. Section 2, (B′)]. Extend the deformation f_t to the complex K^2. In virtue of the remark (C′), Section 2, we have

$$\omega_0(e_1, f_1, K^2) - \omega_0(e_0, f_0, K^2) = \nabla\omega_0(e_t, f_t, K^1) = \nabla x^1 = 0,$$

and since, moreover, $\omega_0(e_0, f_0, K^2) = 0$,

$$\omega_0(e_1, f_1, K^2) = 0,$$

and, consequently, in virtue of (A′), Section 2, the mappings $f_0 = e_1$ and f_1 of the complex K^2 are equivalent with respect to K^1. The continuous deformation of mappings the complex K^2 realizing this equivalence we denote by g_t. The deformation of the mappings of the complex K^2 obtained in the result of successive applications of the deformations f_t and g_t we denote by h_t. The mappings $h_0 = f_0$ and $h_1 = g_1$ of the complex K^2 coincide, and $\omega_0(e_t, h_t, K^1) = x^1$. We extend the deformation h_t to the whole complex K^3 and applying to it Lemma 4 we obtain

$$\omega_1(f_0, h_1, K^3) \mathrel{\tilde{\nabla}} \epsilon 2x^1 \times z^2.$$

Thus $\omega_1(h_1, l, K^3) \mathrel{\tilde{\nabla}} 0$ [cf. Section 2, (D′)], or, in other words, $\omega_1(h_1, l, K^3) = \nabla y^2$.

In virtue of the remark (E), Section 2, there exists a continuous deformation k_t of the complex K^2, transforming the mapping $h_1 = k_0$ into the mapping k_1 and not changing the mapping h_1 on K^1, such that $\omega_1(e'_t, k_t, K^2) = y^2$, where $e'_t = h_1$. Extend the deformation k_t to the whole complex K^3. Then, in virtue of (C′), Section 2, we shall have

$$\omega_1(h_1, k_1, K^3) = \omega_1(e'_1, k_1, K^3) + \nabla y^2.$$

Hence we conclude that $\omega_1(k_1, l, K^3) = 0$ [cf. Section 2, (D′)]. Thus, in virtue of (A′), Section 2, the mappings k_1 and l are equivalent, and consequently so are also the mappings f_0 and l. Theorem 3 is thus proved.

In addition to Theorem 3 we shall prove the following proposition on the existence of mappings.

(A) Let z^2 be a two-dimensional cycle from K^3; then there exists a mapping f of the complex K^3 into S^2 such that

$$\omega_0(f, K^2) = z^2. \tag{38}$$

Further, if f_0 is a certain mapping of the complex K^3 into S^2 and z^3 a certain three-dimensional ∇-cycle from K^3, then there exists a mapping f_1 of the complex K^3 into S^2 coinciding with f_0 on K^2 and satisfying the condition

$$\omega_1(f_0, f_1, K^3) = z^3. \tag{39}$$

Choose on S^2 a certain fixed point q and map every two-dimensional simplex T^2 from K^2 on S^2 with the power $z^2(T^2)$ so that its boundary $\dot{T}^2$ is transformed into the point q. The so arising mapping f of the complex K^2 satisfies the condition (38) [cf. Section 2, (F)]. Since z^2 is a ∇-cycle, the mapping f may be extended to the whole complex K^3. The existence of the mapping f_1 satisfying the condition (39) follows from the proposition (E), Section 2.

Theorem 3 has the following defect: it does not establish the complete system of invariants of the mappings of the complex K^3 into S^2, but only enables us to establish the equivalency or non-equivalency of two mappings. Moreover, in order to establish the equivalency of two mappings already equivalent on K^2, it is necessary to subject one of the mappings to a continuous deformation so as to make it coinciding with the other on K^2.

The contents of Theorem 3 is exposed in more detail by the following proposition, which follows from Theorem 3 and the proposition (A):

(B) Let B^r_∇ be the r-dimensional ∇-Betti group of the complex K^3. If $Z^2 \in B^2_\nabla$, then there exists at least one mapping f of the complex K^3 into S^2, satisfying the condition

$$\omega_0(f, K^2) = z^2 \in Z^2. \tag{40}$$

Denote, further, by $2B^1_\nabla \times Z^2$ the set of all elements of the group B^3_∇, containing cycles of the form $2u^1 \times z^2$, where u^1 is an arbitrary ∇-cycle of dimensionality one. Then to every co-set of the group B^3_∇ with respect to the subgroup $2B^1_\nabla \times Z^2$ corresponds one and only one class of mappings satisfying the condition (40). In order to determine this correspondence it is, however, necessary to choose arbitrarily one definite mapping satisfying the condition (40), to which shall then correspond the co-set $2B^1_\nabla \times Z^2$.

4. Application to Manifolds

In the case when the three-dimensional complex K^3 is an orientable manifold the results of the foregoing paragraph may be formulated by means of the usual homologies, which presents a certain advantage.

We shall understand here under K^3 a three-dimensional, in a definite manner orientated manifold, somehow subdivided into simplexes. By B^r we shall denote the r-dimensional usual Betti group of the manifold K^3 and by B^r — its r-dimensional ∇-Betti group. It is known that between the groups B^r and B^{3-r} there is a quite definite natural isomorphic correspondence (cf. P.C., theorems 3, 4). By S^2 we shall, as above, denote the two-dimensional orientated sphere. By K^r we shall denote the complex composed of all simplexes of the complex K^3, whose dimensionality does not exceed r.

(A) Let f be a simplicial mapping of a certain subdivision of the complex K^3 into S^2 and p a certain point from S^2 lying inside a simplex of the assumed triangulation of the sphere S^2. By $f^{-1}(p)$ we denote the complete orientated original of the point p in K^3 under the mapping f [cf. Section 1, (C′)]. Then $f^{-1}(p)$ is a one-dimensional cycle from K^3; denote the index of its intersection with an arbitrary simplex T^2 of the complex K^3 by $z^2(T^2)$. Then z^2 is a ∇-cycle, and the classes of homologies containing respectively $f^{-1}(p)$ and z^2 correspond to each other (cf. P.C., theorem 3). It is easily seen that $z^2 = \omega_0(f, K^2)$ [cf. Section 2, (F)]. Thus the class of homologies containing $f^{-1}(p)$, as well as the class of homologies containing $\omega_0(f, K^2)$, determines the mapping f up to the equivalency on the complex K^2 [cf. Section 2, (F)].

In view of the fact that the class of homologies, containing $\omega_0(f, K^2)$, may be taken arbitrarily [cf. Section 3, (A)], the class of homologies, containing $f^{-1}(p)$, may be also taken arbitrarily.

Now arises the question of establishing of equivalency or non-equivalency of two mappings f_0 and f_1 in the case, when $f_0^{-1}(p)$ and $f_1^{-1}(p)$ are homologic to each other in K^3.

(B) Let f be a mapping considered in (A) and p^0 and p^1 two points lying inside the simplexes of triangulation of the sphere S^2. It is easily seen that $f^{-1}(p^0)$ and $f^{-1}(p^1)$ are homological cycles from K^3; suppose that they are weakly homologic to zero. Thus, there exists a complex C from K^3 with the boundary $kf^{-1}(p^0)$, where k is a natural number. We define the linkage coefficient $V_{K^3}(f^{-1}(p^0), f^{-1}(p^1))$ by putting

$$V_{K^3}(f^{-1}(p^0), f^{-1}(p^1)) = \frac{1}{k} I_{K^3}(C, f^{-1}(p^1)) = \omega_1(f, K^3). \qquad (1)$$

Thus this linkage coefficient may be also a fraction. Its fractional part is, as is known, an invariant of the class of homologies, containing $f^{-1}(p^0)$, while the linkage coefficient itself is an invariant of the class of mappings, containing f. This is proved in the same way as in Hopf's paper.[2] It turns out that for a given class of homologies, to which the cycle $f^{-1}(p^0)$ belongs, the number $\omega_1(f, K^3)$ is the only invariant of the class of mappings; moreover, the integral part of the number $\omega_1(f, K^3)$ may assume arbitrary values, while the cycle $f^{-1}(p^0)$ belongs to the given class of homologies.

Let us prove the proposition (B). Let f_0 and f_1 be two mappings of the complex K^3 into S^2 coinciding on K^2. Denote by $P_j^i(T^3)$ the complete orientated original of the point p^i in the simplex T^3 from K^3 under the mapping f_j. The complex $P_1^i(T^3) - P_0^i(T^3)$ is a cycle in the simplex T^3 and hence bounds in it a certain complex $Q^i(T^3)$,

$$Q^i(T^3) = P_1^i(T^3) - P_0^i(T^3). \qquad (2)$$

Since the mappings f_0 and f_1 coincide on K^2, the cycles $f_j^{-1}(p^i)$ ($i = 0, 1, j = 0, 1$) belong all to one class of homologies [cf. (A)]; suppose that this class of homologies has a finite order k. Thus there exists a complex C_j^i from K^3 with the boundary $kf_j^{-1}(p^i)$.

Denote the intersection of the complex C_j^i with the simplex T^3 by $C_j^i(T^3)$. Then

$$C_j^i = \sum_{T^2 \in K^3} C_j^i(T^3), \qquad (3)$$

$$f_1^{-1}(p^i) = \sum_{T^2 \in K^3} P_j^i(T^3). \tag{4}$$

Put

$$z^3 = \omega_1(f_0, f_1, K^3) \quad \text{[cf. Section 2, (A')]} \tag{5}$$

and compute $z^3(T^3)$ by means of the introduced complexes. To this end denote by Δ the pair of numbers 0 and 1. Then is defined the mapping f_Δ of the complex $T^3 \cdot \Delta$ (cf. Section 2). This mapping possesses the property that for every $x \in \dot{T}^3$ we have $f_\Delta(x \cdot 0) = f_\Delta(x \cdot 1)$. Identify in one point every pair of points of the form $x \cdot 0$ and $x \cdot 1$ and the sphere so obtained from $T^3 \cdot \Delta$ denote by Σ^3.

The mapping f_Δ may be now considered as the mapping of the sphere Σ^3. It is easily seen that the complete original of the point p^i in Σ^3 under the mapping f_Δ is equal to $P_1^i(T^3) \cdot 1 - P_0^i(T^3) \cdot 0$. Thus

$$z^3(T^3) = V_{\Sigma^3}(P_1^0(T^3) \cdot 1 - P_0^0(T^3) \cdot 0, P_1^1(T^3) \cdot 1 - P_0^1(T^3) \cdot 0) =$$

$$= \frac{1}{k} I_{\Sigma^3}(kQ^0(T^3) \cdot 1 + C_0^0(T^3) \cdot 1 - C_0^0(T^3) \cdot 0, P_1^1(T^3) \cdot 1 - P_0^1(T^3) \cdot 0) =$$

$$= \frac{1}{k} I_{T^3}(kQ^0(T^3) + C_0^0(T^3), P_1^0(T^3)) - \frac{1}{k} I_{T^3}(C_0^0(T^3), P_0^0(T^3)). \tag{6}$$

Since the manifold K^3 is orientated, we may assume that every three-dimensional simplex has a definite orientation, coordinated with the orientations of the whole manifold K^3. Put

$$\sum_{T^2 \in K^3} z^3(T^3) = z^3(K^3), \qquad \sum_{T^2 \in K^3} Q^0(T^3) = Q^0. \tag{7}$$

With these denotations we have

$$(kQ^0 + C_0^0) = kf_1^{-1}(p^0) \quad \text{[cf. (2), (3), (4), (7)]}, \quad (C_0^0) = kf_0^{-1}(p^0). \tag{8}$$

Summing the equality (6) over all $T^3 \in K^3$, we obtain

$$z^3(K^3) = \frac{1}{k} I_{K^3}(kQ^0 + C_0^0, f_1^{-1}(p^1)) - \frac{1}{k} I_{K^3}(C_0^0, f_0^{-1}(p^1)),$$

i.e.,

$$z^3(K^3) = V_{K^3}(f_1^{-1}(p^0), f_1^{-1}(p^1)) - V_{K^3}(f_0^{-1}(p^0), f_0^{-1}(p^1))$$

[cf. (1), (8)], or, which is the same thing,

$$z^3(K^3) = \omega_1(f_1, K^3) - \omega_1(f_0, K^3). \tag{9}$$

The group B^3_∇ is for the three-dimensional orientable manifold K^3 the free cyclic group. If $\overset{*}{T}{}^3$ is a certain simplex from K^3 and $\overset{*}{z}{}^3$ is the ∇-cycle from K^3 assuming the value 1 on $\overset{*}{T}{}^3$ and the value 0 on all other three-dimensional simplexes from K^3, then $\overset{*}{z}{}^3$ may be taken for the basis of ∇-homologies in K^3. It is easy to see that

$$z^3 \underset{\nabla}{\widetilde{\sim}} z^3(K^3)\overset{*}{z}{}^3, \tag{10}$$

and consequently the number $z^3(K^3)$ determines the class of homologies, to which belongs the ∇-cycle z^3. Thus, in the case, when $\omega_1(f_1, K^3) = \omega_1(f_0, K^3)$, we have $z^3 \underset{\nabla}{\widetilde{\sim}} 0$, and hence the mappings f_0 and f_1 are equivalent (cf. Theorem 3).

Let us now show that the integral part of the number $\omega_1(f, K^3)$ may be made arbitrary, the class of homologies, to which $f^{-1}(p^0)$ belongs, being given. In fact, let f_0 be an arbitrary mapping such that $f_0^{-1}(p^0)$ belongs to the given class of homologies; then, in virtue of (A), Section 3, there exists such a mapping f_1 that $z^3 = \omega_1(f_0, f_1, K^3)$ is an arbitrary ∇-cycle from K^3. Thus for a given f_0 we may choose f_1 such that the number $z^3(K^3)$ should have an arbitrary integral value [cf. (10)], and this means that the integral part of the number $\omega_1(f_1, K^3)$ may be chosen arbitrarily [cf. (9)]. The proposition (B) is thus completely proved.

Consider now the case when the cycle $f^{-1}(p)$ [cf. (A)] is weakly not homologic to zero.

(C) Let Z^1 be a fixed free element of the group B^1 and U^2 an arbitrary element of the group B^1. Denote by λ the smallest positive value which the number $I_{K^3}(U^2, Z^1)$ may assume for a given Z^1 and arbitrary U^2. From Poincaré-Veblen's theorem follows that $I_{K^3}(U^2, Z^1)$ admits of positive values, since Z^1 is a free element of the group B^1. It turns out that among the mappings f satisfying the condition

$$f^{-1}(p) \in Z^1 \quad \text{[cf. (A)]} \tag{11}$$

there are exactly 2λ pairwise non-equivalent.

In every class of mappings satisfying the condition (11) we may choose one, so that all chosen mappings should coincide on K^2 [cf. (A)]. Let f_0 and f_1 be two mappings satisfying the condition (11) and coinciding on K^2. Put

$$z^3 = \omega_1(f_0, f_1, K^3), \quad z^2 = \omega_0(f_0, K^2) = \omega_0(f_1, K^2).$$

In virtue of Theorem 3 the mappings f_0 and f_1 are equivalent then and only then, when

$$z^3 \underset{\nabla}{\approx} 2u^1 \times z^2. \tag{12}$$

The class of ∇-homologies, to which the ∇-cycle Z^3 belongs, is determined by the integer $z^3(K^3)$ [cf. (7)]. The class of usual homologies Z^0, corresponding to this class, is also determined by an integer, namely by the index of the nul-dimensional complexes entering into Z^0. Thus we may simply take it that Z^0 is an integer and that it coincides with $Z^3(K^3)$. We denote the class of usual homologies, corresponding to the class of ∇-homologies containing the ∇-cycle u^1, by U^2; since u^1 is an arbitrary ∇-cycle, U^2 is an arbitrary element of the group B^2. Finally, to the class of ∇-homologies containing z^2 corresponds the class of usual homologies Z^1 containing $f_i^{-1}(p)$ [cf. (A)]. In usual homologies the relation (12) may be thus written in the form

$$Z^0 = 2I_{K^3}(U^2, Z^1).$$

In virtue of the arbitrariness of the class of homologies U^2 the right-hand side of the last relation is an arbitrary number divisible by the number 2λ. Thus the mappings f_0 and f_1 are equivalent then and only then, when the number $z^3(K^3)$ is divisible by 2λ; at the same time the number $z^3(K^3)$ may assume, for a given f_0, an arbitrary integral value and, consequently, there exist exactly 2λ non-equivalent among each other mappings f satisfying the condition (11). The proposition (C) is thus proved.

Example. Let S^2 be the metrical two-dimensional sphere and p and q two its diametrically opposite points. Denote by φ_α the mapping of the sphere S^2 on itself, obtained by means of a rotation of the sphere S^2 by the angle α about the axis pq. Denote by S^1 the circumference with the parameter t, $0 \leqslant t \leqslant 2\pi$, introduced on it. The topological product of S^2 and S^1 denote by K^3. Every point $y \in K^3$ is given by a pair $y = x \cdot t, x \in S^2, 0 \leqslant t \leqslant 2\pi$. Define the mapping φ of the manifold K^3 on itself by putting $\varphi(x \cdot t) = \varphi_t(x) \cdot t$. Define, further, the mapping f_0 of the manifold K^3 on the sphere S^2 by putting $f_0(x \cdot t) = x$. Define a second mapping f_1 by putting $f_1 = f_0 \varphi$. Then we obtain two mappings f_0 and f_1 of the manifold K^3 on S^2 such that the complete original of the point p under both mappings is $p \cdot S^1$. The number λ for the cycle $p \cdot S^1$ is easily seen to be equal to 1 [cf. (C)], and hence there exist exactly two non-equivalent mappings satisfying the condition $f^{-1}(p) \sim p \cdot S^1$. These two mappings are precisely f_0 and f_1.

The mappings f_0 and f_1 are completely equipollent, since the mapping φ is an homoeomorphism. Thus there is no possibility to establish in a natural manner a correspondence between the classes of mappings and the co-sets of the group B^3_∇ with respect to the subgroup $2B^1_\nabla \times Z^2$ [cf. Section 3(C)], but it is necessary to choose arbitrarily that mapping, which corresponds to the co-set $2B^1_\nabla \times Z^3$.

From the fact that the mappings f_0 and f_1 are not equivalent follows that the mapping φ is not equivalent to the identical mapping, whereas the mapping φ and the identical mapping are homologically equivalent.

5. The Mappings of the Four-dimensional Complex into the Two-dimensional Sphere

By K^4 we shall denote here the four-dimensional complex and by K^r the sub-complex of the complex K^4 composed of all simplexes of the complex K^4 of dimensionalities not exceeding r. By S^2 we shall denote the two-dimensional sphere.

In Section 2 was shown that the mapping f of the complex K^2 into S^2 may be then and only then extended to the complex K^3, when $\omega_0(f, K^2)$ is a ∇-cycle in K^3 [cf. Section 2, (G)]. Here we shall solve the question on the extension of the mapping f, defined on K^2, to the whole complex K^4.

Theorem 4. Let f be a certain mapping of the complex K^2 into the sphere S^2. The mapping f may be extended to the whole complex K^4 then and only then, when $\omega_0(f, K^2)$ is a ∇-cycle in K^4, satisfying the condition

$$\omega_0(f, K^2) \times \omega_0(f, K^2) \underset{\nabla}{\widetilde{=}} 0 \quad \text{[cf. Section 2, (F)]}. \tag{1}$$

Proof. Suppose that the mapping f is already defined on the complex K^4. Without reducing the generality we may suppose that f is a simplicial mapping of a certain simplicial subdivision of the complex K^4. Let p^0 and p^1 be two inner points of a certain simplex from the assumed triangulation of the sphere S^2. If T^r is a certain orientated r-dimensional simplex of the complex K^4, then we denote by $P^i(T^r)$ the complete original of the point p^i in the simplex T^r under the mapping f. Put, for shortness,

$$\sum_{T^{r-1} \in \dot{T}^r} P^i(T^{r-1}) = P^i(\dot{T}^r). \tag{2}$$

From the fact that the three-dimensional sphere $\dot{T}^4$ bounds in K^4 the simplex T^4 follows $\omega_1(f, \dot{T}^4) = 0$, or, in other words,

$$V_{\dot{T}^4}(P^0(\dot{T}^4), P^1(\dot{T}^4)) = 0. \tag{3}$$

Put

$$\omega_0(f, K^2) = z^3. \tag{4}$$

In virtue of the very definition of the ∇-complex z^2 we have

$$z^2(T^2) = I(P^i(T^2)). \tag{5}$$

Further,

$$P^i(T^3)^{\cdot} = P^i(\dot{T}^3). \tag{6}$$

From the relations (3)–(6) follows the relation (1) (cf. P.C., theorem 2).

Suppose now that f is defined on K^2 and satisfies the conditions of the theorem, i.e. that $\omega_0(f, K^2) = z^2$ is a ∇-cycle in K^4 satisfying the condition (1). Since $\omega_0(f, K^2)$ is a ∇-cycle in K^3, the mapping f may be extended to K^3 [cf. Section 2, (G)], but it may be done in different ways. Suppose that f_0 and f_1 are two such extensions, i.e. two mappings of the complex K^3, coinciding on K^2 with f and, consequently, also one with another. Put

$$\omega_0(f_0, f_1, K^3) = y^3, \quad \omega_1(f_i, \dot{T}^4) = y_i^4(T^4). \tag{7}$$

We shall show that

$$y_1^4 \dot{-} y_0^4 = \nabla y^3. \tag{8}$$

Let Δ be the aggregate of the two numbers 0 and 1. Then is defined the mapping f_Δ of the complex $K^3 \cdot \Delta$ into S^2 (cf. Section 2). Let T^4 be a certain four-dimensional simplex from K^4. Consider the mapping f_Δ in application to the complex $\dot{T}^4 \cdot \Delta$. If $x \in \dot{T}^4, x \in K^2$, then we have $f_\Delta(x \cdot 0) = f_\Delta(x \cdot 1)$. Identify in $K^4 \cdot \Delta$ in one point every pair of points $x \cdot 0$ and $x \cdot 1$ and denote the complex, so obtained from $\dot{T}^4 \cdot \Delta$, by L^3. The mapping f_Δ may be now considered as defined on the complex L^3. It is easily seen that every two-dimensional cycle from L^3 is homologic to zero in L^3 to every modulus, and hence, in virtue of Lemma 3, there exists such a mapping g of the complex L^3 into the three-dimensional sphere S^3 that $f_\Delta = \vartheta g$ [cf. Section 1, (A)].

Denote by $E_k^3, k = 0, 1, 2, 3, 4$, the faces of the simplex T^4. Consider now in the complex L^3 the following orientated three-dimensional spheres:

$$S_k^3 = E_k^3 \cdot 1 - E_k^3 \cdot 0,$$

$$S_5^3 = \sum_{k=0}^{4} E_k^3 \cdot 0 = \dot{T}^4 \cdot 0, \quad S_6^3 = \sum_{k=0}^{4} E_k^3 \cdot 1 = \dot{T}^4 \cdot 1.$$

Evidently we have

$$S_6^3 - S_5^3 = \sum S_k^3.$$

Thus we have

$$\omega_0(g, S_6^3) - \omega_0(g, S_5^3) = \sum_{k=0}^{4} \omega_0(g, S_k^4);$$

further,

$$\omega_0(g, S_n^3) = \omega_1(f_\Delta, S_n^3), \quad n = 0, 1, \ldots, 6 \quad \text{[cf. Section 1, (20)]}.$$

Consequently,

$$\omega_1(f_\Delta, S_6^3) - \omega_1(f_\Delta, S_5^3) = \sum_{k=0}^{4} \omega_1(f_\Delta, S_k^3),$$

and this is the relation (8), written explicitly.

The complete original of the point p^i in the simplex T^r under the mapping f_j we denote by $P_j^i(T^r)$. For shortness introduce the notations:

$$\sum_{T^{r-1} \in \dot{T}^r} P_j^i(T^{r-1}) = P_j^i(\dot{T}^r).$$

Then we have

$$P_j^i(T^3) = P_j^i(\dot{T}^3), \tag{9}$$

$$z^2(T^2) = I(P_j^i(T^2)). \tag{10}$$

From the relations (9), (10) and (7) follows that

$$(z^2 \times z^2)_0(T^4) = V_{\dot{T}^4}(P_0^0(\dot{T}^4), P_0^1(\dot{T}^4)) = y_0^4(T^4), \tag{11}$$

$$(z^2 \times z^2)_1(T^4) = V_{\dot{T}^4}(P_1^0(\dot{T}^4), P_1^1(\dot{T}^4)) = y_1^4(T^4). \tag{12}$$

In the relation (11) the product $z^2 \times z^2$ is calculated by means of the auxiliary complexes $P_0^i(T^r)$, and in the relation (12) the product $z^2 \times z^2$ is calculated by means of the auxiliary complexes $P_1^i(T^i)$. Although the so obtained products are homologic, they need not coincide.

Suppose now that the mapping f defined on K^2 and satisfying the conditions of the theorem is in some way extended by means of the mapping f_0 to K^3. Since $(z^2 \times z^2)_0 \mathrel{\widetilde{\nabla}} 0$,

$$y_0^4 = (z^2 \times z^2)_0 = \nabla u^3.$$

Choose now the mapping f_1, coinciding with f_0 on K^2 and such that $\omega_1(f_0, f_1, K^3) = -u^3$ [cf. Section 2, (A′)]. Then, in virtue of (8), we have $z_1^4 = 0$, and this means that the mapping f of each sphere $\dot{T}^4$ may be extended to the simplex T^4 [cf. (7)], i.e., we obtain an extension of the mapping f to the whole complex K^4. Theorem 4 is thus proved.

Let us now formulate the obtained result in terms of usual homologies in the case when K^4 is a four-dimensional orientable manifold.

(A) Let f be a mapping of the orientated four-dimensional manifold K^4 into the two-dimensional orientated sphere S^2. Without reducing the generality we may assume that the mapping f is a simplicial mapping of a certain simplicial subdivision of the complex K^4. Let p be a certain inner point of a simplex from the assumed triangulation of the sphere S^2 and $f^{-1}(p)$ its complete original in K^4 under the mapping f. It is easily shown that for two equivalent mappings f_0 and f_1 we have $f_0^{-1}(p) \sim f_1^{-1}(p)$. If C^2 is a class of usual two-dimensional homologies, then for the existence of a mapping f satisfying the condition $f^{-1}(p) \in C^2$ it is necessary and sufficient that $I_{K^4}(C^2, C^2) = 0$.

Let c^2 be a certain cycle from C^2. Denote by $z^2(T^2)$ its index of intersection with T^2. Then the class of homologies C^2 corresponds to the class of ∇-homologies containing z^2. In order that there should exist a mapping f satisfying the condition $f^{-1}(p) \in C^2$, it is necessary and sufficient that there should exist a mapping f satisfying the condition $\omega_0(f, K^3) \widetilde{\widetilde{\nabla}} z^2$. But for this it is necessary and sufficient that $z^2 \times z^2 \widetilde{\widetilde{\nabla}} 0$. This last condition has in terms of usual homologies the form: $I_{K^4}(C^2 \times C^2) = 0$. The assertion (A) is thus proved.

It is of interest to note the following:

Let K^{n+2} be a complex of dimensionality $n+2$. Denote by K^r the aggregate of all its simplexes of dimensionalities not exceeding r. Denote by S^n the n-dimensional orientated sphere. If f is a mapping of the complex K^n into S^n, then we introduce, in the same way as in (G), Section 2, a ∇-complex $\omega_0(f_1, K^n)$ of dimensionality n, characterizing the mapping f. Namely, two mappings f_0 and f_1 are then and only then equivalent, when $\omega_0(f_0, K^n) \widetilde{\widetilde{\nabla}} \omega_0(f_1, K^n)$. Further, in order that the mapping f, defined on K^n, could be extended to K^{n+1}, it is necessary and sufficient that $\omega_0(f, K^n)$ should be a ∇-cycle in K^{n+1}. If this condition is satisfied, then there arises the question on the possibility of extension of the mapping f from the complex K^n to the whole complex K^{n+2}. It turns out that the condition of such an extension is expressed by the demand that a certain $(n+2)$-dimensional ∇-cycle from K^{n+2} should be homologic to zero to the modulus 2; this cycle is determined up to homologies by the cycle $\omega_0(f, K^n)$, but its construction can not be carried out by means of the product operations in the complex K^{n+2}, which we possess. Thus we are lead to a new operation of a homological type.

Notes

1. H. Hopf, Die Klassen der Abbildungen der n-dimensionalen Polyeder auf die n-dimensional Sphäre, *Commentaril Mathematici Helvetici,* **5** (1932).

2. H. Hopf, Über die Abbildungen der drei-dimensionalen Sphäre auf die Kugelfläche, *Mathematische Annalen,* **104** (1931).

3. L. Pontrjagin, A classification of continuous transformations of a complex into a sphere. I and II, *Comptes rendus Acad. Sci. U.R.S.S.,* **XIX** (1938).

4. H. Freudenthal, Über die Klassen der Sphärenabbildungen, *Compositio Mathematica,* **5** (1937).

5. L. Pontrjagin, Classification des transformations d'un complexe $(n + 1)$-dimensionnel dans une sphère n-dimensionelle, *Comptes rendus Acad. Sci., Paris,* **208** (1938).

6. W. Hurewicz, Beiträge zur Topologie der Deformationen, *Proc. Kön. Akad. Wet. Amsterdam,* **38,** 112 u. 521; **39,** 117 u. 215 (1935–1936).

7. L. Pontrjagin, Products in complexes, see the preceding paper, p. 321.

8. N. Bruszlinsky, Stetige Abbildungen und Bettische Gruppen, *Mathematische Annalen,* **109** (1934).

9. S. Lefschetz, Intersections and transformations of complexes and manifolds. *Trans. Amer. Math. Soc.,* **28** (1926).

10. H. Whitney, The maps of an n-complex into an n-sphere, *Duke Mathematical Journal,* **3** (1937).

On the Zeros of Certain Elementary Transcendental Functions

Izvestiya Akademii Nauk SSSR, Seriya Matematika (1942)
6(3): 115–34

Let $h(z, t)$ be a polynomial in two variables with real or complex coefficients. The paper deals with conditions for all the zeros of the function $h(z, e^z)$ to have negative real parts.

In the theory of the stability of solutions of ordinary differential equations, the behaviour of the roots of the secular equation $P(z) = 0$ plays an important part. If, to a given equilibrium point p of a system of equations, there corresponds a secular polynomial $P(z)$, then for this equilibrium point to be stable, it is sufficient that all roots of $P(z)$ have negative real parts, and it is necessary that $P(z)$ have no root with positive real part. The behaviour of the roots of the polynomial $P(z)$ therefore plays a decisive role, and Hurwitz has studied this behaviour. In the case of partial differential equations a similar problem sometimes arises, no longer for polynomials, but for a transcendental function. This paper is concerned with the solution of this problem for a transcendental function of the type $H(z) = h(z, e^z)$, where $h(z, t)$ is a polynomial. More precisely, we shall reduce this problem to solving a certain problem for elementary transcendental functions relative to their behaviour in the real domain; then this problem can be solved effectively by the Sturm method.

Let r and s denote the degrees of the polynomial $h(z, t)$ with respect to z and t, respectively. A term of the type $az^r t^s$ is called a *principal term.* If there is no principal term in $h(z, t)$, the function $H(z)$ definitely has an infinite number of zeros with arbitrarily large positive real parts. If, however, $h(z, t)$ contains a principal term, then in order to solve our problem we have to study the behaviour of $H(z)$ on the imaginary axis, i.e., for $z = iy$, where y is a real variable. The function $H(iy)$ can then be separated into real and imaginary parts:

$$H(iy) = F(y) + iG(y)$$

where

$$F(y) = f(y, \cos y, \sin y), \quad G(y) = g(y, \cos y, \sin y),$$

and $f(y, u, v)$, $g(y, u, v)$ are polynomials. For the real parts of all roots of $H(z)$ to be negative, it is necessary and sufficient that the roots of $F(y)$ and $G(y)$ all be real and alternating* and that $G'(y)F(y) - F'(y)G(y) > 0$ for at least one value of y. Whether all roots of $F(z)$ are real or not can be tested by the following criterion: For all roots of the function $F(z)$ to be real, it is necessary and sufficient that in the interval $-2k\pi \leqslant y \leqslant 2k\pi$ it have, for sufficiently large k, exactly $4sk + r$ roots. Here, by roots, we mean real roots. This test is quite similar to a corresponding criterion used for polynomials, namely, the presence of a sufficiently large number of positive roots guarantees the absence of complex roots.

The present paper is a continuation of the work of N. G. Chebotarev. In a communication read at the Moscow Mathmatical Society in the winter of 1941–42, he reported the solution of our problem for a case where the polynomial $h(z, t)$ is linear in t. For this case, he only gave sufficient conditions for the function $H(z)$ to have only zeros with real negative parts, while necessary conditions were not proved. He noticed however the possibility of effectively applying the Strum method to resolve the problem. All his results are published in Doklady Akademii Nauk SSSR.

1. Zeros of the Function $h(z, e^z)$ with no Principal Term

Let $h(z, t)$ be a polynomial in z and t with constant real or complex coefficients

$$h(z, t) = \sum_{m,n} a_{mn} z^m t^n. \qquad (1)$$

A term $a_{rs} z^r t^s$ is called *principal*, if $a_{rs} \neq 0$, and the exponents r and s are maximum, i.e., for every other term $a_{mn} z^m t^n$, if $a_{mn} \neq 0$, either $r > m$, $s > n$, or $r = m$, $s > n$, or $r > m$, $s = n$. Evidently, not every polynomial has a principle term.

---

* See Section 3, Theorem VII, for the explanation of this term (Ed.).

Theorem I. If the polynomials (1) has no principal term, the function

$$H(z) = h(z, e^z) \tag{2}$$

has an infinite number of zeros with arbitrarily large positive real parts.

Proof. Consider a simple polynomial $t - z$ without a principal term. Using it as an example, we shall deduce the solution we seek. Our equation is $e^z - z = 0$. Putting $z = x + iy$, we obtain two equations

$$e^x \cos y = x, \quad e^x \sin y = y. \tag{3}$$

We shall find approximate solutions of these two equations, assuming that x and y are both positive and very large. Under this assumption, from equation (3) it follows that $\cos y = xe^{-x}$ and, consequently, the approximate value of y is $2k\pi + \pi/2$. From equation (3) it also follows that approximately $x = \ln(2k\pi + \pi/2)$. Thus, the solution is of the form

$$z = \ln(2k\pi + \pi/2) + i(2k\pi + \pi/2) + \zeta,$$

where ζ is a small unknown quantity which tends to zero with $1/k$.

In analogy with the solution obtained above, we shall seek the solution of the general equation $H(z) = 0$ containing no principal term in the form:

$$z = \alpha \ln 2k\pi + 2k\pi i + \ln \theta + \zeta. \tag{4}$$

Here α is a positive rational number to be chosen, depending on the polynomial (1), $\theta \neq 0$ is a complex number which is also to be chosen, depending on the polynomial (1), and ζ is an unknown quantity which tends to zero with $1/k$. From (4), we obtain

$$e^z = (2k\pi)^\alpha \theta e^\zeta, \quad z = i2k\pi(1 + \delta_1(\zeta)), \tag{5}$$

where $\delta_1(\zeta)$ is an analytic function of ζ which uniformly tends to zero with $1/k$. Substituting the values found above into $H(z)$, we obtain

$$H(z) = \sum_{m,n} (2k\pi)^{m+\alpha n} a_{mn} i^m \theta^n e^{n\zeta}(1 + \delta_1(\zeta))^m. \tag{6}$$

Thus we have expressed $H(z)$ as a finite sum of fractional powers of $(2k\pi)$. In this expansion, let us pick out the principal terms, i.e., such terms whose exponents $(m + \alpha n)$ attain their maximum for $a_{mn} \neq 0$. Then, (6) can be written as

$$H(z) = \sum_n (2k\pi)^\beta a_{mn} i^m \theta^n e^{n\zeta} + (2k\pi)^\beta \delta_2(\zeta) =$$

$$= \sum_n (2k\pi)^\beta b_n \theta^n e^{n\zeta} + (2k\pi)^\beta \delta_2(\zeta).$$

Here the summation is taken over those n for which $m + \alpha n = \beta$ and $a_{mn} \neq 0$. There is no need to sum over m, because m is uniquely determined by n through the relation $m + \alpha n = \beta$. The quantity $\delta_2(\zeta)$ is an analytic function of ζ which uniformly converges to zero in the circle $|\zeta| \leqslant 1$, as $k \to \infty$.

By the definition of β, there exists at least one n for which $m + \alpha n = \beta$ and $a_{mn} \neq 0$. Below we shall prove that if (1) has no principal term, then for some α, there exist at least two distinct n for which these conditions are satisfied. Under this assumption, the equation

$$\sum_n b_n \theta^n = 0 \tag{7}$$

for the unknown θ will have at least one non-zero solution. Henceforward we shall take this solution as the value of θ. Thus, (7) is no longer an equation, but is an identity satisfied by the θ we have chosen. Instead of the equation $H(z) = 0$, consider the equivalent equation

$$\sum_n b_n \theta^n e^{n\zeta} + \delta_2(\zeta) = 0. \tag{8}$$

As $k \to \infty$, the left-hand side tends uniformly to the function $\Sigma_n \, b_n \theta^n e^{n\zeta}$, but the equation $\Sigma_n \, b_n \theta^n e^{n\zeta} = 0$ has the obvious solution $\zeta = 0$ (see (7)). Because of the uniform convergence, equation (8) has, for sufficiently large k, a solution ζ_k which tends to zero with $1/k$. Thus, the equation $H(z) = 0$ has a solution

$$z = \ln 2k\pi + 2k\pi i + \ln \theta + \zeta_k \tag{9}$$

for sufficiently large k, where ζ_k tends to zero with $1/k$. Since α and θ do not depend on k, this solution evidently has positive real part for sufficiently large k.

It now remains to choose a positive rational number α, for which there exist at least two n for which $m + \alpha n = \beta$ and $a_{mn} \neq 0$. For this purpose, we shall assume that the polynomial (1) has no principal term.

Let s be the maximum n for which $a_{mn} \neq 0$, and r be the maximum m for which $a_{ms} \neq 0$. By our assumption, i.e., that (1) has no principal term, there exist p and q (values of m and n, respectively), such that $p > r, q < s$ and $a_{pq} \neq 0$. Now in (1) substitute z^α for t, where $\alpha > 0$, and consider the resulting expression

$$\sum_{m,n} a_{mn} z^{m+\alpha n} \tag{10}$$

arranged in positive powers of z. In this expansion that term with the maximum exponent $m + \alpha n$ for which $a_{mn} \neq 0$ is called the *principal term.* If α is very large, the principal term in (10) is evidently $a_{rs}z^{r+\alpha s}$. If α is sufficiently close to zero, this term is not the principal term because, for $p > r$, the exponent of $a_{pq}z^{p+\alpha q}$ is evidently greater. Thus, when α is varied continuously from plus infinity to zero, at a certain instant there will occur at least two principal terms in (10). It is precisely this value of α that we denote by α. Clearly, α is rational, since it is determined by an equation of the form $r + \alpha s = m + \alpha n$.

Thus, Theorem I has been demonstrated.

2. Zeros of the function $f(z, \cos z, \sin z)$

Let $f(z, u, v)$ be a polynomial in z, u, v with real constant coefficients. Then

$$f(z, \cos z, \sin z) = F(z) \tag{11}$$

is an entire transcendental function of z and takes real values for real values of its argument. We now state a necessary and sufficient condition for the function $F(z)$ to have only real zeros, moreover, in terms of the behaviour of $F(z)$ in the real domain.

Express $f(z, u, v)$ as

$$f(z, u, v) = \sum_{m,n} z^m \varphi_m^{(n)}(u, v). \tag{12}$$

Here $\varphi_m^{(n)}$ denotes a polynomial of degree n homogeneous in u and v. Henceforth it is assumed that $u = \cos z$, $v = \sin z$, and therefore, without loss of generality, we may suppose that $\varphi_m^{(n)}(u, v)$ is not divisible by $u^2 + v^2$. In other words, we may assume that

$$\varphi_m^{(n)}(1, \pm i) \neq 0 \tag{13}$$

for all the polynomials in (12).

The principal term in (12) is a term $z^r \varphi_r^{(s)}(u, v)$ in the expansion for which the exponents r and s simultaneously attain their maximum, i.e., for every other term $z^m \varphi_m^{(n)}(u, v)$ in (12), we have $r > m$, $s > n$, or $r = m$, $s > n$, or $r > m$, $s = n$. Evidently, not every polynomial (12) has a principal term.

Using the results of Section 1, we can prove

Theorem II. If the polynomial (12) has no principal term, the function $F(z)$ [see (11)] has an infinite number of non-real zeros.

To formulate the theorem in the case that (12) has a principal term, take the coefficient of the highest power of z in

$$f(z,u,v) = z^r\varphi_*^{(s)}(u,v) + \sum_{m<r,n\leqslant s} z^m\varphi_m^{(n)}(u,v). \tag{14}$$

Here $\varphi_*^{(s)}(u,v)$ is no longer a homogeneous polynomial of degree s in u, v:

$$\varphi_*^{(s)}(u,v) = \sum_{n\leqslant s} \varphi_r^{(n)}(u,v). \tag{15}$$

The function $\Phi_*^{(s)}(z) = \varphi_*^{(s)}(\cos z, \sin z)$, where $z = x+iy$, is evidently periodic with a period 2π and, as will be shown below, has only a finite number, i.e., $2s$, of zeros in the strip $a \leqslant x < 2\pi + a$. Consequently, there exist an infinite number of $\alpha = \epsilon$ such that

$$\Phi_*^{(s)}(\epsilon+iy) \neq 0$$

for any y. In the majority of cases, ϵ can be taken to be zero.

Theorem III. Let $f(z,u,v)$ be a polynomial with a principal term $z^r\varphi_r^{(s)}(u,v)$. If ϵ is such that $\Phi_*^{(s)}(\epsilon+iy)$ does not vanish for any real y, then in the strip $-2k\pi+\epsilon \leqslant x \leqslant 2k\pi + \epsilon$ $(z = x+iy)$, the function $F(z)$ has, starting from some sufficiently large k, exactly $4sk+r$ zeros. Thus, for $F(z)$ to have only real zeros, it is necessary and sufficient that in the strip $-2k\pi+\epsilon \leqslant x \leqslant 2k\pi+\epsilon$ it have exactly $4sk+r$ real zeros, for sufficiently large k.

Proof. First we shall show that in the strip $a \leqslant x < 2\pi + a$ $(z = x+iy)$ the function $\Phi_*^{(s)}(z)$ has exactly $2s$ zeros.

Put

$$u = \tfrac{1}{2}(t+1/t), \quad v = \frac{1}{2i}(t-1/t). \tag{16}$$

Then, for $t = e^{iz}$, we have $u = \cos z$, $v = \sin z$. Substituting the expressions (16) for u and v in $\varphi_*^{(s)}(u,v)$, we obtain a finite series for $\varphi_*^{(s)}(t)$ in positive and negative powers of t. The coefficient of the highest positive power s of the variable t, is, as can be easily verified, $\varphi_r^{(s)}(\tfrac{1}{2}, -i/2)$ [see (15)]. Similarly, the coefficient of the lowest negative power $-s$ of t is $\varphi_r^{(s)}(\frac{1}{2}, i/2)$ [(see (15)]. Consequently, both these terms are different from zero [(see (13)]. Thus, the equation $\varphi_*^{(s)}(t) = 0$ has exactly $2s$ roots, and they all are different from zero. Let us denote these roots by $t_1, t_2, \ldots, t_{2s}$. To solve the equation

$$\Phi_*^{(s)}(z) = 0, \tag{17}$$

it is now sufficient to solve all the equations $e^{iz} = t_j$. For a fixed j, each equation has exactly one root in the strip $a \leqslant x < 2\pi + a$. If all the t_j are distinct, then in this strip there are exactly $2s$ roots of equation (17). If, however, there are repeated roots t_j, then equation (17) has a corresponding multiplicity of roots.

Now consider the behaviour of $\Phi_m^{(n)}(z) = \varphi_m^{(n)}(\cos z, \sin z)$ for large y ($z = x + iy$), both positive and negative; we shall show that

$$\left.\begin{aligned} \Phi_m^{(n)}(x+iy) &= e^{ny-nix}(\varphi_m^{(n)}(\tfrac{1}{2}, i/2) + \delta_1); \\ \Phi_m^{(n)}(x+iy) &= e^{-ny+nix}(\varphi_m^{(n)}(\tfrac{1}{2}, -i/2) + \delta_2), \end{aligned}\right\} \tag{18}$$

where δ_1 tends uniformly to zero as $y \to +\infty$, and δ_2 tends uniformly to zero as $y \to -\infty$.

To prove (18), it is sufficient to notice that

$$\cos z = \tfrac{1}{2}(e^{ix-y} + e^{-ix+y}), \quad \sin z = 1/2i(e^{ix-y} - e^{-ix+y})$$

and to consider the behaviour of e^y and e^{-y} for large positive and negative y.

From (18) we directly obtain similar relations for the periodic function $\Phi_*^{(s)}(z)$, i.e.,

$$\left.\begin{aligned} \Phi_*^{(s)}(x+iy) &= e^{sy-six}(\varphi_r^{(s)}(\tfrac{1}{2}, i/2) + \delta_3); \\ \Phi_*^{(s)}(x+iy) &= e^{-sy+six}(\varphi_r^{(s)}(\tfrac{1}{2}, -i/2) + \delta_4), \end{aligned}\right\} \tag{19}$$

where δ_3 tends uniformly to zero as $y \to +\infty$, and δ_4 tends uniformly to zero as $y \to -\infty$.

Now take a $b' > 0$, such that $\Phi_*^{(s)}(x+iy) \neq 0$ for $|y| > b'$. By virtue of (18) and (19),

$$\left|\frac{\Phi_m^{(n)}(x+iy)}{\Phi_*^{(s)}(x+iy)}\right| < c_1 \tag{20}$$

for $|y| > b'$, where c_1 is some constant, depending on the polynomial (12) and on the choice of b'.

Similarly, from (18) and (19), it follows that

$$\left|\frac{\Phi_m^{(n)}(\pm 2k\pi + \epsilon + iy)}{\Phi_*^{(s)}(\pm 2k\pi + \epsilon + iy)}\right| < c_2, \tag{21}$$

where c_2 is some constant, depending on the polynomial (12) and the number ϵ.

On the basis of the preceding estimates, we shall now calculate the number of zeros of the function $F(z)$ inside a rectangular domain in the complex plane of z. Let the rectangle, denoted by P_{kb} be given by the relations:

$$-2k\pi+\epsilon\leqslant x\leqslant 2k\pi+\epsilon, \quad -b\leqslant y\leqslant b.$$

Express $F(z)$ as:

$$F(z) = z^r\Phi_*^{(s)}(z)\left(1+\sum_{m<r,n\leqslant s} z^{m-r}\frac{\Phi_m^{(n)}(z)}{\Phi_*^{(s)}(z)}\right). \tag{22}$$

Here all the exponents $(m-r)$ of z are negative, and therefore, by virtue of (20) and (21), we have on the sides of the rectangle P_{kb}, for sufficiently large k and b, that

$$F(z) = z^r\Phi_*^{(s)}(z)\,(1+\delta_5), \tag{23}$$

where δ_5 tends to zero when k and b tend to zero simultaneously. From (23) it directly follows that $F(z)$ and $z^r\Phi_*^{(s)}(z)$ have the same number of zeros in the rectangle P_{kb}. Elsewhere below we shall prove this assertion in a general form, but here we shall only draw one conclusion following from this result. Take a sufficiently large k and let b tend to infinity. Then, $F(z)$ and $z^r\Phi_*^{(s)}(z)$ have the same number of zeros in the strip $-2k\pi+\epsilon\leqslant x\leqslant 2k\pi+\epsilon$. The second function evidently has $(4sk+r)$ zeros. Thus, the theorem has been demonstrated, with the exception of one gap, which we now fill.

Remark. Let P be a certain closed contour in the plane of the complex variable z, and $g(z)$ an analytic function which has no singularities either inside or on the contour P. Furthermore, assume that it does not vanish on the contour P. Then, by virtue of the known theorem concerning the logarithmic residue, the number of zeros of $g(z)$ inside P is equal to the number of full revolutions made by the coordinate vector $w=g(z)$ about the origin when z traverses the contour P. Now, let $g^*(z)$ be a function, analytic both inside P and on the contour P, and on the contour P let it be connected to $g(z)$ by the relationship $g^*(z)=g(z)\,(1+\delta(z))$, where $|\delta(z)|<1$. On the contour P consider the function $g(z,\tau)=g(z)$ $(1+\tau\delta(z))$, where τ is a real number. For a fixed τ, the vector $w=g(z,\tau)$ describes certain number of full revolutions about the origin when z runs through the contour P. If τ now varies continuously from zero to unity, the vector w will never vanish, and therefore the number of full revolutions cannot change. Thus, $g^*(z)$ and $g(z)$ have the same number of zeros inside the contour P.

This remark fills the gap in the proof of Theorem III.

Now let us prove Theorem II. Incidentally, this theorem is not needed in proving the theorems of Section 3, which demonstrate that the real parts of all the roots of $h(z, e^z)$ are negative.

Proof of Theorem II. Let the polynomial $f(z, u, v)$ have no principal term [see (12)]. Let s denote the largest value that n can take in the sum (12), and let r denote the largest value that m can take for $n = s$. Then, the sum (12) contains the term $z^r \varphi_r^{(s)}(u, v)$. Since there is no principal term in this sum, it contains one more term $z^p \varphi_p^{(q)}(u, v)$ for which $p > r$, and $q < s$. Now substitute the values of u and v from (16), and then multiply the resulting sum by t^s so as to convert it into a polynomial $h(z, t)$. This polynomial will contain the term $z^r t^{2s} \varphi_r^{(s)}(\frac{1}{2}, -i/2)$. It contains the highest power in t, and also the highest power in z that is possible for this power of t. Moreover, the polynomial $h(z, t)$ will also contain the term $z^p t^{q+s} \varphi_p^{(q)}(\frac{1}{2}, -\frac{1}{2})$, where $p > r$, $q + s < 2s$. Thus, $h(z, t)$ will not have a principal term (cf. Section 1). Hence, it immediately follows that the polynomial $h(-iz, t)$ will also not have a principal term, and the equation $h(-iz, e^z) = 0$ therefore has roots with positive real parts (see Theorem I). Consequently, the equation $h(z, e^{zi}) = 0$ has a root with an imaginary part distinct from zero.

Thus, Theorem II has been demonstrated.

Theorems II and III give necessary and sufficient conditions for the function $f(z, \cos z, \sin z)$ to have only real zeros. It is a trivial matter to resolve this question, if the polynomial $f(z, u, v)$ has no principal term. Then it can immediately be asserted that $f(z, \cos z, \sin z)$ has an infinite number of nonreal roots. For functions containing a principal term, there are cases where a simple investigation can reveal an infinite number of nonreal roots:

Theorem IV. Let the polynomial $f(z, u, v)$ have a principal term [see (14)]. If the function $\Phi_*^{(s)}(z) = \varphi_*^{(s)}(\cos z, \sin z)$ has nonreal roots, then the function $F(z)$ has an infinite number of nonreal zeros. If $\Phi_*^{(s)}(z)$ has only real and simple zeros, then $F(z)$ has no more than a finite number of nonreal zeros.

Proof. In place of the equation $F(z) = 0$, consider the equation:

$$\Phi_*^{(s)}(z) + \sum_{m<r, n\leqslant s} z^{m-r}\Phi_m^{(n)}(z) = 0. \tag{24}$$

Assume that $\Phi_*^{(s)}(c) = 0$, where c is a nonreal number. We shall seek a solution of equation (24) in the form $2k\pi + c + \zeta$, where k is large and ζ is small. Equation (24) can be rewritten as

$$\Phi_*^{(s)}(c+\zeta) + \delta(\zeta) = 0, \tag{25}$$

where $\delta(\zeta)$ is an analytic function of ζ which converges uniformly to zero in the circle $|\zeta| \leqslant 1$, as $k \to \infty$. Since the left-hand side of equation (25) converges uniformly to $\Phi_*^{(s)}(c+\zeta)$ as $k \to \infty$, and the equation $\Phi_*^{(s)}(c+\zeta) = 0$ has the obvious solution $\zeta = 0$, equation (25) has a solution ζ_k close to zero for sufficiently large k. As c is nonreal, the solution $2k\pi + c + \zeta_k$, for sufficiently large k, will also be nonreal.

If the equation $\Phi_*^{(s)}(z) = 0$ has all real and simple roots, then in the interval $2k\pi + \epsilon \leqslant x \leqslant 2(k+1)\pi + \epsilon$ the curve $w = \Phi_*^{(s)}(x)$ intersects the axis $w = 0$ at $2s$ distinct points. Since the curve

$$w = \Phi_*^{(s)}(x) + \sum_{m<r, n\leqslant s} x^{m-r}\Phi_m^{(n)}(x), \tag{26}$$

for sufficiently large k, differs very little from the curve $w = \Phi_*^{(s)}(x)$, the curve (26) will also intersect the axis $w = 0$ at $2s$ points in the interval $2k\pi + \epsilon \leqslant x \leqslant 2(k+1)\pi + \epsilon$, for sufficiently large k. Thus, in the interval $-2k\pi + \epsilon \leqslant x \leqslant 2k\pi + \epsilon$ the function $F(z)$ has $(4sk + r')$ real roots for sufficiently large k. By virtue of Theorem III, the function $F(z)$ has $(r - r')$ nonreal zeros. Thus, Theorem IV has been demonstrated.

The question concerning the nature of the roots of the function $\Phi_*^{(s)}(Z)$ in the sense in which we are interested is reduced to solving a similar question for some polynomial. For this purpose it is necessary to express $\cos z$ and $\sin z$ in terms of $\tan z/2$:

$$\cos z = \frac{1 - \tan^2 z/2}{1 + \tan^2 z/2}, \quad \sin z = \frac{2\tan z/2}{1 + \tan^2 z/2},$$

and then take $\tan z/2$ as the unknown variable t. Thus, in $\varphi_*^{(s)}(u, v)$ we have to put

$$u = \frac{1-t^2}{1+t^2}, \quad v = \frac{2t}{1+t^2}$$

and multiply the resulting expression by $(1+t^2)^s$. Denote the polynomial thus obtained by $\varphi^{(s)}(t)$. Its degree is $2s$. If there are no terms of this degree, the polynomial has an infinite root, i.e., the equation $\Phi_*^{(s)}(z)=0$ has the solution $z=\pi$. And this root has a multiplicity equal to the difference of the degree of $\varphi^{(s)}(t)$ and $2s$. To each finite root t_0 of $\varphi^{(s)}(t)$, there corresponds a zero of $\Phi_*^{(s)}(z)$ derived from the equation $\tan z/2 = t_0$. And real zeros correspond to real roots and nonreal zeros to nonreal roots, the only exception being the root $t_0 = \pm i$, since the equation $\tan z/2 = \pm i$ has no solution. But the polynomial $\varphi^{(s)}(t)$ cannot have the roots $\pm i$: this follows directly from the condition (13). Indeed, we have

$$\varphi^{(s)}(t) = \varphi_r^{(s)}(1-t^2, 2t) + \sum_{n<s} \varphi_r^{(n)}(1-t^2, 2t)\,(1+t^2)^{s-n}. \quad (27)$$

Putting $t=\pm i$, we obtain

$$\varphi^{(s)}(\pm i) = \varphi_r^{(s)}(2, \pm 2i) \neq 0 \quad \text{[see (13)]}.$$

3. Zeros of the Function $h(z, e^z)$ with a Principal Term

In Section 1 we have shown that the function $h(z, e^z)$ has an infinite number of zeros with arbitrarily large positive real parts, provided the polynomial $h(z,t)$ has no principal term. Here we shall solve the question concerning the existence of zeros of the function $H(z)=h(z,e^z)$ when $h(z,e^z)$ has a principal term. Let

$$h(z,t) = \sum_{m,n} a_{mn} z^m t^n \quad (28)$$

and $a_{rs}z^r t^s$ be the principal term. In (28) pick out the coefficient of z^r, i.e., put

$$h(z,t) = z^r \chi_*^{(s)}(t) + \sum_{m<r, n\leqslant s} a_{mn} z^m t^n. \quad (29)$$

The function $\chi_*^{(s)}(e^z)$, where $z=x+iy$, is obviously periodic with a period $2\pi i$ and has no more than s zeros in the strip $b \leqslant y < b+2\pi$. Thus, there exists a real number ϵ, such that

$$\chi_*^{(s)}(e^{x+\epsilon i}) \neq 0 \quad (30)$$

for arbitrary x.

Theorem V. Let $h(z, t)$ be a polynomial with the principal term $a_{rs}z^r t^s$ and ϵ be a real number, such that $\chi_*^{(s)}(e^{x+i\epsilon}) \neq 0$ for arbitrary real x [see (29)]. Let N_k denote the number of zeros of the function $H(z)$ in the strip $-2k\pi + \epsilon \leqslant y \leqslant 2k\pi + \epsilon$, $x > 0$ $(z = x + iy)$. Let the function $H(z)$ be non-zero on the imaginary axis, i.e., $H(iy) \neq 0$, and let V_k be the angle which the vector $w = H(iy)$ describes about the origin when y runs through the values from $-2k\pi + \epsilon$ to $2k\pi + \epsilon$. Then

$$V_k = 2\pi(2sk - N_k + \tfrac{1}{2}r) + \delta_k$$

where δ_k tends to zero as $1/k \to 0$.

Proof. Consider the rectangle P_{ka}: $0 \leqslant x \leqslant a$; $-2k\pi + \epsilon \leqslant y \leqslant 2k + \epsilon$. Let us evaluate the angle described by the vector $w = H(z)$ when z traverses counterclockwise three sides of the rectangle P_{ka}, excluding the side $x = 0$, i.e., the lower, right and upper sides.

From (29) and (30), it directly follows that

$$H(z) = z^r \chi_*^{(s)}(e^z)(1 + \delta_1(z)),$$

where $\delta_1(z)$ tends uniformly to zero on these three sides of the rectangle P_{ka}, when k and a simultaneously tend to infinity. Thus, the angle described by the vector w for the function $H(z)$ will differ from the angle for the function $z^r\chi_*^{(s)}(e^z)$ only by a number η which tends to zero as $k \to \infty$, $a \to \infty$.

The angle described by the vector w for the function $z^r\chi_*^{(s)}(e^z)$ is equal to the sum of the angles for the function z^r and for the function $\chi_*^{(s)}(e^z)$. For the function z^r, the angle over the three sides of the rectangle P_{ka} is evidently equal to πr. For the function $\chi_*^{(s)}(e^z)$ the angle over the lower side is cancelled by the angle over the upper side, because the function $\chi_*^{(s)}(e^z)$ is periodic and these two sides are traversed in opposite directions. The angle over the right side for the function $\chi_*^{(s)}(e^z)$ differs very little from the angle over this side for the function $a_{rs}e^{sz}$. The last angle is obviously equal to $4k\pi s$. Thus, the total angle for the function $H(z)$ over the three sides differs very little from $4\pi sk + \pi r$.

Since the number of zeros of the function $H(z)$ inside the rectangle P_{ka} is equal to the number of full revolutions made by the vector $w = H(z)$ when z traverses all sides of the rectangle P_{ka}, the truth of our theorem follows immediately from the calculations made above.

Theorem V suggests that it is important for us now to study the behaviour of the function $H(z)$ on the imaginary axis, i.e., the function $H(iy)$ for real y. Separate the function $H(iy)$ into real and imaginary parts:

$$H(iy) = F(y) + iG(y). \tag{31}$$

It is immediately clear that

$$F(y) = f(y, \cos y, \sin y), \quad G(y) = g(y, \cos y, \sin y),$$

where $f(y, u, v)$ and $g(y, u, v)$ are polynomials. We shall now examine in detail the relationship between the polynomial $h(z, t)$ and the polynomials $f(y, u, v), g(y, u, v)$.

Put

$$\alpha^{(n)}(u, v) + i\beta^{(n)}(u, v) = (u + iv)^n,$$

where $\alpha^{(n)}(u, v)$ and $\beta^{(n)}(u, v)$ are polynomials with real coefficients. Then we have

$$\begin{aligned} \alpha^{(n)}(u, v) &= \tfrac{1}{2}[(u + iv)^n + (u - iv)^n]; \\ \beta^{(n)}(u, v) &= \frac{1}{2i}[(u + iv)^n - (u - iv)^n]. \end{aligned} \tag{32}$$

We now show that $a\alpha^{(n)}(u, v) + b\beta^{(n)}(u, v) = \gamma^{(n)}(u, v)$, where a and b are real and do not simultaneously vanish, satisfies the condition (13), i.e.,

$$\gamma^{(n)}(l, \pm i) \neq 0. \tag{33}$$

From (32), we have $\gamma^{(n)}(1, \pm i) = 2^{n-1}(a \pm ib) \neq 0$.

It is immediately clear that

$$f(y, u, v) + ig(y, u, v) = \sum (a'_{mn} + ia''_{mn}) i^m y^m (\alpha^{(n)}(u, v) + i\beta^{(n)}(u, v)). \tag{34}$$

Here $a'_{mn} + ia''_{mn} = a_{mn}$ [see (28)], and a'_{mn} and a''_{mn} are real. Putting

$$f(y, u, v) = \sum_{m,n} y^m \varphi_m^{(n)}(u, v), \quad g(y, u, v) = \sum_{m,n} y^m \psi_m^{(n)}(u, v), \tag{35}$$

from (34) we find that $\varphi_m^{(n)}(u, v)$ and $\psi_m^{(n)}(u, v)$ are equal to

$$\pm [a'_{mn}\alpha^{(n)}(u, v) - a''_{mn}\beta^{(n)}(u, v)]; \quad \pm [a''_{mn}\alpha^{(n)}(u, v) + a'_{mn}\beta^{(n)}(u, v)], \tag{36}$$

where the sign and order of sequence depend on the remainder obtained after dividing m by four. Let λ and μ be two real numbers that do not simultaneously vanish. Then

$$\lambda f(y, u, v) + \mu g(y, u, v) = \sum_{m,n} y^m [\lambda\varphi_m^{(n)}(u, v) + \mu\psi_m^{(n)}(u, v)].$$

From (36) it follows that $\lambda\varphi_m^{(n)}(u,v)+\mu\psi_m^{(n)}(u,v)=a\alpha^{(n)}(u,v)+b\beta^{(n)}$ (u,v). Since the determinant of the matrix

$$\begin{Vmatrix} a'_{mn} & -a''_{mn} \\ a''_{mn} & a'_{mn} \end{Vmatrix}$$

is distinct from zero for $a_{mn}\neq 0$, under these conditions a and b do not vanish simultaneously. If $a_{rs}z^rt^s$ is the principal term of $h(z,u,v)$, the principal term of $\lambda f(y,u,v)+\mu g(y,u,v)$ is $y^r\gamma^{(s)}(u,v)=y^r[a\alpha^{(s)}(u,v)+b\beta^{(s)}(u,v)]$, where a and b do not vanish simultaneously. Consequently, it satisfies the condition (13).

Just as in Section 2, let us pick out the coefficients $\varphi_*^{(s)}(u,v)$ and $\psi_*^{(s)}(u,v)$ of y^r. Then the coefficient of y^r in $\lambda f(y,u,v)+\mu g(y,u,v)$ is of the type $\lambda\varphi_*^{(s)}(u,v)+\mu\psi_*^{(s)}(u,v)$: and, as in Section 2, there exists a real ϵ, such that $\lambda\Phi_*^{(s)}(\epsilon+iy)+\mu\Psi_*^{(s)}(\epsilon+iy)\neq 0$ for arbitrary real y. Under these conditions, obviously, $\chi_*^{(s)}(e^{x+i\epsilon})\neq 0$ for arbitrary real x.

After these preliminaries, we proceed to prove Theorems VI and VII which give conditions for the function $H(z)$ to have no zeros with positive or zero real parts.

Theorem VI. Let $H(z)=h(z,e^z)$, where $h(z,t)$ is a polynomial with the principal term $a_{rs}z^rt^s$ [see (28)]. Put $H(iy)=F(y)+iG(y)$. If all the zeros of $H(z)$ lie to the left of the imaginary axis, then the vector $w=H(iy)$, as y varies from $-\infty$ to $+\infty$, continuously rotates in the positive direction with a positive velocity: analytically,

$$G'(y)F(y)-F'(y)G(y)>0.$$

Moreover, when y runs through the interval $-2k\pi\leqslant y\leqslant 2k\pi$, the vector w describes an angle equal to $4k\pi s+\pi r+\delta_1$, where $\lim_{k\to\infty}\delta_1=0$. If, conversely, the vector w describes an angle equal to $4k\pi s+\pi r+\delta_1$ when y runs through the interval $-2k\pi\leqslant y\leqslant 2k\pi$, then it continuously rotates in the positive direction with a positive velocity, and all the zeros of $H(z)$ lie to the left of the imaginary axis. (The function $H(z)$ is assumed to have no zeros on the imaginary axis, otherwise it is impossible to determine the angle of rotation of the vector w.)

We now define a concept needed to formulate Theorem VII. Let $p(y)$ and $q(y)$ be two real-valued functions of a real variable. Their zeros are said *to alternate*, if each of these functions has no multiple zeros, and between two consecutive zeros of one function there exists at least one zero of the other. Moreover, it is assumed that the functions $p(y)$ and $q(y)$

do not vanish simultaneously. Under these conditions, the zeros of $p(y)$ and $q(y)$ occur alternately on the y-axis.

Theorem VII. Let $H(z) = h(z, e^z)$, where $h(z, t)$ is a polynomial with a principal term. Separate $H(iy)$ into real and imaginary parts; $H(iy) = F(y) + iG(y)$. If all the zeros of $H(z)$ lie to the left of the imaginary axis, then the zeros of $F(y)$ and $G(y)$ are real and alternate, and

$$G'(y)F(y) - F'(y)G(y) > 0 \tag{37}$$

for every y. Furthermore, for all the zeros of $H(z)$ to lie to the left of the imaginary axis, it is sufficient that one of the following conditions be satisfied.

(1) All the zeros of $F(y)$ and $G(y)$ are real and alternate, and the inequality (37) is satisfied for at least one value of y;

(2) All the zeros of $F(y)$ are real, and for each of its zeros $y = y_0$ the condition (37) is satisfied, i.e., $F'(y_0)G(y_0) < 0$;

(3) All the zeros of $G(y)$ are real, and for each of its zeros $y = y_0$ the condition (37) is satisfied, i.e., $G'(y_0)F(y_0) > 0$.

Proofs of Theorems VI and VII. We divide the proof into separate parts:

(a) Let $v(a, b)$ denote the angle described by the vector $w = H(iy)$ when y runs through the interval $a \leqslant y \leqslant b$. The velocity $(d/dy)v(0, y)$ of rotation of the vector w at the instant y is obviously equal to

$$\frac{d}{dy} v(0, y) = \frac{G'(y)F(y) - F'(y)G(y)}{F^2(y) + G^2(y)} . \tag{38}$$

Thus, the sign of $(d/dy)v(0, y)$ is the same as the sign of $G'(y)F(y) - F'(y)G(y)$.

(b) We claim that

$$v(a + \epsilon, b + \epsilon) = v(a, b) + \delta_2 , \tag{39}$$

where δ_2 tends to zero, for a fixed ϵ, as $a \to \pm\infty$ and $b \to \pm\infty$. To begin with, we have $v(a, b) = v(a, c) + v(c, b)$. Furthermore, from the structure of the function $H(z)$, it is obvious that $v(a, a + \epsilon) = \epsilon + \delta_3$, where $\delta_3 \to 0$, for a fixed ϵ, when $a \to \pm\infty$. Combining these two results, we obtain (39).

(c) Let λ and μ be two real numbers which do not vanish simultaneously. Then there exists a real ϵ, such that, for arbitrary real y, the following four inequalities

$$\begin{aligned} \lambda\Phi_*^{(s)}(\epsilon+iy)+\mu\Psi_*^{(s)}(\epsilon+iy) &\neq 0;\\ \mu\Phi_*^{(s)}(\epsilon+iy)-\lambda\Psi_*^{(s)}(\epsilon+iy) &\neq 0;\\ \Phi_*^{(s)}(\epsilon+iy) &\neq 0;\\ \Psi_*^{(s)}(\epsilon+iy) &\neq 0; \end{aligned} \tag{40}$$

are satisfied. If they are satisfied, then we can apply Theorem III to the functions $\lambda F(y)+\mu G(y)$, $F(y)$ and $G(y)$. From these inequalities it follows that, for sufficiently large k, the point $H(\pm 2k\pi i+\epsilon i)$ does not lie on any of the lines $\lambda w'+\mu w''=0$, $w'=0$, $w''=0$ $(w=w'+iw'')$ in the w-plane.

The fact that all the inequalities (40) are satisfied simultaneously follows from the remark just before Theorem VI.

(d) Assume that $v(-2k\pi, 2k\pi)=\tau(4k\pi s+\pi r)+\delta_4$ where $\tau=\pm 1$, and $\delta_4\to 0$ as $1/k$. Under these conditions, we show that the function $\lambda F(y)+\mu G(y)$ has only real and simple zeros for arbitrary real non-vanishing λ and μ; moreover,

$$\tau[G'(y)F(y)-F'(y)G(y)]>0.$$

To prove the assertion (d), take an ϵ satisfying the condition (c) for given λ and μ. Since the vector w describes an angle equal to $(4k\pi s+\pi r)+\delta_3$ (see assertion b) on the interval $-2k\pi+\epsilon\leqslant y\leqslant 2k\pi+\epsilon$, geometrically it is obvious that the curve $w=H(iy)$ should intersect the line $\lambda w'+\mu w''=0$ at not less than $4ks+r$ distinct values of y in the same interval. By Theorem III, in this interval the function $\lambda F(y)+\mu G(y)$ has no more than $4ks+r$ zeros. Thus, all its zeros are real and simple. That the zeros are simple implies, in particular, that the curve $w=H(iy)$ does not intersect the line $\lambda w'+\mu w''=0$ anywhere, i.e., the vector w rotates continuously with a non-zero rate. Hence, $\tau[G'(y)F(y)-F'(y)G(y)]>0$ (see assertion (a)).

We now proceed to prove the first halves of Theorems VI and VII. Assume that all the zeros of $H(z)$ lie to the left of the imaginary axis. Then, by virtue of Theorem V and assertion b, $v(-2k\pi, 2k\pi)=4k\pi s+\pi r+\delta_6$. Hence, by assertion d, the zeros of $F(y)$ and $G(y)$ are all real, simple, and $[G'(y)F(y)-F'(y)G(y)]>0$. Thus, the vector w rotates continuously counterclockwise with a positive velocity. Hence, it is geometrically obvious that the zeros of $F(y)$ and $G(y)$ occur alternately. The truth of the first halves of Theorems VI and VII has thus been demonstrated.

We now prove the second halves.

To begin with, note that if $v(-2k\pi, 2k\pi) = 4k\pi s + \pi r + \delta_7$, where δ_7 tends to zero as $1/k \to 0$, by virtue of Theorem V and assertion b, all the zeros of $H(z)$ lie to the left of the imaginary axis. This proves the second half of Theorem VI.

If now one of the conditions (1), (2), (3) of Theorem VII is satisfied, we have $v(-2k\pi, 2k\pi) \geqslant 4k\pi s + \pi r + \delta_7$. In case the zeros of $F(y)$ and $G(y)$ are real, simple and, alternate, then, from Theorem III, assertion b, and from geometrical considerations it follows that $v(-2k\pi, 2k\pi) = \tau(4k\pi + \pi r) + \delta_7$. But, by condition (1), the inequality (37) is at least satisfied at one point y, therefore, $\tau = 1$.

If the condition (2) or condition (3) is satisfied, then, by virtue of Theorem III and assertion b, and from simple geometrical considerations, we conclude that $v(-2k\pi, 2k\pi) = 4k\pi s + \pi r + \delta_7$. Consequently, as noted earlier, all the zeros of $H(z)$ lie to the left of the imaginary axis. Thus, Theorem VI and Theorem VII have been proved.

The theorem stated below gives additional conditions for the existence of roots of the function $H(z)$ to the right of the imaginary axis.

Theorem VIII. Let $H(z) = h(z, e^z)$, where $h(z, t)$ is a polynomial with the principal term $a_{rs}z^r t^s$. Let $\chi_*^{(s)}(t)$ denote the coefficient of z^r in $h(z, t)$. If $\chi_*^{(s)}(t)$ has at least one root to the right of the imaginary axis, then $H(z)$ has an infinite number of zeros to the right of the imaginary axis. If all zeros of $\chi_*^{(s)}(t)$ lie to the left of the imaginary axis, then $H(z)$ has no more than a finite number of zeros to the right of the imaginary axis.

This is proved along the same lines as Theorem IV.

The question whether all the zeros of $\chi_*^{(s)}(e^z)$ lie to the left of the imaginary axis can easily be reduced to a similar question concerning polynomials. Note that the zeros of $\chi_*^{(s)}(e^z)$ lie to the left of the imaginary axis, if and only if all the zeros of the polynomial $\chi_*^{(s)}(t)$ lie inside the unit circle $|t| < 1$. The linear fractional transformation $t = (1 + z_*)/(1 - z_*)$ converts the open unit disc in the t-plane into the half-plane lying to the left of the imaginary axis in the z-plane. Thus, in $\chi_*^{(s)}(t)$, we have to substitute $t = (1 + z_*)/(1 - z_*)$, clear the fractions and then solve the problem with the help of Theorems VI and VII.

The V. A. Steklov Mathematical Institute,
The USSR Academy of Sciences

Received 7 March 1942

Mappings of a Three-Dimensional Sphere into an n-Dimensional Complex

Doklady Akademii Nauk SSSR (1942) 34(2): 39–41

This paper gives the classification of the mappings of a three-dimensional sphere into an n-dimensional complex with a trivial fundamental group [1].

A mapping f of a complex L into a complex K is said to be homologically trivial in dimension r, if it transforms each r-dimensional cycle z of the complex L modulo $m \geqslant 2$ into a cycle $f(z)$ modulo m homologous to zero in K.

Lemma. Let f be a homologically trivial mapping in dimension r which transforms an r-dimensional complex L into a complex K with a *trivial fundamental group*. Then there exists a mapping g of the complex L into the complex K, which is equivalent to the mapping f, such that the set $g(L)$ is of dimension $(r-1)$.

Let $S_1, S_2, \ldots, S_k$ be spheres of dimension two and δ_i some point in S_i $(i = 1, 2, \ldots, k)$. Identify the points $\delta_1, \delta_2, \ldots, \delta_k$ with a point δ, and let H_k denote the polyhedron resulting from this identification.

Theorem 1. Let f be a simplicial mapping of a three-dimensional sphere Σ into the polyhedron H_k. Also let p_i be a point inside some triangle in the sphere S_i, and q_j be a similar point in the sphere S_j. Then $f^{-1}(p_i)$ and $f^{-1}(q_j)$ are oriented one-dimensional cycles in Σ; let $p_{ij}(f)$ denote their linking number [2], so that $p_{ij}(f) = p_{ji}(f)$. The two mappings f and g of the sphere into the polyhedron H_k are equivalent, if and only if $p_{ij}(f) = p_{ij}(g)$. Furthermore, for any system of integers p_{ij} satisfying the condition $p_{ij} = p_{ji}$, there exists a mapping f such that $p_{ij}(f) = p_{ij}$.

Theorem 1 gives the classification of the mappings of a three-dimensional sphere Σ into an arbitrary connected two-dimensional complex with a *trivial fundamental group*, because, for some k, any such complex is homotopically equivalent to the polyhedron H_k [3].

Theorem 2. Let K be a three-dimensional connected complex with a *trivial fundamental group*. Take a suitable k and define a mapping φ of the polyhedron H_k into K, so that the spheres $S_1, S_2, \ldots, S_k$ entering into the construction of H_k form a two-dimensional homotopic basis [4] of the complex K. Moreover, assume that all homological relations between the spheres S_i are completely given by the relations $\tau_i\varphi(S_i) \sim 0$ $(i = 1, 2, \ldots, k)$. Here $\tau_1 = \tau_2 = \ldots = \tau_p = 0$, where p is a two-dimensional Betti number of the complex K, and $\tau_{p+1}, \ldots, \tau_k$ are the two-dimensional torsion coefficients of K. Let d_{ij} denote the greatest common divisor of the numbers τ_i and τ_j, for $i \neq j$. And let d_{ii} stand for τ_i when τ_i is odd. Finally, if τ_i is even, take $d_{ii} = 2\tau_i$. Let f and g be two mappings of the three-dimensional sphere Σ into the polyhedron H_k. Then the mappings $f' = \varphi f$ and $g' = \varphi g$ of the sphere Σ into the complex K are equivalent in K, if and only if $p_{ij}(f) \equiv p_{ij}(g) \pmod{d_{ij}}$ (see Theorem 1).

By virtue of the lemma, Theorem 2 gives the complete classification of the mappings of the sphere Σ into the complex K for homologically trivial mappings.

In order to formulate this result for an n-dimensional complex K, we have to introduce a new operation over ∇-complexes. We shall precede the introduction of this operation by intutive considerations.

Let M be an orientable manifold and X some homology class modulo $\tau = 2m$ in M. Then the intersection $X \cdot X$ in the sense of Lefschetz [5] is defined as a homology class modulo 2τ. This is done as follows: let x and y be two cycles modulo τ in the class X in general position: $\Delta x = \tau a$, $\Delta y = \tau b$. Let d be a complex which realises the homology between x and y, then $\Delta d = x - y - \tau r$. We have $\Delta r = a - b$. Further, let c be a cycle homologous to the cycles a and b, such that it is in general position with d. Then there exist complexes p and q, such that $\Delta p = a - c$, $\Delta q = b - c$. Moreover, we shall assume that the complex p is in general position with y, and the complex q in general position with x. In addition, assume that the cycle $r - p + q$ is homologous to zero in M. Now let us form the expression:

$$x \cdot y + \tau(py + qx + c \cdot d) = z.$$

By direct calculations it can be verified that z is a cycle modulo 2τ, and the homology class Z containing the cycle z is uniquely defined by the homology class X.

Let K be a complex, its vertices being taken in a definite order, and let u and v be two ∇-complexes [6] in K modulo 2 of dimensions r and s, respectively. Let us define a ∇-complex $(u*v)$ mod 2 of dimension $r+s-1$ dependent on the complexes u and v as well as on the order chosen for the vertices of the complex K. Let T be some simplex of dimension $r+s-1$ in K with vertices $a_1, a_2, \ldots, a_{r+s}$ taken in increasing order of the numeration used for K. The value of the complex $(u*v)$ on T is given by the expression:

$$(u*v)(T) = \sum_{i=1}^{s} u(a_i, a_{i+1}, \ldots, a_{i+r})\, v(a_1, \ldots, a_i, a_{i+r}, \ldots, a_{i+s}).$$

It can be directly verified that

$$\nabla(u*v) = (\nabla u * v) + (u * \nabla v) + u \times v + v \times u \pmod 2. \qquad \text{(A)}$$

Here $u \times v$ and $v \times u$ are the products in the sense of Alexander calculated using the order of vertices taken in K.

Now, in the complex K let there be given a class of ∇-homology X modulo $\tau = 2m$, and let x be some ∇-cycle modulo τ in X. Then $\nabla x = \tau a$. Let us define the $x \dot{\times} x$ operation as follows:

$$x \dot{\times} x = x \times x + \tau(a*x).$$

Calculating with the help of (A), we find that $x \dot{\times} x$ is a ∇-cycle modulo 2τ and that the ∇-homology class Z containing $x \dot{\times} x$ is uniquely defined by the homology class X. Thus, for ∇-homology, also, the homology class modulo $\tau = 2m$ uniquely defines a homology class modulo 2τ, and we simply write $Z = X \times X$.

Theorem 3. Let K be a connected n-dimensional complex with a trivial fundamental group. For a suitable k, define a mapping φ of the polyhedron H_k into K, such that the spheres $S_1, S_2, \ldots, S_k$, constituting the polyhedron H_k give a two-dimensional basis of homotopy of the complex K with the relations $\tau_i \varphi(S_i) \sim 0$ $(i = 1, 2, \ldots, k)$ (see Theorem 2). If we assume that the spheres S_i are oriented, then $\varphi(S_i)$ will be a cycle in K. Now take two-dimensional ∇-cycles $x_1, x_2, \ldots, x_k$ (x_i are ∇-cycles modulo τ_i) for which the scalar product $\varphi(S_i)x_j = \delta_{ij}$, where δ_{ij} is the Kronecker delta. If X_i is a homology class containing x_i, the product $X_i \times X_j$ is defined with respect to the modulus d_{ij} (see Theorem 2). Let w be some four-dimensional Δ-cycle in K, $\Delta w = 0$, and let $q_{ij}(w)$ denote the scalar product of the ∇-homology class $X_i \times X_j$ with the cycle w. The number $q_{ij}(w)$ is defined with respect to the modulus d_{ij}. Now let f and g be two mappings of the three-dimensional sphere Σ into the polyhedron H_k. Then the mappings $f' = \varphi f$ and $g' = \varphi g$ of the sphere Σ into the complex K are homotopic in K, if and only if there exists a four-dimensional Δ-cycle w for which

$$p_{ij}(f) - p_{ij}(g) \equiv q_{ij}(w) \pmod{d_{ij}}.$$

By virtue of the lemma, we see that Theorem 3 gives a complete solution of the classification problem for homologically trivial mappings of a three-dimensional sphere Σ into the complex K. For homologically nontrivial mappings, we have to take recourse to homology.

References

1. For the formulation of the classification problem, see Pontryagin, *Matem. Sborn.* 9 (51) (1941).
2. Hopf, *Math. Ann.* **104** (1931).
3. Hurewicz, *Proced. Akad., Amsterdam,* 38–39 (1935–36).
4. *Ibid.*
5. Lefschetz, *Trans. Amer. Math. Soc.* 28 (1926). (We have denoted the intersection of two complexes X and Y by $X \cdot Y$).
6. Alexander, *Annals of Math.* 37 (1936).
7. *Ibid.* (We have used the symbol $X \times X$ to denote the product of X and Y in the sense of Alexander.)

Received 6 January 1942

Characteristic Cycles of Manifolds

Doklady Akademii Nauk SSSR (1942) 35(2): 35–39

Let R^{k+l} be $(k+l)$-dimensional Euclidean space, and $H(k,l)$ the manifold of all oriented k-dimensional planes in the space R^{k+l} that pass through a fixed point O. Let M be a k-dimensional oriented manifold with a continuous tangent in R^{k+l}. To each point x of the manifold M^k let us put into correspondence that plane $T(x)$ in $H(k,l)$ which is parallel to the tangent to the manifold M at the point x. Evidently, under a continuous differentiable deformation of M^k in R^{k+l}, the mapping T changes continuously, and consequently, the homotopy type of this mapping remains invariant.

Hence it follows that for $l \geqslant k+2$, the homotopy type of the mapping T is a topological invariant of the oriented manifold M^k (by a topological transformation of a manifold M we mean here a topological differentiable transformation). Thus, in particular, the homomorphisms of the Betti group of the manifold M^k into the Betti groups of the manifold $H(k,l)$ under the mapping T are certain topological invariants of the manifold M^k. These homomorphisms, by virtue of the Poincare–Veblen theorem, can be characterized as follows. Let z_j^{m-r} $(j=1,2,\ldots,p^r)$ be an $(m-r)$-dimensional basis of homology of the manifold $H(k,l)$, where $m=kl$ is the dimension of the manifold $H(k,l)$. The intersection of the manifold $T(M^k)$ with the cycle z_j^{m-r} is a cycle y_j^{k-r} of dimension $(k-r)$ in the manifold $T(M^k)$. Then $T^{-1}(y_j^{k-r})=x_j^{k-r}$ is a cycle in the manifold M^k. It can be readily verified that for $l \geqslant k+2$, the homology class X_j^{k-r} containing the cycle x_j^{k-r} is uniquely defined by the homology class of the cycle z_j^{m-r} and does not depend on the location of M^k in R^{k+l}. Thus, each homology class X_j^{k-r} is a topological invariant of the oriented manifold M^k. Moreover, a knowledge of the homology classes X_j^{k-r} ($j = 1, 2, \ldots, p^r$) is equivalent to knowledge of the homomorphism of the r-dimensional Betti group of the manifold M^k into the r-dimensional Betti group of the manifold $H(k, l)$ under the mapping T. The cycles x_j^{k-r} are called the *characteristic cycles* of the manifold M^k. Characteristic cycles are rather typical features of the topological structure of a manifold and are encountered, as we shall show below, in different problems. Study of characteristic cycles requires, first of all, a knowledge of the homology of $H(k, l)$.

Let $\omega(p)$ be an integral-valued non-decreasing function of p ($p = 1, 2, \ldots, k$), satisfying the condition $0 \leqslant \omega(p) \leqslant l$. In the space R^{k+l}, take a sequence of planes $0 \subset R_1 \subset R_2 \subset \ldots \subset R_k$, the dimension of the plane R_p being $p + \omega(p)$. Let $Z(\omega(p))$ denote the set of all planes R_ξ^k in $H(k, l)$ satisfying the condition that the intersection of R_ξ^k with R_p is of dimension not less than p. The set $Z(\omega(p))$ is a pseudomanifold of dimension $s = \Sigma_{p=1}^k \omega(p)$. If it is orientable, it can be regarded as a cycle in $H(k, l)$, and if not, as a cycle modulo 2. The cycle $Z(\omega(p))$ depends on the function $\omega(p)$; its dependence on the planes R_p is unimportant from the viewpoint of homology because by rotating the space R^{k+l}, any system of planes R_p can be converted into any other system with the same function $\omega(p)$.

The number p is called *a jump* of the function ω, if $\omega(p+1) - \omega(p) > 0$. Let $p_1, p_2, \ldots, p_{n-1}$ be an increasing sequence of jump points. Put

$$p_0 = 0, \quad p_n = k, \quad \alpha_i = p_i - p_{i-1} \quad (i = 1, 2, \ldots, n);$$
$$\beta_1 = \omega(1), \quad \beta_{i+1} = \omega(p_i + 1) - \omega(p_i) \quad (i = 1, 2, \ldots, n-1).$$

The numbers α_i and β_i uniquely define the function $\omega(p)$. Furthermore, they are essential because the set $Z(\omega(p))$ is uniquely defined by the planes R_p for $p = p_i$ ($i = 1, 2, \ldots, n$). Applying the aforementioned condition on the dimension of intersection to the intersection of R_ξ^k with the planes R_{p_i} only, we obtain the set $Z(\omega(p))$. The following two sequences

$$\alpha_1, \beta_2, \alpha_2, \beta_3, \ldots, \alpha_{n-1}, \beta_n; \quad \alpha_1, \beta_2, \alpha_2, \beta_3, \ldots, \beta_n, \alpha_n \tag{1}$$

play an important part; if $\omega(k) = l$, the first sequence is to be taken, and if $\omega(k) < l$, the second sequence. If $\alpha_1 \geqslant 2$ and $\Sigma_{p=1}^k (p) \geqslant m - (l-2)$, denote the function $\omega(p)$ by $\omega^*(p)$. Thus, the dimension of $Z(\omega^*(p))$ is equal to $(m - r)$, where $r \leqslant l - 2$. Let $\omega'(p)$ denote those functions $\omega^*(p)$ for which all the numbers in the sequence (1) are even, and let $\omega''(p)$ stand for those functions $\omega^*(p)$ for which the sequence (1) contains odd numbers, and let β be the first odd number in this sequence.

Theorem 1. All pseudomanifolds $Z(\omega'(p))$ of dimension $(m-r)$ are orientable and when each is taken with a certain orientation, they form an $(m-r)$-dimensional independent basis of the weak homology of the manifold $H(k, l)$. The pseudomanifold $Z(\omega''(p))$ is not orientable, and if its boundary is taken in a certain triangulation, we obtain a cycle $2Y(\omega''(p))$. The set of all cycles $Z(\omega'(p))$ and $Y(\omega''(p))$ of dimension $(m-r)$ form an $(m-r)$-dimensional canonical basis of homology in the manifold $H(k, l)$. Hence, all torsion coefficients of dimension $(m-r)$, where $r \leqslant l-1$, of the manifold $H(k, l)$ are equal to two, and the pseudomanifolds $Z(\omega''(p))$, which are cycles modulo 2, are independent in the sense of homology modulo 2.

Proof. From the set $Z(\omega(p))$ take a plane R_0^k whose intersection with R_{p_i} is of dimension p_i $(i = 1, 2, \ldots, n)$. Let $U(\omega(p))$ denote the closure of the set of those R_ξ^k in $Z(\omega(p))$ for which the orthogonal projection of R_ξ^k on R_0^k is positive, and let $\bar{U}(\omega(p))$ denote a similar set with a negative projection. Now let us impose an additional condition for the choice of the planes R_{p_i} and R_0^k, so that U and $\bar{U}$ should depend only on the function $\omega(p)$, but not on the random choice of these planes. For this purpose, in the space R^{k+l} fix an orthogonal system of coordinate axes taken in a definite order: $(1), (2), \ldots, (k+l)$. The plane defined by the first $p_i + \omega(p_i)$ coordinates axes shall always be taken as the plane R_{p_i}. And R_0^k is always taken to be the plane defined by the axes:

$$(\beta_1 + 1), \ldots, (\beta_1 + \alpha_1), (\beta_1 + \beta_2 + \alpha_1 + 1), \ldots, (\beta_1 + \beta_2 + \alpha_1 + \alpha_2),$$
$$(\beta_1 + \beta_2 + \beta_3 + \alpha_1 + \alpha_2 + 1), \ldots, (\beta_1 + \ldots + \beta_n, \alpha_1 + \ldots + \alpha_n),$$

with an orientation given by the order already specified. In the pseudomanifolds U and $\bar{U}$ with boundaries thus uniquely defined by the function $\omega(p)$, let us introduce a certain definite orientation. If $\omega(p)$ is a given function, then let $\omega_i(p)$ be a function that coincides everywhere with $\omega(p)$, except at the point $p_{i-1} + 1$, $(i = 1, 2, \ldots, n)$ where it is equal to $\omega(p) - 1$; if $\omega_1(p) < 0$, the corresponding term in (2) is discarded. Thus, we have

$$\Delta U(\omega(p)) = \sum_{i=1}^{n} [(-1)^{\beta_1 + \ldots + \beta_i + \alpha_i + \ldots + \alpha_n} U(\omega_i(p)) - \bar{U}(\omega_i(p))](-1)^{q_i(\omega)};$$

$$\Delta \bar{U}(\omega(p)) = \sum_{i=1}^{n} [(-1)^{\beta_1 + \ldots + \beta_i + \alpha_i + \ldots + \alpha_n} \bar{U}(\omega_i(p)) - U(\omega_i(p))](-1)^{q_i(\omega)}.$$

Here $q_i(\omega) = \Sigma_{p=1}^{n} \omega(p)$. Each linear form in various U and $\bar{U}$ is called a *U-complex*. It turns out that every cycle in $H(k, l)$ is homologous to some

U-cycle, and each U-cycle which is homologous to zero in $H(k, l)$ bounds some U-complex. From this, by virtue of (2), the proof of Theorem 1 can be obtained.

Now put $Z(\omega(p)) = U(\omega(p)) + (-1)^{\beta_1 + k}\, \bar{U}(\omega(p))$. Thus, we assign a definite orientation to each pseudomanifold $Z(\omega(p))$, so that $Z(\omega'(p))$ turns out to be a cycle, while $Z(\omega''(p))$ acquires a definite boundary. In particular, the manifold $H(k, l) = Z(l)$ itself acquires an orientation. It can be readily found that the intersection of $H(k, l)$ with $Z(1 + \omega(p))$ in $H(k, l+1)$ is $\pm Z(\omega(p))$. Thus we eliminate the apparent dependence of the characteristic cycle on the number l. In order to emphasize this independence, let us introduce a function $\varphi(p) = l - \omega(p)$ in place of the function $\omega(p)$, and let $X(\varphi(p))$ denote the intersection of the manifold $T(M^k)$ with $Z(l - \varphi(p))$ transferred into M^k with the help of the mapping T^{-1}. The dimension of $X(\varphi(p))$ is equal to $k - r(\varphi)$, where $r(\varphi) = \Sigma_{p=1}^{k} \varphi(p)$. Special mention should be made of the case where $r(\varphi) = k$. The cycle $X(\varphi(p))$ in this case is zero dimensional, and the number of its points is a numerical invariant of the manifold M^k. In particular, $X(1)$, i.e., $X(\varphi(p))$ when $\varphi(p) \equiv 1$, gives the Euler characteristic for even k. Other types of invariants also exist, for instance, when $k = 4$, the function defined by the conditions

$$\varphi(1) = \varphi(2) = 2, \quad \varphi(3) = \varphi(4) = 0$$

gives a zero-dimensional cycle $X(\varphi(p))$.

We now give examples where characteristic cycles are encountered in certain problems.

Let f be a differentiable mapping of the oriented manifold M^k into a q-dimensional Euclidean space R^q. Using the local coordinates $t_1, t_2, \ldots, t_k$ in M^k, we can analytically express it as follows:

$$y_i = f_i(t_1, \ldots, t_n) \quad (i = 1, \ldots, q).$$

The functional matrix $\|\partial y_i/\partial t_j\|$ has, in general, a rank equal to the lesser of the two numbers k and q. Let $C(k, q, \rho)$ denote the set of those points in M^k for which this rank is less than the number just indicated or less than or equal to ρ.

Theorem 2. The critical set $C(k, q, \rho)$ is a characteristic cycle $X(\varphi(p))$, where $\varphi(p)$ is given by the conditions: $\varphi(1) = \ldots = \varphi(k - \rho) = q - \rho$, $\varphi(k - \rho + 1) = \ldots = \varphi(k) = 0; r(\varphi) = (k - \rho)(q - \rho)$.

Proof. Let R^{k+l} be a Euclidean space with sufficiently large l and a fixed orthogonal coordinate system. Take the subspace of R^{k+l} given by the last q axes as R^q, and the normal to R^q as R^{k+l-q}. Now imbed M^k into R^{k+l} in such a manner that its orthogonal projection in R^q coincides with the mapping f. It is easily seen that a point x in M belongs to $C(k, q, \rho)$, if and only if the plane $T(x)$ parallel to the tangent intersects R^{k+l-q} along a plane of dimension $\leqslant (k-\rho)$. Theorem 2 follows directly from this assertion as a corollary.

Let M^k be a differentiable oriented manifold. In each of its tangential planes T_x, take $s < k$ vectors $v_1(x), \ldots, v_s(x)$ continuously dependent on x. And let $V(k, s)$ denote the set of all those points for which these vectors are linearly dependent. Stiefel investigated this problem.

Theorem 3. The set $V(k, s)$ is a characteristic cycle $X(\varphi(p))$, where $\varphi(p)$ is given by the conditions:

$$\varphi(1) = \ldots = \varphi(k-s+1) = 1; \quad \varphi(k-s+2) = \ldots = \varphi(k) = 0;$$
$$k - r(\varphi) = (s-1).$$

Proof. Assume that M^k sits in R^{k+l}. Let R^s denote a plane in R^{k+l} spanned by the last s axes, and R^{k+l-s} the normal to this plane. In the plane R^s take mutually orthogonal vectors $v_1, \ldots, v_s$. If the tangent T_x is in general position with R^{k+l-s}, the intersection of $T(x)$ with R^{k+l-s} is of dimension equal to $(k-s)$. Let $R^s(x)$ denote the normal to this intersection in $T(x)$. Projection of the plane $R^s(x)$ on the plane R^s is a one-to-one correspondence. Consequently, the vectors $v_1, \ldots, v_s$ uniquely define the vectors $v_1(x), \ldots, v_s(x)$ in $R^s(x)$. Thus, x in M^k will belong to $V(k, s)$, if and only if the intersection of $T(x)$ with R^{k+l-s} is of dimension greater than or equal to $(k-s+1)$. Theorem 3 follows directly from this assertion as a corollary.

Received 16 March 1942

A Method of Calculation of Homology Groups

Matematicheskii Sbornik, Novaya Seriya (1942) 11(1/2): 3–14

In the present paper I generalize the method of calculation of homology groups of manifolds, which was already used by me in the investigation of group manifolds.[1] The essence of this method consists in the following.

Let M be an orientable differentiable compact manifold, in which there is given a non-degenerated Riemannian metric. Let, further, f be a real numerical differentiable function, defined on M and possessing the property that it attains its maximum on an orientable sub-manifold P of the manifold M and its minimum – on a sub-manifold Q of the manifold M and that its total differential does not vanish anywhere, except the points of the manifolds P and Q. Under these conditions through every point x from M, not belonging to P and Q, there passes one definite trajectory orthogonal to the level surfaces of the function f. Using a well known device of Morse,[2] we may subject every compact set $F \subset M$, not intersecting P, to a continuous deformation along the said trajectories in such a way that in the result of this deformation the set F will be transformed into a beforehand given neighbourhood of the manifold Q. Using this remark and certain considerations of the theory of homologies and intersections, we may, to a certain extent, reduce the study of topological properties of the manifold M to that of the topological properties of the manifolds P and Q (cf. Theorems 1 and 2). In the case, when we have to consider the topological properties of a whole series of manifolds, we may, using the outlined transition from P and Q to M, carry out a complete mathematical induction over the whole series of manifolds (cf.

Theorem 3). When all manifolds of the considered series are of even dimensionality, the calculation of homology groups is particularly simple (cf. Theorem 4).

1. The General Method

The present paragraph contains the chief results of the paper. In order to avoid misunderstandings in the terminology, we begin by recollecting the fundamental notions of the theory of homologies.

Let $a_0, a_1, \ldots, a_r$ be a system of $r+1$ points of a certain Euclidean space, not lying in any linear space of dimensionality less than r. The minimal convex body T containing the points $a_0, a_1, \ldots, a_r$ is called a polyhedral r-dimensional simplex. The aggregate consisting of the simplex T and of its continuous mapping f into a certain manifold M is called an r-dimensional simplex of the manifold M, $E=(T,f)$. If the simplex T is orientated, then the simplex E is also said to be orientated. Two simplexes $E=(T,f)$ and $E'=(T',f')$ of the manifold M are considered to be coinciding, if there exists such a nondegenerated affine superposition of the simplexes T and T' on each other that the functions f and f' coincide. If the simplexes E and E' are orientated, then they are considered to be coinciding, if the described superposition preserves the orientation. A linear form of the type

$$C = c_1E_1 + c_2E_2 + \ldots + c_nE_n,$$

where $c_1, c_2, \ldots, c_n$ are integers and $E_1, E_2, \ldots, E_n$ are r-dimensional orientated simplexes from M, is called an r-dimensional algebraical complex from M. By the boundary of an oriented r-dimensional simplex E is meant the sum of its $(r-1)$-dimensional orientated faces, mapped into M by means of f. Thus to each oriented r-dimensional simplex from M and hence, in virtue of additiveness, to each algebraic r-dimensional complex C from M is correlated an algebraical $(r-1)$-dimensional complex $\dot{C}$ called the boundary of the original complex. An algebraic complex from M is called a cycle, if its boundary is equal to zero. As is known, the boundary of every algebraical complex from M is a cycle. A cycle Z, which is the boundary of a certain complex, is said to be homologic to zero, in symbols: $Z \sim 0$. Two cycles X and Y, the difference of which is homologic to zero, are said to be homologic to each other, in symbols: $X \sim Y$. A cycle W is said to be weakly homologic to zero, if there exists a positive integer, q such that $qW \sim 0$; then we write: $W \approx 0$. The cycles U and V are said to be weakly homologic to each other, if their difference is weakly homologic to zero; in symbols: $U \approx V$.

If the manifold M is divided into simplexes forming the complex K, then the simplexes of the complex K may be considered as affine non-degenerated images of polyhedral simplexes and are thus simplexes from M. We shall call the algebraical complexes of the manifold M, formed from such simplexes, the polyhedral algebraic complexes from K.

Let C be an r-dimensional algebraical complex from M. Then there exists such an r-dimensional polyhedral complex C^* from K and such an r-dimensional complex C_* from M that C^* is situated in the neighbourhood of C and C_* – in the neighbourhood of $\dot{C}$ and, moreover, the boundary of the complex C_* is equal to $\dot{C}-\dot{C}^*$ and the cycle $C-C_*-C^*$ is homologic to zero in the neighbourhood of C. The proximity, which may be here achieved, depends on the smallness of the simplexes of the complex K. The complex C^* is said to be polyhedrally approximating the complex C.

Let K' be the barycentric subdivision of the complex K. Every barycentric star of the complex K is a polyhedral algebraical complex from K'. We shall call a linear form of the type $c_1B_1+c_2B_2+\ldots+c_nB_n$, where $c_1,c_2,\ldots,c_n$ are integers and $B_1,B_2,\ldots,B_n$ are r-dimensional barycentric stars of the complex K, the r-dimensional star complex from K.

If C is an r-dimensional polyhedral complex from K', then there exist such an r-dimensional star complex C^* from K and such an r-dimensional polyhedral complex C_* from K' that C^* is situated in the neighbourhood of C and C_* – in the neighbourhood of C and, moreover, the boundary of the complex C_* is equal to $\dot{C}-\dot{C}^*$ and the cycle $C-C^*-C_*$ bounds a polyhedral complex from K' situated in the neighbourhood of C. The proximity, which may be achieved here, depends on the smallness of the simplexes of the complex K. The complex C' is said to be star-approximating the complex C.

If C is a complex from M, then approximating it first by a polyhedral complex from K' and then by a star complex from K we obtain a star approximation of an arbitrary complex C from the manifold M by means of a star complex from K.

Lemma 1. Let M be an orientable compact manifold and N its orientable closed sub-manifold situated in it without self-intersections. Suppose that there exists such a division of the manifold M into simplexes forming a complex K that the sub-manifold N represents a sub-complex L of the complex K. Let further X be a certain r-dimensional algebraical complex from M, whose boundary does not meet N and whose simplexes are sufficiently small. If the algebraical intersection[3] of the complex X with the manifold N is homologic to zero in N, then there exists in any given neighbourhood U of the manifold N such an $(r+1)$-dimensional complex Y from M that the complex $X - \dot{Y}$ does not already meet N.

Proof. Let V be such a neighbourhood of the manifold N, the closure of which lies in U. Denote by C the algebraic complex formed of all simplexes of the complex X, lying in V. If the simplexes of the complex X are sufficiently small, then the complex $X - C$ already does not meet N. Denote by C^* the star approximation of the complex C. If the simplexes of the complex K are sufficiently small, then the complex C_*, with the boundary $\dot{C} - \dot{C}^*$, lying in the neighbourhood of $\dot{C}$, does not meet N and the complex D, with the boundary $C - C^* - C_*$, lies in U. The complex $X' = X - \dot{D}$ is representable in the form

$$X' = X - \dot{D} - C^* + C^*.$$

It is easily seen that the complex $X - \dot{D} - C^*$ does not meet N. Denote by X_2 the star complex formed of all stars of the complex C^* meeting N; then $X' = X_1 + X_2$, X_1 does not meet N and X_2 consists of stars meeting N.

Put

$$X_2 = d_1 B_1 + d_2 B_2 + \ldots + d_n B_n,$$

where $d_1, d_2, \ldots, d_n$ are integers and $B_1, B_2, \ldots, B_n$ – barycentric stars of the complex K, which meet L. Denote by T_j the simplex from K dual to the star B_j, i.e., such that the index of intersection of T_j with B_j in M is equal to $+1$. Since B_j intersects with L, T_j must belong to L. Denote by $\bar{B}_j$ the barycentric star of the complex L dual to the simplex T_j, i.e., such that the index of intersection of T_j with $\bar{B}_j$ in N is equal to $+1$. The cycle

$$\bar{X} = d_1 \bar{B}_1 + d_2 \bar{B}_2 + \ldots + d_n \bar{B}_n$$

is homologic in N to the algebraic intersection of X with N and hence is homologic to zero in N. Let

$$\bar{Y} = c_1 \bar{A}_1 + c_2 \bar{A}_2 + \ldots + c_m \bar{A}_m$$

be that star complex from L, the boundary of which is equal to $\bar{X}$. Denote by S_i the simplex from L dual to the star $\bar{A}_i$ in L and by A_i the barycentric star of the complex K dual to the simplex S_i. Put

$$Y' = c_1 A_1 + c_2 A_2 + \ldots + c_m A_m.$$

It is easily seen that the complex formed from all stars entering into the boundary Y' of the complex Y' and meeting L coincides with X_2. Thus the complex $X' - \dot{Y}' = X_1 + X_2 - \dot{Y}'$ does not meet N. Passing from the complex X' to the complex X, we obtain that

$$X - \dot{D} - \dot{Y}' = X' - \dot{Y}' = X - \dot{Y}$$

does not meet N, while the complex $Y = D + Y'$ lies completely in U. Lemma is thus proved.

Lemma 2. Let M be a compact orientable differentiable manifold, in which it is possible to introduce a positive nowhere degenerated Riemannian metric, and $f(x)$ – a real numerical differentiable function defined on M, $x \in M$. Suppose that the function f attains its maximum on a differentiable orientable sub-manifold P of the manifold M and its minimum on a differentiable orientable sub-manifold Q of the manifold M. Suppose, further, that the total differential of the function f vanishes only on the manifolds P and Q. We shall, moreover, assume that M admits of such a triangulation, in which P and Q consist of its simplexes. If X is an algebraical complex from M, the boundary of which lies in Q and the algebraical intersection of which with P is homologic to zero in P, then there exists such a complex X', lying in Q, that $X - X'$ is a cycle homologic to zero in M.

Proof. We shall assume that a positive nowhere degenerated Riemannian metric is already introduced in M. Denote by W a neighbourhood of the manifold Q in the manifold M so small that through every point $x \in W - Q$ there passes one and only one geodetic normal to the manifold Q. Let now the point x uniformly travel along the geodetic from the position x to the foot of the normal on Q with such a velocity that the whole way is traversed in a unit of the time. We have thus defined a continuous deformation φ of the neighbourhood W in Q, under which every point belonging to Q remains fixed.

Let d be the minimum of the function f. Denote by ϵ a positive number so small that from $f(x) \leqslant d + \epsilon$ follows $x \in W$.

In virtue of Lemma 1 there exists such a complex Y situated in $M - Q$ that $X - Y = X_1$ does not already meet P. Let c be the maximum of the function f. Denote by δ a positive number so small that from $f(x) > c - \delta$ follows that x does not belong to X_1. In virtue of the known Morse's construction[4] there exists a continuous deformation ψ of the set of all x, for which $f(x) \leqslant c - \delta$, into the set of all x, for which $f(x) \leqslant d + \epsilon$, i.e. into W, such that under this deformation all points from Q remain fixed. The defrormation ψ transforms the complex X_1 into a complex X_2 situated in W. Applying to the complex X_2 the deformation φ, we obtain a complex X' satisfying the conditions of Lemma 2. Lemma 2 is thus proved.

Lemma 3. Let M, P and Q be the manifolds defined in Lemma 2. Then for every cycle $\bar{Z}$ from P there exists such a complex U from M with the boundary belonging to Q that the algebraical intersection of the manifold P with U is homologic in P to the cycle $\bar{Z}$. If this condition is satisfied, then, moreover, the class of homologies of the cycle $\dot{U}$ in Q is uniquely determined by the class of homologies of the cycle $\bar{Z}$ in the manifold P.

Proof. We shall assume that in the manifold M is realized the triangulation mentioned in the conditions of Lemma 2 and denote the complexes obtained from the manifolds M, P, Q again by M, P, Q.

Without restricting the generality we may suppose that the cycle $\bar{Z}$ is an algebraic star complex from the complex P, i.e.

$$\bar{Z} = c_1\bar{B}_1 + c_2\bar{B}_2 + \ldots + c_n\bar{B}_n,$$

where $c_1, c_2, \ldots, c_n$ are integers and $\bar{B}_1, \bar{B}_2, \ldots, \bar{B}_n$ – barycentrical stars of the complex P. Let T_j be a simplex from the complex P dual in P to the star $\bar{B}_j$ and B_j – the barycentrical star of the complex M dual to the simplex T_j in the complex M. Put

$$Z = c_1B_1 + c_2B_2 + \ldots + c_nB_n;$$

then the algebraical intersection of the manifold P with the complex Z is equal to $\bar{Z}$ and the boundary $\dot{Z}$ of the complex Z does not intersect with P. By a continuous deformation described in Lemma 2 the cycle $\dot{Z}$ may be transformed into the cycle Y lying in Q so that in the process of transformation the cycle $\dot{Z}$ will not intersect P. Thus the cycle $Y - \dot{Z}$ bounds in $M - P$ a complex U'. It is easily seen that the complex $U = Z + U'$ satisfies the conditions of Lemma 3.

Suppose now that there are two complexes U and V satisfying the conditions of Lemma 3; then their difference $U - V$ satisfies the conditions of Lemma 2 and hence there exists a complex C from Q, the boundary of which is equal to $\dot{U} - \dot{V}$, i.e., $\dot{U} \sim \dot{V}$ in Q. Lemma 3 is thus proved.

Theorem 1. Let M be a compact orientable differentiable manifold admitting of a non-degenerated Riemannian metric and f – a real numerical differentiable function defined on M. Suppose that f assumes its maximum on a differentiable orientable sub-manifold P of the manifold M and its minimum – on a differentiable sub-manifold Q of the manifold M and that the total differential of the function f vanishes only on the manifolds P and Q; suppose, besides, that M admits of a triangulation, for which the manifolds P and Q are composed of its simplexes. The dimensionalities of the manifolds M, P, Q we denote respectively by m, p, q. The Betti groups of dimensionality r of the manifolds M, P, Q we denote respectively by M^r, P^r, Q^r. Put $s = m - p - 1$, $t = m - p$. Define a homomorphism φ^r of the group P^r into the group Q^{r+s} in the following way: let $x \in P^r$ and X be a certain cycle from x; in virtue of Lemma 3 there exists a complex U from M with the boundary in Q such that the algebraical intersection of P with U is homologic in P to the cycle X; if $y \in Q^{r+s}$ and $\dot{U}$ belongs to y, we put $\varphi^r(x) = y$. The nucleus of the homomorphism φ^r we denote by P_1^r and put

$$P_2^r = P^r/P_1^r, \quad Q_1^{r+s} = \varphi^r(P^r), \quad Q_2^{r+s} = Q^{r+s}/Q_1^{r+s}.$$

Denote the natural homomorphic mapping of the group Q^r into the group M^r, arising due to the fact that $Q \subset M$, by ψ^r. It turns out that the nucleus of the homomorphism ψ^r is equal to Q_1^r. Put

$$\psi^r(Q^r) = M_1^r, \quad M_2^r = M^r/M_1^r.$$

Then the groups M_2^{r+t} and P_1^r are isomorphic.

Thus, knowing the homomorphisms φ^r $(r = 0, 1, \ldots, p)$, we know the subgroup M_1^r and the factor group $M_2^r = M^r/M_1^r$ of the group M^r, which gives us at least the possibility to calculate the rank of the group M^r.

Proof. Let us show that the nucleus Q^{*r} of the homomorphism ψ^r is equal to

$$Q_1^r = \varphi^{r-s}(P^{r-s}).$$

In virtue of the very definition of the homomorphism φ^{r-s} it is evident that $\varphi^{r-s}(P^{r-s}) \subset Q^{*r}$. Let now $y \in Q^{*r}$, and Y be a certain cycle from y; then $Y \sim 0$ in M, and consequently there exists a complex U from M with the boundary Y, $\dot{U} = Y$. The algebraical intersection of the manifold P with the complex U we denote by X and that element from P^{r-s}, to which X belongs, – by x. In virtue of the definition of φ^{r-s} we have $\varphi^{r-s}(x) = Y$. Thus $\varphi^{r-s}(P^{r-s}) \supset Q^{*r-s}$, and consequently

$$Q^{*r} = \varphi^{r-s}(P^{r-s}).$$

Let us now show that the group $M_2^r = M^r/M_1^r$ is isomorphic to the group P_1^{r-t}. To this end introduce the homomorphism ξ^r of the group M^r into the group P^{r-t}; we shall show that the nucleus of the homomorphism ξ^r coincides with M_1^r and that $\xi^r(M^r) = P_1^{r-t}$.

We define the homomorphism ξ^r in the following way: let $u \in M^r$ and let U be a certain cycle from u; denote by X the algebraic intersection of the manifold P with the cycle U and by x that element of the group P^{r-t}, which contains the cycle X. Then $\xi^r(u) = x$.

If x is obtained in this manner, i.e., $x = \xi^r(u)$, then in virtue of the definition of the homomorphism φ^{r-t} it is clear that $\varphi^{r-t}(x) = 0$, since $\dot{U} = 0$ and $\dot{U} \subset Q$. Thus $\xi^r(M^r) \subset P_1^{r-t}$. If, conversely, $x \in P_1^{r-t}$, then there exists a complex U such that the algebraical intersection of P with U, which is equal to X, enters into x and $\dot{U} \subset Q$, $\dot{U} \sim 0$ in Q. In this case there is a complex C from Q such that $U - C = U'$ is a cycle, and consequently the element $u' \in M^r$ containing U' possesses the property that $\xi^r(u') = x$. Thus $\xi^r(M^r) = P_1^{r-t}$.

In order to prove the fact that the nucleus M^{*r} of the homomorphism ξ^r coincides with M_1^r, we observe in the first place that M_1^r consists of those classes of homologies in M, which have cycles lying in Q. If now $u \in M_1^r$, then there exists a cycle U from u lying in Q, and then the algebraical intersection of P with U is evidently equal to zero, and consequently $\xi^r(u) = 0$. Thus $M_1^r \subset M^{*r}$. Let now $u \in M^{*r}$ and $U \in u$; this means that the algebraical intersection of P with U is homologic to zero in P. Then, in virtue of Lemma 2, there exists such a complex C from Q that $U - C$ is a cycle homologic to zero in M. Thus C belongs to the class u and lies in Q, and so $u \in M_1^r$. Consequently we have $M_1^r \supset M^{*r}$ and finally obtain $M^{*r} = M_1^r$. Theorem 1 is proved.

The essence of Theorem 1 consists in that it enables us to a certain extent to calculate the Betti groups of the manifold M, if we know the Betti groups of the manifolds P and Q and the homomorphisms φ^r. On ground of these data we can, of course, find only the subgroup M_1^r and the factor group $M_2^r = M^r/M_1^r$ of the group M^r, which, as is known, is not sufficient for the determination of the group M^r itself. In certain particular cases, however, we can nevertheless find also the group M^r itself (cf. Section 2). In order to determine the group M^r in the general case certain complementary considerations are necessary, which I shall not develop here. The results of Theorem 1 fully enable us to find the Betti number of the manifold M (cf. Theorem 2).

In the proof of Theorem 2 we use the following known proposition:

If G is a certain commutative group, H its subgroup and $K = G/H$ the factor group, then the ranks of these groups are connected by the following relation:

$$R(G) = R(H) + R(K).$$

Theorem 2. Let M, P, Q be the manifolds of the Theorem 1. If we denote the ranks of the groups introduced in Theorem 1 by corresponding small letters, then we have

$$m^r = q^r - p^{r-s} + p_1^{r-s} + p_1^{r-t}.$$

Proof. In virtue of Theorem 1 and the foregoing remark on the ranks or groups we have

$$p^r = p_1^r + p_2^r,\ q^r = q_1^r + q_2^r,\ m^r = m_1^r + m_2^r,\ p_2^r = q_1^{r+s},$$
$$q_2^r = m_1^r,\ m_2^r = p_1^{r-t}.$$

From these inequalities immediately follows the assertion of Theorem 2.

The best application Theorems 1 and 2 find in the case when a whole series of manifolds satisfying certain conditions is considered (cf. Theorem 3).

Theorem 3. Let Ω be a certain series of orientable differentiable manifolds, each of which admits of a non-degenerated Riemannian metric. With respect to Ω we assume that in it there is a manifold M_0, the topological structure of which is known, and that on every other manifold $M \neq M_0$ from Ω is given a non-constant function f satisfying the conditions of Theorem 1 and such that the sub-manifolds P and Q determined by this function belong to the series Ω. Under these assumptions the topological properties of the manifolds of the series Ω may be studied inductively on ground of Theorems 1 and 2.

Proof. Let M be an arbitrary manifold of the series Ω. Denote by $M(1)$ and $M(2)$ the manifolds P and Q determined in M by the corresponding function f. Suppose now that the manifolds

$$M(i_1, i_2, \ldots, i_k) \quad (i_1 = 1, 2,\ i_2 = 1, 2, \ldots,\ i_k = 1, 2)$$

are already determined and define the manifolds

$$M(i_1, i_2, \ldots, i_k, 1) \quad \text{and} \quad M(i_1, i_2, \ldots, i_k, 2)$$

as determined in $M(i_1, i_2, \ldots, i_k)$ by the corresponding function f. If the dimensionality of the manifold M is equal to m, then the dimensionality of the manifold $M(i_1, i_2, \ldots, i_k)$ clearly does not exceed $m-k$, and consequently $k \leqslant m$. At the same time the described inductive process can break up only then, when we arrive at the manifold M_0, since on every other manifold from Ω the function f is defined. From the above immediately follows:

If any assertion holds for the manifold M_0 and if, under the assumption that it is true for the manifolds P and Q, it may be proved for M, then this assertion is also true for all manifolds of the series Ω.

The possibility of the transition from the manifolds P and Q to the manifold M is given by Theorems 1 and 2.

2. Series of Manifolds of Even Dimensionality

Complex analytical manifolds have even dimensionality. In the case, when it is possible to compose a series Ω of some type of these manifolds satisfying the conditions of Theorem 3, this series admits of a complete and simple study on ground of Theorems 1 and 3. The general result obtained here is given in Theorem 4. Theorem 4 has numerous applications. In the present paper I give only one concrete example.

For the formulation of Theorem 4 recollect that by the Poincaré polynomial $m(x)$ of the manifold M is meant the polynomial $m^0 + m^1x + m^2x^2 + \ldots + m^m x^m$ where m^i is the i-dimensional Betti number of the manifold M $(i = 0, 1, \ldots, m)$.

Theorem 4. Let Ω be a series of manifolds of even dimensionality, satisfying the conditions of Theorem 3, and let M_0 consist of one point. Then for every manifold M from Ω all Betti groups of odd dimensionality are trivial and that of even dimensionality have no torsions and so are determined by the Betti numbers. If now P and Q are manifolds determined in M by the corresponding function f and $m(x), p(x), q(x)$ are the Poincaré polynomials of the manifolds M, P, Q, then we have

$$m(x) = x^{m-p}p(x) + q(x) = p(x) + x^{m-q}q(x).$$

Thus, in order to calculate the Betti groups of the manifolds of the series Ω, it suffices to know only the dimensionalities of the manifolds and the correspondence referring to every manifold M its sub-manifolds P and Q.

Proof. The assertion of Theorem 4 is evident for the manifold M_0; thus, in virtue of Theorem 3, it suffices to prove it for the manifold M under the assumption that it is true for the manifolds P and Q.

In proving the theorem we shall use the notations of Theorems 1 and 2. Since the numbers m, p, q are even, the number $s = m - p - 1$ is odd. If now r is even, then the homomorphism φ^r maps the Betti group P^r of even dimensionality into the Betti group Q^{r+s} of odd dimensionality; this last is by assumption trivial, and hence the nucleus P_1^r of the homomorphism φ^r coincides with P^r. Thus $P_1^r = P^r$. In the case of an odd r the group P^r is trivial, and consequently we have again $P_1^r = P^r$. Since the group Q_1^{r+s} is isomorphic to the group P^r/P_1^r and this last is trivial, $Q_1^{r+s} = \{0\}$. Hence follows that $Q_2^r = Q^r$. Further we have that M_1^r is isomorphic to $Q_2^r = Q^r$, and consequently M_1^r has no torsions. On the other hand, M_2^r is isomorphic to $Q_1^{r-t} = Q^{r-t}$, and this last has no torsions. Thus the group M possesses a subgroup M_1^r and a factor group $M_2^r = M^r/M_1^r$, both having no torsions; hence immediately follows that M^r also has no torsions and is isomorphic to the direct sum of the groups P^{r-t} and Q^r. If r is odd, then both summands are by assumption trivial, and consequently the group M^r itself is trivial.

From what has been established for the Betti numbers follows that $m^r = p^{r-t} + q^r$. Multiplying the obtained equality by x^r and summing over r ($r = 0, 1, \ldots, m$) we obtain

$$m(x) = x^{m-p}p(x) + q(x).$$

Since the assumptions are completely symmetric in the manifolds P and Q, we also have

$$m(x) = p(x) + x^{m-q}q(x).$$

Theorem 4 is thus proved.

As an example of application of Theorem 4 consider the following manifolds.

Let K^n be the complex projective space of the complex dimensionality n and

$$\xi = (l_0, l_1, \ldots, l_k) \tag{1}$$

the sequence of its linear subspaces of respective dimensionalities $0, 1, \ldots, k$, where all of them contain the foregoing one. The set of all these sequences, taken with the natural topology, forms a manifold, which we shall denote by H_k^n. Denote the polynomial $1 + x^2 + x^4 + \ldots + x^{2i}$ by $a_i(x)$. Then the Poincaré polynomial of the manifold H_k^n has the form

$$a_{n-k}(x)a_{n-k+1}(x)\ldots a_n(x). \tag{2}$$

We shall compute the polynomial (2) on ground of Theorem 4. In order to construct the series Ω satisfying the conditions of Theorem 3, we extend the class of considered manifolds.

Let K^α be an α-dimensional linear subspace of the space K^n. The set of all elements $\xi \in H_k^n$ satisfying the condition $l_0 \in K^\alpha$ we denote by $H_{k,\alpha}^n$. It is easily verified that $H_{k,\alpha}^n$ is an orientable differentiable manifold admitting of a non-degenerated Riemannian metric. The dimensionality of the manifold $H_{k,\alpha}^n$ is, as may be easily computed,

$$2nk + k(k+1) + 2\alpha. \tag{3}$$

Thus it is an even number.

Observe that

$$H^n_{k,n} = H^n_k \tag{4}$$

and $H^n_{k,0}$ is homeomorphic to $H^{n-1}_{k-1,n-1}$. We shall simply write

$$H^n_{k,0} = H^{n-1}_{k-1,n-1}. \tag{5}$$

This follows thus: if $l_0 \in K^0$, then l_0 is a fixed point. Let K^{n-1} be a certain linear $(n-1)$-dimensional subspace from K^n not passing through l_0; then the sequence $\xi = (l_0, l_1, \ldots, l_k)$ determines on K^{n-1} the sequence $\xi' = (l'_0, l'_1, \ldots, l'_{k-1})$ and this establishes a homomorphism between $H^n_{k,0}$ and $H^{n-1}_{k-1,n-1}$.

In order to introduce on $H^n_{k,\alpha}$ $(\alpha > 0)$ the function $f(\xi)$, introduce in K^α linear complex coordinates $z_0, z_1, \ldots, z_\alpha$ and assume that they are normed by the condition

$$z_0\bar{z}_0 + z_1\bar{z}_1 + \ldots + z_\alpha\bar{z}_\alpha = 1.$$

Then the coordinates of a point shall be determined not up to an arbitrary factor, as it usually occurs in a projective space, but to a factor, whose modulus is equal to one. Thus the number $z_0\bar{z}_0$ is uniquely determined by the point and the chosen system of coordinates. If now $z_0, z_1, \ldots, z_\alpha$ are the coordinates of the point l_0 [cf. (1)], then $f(\xi)$ shall be defined by the equality $f(\xi) = z_0\bar{z}_0$. It is easily verified that the so defined function $f(\xi)$ is differentiable and that it attains its maximum for $z_0 = 1$, $z_1 = 0, \ldots,$ $z_\alpha = 0$ and its minimum for $z_0 = 0$. The fact that the total differential of the function $f(\xi)$ vanishes only under the formulated conditions may be easily verified.

If we denote by M the manifold $H^n_{k,\alpha}$, then we see directly that

$$P = H^n_{k,0} = H^{n-1}_{k-1,n-1}, \quad Q = H^n_{k,\alpha-1}. \tag{6}$$

We define the function $f(\xi)$ on the manifold $H^n_{k,0}$, interpreting it as $H^{n-1}_{k-1,n-1}$ [cf. (5)]. This transition is impossible only in the case when $k = 0$, but it is evident that all manifolds $H^n_{0,0}$ are homeomorphic to each other and contain only one point; thus they play here the role of M_0.

Thus all conditions of Theorem 4 are realized for the series of manifolds $\Omega = \{H^n_{k,\alpha}\}$. The application of Theorem 4 will now proceed only on grounds of the relations (3) and (6).

The Poincaré polynomial of the manifold $H^n_{k,\alpha}$ we denote by $h^n_{k,\alpha}(x)$. In virtue of Theorem 4 we have

$$h^n_{k,\alpha}(x) = h^{n-1}_{k-1,n-1}(x) + x^2 h^n_{k,\alpha-1}(x). \tag{7}$$

Applying this formula repeatedly, we obtain

$$h^n_{k,\alpha}(x) = h^{n-1}_{k-1,n-1}(x)a_\alpha(x). \tag{8}$$

Denoting the Poincaré polynomial of the manifold H^n_k by $h^n_k(x)$ and putting $\alpha = n$, we obtain

$$h^n_k(x) = h^{n-1}_{k-1}(x)a_n(x). \tag{9}$$

For $k = 0$ the relation (7) assumes the form

$$h^n_{0,\alpha}(x) = h^n_{0,0}(x) + x^2 h^n_{0,\alpha-1}(x). \tag{10}$$

Applying it successively, we obtain

$$h^n_0(x) = a_n(x). \tag{11}$$

Comparing (9) and (11), we obtain (2).

Notes

1. L. Pontrjagin, Homologies in compact Lie groups, *Recueil math.*, new series, 8(48): 3 (1939).
2. Morse, Relations between the critical points of a real function of n independent variables, *Trans. Amer. Math. Soc.*, **27** (1925).
3. Cf. S. Leischetz, *Topology*.
4. Cf. Note 2.

Received 4 February 1941

Hermitian Operators in a Space With an Indefinite Metric*

Izvestiya Akademii Nauk SSSR, Seriya Matematika (1944) 8(6): 243–80

The linear space H_k, like the Hilbert space, consists of all sequences $x_1, x_2, \ldots, x_k, \ldots$ of complex numbers, the sum of the squares of their moduli being convergent; but the scalar product in this space is defined as:

$$(x,y) = -x_1\bar{y}_1 - \ldots - x_k\bar{y}_k + x_{k+1}\bar{y}_{k+1} + \ldots .$$

The present paper is concerned with Hermitian operators in H_k. Every Hermitian operator in H_k is shown to have an invariant finite-dimensional subspace of dimension k.

Section 1 deals with the elementary properties of the space H_k which are somewhat more complicated than the corresponding properties of the Hilbert space. A central role is here played by the subspaces on which the quadratic form (x,x) is non-degenerate. It is shown that if the quadratic form (x,x) is non-degenerate on a subspace Q, then

(a) Q is isomorphic to some subspace $H_l, l \leqslant k$;

(b) the quadratic form (x,x) is also non-degenerate on the orthogonal complement Q' of Q;

---

* That Hermitian operators in a space with an indefinite metric may be of some interest I learned from S. L. Sobolev who encountered such an operator in a problem of mechanics.

The basic concepts of the theory of Hermitian operators in a Hilbert space are given in "Plessner: Spectral Theory of Linear Operators", *Uspekhi Matematicheskikh Nauk* **IX**, 3–125 (1941).

(c) the orthogonal complement of Q' is Q itself, and H_k is the direct sum of Q and Q', so that there exist orthogonal projections on both of these subspaces.

Section 2 is devoted to operators acting in H_k. Unlike in a Hilbert space, a symmetric operator in H_k may have complex eigenvalues and non-simple elementary divisors. Their number is greatly restricted by the dimension k.

Section 3 establishes a nontrivial specific property of a Hermitian operator A in H_k, namely, that the operator A has an invariant subspace I of dimension k, and $(x,x) \leqslant 0$ for each vector x in I. Moreover, each eigenvalue of the operator A in I has a non-negative imaginary part. Thus, a Hermitian operator in H_k, unlike in a Hilbert space, always has eigenvalues for $k > 0$.

1. The Linear Space H_k

Definition 1. The linear space H_k consists of all sequences

$$x_1, x_2, \ldots, x_k, \ldots \tag{1}$$

of complex numbers, the number of terms being equal to $n \geqslant k$ (finite dimensional case) or infinite (infinite dimensional case), for which the series

$$x_1\bar{x}_1 + x_2\bar{x}_2 + \ldots + x_k\bar{x}_k + \ldots \tag{2}$$

converges. The convergence and linear operations in H_k are defined in the same way as in a finite-dimensional affine or Hilbert space, respectively. The scalar product (x, y) of two vectors x and y is defined to be

$$(x,y) = -x_1\bar{y}_1 - x_2\bar{y}_2 - \ldots - x_k\bar{y}_k + x_{k+1}\bar{y}_{k+1} + \ldots . \tag{3}$$

In order to consider H_k as a Banach space, it is sufficient to define a norm in H_k as:

$$\|x\|^2 = x_1\bar{x}_1 + x_2\bar{x}_2 + \ldots + x_k\bar{x}_k + \ldots . \tag{4}$$

A linear manifold and subspace are defined in the usual way.

The scalar product satisfies the usual conditions:

$$(\lambda x, y) = \lambda(x,y); \quad (x + x', y) = (x,y) + (x',y); \quad (x,y) = (\bar{y}, \bar{x}).$$

Two vectors x and y are said to be *orthogonal*, if $(x, y) = 0$. Orthogonality of subspaces is defined as usual by this relationship.

If P is a subspace of H_k, its orthogonal complement P' is defined to be the subspace consisting of all vectors orthogonal to P. Unlike the usual case, the intersection of P and P' may contain non-zero vectors. In this case, we say the quadratic form (x,x) is degenerate on P, i.e., P contains a vector $x \neq 0$ orthogonal to all other vectors in P. If there is no such vector, we say the quadratic form (x,x) is non-degenerate on P. Evidently, the quadratic form (x,x) is non-degenerate on the linear space H_k itself.

Definition 2. Let H_k and H_l be two linear spaces, and P, Q their subspaces. A mapping f of P onto Q is said to be *isomorphic,* if it is bijective, bicontinuous and preserves linear operations and the scalar product. Two subspaces P and Q are said to be *isomorphic,* if there exists an isomorphic mapping f of P onto Q.

Below, we shall prove that if two spaces H_k and H_l are isomorphic, then the numbers k and l are equal:

$$k = l. \tag{5}$$

An isomorphic mapping of a subspace P onto itself is called an *automorphism*.

In Definition 1 the space H_k was defined by the coordinate method – each vector x was specified by its coordinates $x_1, x_2, \ldots, x_k, \ldots$. By means of any automorphism f of the space H_k, we can introduce a new system of coordinates in H_k. If $x'_1, x'_2, \ldots, x'_k, \ldots$ are the coordinates of the vector $f(x)$ then the numbers $x'_1, x'_2, \ldots, x'_k, \ldots$ can be taken as the new coordinates of the vector x.

(A) Let H_+ be the subspace of H_k consisting of all vectors x for which

$$x_1 = x_2 = \ldots = x_k = 0$$

and let H_- be the subspace of H_k consisting of all vectors for which

$$x_{k+1} = x_{k+2} = \ldots = 0.$$

Evidently, the quadratic form (x,x) is non-degenerate on H_+ and H_-. It is also obvious that every vector x in H_k can be represented uniquely as $x = x_- + x_+$, where $x_- \in H_-$ and $x_+ \in H_+$. Put $x_- = \varphi_-(x)$ and $x_+ = \varphi_+(x)$. The mappings φ_- and φ_+ are linear and continuous, and their kernels are H_+ and H_-, respectively.

(B) Let P be a subspace of H_k on which the quadratic form (x,x) is non-positive, i.e., $(x,x) \leqslant 0$ for every $x \in P$. The dimension of P is then not greater than k. Equality (5) follows directly from this assertion.

Since $(x,x) \leqslant 0$, the intersection of P with H_+ contains only zero. Consequently, P is mapped by φ_- onto a certain subspace of H_- linearly, continuously and bijectivly. Thus, the dimension of P is not greater than the dimension of H_-.

(C) Let P_- be some subspace of H_-, and P_+ be some subspace of H_+. Then the direct sum P of P_- and P_+ $(P = P_- \oplus P_+)$ is isomorphic to some linear space $H'_{k'}$ (Definition 1).

Construct the space H'_k, so that the dimension of H'_- is equal to the dimension of P_- and the dimension of H'_+ to the dimension of P_+. Then the spaces P_- and H'_- are isomorphic; and so are also the spaces P_+ and H'_+. These two isomorphisms taken together generate an isomorphic mapping of P onto $H'_{k'}$.

Lemma 1. Let P be a subspace of the space H_k on which the quadratic form (x,x) is non-degenerate and has the same sign ϵ for all its vectors; then there exists an automorphism f of H_k under which the subspace P is mapped into H_ϵ, $f(P) \subset H_\epsilon$.

Proof. We first reduce an infinite-dimensional P to a finite-dimensional P. Evidently, the infinite-dimensional case is true only when $\epsilon = +$ (see (B)).

Let Q be the intersection of P and H_+. Since H_+ is an ordinary Hilbert space, it can be expressed as the direct sum of two orthogonal subspaces Q and H'_+. Let H'_k be the direct sum of the subspaces H_- and H'_+, and P' the intersection of P and H'_k. Obviously, H_k is the direct sum of the orthogonal subspaces H'_k and Q. Hence, P is also the direct sum of the orthogonal subspaces P' and Q. Since zero is the only element in the intersection of P' with H_+, the dimension of P' does not exceed k, i.e., it is finite. The quadratic form (x,x) on P' is non-degenerate and positive. Assume now that there exists an automorphism f' of H'_k, such that $f'(P') \subset H'_+$. Define an automorphism f of the whole space H_k such that it coincides with f' on H'_k and is the identity on Q. Then $f(P) \subset H_+$, which proves the lemma.

It now remains to prove the lemma for a finite dimensional P. Assume that P is finite-dimensional and denote by H''_+ the subspace $\varphi_+(P)$ of the Hilbert space H_+. Let R denote the orthogonal complement of H''_+ in H_+, and H''_k the direct sum of H_- and H''_+. Then H_k is the direct sum of orthogonal subspaces H''_k and R, where H''_k is finite-dimensional and contains P. Now suppose that there exists an automorphism f'' of the finite-dimensional space H''_k such that P is mapped into H''_ϵ. Define an automorphism

f of the whole space H_k such that it is f'' on H_k'' and is the identity on R. Then $f(P) \subset H_\epsilon$, which proves the lemma.

Thus, it suffices to demonstrate the lemma for a finite-dimensional H_k.

Assume that H_k is finite-dimensional. Under this assumption, the cases $\epsilon = +$ and $\epsilon = -$ are equivalent, and as such we can confine ourselves to the case $\epsilon = +$.

Put $\varphi_-(P) = H_-'''$ and $\varphi_+(P) = H_+'''$. Let S_- and S_+ be the orthogonal complements of H_-''' and H_+''' in H_- and H_+, respectively. The space H_k is the direct sum of two orthogonal subspaces $H_{k'}''' = H_-''' \oplus H_+'''$ and $S = S_- \oplus S_+$.

As in the previous two cases, we first define a proper automorphism in $H_{k'}'''$ and then complement it with the identity autmorphism of S. Thus, it is sufficient to consider the case where

$$\varphi_-(P) = H_-, \quad \varphi_+(P) = H_+. \tag{6}$$

In P the quadratic form (x, x) is positive-definite; taking this quadratic form as the metric, we find that the form $(\varphi_+(x), \varphi_+(x))$ is a positive-definite Hermitian form in P.

In the space P, choose an orthonormal basis $e^{(1)}, e^{(2)}, \ldots, e^{(r)}$, in which the Hermitian form $(\varphi_+(x), \varphi_+(x))$ is canonical. If $i \neq j$, then $(e^{(i)}, e^{(j)}) = 0$, $(e_+^{(i)}, e_+^{(j)}) = 0$; hence $(e_-^{(i)}, e_-^{(j)}) = 0$. Here $e^{(i)} = e_-^{(i)} + e_+^{(i)}$ (see Proposition (A)). By virtue of the condition (6), the vectors $e_+^{(1)}, \ldots, e_+^{(r)}$ form an orthogonal basis in H_+. By virtue of the same condition those vectors among $e_-^{(1)}, \ldots, e_-^{(r)}$ that are non-zero form an orthogonal basis in H_-. Let $H^{(i)}$ denote the linear span of the vectors $e_-^{(i)}$ and $e_+^{(i)}$. Then the space H_k is a direct sum of the mutually orthogonal subspaces $H^{(1)}, \ldots, H^{(r)}$. Now assume that for each $H^{(i)}$, there exists an automorphism $f^{(i)}$ such that $f^{(i)}(e^{(i)}) = \alpha^{(i)} e_+^{(i)}$, where $\alpha^{(i)}$ is some number. Define an automorphism f such that it coincides with $f^{(i)}$ on $H^{(i)}$. Then, $f(P) \subset H_+$. This completes the proof of the lemma.

Hence it is sufficient to demonstrate the lemma for a space H_k of dimension not greater than two. The one-dimensional case is trivial, so we take up the two-dimensional case.

Thus, let H_1 have a basis consisting of two vectors **a** and **b**, where $(a, a) = -1$, $(b, b) = 1$, $(a, b) = 0$. Let q be the unit vector in P, so that $(q, q) = 1$. Let p be a vector orthogonal to q such that $(p, q) = -1$. Evidently, p and q form a basis of H_1. The desired automorphism f of H_1 is defined by the conditions

$$f(p) = a, \quad f(q) = b.$$

Lemma 1 has been demonstrated in full.

Theorem 1. Let Q be some subspace of the space H_k on which the quadratic form (x,x) is non-degenerate; then there exists an automorphism f of H_k which maps Q into a direct sum of the subspaces P_- and P_+, where $P_- \subset H_-, P_+ \subset H_+$.

Theorem 1 implies that any subspace Q of the space H_k with the non-degenerating quadratic form (x,x) is isomorphic to some space $H'_{k'}$ (see Definition 2).

Proof. If the quadratic form (x,x) has the same sign for all the vectors in Q, the assertion of Theorem 1 follows as a corollary from Lemma 1. Discarding this particular case, let us construct in Q a maximal system of vectors $e^{(1)}, e^{(2)}, \ldots, e^{(r)}$, satisfying the condition:

$$(e^{(i)}, e^{(j)}) = -\delta_{ij}. \tag{7}$$

Since the subspace Q contains vectors x, for which $(x,x) < 0$, there always exists one vector $e^{(1)}$ satisfying the condition (7). If, amongst the vectors $x \in Q$, orthogonal to $e^{(1)}$, there exist such vectors that $(x,x) < 0$, there also exists a second vector $e^{(2)}$. By Assumption (B), this process cannot be continued *ad infinitum*, and there exists therefore a maximal system of vectors satisfying the condition (7).

Let Q_- denote the linear span of the maximal system of vectors $e^{(1)}, \ldots, e^{(r)}$. The quadratic form (x,x) on Q_- is non-degenerate and negative. Thus, by virtue of Lemma 1, there exists an automorphism f' such that $f'(Q_-) = P_- \subset H_-$.

Put $f'(Q) = Q'$ and let P'_+ be the intersection of Q' with H_+. Let H'_- and H'_+ be the orthogonal complements of the subspaces P_- and P'_+ in the spaces H_- and H_+, respectively. Then the space H_k is the direct sum of the orthogonal subspaces $P' = (P_- \oplus P'_+)$ and $H' = (H'_- \oplus H'_+)$. The subspace P' is contained in Q' and therefore Q' can be represented as the direct sum of P' and the intersection R of Q' with H'. Obviously, on R the quadratic form (x,x) is non-degenerate and positive. Thus, by virtue of Lemma 1, there exists an automorphism f'' of the space H' which maps R into H'_+. Define an automorphism f''', such that it is the same as f'' on H' and is the identity map on P'. The map $f'''f' = f$ is the desired automorphism, and Theorem 1 has been proved.

Theorem 2. Let Q be a subspace of the space H_k on which the quadratic form (x,x) is non-degenerate, and Q' the orthogonal complement of Q in H_k; then the quadratic form (x,x) is also non-degenerate on Q' and H_k is the direct sum of Q and Q'.

This means, in particular, that the space H_k can be projected orthogonally onto Q. Moreover, Q is the orthogonal complement of Q'.

Proof. By virtue of Theorem 1, there exists an automorphism f of the space H_k such that $f(Q) = P = P_- \oplus P_+$, where $P_- \subset H_-$, $P_+ \subset H_+$. Let P'_- and P'_+ be the orthogonal complements of P_- and P_+ in the subspaces H_- and H_+, respectively. Then, the space H_k is the direct sum of the orthogonal subspaces P and $P' = P'_- \oplus P'_+$. Evidently, the quadratic form (x, x) is non-degenerate on P'. The automorphism f^{-1} maps the subspace P onto Q, and the subspace P' into some Q' where $H_k = Q \oplus Q'$. The theorem is proved.

2. Linear Operators in the Space H_k

As H_k is a Banach space, there is no need for defining a linear functional or a linear operator on it; we have only to consider anew those concepts in the theory of operators in a Hilbert space that are related to the concept of scalar product. Here we come across certain essential differences of fundamental interest.

Let E denote the identity operator, $Ex = x$. The domain of definition Ω of a linear operator A is always assumed to be everywhere dense in H_k.

A linear operator A is called *symmetric*, if for any two vectors x and y in its domain of definition,

$$(Ax, y) = (x, Ay). \tag{1}$$

Unlike the ordinary theory, a symmetric operator in H_k may have complex eigenvalues, and also non-simple elementary divisors. The numbers of both the former and the latter are related to the number k by stringent conditions due to (B) of Section 1.

A vector x is said to belong to the eigenvalue λ of the operator A, if there exists a natural integer r such that

$$(A - \lambda E)^r x = 0. \tag{2}$$

The set of all vectors belonging to the eigenvalue λ of the operator A forms a linear manifold in H_k called the eigenspace of λ for the operator A.

(A) If $\lambda \neq \bar{\mu}$, two vectors x and y belonging to the eigenvalues λ and μ of a symmetry operator A are orthogonal or, which is the same thing, the eigenspaces S_λ and S_μ are orthogonal. If $\lambda = \mu$, where λ is complex, the eigenspace S_λ is orthogonal to itself, i.e., the quadratic form (x, x) is identically zero on this manifold. Consequently, the dimension of S_λ is not greater than k ((B) Section 1).

By assumption,

$$(A - \lambda E)^r x = 0, \quad (A - \mu E)^s y = 0 \tag{3}$$

It is required to prove that $(x, y) = 0$.

We prove by induction on $r + s$, assuming $r \geqslant 0$, $s \geqslant 0$. For $r + s = 0$, the proposition is obvious. Put

$$(A - \lambda E)x = x', \tag{4}$$

$$(A - \mu E)y = y'. \tag{5}$$

Then, by virtue of (3)

$$(A - \lambda E)^{r-1} x' = 0, \quad (A - \mu E)^s y = 0, \tag{6}$$

$$(A - \lambda E)^r x = 0, \quad (A - E)^{s-1} y' = 0. \tag{7}$$

By induction, from (6) it follows that $(x', y) = 0$ and from (7) that $(x, y') = 0$. Taking the scalar product of both sides of (4) with y, and both sides of (5) with x, we obtain

$$((A - \lambda E)x, y) = (x', y) = 0, \tag{8}$$

$$(x, (A - \mu E)y) = (x, y') = 0. \tag{9}$$

Subtracting (9) from (8), and taking into account that A is symmetric, we obtain

$$(\lambda - \bar{\mu})(x, y) = 0, \quad \text{i.e.} \ (x, y) = 0.$$

Proposition (A) asserts that the eigenspace of a complex eigenvalue λ is a finite-dimensional manifold. This cannot be said of a real eigenvalue α, although nonsimple elementary divisors are clustered in a finite-dimensional subspace of the manifold S_α, and their number and order are bounded, depending on the number k.

(B)* Let S_α be an eigenspace of a real eigenvalue α of a closed symmetric operator A in H_k. Then S_α is closed in H_k and is the direct sum of two orthogonal subspaces S and S' invariant relative to A, where the dimension of S is finite and S' consists of the eigenvectors of the operator A. Furthermore, the quadratic form (x, x) is non-degenerate and positive

---

* Proposition (B) and Theorem 3 are not used in what follows.

on S', so that S' is a Hilbert space. (If S_α itself is finite-dimensional, then we may take $S = S_\alpha$.)

On a finite-dimensional subspace S the operator A has the following properties. Let $D_1, D_2, \ldots, D_r$ be the elementary divisors of the operator A on S, and $d_1, d_2, \ldots, d_r$ their orders. The number of nonsimple elementary divisors of a given order is an *invariant* of the operator A and of the number α. Furthermore, if

$$\rho(\alpha) = \sum_{i=1}^{r} \left[\frac{d_i}{2}\right], \tag{10}$$

where the square bracket stands for the integral part of the number inside the bracket, there exists in S a subspace P_α of dimension $\rho(\alpha)$ on which the quadratic form (x, x) vanishes identically. Consequently, $\rho(\alpha) \leqslant k$ (see (B) Section 1).

In proving this proposition we put $\alpha = 0$. This in no way imposes any limitation, but in place of the symmetric operator A we then have to take the symmetric operator $A - \alpha E$.

If $S_\alpha = S_0$ is finite-dimensional, take $S = S_0$. If S_0 is infinite-dimensional, let S_0^i be the set of all vectors $\mathbf{x}$ in S_0 for which $A^i x = 0$. Then $S_0^i \subset S_0^{i+1}, i = 1, 2, \ldots$.

Let $P \subset Q$ be two linear manifolds. The maximum number of vectors in Q that are linearly independent relative to P is called the *index of P in Q*. If the index of P in Q is finite, and equal to r, there exists a linear manifold R of dimension r, such that Q is the direct sum of P and R: $Q = P \oplus R$.

We shall now show that the index of S_0^i in S_0^{i+1} is not greater than k, and consequently, is finite.

Let R be a linear manifold in S_0^{i+1} such that the intersection of S_0^i with R contains only zero. Then $A^i R \subset S_0^1$, where the mapping A^i is linear and bijective on R, so that $A^i R$ is of the same dimension as R. Further, let $y \in R$ and $x \in S_0^1$. Then

$$(x, A^i y) = (A^i x, y) = (0, y) = 0$$

and, consequently, the space S_0^i is orthogonal to $A^i R$. Since $A^i R \subset S_0^1$, $A^i R$ is orthogonal to itself and the dimension of $A^i R$ is therefore not greater than k (see (B) Section 1).

Let t be a natural integer such that S_0^{t-1} is distinct from S_0^t. Let R_0^t be a finite-dimensional space in S_0^t, such that $S_0^t = S_0^{t-1} \oplus R_0^t$. Then the mapping A is one-to-one on R_0^t and the intersection of S_0^{t-2} with AR_0^t contains only zero. Consequently, S_0^{t-1} is a direct sum: $S_0^{t-1} =$

$S_0^{t-2} \oplus R_0^{t-2}$, where $AR_0^t \subset R_0^{t-1}$. Continuing this process further, and assuming that $R_0^1 = S_0^1$, we obtain a sequence

$$R_0^1, R_0^2, \ldots, R_0^t \tag{11}$$

such that S_0^t is a direct sum of the terms in (11), where $AR_0^i \subset R_0^{i-1}$ ($i = 2, 3, \ldots, t$).

Now take $A^i R_0^{2i} = Q_0^i$ ($i = 1, 2, \ldots, [t/2]$). Then $Q_0^i \subset R_0^i$. Hence it follows that we can form the direct sum

$$P_0^t = Q_0^1 \oplus Q_0^2 \oplus \ldots \oplus Q_0^{[t/2]}.$$

We shall show that the quadratic form (x, x) vanishes identically on P_0^t. For this, it suffices to show that the spaces Q_0^i and Q_0^j are orthogonal for arbitrary i and j not greater than $[t/2]$.

Assume that $i \leqslant j$, and let $x \in Q_0^i, y \in Q_0^j$. Then $x = A^i x', y = A^j y'$ and

$$(x, y) = (A^i x', A^j y') = (A^{i+j} x', y') = (0, y') = 0.$$

Thus, the dimension of P_0^t is not greater than k, and, in particular, $t \leqslant 2k + 1$. Hence the sequence $S_0^1, S_0^2, \ldots, S_0^t, \ldots$ stabilizes after a finite number of terms, and we may assume that t is the largest number for which S_0^t does not coincide with S_0^{t-1}. If r_0^i is the index of S_0^{i-1} in S^i, then the dimension of R_0^i is r_0^i, and there exists in the manifold S^0 a subspace P_0^t of dimension equal to $\rho = r_0^2 + r_0^4 + \ldots + r_0^{2i} + \ldots + r^{2[t/2]}$, and the quadratic form (x, x) on P_0^t is identically zero.

Let x^i be a vector in R_0^i that does not belong to AR_0^{i+1}. Then the vectors $x^i, Ax^i, \ldots, A^{i-1}x^i$ are linearly independent. Let $N_x i$ be the linear span of these vectors. Evidently, the space $N_x i$ is invariant relative to A, and on $N_x i$ the operator A has only one elementary divisor of order i. Let x^i, y^i, z^i be a maximal system of vectors in R_0^i linearly independent relative to AR_0^{i+1}; then the space S_0 is the direct sum of the totality of all spaces of the type $N_x i, N_y i, N_z i, \ldots$ ($i = 1, 2, \ldots, t$). Hence the number of elementary divisors of order i of the operator A on S_0 is equal to $r_0^i - r_0^{i+1} = p_0^i$. Consequently,

$$r_0^i = p_0^i + p_0^{i+1} + \ldots + p_0^t.$$

Thus,

$$\rho = \sum_{i=1}^{t} [i/2]\, p_0^i,$$

which is the same as (10).

Up to this point we have not used that A is closed. Since the dimension of S_0 may be infinite, we are not fully justified in using the concept of elementary divisors, though for the comparatively simple structure which the operator A has on S_0, we could easily develop a theory of elementary divisors.

We now assume that the operator A is closed. Since S_0^1 is composed of the eigenvectors of A corresponding to the eigenvalue 0, the linear manifold S_0^1 is closed and is a subspace of the space H_k. And since the index of S_0^1 in S_0 is finite, S_0 is also a subspace of H_k.

In H_k choose coordinates and let S' denote the intersection of S_0^1 with H_+. Evidently, the quadratic form (x,x) is non-degenerate on S' and is always positive, because it is so on H_+. By virtue of Theorem 2, the space H_k can be represented as the direct sum of S' and some other subspace H'_k. If S is the intersection of S_0 with H'_k, obviously $S_0 = S \oplus S'$. As S' is invariant under A, namely $AS' = \{0\}$, the space H'_k is also invariant under A. Hence, S is invariant. For S we can now make all the constructions that were made earlier for S_0, the only important difference being that S^1 is finite dimensional, because the intersection of S^1 with H_+ contains only zero. Since the index of S^1 in S is finite, the dimension of S is therefore also finite. Hence, no further explanation is needed in applying elementary divisors to the space S. And the relations derived above for S_0, when applied to S, demonstrate the truth of (B).

Thus, Proposition (B) has been demonstrated in full.

Combining the results stated in (A) and (B), we formulate:

Theorem 3. Let λ be an eigenvalue of a symmetric operator A in the space H_k. Let $\rho(\lambda)$ denote the dimension of the eigenspace of λ for non-real λ, and the number introduced in (B) for real λ. For real λ, $\rho(\lambda)$ is evidently zero, if its eigenspace is composed of eigenvectors, i.e., if all the elementary divisors belonging to the eigenvalue λ are simple. For a system of eigenvalues $\lambda_1, \lambda_2, \ldots, \lambda_n$ not containing any pair of conjugate numbers (the set of all eigenvalues with non-negative imaginary parts may be taken as this system), we have

$$\rho(\lambda_1) + \rho(\lambda_2) + \ldots + \rho(\lambda_n) \leqslant k.$$

Proof. For complex λ, let P_λ denote its eigenspace and for real λ, the space on which the quadratic form (x,x), by virtue of (B), identically vanishes. Then the dimension of P_λ is equal to $\rho(\lambda)$. The direct sum of all spaces P_{λ_i} $(i = 1, 2, \ldots, n)$ is some space P on which the quadratic form (x,x) vanishes identically, because, by virtue of (A) and (B), each P_{λ_i} is orthogonal to every other P_{λ_j}, even for $i = j$.

Hence, by Proposition (B) of Section 1, the dimension of P is not greater than k.

Thus, Theorem 3 has been demonstrated.

For the space H_k, Theorem 3 replaces the assertion that a symmetric operator in a Hilbert space has only real eigenvalues and admits only simple elementary divisors.

We now define a Hermitian operator in the linear space H_k, after certain preliminary remarks.

(C) If L is a linear functional in H_k, there exists a vector l in H_k such that $Lx = (x, l)$ for any $x \in H_k$.

Since linear operations and convergence in H_k are defined exactly in the same way as in a Hilbert space, there exists a sequence of numbers l'_1, $\ldots, l'_k$, $l'_{k+1}, \ldots$ the sum of the squares of whose moduli converges, such that

$$Lx = x_1\bar{l}'_1 + \ldots + x_k\bar{l}'_k + x_{k+1}\bar{l}'_{k+1} + \ldots \tag{12}$$

for any $x \in H_k$. Take the coordinates of the vector l to be the numbers $-l'_1, \ldots, -l'_k, +l'_{k+1}, \ldots$. Then, (12) can be rewritten as $Lx = (x, l)$.

(D) A sequence $x^1, x^2, \ldots, x^n, \ldots$ of elements of H_k is said to *converge weakly* to x, $x^n \to x$, if $\lim_{n\to\infty}(x^n, y) = (x, y)$ or which is the same, if $\lim_{n\to\infty}(y, x^n) = (y, x)$, for any $y \in H_k$. Just as in the case of a Hilbert space, for a sequence to converge weakly, it is sufficient that either $\lim_{n\to\infty}(y, x^n) = (y, x)$ or $\lim_{n\to\infty}(x^n, y) = (x, y)$, for $y \in M$, where M is a set everywhere dense in H_k.

Definition 3. Let A be a linear operator in H_k with an everywhere dense domain of definition Ω and let $y \in H_k$. Suppose that for every $x \in \Omega$ the following relation holds

$$(Ax, y) = (x, z).$$

Denote the set of all y satisfying (13) by Ω^* and put $z = A^*y$.

As Ω is everywhere dense in H_k, the relation (13) uniquely defines z and therefore A^* is a linear operator with Ω^* as its domain of definition. The operator A^* is called the adjoint of A. If $A^* = A$, i.e., $\Omega^* = \Omega$ and $A^*x = Ax$ for every $x \in \Omega$, the operator A is called *Hermitian.*

(E) Let A be a Hermitian operator with $\Omega \subset H_k$ as its domain of definition, and $y^1, y^2, \ldots, y^n, \ldots$ a sequence of elements in Ω, such that $y^n \to y$, $Ay^n \to z$ (see Proposition (D)); then $y \in \Omega$ and $Ay = z$.

For every $x \in \Omega$,

$$(Ax, y) = \lim_{n\to\infty} (Ax, y^n) = \lim_{n\to\infty} (x, Ay^n) = (x, z),$$

i.e., $(Ax, y) = (x, z)$, and as A is Hermitian, this implies that $y \in \Omega$, $Ay = z$.

As usual, a linear operator A in H_k is said to be continuous, if its domain of definition is H_k, and $x^n \Rightarrow x$ implies $Ax^n \Rightarrow Ax$ (where the symbol $\Rightarrow$ stands for strong convergence).

(F) A continuous Hermitian operator P is called a *projection,* if $P^2 = P$. For a projection operator P, let H' denote the set of all vectors of the type Px, where $x \in H_k$. Then, H' is a subspace of H_k on which the quadratic form (x, x) is non-degenerate. Let H'' be the orthogonal complement of H' in H_k; then for $x' \in H', x'' \in H''$,

$$Px' = x', \quad Px'' = 0. \tag{14}$$

Obviously, the conditions (14) uniquely define the operator P throughout the space H'. For a given subspace H' with the non-degenerate quadratic form (x, x) and its orthogonal complement H'', the relations (14) define a projection P. A projection P and the subspace H' are called *mutually corresponding*.

The set H' defined as the totality of all vectors of the form Px is evidently a linear manifold in H_k; it is closed as well. Indeed, if $x \in H'$, then $Px = x$, and conversely, if $Px = x$, then $x \in H'$. Thus, H' can be defined as the set of all vectors $x \in H_k$ for which $Px = x$. And this relationship, by virtue of the continuity of the operator P, defines a closed set in H_k. We now show that the quadratic form (x, x) is non-degenerate on H'. If $h \in H'$ is a vector, such that $(x, h) = 0$ for any $x \in H'$, then

$$(y, h) = (y, Ph) = (Py, h) = (x, h) = 0$$

for any $y \in H_k$. As the quadratic form (x, x) is non-degenerate on H_k, we have $h = 0$. Let H'' be the orthogonal complement of H' in H_k. If $x'' \in H''$ and $x \in H_k$, then $(Px'', x) = (x'', Px) = 0$, i.e., $Px'' = 0$. Conversely, if $Px = 0$ and $x' \in H'$, then $(x, x') = (x, Px') = (Px, x') = 0$, i.e., $x \in H''$. Thus, the subspace H'' is defined as the set of all vectors x in H_k for which $Px = 0$.

If H' is a subspace of H_k on which the quadratic form (x, x) is non-degenerate, and H'' is the orthogonal complement of H' in H_k, then the relations (14) evidently define a projection operator.

Theorem 4. Let A be a Hermitian operator with $\Omega \subset H_k$ as its domain of definition. Let $H'' \subset \Omega$ be some finite-dimensional subspace of H_k with the non-degenerate quadratic form (x,x), and H' the orthogonal complement of H'' in H_k. Let P be the projection operator corresponding to the subspace H'. Then the operator $B = PAP$ is a Hermitian and its domain of definition is Ω. It is also Hermitian on H' and has the domain of definition $\Omega' = \Omega \cap H'$.

Proof. We affix one prime to all the quantities relating to H' and two primes to those relating to H''. Thus, an operator B acting on H' is denoted by B'.

Since $H'' \subset \Omega$, we have $\Omega = H'' + \Omega'$. Hence, as Ω is everywhere dense in H_k, it follows that Ω' is everywhere dense in H'. The domain of definition of the operator B' is the set Ω'. We now show that the adjoint of B', namely, the operator B'^* has the same domain of definition as B' and coincides with B' on this domain. It would then follow that the operator B' is Hermitian in H'.

Let y' be a vector in H', such that

$$(B'x', y') = (x', z') \tag{15}$$

for any $x' \in \Omega'$. From this relationship it follows that

$$(PAPx', y') = (Ax', y') = (x', z'). \tag{16}$$

If $x'' \in H''$, then (A'', y') is a linear functional of x'' in H'', because $H'' \subset \Omega$. Thus,

$$(Ax'', y') = (x'', z'') \quad \text{(see (C))}. \tag{17}$$

Adding (16) and (17), we obtain

$$(A(x' + x''), y') = (x' + x'', z' + z'').$$

This can also be rewritten as $(Ax, y') = (x, z)$, where x is an arbitrary vector in Ω. Since A is a Hermitian operator, from the last relation it follows that $y' \in \Omega'$. Then from (15) we find $(B'x', y') = (PAPx', y') = (x', PAPy') = (x'B'y')$. Consequently, the operator B' is Hermitian.

Now consider the operator B. Its domain of definition, as can be easily verified, is Ω. Let y be a vector in H_k, such that $(Bx, y) = (x, z)$ for any $x \in \Omega$. This can be rewritten as

$$(PAP(x' + x''), y' + y'') = (Ax', y') = (x', z') + (x'', z'').$$

For $y'' = 0$ we obtain the relation (16), from which we have already derived that $y' \in \Omega'$; but as $H'' \subset \Omega$, it follows that $y = y' + y'' \in \Omega$. Hence

$$(Bx,y) = (PAPx,y) = (x,PAPy) = (x,By).$$

This completes the proof of our theorem.

3. Hermitian Operators in the Space H_k

In this section we shall demonstrate the main property of a Hermitian operator in H_k $(k>0)$, i.e., the existence of eigenvalues (see the fundamental theorem).

Certain details in the proof of the fundamental theorem are given after the proof of lemmas. This is done for the sake of clarity.

Fundamental Theorem. Let A be a Hermitian operator with $\Omega \subset H_k$ as its domain of definition. Then in H_k there exists an invariant subspace I of dimension k, on which the quadratic form (x,x) is non-positive, i.e., for every $x \in I$, $(x,x) \leqslant 0$ (see Definitions 1 and 3). All eigenvalues of A in I have non-negative imaginary parts.

Proof. In H_k choose coordinates so that H_- is contained in Ω. This is always possible. Indeed, suppose that certain arbitrary coordinates are taken in H_k and that $e_1', e_2', \ldots, e_k'$, is the basis of H_-' in these coordinates. As Ω is everywhere dense in H_k, for every vector e_p' we can always find a vector e_p'' arbitrarily close to e_p'. Let H_- be the linear span of the vectors $e_1'', e_2'', \ldots, e_k''$. Since each vector e_p'' is close to e_p', the quadratic form (x,x) is non-degenerate on H_- and is negative. Let H_+ denote the orthogonal complement of H_-. Choosing arbitrary orthogonal coordinates in H_- and H_+, we obtain the coordinate system in H_k we need. We then construct a normal orthogonal basis in H_- of certain vectors $e_1, e_2, \ldots, e_k$. Thus,

$$(e_p, e_q) = -\delta_{pq}, \quad e_p \in \Omega, \quad p,q = 1,2,\ldots,k. \tag{1}$$

Splitting of the space H_k into the direct sum $H_- \oplus H_+$ is the technique that underlies all the constructions made in what follows. Accordingly, express each vector in H_k as

$$c + x = \sum_{p=1}^{k} c_p e_p + x, \quad c \in H_-, \quad x \in H_+. \tag{2}$$

Let P' and P be the projection operators corresponding to the subspaces H_- and H_+, respectively (see Section 2). Let

$$F = P'AP', \quad V = PAP, \tag{3}$$

where F is a Hermitian operator in H_-, and V a Hermitian operator in H_+, its domain of definition being $\Omega' = \Omega \cap H_+$ (see Theorem 4). We have

$$A = (P+P')A(P+P') = V+F+PAP'+P'AP. \tag{4}$$

By virtue of (4),

$$Ae_p = Fe_p - u_p = \sum_{q=1}^{k} f_{pq}e_q - u_p, \quad u_p \in H_+, \tag{5}$$

$$Ax = Vx + P'Ax, \quad P'Ax = y \in H_-, \quad x \in H_+. \tag{6}$$

A vector y belonging to H_- can be written as

$$y = \sum_{p=1}^{k} -(y, e_p)e_p. \tag{7}$$

This can be readily verified by multiplying on the right by e_q (see (1)). The Hermitian nature of the operator A implies that $(Ax, e_p) = (x, Ae_p)$. By virtue of (5) and (6), we can write

$$(y, e_p) = -(x, u_p).$$

Thus, (6) takes the form

$$As = Vx + \sum_{p=1}^{k} (x, u_p)e_p. \tag{8}$$

Combining (5) and (8), and using (2), we finally obtain

$$A(c+x) = \sum_{p,q=1}^{k} f_{pq}c_pe_q + \sum_{q=1}^{k} (x, u_q)e_q + Vx - \sum_{p=1}^{k} c_pu_p. \tag{9}$$

We shall now find an eigenvector $c+x$ of the operator A, $A(c+x) = \lambda(c+x)$. By virtue of (9), we can write this relation as

$$\sum_{p=1}^{k} f_{pq}c_p + (x, u_p) = \lambda c_q, \tag{10}$$

$$Vx - \sum_{p=1}^{k} c_pu_p = \lambda x. \tag{11}$$

We now solve equations (10) and (11). For this purpose, introduce the resolvent R_λ of the Hermitian operator V in H_+. This resolvent is defined for all complex (non-real) λ, as H_+ is an ordinary Hilbert space and V is an ordinary Hermitian operator in H_+. The calculations made below are purely heuristic, and there is therefore no need to worry whether they have meaning for all values of the quantities contained in them. Rewrite (11) as follow:

$$x = \sum_{p=1}^{k} c_p R_\lambda u_p. \tag{12}$$

Substituting this expression into (10), we obtain

$$\sum_{p=1}^{k} f_{pq} c_p + \sum_{p=1}^{k} (R_\lambda u_p, u_q) c_p = \lambda c_q. \tag{13}$$

The unknown quantities in this equation are λ and the coordinates $c_1, c_2, \ldots, c_k$ of the vector c. The operator function R_λ of the parameter λ, the vectors $u_1, u_2, \ldots, u_k$ and the matrix $\|f_{pq}\|$ depend on the operator A and the choice of the coordinates in H_k. With respect to the numbers $c_1, c_2, \ldots, c_k$, the equation (13) is a system of homogeneous linear equations. For a nontrivial solution $c \neq 0$ to exist, it is necessary that the determinant of this system be zero, i.e.,

$$|f_{pq} + (R_\lambda u_p, u_p) - \lambda \delta_{pq}| = 0. \tag{14}$$

If this equation has a non-real solution λ, then the operator R_λ is well defined for every vector in H_+, and in particular for all vectors $u_1, u_2, \ldots, u_k$. In this case we can reverse all calculations, and thus we obtain an eigenvector $c + x$, $c \neq 0$, of the operator A with non-real eigenvalue λ. Conversely, if the operator A has an eigenvector $c + x$ with non-real eigenvalue λ, then all our calculations have meaning and the vector c is non-zero, because the vector x, having a positive square (x, x), cannot be an eigenvector for the complex eigenvalue λ (see (A) Section 2). Thus, if the operator A has a non-real eigenvalue, it satisfies the equation (14).

It is a simple matter to give examples of Hermitian operators, even in a finite-dimensional space H_k, all eigenvalues of which are real. Thus, equation (14) may have no complex solutions at all.

In place of (14), consider the equation:

$$|f_{pq} + (R_\lambda u_p, u_q) + i\epsilon_{pq} - \lambda \delta_{pq}| = 0, \quad \|\epsilon_{pq}\| = E, \tag{15}$$

where E is a positive-definite Hermitian matrix.

In Lemma 1 (see below) we shall prove that equation (15) has exactly k roots with positive imaginary parts. The roots may, however, be multiple ones, and therefore, instead of equation (15), consider the equation:

$$|g_{pq} + (R_\lambda u_p, u_q) + i\epsilon_{pq} - \lambda \delta_{pq}| = 0, \tag{16}$$

where $G = \|g_{pq}\|$ and, as before, F is a Hermitian matrix. In (A) (see below) we shall prove that it is possible to find a matrix G, arbitrarily close to F, and an arbitrarily small matrix E, so that equation (16) has no multiple roots with positive imaginary part. Denote the roots of (16) by

$\lambda_1, \lambda_2, \ldots, \lambda_k$; they are pairwise distinct and have positive imaginary parts. Now consider the following system of equations similar to (13):

$$\sum_{p=1}^{k} g_{pq}c_p + \sum_{p=1}^{k} (R_{\lambda_r}u_p, u_q)c_p + i \sum_{p=1}^{k} \epsilon_{pq}c_p = \lambda_r c_q. \tag{17}$$

This system has a nontrivial solution in c. Let us denote this solution by f_r. Let the coordinates of f_r be $f'_{r_1}, f'_{r_2}, \ldots, f'_{r_k}$. Furthermore, introduce the following vector

$$y_r = R_{\lambda_r}v_r, \quad v_r = \sum_{p=1}^{k} f'_{r_p}u_p. \tag{18}$$

This vector makes sense since λ_r is non-real. The relationship between these quantities can now be expressed as follows:

$$Gf_r + \sum_{q=1}^{k} (R_{\lambda_r}v_r, u_q)e_q + iEf_r = \lambda_r f_r, \tag{19}$$

$$Gf_r + \sum_{q=1}^{k} (y_r, u_q)e_q + iEf_r = \lambda_r f_r, \tag{20}$$

$$Vy_r + \sum_{q=1}^{k} (f_r, e_q)u_p = \lambda_r y_r. \tag{21}$$

The systems (19) and (20), by virtue of (21), are equivalent.

The problem now is, by taking limits as $E \to 0$ and $G \to F$, to derive an invariant sub-space I from the vectors $f_r + y_r$ $(r = 1, 2, \ldots, k)$.

In our calculations we need the following basic properties of the resolvent of a Hermitian operator in Hilbert space:

$$R_\lambda - R_\mu = (\lambda - \mu)R_\lambda R_\mu, \tag{22}$$

$$(R_\lambda x, y) = (x, R_{\bar{\lambda}}y), \text{ because } \overline{(R_\lambda x, y)} = (R_{\bar{\lambda}}y, x). \tag{23}$$

Taking the scalar product of (19) on the right with the vector f_s, we get

$$(Gf_r, f_s) - (R_{\lambda_r}v_r, v_s) = \lambda_r(f_r, f_s) - i(Ef_r, f_s). \tag{24}$$

Substituting s for r in (19), and then multiplying by f_r, we find using (23) that

$$(f_r, Gf_s) - (R_{\bar{\lambda}_s}v_r, v_s) = \bar{\lambda}_s(f_r, f_s) + i(f_r, Ef_s). \tag{25}$$

Subtracting (24) from (25), we get

$$((R_{\lambda_r} - R_{\bar{\lambda}_s})v_r, v_s) = -(\lambda_r - \lambda_s)(f_r - f_s) + 2i(Ef_r, f_s). \tag{26}$$

By virtue of (22) and (23), we have

$$(R_{\lambda_r}v_r, R_{\lambda_s}v_s) = -(f_r, f_s) + \frac{2i}{\lambda_r - \bar{\lambda}_s}(Ef_r, f_s) \tag{27}$$

which can be rewritten as

$$(y_r, y_s) = -(f_r, f_s) + \frac{2i}{\lambda_r - \bar{\lambda}_s}(Ef_r, f_s). \tag{28}$$

Henceforth this expression plays a central role. We shall write it in a slightly different form. Let $d_1, d_2, \ldots, d_k$ be any complex numbers. Let

$$d = \sum_{r=1}^{k} d_r f_r, \quad Yd = \sum_{r=1}^{k} d_r y_r. \tag{29}$$

Here Y is not yet an operator, because Yd is not determined by the vector d itself, but is defined by the numbers $d_1, d_2, \ldots, d_k$; the d's may not be determined by the vector d because as yet we do not know whether the vectors $f_1, f_2, \ldots, f_k$ are linearly independent. Multiplying (28) by $d_r\bar{d}_s$, and then summing over r and s, we obtain

$$(Yd, Yd) = -(d, d) + \sum_{r,s=1}^{k} \frac{2i}{\lambda_r - \bar{\lambda}_s}(Ed_r f_r, d_s f_s). \tag{30}$$

Below it will be proved (see (B)) that under these assumptions

$$\sum_{r,s=1}^{k} \frac{i}{\lambda_r - \bar{\lambda}_s}(Ed_r f_r, d_s f_s) < -\delta \sum_{r=1}^{k} d_r \bar{d}_r \ (\delta > 0). \tag{31}$$

We now prove that the vectors $f_1, f_2, \ldots, f_k$ are linearly independent. Assume the contrary. Then there exists numbers $d_1, d_2, \ldots, d_k$, which do not vanish simultaneously, such that $d = 0$. Then, (30) and (31) yield

$$(Yd, Yd) < -2\delta \sum_{r=1}^{k} d_r \bar{d}_r \tag{32}$$

which is impossible.

Since the vectors $f_1, f_2, \ldots, f_k$ are linearly independent, any vector d in H_- can be expressed in the form (29), where $d_1, d_2, \ldots, d_k$ are uniquely determined by the vector d. Hence, Y is a linear operator defined on H_-.

As $Yd \in H_+$ and H_+ is an ordinary Hilbert space, we have $\|Yd\|^2 = (Yd, Yd)$. Since $d \in H$ and H_- is a finite-dimensional space with non-degenerate negative quadratic form (d, d), we have $\|d\|^2 = -(d, d)$. From (30) and (31), we obtain

$$\|Yd\| < \|d\|. \tag{33}$$

Using the operator Y, we rewrite (20) in a different form: multiply (20) by d_r and then sum over r:

$$Gd + \sum_{q=1}^{k} (Yd, u_q)e_q + iEd = \sum_{r=1}^{k} \lambda_r d_r f_r. \tag{34}$$

The right-hand side depends linearly on $d_1, d_2, \ldots, d_k$. We interpret it as the result obtained by the application of the operator Λ to the vector d:

$$\Lambda d = \sum_{r=1}^{k} \lambda_r d_r f_r, \quad \Lambda f_r = \lambda_r f_r. \tag{35}$$

The operator Λ maps the space H_- into itself. Now rewrite (34) as

$$Gd + \sum_{q=1}^{k} (Yd, u_q)e_q + iEd = \Lambda d. \tag{36}$$

From (33) and (36), we obtain an important inequality:

$$\|\Lambda d\| < C\|d\|. \tag{37}$$

Here the positive constant C depends, for a given operator A and the coordinates chosen in H_k, only on the matrices G and E. This constant is bounded, if the elements of these matrices are bounded.

From (35) it follows that the eigenvalues of Λ are equal to $\lambda_1, \lambda_2, \ldots, \lambda_k$ which all have positive imaginary parts.

Using the operators Y and Λ, we can write (21) as follows:

$$VYf_r + \sum_{p=1}^{k} (f_r e_q)u_p = Y\Lambda f_r. \tag{38}$$

Multiplying this expression by d_r, and then summing over r, we obtain

$$VYd + \sum_{p=1}^{k} (d, e_p)u_p = Y\Lambda d. \tag{39}$$

Recall that the operators Y and Λ depend on the matrices G and E; almost no contraints have so far been imposed on these matrices. Choose a sequence of values $E^{(1)}, E^{(2)}, \ldots, E^{(n)}$ of the matrix E which converge to zero, and a sequence of matrices $G^{(1)}, G^{(2)}, \ldots, G^{(n)}, \ldots$ that converge to F, such that equation (16) has no multiple zeros in the upper half-plane of λ for any n. Let $\Lambda^{(n)}$ and $Y^{(n)}$ denote the corresponding values of the operators Λ and Y.

The inequality (33) implies that the norm of the vector $Y^{(n)}e_p$ is less than one, therefore we can choose a weakly convergent subsequence from $Y^{(n)}e_p$, $n = 1, 2, \ldots$. Similarly, by virtue of (37), we can choose a convergent subsequence from the sequence of vectors $\Lambda^{(n)}e_p, n = 1, 2, \ldots$. As there are exactly k vectors e_p, making the selection one after another, we obtain a subsequence

$$n_1, n_2, \ldots, n_m, \ldots$$

of natural numbers $1, 2, \ldots$, for which

$$Y^{(n_m)}e_p \to Ze_p, \quad \Lambda^{(n_m)}e_p \to Me_p. \tag{40}$$

As the vectors $e_1, e_2, \ldots, e_k$ form a basis of H_-, (40) can be rewritten as

$$Y^{(n_m)}d \to Zd, \quad \Lambda^{(n_m)}d \to Md. \tag{41}$$

Since the eigenvalues of Λ all have positive imaginary parts, the limiting operator M has eigenvalues with non-negative imaginary parts. As Y satisfies (33), the limiting operator Z satisfies one of the inequalities:

$$\|Zd\| \leqslant \|d\| \quad \text{or} \quad -(d,d) + (Zd, Zd) \leqslant 0, \tag{42}$$

although its defining sequence is only weakly convergent.

We can now take limits in (36) and (39). The scalar product in (36) converges, given the weak convergence of the first vector; the other terms do not give rise to any doubt. In (39) all terms, except VYd, evidently converge, and VYd therefore converges weakly to some vector. Thus, by virtue of (E) Section 2, the operator V is defined on the vector Zd and the weak limit of VYd is VZd.

Thus,

$$Fd + \sum_{q=1}^{k} (Zd, u_q)e_q = Md, \tag{43}$$

$$VZd + \sum_{p=1}^{k} (d, e_p)u_p = ZMd. \tag{44}$$

By virtue of (9), we have

$$A(d + Zd) = Md + ZMd. \tag{45}$$

The set I of all vectors of the type $d + Zd$, where $d \in H_-$ forms, as can be readily verified, a k-dimensional subspace of H_k. From (42) it follows that

$$(d + Zd, d + Zd) \leqslant 0. \tag{46}$$

Thus, the quadratic form (x, x) on I is non-positive. The relation (45) shows that the subspace I is invariant relative to the operator A and that the operator A on I is equivalent to the operator M. Consequently, all eigenvalues of the operator A on I have non-negative imaginary parts.

Thus, we have demonstrated the fundamental theorem, except for three lemmas involved which we now proceed to prove.

Lemma 1. Let R_λ be the resolvent of some Hermitian operator V acting in the Hilbert space $H; u_1, u_2, \ldots, u_k$ an arbitrary finite system of vectors in H; $G = \|g_{pq}\|$, $E = \|\epsilon_{pq}\|$, $(p, q) = 1, 2, \ldots, k$ Hermitian matrices, the last of which is positive-definite. Then the equation in λ

$$|g_{pq} + (R_\lambda u_p, u_q) + i\epsilon_{pq} - \lambda\delta_{pq}| = 0 \tag{47}$$

has exactly k roots with positive imaginary parts.

Proof. Let α be a real number such that $0 \leqslant \alpha \leqslant 1$. Consider the equation:

$$|g_{pq}\alpha + (R_\lambda u_p, u_q)\alpha + i\epsilon_{pq} - \lambda\delta_{pq}| = 0. \tag{48}$$

For $\alpha = 1$, this equation coincides with (47), and for $\alpha = 0$, it takes the form $|i\epsilon_{pq} - \lambda\delta_{pq}| = 0$. This last equation has k zeros in the upper half-plane.

The proof will consist of passing continuously from the last equation to equation (47).

Put $\lambda = \mu + i\nu$. Using functions of the operator V, we can express the resolvent as follows:

$$R_\lambda = S_\lambda + iT_\lambda, \quad S_\lambda = \frac{V - \mu E}{(V - \mu E)^2 + \nu^2 E}, \quad T_\lambda = \frac{\nu E}{(V - \mu E)^2 + \nu^2 E}. \tag{49}$$

For $\nu > 0$, the operators S_λ and T_λ are bounded, and the latter is positive-definite.

Let us consider the system of linear equations in the numbers $c_1, c_2, \ldots, c_k$,

$$\alpha \sum_{p=1}^{k} g_{pq}c_p + \alpha \sum_{p=1}^{k} (S_\lambda u_p, u_q)c_p + i\alpha \sum_{p=1}^{k} (T_\lambda u_p, u_q)c_p + i \sum_{p=1}^{k} \epsilon_{pq}c_p = \mu c_q + i\nu c_q. \tag{50}$$

If λ is a solution of (48), then there exists a nontrivial solution to the system (50) – the vector c with coordinates $c_1, c_2, \ldots, c_k$ satisfying the condition $\|c\| = 1$. Put

$$u = \sum_{p=1}^{k} c_p u_p. \tag{51}$$

Multiplying (50) by $\bar{c}_q$ and then summing over q, we obtain

$$\alpha(Gc, c) + \alpha(S_\lambda u, u) + \alpha i(T_\lambda u, u) + i(Ec, c) = \mu + i\nu. \tag{52}$$

Separating the imaginary part from (52), we find

$$\nu = \alpha(T_\lambda u, u) + (Ec, c). \tag{53}$$

Thus, for the root of equation (48), we obtain

$$\nu \geqslant (Ec, c) > \epsilon > 0. \tag{54}$$

By virtue of the properties of resolvents, for $\nu > \epsilon$, we have

$$(R_\lambda u, u) < \frac{C}{\rho} \quad \text{for } |\lambda| > \rho. \tag{55}$$

Here C is some positive constant, and ρ is a sufficiently large number.

Now in the λ-plane, take a domain U as follows: $\nu > \epsilon$, $|\lambda| < \rho'$. From (52) and (55), it is clear that for sufficiently large ρ', all roots of (48) in the upper half-plane belong to U; In particular, they do not occur on the boundary of this domain. This is true of any α that satisfies the condition $0 \leqslant \alpha \leqslant 1$. Therefore the number of roots of equation (48) inside U does not depend on α. But the number of roots in the upper half-plane that lie outside U is always equal to zero. Hence, the number of roots of equation (48) in the upper half-plane is the same as the number of roots of the equation $|i\epsilon_{pq} - \lambda\delta_{pq}| = 0$, which is k.

Thus, Lemma 1 has been demonstrated.

Lemma 2. Let R_λ be the resolvent of a Hermitian operator in the Hilbert space H, and $u_1, u_2, \ldots, u_k$ an arbitrary system of vectors in H; $M = \|m_{pq}\|$, $(p, q) = 1, 2, \ldots, k$ an arbitrary matrix with complex elements. In the space of matrices M, there exists an everywhere dense set of matrices, for which the equation

$$f(\lambda) = |m_{pq} + (R_\lambda u_p, u_q) - \lambda\delta_{pq})| = 0 \tag{56}$$

has no multiple non-real roots.

Proof. Put

$$T(\lambda) = \|(R_\lambda u_p, u_q) - \lambda\delta_{pq}\|. \tag{57}$$

The matrix $T(\lambda)$ is an analytic function of λ for all non-real λ. This matrix cannot be a constant, because for some value of the matrix M, equation (56) has exactly k roots in the upper half-plane (see Lemma 1).

Let λ' be a multiple root of equation (56) for $M=M'$. Expand $T(\lambda)$ as a power series in $\mu=(\lambda-\lambda')$:

$$T(\lambda) \ = \ A+B\mu^r+\ldots+, \quad A \ = \ \|a_{pq}\|, \quad B \ = \ \|b_{pq}\|. \qquad (58)$$

Here B is the first non-zero coefficient of a power of μ. Let M_{pq} denote the minor of the element $(m_{pq}+a_{pq})$ of the matrix $M+A$. Then the expansion of $f(\lambda)$ in powers of μ takes the form:

$$f(\lambda) \ = \ D(M)+(M)\mu^r+\ldots, \quad D(M) \ = \ |m_{pq}+a_{pq}|,$$

$$\Lambda(M) \ = \ \sum_{(p,q=1)}^{k} b_{pq}M_{pq}. \qquad (59)$$

Here we have omitted the powers of μ greater than r. $D(M)$ and $\Lambda(M)$ are polynomials of complex variables – the elements of the matrix M. The polynomial $D(M)$, being a determinant, is irreducible. And as the polynomial $\Lambda(M)$ is not divisible by $D(M)$, the intersection of the algebraic manifolds $D(M)=0$ and $\Lambda(M)=0$ is nowhere dense in the manifold $D(M)=0$.

We have $D(M')=0$. If $\Lambda(M')=0$, choose a matrix M'' arbitrarily close to M' and such that

$$D(M'') \ = \ 0 \quad \text{and} \quad \Lambda(M'') \ \neq \ 0. \qquad (60)$$

From (59) it follows that for $M=M''$, equation (56) has a root λ' of multiplicity r. For $M=M'$, the multiplicity of λ' can be only greater than r. If the multiplicity were smaller, this would signify that some roots were displaced from λ'.

Now there exists a small positive ϵ, such that for $M=M''$ the equation (56) has exactly r roots (counted with multiplicities) in the circle $|\lambda-\lambda'|<\epsilon$, but the equation

$$\frac{d^{r-1}f(\lambda)}{d^{r-1}} \ = \ 0 \qquad (61)$$

has exactly only one root. The roots of equation (56) all coincide with λ'; similarly, the root of (61) is also equal to λ'. This assertion, especially concerning the number of roots, holds true also for all matrices M in a certain neighbourhood U of the matrix M'', because $f(\lambda)$ depends continuously on the matrix M. The roots of (56) may shift, always remaining within the circle $|\lambda-\lambda'|\leqslant\epsilon$, while the root of (61) remains equal to λ'. The last fact is important.

Let now M''' be a matrix in U, such that $D(M''')\neq 0$. If $M=M'''$, equation (56) no longer has roots coincident with λ', and, consequently, the derivative $d^{r-1}f(\lambda)/d\lambda^{r-1}$ at those roots does not vanish. This means that each of the corresponding roots of equation (56) now has a multiplicity less than r.

Let G be some finite domain in the λ-plane, whose closure does not intersect the real axis. The domain G may contain only a finite number of the roots of (56). If, for $M = M'$, the greatest multiplicity of the roots of equation (56) contained in G is equal to s, this multiplicity will not be greater for any of the matrices within some neighbourhood V of the matrix M'. In particular, the set $W(G)$ of all the values of the matrix M for which the domain G contains only simple roots of (56) is a domain in the space of all matrices M. Applying this procedure of shifting the roots a finite number of times to all those roots of equation (56) that are inside G, we find that the domain $W(G)$ is everywhere dense in the space of matrices M.

Let us cover the λ-plane, except for the real axis, with a countable number of domains G_n ($n = 1, 2, \ldots$) of the type used above. The intersection W of the domains $W(G_n)$ is everywhere dense in the space of matrices M and for every matrix $M \in W$, all the roots of (56) are simple. This completes the proof of Lemma 2.

(A) If F is an arbitrary Hermitian matrix, there exists a Hermitian matrix G arbitrarily close to F, and an arbitrarily small positive-definite Hermitian matrix E for which equation (16) has only simple roots.

Let E' be a small positive-definite Hermitian matrix. By virtue of Lemma 2, there exists a matrix M arbitrarily close to the matrix $F + iE'$, such that equation (56) has only simple roots. Take

$$G = \tfrac{1}{2}(M + M^*), \quad E = -\tfrac{1}{2}(M - M^*)i,$$

where M^* is the adjoint of M. The Hermitian matrix G is close to F, and E is a Hermitian matrix close to E'. Since E is close to E', E is small and positive-definite.

Lemma 3. Let H be a unitary space with a normal orthogonal basis $e_1, e_2, \ldots, e_k$, $(e_p, e_q) = \delta_{pq}$, let A and B be Hermitian positive-definite operators:

$$Ae_p = \sum_{q=1}^{k} a_{pq}e_q, \quad Be_p = \sum_{q=1}^{k} b_{pq}e_q$$

and let $f_1, f_2, \ldots, f_k$ be non-zero vectors in H (they may be linearly dependent). Then the matrix $R = \|q_{pq}(Bf_p, f_q)\|$ is Hermitian and positive-definite.

Proof. By our assumption, the eigenvalues of the operator B can be taken to be $b_1^2, b_2^2, \ldots, b_k^2$, where $b_1, b_2, \ldots, b_k$ are real positive numbers. On changing over from coordinates which are canonical for the operator B to coordinates initially defined in H, we obtain

$$b_{pq} = \sum_{\alpha=1}^{k} d'_{p\alpha}\bar{d}'_{q\alpha} b_\alpha^2. \tag{62}$$

Here the matrix $\|d'_{p\alpha}\|$, which effects the coordinate transformation, is non-degenerate and therefore the matrix $\|d_{p\alpha}\|$, where $d_{p\alpha} = d'_{p\alpha} b_\alpha$, is also non-degenerate. From (62), we get

$$b_{pq} = \sum_{\alpha=1}^{k} d_{p\alpha}\bar{d}_{q\alpha}, \quad B = D^*D, \tag{63}$$

where D is non-degenerate operator defined by the condition

$$De_p = \sum_{q=1}^{k} d_{pq} e_q.$$

Putting $g_p = Df_p$, we find

$$(Bf_p, f_q) = (Df_p, Df_q) = (g_p, g_q). \tag{64}$$

Thus, we have to show that the matrix $R = \|a_{pq}(g_p, g_q)\|$ is positive-definite, bearing in mind that none of the vectors $g_1, g_2, \ldots, g_k$ vanishes.

Let the coordinates of the vector g_p be $g_{1p}, g_{2p}, \ldots, g_{kp}$. We shall express the matrix A, just as we did the matrix B earlier, in the form $a_{pq} = \Sigma_{\beta=1}^{k} c_{p\beta}\bar{c}_{q\beta}$. Let the coordinates of an arbitrary vector x in H be $x_1, x_2, \ldots, x_k$. With this notation,

$$(Rx, x) = \sum_{\alpha,\beta=1}^{k} s_{\alpha\beta}\bar{s}_{\alpha\beta}, \quad s_{\alpha\beta} = \sum_{p=1}^{k} g_{\alpha p} x_p c_{p\beta}. \tag{65}$$

If $(Rx, x) = 0$, then from (65) it follows that $\Sigma_{p=1}^{k} g_{\alpha p} x_p c_{p\beta} = 0$. Therefore

$$g_{\alpha p} x_p = 0, \tag{66}$$

because the matrix $\|c_{p\beta}\|$ is non-degenerate. Now (66) can be rewritten as $x_p g_p = 0$ $(p = 1, 2, \ldots, k)$; and since none of the vectors $g_1, g_2, \ldots, g_k$ vanishes, we have $x_1 = x_2 = \ldots = x_k = 0$. Thus, the matrix R is definite, and its positiveness follows from (65). This completes the proof of Lemma 3.

Lemma 4.* Let $a_1, a_2, \ldots, a_k$ be complete numbers, all distinct and with positive real parts. The matrix $M = \|m_{pq}\|$, where

* The proof of this lemma is due to V. V. Morozov.

$$m_{pq} = \frac{1}{a_p + \bar{a}_q}, \quad (p, q = 1, 2, \ldots, k),$$

is Hermitian and positive-definite.

Proof. It is obvious that the matrix M is Hermitian. We show it is positive. For this purpose, let us calculate its determinant $|M|$.

Let m_{pq} belong to the p-th row and q-th column. To calculate the determinant $|M|$, subtract the last column from all preceding columns, after multiplying with suitable coefficients, so that the last row vanishes. Let $N = \|n_{pq}\|$ $(p, q = 1, 2, \ldots, k-1)$ denote the minor of the element m_{kk} in the new matrix thus obtained. In this way we have

$$|M| = m_{kk}|N|.$$

In subtracting the k-th column from the q-th column, the appropriate coefficient is the number m_{kq}/m_{kk}. Hence

$$n_{pq} = m_{pq} - \frac{m_{pk}m_{kq}}{m_{kk}} = \frac{m_{pq}m_{kk} - m_{pk}m_{kq}}{m_{kk}}$$

$$= m_{pq}m_{pk}m_{kq}\left((a_p + \bar{a}_k)(a_k + \bar{a}_q) - (a_p + \bar{a}_q)(a_k + \bar{a}_k)\right)$$

$$= m_{pq}m_{pk}m_{kq}\,(a_p - a_k)(\bar{a}_q - \bar{a}_k).$$

Thus,

$$n_{pq} = m_{pq}m_{pk}m_{kq}\,(a_p - a_k)(\bar{a}_q - \bar{a}_k).$$

In calculating the determinant of the matrix N, the number $m_{pk}(a_p - a_k)$ can be taken outside the p-th row, and the number $m_{kq}(\bar{a}_q - \bar{a}_k)$ outside the q-th column. If now M' denotes the minor of the element m_{kk} in the matrix M, we obtain

$$|M| = m_{kk}|M'|\left(\prod_{r=1}^{k-1} m_{rk}\bar{m}_{rk}\right) \times \left(\prod_{s=1}^{k-1} (a_s - a_k)(\bar{a}_s - \bar{a}_k)\right).$$

The matrix M' here is of the same form as the matrix M, but its order is $(k-1)$. Its determinant can also be calculated in the same way. By an obvious induction, we have

$$|M| = \prod_{r=1}^{k} m_{rr} \prod_{(p<q)=1}^{k} m_{pq}\bar{m}_{pq}\,(a_p - a_q)(\bar{a}_p - \bar{a}_q).$$

This expression shows that $|M|$ is positive. Since each diagonal minor of M is of the same form as the matrix M, all these minors are positive, and, as is known, this implies that the Hermitian matrix is positive definite.

This completes the proof of Lemma 4.

(B) The inequaltiy (31) is true. To prove it, take $a_p = -i\lambda_p$, so that $a_1, a_2, \ldots, a_k$ are all distinct and have positive real parts. Moreover,

$$\frac{i}{\lambda_p - \bar{\lambda}_p} = \frac{1}{a_p + \bar{a}_q}$$

and, consequently, the matrix $\|i/(\lambda_p + \bar{\lambda}_p)\|$ is positive-definite (Lemma 4). The matrix E is positive-definite, and as the quadratic form (x,x) is non-degenerate and negative in H_-, by virtue of Lemma 3, we find that the matrix

$$\left\| \frac{i}{\lambda_p - \bar{\lambda}_q} (Ef_p, f_p) \right\|$$

is negative-definite. Hence (31) follows directly as a consequence.

Now the gaps in the proof of the Fundamental Theorem 1 have been filled, and consequently, it has been demonstrated in full.

Now we state certain supplements and corollaries of this fundamental theorem.

(C) Besides the invariant subspace I stated in the fundamental theorem, there exists an analogous invariant subspace I' of dimension k on which the quadratic form (x,x) is non-positive, and in which the eigenvalues of the operator A all have non-positive imaginary parts.

To construct the subspace I' we have to proceed as in the fundamental theorem, using in place of equation (16), the following equation

$$|g_{pq} + (R_\lambda u_p, u_q) - i\epsilon_{pq} - \lambda\delta_{pq}| = 0.$$

In the lower half-plane of the variable λ this equation has exactly k roots.

(D) The space I contains all the eigenspaces S_λ of the operator A for the eigenvalues λ with positive imaginary parts. The space I' contains all the eigenspaces S_λ of the operator A for the eigenvalues λ with negative imaginary parts (cf. the fundamental theorem and (C)).

We prove this assertion for I; for the space I' it can be demonstrated in a similar manner.

Let S_λ be an eigenspace of the operator A for the eigenvalue λ with a positive imaginary part. The space S_λ is then orthogonal to I, because I is the direct sum of the eigenspaces for eigenvalues with non-negative imaginary parts (see, (A) Section 2).

By virtue of Proposition (A) Section 2, the quadratic form (x,x) on S_λ is identically zero. Moreover, as the quadratic form (x,x) is non-positive on I, so it is on the linear span $S_\lambda + I$ of the subspaces S_λ and I. Hence the dimension of $S_\lambda + I$ is not greater than k. But the dimension of I is k, therefore $S_\lambda \subset I$. This completes the proof of Proposition (D).

Theorem 5. Let A be a Hermitian operator with $\Omega \subset H_k$ as its domain of definition, and let K be the direct sum of all eigenspaces S_λ for non-real eigenvalues λ. Then the quadratic form (x,x) is non-degenerate on K. The space H_k is thus the direct sum of orthogonal spaces K and H'_k, where the eigenvalues of the operator A in H'_k are all real. (The structure of the operator A in the finite-dimensional space K is given by Proposition (E).)

Proof. Let J be the intersection of the subspaces K and I, and J' the intersection of the subspaces K and I'. By virtue of Proposition (D), the space K is then the direct sum of its subspaces J and J'. Assume that in J' there exists a vector $x' \neq 0$ orthogonal to J. By virtue of (A) Section 2 this vector is orthogonal to the whole space I. Thus, on the linear span of the space I and the vector x', which is of dimension $k+1$, the quadratic form (x,x) is non-positive. This is impossible, by Proposition (B) Section 1. Hence, none of the vectors in J' different from zero is orthogonal to J. Similarly, none of the vectors in J different from zero is orthogonal to J'. And, as the quadratic form (x,x) is identically zero on J and J' (Proposition (A) Section 2), the quadratic form (x,x) is non-degenerate on K. Theorem 5 has thus been demonstrated.

(E) Let K be a finite-dimensional space of the type H_k, and A a Hermitian operator of all whose eigenvalues in K are non-real. Decompose K as a direct sum of the subspaces J and J', where J consists of all eigenspaces S_λ of the eigenvalues λ with positive imaginary parts, and J' consists of all eigenspaces of the operator A for the eigenvalues with negative imaginary parts. For coordinates chosen arbitrarily in J, it is possible to choose coordinates in J' so that the operator A is expressed in the matrix form:

$$A = \left\| \begin{matrix} B & 0 \\ 0 & B^* \end{matrix} \right\|, \tag{67}$$

where B is a matrix in J and B^* is the adjoint of B (complex conjugate transpose) acting on J'. Hence, it follows that the elementary divisors of the operator A for the eigenvalues λ may be put into one-to-one correspondence with the elementary divisors of the operator A for the eigenvalues $\bar{\lambda}$, so that the orders of the corresponding elementary divisors are the same.

To prove Proposition (E), introduce a linear basis $e_1, e_2, \ldots, e_r$ in J, and a linear basis $f_1, f_2, \ldots, f_s$ in J'. Put

$$Ae_p = \sum_{\alpha=1}^{r} b_{p\alpha} e_\alpha, \quad Af_q = \sum_{\beta=1}^{s} c_{q\beta} f_\beta, \quad B = \|b_{p\alpha}\|, \quad C = \|c_{q\beta}\|.$$

As the subspaces J and J' are invariant under A, it takes the matrix form:

$$A = \begin{Vmatrix} B & 0 \\ 0 & C \end{Vmatrix}. \tag{68}$$

Furthermore, $(e_p, e_\alpha) = 0$ and $(f_q, f_\beta) = 0$. Hence, since the quadratic form (x, x) is non-degenerate on K, $\|e_p, f\|$ is a square matrix with a non-zero determinant. Properly choosing an appropriate basis in J', we can arrange that $(e_p, f_q) = \delta_{pq}$. Thus,

$$(e_p, f_q) = \delta_{pq}, \quad (e_p, e_\alpha) = 0, \quad (f_q, f_\beta) = 0. \tag{69}$$

The condition for the operator A to be Hermitian in the chosen bases is:

$$(Ae_p, f_q) = b_{pq} = (e_p, Af_q) = \bar{c}_{qp}.$$

Thus, $C = B^*$, and this completes the proof of Proposition (E).

(F) Let A be a Hermitian operator with $\Omega \subset H_k$ as its domain of definition, J its invariant subspace consisting of all eigenspaces S_λ for eigenvalues λ with positive imaginary parts. If the dimension of J is k, the space H_k is the direct sum of an invariant subspace K of dimension $2k$ and an invariant Hilbert subspace H.

Under the assumptions of (F), we have $I = J$, and $I' = J'$. Take $K = I + I'$. The quadratic form (x, x) is non-degenerate on K (Theorem 5). In K take the basis $e_1, e_2, \ldots, e_k, f_1, f_2, \ldots, f_k$ stated in (E). Then $x \in K$ can be expressed as

$$x = x_1 e_1 + \ldots + x_k e_k + y_1 f_1 + \ldots + y_k f_k$$

and the quadratic form (x, x) takes the form

$$(x, x) = \sum_{p=1}^{k} x_p \bar{y}_p + \bar{x}_p y_p \quad \text{(see (69))}.$$

In K now introduce new coordinates as follows:

$$x_p = u_p + v_p, \quad y_p = u_p - v_p.$$

The quadratic (x,x) on K is then equal to

$$(x,x) = \sum_{p=1}^{k} 2(u_p\bar{u}_p - v_p\bar{v}_p).$$

Thus, the quadratic form (x,x) on K has exactly k negative squares and, consequently has no negative squares on the orthogonal complement H_0 of K. This implies that H_0 is an ordinary Hilbert space. This completes the proof of Proposition (F).

An estimate for the dimension of the eigenspace S_λ for a complex eigenvalue λ is given by the following proposition.

(G) Let A be a Hermitian operator with $\Omega \subset H_k$ as its domain of definition. A non-real λ is an eigenvalue of the operator A, if and only if it is a root of the equation

$$|f_{pq} + (R_\lambda u_p, u_p) - \lambda\delta_{pq})| = 0. \tag{70}$$

The multiplicity of the root λ is equal to the dimension of the eigenspace S_λ of the operator A.

In demonstrating this proposition we use some of the steps used in proving the fundamental theorem. Here we only take up the case in which λ has a positive imaginary part; the proof is similar for the case in which λ has a negative imaginary part.

Besides equations (14) and (70), consider equation (16)

$$|g_{pq}^{(n)} + (R_\lambda u_p, u_q) + i\epsilon_{pq}^{(n)} - \lambda\delta_{pq}| = 0, \tag{71}$$

where the sequences $G^{(n)}$ and $E^{(n)}$ $(n = 1, 2, \ldots)$ are chosen as in the proof of the fundamental theorem, i.e., in particular the operator $\Lambda^{(n)}$, acting on H_-, converges to the operator M.

The roots of equation (71) that lie in the upper half-plane, as already demonstrated in the proof of the fundamental theorem, are all simple and their number is equal to k. These roots

$$\lambda_1^{(n)}, \lambda_2^{(n)}, \ldots, \lambda_k^{(n)}. \tag{72}$$

are the eigenvalues of the operator $\Lambda^{(n)}$ and, consequently, satisfy the equation:

$$|\Lambda^{(n)} - \lambda E| = 0. \tag{73}$$

As the operator $\Lambda^{(n)}$ converges to the operator M, the numbers (72) converge to the roots of the equation

$$|M-\mu E| = 0, \tag{74}$$

and, in particular, $\lim \lambda_p^{(n)} = \mu_p$, exists, where μ_p is a root of (74). The numbers

$$\mu_1, \mu_2, \ldots, \mu_k$$

are no longer distinct. The multiplicity of any root μ of equation (74) is equal to the dimension of the eigenspace T_μ of the operator M. Since the operator M in H_- is equivalent to the operator A on I, the dimension of the subspace T_μ is equal to the dimension of the eigenspace S_μ of the operator A in the space I. If μ is a complex number with positive imaginary part, the eigenspace S_μ of the operator A on all of H_k is the same as the eigenspace S'_μ of the operator A in I (see Proposition (D)). Thus, the dimension of S_μ is equal to the multiplicity of the root μ of equation (74), if μ has a positive imaginary part.

Since the left-hand side of (73) converges to the left-hand side of (74), the multiplicity of a root μ of (74) is equal to the number of roots of (73) that tend to μ as $n \to \infty$. The roots of equation (73) coincide with the roots of equation (71) in the upper half-plane of the unknown. Since, in the upper half-plane, the left-hand side of (71) converges uniformly to the left-hand side of (70), the number of roots of the system (72) that tend to the complex number μ (when μ has a positive imaginary part) is equal to the multiplicity of the root μ of equation (70). Thus, the multiplicity of the root μ of equation (70) is equal to the dimension of the eigenspace.

Thus, Proposition (G) has been demonstrated.

Theorem 6. Let A be a Hermitian operator with $\Omega \subset H_k$ as its domain of definition. Consider the equation (14) = (70):

$$|f_{pq} + (R_\lambda u_p, u_q) - \lambda\delta_{pq}| = 0. \tag{75}$$

If the number of roots of this equation (including multiplicity) with a positive imaginary part is k (there cannot be more than k roots), the space H_k is demonstrated into the direct sum of an invariant subspace K of dimension $2k$ and an orthogonal invariant Hilbert space H_0.

The proof follows directly from Propositions (F) and (G).

Received 1 December 1943

The Classification of Certain Skew Products

Doklady Akademii Nauk SSSR (1945) 47(5): 327–30

The present communication deals with the classification of certain fibre-bundles or skew products as I prefer to call them. This classification is based on my previous papers [1, 2]. Here we give the classification (see Definition 2 in [2]) of skew products $P(R^k, B)$ with the group Γ_k, where R^k is a k-dimensional Euclidean space and Γ_k is the group of all its isometric transformations which preserve the origin of coordinates and orientation, while B is a complex of dimension $n \leqslant 4$. By virtue of Remark A given below, this classification depends only on the group Γ_k and the complex B, but does not depend on the space R^k. Thus, the results can be generalized to the case studied by Whitney, i.e., to the skew product $P(S^{k-1}, B)$ with the group Γ_k, where S^{k-1} is a $(k-1)$-dimensional sphere and Γ_k is its group of rotations. Whitney has also investigated the cases $n = 2$ and 3 in detail in [3]. In the same paper he gives a result concerning the case $k = 3$, $n = 4$ without proof. As demonstrated by Theorems 1 and 2 given below, this result is, however, erroneous.

(A) By virtue of Propositions A and B of [2] a skew product $P(A, B)$ with the group Γ is uniquely defined up to isomorphism by the defining functions $\zeta_{\alpha\beta y} \in \Gamma$, $y \in T_\beta$. But, by Proposition B of [2], two skew products $P(A, B)$ and $P'(A, B)$ with the same group Γ are isomorphic, if and only if the defining functions satisfy the condition: $\zeta_{\alpha\beta y} = \tau_{\alpha y}^{-1} \zeta_{\alpha\beta y} \tau_{\beta y}$, $\tau_{\alpha y} \in \Gamma$. Thus, classification of skew products depends only on the complex B and on the group Γ, but not on the space A.

To classify $P(R^k, B)$, we shall use the univeral skew product $P(R^k, H_l)$ (see Definition 3 and Theorem 1 in [2]). Since the dimension n of the complex B does not exceed 4, we are only interested in cycles of dimension $(h-r)$ in the space $H_l = H(k, l)$, where h is the dimension of H_l and $r \leqslant 4$. These cycles are defined in [1], nonetheless, we recall them here.

(B) Let R^{k+l} be a Euclidean space of dimension $k+l$, and $H(k, l)$ the set of all oriented k-dimensional linear spaces in R^{k+l} that pass through the origin. Let R^{l+1}, R^l, R^{l+3} be three fixed linear spaces in R^{k+l} of dimensions $l+1$, l, $l+3$, respectively, that pass through the origin. Let Z^{h-2} denote the set of all R^k in $H(k, l)$ whose intersection with R^{l+1} is of dimension not less than 2. Let Z_1^{h-4} denote the set of all R^k in $H(k, l)$ whose intersection with R^l is of dimension not less than two, and Z_2^{h-4} the set of all R^k in $H(k, l)$ whose intersection with R^{l+3} is of dimension not less than four. In [1] it was proved that:

(a) Z^{h-2} is a cycle of dimension $(h-2)$ modulo 2 if $k > 2$ and an oriented cycle if $k = 2$;

(b) Z_1^{h-4} is an oriented cycle of dimension $(h-4)$;

(c) Z_2^{h-4} is a cycle of dimension $(h-4)$ modulo 2 if $k > 4$, is an oriented cycle if $k = 4$; and $Z_2^{h-4} = 0$, if $k < 4$.

In [4] it was shown that the Lefschetz intersection of a homology class modulo 2 with itself can be defined as a homology class modulo 4. In this sense

$$Z^{h-2} \times Z^{h-2} \sim Z_1^{h-4} - 2Z_2^{h-4} \pmod 4. \tag{1}$$

(C) If f is some mapping of the complex B into H_l let u_f^2, u_{1f}^4, and u_{2f}^4 denote the characteristic ∇-cycles (see Definition 4 in [2]) corresponding to the cycles Z^{h-2}, Z_1^{h-4}, and Z_2^{h-4}, respectively. From (1), it follows that

$$u_f^2 \times u_f^2 \sim u_{1f}^4 - 2u_{2f}^4 \pmod 4, \tag{2}$$

where $u_f^2 \times u_f^2$ is the product defined in [4].

In [2] it was demonstrated that the homotopy group $\pi_r(H_l)$ is isomorphic to the homotopy group $\pi_{r-1}(\Gamma_k)$ (see Theorem 2 in [2]). Hence, the homotopy groups $\pi_r(H_l)$ of dimensions $r = 1, 2, 3, 4$ can be easily constructed.

(D) Since Γ_k is connected, $\pi_1(H_l)$ is trivial. As $\Gamma_2 \subset \Gamma_k$ is a one-dimensional basis of homotopy in Γ_k, the two-dimensional sphere $S^2 \subset H_l$ corresponding to Γ_2 is a basis of the two-dimensional homotopy in H_l, where S^2 is free if $k=2$ and is of order two, if $k>2$. As $\pi_2(\Gamma_k)$ is trivial, $\pi_3(H_l)$ is also trivial. And $\Gamma_3 \subset \Gamma_k$ is homeomorphic to a three-dimensional projective space. Mapping some three-dimensional sphere with degree two onto Γ_3, we obtain a homotopy basis of the space Γ_3. To this generator there corresponds a sphere S_1^4 of dimension 4 in $H_l = H(k,l)$, which forms a homotopy basis in $H(k,l)$ for $k=3$, and is free. If $k>3$, S_1^4 is still free but is no longer a homotopy basis of $H(k,l)$; it can only be embedded in such a basis. There exists a sphere S^3 in Γ_4, which, together with the generator in Γ_3 form a homotopy basis in Γ_4. To the sphere S^3 there corresponds a sphere $S_2^4 \subset H(k,l)$, which together with S_1^4, forms a homotopy basis in $H(k,l)$ if $k \geqslant 4$. If $k=4$, the spheres S_1^4 and S_2^4 are independent; if $k>4$,

$$2S_2^4 - S_1^4 \text{ is homotopic to zero.} \tag{3}$$

The spheres S^2, S_1^4, and S_2^4 can be regarded as cycles in H_l. We easily find that the indices of their intersections with the cycles stated in (B) are

$$I(S^2, Z^{h-2}) = 1, \tag{4}$$

$$I(S_1^4, Z_1^{h-4}) = 4; \quad I(S_1^4, Z_2^{h-4}) = 0; \quad I(S_2^4, Z_1^{h-4}) = 2;$$
$$I(S_2^4, Z_2^{h-4}) = 1. \tag{5}$$

By virtue of Definitions 3 and Theorem 1 given in [2], the classification of skew products $P(R^k, B)$ reduces to the classification of the continuous mappings of B into $H(k,l)$. For $n=4$ the latter is solved by

Theorem 1. Let B be a four-dimensional complex and $H(k,l)$, $k \geqslant 3$, $l \geqslant 5$ be a space as defined in Remark B. Assume that ∇-cycles u^2, u_1^4, u_2^4 are given in B, where u^2 is a two-dimensional ∇-cycle modulo 2, u_1^4 is a four-dimensional integral ∇-cycle, u_2^4 is a four-dimensional ∇-cycle modulo 2 if $k>4$, or integral if $k=4$. Furthermore, let these cycles satsify the condition

$$u^2 \times u^2 \sim u_1^4 - 2u_2^4 \pmod 4. \tag{6}$$

Then there exists a mapping f of the complex B into $H(k,l)$ such that (see C)),

$$u_f^2 \sim u^2, \quad u_{1f}^4 \sim u_1^4, \quad u_{2f}^4 \sim u_2^4. \tag{7}$$

If, in addition, the complex B has no four-dimensional even ∇-torsions, and f, g are two mappings of B into $H(k,l)$ such that

$$u_f^2 \sim u_g^2, \quad u_{1f}^4 \sim u_{1g}^4, \quad u_{2f}^4 \sim u_{2g}^4, \tag{8}$$

then the mappings f and g are equivalent.

If the complex B has four-dimensional even ∇-torsions, this problem is not so simple: the results are complicated, and the proof, in particular, is rather involved. We therefore avoid them here.

Proof. We shall only prove the second part of the theorem, i.e., the equivalence of the mappings f and g when the conditions (8) are satisfied. Let B^n denote the n-dimensional part of the complex B. Let $p \in S^2 \subset H(k,l)$ (see (D)). Since the fundamental group of $H(k,l)$ is trivial, we may assume that $f(B^1)=g(B^1)=p$. As the two-dimensional homotopy basis of $H(k,l)$ consists of only S^2, we may assume that $f(B^2) \subset S^2$ and $g(B^2) \subset S^2$. Let T^2 be an arbitrary oriented two-dimensional simplex in B. Let $v_f^2(T^2)$ and $v_g^2(T^2)$ denote the degrees of the mappings f and g of the simplex T^2 onto S^2. Then, we find v_f^2 and v_g^2 are integral two-dimensional ∇-chains in B. From (4) and (8), it follows that $v_f^2 \sim v_g^2$ (mod 2). Deforming the map g as usual, we can obtain $v_f^2 \equiv v_g^2$ (mod 2). Since S^2 is of order 2, we can, on the basis of the last identity, make the mappings f and g coincident on B^2. Since the three-dimensional homotopy group of $H(k,l)$ is trivial, we may assume that f and g coincide on B^3. We now deal at length with the case of $k=3$. In this case the four-dimensional homotopy basis of $H(k,l)$ consists of only S_1^4; if $k>3$, we have to consider S_2^4 as well. Let T^4 be an arbitrary oriented four-dimensional simplex in B. Since the mappings f and g coincide on the boundary of this simplex, both of them taken together generate a mapping of an oriented four-dimensional sphere into $H(k,l)$. This map is homotopic to some multiple of the sphere S_1^4. Let $v_{fg}^4(T^4)$ denote the multiplicity, where v_{fg}^4 is a four-dimensional ∇-chain in B. From (5) it follows that $4v_{fg}^4 = u_{1f}^4 - u_{1g}^4$, and from (8), $4v_{fg}^4 \sim 0$. If B has no even four-dimensional ∇-torsions, the last relation implies that $v_{fg}^4 \sim 0$. Hence the equivalence of f and g follows immediately. If $k>3$, the proof proceeds along similar lines and here the sphere S_2^4 no longer presents any difficulties associated with torsion, because $I(S_2^4, Z_2^{h-4})=1$.

Theorem 1 demonstrates that the characteristic cycles of the skew product $P(R^k, B^4)$ are connected only by the condition (6). If B^4 is an oriented manifold M^4, and the skew product $P(R^4, M^4)$ consists of tangents then, in addition, we have

$$u_1^4 \sim u_2^4 \text{ (mod 2)}. \tag{9}$$

Whether all the relations between the characteristic cycles of the manifold M^4 are given by (6) and (9) is a question that has an important bearing on solving certain geometric problems. The characteristic cycles

u_2^4 and u_1^4 of the manifold M^4 are defined, up to an isomorphism, by integers, where the Euler characteristic corresponds to the cycle u_2^4, and the cycle u_1^4 is a new invariant. It is rather difficult to calculate this invariant in concrete cases. It is equal to 3 for the two dimensional complex projective space. Interestingly, if the manifold M^4 is the boundary of a five-dimensional oriented manifold, then $u_1^4 \sim 0$. I do not know whether the converse is true.

We shall state here an elementary theorem which can be proved without the help of characteristic cycles.

Theorem 2. Let Γ be a connected topological group and $P(A, B)$ a skew product with the group Γ (see Definition 1 in [2]), which is a direct product on the n-dimensional part B^n of the complex B. This means that there exists a homeomorphic mapping $f(x, y)$ of the direct product $A.B^n (x \in A,\ y \in B)$ onto the skew product $P(A, B^n)$, where $f(x, y) = f_y(x) \in A_y$ and $f_y^{-1}\,\mathscr{E}_{\alpha y} \in \Gamma$, $y \in T_\alpha \subset B^n$ (see Definition 1 in [2]). Let h be a continuous mapping of some oriented n-dimensional sphere S^n into Γ, then $h(S^n)$ is an n-dimensional cycle in Γ. Let Σ_n denote the group of all homology classes containing cycles of this kind, and let T_α^{n+1} be an arbitrary oriented $(n+1)$-dimensional simplex in B and S_α^n, its boundary. The mapping $f_y^{-1}\,\mathscr{E}_{\alpha y} \in \Gamma$, $y \in S_\alpha^n$ is well defined on S_α^n and it generates an element $w_f^{n+1}(T_\alpha^{n+1})$ of the group Σ_n. Then w_f^{n+1} is an $(n+1)$-dimensional ∇-cycle in B with respect to the field of Σ_n, where the homology class containing this cycle is an invariant of the skew product $P(A, B)$, i.e., it does not depend on the choice of the mapping f.

The proof is based on the fact that if g is a mapping similar to f, then $f_y^{-1} g_y = f_y^{-1}\,\mathscr{E}_{\alpha y}(g_y^{-1}\,\mathscr{E}_{\alpha y})^{-1} \in \Gamma$ is a continuous mapping of B^n into Γ.

References

1. L. Pontryagin, *Doklady AN SSSR* **XXXV**, N 2 (1942).
2. L. Pontryagin, *Doklady AN SSSR* **XLVII**, N 4 (1945).
3. H. Whitney, *Proc. Nation. Acad.* **26**, 148 (1940).
4. L. Pontryagin, *Doklady AN SSSR* **XXXIV**, N 2 (1942).

Received 24 February 1945

Characteristic Cycles

Doklady Akademii Nauk SSSR (1945) 47(4): 246–49

Whitney, in his papers [1, 2], introduced the concept of a fibre-bundle and studied mainly the case where fibres are spheres. My paper [3] also deals with a fibre-bundle for a case where a bundle consists of tangents to an oriented manifold. My technique is stronger than the Whitney method in that I succeeded in constructing more characteristic cycles than Whitney did. In the present note I demonstrate that my method is applicable to the general case as well.

Definitions 1 and 2, due to Whitney, will be supplemented by several remarks.

Definition 1. Let A be a topological space with a given topological group Γ of homeomorphisms, B a complex of simplices $T_1, T_2, \ldots$, and P a topological space. The space P is called a *fibre-bundle* or *skew product*, $P = P(A, B)$ of A and B, if:

(1) to every point $y \in B$, there corresponds some subset $A_y \subset P$; to distinct points y and z in B there correspond disjoint subsets A_y and A_z;

(2) to each simplex T_α, there corresponds some homeomorphic mapping ξ_α of the topological product $A \cdot T_\alpha$, i.e., a function $\xi_\alpha(x, y) = \xi_{\alpha y}(x)$ $(x \in A, y \in T)$ is defined so that $\xi_{\alpha y}$ is the homeomorphic mapping of the space A onto A_y;

(3) if $y \in T_\alpha \cap T_\beta$, then $\xi_{\alpha y}^{-1} \xi_{\beta y} = \eta_{\alpha \beta y} \in \Gamma$ and $\eta_{\alpha\beta y}$ is a continuous mapping of $T_\alpha \cap T_\beta$ into the group Γ.

Evidently, if $y \in T_\alpha \cap T_\beta \cap T_\gamma$, then

$$\eta_{\alpha\beta y}\eta_{\beta\gamma y} = \eta_{\alpha\gamma y}. \tag{1}$$

Assume that all the simplices $T_1, T_2, \ldots$ are oriented and let $\epsilon_{\alpha\beta}$ denote the coefficient with which the simplex T_β enters into the boundary of the simplex T_α. If $\epsilon_{\alpha\beta} \neq 0$, take $\eta_{\alpha\beta y} = \zeta_{\alpha\beta y}$. Let $\epsilon_{\alpha\beta} \neq 0$, $\epsilon_{\alpha\gamma} \neq 0$, $\epsilon_{\beta\delta} \neq 0$, $\epsilon_{\gamma\delta} \neq 0$; then for $y \in T_\delta$, from (1) it follows that

$$\zeta_{\alpha\beta y}\zeta_{\beta\delta y} = \zeta_{\alpha y\gamma}\zeta_{\gamma\delta y}. \tag{2}$$

If we put $\Delta_\Gamma T_\alpha = \Sigma_\beta\, \epsilon_{\alpha\beta}\zeta_{\alpha\beta y} T_\beta$, then (2) can be rewritten as

$$\Delta_\Gamma \Delta_\Gamma T_\alpha = 0.$$

It is a simple matter to verify that $\eta_{\alpha\beta y}$ satisfying (1) is uniquely determined by $\zeta_{\alpha\beta y}$ satisfying (2). The following existence theorem is easily proved.

(A) Let A be a topological space with a given topological group Γ of homeomorphisms, B a complex consisting of the simplices $T_1, T_2, \ldots,$ where $\Delta T_\alpha = \Sigma_\beta\, \epsilon_{\alpha\beta} T_\beta$ and $\zeta_{\alpha\beta y} \in \Gamma$ are continuous functions of $y \in T_\beta$ satisfying (2), defined for α, β for which $\epsilon_{\alpha\beta} \neq 0$. Then there exists a skew product $P(A, B)$ for which $\xi_{\alpha y}^{-1}\xi_{\beta y} = \zeta_{\alpha\beta y}$.

Definition 2. Let $P(A,B)$ and $P'(A,B)$ be two skew products with the same group Γ. A homeomorphic mapping f of the space P' onto the space P is called an *isomorphic mapping* of $P'(A, B)$ onto $P(A, B)$, if $f(A'_y) = A_y$ and for $y \in T_\alpha$, $\xi_{\alpha y}^{-1} f \xi'_{\alpha y} = \tau_{\alpha y} \in \Gamma$.

From the last assertion, it follows that for $y \in T_\beta$,

$$\zeta_{\alpha\beta y} = \tau_{\alpha y}^{-1}\zeta'_{\alpha\beta y}\tau_{\beta y}. \tag{3}$$

The following theorem is easily proved.

(B) Two skew products $P(A,B)$ and $P'(A,B)$ with defining functions $\zeta_{\alpha\beta y}$ and $\zeta'_{\alpha\beta y}$ are isomorphic, if and only if there exist continuous functions $\tau_{\alpha y} \in \Gamma$ of $y \in T_\alpha$ satisfying the conditions (3).

Whitney noted that if $P(A, B)$ is a skew product and f is a continuous mapping of some complex C into B, there exists a skew product $P(A, C, f)$ of A and C with the same group Γ as the initial product $P(A, B)$, where $P(A, C, f)$ is defined uniquely up to an isomorphism. It is a simple matter to demonstrate that if two mappings f and g are equivalent, then $P(A, C, f)$

and $P(A,C,g)$ are isomorphic. But, often isomorphism of the products $P(A,C,f)$ and $P(A,C,g)$ implies equivalence of the mappings f and g. Thus we arrive at the following

Definition 3. A skew product $P(A,B)$ with the group Γ is called *n-dimensionally universal* for the space A with the group Γ, provided:

(1) the skew product $P(A,C)$ with the group Γ, where C is of a dimension not greater n, is always isomorphic to $P(A,C,f)$, where f is a continuous mapping of C into B;

(2) two skew products $P(A,C,f)$ and $P(A,C,g)$ are isomorphic, if and only if the mappings f and g of C into B are equivalent.

Definition 4. Using the notation of Definition 3, assume that B is a b-dimensional oriented manifold and let z denote some $(b-r)$-dimensional cycle in B. Let f be a continuous mapping of C into B, and T some r-dimensional oriented simplex in C. Let $u_z^r(T)$ denote the index of intersection of $f(T)$ with z. We can easily show that u_z^r is an r-dimensional ∇-cycle in C, whose homology class is defined by the homology class of the cycle z and the class of the mapping f. Thus, since $P(A,B)$ is universal, the ∇-cycle u_z^r is an invariant of the skew product $P(A,C,f)$. The cycle u_z^r is called a *characteristic cycle* of the skew product $P(A,C,f)$.

Let $H_l = H(k,l)$ be the manifold of all k-dimensional oriented planes of $(k+l)$-dimensional Euclidean space R^{k+l} that pass through the origin of coordinates in R^{k+l}. Let A be some k-dimensional oriented Euclidean space R^k, and let Γ be the group of all isometric mappings of R^k onto itself that preserve the origin and orientation in R^k. We now define the skew product $P(R^k, H_l)$: a point in the space P is a pair (R_y^k, x), where $R_y^k \in H_l$, and $x \in R_y^k$. Then the set of all pairs (R_y^k, x) with a fixed R_y^k is a fibre A_y. Whitney noted that $P(R^k, H_2)$ satisfies the condition 1 of Definition 3 when $l=n$. In Theorem 1 we give another proof of this assertion, demonstrating at the same time that $P(R^k, H_2)$ is universal.

Theorem 1. The skew product $P(R^k, H_2)$ is an $(l-1)$-dimensional universal product for the space R^k with the rotation group Γ.

Proof. We show that $P(R^k, H_l)$ is universal for sufficiently large l; the dimension can be lowered without any difficulty. Let C be a complex consisting of the simplices $T_1, T_2, \ldots,$ and C^r a subcomplex of C that consists of all simplices of dimension not greater r. Let $P(R^k, C)$ be some skew product with defining functions $\zeta_{\alpha\beta y} \in \Gamma$, $y \in T_\beta$ (see (A)). We shall construct a continuous mapping f of the complex C into H_l, such that $P(R^k, C, f)$ is isomorphic to $P(R^k, C)$, $f(y) = \bar{R}^k_y \in H_l$. We shall also construct an isometric mapping $\bar{\xi}_{\alpha y}$, $y \in T_\alpha$ of the space R^k onto the space $\bar{R}^k_y$ so that $\bar{\xi}^{-1}_{\alpha y} \bar{\xi}_{\beta y} = \zeta_{\alpha\beta y}$. Suppose that we have already constructed f and $\bar{\xi}_{\alpha y}$ for C^r, $f(y) = \bar{R}^k_y \subset R^{k+l}$ for $y \in C^r$. We assume that $R^{k+l} \subset R^{2k+l}$ and that $\bar{R}^k$ is the orthogonal complement of R^{k+l} in R^{2k+l}. Let T_α be an arbitrary $(r+1)$-dimensional simplex in C with center p. Put $f(p) = \bar{R}^k$, and arbitrarily define an isometric mapping $\bar{\xi}_{\alpha p}$ of the space R^k onto $\bar{R}^k$. Let S_α be the boundary of the simplex T_α, then the point $z \in T_\alpha$ in the simplex T_α can be defined as the pair (y, t), where $y \in S_\alpha$, and t is a real parameter $0 \leqslant t \leqslant \pi/2$, where $(y, 0) = p$, $(y, \pi/2) = y$, $z = (y, t)$. Let T_β be an arbitrary r-dimensional face of the simplex T_α, then $\bar{\xi}_{\beta y}$ is well defined. Let us define the function $\bar{\xi}_{\alpha z}$, $z = (y, t)$, $y \in T_\beta$ as follows:

$$\bar{\xi}_{\alpha z}(x) = \bar{\xi}_{\alpha p}(x) \cos(t) + \bar{\xi}_{\beta y}(\zeta^{-1}_{\alpha\beta y}(x)) \sin(t), \quad x \in R^k, \tag{4}$$

and the function f as follows

$$f(z) = \bar{\xi}_{\beta z}(R^k). \tag{5}$$

If y belongs simultaneously to two r-dimensional faces T_β and T_γ of the simplex T_α, putting $T_\delta = T_\beta \cap T_\gamma$, we find that, by virtue of (2), the relation (4) uniquely defines the mapping $\bar{\xi}_{\alpha z}$, irrespective of the choice of T_β or T_γ.

Now let f and g be two continuous mappings of C into H_l, such that the products $P(R^k, C, f)$ and $P(R^k, C, g)$ are isomorphic. This implies that there exists an isometric mapping h_y of the plane $f(y)$ onto the plane $g(y)$ which depends continuously on $y \in C$. Let $R^{k+l} \subset R^{2k+2l}$, $\bar{R}^{k+l}$ the orthogonal complement of R^{k+l} in R^{2k+2l}, and let φ be some isometric mapping of R^{k+l} onto $\bar{R}^{k+l}$. Then φg is a mapping of C into H_{k+2l} isotopic to g, because φ can be realized by continuously rotating R^{k+l} in R^{2k+2l}. If $x \in f(y)$, take

$$\psi_{yt}(x) = \varphi(h_y(x)) \cos(t) + x \sin(t), \quad \psi_t(y) = \psi_{ty}(f(y)). \tag{6}$$

Clearly, $\psi_t(y)$ is a k-dimensional plane in R^{2k+2l}; therefore ψ_t is a mapping of C into H_{k+2l}. Thus, from (6), we have $\psi_0 = \varphi g$, $\psi_{\pi/2} = f$, and as g and φg are equivalent, the mappings f and g are equivalent in H_{k+2l}. This completes the proof of Theorem 1.

The homology of the manifold $H(k, l)$ is discussed in detail in [3]. In studying the homotopic properties of $H(k, l)$ the following theorem is helpful.

Theorem 2. For $n \leqslant l-1$, the n-dimensional homotopy group of the manifold $H(k, l)$ is isomorphic to the $(n-1)$-dimensional homotopy group of the manifold Γ_k, where Γ_k is the group of positive rotations of a k-dimensional Euclidean space.

Proof. Just as in Theorem 1, we give the proof for sufficiently large l. Let R_0^k and R_1^k be two orthogonal k-dimensional subspaces of R^{k+l}. Assume that the group Γ_k acts in R_1^k and let φ be some isometric mapping of R_1^k onto R_0^k. Let S^{n-1} be an oriented $(n-1)$-dimensional sphere and ψ_y be an element of the group Γ_k which depends continuously on $y \in S^{n-1}$. For $x \in R_1^k$, suppose that

$$\theta(x, y, t) = \varphi(x) \cos(t) + \psi_y(x) \sin(t); \quad \theta(y, t) = \theta(R_1^k, y, t). \tag{7}$$

Obviously, $\theta(y, t)$ is a k-dimensional plane in R^{k+l}, and $\theta(y, 0) = R_0^k$, $\theta(y, \pi/2) = R_1^k$. Thus, $\theta(y, t)$ is a continuous mapping of the direct product $S^{n-1} \cdot T$ into H_l, where T is the interval $0 \leqslant t \leqslant \pi/2$. Under this mapping both the bases $S^{n-1} \cdot 0$ and $S^{n-1} \cdot \pi/2$ of the cylinder $S^{n-1} \cdot T$ are mapped into points, and, consequently, we have a continuous mapping of the sphere S^n into H_l. The correspondence thus established between the mapping of S^{n-1} into Γ_k and the mapping of S^n into H_l gives an isomorphism of the corresponding homotopy groups. The proof proceeds in a manner similar to the proof of Theorem 1.

It should be noted that universal skew products can be easily constructed for any compact Lie group Γ, so $P(R^k, H_2)$ is no exception in this respect.

References

1. H. Whitney, *Bull. Am. Math. Soc.* **43,** 785 (1937).
2. H. Whitney, *Proc. Nation. Acad.* **26,** 148 (1940).
3. L. Pontryagin, *Doklady AN USSR*, **XXXV,** N 2 (1942).

Received 9 February 1945

Topological Duality Theorems

Uspekhi Matematicheskikh Nauk (1947) 2(2): 21–44

Duality theorems play a prominent role in topology. Historically the first duality theorem is due to Pioncare, and reads as follows:

Let M^n be a closed orientable n-dimensional manifold and p^r its r-dimensional Betti number. Then $p^r = p^{n-r}$ $(r = 0, 1, \ldots, n)$ and the r-dimensional torsion coefficients of the manifold M^n coincide with its $(n - r - 1)$-dimensional torsion coefficients [1].

Veblen refined this theorem as follows:

Let z_i^r $(i = 1, \ldots, p^r)$ be an r-dimensional basis of the weak homology of a manifold M^n. Denote the index of intersection of the cycles z_i^r and z_j^{n-r} by $a_{ij}^r = I(z_i^r, z_j^{n-r})$. Then the determinant of the matrix $\|a_{ij}^r\|$ is equal to ± 1 (see [2]).

That the determinant of the matrix $\|a_{ij}^r\|$ $(r = 0, 1, 2, \ldots, n)$ is non-zero is equivalent to the following assertion:

An r-dimensional cycle z^r in the manifold M^n is weakly not homologous to zero, if and only if there exists in M^n a cycle u^{n-r} of complementary dimension such that $I(z^r, u^{n-r}) \neq 0$.

Thus, the Veblen theorem gives the necessary and sufficient condition for a cycle to be weakly not homologous to zero in a manifold. This condition is quite effective, and is often used in calculating the homology in a manifold.

Independently of Poincare and Veblen, Alexander [3] ennunciated his duality theorem generalizing the Jordan theorem on the separation of a plane by a closed curve as follows:

Let K be a polyhedron in an n-dimensional Euclidean space R^n. The r-dimensional Betti number modulo 2 of the space $R^n - K$ is equal to the $(n - r - 1)$-dimensional Betti number modulo 2 of the polyhedron K [3].

This theorem greatly stimulated the development of topological duality theorems. To begin with, it can be improved along the same lines as Veblen refined the Poincare theorem. The linking number plays here the part of the intersection index. Thus, the following theorem is obtained.

Let z_i^r $(i = 1, 2, \ldots, p^r)$ be an r-dimensional basis of the homology modulo 2 in the space $R^n - K$, and u_i^{n-r-1} $(i = 1, 2, \ldots, p^r)$ an $(n - r - 1)$-dimensional basis of the homology modulo 2 of the polyhedron K. Then the determinant of the matrix $\|b_{ij}^r\|$ is equal to unity, where b_{ij}^r is the linking number of the cycles z_i^r and u_j^{n-r-1} [4, 5].

Thus, the improved Alexander duality theorem turned out to be an effective tool for generalizing the original theorem itself as well as for extending it to any compact subset K of the Euclidean space R^n [5, 6].

Subsequently, the duality theorems were extended to the case where the polyhedron or the set K belongs to an arbitrary closed manifold M^n. The problem here was to study the homology in the space $M^n - K$ starting from the homology in K and its imbedding in M^n, which was also expressed in terms of homology. Thus a close relationship, both algebraic and geometric [7], was detected between the duality theorems of Poincare and Alexander. Furthermore, it became clear that in constructing a homology theory for the purposes of duality theorems, it was inexpedient, and even impossible, to restrict the coefficient realm to be the domain of integers or residues modulo 2. On the contrary, it is helpful to use more general groups, even topological groups [8]. The reasons for this are discussed in detail in [8]. Owing to the use of more general groups and the theory of characters [9], duality theorems have to some degree acquired a more refined and finished form.

The ∇-homology theory developed by Kolmogorov [10] and Alexander [11] promoted the rapid advancement of the duality theorems, which found its final culmination in the work of Alexandroff [12].

The first three sections in this paper are devoted to the preliminaries needed in the formulation and proof of the duality theorems. Therefore, a knowledge of the first four chapters in [13] and the first chapter in [9] is sufficient to understand this paper. The fourth section presents the results in [7], which have been modernized by using the theory of characters [9].

1. The Manifold and its Orientation

The manifold is a fundamental concept in topology: but its definition is no trivial problem. There are several definitions, and the relationship between them is so far not completely clear. On the other hand, the invariance of certain definitions has not been demonstrated.

Except when considering combinatorial analysis, use is always made of the following

Definition 1. A topological space M is called an *n-dimensional manifold*, if it is connected and satisfies the second axiom of countability, and if each point of this space admits a neighbourhood homeomorphic to an n-dimensional Euclidean space.

In combinatorial topology we have often to take recourse to other definitions, because Definition 1 does not agree well with the combinatorial properties of a complex, in case the space M is a polyhedron. I say "in case" because the speculation that every manifold is, in the sense of Definition 1, a polyhedron has neither been proved, nor disproved. To formulate the combinatorial definition, we need the following preliminary constructions:

(A) Let K be some complex and T a simplex of K. The set of all open simplices in K which have T as a face, including T itself, is called a star of the simplex T in the complex K and is denoted by $U(T,K)$ or simply $U(T)$. The set of all simplices in K that have no common vector with T and are the faces of the simplices contained in $U(T)$ is denoted by $P(T,K)$ or $P(T)$. In other words, $P(T)$ consists of all closed simplices of the complex $\overline{U(T)}$, ($\overline{U(T)}$ is the complex consisting of all simplices in $U(T)$ and their faces) that do not intersect the closed simplex T. The collection $P(T)$ is called a *representative* of the simplex T in the complex K. Note that $U(T)$ is an open, while $P(T)$ is a closed, subcomplex of the complex K. In combinatorial definitions of a manifold, either $U(T)$ or $P(T)$ can be used with equal success, but I prefer the latter.

Definition 2. A connected complex K is called an *n-dimensional manifold*, if the representative $P(T^r,K)$ of each r-dimensional simplex T^r of K is homeomorphic to an $(n-r-1)$-dimensional sphere S^{n-r-1} $(r = 0, 1, \ldots, n-1)$. A zero-dimensional sphere is a pair of points.

This definition has two drawbacks: first, its topological invariance is obscure, and second, its relationship with Definition 1 has so far not been established. If a complex K satisfies Definition 2, we can easily show that the polyhedron $|K|$ satisfies Definition 1. But the converse is so far unproved, i.e., it is not known whether the complex K satisfies Definition 2, if the polyhedron K satisfies Definition 1. These are difficult and unsolved problems of contemporary topology.

Henceforth we shall use the following combinatorial definition of a manifold, which is much more general than Definition 2.

Definition 3. A connected complex K is called an *n-dimensional homological manifold* or simply *a manifold*, if the representative $P(T^r, K)$ of any of its r-dimensional simplices T^r has the same homology groups as a sphere S^{n-r-1} of dimension $(n-r-1)$ $(r = 0, 1, 2, \ldots, n-1)$.

The logical meaning of this definition is quite clear. First, it is relatively easy, though not entirely simple, to demonstrate its topological invariance. Further one can show the equivalence of Definitions 2 and 3 for $n = 1, 2$ and 3; beginning with $n = 4$, Definition 3 gives a wider class of objects than Definition 2. The relationship between Definitions 1 and 3 is also quite clear: if a polyhedron K satisfies Definition 1, it also satisfies Definition 3. Furthermore, if a complex K satisfies Definition 3, it also satisfies Definition 1 for $n = 1, 2, 3$. Starting with $n = 4$, there exist complexes satisfying Definition 3, such that their corresponding polyhedra do not satisfy Definition 1. We leave all these assertions without proof. Due to logical clarity and adequacy, Definition 3 finds successful application in duality theorems. Henceforth we shall only make use of this definition.

(B) Let K be an arbitrary complex and T_0 a simplex in K, T_1 some face of this simplex, T_2 a face of the simplex T_0 opposite to T_1. Obviously, $T_2 \in P(T_1, K)$, and

$$P(T_2, P(T_1, K)) = P(T_0, K). \tag{1}$$

In the proof we use the following notation: Let A and B be two simplices in K. If A and B do not have common vertices and there exists in K a simplex C spanned by all the vertices belonging to A and B, then the simplex $C = AB$ is said to exist in K.

Let $T \in P(T_0, K)$: this implies that the simplex $TT_0 = TT_2T_1$ exists in K; but then the simplex TT_2 also exists in K and obviously belongs to $P(T_1, K)$, i.e., $T \in P(T_2, P(T_1, K))$.

Now let $T \in P(T_2, P(T_1 K))$. This means that the simplex TT_2 exists in

$P(T_1, K)$; then the simplex $TT_2T_1 = TT_0$ exists in K which implies that $T \in P(T_0, K)$.

From (B) follows easily an important corollary:

(C) Let K be an n-dimensional manifold and $T^r(0 \leqslant r \leqslant n-2)$, an r-dimensional simplex of K; then $P(T^r, K)$ is an $(n-r-1)$-dimensional manifold (see Definition 3).

Let $T^r = T_1$ and $T_2 \in P(T_1, K)$; then the simplex $T_0 = T_1T_2$ exists in K. Since K is a manifold, the homology groups of $P(T_0, K)$ coincide with those of a sphere of the proper dimension, and by virtue of (1), the same is true for $P(T_2, P(T_1, K))$. Thus, $P(T_1, K)$ satisfies the requirements of Definition 3, except, possibly, the condition of connectedness. But connectedness is implied as the homology groups of $P(T_1, K)$ are the same as the homology groups of a sphere.

Definiton 4. A complex K is called an *n-dimensional pseudomanifold* if it satisfies the following conditions:

(1) The complex K is dimensionally homogeneous, i.e., each of its simplices is a face of one of its n-dimensional simplices;

(2) To each $(n-1)$-dimensional simplex of the complex K, there adjoin exactly two n-dimensional simplices; and

(3) If A and B are two n-dimensional simplices of K, there exists in K a chain of n-dimensional simplices

$$A = T_1, T_2, \ldots, T_1 = B, \tag{2}$$

such that any two adjacent members in this chain have a common $(n-1)$-dimensional face.

(D) Every manifold is a pseudo-manifold.

Let K be an n-dimensional manifold; we shall demonstrate that the conditions (1), (2), and (3) of Definition 4 are satisfied for this manifold.

If condition 1 is not satisfied, then there exists in K some r-dimensional simplex T^r $(r < n)$, which is not a face of any simplex of higher dimension in K. Then, evidently, $P(T^r, K)$ is empty; but, by virtue of Definition 3, the homology groups of $P(T^r, K)$ are the same as those of S^{n-r-1}. The zero-dimensional homology group of this sphere is not trivial, while for an empty set it is trivial.

If T^{n-1} is an $(n-1)$-dimensional simplex of K, then by virtue of

Definition 3, $P(T^{n-1}, K)$ consists two points. This means that exactly two n-dimensional simplices of K are adjacent to T^{n-1}, i.e., condition 2 is fulfilled.

We prove condition 3 by induction. We assume that this condition is satisfied for all $(n-1)$-dimensional manifolds and we show that it holds true for an n-dimensional manifold K. Let A be an arbitrary n-dimensional simplex of K, and let L denote the set of all n-dimensional simplices in K which may be adjoined to A by means of a chain of the type (2). Let M stand for the set of all remaining n-dimensional simplices of K. Since K is a connected complex, there exists in K a vertex a common to the simplex $C \in L$ and to the simplex $D \in M$. It now suffices to show that C and D can be connected in K by means of a chain of the type (2). Let $K^* = P(a, K)$. Then, $C = aC^*$, $D = aD^*$, where C^* and D^* are contained in K^*. Since K^* is a manifold (see (C)), by induction, there exists in K^* a chain

$$C^* = T_1^*, T_2^*, \ldots, T_1^* = D^* \tag{3}$$

which connects the simplices C^* and D^*. The chain we need is obtained from (3) by adding the vertex a to each simplex. This completes the proof of assertion (D).

(E) Let K be an n-dimensional pseudo-manifold. Two n-dimensional oriented simplices T and T' of K having a common $(n-1)$-dimensional face S are said to be *coherently oriented*, if their orientations induce opposite orientations on S. Let A be an arbitrary oriented n-dimensional simplex in K. To all simplices in the chain (2) we can assign orientations such that any two neighbouring simplices will be coherently oriented. Thus, the orientation defined in A can be extended to B. It may happen that in passing from an oriented simplex A to a simplex B via two different chains, we obtain two different orientations in B. In this case there obviously exists a closed chain starting from A and returning to A that inverts the orientation. Clearly, if there exists a closed chain that inverts the orientation of some simplex A, then such a chain exists for any other simplex as well. A pseudo-manifold K, in which there exist closed chains inverting the orientation, is called a *non-orientable manifold.* A pseudo-manifold K, in which there are no closed chains inverting the orientation, is called an *orientable manifold*. By giving an orientation to any one of the n-dimensional simplices in an orientable pseudo-manifold K, we can extend this orientation via chains of the type (2) to every other n-dimensional simplex $B \in K$.

This definition can be rephrased in a slightly different form as

Definition 5. An n-dimensional pseudo-manifold K is called *orientable*, if all its n-dimensional simplexes can be so oriented that the sum of their boundaries is zero; otherwise, the pseudo-manifold is called *non-orientable.* If, all the n-dimensional simplices of a pseudo-manifold K are assigned an orientation such that the sum of their boundaries is zero, then an *orientation* is said to have been defined in K or K has been *oriented.* From (E), it follows that there exist only two distinct orientations in a pseudo-manifold K.

Definition 5, by virtue of (D), can be applied to any manifold.

(F) Let K be an n-dimensional orientable manifold, and $\vec{z}$ the sum of its n-dimensional simplices so oriented that the sum of their boundaries is zero (see Definition 5); then $\vec{z}$ is an n-dimensional cycle in K and it defines the orientation of K. Any n-dimensional cycle x in K is equal to $\alpha\vec{z}$, where α is an integer.

To all the n-dimensional simplices in K let us assign the orientations which they are given in $\vec{z}$. Let A be an n-dimensional oriented simplex and α its coefficient in x. Then, every simplex in K having an $(n-1)$-dimensional common face with A is evidently contained in x with the coefficient α, because x is a cycle. By virtue of the chain (2), we find that each n-dimensional simplex B in K is contained in x with the coefficient α and, consequently, $x = \alpha\vec{z}$.

2. Barycentric Stars

Barycentric stars play an important role in almost all the proofs of duality theorems. In formulating certain properties of barycentric stars, it is convenient to introduce the concept of ∇-boundary.

(A) Let K be some complex and T^r some oriented r-dimensional simplex in K. The sum of all oriented $(r+1)$-dimensional simplices T^{r+1} of the complex K such that ΔT^{r+1} contains $+T^r$ is called the ∇-boundary of the simplex T^r and is denoted by ∇T^r. The ∇-operation is extended to any r-dimensional chain x by additivity. It is easy to verify that

$$\nabla\nabla x = 0. \tag{1}$$

We now proceed to construct barycentric stars.

(B) Let T^n be an n-dimensional oriented simplex whose vertices we denote by $0, 1, \ldots, n$, so that

$$T^n = \epsilon(0, 1, \ldots, n) \quad (\epsilon = \pm 1),$$

and let T^r be some oriented r-dimensional face of the simplex T^n:

$$T^r = \delta(0, 1, \ldots, r) \quad (\delta = \pm 1).$$

Let us rewrite the sequence of numbers

$$r+1, r+2, \ldots, n \tag{2}$$

in some arbitrary order:

$$i_1, i_2, \ldots, i_s \quad (s = n-r). \tag{3}$$

Let $\alpha(i_1, i_2, \ldots, i_s)$ be $+1$, if the sequence (3) can be obtained from the sequence (2) by an even number of permutations, and -1 otherwise. Let a_k $(k = 0, 1, \ldots, s)$ denote the center of the simplex $(0, 1, \ldots, r, i_1, i_2, \ldots, i_k)$.

Put

$$T^s(i_1, \ldots, i_s) = \delta\epsilon\alpha(i_1, \ldots, i_s)(a_0, a_1, \ldots, a_s). \tag{4}$$

It defines an s-dimensional oriented simplex of the barycentric subdivision of the simplex T^n. Let $B(T^r, T^n)$ denote the sum of all simplices of the type (4), when the sequence (3) runs through all permutations of the sequence (2):

$$B(T^r, T^n) = \sum T^s(i_1, \ldots, i_s). \tag{5}$$

It is a simple matter to verify that the chain $B(T^r, T^n)$ does not depend on the numeration used to number the vertices of the simplex T^n, it only depends on the oriented simplex T^n and its oriented face T^r, and

$$B(-T^r, T^n) = B(T^r, -T^n) = -B(T^r, T^n). \tag{6}$$

Note that $B(T^r, T^n)$ is well defined even for $r = 0$ and $r = n$; in the latter case, $B(T^r, T^n) = \delta\epsilon(a_0)$, where a_0 is the center of the simplex T^n.

(C) Let K be a complex, and T^r and T^n two oriented simplices in K of dimensions r and n $(r \leqslant n)$, respectively. If T^r is a face of the simplex T^n, then $B(T^r, T^n)$ is defined by the relation (5); if not, take $B(T^r, T^n) = 0$. If $x = \Sigma_i \alpha_i T_i^r$, $y = \Sigma_j \beta_j T_j^n$ are two chains of dimensions r and n, respectively, in K, let

$$B(x, y) = \sum_{i,j} \alpha_i \beta_j B(T_i^r, T_j^n). \tag{7}$$

Then

$$\Delta B(x,y) = (-1)^{r+1} B(\nabla x, y) + (-1)^r B(x, \Delta y). \tag{8}$$

It is sufficient to demonstrate the truth of (8) for the case $x = T^r$, $y = T^n$, where T^r is a face of T^n. If T^r is not a face of T^n, the relation (8) is self-evident. It can obviously be extended to arbitrary chains x and y because of the additivity of the ∇ and Δ operations.

Thus, let T^r and T^n be two oriented simplices in K, where T^r is a face of the simplex T^n. Hereafter we use the notation adopted in (B).

The number $r + k$ is called the weight of the vertex a_k of a simplex $T^s\ (i_1, \ldots, i_s)$ because a_k is the center of a simplex of dimension $r + k$ in K. Thus, the simplex T contains vertices of all weights $r, r+1, \ldots, n$. If S is some $(s-1)$-dimensional face of the simplex T obtained by omitting the vertex a_k, then S contains vertices of weights $r, \ldots, r+k-1$, $r+k+1, \ldots, n$.

Let $T' = T^s\ (i'_1, \ldots, i'_s)$ be a simplex different from T with the vertices $a'_0, a'_1, \ldots, a'_s$. Note that

$$a'_0 = a_0, \quad a'_s = a_s \tag{9}$$

(see (B)). If S' is an $(s-1)$-dimensional face of the simplex T' obtained by omitting the vertex a'_1, then S' contains vertices of weights

$$r, \ldots, r+l-1, r+l+1, \ldots, n.$$

We now find the conditions under which the non-oriented simplices S and S' coincide, taking in account that the sequence (3) and the sequence

$$i'_1, i'_2, \ldots, i'_s \tag{10}$$

defining the simplices T and T', are distinct. Since the simplex S (and consequently, S') contains no more than one vertex of each weight, the weights of the vertices in S and S' should be the same; therefore, $k = l$. Moreover, vertices of identical weights in the simplexes S and S' should be the same, therefore

$$\begin{aligned} &a'_0 = a_0, \quad a'_1 = a_1, \ldots, a'_{k-1} = a_{k-1}, \\ &a'_k \neq a_k, \quad a'_{k+1} = a_{k+1}, \ldots, a'_s = a_s \end{aligned} \tag{11}$$

From (9) and (11), it follows that $0 < k < 1$. Furthermore, from (11) we find that the sets $\{0, \ldots, n, i_1, \ldots, i_m\}$ and $\{0, \ldots, r, i'_1, \ldots, i'_m\}$ coincide for any $m \neq k$, and this is possible only when

$$\begin{aligned} i'_1 = i_1, \ldots, i'_{k-1} = i_{k-1}, \quad i'_k = i_{k+1}, \quad i'_{k+1} = i_k, \\ i'_{k+2} = i_{k+2}, \ldots, i'_s = i_s. \end{aligned} \tag{12}$$

Hence from (12), it follows that the sequences (3) and (10) differ only by a permutation of two neighbouring terms.

Now suppose that S and S' are oriented faces of the simplices T and T', respectively; then

$$S = \delta\epsilon\alpha(i_1, \ldots, i_s)(-1)^k (a_0, \ldots, a_{k-1}, a_{k+1}, \ldots, a_s),$$
$$S' = \delta\epsilon\alpha(i'_1, \ldots, i'_s)(-1)^k (a'_0, \ldots, a'_{k-1}, a'_{k+1}, \ldots, a'_s).$$

Since the sequences (3) and (10) differ only by a permutation of two neighbouring terms, by (11) we have $S = -S'$.

From the foregoing it follows that in calculating the boundary of $B(T^r, T^n)$ (see (5)), all faces obtained from the simplices $T^s\ (i_1, \ldots, i_s)$ by omitting vertices of weight $r+k$, $0<k<s$, are cancelled. Hence we have to consider only those terms in the boundary of $B(T^r, T^n)$ that are obtained by omitting vertices of weights r and n.

Consider those terms in the boundary of $B(T^r, T^n)$ that are obtained by omitting vertices of weight r. Separating the corresponding term in the boundary of the simplex $T^s\ (i_1, \ldots, i_s)$, we obtain:

$$\Delta T^s(i_1, \ldots, i_s) = \delta\epsilon\alpha(i_1, \ldots, i_s)(a_1, \ldots, a_s) + \ldots . \tag{13}$$

Note that

$$T^n = \epsilon\alpha(i_1, \ldots, i_s)(0, \ldots, r, i_1, \ldots, i_s). \tag{14}$$

Further, put

$$T^{r+1}_{i_1} = \delta(0, \ldots, r, i_1). \tag{15}$$

By virtue of (B), we have

$$B(T^{r+1}_{i_1}, T^n) = \delta\epsilon\alpha(i_1, \ldots, i_s)(a_1, \ldots, a_s) + \ldots . \tag{16}$$

Here only one term has been written corresponding to the numeration $0, \ldots, r, i_1, \ldots, i_s$ of the vertices of the simplex T^n. Note that $\Delta T^{r+1}_{i_1}$ contains the simplex T^r with the sign $(-1)^{r+1}$, hence we have

$$\nabla T^r = (-1)^{r+1} T^{r+1}_{i_1} + \ldots . \tag{17}$$

Here only one term in the ∇-boundary of the simplex T^r has been written.

Summing (13) and (16) over all possible permutations (3) of the sequence (2), and using (17), we obtain

$$\Delta B(T^r, T^n) = (-1)^{r+1} B(\nabla T^r, T^n) + \ldots . \tag{18}$$

On the right-hand side of this expression, we have written only those terms that are obtained by omitting vertices of weight r. Moreover, the expression (18) should contain only those simplices in ∇T^r that are faces of the simplex T^n, because only these $T^{r+1}_{i_1}$ are contained in (16).

But if ∇T^r is interpreted in the usual way in K, then (18) remains valid since, by virtue of (C), all superfluous terms vanish.

Now consider those terms in the boundary of $B(T^r, T^n)$ that are obtained by omitting vertices of weight n. Let us express the boundary of $T^s(i_1, \ldots, i_s)$ by writing out a corresponding term:

$$\Delta T^s(i_1, \ldots, i_s) = \delta\epsilon\alpha(i_1, \ldots, i_s)(-1)^s(a_0, \ldots, a_{s-1}) + \ldots . \tag{19}$$

Put

$$T_{i_s}^{n-1} = \epsilon\alpha(i_1, \ldots, i_s)(0, \ldots, r, i_1, \ldots, i_{s-1}); \tag{20}$$

then we have

$$B(T^r, T_{i_s}^{n-1}) = \delta\epsilon\alpha(i_1, \ldots, i_s)(a_0, \ldots, a_{s-1}) + \ldots . \tag{21}$$

Further, the simplex $T_{i_s}^{n-1}$ is contained in the boundary of the simplex T^n with the sign $(-1)^n$, i.e.,

$$\Delta T^n = (-1)^n T_{i_s}^{n-1} + \ldots . \tag{22}$$

Adding (19) to (21), and using (22), we obtain

$$\Delta B(T^r, T^n) = (-1)^r B(T^r, \Delta T^n) + \ldots . \tag{23}$$

On the right-hand side of this expression we have written only those terms that are obtained by omitting vertices of weight n. Note that (23) should contain only those faces of T^n that contain T^r; but, by virtue of (C), the equality (23) remains valid, if ΔT^n is interpreted in the usual way.

The relations (23) and (18) taken together yield (8).

Definition 6. Let K be an n-dimensional oriented manifold and $\overset{\circ}{z}$ a cycle defining its orientation (see Definition 5 and (F), Section 1). Let $B(x, \overset{\circ}{z}) = B(x)$. In case x is an oriented r-dimensional simplex T^r in K, we call $B(T^r)$ a *barycentric star* of the simplex T^r in the oriented manifold K. If K is a non-orientable manifold, then the whole construction has to be made modulo 2, and the barycentric star is defined to be a chain modulo 2. From (8), it follows that

$$\Delta B(x) = (-1)^{r+1} B(\nabla x), \quad \Delta B(T^r) = (-1)^{r+1} B(\nabla T^r); \tag{24}$$

if K is non-orientable, then the latter expressions are true only modulo 2, the chain x is also to be considered modulo 2, and T^r can be considered without any orientation.

Note that the barycentric star of an r-dimensional simplex in an n-dimensional manifold K is an $(n-r)$-dimensional chain of the barycentric subdivision K' of the manifold K.

If $T_i^r (i = 1, \ldots, k)$ is the set of all arbitrarily oriented r-dimensional simplices of the manifold K, any linear form $x^* = \Sigma_{i=1}^k \alpha_i B(T_i^r)$ with integral coefficients is an $(n-r)$-dimensional chain in K', called a *star chain* of the manifold K. Evidently, it can be expressed as

$$x^* = B(x) \tag{25}$$

where

$$x = \sum_{i=1}^{k} \alpha_i T_i^r.$$

Since the barycentric stars of two distinct simplices T_i^r and T_j^r cannot have a common $(n-r)$-dimensional simplex, the relation (25) enables us to determine x by x^* uniquely. In other words, barycentric stars of a manifold K are linearly independent chains in K'. The relation (24) shows that the boundary of a star chain is also a star chain. Below we shall demonstrate (see the lemma) that star chains can be used to calculate the homology in K'.

(D) Let K be a manifold, and K' its barycentric subdivision. Let $B[x]$ denote the carrier of the chain $B(x)$, i.e., the smallest subcomplex of the complex K' containing all the simplices in $B(x)$ with non-zero coefficients. From (B), (C) and (A) Section 1, it immediately follows that

$$B[T^r] \subset U(T^r). \tag{26}$$

If P is some subcomplex of the complex K and T^r does not belong to P, then $B[T^r]$ does not intersect P, because $U(T^r)$ does not intersect P.

Lemma. Let K be an n-dimensional orientable manifold, and K' its barycentric subdivision, let P be some subcomplex of the complex K and Q some subcomplex of the complex K' formed as the sum of all $B[T]$ (see (D)), where T is not contained in P. Evidently, $Q \subset K-P$. Let x be a general (possibly curvilinear) chain in $K-P$, whose boundary Δx is a star chain in Q. Then there exists in Q a star chain y such that $x-y$ is a cycle homologous to zero in $K-P$. For a non-orientable complex K, the lemma is to be understood modulo 2.

Thus, to calculate the homology of $(K-P)$, it is sufficient to consider the star cycles in Q and the star homology between them.

Proof. We shall construct the proof in several steps.

(a) Let T be an arbitrary simplex in K; it is natural to call the complex

$B[\nabla T]$ the edge of the star $B[T]$. We shall show that this edge is homeomorphic to a representative $P(T)$ of the simplex T in K (see (A) Section 1 and (D) Section 2). Call the center σ of the simplex T the center of the star $B[T]$. Let a be an arbitrary point in $P(T)$. Join the points σ and a by a linear segment (σ, a). Elementary geometrical considerations show that the segment (σ, a) intersects $B[\nabla T]$ at only one point $b = \varphi(a)$. Thus, we obtain a homeomorphic mapping φ of the complex $P(T)$ onto the complex $B[\nabla T]$. It is a simple matter to verify that φ is a combinatorial isomorphic mapping of the barycentric subdivision $P(T)'$ of the complex $P(T)$ onto the complex $B[\nabla T]$. Furthermore, we can easily verify that two segments (σ, b) and (σ, b'), where b and b' are distinct points in $B[\nabla T]$, intersect only at σ, and that the polyhedron $B[T]$ is covered by intervals of the type (σ, b). Thus, $B[T]$ is a cone with vertex at σ and base $B[\nabla T]$ homeomorphic to $P(T)$.

Now if F is a closed subset of the polyhedron $B[T]$ not containing the point σ, then by sliding each point $c \in F$ along the straight segment (σ, b) containing c, from c to b, we continuously deform the set F into the edge $B[\nabla T]$. In this deformation ψ, the points of F belonging to the edge $B[\nabla T]$ remain fixed.

(b) Let F be a closed subset (in K) of the domain $K - P$; then there exists a continuous deformation ψ of the set F, occurring inside K–P, which transforms F into Q, such that the points of the set F belonging to Q remain fixed.

Let T^0 be an arbitrary zero-dimensional simplex of P. By deforming, as in (a), that part of the set F which lies inside $B[T^0]$, we drive the set F out from the interior of the star $B[T^0]$. Applying this operation to all the zero-dimensional simplices of P, we obtain from F a new set F_1 which has no longer any points inside the star $B[T^0]$, for $T^0 \in P$. Now we can apply this sweeping operation to the stars $B[T^1]$, where T^1 is a one-dimensional simplex in P. By applying it to simplices of all dimensions, we arrive at the desired result.

(c) Let $B[T^r]$ be a barycentric star of some simplex $T^r \in K$ and $S^t = \epsilon(b_0, \ldots, b_t)$ an arbitrary oriented t-dimensional simplex in the edge $B[\nabla T^r]$. Let $\sigma(S^t)$ denote a $(t+1)$-dimensional simplex $\epsilon(\sigma, b_0, \ldots, b_t)$ (σ is the center of T^r) in $B[T^r]$. It is called a *cone* constructed on the base S^t. The operation σ of cone construction can be extended to any t-dimensional chain u in $B[\nabla T^r]$ by additivity. It is easy to see that

$$\Delta\sigma(u) = u - \sigma(\Delta u). \tag{27}$$

Obviously, every $(t+1)$-dimensional chain v in $B[T^r]$ formed from simplices with the vertex σ can be expressed in the form $v = \sigma(u)$. If v

has a boundary belonging to $B[\nabla T^r]$, then from (27) it follows that $\Delta u = 0$. Now if the dimension $(t+1)$ of the chain v is less than $n-r$, the dimension of u is less than $n-r-1$. Therefore, by virtue of (a), the cycle u is homologous to zero in $B[\nabla T^r]$, i.e., $u = \Delta w$, $w \subset B[\nabla T^r]$. Then, we have $\Delta\sigma(w) = w - v$, i.e., for the chain v, there exists a chain $w \subset B[\nabla T^r]$ such that $v - w \sim 0$ in $B[T^r]$. If the dimension of v is equal to $n-r$, the cycle u has a dimension equal to $n-r-1$, therefore $u = \alpha\sigma(B(\nabla T^r))$, where α is an integer. Hence,

$$v = \sigma(u) = \alpha\sigma(B(\nabla T^r)) = \alpha(-1)^{r+1}\sigma(\Delta B(T^r)) = \alpha(-1)^{r+1}B(T^r). \tag{28}$$

(d) Let $\bar{x}$ be an arbitrary chain in the complex Q, whose boundary $\Delta\bar{x}$ is a star chain; then there exists a star chain y in Q such that $\bar{x} - y \sim 0$ in Q.

Let T^0 be an arbitrary zero-dimensional simplex in K not belonging to P. Let v denote the part of the chain $\bar{x}$ that is formed from simplices having a vertex at the center σ of the simplex T^0. Then the chain v can be expressed as $\sigma(u)$, where u belongs to $B[\nabla T^0]$ (see (c)). Since the boundary of $\bar{x}$ is a star chain, v has a boundary lying on $B[\nabla T^0]$, and u is therefore a cycle (see (c)). Thus, we have $v = \beta B[T^0]$, if the dimension of the chain $\bar{x}$ is equal to n. If, however, the dimension of the chain $\bar{x}$ is less than n, a part v of the chain $\bar{x}$ can be replaced by a chain w lying on the edge $B[\nabla T^0]$, so that the new chain $\bar{x}_1 = \bar{x} - v + w$ is homologous to the chain $\bar{x}$ in Q, and no longer has any simplices inside $B[T^0]$. Applying this procedure to every zero-dimensional simplex T^0 not contained in P, we obtain from $\bar{x}$ a new chain having no simplices inside an n-dimensional star. Applying this procedure to one-dimensional simplices T^1 not belonging to P, we obtain from $\bar{x}$ a new chain whose simplices lie neither in an n-dimensional star nor in an $(n-1)$-dimensional star. After a finite number of steps of this kind, we obtain a star chain y.

(e) Now if x is a general (curvilinear) chain with a star boundary z in $K - P$, then deforming as in (b), we obtain a general chain in the polyhedron Q. This chain can be approximated by a chain $\bar{x}$ consisting of the simplices in the polyhedron Q, and then replaced by a star chain y as described in (d). In the course of all these reconstructions, the boundary z of the chain x remains unchanged.

This completes the proof of our lemma.

3. Intersection Indices and Linking Numbers

In almost all formulations of duality theorems, either intersection indices or linking numbers play an important part. This section concerns itself with these concepts.

For a long time now in homology theory, not only integers (as were used by Poincare), but also residues modulo 2 and residues modulo arbitrary m have been used as coefficients. Later, it was found that the elements of a quite arbitrary commutative group (in additive notation) can be, and should preferably be, taken as the coefficient realm.

(A) Let G be a commutative group in additive notation, where G is either a discrete group and has no more than a denumerable number of elements, or G is a topological compact group and satisfies the second axiom of countability.

In both cases, the group G can be used in constructing the homology theory. Let K be a complex and T_i^r $(i = 1, \ldots, \alpha^r)$ the set of all of its r-dimensional (arbitrarily) oriented simplices. Any linear form of the type $\mathbf{x} = \Sigma_{i=1}^{\alpha_r} g_i T_i^r$, where $g_i \in G$ $(i = 1, \ldots, \alpha^r)$ is called an r-dimensional chain in K over the coefficient realm G. The set of all such chains is denoted by L_G^r. Being a collection of linear forms, this set is naturally an additive group. Obviously, the group L_G^r is isomorphic to the direct sum of α^r copies of the group G. If G is a topological group, and the product topology is introduced in L_G^r, then L_G^r is a compact topological group satisfying the second axiom of countability. If the boundary ΔT_i^r of the simplex T_i^r is given by the relationship $\Delta T_i^r = \Sigma_{j=1}^{\alpha_{r-1}} \epsilon_{ij}^r T^{t-1}$, where $\epsilon_{ij}^r = 0, \pm 1$, then the boundary Δx of the chain $\mathbf{x}$ is given by the relationship

$$\Delta x = \sum_{i=1}^{\alpha^r} \sum_{j=1}^{\alpha^{r-1}} g_i \epsilon_{ij}^r T^{r-1} .$$

Evidently, $\Delta(x + y) = \Delta x + \Delta y$; thus Δ is a homomorphic mapping of the group L_G^r into the group L_G^{r-1}. If G is a topological group, this homomorphism is continuous. The kernel of the homomorphism Δ is denoted by Z_G^r, and the image of L_G^r into L_G^{r-1} by H_G^{r-1}. It is easy to verify that

$$H_G^r \subset Z_G^r. \tag{1}$$

The elements of the group Z_G^r are called *cycles*, and the elements of the group H_G^r, *cycles homologous to zero.* In the case of a topological group G, the groups Z_G^r and H_G^{r-1} are compact due to the continuity of the homomorphism Δ, and are therefore closed. The factor group $B_G^r = Z_G^r/H_G^r$ is called the *r-dimensional Betti group* of the complex K over the coefficient realm G.

If the complex K is of dimension n, the group H_G^n is not defined, so we take it to be zero. Under this supposition, the relationship (1) is satisfied. For the group L_G^0 the homomorphism Δ is not defined, because the group L_G^{-1} does not exist. Hence the group Z_G^0 is also not defined. This group is usually defined in two different ways: (1) $Z_G^0 = L_G^0$, or (2) the chain $x = \Sigma_{i=1}^{\alpha^0} g_i T_i^0$ is contained in Z_G^0, if and only if

$$\sum_{i=1}^{\alpha^0} g_i = 0. \qquad (2)$$

In order to make a distinction between the definitions (1) and (2), we denote the group given by the latter by $Z_G^{0'}$. For both definitions, the relationship (1) is satisfied. In the second case we denote the Betti group by $B_G^{0'}$. The condition (2) is meaningful only for zero-dimensional chains, because in this case all zero-dimensional simplices can be assigned a positive orientation, which we assume has been done. For a connected complex K, the group H_G^0 is the same as the group $Z_G^{0'}$. In general, the relationship between the groups B_G^0 and $B_G^{0'}$ is quite simple: B_G^0 is isomorphic to the direct sum of the groups $B_G^{0'}$ and G.

(B) Duality theorems are generally concerned with two groups, a discrete group G and its compact character group Γ; or (which comes to the same thing) the orthogonal pair of groups G, Γ (see Ref. [9], the definitions 1, 3 and Theorem 5), for which multiplication of $g \in G$ and $\gamma \in \Gamma$ is defined: $\gamma g = \gamma(g) \in \mathrm{K}$, where K is a topological group of real numbers reduced modulo one. If G is the group of integers, then Γ isomorphic to K; if G is a finite group, Γ is isomorphic to G. This is true, in particular, when G is the group of integer residues modulo m. Then Γ is also to be interpreted as a group of residues modulo m. We obtain the usual theory modulo m for these two groups G and Γ.

(C) Let K be a complex, G a discrete group, and Γ its character group (see (B)). Let

$$x^r = \sum_{i=1}^{\alpha^r} g_i T_i^r \in L_G^r, \quad \xi^r = \sum_{i=1}^{\alpha^r} \gamma_i T_i^r \in L_\Gamma^r.$$

The scalar product (ξ^r, x^r) of the chains ξ^r and x^r is defined to be

$$(\xi^r, x^r) = \sum_{i=1}^{\alpha^r} \gamma_i g_i \in K. \tag{3}$$

Then, we find that

$$(\Delta\xi^{r+1}, x^r) = (\xi^{r+1}, \nabla x^r) \tag{4}$$

(see (A), Section 2 and (A), Section 3). If K is an n-dimensional oriented manifold, then $\overset{*}{x}{}^{n-r} = B(x^r)$ is an $(n-r)$-dimensional star chain in K over the group G, is an arbitrary chain because x^r is arbitrary (see Definition 6). The intersection index of the chains ξ^r and $\overset{*}{x}{}^{n-r}$ is defined to be

$$I(\xi^r, \overset{*}{x}{}^{n-r}) = (\xi^r, x^r). \tag{5}$$

From (4) of Section 3 and (24) of Section 2, it follows that

$$I(\Delta\xi^{r+1}, \overset{*}{x}{}^{n-r}) = (-1)^{r+1} I(\xi^{r+1}, \Delta\overset{*}{x}{}^{n-r}). \tag{6}$$

We shall prove (4) only for chains with single terms $\xi^{r+1} = \gamma T_i^{r+1}$ and $x^r = gT_j^r$. It is then extended to general chains by additivity. We have

$$\Delta T_i^{r+1} = \sum_{k=1}^{\alpha^r} \epsilon_{ik}^{r+1} T_k^r, \quad \nabla T_j^r = \sum_{l=1}^{\alpha^{r-1}} \epsilon_{lj}^{r+1} T_l^{r+1}.$$

Hence

$$(\Delta\gamma T_i^{r+1}, gT_j^r) = \epsilon_{ij}^{r+1}\gamma g, \quad (\gamma T_i^{r+1}, \nabla g T_j^r) = \epsilon_{ij}^{r+1}\gamma g,$$

i.e., (4) has been demonstrated. And we obtain (6) from:

$$\begin{aligned} I(\Delta\xi^{r+1}, \overset{*}{x}{}^{n-r}) &= (\Delta\xi^{r+1}, x^r) = (\xi^{r+1}, \nabla x^r) = I(\xi^{r+1}, B(\nabla x^r)) \\ &= (-1)^{r+1} I(\xi^{r+1}, \Delta\overset{*}{x}{}^{n-r}). \end{aligned}$$

This completes the proofs of (4) and (6).

For a non-orientable manifold K, relationship (6) is true only modulo 2, i.e., G and Γ are assumed to be groups of residues modulo 2.

Prior to taking up the general definition of the intersection index, let us note that if K_1 and K_2 are homeomorphic manifolds with the homeomorphism φ, $\varphi(K_1) = K_2$, the orientation of K_1 induces a definite orientation of K_2. If $\vec{z}_1$ is a cycle defining the orientation in K_1, then $\varphi(\vec{z}_1)$ is a general (curvilinear) cycle in K_2 which, under simplicial approximation, is transformed into a cycle $\vec{z}_2$ defining the orientation of the manifold K_2. By virtue of this remark, we may speak of the orientation of a polyhedron R homeomorphic to the manifold K.

Definition 7. Let R be an oriented polyhedron homeomorphic to an n-dimensional manifold, G and Γ an orthogonal pair of groups (see (B)), ξ^r and x^{n-r} two general (curvilinear) chains of dimensions r and $n-r$, respectively, in R, where the coefficients of the former are taken from Γ and the coefficients of the latter are taken from G. The chains ξ^r and x^{n-r} are assumed to be such that one does not intersect the boundary of the other. Let ϵ denote the lesser of the two distances of one chaine from the boundary of the other. Since R is homeomorphic to an n-dimensional manifold K, it may be assumed that both ξ^r and x^{n-r} are located in K. Moreover, as K can be chosen arbitrarily at least up to subdivision, the simplices of K may be assumed to be small compared to ϵ, say, they do not exceed $\frac{1}{10}\epsilon$ in diameter. Approximating the chain ξ^r by a simplicial chain ξ_1^r in K, and the chain x^{n-r} by a star chain x_1^{n-r} in K (details are given below), we define the intersection index as follows:

$$I(\xi^r, x^{n-r}) = I(\xi_1^r, x_1^{n-r}). \tag{7}$$

It can be proved that the intersection index thus defined does not depend on the choice of the complex K, nor on the approximation method. The first assertion is rather a complicated theorem, which we leave without proof, while we demonstrate the second assertion later below. From (6), it follows that for general chains

$$I(\Delta\xi^{r+1}, x^{n-r}) = (-1)^{r+1} I(\xi^{r+1}, \Delta x^{n-r}). \tag{8}$$

The simplicial chain ξ_1^r used to approximate ξ^r is obtained by sufficiently subdividing the chain ξ^r and then simplicially approximating it by a chain ξ_1^r in K in the usual way. The star chain x_1^{n-r} approximating the chain x^{n-r} is derived in a more complicated manner. Let $\Delta x^{n-r} = z$. Let P_1 denote the set of all closed simplices in K that do not intersect z, and let P_2 denote the set of all closed simplices in K that do not intersect x^{n-r}. By virtue of the lemma stated in Section 2, there exists in $K - P_1$ a star cycle z_1 homologous to z in $K - P$:

$$z_1 - z = \Delta y, \quad y \subset K - P_1.$$

The chain $x^{n-r} + y$ in $K - P_2$ has the star boundary z_1, therefore there exists in $K - P_2$ a star chain x_1^{n-r} such that

$$x^{n-r} + y - x_1^{n-r} \sim 0 \text{ in } K - P_2.$$

The star chain x_1^{n-r} thus obtained is precisely the star approximation of the general chain x^{n-r}.

We now prove that, for a fixed K, the definition (7) of the intersection index is invariant with respect to the choice of the approximating chains ξ_1^r and x_1^{n-r}. Substitute ξ_1^n by some other approximating chain ξ_2^r. We pass from the chain x_1^{n-r} to the chain x_2^{n-r} in exactly the same way. As the boundaries $\Delta\xi_1^r$ and $\Delta\xi_2^r$ are homologous to the boundary $\Delta\xi^r$ in some neighbourhood of the latter, the cycles $\Delta\xi_1^r$ and $\Delta\xi_2^r$ are homologous to each other in some neighbourhood of $\Delta\xi^r$, $\Delta\xi_2^r - \Delta\xi_1^r = \Delta\eta$, where η is a chain in K located in a neighbourhood of the cycle $\Delta\xi^r$, the size of which depends on the size of the simplices of the complex K. Similarly, the cycle $\xi_1^r - \xi_2^r + \eta$ is homologous to zero in some neighbourhood of the chain ξ^r, $\xi_1^r - \xi_2^r + \eta = \Delta\zeta$, where ζ is a chain in K located in a neighbourhood of the chain ξ^r, whose size depends on the size of the simplices of the complex K. Since neither of the chains ξ^r and x^{n-r} intersects the boundary of the other, from what has been said regarding the locations of the chains η and ζ, we find that η does not intersect x_1^{n-r}, i.e., $I(\eta, x_1^{n-r}) = 0$. Similarly, the chains ζ and Δx_1^{n-r} do not intersect and therefore $I(\zeta, \Delta x_1^{n-r}) = 0$. Hence we have

$$\begin{aligned} I(\xi_1^r, x_1^{n-r}) - I(\xi_2^r, x_1^{n-r}) &= I(\xi_1^r - \xi_2^r + \eta, x_1^{n-r}) = I(\Delta\zeta, x_1^{n-r}) \\ &= (-1)^{r+1} I(\zeta, \Delta x_1^{n-r}) = 0. \end{aligned}$$

Thus, we have demonstrated that the intersection index is invariant with respect to the choice of the approximating chains ξ_1^r and x_1^{n-r}.

(D) If ξ^r and x^{n-r} are two cycles in an n-dimensional oriented manifold K, the intersection index $I(\xi^r, x^{n-r})$ is an invariant of the homology class to which the cycles ξ^r and x^{n-r} belong. If one of these cycles is homologous to zero, the intersection index is zero.

Let $\xi_1^r \sim \xi^r$ and $\xi_1^r - \xi^r = \Delta\eta$, then by virtue of (8)

$$I(\xi_1^r, x^{n-r}) - I(\xi^r, x^{n-r}) = I(\Delta\eta, x^{n-r}) = (-1)^{r+1} I(\eta, \Delta x^{n-r}) = 0.$$

Definition 8. Let ξ^r and x^{n-r-1} be two general non-intersecting cycles in an oriented n-dimensional manifold K, where the former is of dimension r and is taken over the coefficient realm Γ, and the latter is of dimension $n-r-1$ and is taken over the coefficient realm G (see (B)). Moreover, assume that each of the cycles ξ^r and x^{n-r-1} is homologous to zero in K. If $x^{n-r-1} = \Delta y^{n-r}$, the linking number $V(\xi^r, x^{n-r-1})$ of the cycles ξ^r and x^{n-r-1} is defined to be

$$V(\xi^r, x^{n-r-1}) = I(\xi^r, y^{n-r}). \tag{9}$$

It turns out that the linking number does not depend on the choice of the chain y^{n-r}. We prove this assertion.

Let $\xi^r = \Delta\eta^{r+1}$. Then, by virtue of (8), we have

$$V(\xi^r, x^{n-r-1}) = I(\xi^r, y^{n-r}) = I(\Delta\eta^{r+1}, y^{n-r})$$
$$= (-1)^{r+1} I(\eta^{r+1}, x^{n-r-1}).$$

The chain y^{n-r} is not contained in the last term of this equality, and therefore $V(\xi^r, x^{n-r-1})$ does not depend on this chain.

(E) Retaining the notation of Definition 8, assume that ξ_1^r is a cycle homologous to the cycle ξ^r in the complement of x^{n-r-1} and that x_1^{n-r-1} is a cycle homologous to the cycle x^{n-r-1} in the complement of ξ^r. Then

$$V(\xi_1^r, x^{n-r-1}) = V(\xi^r, x_1^{n-r-1}) = V(\xi^r, x^{n-r-1}). \tag{10}$$

Let $\xi_1^r - \xi^r = \Delta\zeta^{r+1}$, $\zeta^{r+1} \subset K - x^{n-r-1}$. By virtue of (8) and $x^{n-r-1} = \Delta y^{n-r}$, we have

$$V(\xi_1^r, x^{n-r-1}) - V(\xi^r, x^{n-r-1}) = I(\xi_1^r - \xi^r, y^{n-r}) = I(\Delta\zeta^{r+1}, y^{n-r})$$
$$= (-1)^{r+1} I(\zeta^{r+1}, x^{n-r-1}) = 0.$$

Let $x_1^{n-r-1} - x^{n-r-1} = \Delta z^{n-r}$, $z^{n-r} \subset K - \xi^r$, then we have $x_1^{n-r-1} = \Delta(y^{n-r} + z^{n-r})$. Therefore, by virtue of (9)

$$V(\xi^r, x_1^{n-r}) = I(\xi^r, y^{n-r} + z^{n-r}) = I(\xi^r, y^{n-r}) = V(\xi^r, x^{n-r-1}).$$

This completes the proof of Proposition (E).

4. Duality Theorems

In this section we formulate and prove a general duality theorem (Theorem 4) for a subcomplex P of a manifold K. The formulation of this theorem presupposes that the Poincare–Veblen theorem has already been established (Theorem 3). Alexander's duality theorem and the cycle removal theorem follow as corollaries from the general duality theorem and are given immediately after it.

In demonstrating the general duality theorem, it is convenient, though not necessary, to use ∇-homology. Therefore at the beginning of this section we give the definition of the Betti ∇-group (see (A)). Theorem 1 establishes a relationship between the Betti ∇-group and the Betti Δ-group of the same dimension of a complex. It also serves as an auxiliary tool in proving Theorem 4. Theorem 2 relates the Betti ∇-group and the Betti Δ-group of complementary dimension in a manifold. Theorems 1 and 2 taken together give the Poincare–Veblen duality theorem.

Using the ∇ operation (see Proposition (A), Section 2), we can construct a theory of ∇-homology in the same way that we did with the help of the Δ operation (see Proposition (A), Section 3).

(A) Let G be a commutative group (in additive notation) for which we assume: (a) either G is a discrete group and has no more than a countable number of elements, or (b) G is a compact topological group and satisfies the second axiom of countability. Furthermore, assume that K is a complex and $T_i^r (i = 1, \ldots, \alpha^r)$ is the set of all of its (arbitrarily) oriented r-dimensional simplexes. Let $\bar{L}_G^r = L_G^r$ denote the collection of all r-dimensional chains in K over the coefficient realm G, i.e., the set of all linear forms of the type

$$x = \sum_{i=1}^{\alpha^r} g_i T_i^r \quad \text{(see (A), Section 3).}$$

If

$$\nabla T_i^r = \sum_{j=1}^{\alpha^{r+1}} \epsilon_{ji}^{r+1} T_j^{r+1}, \tag{1}$$

the ∇-boundary of a chain $\mathbf{x}$ is defined as follows:

$$\nabla x = \sum_{i=1}^{\alpha^r} \sum_{j=1}^{\alpha^{r+1}} g_i \epsilon_{ji}^{r+1} T_j^{r+1}. \tag{2}$$

Thus, we have a homomorphic mapping ∇ of the group $\bar{L}_G^r$ into the group $\bar{L}_G^{r+1}$. We denote the kernel of this homomorphism ∇ by $\bar{Z}_G^r$, and the image of the group $\bar{L}_G^r$ in the group $\bar{L}_G^{r+1}$ by $\bar{H}_G^{r+1}$. We can easily verify that $\bar{H}_G^r \subset \bar{Z}_G^r$. Therefore it is possible to define the factor group $\bar{B}_G^r = \bar{Z}_G^r / \bar{H}_G^r$, which is called the r-dimensional Betti ∇-group of the complex K over the coefficient realm G. A chain in $\bar{Z}_G^r$ is called a *∇-cycle*, and a chain in $\bar{H}_G^r$, a *∇-cycle ∇-homologous to zero*. If G satisfies the conditions (b), all these groups also satisfy these conditions. Henceforth, we shall however only consider ∇-homology for a G satisfying the conditions (a).

Theorem 1 stated below establishes a relationship between the ∇-homology and Δ-homology of the same dimension of a complex K, where a discrete group G is used to constructing the ∇-homology and a compact topological group Γ in constructing the Δ-homology; G and Γ are orthogonal (see (E) Section 3). A similar theorem is valid when the groups G and Γ are interchanged, but this is true only if the complex K is finite.

Theorem 1. Let K be a complex, let G and Γ be an orthogonal pair of groups (see (B), Section 3). Let $L^r_\Gamma, Z^r_\Gamma, H^r_\Gamma$ denote the groups introduced in (A), Section 3, and $\bar{L}^r_G, \bar{Z}^r_G, \bar{H}^r_G$ the groups introduced in (A), Section 4. Let the multiplication of an element $\xi \in L^r_\Gamma$ by an element $x \in \bar{L}^r_G$ be defined as follows: $\xi x = (\xi, x)$ (see Proposition (C), Section 3). Then the groups L^r_Γ and $\bar{L}^r_G$ form an orthogonal pair (see Definition 3 in Ref. [9]). Furthermore, the annihilators obey the conditions

$$(\bar{L}^r_G, H^r_\Gamma) = \bar{Z}^r_G, \quad (L^r_\Gamma, \bar{H}^r_G) = Z^r_\Gamma. \tag{3}$$

It follows from these relations (see Theorems 2 and 4 in Ref. [9]) that

$$(\bar{L}^r_G, Z^r_\Gamma) = \bar{H}^r_G, \quad (L^r_\Gamma, \bar{Z}^r_G) = H^r_\Gamma, \tag{4}$$

from which we obtain

$$(\bar{Z}^r_G, Z^r_\Gamma) = \bar{H}^r_G, \quad (Z^r_\Gamma, \bar{Z}^r_G) = H^r_\Gamma. \tag{5}$$

This means that the factor groups $B^r_\Gamma = Z^r_\Gamma / H^r_\Gamma$ and $\bar{B}^r_G = \bar{Z}^r_G / \bar{H}^r_G$ form an orthogonal pair, i.e., each of them is the character group of the other (see Theorem 5 in Ref. [9]).

Proof. It is obvious that the multiplication law $\xi x = (\xi, x)$ is continuous and distributive. We shall show that $(\bar{L}^r_G, L^r_\Gamma) = (0)$. Let $x \neq 0$, then one of the coefficients in the linear form x is non-zero. Let $g_k \neq 0$. As G and Γ are orthogonal, there exists an element $y \in \Gamma$, such that $\gamma g_k \neq 0$, hence $(\gamma T^r_k, x) = \gamma g_k \neq 0$. Therefore, $(\bar{L}^r_G, L^r_\Gamma) = (0)$. In the same way we establish that $(L^r_\Gamma, \bar{L}^r_G) = (0)$. Thus, the groups $\bar{L}^r_G$ and L^r_Γ, by virtue of our multiplication operation, are orthogonal.

Let $x \in \bar{Z}^r_G$ and $\xi \in H^r_\Gamma$, i.e., $\nabla x = 0$, $\xi = \Delta\eta$; then $(\xi, x) = (\Delta\eta, x) = (\eta, \nabla x) = 0$ (see (4), Section 3), and, consequently, $\bar{Z}^r_G \subset (\bar{L}^r_G, H^r_\Gamma)$. Assume that $x \in \bar{L}^r_G$, but does not belong to $\bar{Z}^r_G$, i.e., $\nabla x \neq 0$. As $\bar{L}^{r+1}_G$ and $\bar{L}^{r+1}_\Gamma$ are orthogonal there exists an element $\eta \in L^{r+1}$ such that $(\eta, \nabla x) \neq 0$. By virtue of (4) Section 3, we have $(\xi, x) = (\Delta\eta, x) = (\eta, \nabla x) \neq 0$ for $\xi = \Delta\eta \in H^r_\Gamma$. Therefore $(\bar{L}^r_G, H^r_\Gamma) = \bar{Z}^r_G$. The second equality in (3) is proved in a similar manner.

The relations (4) follow from (3) (see Theorems 2 and 4 in Ref. [9]), from (4) we obtain the relations (5), and consequently, the orthogonality of the groups $\bar{B}^r_G$ and B^r_Γ. This completes the proof of Theorem 1.

If K is a manifold, Theorem 2 gives another relationship between the ∇-homology and the Δ-homology in K.

Theorem 2. Let K be an n-dimensional oriented manifold, $\bar{B}^r_G$ the

r-dimensional Betti ∇-group of the complex K and B_G^{n-r} the $(n-r)$-dimensional Betti Δ-group of the complex K. Then, the groups $\bar{B}_G^r$ and B_G^{n-r} are isomorphic. In more detail, let $X \in \bar{B}_G^r$ and let x be some ∇-cycle in the homology class X; then $\overset{*}{x} = B(x)$ (see Definition 6) is an $(n-r)$-dimensional star Δ-cycle and belongs to some homology class $\overset{*}{X} \in B_G^{n-r}$, which depends only on the element X, and does not depend on the choice of the ∇-cycle $x \in X$, so that we can put $\overset{*}{X} = \varphi(X)$. The mapping φ of the group $\bar{B}_G^r$ onto the group B_G^{n-r} is isomorphic. For a non-orientable manifold K, this theorem is true modulo 2, i.e., if G is a cyclic group of order two.

Proof. Let $\bar{L}_G^r, \bar{Z}_G^r, \bar{H}_G^r$ be the groups defined in (A). Let L_G^{n-r} denote the group of all $(n-r)$-dimensional star chains in K over the field G. Let Z_G^{n-r} denote the subgroup of the group L_G^{n-r} that consists of all cycles, and H_G^{n-r} the subgroup of the group Z_G^{n-r} that consists of all cycles homologous to zero. By virtue of the lemma stated in Section 2 (the empty complex is to be taken for P) the group B_G^{n-r} is isomorphic to the group Z_G^{n-r}/H_G^{n-r}. Let $x \in \bar{L}_G^r$, then $B(x) = \overset{*}{x}$ is an $(n-r)$-dimensional star chain over the realm G, where the relationship $\overset{*}{x} = B(x)$ uniquely defines x by $\overset{*}{x}$ (see Definition 6). Thus, B is an isomorphic mapping of the group $\bar{L}_G^r$ into the group L_G^{n-r}. By virtue of (24) Section 2, we have $B(\bar{Z}_G^r) = Z_G^{n-r}$, $B(\bar{H}_G^r) = H_G^{n-r}$. Hence it immediately follows that $\bar{Z}_G^r/\bar{H}_G^r = \bar{B}_G^r$ and $Z_G^{n-r}/H_G^{n-r} = B_G^{n-r}$ are isomorphic.

This completes the proof of Theorem 2.

Combining Theorems 1 and 2 we obtain the Poincare–Veblen duality theorem:

Theorem 3. Let K be an n-dimensional oriented manifold, and G and Γ an orthogonal pair of groups (see (B), Section 3). Let B_Γ^r denote the r-dimensional Betti group of the manifold K over the field Γ, and B_G^{n-r} the $(n-r)$-dimensional Betti group of the manifold K over the field G. Define multiplication of an element $\Xi \in B_\Gamma^r$ and $\overset{*}{X} \in B_G^{n-r}$ as follows: $\Xi\overset{*}{X} = I(\xi, \overset{*}{x})$ (see Definition 7), where $\xi \in \Xi$, $\overset{*}{x} \in \overset{*}{X}$ (see Proposition (D), Section 3). Then, with this definition of multiplication, the groups B_Γ^r and B_G^{n-r} form an orthogonal pair (see Definition 3 in [9]).

Proof. Let $\xi \in Z_\Gamma^r, x \in \bar{Z}_G^r$ (see Theorem 1). Let Ξ denote that Δ-homology class to which ξ belongs, and X that ∇-homology class to which x belongs. Hence $\xi \in \Xi \in B_\Gamma^r, x \in X \in \bar{B}_G^r$.

Take $\Xi X = \xi x = (\xi, x)$. By virtue of the relations (5), this multiplication operation defined on the elements Ξ and X does not depend on the choice of ξ and x. Theorem 1 asserts that this multiplication turns the groups B^r and $\bar{B}^r_G$ into an orthogonal pair. Now put $\overset{*}{x} = B(x)$ and let $\overset{*}{X}$ denote that homology class to which $\overset{*}{x}$ belongs: $\overset{*}{x} \in \overset{*}{X} \in B^{n-r}_G$. We have $\overset{*}{X} = \varphi(X)$, and φ is an isomorphic mapping of the group $\bar{B}^r_G$ onto the group B^{n-r}_G (see Theorem 2). Furthermore, $\Xi\overset{*}{X} = I(\xi, \overset{*}{x}) = (\xi, x) = \Xi X$. Thus, the orthogonality of the groups B^r_Γ and $\bar{B}^r_G$, under the multiplication operation defined in Theorem 1, implies the orthogonality of the groups B^r_Γ and B^{n-r}_G under the multiplication operation defined in Theorem 3.

Thus, Theorem 3 has been demonstrated.

(B) Henceforth we deal with different complexes simultaneously. Therefore, to avoid confusion, we shall indicate the corresponding complex in the notation of the groups introduced in (A), Section 3 and in (A), Section 4. Thus, $B^r_\Gamma(K)$ stands for the Betti group of the complex K and $B^r_\Gamma(P)$, the Betti group of the complex P. If P is a subcomplex of the complex K, each cycle in P is a cycle in K; cycles homologous in P are also homologous in K. Thus, if $\Xi \in B^r_\Gamma(P)$, there exists an element $H \in B^r_\Gamma(K)$, such that $\Xi \subset H$. The correspondence $\Xi \to H$ is clearly a homomorphism; we shall call it the *natural homomorphism* of the group $B^r_\Gamma(P)$ into the group $B^r_\Gamma(K)$. There is no analogue for ∇-homology.

The theorem ennunciated below is a general duality law for a complex in an orientable manifold. For a non-orientable manifold this theorem is true modulo 2.

Theorem 4. Let K be an n-dimensional oriented manifold, P a subcomplex of the complex K, G and Γ an orthogonal pair of groups (see (B), Section 3). As $P \subset K$, there exists a natural homomorphism (see (A)) of the group $B^r_\Gamma(P)$ into the group $B^r_\Gamma(K)$; let $A^r_\Gamma(P)$ denote the kernel of this homomorphism and $C^r_\Gamma(P)$ the image of the group $B^r_\Gamma(P)$ in the group $B^r_\Gamma(K)$. Similarly, as $K-P \subset K$, there exists a natural homomorphism of the group $B^s_G(K-P)$ into the group $B^s_G(K)$. Let $A^s_G(K-P)$ denote the kernel of this homomorphism, and $C^s_G(K-P)$ the image of the group $B^s_G(K-P)$ in the group $B^s_G(K)$. Then the following relations hold.

(1) By virtue of Theorem 3, the groups $B^r_\Gamma(K)$ and $B^{n-r}_G(K)$ form an orthogonal pair; and by the multiplication operation defined, the subgroups $C^r_\Gamma(P) \subset B^r_\Gamma(K)$ and $C^{n-r}_G(K-P) \subset B^{n-r}_G(K)$ are mutual annihilators (see Definition 3 in Ref. [9], i.e.,

$$(B^{n-r}_G(K),\ C^r_\Gamma(P)) = C^{n-r}_G(K-P),\ (B^r_\Gamma(K),\ C^{n-r}_G(K-P)) = C^r_\Gamma(P). \quad (6)$$

(2) Let $\Xi \in A^r_\Gamma(P)$ and $\overset{*}{U} \in A^{n-r-1}_G(K-P)$. Define a multiplication operation on the elements Ξ and $\overset{*}{U}$ as follows: $\Xi\overset{*}{U} = V(\xi, \overset{*}{u})$, where $\xi \in \Xi$, $\overset{*}{u} \in \overset{*}{U}$ (see Definition 8 and (E), Section 3). With this definition of multiplication, the groups $A^r_\Gamma(P)$ and $A^{n-r-1}_G(K-P)$ form an orthogonal pair.

For a non-orientable manifold K, this theorem is true modulo 2, i.e., assuming the groups G and Γ are second order cyclic groups.

Proof. If x is an arbitrary chain in K, it can be uniquely expressed as the sum

$$x = x' + x'', \tag{7}$$

where x' is a linear form in the simplices belonging to P and x'' a linear form in the simplices that do not belong to P.

If y is an arbitrary chain in P, then, by virtue of (7), the chain ∇y in K can be expressed as the sum:

$$(\nabla y) = \nabla_p y + \bar{\bar{\nabla}} y, \quad (\nabla y)' = \nabla_p y, \quad (\nabla y)'' = \bar{\bar{\nabla}} y. \tag{8}$$

Here $\nabla_p y$ is the ∇-boundary of the chain y in P, and $\bar{\bar{\nabla}} y$ is a chain which has no simplices in P.

If x is some chain in K, the star chain $B(x)$ does not intersect P, if and only if $x' = 0$, i.e., if and only if the chain x consists of simplices that do not belong to P (see (D), Section 2). Thus,

$$B(x) \subset K - P \text{ is equivalent to } x' = 0. \tag{9}$$

We now proceed to prove assertion (1) of Theorem 4. We shall only demonstrate the first relation in (6); the second follows from the first by purely algebraic considerations (see Theorem 4 in Ref. [9]).

Let $\Xi \in C^r_\Gamma(P)$, $\overset{*}{X} \in C^{n-r}_G(K-P)$. This means that the homology class Ξ contains a cycle ξ, belonging to P, and the homology class $\overset{*}{X}$ contains a cycle $\overset{*}{x}$ belonging to $K-P$; thus, we have $\Xi\overset{*}{X} = I(\xi, \overset{*}{x}) = 0$. This means that

$$C^{n-r}_G(K-P) \subset (B^{n-r}_G(K), C^r_\Gamma(P)). \tag{10}$$

Now let $\overset{*}{X} \in (B^{n-r}_G(K), C^r_\Gamma(P))$, i.e., for any $\Xi \in C^r_\Gamma(P)$, we have $\Xi\overset{*}{X} = 0$. If now $\xi \in \Xi$, $\overset{*}{x} \in \overset{*}{X}$, from the last equality we deduce $I(\xi, \overset{*}{x}) = 0$. Since Ξ is an arbitrary element in $C^r_\Gamma(P)$, we can consider ξ to be an arbitrary cycle in P, i.e.,

$$I(\xi, \overset{*}{x}) = 0 \text{ for } \xi \in Z^r_\Gamma(P). \tag{11}$$

By virtue of the lemma of Section 2, we may assume that $\overset{*}{x}$ is a star cycle and therefore $\overset{*}{x} = B(x)$, where x is a chain in K. The relationship (11) gives

$$I(\xi, \overset{*}{x}) = (\xi, x) = (\xi, x' + x'') = (\xi, x') = 0. \tag{12}$$

Since the relation is satisfied for any $\xi \in Z^r_\Gamma(P)$, we find, by virtue of (4), that $x' \in H^r_G(P)$, i.e., $x' = \nabla_P y$, $y \in \bar{L}^{r-1}_G(P)$. Now consider the chain $x - \nabla y$. We have

$$(x - \nabla y)' = x' - \nabla_P y = 0.$$

This implies (see (9)) that $B(x - \nabla y) = \overset{*}{z}$ does not intersect P. Furthermore, we have $\overset{*}{z} = \overset{*}{x} - (-1)^r \Delta B(\overset{*}{y})$ (see (24) in Section 2); this means that $\overset{*}{z} \sim \overset{*}{x}$ in K and therefore $\overset{*}{z} \in \overset{*}{X}$. Thus, in $\overset{*}{X}$ there exists a cycle $\overset{*}{z}$ that belongs to $K - P$, i.e., $\overset{*}{X} \in C^{n-r}_G(K - P)$.

This completes the proof of assertion (1) of Theorem 4.

We now take up the proof of assertion (2) of Theorem 4.

Let $0 \neq \Xi \in A^r_\Gamma(P)$. We show that there exists an $\overset{*}{U} \in A^{n-r-1}_G(K - P)$ such that $\Xi\overset{*}{U} \neq 0$. For this purpose it is sufficient to show that if $\xi \in \Xi$, there exists in $K - P$ a cycle $\overset{*}{u}$ such that $\overset{*}{u} \sim 0$ in K and $V(\xi, \overset{*}{u}) \neq 0$.

Since $\Xi \neq 0$, we have $\xi \in Z^r_\Gamma(P)$, but ξ is not contained in $H^r_\Gamma(P)$. By virtue of (5), there exists an $x \in \bar{Z}^r_G(P)$ such that $(\xi, x) \neq 0$ (see (5)). Since $x \in \bar{Z}^r_G(P)$, we have $\nabla_P x = 0$, and therefore $\overset{*}{u} = (-1)^{r+1} B(\nabla x) = \Delta B(x) = \Delta \overset{*}{x} \subset K - P$, $\overset{*}{x} = B(x)$ (see (24) in Section 2). On the other hand, $(\xi, x) \neq 0$ implies $V(\xi, \overset{*}{u}) = I(\xi, \overset{*}{x}) = (\xi, x) \neq 0$. Hence

$$(A^r_\Gamma(P), A^{n-r-1}_G(K - P)) = (0). \tag{13}$$

It now remains to show (see Definition 3 in Ref. [9]), that

$$(A^{n-r-1}_G(K - P), A^r_\Gamma(P)) = (0). \tag{14}$$

Let $\overset{*}{U}$ be an element in $A^{n-r-1}_G(K - P)$ such that $\Xi\overset{*}{U} = 0$ for every $\Xi \in A^r_\Gamma(P)$. We now show that $\overset{*}{U} = 0$, from which will immediately follow the truth of (14).

Let $\overset{*}{u}$ be some cycle in $\overset{*}{U}$; by virtue of the lemma stated in Section 2, we may assume that $\overset{*}{u}$ is a star cycle that does not intersect P. Since $\overset{*}{U} \in A^{n-r-1}_G(K - P)$, we have $\overset{*}{u} \sim 0$ in K. Therefore there exists a star chain $\overset{*}{v}$ for which $\overset{*}{u} = \Delta \overset{*}{v}$.

Let $\overset{*}{v} = B(v)$, then the relation $\alpha(\xi) = (\xi, v)$, $\xi \in Z^r(P)$ defines a character α of the group $Z^r_\Gamma(P)$. In this group let us take the subgroup $K^r_\Gamma(P)$ consisting of all cycles that are homologous to zero in K. If $\xi \in K^r_\Gamma(P)$, then $\alpha(\xi) = (\xi, v) = I(\xi, \overset{*}{v}) = V(\xi, \overset{*}{u}) = \Xi\overset{*}{U} = 0$. This means that the character α vanishes on the subgroup $K^r_\Gamma(P)$, and is therefore defined on the factor group $Z^r_\Gamma(P)/K^r_\Gamma(P)$. This factor group is evidently the same as the group $C^r_\Gamma(P)$. Thus, the character α is defined on the subgroup $C^r_\Gamma(P)$ of the group $B^r_\Gamma(K)$ and can therefore be extended to the whole group $B^r_\Gamma(K)$ (see Ref. [9]). Consequently, there exists an element $\overset{*}{W} \in B^{n-r}_G(K)$ such that $\alpha(\Xi) = \Xi\overset{*}{W}$ for any $\Xi \in B^r(K)$ (see Theorem 3). If $\overset{*}{w}$ is some star cycle in $\overset{*}{W}$, then we obtain $I(\xi, \overset{*}{v}) = \alpha(\xi) = I(\xi, \overset{*}{w})$ for any $\xi \in Z^r_\Gamma(P)$. Thus, for $\overset{*}{x} = \overset{*}{v} - \overset{*}{w}, \overset{*}{x} = B(x)$, we have $\Delta\overset{*}{x} = \overset{*}{u}$ and

$$(\xi, x) = I(\xi, \overset{*}{x}) = 0, \tag{15}$$

where ξ is an arbitrary element in $Z^r_\Gamma(P)$.

From this point on we argue as we did in demonstrating assertion (1). From (15), we obtain

$$(\xi, x) = (\xi, x' + x'') = (\xi, x') = 0.$$

By virtue of (4), the last equality shows that $x' \in \bar{H}^r_\Gamma(P)$, and therefore $x' = \nabla_p y, y \in \bar{L}^{r-1}_G(P)$. Furthermore,

$$z = x - \nabla y \quad \text{and} \quad z' = x' - \nabla_p y = 0,$$

i.e., $\overset{*}{z} = B(z)$ does not intersect P. Furthermore we have

$$\overset{*}{z} = \overset{*}{x} - (-1)^r \Delta B(y), \text{consequently, } \Delta\overset{*}{z} = \Delta\overset{*}{x} = \overset{*}{u}.$$

This completes the proof of assertion 2).

In the particular case where K is homeomorphic to an n-dimensional sphere or more generally, if K has the same homology groups as an n-dimensional sphere, we obtain the following corollary of theorem 4, known as Alexander's law of duality.

Alexander's Duality Theorem. Let K be an n-dimensional Poincare space, i.e., a manifold with the same homology groups as an n-dimensional sphere, P a subcomplex of the complex K, G and Γ an orthogonal pair of groups (see (B), Section 3). Define a multiplication operation of the elements $\xi \in B^r_\Gamma(P)$ and $\overset{*}{U} \in B^{n-r-1}_G(K-P)$ as follows: $\Xi\overset{*}{U} = V(\xi, \overset{*}{u})$, where $\xi \in \Xi$, $\overset{*}{u} \in \overset{*}{U}$ (see Definition 8). By virtue of this definition of multiplication, the groups $B^r_\Gamma(P)$ and $B^{n-r-1}_G(K-P)$ form an orthogonal pair, i.e., each is a character group of the other. (By zero-dimensional Betti groups of the complex P and the domain $K-P$, we mean the groups $B^{0'}_\Gamma(P)$ and $B^{0'}_G(K-P)$, respectively (see (A), Section 3).

Proof. This theorem follows as a corollary directly from Theorem 4.

The assertion given below, known also as the cycle removal theorem, is equivalent to assertion (1) of Theorem 4.

Cycle Removal Theorem. Let K be an n-dimensional oriented manifold, P a subcomplex of the complex K, G and Γ an orthogonal pair of groups (see (B), Section 3). Let $\overset{*}{x}$ be a cycle in K over the coefficient realm G such that for any cycle ξ in P over the coefficient realm Γ, we have $I(\xi, \overset{*}{x}) = 0$. Then there exists a cycle $\overset{*}{z}$ in $K-P$ homologous to $\overset{*}{x}$. For a non-orientable manifold K, this theorem is true modulo 2.

Proof. The homology class $\overset{*}{X}$ containing the cycle $\overset{*}{x}$ evidently belongs to $(B_G^{n-r}(K), C_\Gamma^r(P))$ (see Theorem 4). Therefore, by virtue of Theorem 4, it is contained in $C_G^{n-r}(K-P)$. This means that there exists in $\overset{*}{X}$ a cycle $\overset{*}{z}$ belonging to $K-P$.

References

1. Poincare, Analysis Situs, *J. Ec. Pol.* (1985).
2. Veblen, *Trans. Amer. Mathem. Soc.* **25**, 540 (1923).
3. Alexander, A proof and extension of the Jordan–Brouwer theorem, *Trans. Amer. Mathem. Soc.* **23**, 333–339 (1922).
4. Pontrjagin, Zum Alekanderschen Dualitätssatz, *Gött. Nachr. Math. Phys. Kl.* **25**, November 1927.
5. Alexandroff, *Gött. Nachr. Math. Phys. Kl.* (25 November 1927).
6. Frankl, *Wien. Ber., Dez.* 689 (1927).
7. Pontrjagin, Uber den algebraischen Inhalt topologischer Dualitätssätz, *Math. Ann.* **105** (1931).
8. Pontrjagin, The general topological theorem of duality for closed sets, *Ann. of Math.* **35** (1934).
9. Pontrjagin, The theory of topological commutative groups, *Ann. of Math.* **35** (1934).
10. Kolmgorov, Uber die Dualität im Aufbau der Kombinatorischen Topologie, *Matem. Sborn.* **1** (1936).
11. Alexander, On the connectivity ring of an abstract space, *Ann. of Math.* **37** (1936).
12. Alexandroff, On the homological properties of location of complexes and closed sets, *Izv. AN SSSR*, ser Matem. **6** (1942).
13. Seifert and Threlfall, *Topology*, Moscow–Leningrad (1928) (Russian translation).

Characteristic Cycles of Differentiable Manifolds

Matematicheskii Sbornik, Novaya Seriya (1947)
21(2): 233–84

This paper presents a full exposition of some of my results published in *Doklady AN SSSR* [1].

In geometry an important role is played by the spherical mapping method, which consists in the following.

Let M^k be a differentiable oriented manifold of dimension k with a continuous tangent in a Euclidean space R^{k+1} of dimension $k + 1$. At the point $x \in M^k$ draw a unit normal N_x to M^k consistent with the orientation of M^k. Translate the normal N_x in R^{k+1}, so that its origin coincides with a fixed point 0 in the space R^{k+1}. Then the end of the normal will be at a point $N(x)$ on the unit sphere S^k with center at 0.

Thus, we obtain a spherical mapping N of the manifold M^k which puts each point $x \in M^k$ into correspondence with a point $N(x) \in S^k$.

A study of the mapping N reveals certain invariants of the manifold M^k, both differential geometric and topological. In particular, if the manifold M^k is closed, and k is even, the degree of the mapping N is a topological invariant equal to one half of the Euler characteristic [2] of the manifold M^k.

Not every manifold M^k of dimension k can be properly embedded in a $(k + 1)$-dimensional Euclidean space. Hence, the idea naturally arises, to give a construction similar to the spherical mapping of a manifold M^k in a Euclidean space R^{k+l} of dimension $(k + l)$, where l is an arbitrary integer.

This paper is devoted to construction of certain invariants of closed differentiable oriented manifolds on the basis of such a mapping. Whether these invariants are new, or whether they can be calculated in terms of already known invariants, say, the intersection ring, are unresolved questions. Undoubtedly these invariants are closely related to many geometrical problems; they are in particular related to the classification of the mappings of a sphere of large dimension into a sphere of small dimension.

The invariants which I construct here also include the Stiefel cycles [3]. As we shall deal with these cycles, this paper touches on Whitney's paper [4] which gives a detailed account of Stiefel's cycles. Stiefel's cycles constitute, however, only a part of the whole range of invariants discussed here. The relationship with Stiefel's cycles and other similar invariants will be dealt with in [5], to be published later.

The subject matter is treated along the following lines.

Two differentiable manifolds are said to be *equivalent* or simply *homeomorphic*, if there exists a homeomorphic continuously differentiable mapping of one onto the other, with nowhere degenerating Jacobian. The invariants of a differentiable manifold discussed here are invariants precisely from the point of view of this equivalence.

Henceforth the dimension of a set P is always denoted by $D(P)$.

Let R^{k+l} be a Euclidean space of dimension $k+l$, and 0 some fixed point in this space.

Let $H(k,l)$ denote the manifold of all oriented k-dimensional linear subspaces of the space R^{k+l} containing the fixed point 0.

The dimension of the manifold $H(k,l)$ is given by the following relationship:

$$D(H(k,l)) = k\cdot l. \tag{1}$$

Evidently, if $R^{k+l'} \subset R^{k+l}$, then

$$H(k,l') \subset H(k,l). \tag{2}$$

This inclusion defines the relationship between various manifolds $H(k,l)$ for a fixed k.

Now let M^k be an abstractly given oriented differentiable manifold, and f homeomorphic mapping of M^k onto the manifold $f(M^k) \subset R^{k+l}$ with a continuous tangent.

At a point $f(x) \in f(M^k)$, draw an oriented linear tangent space T_x of dimension k to $f(M^k)$ and let $T(x)$ denote an element of the manifold $H(k,l)$ that is parallel to T_x.

In this way we obtain a mapping T of the manifold M^k into the manifold $H(k,l)$ that puts a point $x \in M^k$ into correspondence with the point $T(x) \in H(k,l)$. This mapping T is called the *tangential mapping* of the manifold M^k; it depends on the embedding f of the manifold M^k into the space R^{k+l} and plays the role of the spherical mapping N.

If f_0 and f_1 are two different embeddings of the manifold M^k into R^{k+l}, there exist two different tangential mappings T_0 and T_1. It is easy to show that for $l \geqslant k+1$, the mappings T_0 and T_1 are always homotopic in $H(k, l)$.

Thus, up to a homotopy, a tangential mapping does not depend on the embedding of the manifold M^k into the Euclidean space R^{k+l}. By virtue of the inclusion relation (2), the tangential mapping also does not depend on the number l.

In this paper we shall construct a complete system of homological invariants of a tangential mapping of a closed manifold M^k into $H(k, l)$. These invariants are constructed as follows.

Let Z be some $(kl-r)$-dimensional cycle of the manifold $H(k, l)$, where $r \leqslant k$ (see [1]). As the dimension of the manifold $H(k, l)$ is large compared with the dimension of the manifold M^k, there exists a homeomorphic mapping T_1 of the manifold M^k into $H(k, l)$ close to the tangential mapping T such that the cycle Z and the manifold $T_1(M^k)$ are in general position in $H(k, l)$.

The algebraic intersection [6] $Z \times T_1(M^k)$ of the cycle Z with the manifold $T_1(M^k)$ is a cycle of dimension $k-r$ in $T_1(M^k)$. Let X denote the image of this cycle in M^k under the mapping T_1^{-1}. Since the cycle X does not depend, up to homology, on the approximation T_1 of the mapping T, we write

$$X \sim T^{-1}\,(Z \times T(M^k)), \tag{3}$$

although this expression is devoid of immediate meaning.

Since the tangential mapping T is uniquely determined, up to homology, by the manifold M^k itself, the homology class of the cycle X depends only on the homology class of the cycle Z and on the manifold M^k. It is to be assumed that the orientation of the manifold $H(k, l)$ is already given.

By choosing the cycle Z in different ways, we obtain a complete system of homological invariants of the tangential mapping T. By virtue of the inclusion (2), we avoid dependence on the number l as well.

In order to decide which of the cycles Z are to be taken in the manifold $H(k, l)$, we have to study the homology of dimension $kl-r$, $(r \leqslant k)$ of the manifold $H(k, l)$. This study is an important point in this paper and is carried out using Ehresmann's method [7]. Since it is not possible to quote Ehresmann's final results here we give an independent presentation.

A sufficient supply of cycles is obtained as follows.

Let ω be a monotone non-decreasing integral-valued function of an integral valued argument $i = 1, 2, \ldots, k$ satisfying the condition

$$0 \leqslant \omega(i) \leqslant l. \tag{4}$$

In the space R^{k+l} consider the sequence

$$R_1 \subset R_2 \subset \ldots \subset R_k \tag{5}$$

of linear subspaces, containing 0, of dimension satisfying the condition

$$D(R_i) = \omega(i) + i, \quad i = 1, 2, \ldots, k. \tag{6}$$

Let $Z(\omega)$ denote the set of all elements $R^k \in H(k, l)$ satisfying the condition:

$$D(R^k \times R_i) \geqslant i, \quad i = 1, 2, \ldots, k. \tag{7}$$

We find that $Z(\omega)$ is a closed pseudo-manifold whose dimension is

$$D(Z(\omega)) = r(\omega) = \sum_{i=1}^{k} \omega(i). \tag{8}$$

In case $Z(\omega)$ is orientable, when taken with a definite orientation, it can be regarded an integral cycle in $H(k, l)$.

If $Z(\omega)$ is non-orientable, it can be regarded a cycle modulo 2. In this case an integral cycle can be extracted from it by the following general technique. Let Y be a cycle modulo 2. Taking all simplices of the cycle Y with a definite orientation, we obtain a chain Y' with integral coefficients, for which

$$\Delta Y' = 2\Gamma Y, \tag{9}$$

and ΓY is a cycle with integral coefficients, whose homology class is uniquely determined by the homology class of the cycle Y.

In place of the function ω, it is more convenient to introduce a function χ defined as follows:

$$\omega + \chi \equiv l, \quad \text{i.e. } \omega(i) + \chi(i) = l, \quad i = 1, 2, \ldots, k. \tag{10}$$

χ is a monotone non-increasing integral-valued function of an integral-valued argument i satisfying the condition

$$l \geqslant \chi(i) \geqslant 0. \tag{11}$$

Using (10), from χ we can always derive a function ω satisfying the conditions (4).

The pseudo-manifold $Z(\omega)$, and the corresponding cycle, will be denoted by Z_χ; its dimension, by virtue of (8) and (10), is

$$D(Z_\chi) = kl - r(\chi), \quad r(\chi) = \sum_{i=1}^{k} \chi(i). \tag{12}$$

We now strengthen the condition (11) imposed on the function χ; we assume that

$$l-1 \geqslant \chi(i) \geqslant 0. \tag{13}$$

When this condition is satisfied, the orientability of the pseudo-manifold Z_χ depends only on the properties of the function χ and does not depend on the number l.

If, in (3), in place of an arbitrary cycle Z, we put the cycle Z_χ, then we come to the following:

Definition. Let M^k be some oriented differentiable closed manifold of dimension k, T some tangential mapping of M^k into $H(k, l)$, and χ a function satisfying the conditions (13) and the condition $r(\chi) \leqslant k$. Put

$$X_\chi(M^k) = X_\chi \sim T^{-1}(Z_\chi \times T(M^k)), \quad (D(X_\chi) = k - r(\chi)). \tag{14}$$

If the pseudo-manifold Z_χ is non-orientable, the intersection is taken modulo 2, then X_χ is a cycle modulo 2.

If the pseudo-manifold Z_χ is orientable, and certain definite orientations are assigned to Z_χ and to the manifold $H(k, l)$, then X_χ is a cycle with integral coefficients.

It turns out that the homology class of the cycle X_χ in the manifold M^k is uniquely determined by the oriented differentiable manifold M^k and the function χ; the cycle X_χ is called a *characteristic cycle of the type* χ of the oriented manifold M^k.

If Z_χ is a non-orientable cycle, the homology class of the cycle

$$\Gamma X_\chi \sim T^{-1}(\Gamma Z_\chi \times T(M^k)) \quad (D(\Gamma X_\chi) = k - r(\chi) - 1); \tag{15}$$

is also defined and it is called a *secondary characteristic cycle of the type* χ of the oriented manifold M^k.

If $r(\chi) = k$, then $D(X_\chi) = 0$ and the homology class of the characteristic cycle X_χ is related to a unique integer or residue modulo 2, depending on whether the pseudo-manifold Z_χ is orientable or not. In this case we use the symbol X_χ to denote not only the zero-dimensional cycle, but also the corresponding integer or residue. The characteristic number or residue X_χ is an invariant of the oriented manifold M^k.

To study the characteristic cycles in greater detail, let us examine some properties of the function χ.

A jump point of the function χ is a value i of its argument where $\chi(i+1) < \chi(i)$. Let $i_1, i_2, \ldots, i_{n-1}$ be the set of all jump points of the function χ written in increasing order of magnitude. Moreover, assume that $i_0 = 0, i_n = k$, and

$$\left.\begin{aligned} \alpha_h &= i_h - i_{h-1}, \quad h = 1, 2, \ldots, n; \\ \beta_h &= \chi(i_h) - \chi(i_h - 1), \quad h = 1, 2, \ldots, n-1; \quad \beta_n = \chi(k). \end{aligned}\right\} \tag{16}$$

Consider the relations:

$$\alpha_1 + \beta_1 \equiv \alpha_2 + \beta_2 \equiv \ldots \equiv \alpha_{n-1} + \beta_{n-1} \equiv 0 \ (\mathrm{mod}\ 2). \tag{17}$$

Let X_0 denote the set of all functions χ satisfying the conditions (13) for which the relations (17) hold.

Let X_2 denote the set of all functions χ satisfying the conditions (13) for which the relations (17) do not hold.

It turns out that the pseudo-manifold Z_χ is orientable if $\chi \in X_0$ and non-orientable if $\chi \in X_2$.

Let X' denote the set of all such functions χ satisfying the conditions (13) for which

$$\alpha_1 \geqslant 2. \tag{18}$$

For the function χ consider two sequences of numbers:

$$\left.\begin{aligned} &\alpha_1, \beta_1, \alpha_2, \beta_2, \ldots, \alpha_{n-1}, \beta_{n-1}, \alpha_n; \\ &\alpha_1, \beta_1, \alpha_2, \beta_2, \ldots, \alpha_{n-1}, \beta_{n-1}. \end{aligned}\right\} \tag{19}$$

If $\beta_n > 0$, we put the first sequence into correspondence with the function χ, and if $\beta_n = 0$, the second sequence.

Put the function $\chi \in X'$ in the set X, if all the terms in the corresponding sequence (19) are even.

Put the function $\chi \in X'$ in the set X_α, if the corresponding sequence (19) contains odd numbers of which the first is some α.

Put the function $\chi \in X'$ in the set X_β, if the corresponding sequence (19) contains odd numbers of which the first is some β.

Thus the set X' is the sum of three disjoint sets X, X_α, and X_β. Evidently, $X \subset X_0$ and $X_\beta \subset X_2$.

Furthermore, if $\chi \in X_0 - X$, we have

$$2Z_\chi \sim 0. \tag{20}$$

Theorem 1, stated below, gives a homology basis of the manifold $H(k, l)$ for the dimensions of interest to us.

Theorem 1. A canonical basis of homology of dimension $kl - r$, $r \leqslant l - 1$, of the manifold $H(k, l)$ can be formed from the cycles:

$$Z_\chi, \chi \in X, r(\chi) = r; \quad \Gamma Z_{\chi'}, \chi' \in X_\beta, \quad r(\chi') = r - 1. \tag{21}$$

Here the cycles Z_χ are free, and the cycles ΓZ_χ are of order two.

From Theorem 1 it follows that all characteristic cycles of a manifold M^k can be expressed by the cycles X_χ, $\chi \in X$ and $X_{\chi'}$, $\chi' \in X_\beta$. These characteristic cycles are called *fundamental cycles.* Here the X_χ, with $\chi \in X$ are cycles with integral coefficients, and the $X_{\chi'}$, with $\chi' \in X_\beta$ are cycles modulo 2. If $\chi^* \in X$, X_{χ^*} is a fundamental cycle. If $\chi \in X_0 - X$, then

$$X_{\chi^*} \sim \sum_{\chi \in X_\beta} b_\chi \Gamma X_\chi, \tag{22}$$

where b_χ are residues modulo 2 uniquely determined by the function χ^*. If $\chi^* \in X_2$, then

$$X_{\chi^*} \sim \sum_{\chi \in X} a_\chi X_\chi + \sum_{\chi \in X_\beta} b_\chi \Gamma X_\chi + \sum_{\chi \in X_\beta} c_\chi X_\chi \pmod 2, \tag{23}$$

where a_χ, b_χ, and c_χ are residues modulo 2 uniquely determined by the function χ^*.

Further, if $\chi^* \in X_2$, we have

$$\Gamma X_{\chi^*} \sim \sum_{\chi \in X_\beta} c_\chi \Gamma X_\chi, \tag{24}$$

where c_χ are residues modulo 2 uniquely determined by the function χ^*.

These results imply that a characteristic number X_χ, $\chi \in X_0$ may be non-zero only when $\chi \in X$. Indeed, the zero-dimensional Betti group over the integers has no torsion; therefore, if a zero-dimensional cycle X_χ satisfies the condition $2X_\chi \sim 0$ (see (20)), then $X_\chi \sim 0$, i.e., the characteristic number X_χ vanishes. For the same reason, zero-dimensional cycles with integral coefficients of the type ΓX_χ do not give non-zero characteristic numbers.

Therefore, it makes sense to consider only characteristic numbers X_χ for $\chi \in X$. For a characteristic residue we obtain from (23)

$$X_{\chi^*} = \sum_{\chi \in X} a_\chi X_\chi + \sum_{\chi \in X_\beta} c_\chi X_\chi \pmod 2, \tag{25}$$

where a_χ and c_χ are residues modulo 2 uniquely determined by the function χ.

From among the characteristic numbers, let us choose one particular number, namely, $X_\chi = X_1$, where $\chi \equiv 1$. The function χ belongs to the set X_0; and it belongs to X only when k is even. Thus, for an odd k, the number X_1 is zero. In [5] it was shown that the characteristic number X_1 is the Euler characteristic of the manifold under consideration. This fact corresponds to Hopf's theorem that the degree of a spherical mapping of an even-dimensional manifold is half of its Euler characteristic.

If the orientation of the manifold M^k is inverted, i.e., if we are dealing with the manifold $-M^k$, the tangential mapping of the manifold $-M^k$ is essentially different, because we are dealing with oriented tangents.

Theorem 2, given below, establishes a relationship between the characteristic cycles of a manifold M^k and its inverse $-M^k$.

Theorem 2. If $-M^k$ is a manifold geometrically coinciding with M^k but having an opposite orientation, then

$$X_1(-M^k) \sim X_1(M^k). \tag{26}$$

If $\chi \not\equiv 1$, but $\chi \in X$,

$$X_\chi(-M^k) \sim -X_\chi(M^k). \tag{27}$$

For the remaining characteristic cycles

$$X_\chi(-M^k) \sim X_\chi(M^k). \tag{28}$$

This theorem shows that if a characteristic cycle X_χ, $\chi \not\equiv 1$, $\chi \in X$, is such that $2X_\chi$ is not homologous to zero, the manifold is asymmetric, i.e., it cannot be mapped onto itself by an orientation-inverting homeomorphism.

Theorem 3 gives a particular geometric property of characteristic cycles and residues.

Theorem 3. If an oriented closed manifold M^k is the boundary of an oriented bounded manifold M^{k+1}, all characteristic numbers and residues of the manifold M^k are zero, except, possibly, the characteristic number X_1, which is even.

By doubling the manifold M^{k+1} we can easily demonstrate that the Euler characteristic X_1 of M^k is even.

Thus, an interesting geometric fact comes to light: not every oriented closed manifold can be the boundary of an oriented manifold. A simple example is the complex projective plane which, when regarded as a four-dimensional manifold, is orientable and has an Euler characteristic equal to three.

Theorem 3 does not say anything about two- and three-dimensional manifolds.

It is known that any oriented surface may serve as the boundary of a three-dimensional oriented manifold. For certain reasons, it would be worthwhile to demonstrate that this assertion holds true for three-dimensional manifolds as well.

One more special property of characteristic cycles deserves mention.

If a manifold M^k can be embedded in a Euclidean space of dimension $k + 1$, all its characteristic cycles are homologous to zero, except, possibly, X_1; the characteristic number X_1 in this case is even.

For two- and three-dimensional manifolds, the characteristic cycles do not give any new invariants.

For a two-dimensional manifold M^2, we only obtain the characteristic number X_1 which is equal to the Euler characteristic.

For a three-dimensional manifold M^k, all its characteristic cycles reduce to Stiefel cycles which are, as has been proved by Stiefel, homologous to zero.

A four-dimensional manifold M^4 has three fundamental characteristic cycles. The first X_1, which has already been considered, gives the Euler characteristic. The second X_{22}, which is also zero-dimensional, is given by the function χ with the values $\chi(1) = \chi(2) = 2$, and $\chi(3) = \chi(4) = 0$.

The characteristic numbers X_1 and X_{22} are congruent modulo 2, but are, as is established in [5], quite independent. The characteristic cycle X_{22} is not a Stiefel cycle.

Besides these two cycles, there is one more fundamental characteristic cycle X_{21} of a four-dimensional manifold, defined by the function χ which takes the values $\chi(1) = \chi(2) = 1$, $\chi(3) = \chi(4) = 0$. The cycle X_{21} is defined modulo two, has dimension two, and is a Stiefel cycle.

If it were possible to establish all relationships between the three fundamental characteristic cycles X_1, X_{22}, and X_{21} of a four-dimensional manifold M^4, considerable advancement would be made in the classification of mappings of an $(n + 3)$-dimensional sphere S^{n+3} into an n-dimensional sphere S^n. For instance, if it were known that $X_{21} \sim 0$ implies $X_{22} \sim 0$, then we could find a denumerable number of distinct homotopy classes of mappings of the sphere S^{n+3} onto the sphere S^n.

In conclusion, we note that it is rather very difficult to calculate characteristic cycles on the basis of the definitions given here. So it is worth exploring other constructions that give characteristic cycles. Such constructions are considered in [5] where characteristic cycles are derived from systems of certain vector fields on manifolds. But, how far this method aids in calculating characteristic cycles, is still obscure.

It would be important to give a definition of characteristic cycles that would be suitable for combinatorial manifolds rather than for differentiable manifolds, so that we could find an algorithm for calculating characteristic cycles from the combinatorial scheme of a manifold.

1. The Manifold $H(k, l)$

We shall construct the manifold $H(k, l)$ and certain pseudo-manifolds needed in constructing homology bases in $H(k, l)$.

We confine ourselves to finite-dimensional real vector spaces.

The orientation of an n-dimensional vector space R^n can be defined by a sequence $e_1, \ldots, e_n$ of its basis vectors, meaning that the positive orientation of R^n is given by the orientation of the simplex $+(0, e_1, \ldots, e_n)$.

We denote the dimension of a set M by $D(M)$.

Definition 1. Let R^{k+l} be a vector space of dimension $k+l$ $(k \geqslant 1, l \geqslant 1)$. Let $H(k, l)$ denote the manifold of all oriented k-dimensional vector subspaces of the space R^{k+l}. In (A) we shall show that $D(H(k, l)) = kl$ and that by a proper choice of local coordinates, the manifold $H(k, l)$ can be converted into an analytical manifold.

If R^k is an arbitrary element in $H(k, l)$, let $\hat{R}^k$ denote an element of $H(k, l)$ differing from R^k only in orientation, i.e., $\hat{R}^k = -R^k$. Evidently, the mapping $R^k \rightleftharpoons \hat{R}^k$ is homeomorphic and from (A) it will be clear that it is analytic as well. If M is an arbitrary set of elements in $H(k, l)$, let $\hat{M}$ denote the set which is the image of M under the mapping $R^k \to \hat{R}^k$.

Let $R^{k+l'}$ be a vector space of dimension $k+l'$, and $H'(k, l')$ the manifold of all k-dimensional oriented vector subspaces of $R^{k+l'}$. If a is a non-degenerate linear mapping of $R^{k+l'}$ into R^{k+l}, then we can put each element R^k of $H'(k, l')$ into correspondence with an element $a(R^k)$ of $H(k, l)$. Thus we obtain a mapping a of the manifold $H'(k, l')$ into the manifold $H(k, l)$.

(A) Let $R_0^k \in H(k, l)$ and $(e_1, \ldots, e_k)$ be a basis in R_0^k defining its orientation. Let $f_1, \ldots, f_1$ be a system of vectors in R^{k+l} such that the vectors

$$e_1, \ldots, e_k, \quad f_1, \ldots, f_l \tag{1}$$

form a basis of the space R^{k+l} and let P be the linear span of the vectors $f_1, \ldots, f_l$. Then R^{k+l} is the direct sum of its subspaces R_0^k and P, so that $x = u + v$ for each $x \in R^{k+l}$, where $u \in R_0^k$ and $v \in P$.

Put $u = \varphi(x)$, where φ is the projection of the space R^{k+l} onto R_0^k in the direction of P. Henceforth this mapping plays an important part.

Let U denote the set of all elements in $H(k, l)$ which under the projection φ are mapped onto R_0^k without degenerating, with their orientation preserved. Evidently, U is a domain in $H(k,l)$ that contains R_0^k. The domains U and $\hat{U}$ do not intersect; let V denote the complement of their sum in $H(k,l)$.

In U let us now introduce coordinates that depend on the system (1). If $R_\xi^k \in U$, then in R_ξ^k, there exists a basis $e_1', \ldots, e_k'$ defining the orientation of R_ξ^k and such that

$$\varphi(e_i') = e_i, \quad i = 1, \ldots, k.$$

We have

$$e_i' = e_i + \sum_{j=1}^{l} \xi_i^j f_j, \quad i = 1, \ldots, k; \tag{2}$$

where $\|\xi_i^j\| = \xi$ is a matrix of real numbers. We take the kl elements of this matrix as the coordinates of $R_\xi^k \in U$. If this matrix is given arbitrarily, then after finding the vectors $e_1', \ldots, e_k'$ from (2), and letting R_ξ^k be the span of these vectors, we find that R_ξ^k is an element of U.

Thus, we have introduced coordinates in the neighbourhood U of the element R_0^k, where R_0^k is the origin of this system. Consequently, U is homeomorphic to a kl-dimensional Euclidean space. This coordinate system in U depends on the basis (1) of the space R^{k+l}. It can be demonstrated without any difficulty that one coordinate system is analytically transformed into any other.

Definition 2. Let $\omega(i)$ be an integral-valued monotone non-decreasing function of an integral-valued argument $i = 1, \ldots, k; 0 \leqslant \omega(i) \leqslant l$. In the space R^{k+l} (see Definition 1), choose an increasing sequence

$$R_1 \subset R_2 \subset \ldots \subset R_k \tag{3}$$

of vector subspaces, such that

$$D(R_i) = i + \omega(i), \quad i = 1, \ldots, k.$$

Let $Z(\omega)$ denote the set of all elements R^k in $H(k,l)$ such that

$$D(R^k \cap R_i) \geqslant i, \quad i = 1, \ldots, k.$$

It is clear that $Z(\omega)$ is compact, and that $\hat{Z}(\omega) = Z(\omega)$. Below we shall show (see (G)) that $Z(\omega)$ is a pseudo-manifold of dimension $r(\omega) = \Sigma_{i=1}^k \omega(i)$. If $Z(\omega)$ is orientable, by assigning one of the two possible orientations to it, we obtain a cycle in $H(k,l)$ over the integers. If $Z(\omega)$ is non-orientable, it can be regarded as a cycle modulo 2 in $H(k,l)$.

In the space R^{k+l}, in place of the sequence (3), let us now consider a sequence

$$R_1' \subset R_2' \subset \ldots \subset R_k' \tag{4}$$

of vector subspaces of the same dimensions

$$D(R_i') = D(R_i), \quad i = 1, \ldots, k.$$

In the same way that we defined the pseudo-manifold $Z(\omega)$ with the help of the sequence (3), let us now define a pseudo-manifold $Z'(\omega)$ with the help of the sequence (4). Obviously, there exists a continuous rotation of the space R^{k+l} that transforms the sequence (3) into the sequence (4). Since, in the course of rotation, the manifold $H(k, l)$ is sent into itself, the pseudo-manifold $Z(\omega)$ moves inside $H(k, l)$ and occupies the position of $Z'(\omega)$ at the end of the rotation.

Thus, the pseudo-manifolds $Z(\omega)$ and $Z'(\omega)$, interpreted as cycles and taken with proper orientations, are homologous to each other in $H(k, l)$. This explains why only the function ω is depicted in the notation $Z(\omega)$ instead of the sequence (3).

(B) A *jump point* of a function ω (see Definition 2) is a value of its argument i for which $\omega(i+1) \neq \omega(i)$.

Let $i_1, \ldots, i_{n-1}$ be the set of all jump points of the function ω taken in increasing order of magnitude, where $i_0 = 0$ and $i_n = k$. Let S_h denote the space R_{i_h} of the sequence (3). Now consider the increasing sequence

$$S_1 \subset S_2 \subset \ldots \subset S_n \tag{5}$$

of vector subspaces of the space R^{k+l}. Imposing the condition

$$D(R^k \cap S_h) \geqslant i_h, \quad h = 1, \ldots, n,$$

on an element R^k in $H(k, l)$, we once again obtain the set $Z(\omega)$. Thus the set $Z(\omega)$ is uniquely determined by the subsequence (5) of the sequence (3).

We now prove this assertion. Assume that i is not a jump point of the function ω and that i is not equal to k. We show that if $D(R^k \cap R_{i+1}) \geqslant i+1$, then $D(R^k \cap R_i) \geqslant i$; this will demonstrate that the spaces R_i can be eliminated from (3) without any loss. Thus, we can use the sequence (5) instead of (3).

Put

$$R^k \cap R_{i+1} = R^s \quad \text{and} \quad R^k \cap R_i = R^r.$$

The dimensions of the spaces R^s and R^r are s and r, respectively. Obviously, $R^r = R^s \cap R_i$, and as both R^s and R_i are located in R_{i+1}, we have

$$r \geqslant s + (i + \omega(i)) - (i + 1 + \omega(i+1)) = s - 1.$$

But $s \geqslant i + 1$, consequently, $r \geqslant i$.

This completes the proof of assertion (B).

(C) We say an element R^k of the set $Z(\omega)$ is *in general position* in $Z(\omega)$, if

$$D(R^k \cap S_h) = i_h, \quad h = 1, \ldots, n \text{ (see (B))}.$$

If R_0^k is in general position in $Z(\omega)$, then there exists a basis

$$e_1, \ldots, e_k, \quad f_1, \ldots, f_l \tag{6}$$

of the space R^{k+l}, such that the vectors

$$e_1, \ldots, e_k \tag{7}$$

form a basis of the space R_0^k that defines the orientation of R_0^k, and the vectors

$$e_1, \ldots, e_{i_h}, \quad f_1, \ldots, f_{\omega(i_h)}, \quad (h = 1, \ldots, n), \tag{8}$$

form a basis of the space S_h.

Now let U be a coordinate domain in $H(k, l)$ constructed with the help of the basis (6) as in (A). Then every element of $U \cap Z(\omega)$ is in general position in $Z(\omega)$. And $R_\xi^k \in U$ belongs to $Z(\omega)$ if and only if the matrix $\xi = \|\xi_i^j\|$ satisfies the condition

$$\xi_i^j = 0 \text{ for } j \geqslant \omega(i), \quad i = 1, \ldots, k. \tag{9}$$

Thus, the neighbourhood $U \cap Z(\omega)$ of the element R_0^k in the set $Z(\omega)$ is homeomorphic to a Euclidean space of dimension $r(\omega) = \Sigma_{i=1}^k \omega(i)$.

Let us now choose the basis (6) defined in (C) in the space R^{k+l}. Put $R_0^k \cap S_h = R^{i_h}$. Since the spaces R^{i_h}, $h = 1, \ldots, n$, form an increasing system, the vectors $e_1, \ldots, e_k$ can be so chosen that the system $e_1, \ldots, e_{i_h}$ forms a basis of the space R^{i_h}, $h = 1, \ldots, n$. As the vectors $e_1, \ldots, e_{i_1}$ are independent and lie in S_1, a basis of the space S_1 can be constructed by supplementing the system $e_1, \ldots, e_{i_1}$ with arbitrary independent vectors $f_1, \ldots, f_{\omega(i_1)}$ in S_1. Thus we have constructed the system (8) for $h = 1$, so that it is a basis of the space S_1.

Now suppose that the system (8) has already been constructed for a given h, so that it is a basis of the space S_h. The vectors of the system

$$e_1, \ldots, e_{i_h}, \quad f_1, \ldots, f_{\omega(i_h)}$$

are linearly independent, and the vectors $e_{i_h+1}, \ldots, e_{i_{h+1}}$ are independent of the former system because $D(R_0^k \cap S_h) = i_h$. And as the vectors $e_1, \ldots, e_{i_{k+1}}, f_1, \ldots, f_{(i_h)}$ are independent and lie in S_{h+1}, this sytem can be supplemented with the vectors $f_{\omega(i_h)+1}, \ldots, f_{\omega(i_{h+1})}$ to form a basis of S_{h+1}. Thus we obtain a system (8) for higher values of h.

Thus we have constructed the system (8) by induction. If $\omega(k) < l$, we supplement the system (8) for $h = n$ to the system (6).

Let ξ be a matrix satisfying the condition (9). We show that R_ξ^k belongs to $Z(\omega)$ and is in general position in $Z(\omega)$. Recall that R_ξ^k is spanned by the vectors $e'_1, \ldots, e'_k$, and the vector e'_i is expressed by formula (2). By virtue of the condition (9), the vector e'_i belongs to S_h when $i \leqslant i_h$; consequently,

$$D(R_\xi^h \cap S_h) \geqslant i_h.$$

Moreover, we readily find that

$$D(R_\xi^k \cap S_h) = i_h.$$

Thus, R_ξ^k belongs to $Z(\omega)$ and is in general position in $Z(\omega)$.

Let us now suppose that $R_\xi^k \in Z(\omega)$. Then we show that the matrix ξ satisfies the condition (9).

Let x be an arbitrary vector in $R_\xi^k \cap S_h$. Since $x \in R_\xi^h$, we have

$$x = \sum_{i=1}^{k} x^i e'_i.$$

Since $x \in S_h$, it follows that $x^i = 0$ for $i > i_h$, so that

$$x = \sum_{i=1}^{i_h} x^i e'_i. \tag{10}$$

Since $D(R_\xi^k \cap S_h) \geqslant i_h$, and the expression (10) contains exactly i_h parameters, these parameters can take arbitrary values. In particular, we find that the vector e'_i belongs to S_h for $i \leqslant i_h$. Since the vector e'_i is expressed by (2), and belongs to the space S_h, whose basis is the system (8), we find

$$\xi_i^j = 0 \text{ for } i \leqslant i_h, \quad j > \omega(i_h). \tag{11}$$

We assume that $i_{h-1} < i \leqslant i_h$; then $\omega(i) = \omega(i_h)$ and (11) can be expressed as

$$\xi_i^j = 0 \text{ for } i_{h-1} < i \leqslant i_h, \quad j > \omega(i), \quad h = 1, \ldots, n.$$

This means that the matrix ξ satisfies the condition (9).

Thus, assertion (C) has been demonstrated.

(D) Let $\omega \not\equiv 0$. We shall assume that the space R^{k+l} and the spaces of the system (5) are oriented. If $\omega(k) = l$, the spaces R^{k+l} and S_n coincide. In this case we assume that their orientations also coincide. Then the basis (6) of the space R^{k+l} constructed in (C) can be chosen so that the orientations of the spaces R^{k+l}, R_0^k, and S_k are determined by the bases (6), (7), (8), respectively.

In order to satisfy the requirements imposed here, it is sufficient to change the sign of certain vectors in the basis (6). If $\omega \equiv 0$, this cannot be done because the spaces R_0^k and S_1 will then coincide and their pre-assigned orientations may be conflicting. If, however, their orientations coincide, the requirements may be satisfied even when $\omega \equiv 0$.

(E) If $\omega \not\equiv 0$, the set M of all elements in general position in $Z(\omega)$ is connected, and therefore M is a manifold of dimension $r(\omega)$ (see (C)).

If $\omega \equiv 0$, the set $Z(\omega)$ consists of only two elements R_0^k and $\hat{R}_0^k$, and both these elements are in general position in $Z(\omega)$.

We now prove that the set M is connected if $\omega \not\equiv 0$. Assume that the space S_h of the system (5) and the space R^{k+l} are orientated as described in (D). Take

$$S_0 = \{0\} \quad \text{and} \quad S_{n+1} = R^{k+l}.$$

Let T_{h+1} denote a vector subspace of the space S_{h+1} such that S_{h+1} is the direct sum of S_h and T_{h+1}, $h = 0, 1, \ldots, n$.

Let A stand for the group of all linear transformations of the space R^{k+l} that map each space S_h into itself, and preserve orientation. We prove that A is connected.

Let A_1 be the group of all linear transformations of the space R^{k+l} that map every space T_h into itself, and preserve orientation; then $A_1 \subset A$. Since R^{k+l} is the direct sum of all the spaces T_h, the group A_1 is the direct product of the group of positive (i.e., those which preserve orientation) transformations of the spaces T_h.

Thus, the group A_1 is connected. Let

$$a \in A, \quad x = \sum_{h=1}^{n+1} x_h, \quad x_h \in T_h.$$

Put

$$a(x_h) = b(x_n) + a_1(x_h),$$

where $b(x_h) \in S_{h-1}, a_1(x_h) \in T_h$. Let t $(0 \leqslant t \leqslant 1)$ be a real number and

$$a_t(x) = \sum_{h=1}^{n+1} (1-t)b(x_h) + a_1(x_h).$$

Then we have

$$a_t \in A, \quad a_0 = a, \quad a_1 \in A_1.$$

Thus, A is a connected group.

Let R_0^k and $\bar{R}_0^k$ be two elements in M, and $e_1, \ldots, e_k, f_1, \ldots, f_l$ and $\bar{e}_1, \ldots, \bar{e}_k; \bar{f}_1, \ldots, \bar{f}_l$ be two bases of the space R^{k+l} constructed for these elements by the method described in (C) and (D).

Define a linear transformation a as follows:

$$a(e_i) = \bar{e}_i, \quad i = 1, \ldots, k, \quad a(f_j) = \bar{f}_j, \quad j = 1, \ldots, l.$$

Then $a \in A$, where $a(R_0^k) = \bar{R}_0^k$.

Thus, M admits a connected transitive group A of transformations and is therefore connected.

(F) Let $N = Z(\omega) - M$ (see (E)). We prove that

$$D(N) \leqslant r(\omega) - 3, \quad D(Z(\omega)) = r(\omega).$$

Assume that R^k is not in general position in $Z(\omega)$, then for some q $(1 \leqslant q \leqslant n-1)$, we have

$$D(R^k \cap S_q) \geqslant i_q + 1.$$

Let us introduce a new function $\omega_{(q)}(i)$ defined as follows:

$$\omega_{(q)}(i) = \omega(i) \quad \text{for } i \leqslant i_{q-1},$$

$$\omega_{(q)}(i) = \omega(i_q) - 1 \quad \text{for } i_{q-1} < i \leqslant i_q + 1,$$

$$\omega_{(q)}(i) = \omega(i) \quad \text{for } i > i_q + 1.$$

This function no longer has a jump at the point i_q but it has a jump at the point $i_q + 1 = i'_q$. For the other values of h, put $i'_h = i_h$. Then R^k satisifes the condition

$$D(R^k \cap S_h) \geqslant i'_h, \quad h = 1, \ldots, n,$$

and therefore $R^k \subset Z(\omega_{(q)})$.

We have

$$r(\omega) - r(\omega_{(q)}) = \sum_{i=1}^{k} (\omega(i) - \omega_{(q)}(i)) =$$

$$= i_q - i_{q-1} + 1 + \omega(i_q + 1) - \omega(i_q) \geqslant 3. \quad (12)$$

Furthermore, it is obvious that

$$N \subset \sum_{q=1}^{n-1} Z(\omega_{(q)}). \quad (13)$$

Now if we assume that $D(Z(\omega')) = r(\omega')$ for every function ω' satisfying the condition $r(\omega') < r(\omega)$, then the relations (12) and (13) give

$$D(N) \leqslant r(\omega) - 3.$$

Since $Z(\omega) = M + N$, we find that $D(Z(\omega))$ is equal to the larger of the two numbers $D(M)$ and $D(N)$, i.e., equal to $r(\omega)$.

This completes the proof of (F).

(G) From (E) and (F) it immediately follows that the space $Z(\omega)$ represented as a complex is indeed a pseudo-manifold of dimension $r(\omega)$.

(H) If $kl > 1$, the manifold $H(k, l)$ is simply connected, i.e., has a trivial fundamental group. And as $H(1, 1)$ is homeomorphic to a circle, the manifold $H(k, l)$ is always orientable. Furthermore, for $k \geqslant 2$, $l \geqslant 2$ the domain $H(k, l) - Z(\omega)$ is simply connected.

Let R^{k+t} be an oriented Euclidean vector space. Then to each oriented k-dimensional subspace R^k of the space R^{k+l} there corresponds a unique oriented complement R^l, $R^k \to R^l$. This one-to-one correspondence is obviously a homeomorphic mapping of the manifold $H(k, l)$ onto the manifold $H(l, k)$. We shall denote this mapping by ψ.

The manifold $H(1, l)$ is obviously homeomorphic to an l-dimensional sphere, and applying the mapping ψ, we find that $H(k, 1)$ is homeomorphic to a k-dimensional sphere.

Thus, if $kl > 1$, and if k or l is equal to one, the manifold $H(k, l)$ is simply connected.

Let P^s be an s-dimensional subspace of the space R^{k+l}, and Q^{k+l-s} its orthogonal complement in R^{k+l}. Put

$$\omega_1(1) = l-1, \quad \omega_1(2) = \ldots = \omega_1(k) = l;$$

$$\omega_2(1) = l-2, \quad \omega_2(2) = \ldots = \omega_2(k) = l;$$

$$\omega_3(1) = \omega_3(2) = l-1, \quad \omega_3(3) = \ldots = \omega_3(k) = l.$$

The pseudo-manifolds $Z(\omega_1)$, $Z(\omega_2)$, and $Z(\omega_3)$ are defined by the conditions:

$$D(R^k \cap P^l) \geqslant 1, \quad D(R^k \cap P^{l-1}) \geqslant 1, \quad \text{and} \quad D(R^k \cap P^{l+1}) \geqslant 2$$

(see (B)),

respectively. We have

$$r(\omega_1) = kl - 1, \quad r(\omega_2) = r(\omega_3) = kl - 2.$$

It is easily verified that ω_1, ω_2, and ω_3 are the only functions ω for which $r(\omega) \geqslant kl - 2$, except for $\omega \equiv l$.

It is a simple matter to verify that under the mapping ψ, the pseudo-manifold $Z(\omega_2) \subset H(k, l)$ is transformed into the pseudo-manifold $Z(\omega_3) \subset H(l, k)$.

The domain $H(k, l) - Z(\omega_1)$ consists of two regions U and $\hat{U}$ (see (A)) and is therefore simply connected.

The domain $H(k,l)-Z(\omega_2)$ consists of all those $R^k \in H(k,l)$ whose intersection with P^{l-1} contains only the zero point. Thus the orthogonal projection of R^k into Q^{k+1} is nondegenerate. Making R^k move into its projection $R_0^k \subset Q^{k+1}$ according to some definite rule, we obtain a continuous deformation of the domain $H(k,l)-Z(\omega_2)$ into the manifold $H(k,1)$ consisting of all R_0^k lying in Q^{k+1}. Thus the fundamental group of the domain $H(k,l)-Z(\omega_2)$ is isomorphic to the fundamental group of the manifold $H(k,1)$. And if $k>1$, this fundamental group is trivial.

Since $r(\omega_2)=kl-2$, any curve in $H(k,l)$ can be deformed into a curve lying outside $Z(\omega_2)$; if $k>1$ this curve can be contracted into a point in the domain $H(k,l)-Z(\omega_2)$. Hence the fundamental group of the manifold $H(k,l)$ in this case is trivial.

The domain $H(k,l)-Z(\omega_3)$ is transformed by the mapping ψ into the domain $H(l,k)-Z(\omega_2)$ and is therefore simply connected when $l>1$.

Now if $r(\omega)\leqslant kl-3$, the domain $H(k,l)-Z(\omega)$ is simply connected as the manifold $H(k,l)$ is simply connected. Indeed, not only any curve, but also any two-dimensional film in $H(k,l)$ can be removed from $Z(\omega)$ by a continuous deformation.

Thus, the assertion (H) has been demonstrated.

2. Orientations of the Pseudo-Manifolds $Z(\omega)$

In this section we shall assign a definite orientation to each orientable pseudo-manifold $Z(\omega)$.

(A) Let R^n be an n-dimensional vector space with some system of Cartesian coordinates, i.e., a sequence of numbers $x^1,\ldots,x^n$ is put into correspondence with each $x\in R^n$: $x=(x^1,\ldots,x^n)$. A definite basis $e_1,\ldots,e_n$ corresponds to the coordinate system chosen, so that $x=\Sigma_{i=1}^n x^i e_i$. And this basis defines a definite orientation of the space R^n as stated in the beginning of Section 1. Thus, to every definite coordinate system in R^n there corresponds a definite orientation of the space R^n.

Let $\{R^n;\ x^{j_1}=0,\ldots,x^{j_r}=0\}$ denote the coordinate plane of the space R^n defined by the equations:

$$x^{j_1}=0,\ldots,x^{j_r}=0.$$

It has its own coordinates $x^{i_1},\ldots,x^{i_{n-r}}$, and therefore a definite orientation is given.

Let $\{R^n, x^m < 0\}$ denote the half-space $x^m < 0$ of the space R^n. This half-space may be regarded as oriented, because R^n is oriented. It is readily seen that

$$\Delta\{R^n, x^m < 0\} = (-1)^{m-1}\{R^n, x^m = 0\}. \tag{1}$$

In the coordinate neighbourhood U (see (A) Section 1) the coordinates ξ_i^j are given as the elements of the matrix ξ. In order that these coordinates may define the orientation in U and in the coordinate planes in U, it is sufficient to number them in some order. We use a subscript to denote the number of the row, and a superscript to denote the number of the column. We number the elements of the matrix ξ with respect to the columns, i.e., write them in the following sequence:

$$\xi_1^1, \xi_2^1, \ldots, \xi_k^1, \quad \xi_1^2, \xi_2^2, \ldots, \xi_k^2, \ldots, \xi_1^l, \xi_2^l, \ldots, \xi_k^l. \tag{2}$$

Thus, the coordinate plane $U \cap Z(\omega)$ defined by the equations (9) of Section 1 is assigned, according to our condition, a definite orientation. Henceforth we shall assign this orientation to the neighbourhood $U \cap Z(\omega)$. This orientation thus depends on the basis (6) Section 1 of the space R^{k+l}.

(B) Let $\omega \not\equiv 0$ be a function for which the pseudo-manifold $Z(\omega)$ is orientable. We assume that all the spaces $S_1, \ldots, S_n$ of the system (5) Section 1 defining $Z(\omega)$ are oriented, and assign $Z(\omega)$ an orientation, depending on the orientations chosen for the spaces S_h. For this purpose, assign the space R^{k+l} some orientation so that it coincides with the orientation of S_n when $\omega(k) = l$. Choose the basis (6) Section 1 of the space R^{k+l} in such a manner that the conditions (D) Section 1 are satisfied. This basis defines the orientation of the neighbourhood $U \cap Z(\omega)$ of the element $R_0^k \in Z(\omega)$ as described in (A). Since the pseudo-manifold $Z(\omega)$ is orientable, the orientation of the neighbourhood $U \cap Z(\omega)$ induces the orientation of the whole set $Z(\omega)$. The orientation thus obtained for $Z(\omega)$ depends only on the orientations chosen for the spaces of the system (5) Section 1. We now examine the nature of this dependence when $\omega(1) > 0$.

Let us introduce the following notation:

$$\alpha_h = i_h - i_{h-1}, \quad h = 1, \ldots, n;$$

$$\beta_h = \omega(i_h + 1) - \omega(i_h), \quad h = 1, \ldots, n-1; \quad \beta_n = l - \omega(k) \tag{3}$$

(see Section 1, (B));

where the numbers β_h will be used later. Now let

$$S'_h = \epsilon_h S_h, \quad \epsilon_h = \pm 1, \quad h = 1, \ldots, n.$$

The pseudo-manifold $Z'(\omega)$ constructed with the help of the spaces S'_h can differ from the space $Z(\omega)$ only in orientation. Thus, we find

$$Z'(\omega) = \epsilon_1^{\alpha_1} \cdot \epsilon_2^{\alpha_2} \ldots \cdot \epsilon_n^{\alpha_n} \dot{Z}(\omega) \quad \text{for } \omega(1) > 0. \tag{4}$$

Let R_0^k and $\bar{R}_0^k$ be two elements in general position in $Z(\omega)$ and

$$e_1, \ldots, e_k, \quad f_1, \ldots, f_l \tag{5}$$

and

$$\bar{e}_1, \ldots, \bar{e}_k, \quad \bar{f}_1, \ldots, \bar{f}_l \tag{6}$$

two bases of the space R^{k+l} constructed for these elements by the method described in (C) and (D) Section 1, i.e., with due regard for the orientations of the spaces S'_h and R^{k+l}. We show that the bases (5) and (6) induce the same orientation in $Z(\omega)$.

Define a linear transformation a of the space R^{k+l} as follows:

$$a(e_i) = \bar{e}_i, \quad i = 1, \ldots, k; \quad a(f_j) = \bar{f}_j, \quad j = 1, \ldots, l.$$

By virtue of the constructions given in (E) Section 1, there exists one-parameter family a_t of linear transformations of the space R^{k+l} such that $a_t(S_h) = S_h, h = 1, \ldots, n$, where $a_1 = a, a_0$ is the identity transformation. The basis

$$a_t(e_1), \ldots, a_t(e_k), \quad a_t(f_1), \ldots, a_t(f_l)$$

of the space R^{k+l} determines the orientation in the neighbourhood $a_t(U \cap Z(\omega)) \subset Z(\omega)$. By continuity it follows that the orientation of the pseudo-manifold $Z(\omega)$ thus determined always coincides with the orientation induced by the basis (5). For $t = 1$, we obtain the desired result.

We now prove (4). First, let us note that if the sign of the vector f_q in the basis (5) is changed, the orientation of the neighbourhood $U \cap Z(\omega)$ defined in (B) acquires the factor $(-1)^\gamma$, where γ is the number of coordinates ξ_i^q that may be non-zero in $U \cap Z(\omega)$ (see (9) Section 1). If $q = \omega(i_{h-1} + 1)$, then $\gamma = k - i_{h-1}$; if $q = \omega(i_{h-1} + 1) + 1$, we have $\gamma = k - i_h$. Now simultaneously change the signs of two of the indicated vectors, then the orientation of the neighbourhood $U \cap Z(\omega)$ acquires the factor $(-1)^{\alpha_h}$ (see (3)). On the other hand, when the signs of two vectors are simultaneously changed, the orientation of the space S_h changes, but the orientations of the remaining spaces in the sequence (5) Section 1 do not change. The orientation of the space R^{k+l} changes, under this change of the sign of vectors, only if $h = n$ and $S_n = R^{k+l}$. From the foregoing follows the truth of the formula (4). If $\omega(k) < l$, the orientation of $Z(\omega)$

does not depend on the orientation of R^{k+l}, because, on changing only the sign of the vector f_q when $q = \omega(i_{n-1} + 1) + 1$, we have $\gamma = k - i_n = 0$.

(C) Let ω be a function such that $\omega(1) > 0$, $Z(\omega)$ is orientable and $2Z(\omega)$ is not homologous to zero in $H(k, l)$. Then we have

$$\alpha_1 \equiv \alpha_2 \equiv \ldots \equiv \alpha_n \equiv 0 \ (\mathrm{mod}\ 2) \text{ for } \omega(k) < l, \tag{7}$$

$$\alpha_1 \equiv \alpha_2 \equiv \ldots \equiv \alpha_{n-1} \equiv 0 \ (\mathrm{mod}\ 2) \text{ for } \omega(k) = l. \tag{8}$$

Thus, if $\omega(k) < l$, the orientation of $Z(\omega)$ defined in (B) does not depend on the orientations of the spaces in the system (5) Section 1 (see (4)). If $\omega(k) = l$, the orientation of $Z(\omega)$ depends only on the orientation of the space $S_n = R^{k+l}$. Both of these cases can be combined into one expression:

$$Z'(\omega) = \epsilon^k Z(\omega), \quad \omega(1) > 0, \tag{9}$$

where $Z(\omega)$ is constructed from the oriented space R^{k+l}, and $Z'(\omega)$ from the oriented space ϵR^{k+l}, $\epsilon = \pm 1$. If, however, we fix the orientation of the space R^{k+l}, and construct an oriented pseudo-manifold $Z(\omega)$ from the system of spaces (5) Section 1, and an oriented pseudo-manifold $Z^*(\omega)$ from some other system of spaces S_h^*, $h = 1, \ldots, n$, then we find

$$Z^*(\omega) \sim Z(\omega) \text{ in } H(k, l). \tag{10}$$

We now proceed to prove (7) and (8). Let

$$S_h' = \epsilon_h S_h, \quad \epsilon_h = \pm 1, \quad h = 1, \ldots, n.$$

If $\omega(k) = l$, we take $\epsilon_n = \pm 1$. Then we can easily construct a linear transformation a of R^{k+l}, preserving orientation, such that

$$a(S_h) = S_h', \quad h = 1, \ldots, n.$$

Since the transformation a preserves the orientation of R^{k+l}, there exists a one-parameter family of linear transformations a_t of the space R^{k+l}, such that $a_1 = a$, and a_0 is the identity transformation. The pseudo-manifold $a_t(Z(\omega))$ depends on the parameter t. For $t = 0$, this manifold becomes the manifold $Z(\omega)$, and for $t = 1$, the manifold $Z'(\omega)$. Thus, $Z'(\omega) \sim Z(\omega)$. Hence, from (4), we obtain

$$(1 - \epsilon_1^{\alpha_1} \cdot \epsilon_2^{\alpha_2} \ldots \epsilon_n^{\alpha_n}) Z(\omega) \sim 0.$$

Since, $2Z(\omega)$, by assumption, cannot be homologous to zero, we find that (7) and (8) are indeed true.

We now prove (9). By virtue of (3), we have $k = \alpha_1 + \alpha_2 + \ldots + \alpha_n$. Hence, by virtue of (4), (7) and (8), we obtain (9).

We now prove (10). Let a be a linear transformation of the space R^{k+l} preserving orientation, such that

$$a(S_h) = S_h^*, \quad h = 1, \ldots, n.$$

Here we assume that spaces S_h and S_h^* are non-oriented, because the orientations of the pseudo-manifolds $Z(\omega)$ and $Z^*(\omega)$ do not depend on the orientations of these spaces. Since the transformation a preserves the orientation of the space R^{k+l}, there exists a one-parameter a_t of linear transformations of the space R^{k+l}, such that $a_1 = a$ and a_0 is the identity transformation. The oriented pseudo-manifold $a_t(Z(\omega))$ depends on the parameter t. For $t = 0$, it becomes the manifold $Z(\omega)$, and for $t = 1$, the manifold $Z^*(\omega)$. Thus, $Z(\omega) \sim Z^*(\omega)$.

This completes the proof of (C).

(D) Let $\omega \equiv c$, where c is an integer. It is easy to see that the pseudo-manifold $Z(c)$ is homeomorphic to the manifold $H(k, c)$ which is orientable (see (H), Section 1). Hence $Z(c)$ is also orientable.

Since the function $\omega \equiv c$ has no jumps, for this function we have $n = 1$, $\alpha_1 = k$, and the system (5) Section 1 consists of only one space $S_1 = R^{k+c}$, whose orientation determines the orientation of $Z(c)$ (see (B)). In this case

$$Z'(c) = \epsilon_l^k Z(c) \quad \text{(see (4)).} \tag{11}$$

(E) Let R^{k+l} and $R^{k+l''} \subset R^{k+l}$, $l - l'' = m > 0$ be two oriented vector spaces; then $H(k, l'') \subset H(k, l)$ and both these manifolds are, by virtue of (B) (see (D)), oriented. Furthermore, let $Z(\omega)$ be a pseudo-manifold defined by the system $\{S_h\}$, $h = 1, \ldots, n$. If $Z(\omega)$ is orientable, we may consider the spaces S_h to be oriented and to generate the orientation of $Z(\omega)$.

We now proceed to study the algebraic intersection $Z(\omega) \times H(k, l'')$ modulo 2, when $Z(\omega)$ is non-orientable, also with regard to orientation when it is orientable.

Put $\omega'' = \omega - m$; the function ω'' is useful in determining the pseudo-manifold $Z(\omega_2)$ only when $\omega(1) \geqslant m$.

We assume that the spaces S_h and $R^{k+l''}$ are in general position in R^{k+l}, and put $S_h'' = S_h \times R^{k+l''}$ taking orientation into account when $Z(\omega)$ is orientable. We have

$$D(S_h'') = i_h + \omega(i_h) - m.$$

Thus, if $\omega(1) \geqslant m$, the spaces S''_h, $h = 1, \ldots, n$ may serve as a basis in constructing the pseudo-manifold $Z(\omega'') \subset H(k, l'')$. We find that

$$Z(\omega) \cap H(k, l'') \text{ is empty when } \omega(1) < m, \tag{12}$$

$$Z(\omega) \times H(k, l'') = Z(\omega'') \text{ when } \omega(1) \geqslant m, \omega \not\equiv m, \tag{13}$$

$$I(Z(\omega), H(k, l'')) = 1 + (-1)^k, \text{ when } \omega \equiv m. \tag{14}$$

The expression (14) gives the intersection index, while (13) gives the intersection modulo 2 for a non-orientable $Z(\omega)$, and takes orientation into account for an orientable $Z(\omega)$.

We now take up the proof. If $R^k \in Z(\omega)$, $R^k \in H(k, l'')$, then

$$D(R^k \cap S_h) \geqslant i_h, \quad R^k \subset R^{k+l''},$$

and therefore $D(R^k \cap S''_h) \geqslant i_h$. If $\omega(1) < m$, we thus have $i_1 \leqslant i_1 + \omega(i_1) - m \leqslant i_1 - 1$. This is impossible, and therefore (12) is true. If $\omega(1) \geqslant m$, we obtain $Z(\omega) \cap H(k, l'') \subset Z(\omega'')$. The converse embedding is obvious, and therefore

$$Z(\omega) \cap H(k, l'') = Z(\omega'') \text{ when } \omega(1) \geqslant m. \tag{15}$$

We now assume that $\omega(1) \geqslant m$. If $\omega \equiv m$, we take $R_0^k = S''_1$. If $\omega \not\equiv m$, for R_0^k we take any element in general position in $Z(\omega'')$. Let

$$e_1, \ldots, e_k, \quad f''_1, \ldots, f''_{l''}, \tag{16}$$

be a basis of the space $R^{k+l''}$ constructed as in (C) Section 1 for $R_0^k \in Z(\omega)$, i.e., such that the vectors

$$e_1, \ldots, e_k \tag{17}$$

form a basis of the space R_0^k, and the vectors

$$e_1, \ldots, e_k, \quad f''_1, \ldots, f''_{\omega''(i_h)}, \quad h = 1, \ldots, n, \tag{18}$$

form a basis of the space S''_h. In this case the orientations of the spaces R^{k+l}, R_0^k, and S''_h are determined by the bases (16), (17), and (18), respectively. In the space S_1, choose the vectors $f_1, \ldots, f_m$ linearly independent of S''_1. The vectors

$$e_1, \ldots, e_k, \quad f_1, \ldots, f_m, \quad f_{m+1} = f''_1, \ldots, f_l = f''_{l''}, \tag{19}$$

form a basis for the space R^{k+l}, and a basis of the space S_h will consist of the vectors

$$e_1, \ldots, e_k, \quad f_1, \ldots, f_{\omega(i_h)}. \tag{20}$$

Choose the vectors $f_1, \ldots, f_m$ in such a manner that the basis (19) determines the orientation of the space R^{k+l}. Then, from the condition $S''_h =$

$S_h \times R^{k+l''}$, it readily follows that the basis (20) gives an orientation of the space S_h.

The basis (19) defines a coordinate neighbourhood U of the element R_0^k in $H(k, l)$ (see (A) Section 1). The neighbourhood $U' = U \cap Z(\omega)$ of the element R_0^k in $Z(\omega)$ is given in U by the equations:

$$\xi_i^j = 0 \text{ for } j > \omega(i), \quad i = 1, \ldots, k. \tag{21}$$

The neighbourhood $U'' = U \cap H(k, l'')$ of the element R_0^k in $H(k, l'')$ is given by the equations:

$$\xi_i^j = 0 \text{ for } j \leqslant m. \tag{22}$$

The neighbourhood $U''' = U \cap Z(\omega'')$ of the element R_0^k in $Z(\omega'')$ is given by the systems (21) and (22). The orientations of the coordinate spaces U, U', U'', and U''' are determined by the rule described in (A), i.e., by the numeration of elements of the matrix ξ by columns.

Let A be the sequence of all coordinate unit vectors in U written in this ordering. Let A', A'', and A''' denote those parts of the sequence A that are contained in U', U'', and U''', respectively. Then the orientations of the neighbourhoods U, U', U'', and U''' are given by the sequences A, A', A'', and A''', respectively. Let B' denote the sequence that is obtained from A' on omitting the elements contained in A'''. Similarly, let B'' denote the sequence that is obtained from A'' on omitting the elements contained in A'''. Since numeration has been made with respect to the matrix columns, in the sequence A every element from B' precedes every element from B''. Let

$$A = \epsilon(A'''B'B''), \quad A' = \epsilon'(A''', B'), \quad A'' = \epsilon''(A''', B''). \tag{23}$$

Here $\epsilon = \pm 1$, $\epsilon' = \pm 1$, $\epsilon'' = \pm 1$, and (23) gives the relationship between the orientations of the spaces considered, defined by various vector sequences. Since in the sequence A every element from B' precedes every element from B'', we have $\epsilon = \epsilon'' \cdot \epsilon'$. This implies, by virtue of the Lefschetz rule, that $U''' = U' \times U''$, the orientations being taken into account.

Thus we have demonstrated the relation (13).

When $\omega \equiv m$, this result gives $I(U', U'') = +1$. Thus, the intersection index of $Z(\omega)$ and $H(k, l'')$ at the point R_0^k is $+1$. If $\omega \equiv m$, the intersection $Z(\omega) \cap H(k, l'') = Z(\omega'') = Z(0)$ (see (15)) contains only two points, namely, the point R_0^k already considered and some other $\hat{R}_0^k$. In order to calculate the intersection index of $Z(\omega)$ and $H(k, l'')$ at the point $\hat{R}_0^k$, let us invert the orientation of S_1, i.e., take $S_1' = -S_1$. Let $Z'(\omega)$ denote the pseudo-manifold given by the space S_1'; then

$$Z'(\omega) = (-1)^k Z(\omega) \quad \text{(see (11))}. \tag{24}$$

Here the roles of the elements R_0^k and $\hat{R}_0^k$ have been interchanged: R_0^k is defined as $S_1'' = S_1 \cap R^{k+l''}$. Thus, the intersection index of $Z'(\omega)$ and $H(k, l'')$ at the point $\hat{R}_0^k$ is equal to $+1$, and from (24) we find that the intersection index of $Z(\omega)$ and $H(k, l'')$ at the point $\hat{R}_0^k$ is $(-1)^k$. Consequently, (14) is indeed true.

This completes the proof of (E).

The pseudo-manifolds $Z(\omega)$ give a significant supply of cycles of the manifold $H(k, l)$. In Section 5 we shall consider the construction of homology bases with the help of these cycles. Here we give a general method of constructing cycles.

(F) Let Z be an r-dimensional cycle modulo 2. Assigning some orientation to each of its simplices, we obtain a chain Y with integral coefficients such that $\Delta Y = 2X$, where X is an $(r-1)$-dimensional chain with integral coefficients, whose homology class is uniquely determined by the homology class of the cycle Z. Put $X = \Gamma Z$. The operation Γ, defined up to homology will henceforth be applied to non-orientable pseudo-manifolds $Z(\omega)$.

If Z is an r-dimensional ∇-cycle modulo 2, by assigning some arbitrary orientation to each of its simplices, we obtain a chain Y with integral coefficients such that $\nabla Y = 2X$, where X is an $(r+1)$-dimensional ∇-cycle with integral coefficients, whose homology class is uniquely determined by the homology class of the ∇-cycle Z. We put $X = \Gamma Z$.

3. Tangential Mappings and Characteristic Cycles

In this section we introduce the concept of a characteristic cycle of a closed differentiable oriented k-dimensional manifold M^k – a concept which is basic for the whole paper.

Let U^k be some coordinate neighbourhood of a point a of the differentiable manifold M^k with the local coordinates $x^1, \ldots, x^k$. A mapping f of the manifold M^k into a vector space R^{k+l} with the coordinates $y^1, \ldots, y^{k+l}$ is called *regular at the point a* by Whitney, if near the point a this mapping can be expressed in the coordinate form:

$$y^j = y^j(x^1, \ldots, x^k) = f^j(k), \quad j = 1, \ldots, k+l, \tag{1}$$

where the functional matrix

$$F = \left\| \frac{\partial j^j(x)}{\partial x^i} \right\|, \quad i = 1, \ldots, k, \quad j = 1, \ldots, k+l,$$

at the point a is of rank k and continuously depends on x. If the mapping f is regular at each point $a \in M^k$, it is called *regular.* Whitney has demonstrated that a differentiable manifold M^k can be mapped regularly into a vector space R^{2k}. The manifold M^k can be mapped regularly and at the same time homeomorphically into the vector space R^{2k+1}. Such a mapping we call an *embedding* of a differential manifold M^k into a vector space R^{k+l}. Often we take the image $f(M^k)$ to be the initial manifold $M^k \subset R^{k+l}$ itself.

Henceforth we have to deal with mappings of the manifold M^k into various manifolds $H(k,l)$ (see Definition 1); consequently, it is desirable to define the concept of equivalence of these mappings.

(A) Let $R'^{k+l'}$ and $R''^{k+l''}$ be two vector spaces, $H'(k,l')$ and $H''(k,l'')$ be their corresponding manifolds (see Definition 1).

Two mappings θ' and θ'' of a p-dimensional complex K^p into $H'(k,l')$ and $H''(k,l'')$, respectively, are said to be *equivalent,* if there exist linear nondegenerate mappings a' and a'' of the vector spaces $R'^{k+l'}$ and $R''^{k+l''}$ into some vector space R^{k+l}, such that the mappings $a'\theta'$ and $a''\theta''$ of the complex K^p into $H(k,l)$ are homotopic in $H(k,l)$. Here we keep in mind that the mappings a' and a'' of the manifolds $H'(k,l')$ and $H''(k,l'')$ into $H(k,l)$ correspond to the linear mappings a' and a'', respectively. Reflexivity and symmetry of this definition are quite obvious, and therefore we shall only demonstrate its transitivity. It turns out that each mapping θ of the complex K^p into some $H(k,l)$ is equivalent to some mapping of it into $H(k,p)$, and, consequently, is equivalent to some mapping into $H(k,p+1)$ because $H(k,p) \subset H(k,p+1)$. Furthermore, two mappings θ' and θ'' of the complex K^p into the manifold $H(k,p+1)$ are equivalent in this sense, if and only if they are homotopic in $H(k,p+1)$. Thus, to classify the mappings of the complex K^p from the viewpoint of this equivalence principle, it is sufficient to classify all the mappings of this complex into the manifold $H(k,p+1)$ from the standpoint of the ordinary homotopy theory. Hence follows the transitivity of our equivalence relation.

To demonstrate the truth of the assertions made in (A), we first prove (B).

(B) Let R^{k+l} and $R^{k+l-1} \subset R^{k+l}$ be two vector spaces, and $H(k,l-1) \subset$

$H(k,l)$ their corresponding manifolds (see Definition 1). If $p \leqslant l-1$, every mapping θ of the complex K^p into $H(k,l)$ is homotopic to some mapping θ' of the complex K^p into $H(k,l-1)$. If $p \leqslant l-2$, and θ_0 and θ_1 are two mappings of the complex K^p into $H(k,l-1)$ homotopic in $H(k,l)$, they are also homotopic in $H(k,l-1)$.

We assume that R^{k+l} is Euclidean and let e denote the unit vector in R^{k+l} orthogonal to R^{k+l-1}. The set of all elements $R^k \in H(k,l)$ containing e we denote by H. It is easily seen that H is homeomorphic to the manifold $H(k-1,l)$, and is therefore of dimension equal to $(k-1)l$.

Now let R_0^k be an element of $H(k,l)$ belonging neither to H, nor to $H(k,l-1)$. Let g denote the projection of the vector e on R_0^k and h the projection of the vector g on R^{k+l-1}. We denote the unit vectors in the directions of g and h by e_0 and e_1, respectively. Let R_0^{k-1} be a linear subspace of the space R_0^k orthogonal to e_0. Let e_t $(0 \leqslant t \leqslant 1)$ be the unit vector that uniformly rotates from the position e_0 to the position e_1 in the (e_0, e_1)-plane. The linear span of the vector e_t and the space R_0^{k-1}, we denote by R_t^k.

Now if $R_0^k \in H(k,l-1)$, we take $R_t^k = R_0^k$. Thus, for each element R_0^k that does not belong to H, we have defined a procedure of moving R_t^k from the position R_0^k into the position $R_1^k \in H(k,l-1)$, during which the elements in $H(k,l-1)$ remain fixed. Denote the deformation thus obtained by ψ.

Let $p \leqslant l-1$ and θ be a mapping of the complex K^p into $H(k,l)$. Since the difference in the dimensions of $H(k,l)$ and H is $l > p$, by slightly modifying the mapping θ, we can obtain a mapping θ'' such that $\theta''(K^p)$ no longer intersects H. Applying the deformation ψ to the mapping θ'', we obtain the mapping θ' that we need.

Now let $p \leqslant l-2$, and $\theta_t (0 \leqslant t \leqslant 1)$ be a family of mappings of the complex K^p into $H(k,l)$ such that

$$\theta_0(K^p) \subset H(k,l-1), \quad \theta_1(K^p) \subset H(k,l-1).$$

This means that the mappings θ_0 and θ_1 of the complex K^p into $H(k,l-1)$ are homotopic in $H(k,l)$. Let K^{p+1} denote the product of the complex K^p and the unit segment $[0,1]$. The mapping θ of the complex K^{p+1} into $H(k,l)$ corresponds to the family θ_t in a well known way. As before, by slightly modifying the mapping θ, we can change it into a mapping θ'' such that the image $\theta''(K^{p+1})$ does not intersect H. Applying the deformation ψ to the mapping θ'', we obtain the mapping θ' of the complex K^{p+1} into $H(k,l-1)$ which realises the homotopy between θ_0 and θ_1 in $H(k,l-1)$.

Thus, we have proved Proposition (B).

We now proceed to prove (A). From (B) it immediately follows that every mapping θ of the complex K^p into some $H(k,l)$ is equivalent to some mapping of the complex K^p into the manifold $H(k,p)$.

Now let θ' and θ'' be two equivalent mappings of the complex K^p into $H(k,p+1)$. This means that there exist linear nondegenerate mappings a' and a'' of the space R^{k+p+1} into some space R^{k+l} such that the mappings $a'\theta'$ and $a''\theta''$ of the complex K^p into $H(k,p)$ are homotopic in $H(k,l)$. Without loss of generality, we may assume that $p+1<l$, so that there exists a one-parameter family a_t $(0 \leqslant t \leqslant 1)$ of linear nondegenerate mappings of the space R^{k+p+1} into the space R^{k+l} such that $a_0 = a'$ and $a_1 = a''$. Thus, the mappings $a_0\theta'$ and $a_1\theta'$ are homotopic in $H(k,l)$. Since, by assumption, the mappings $a_0\theta'$ and $a_1\theta''$ are homotopic in $H(k,l)$, the mappings $a_1\theta'$ and $a_1\theta''$ are also homotopic in $H(k,l)$. Hence, by virtue of (B), it directly follows that the mappings $a_1\theta'$ and $a_1\theta''$ of the complex K^p into $a_1(H(k,p+1))$ are homotopic in the manifold $a_1(H(k, p+1))$ itself. Taking the inverse mapping a_1^{-1}, we find that the mappings θ' and θ'' are homotopic in $H(k,p+1)$. Thus, we have proved Proposition (A).

Definition 3. Let f be a regular (possibly not homeomorphic) mapping of a differentiable oriented manifold M^k into the vector space R^{k+l} (see (4)). Then, to a point $x \in M^k$ there corresponds a definite oriented tangent T_x of $f(U^k)$ at the point $f(x)$, where U^k is a small neighbourhood of the point x in M^k. Let $T(x)$ denote an oriented k-dimensional vector subspace of the space R^{k+l} parallel to T_x. Since $T(x) \in H(k,l)$, we obtain a continuous mapping T of the manifold M^k into $H(k,l)$ called the *tangential mapping.* It turns out that all tangential mappings of a closed oriented manifold M^k are equivalent in the sense defined in (A).

Let f' and f'' be two regular mappings of the manifold M^k into vector spaces $R'^{k+l'}$ and $R''^{k+l''}$, respectively, and T' and T'' their corresponding tangential mappings of M^k into the manifolds $H'(k,l')$ and $H''(k,l'')$. We shall prove that the mappings T' and T'' are equivalent (see (A)).

Let R^{k+l} $(l = k + l' + l'')$ denote the direct sum of the spaces $R'^{k+l'}$ and $R''^{k+l''}$. We assume that $R'^{k+l'}$ and $R''^{k+l''}$ are subspaces of the space R^{k+l}. The inclusions $H'(k,l') \subset H(k,l)$ and $H''(k,l'') \subset H(k,l)$ correspond to the inclusions $R'^{k+l'} \subset R^{k+l}$ and $R''^{k+l''} \subset R^{k+l''}$. It is now sufficient to show that the mappings T' and T'' of the manifold M^k into $H(k,l)$ are homotopic in $H(k,l)$. Let us put

$$f_t(x) = (1-t)f'(x) + tf''(x), \quad x \in M^k.$$

From the regularity of the mappings f' and f'' it readily follows that the mapping f_t is regular. Let T_t denote the tangential mapping that corresponds to the regular mapping f_t. We thus have a continuous deformation T_t which deforms the mapping T' into the mapping T''; these mappings are therefore homotopic in $H(k, l)$.

Thus, the equivalence of all tangential mappings of a closed oriented differentiable manifold has been demonstrated.

We should, however, bear in mind that a tangential mapping T depends on the orientation of the manifold M^k. When the orientation of the manifold is changed, we obtain some other tangential mapping which, in general, is not equivalent to the original mapping.

From among all possible mappings of an oriented manifold M^k into a manifold $H(k, k+1)$ (see (A)), let us distinguish the special class of mappings homotopic to tangential mappings. This tangential class of mappings undoubtedly reflects certain deep properties of the differentiable manifold M^k. We believe that the tangential class of mappings does not depend on the procedure by which a topological manifold M^k is converted into a differentiable manifold; and that possibly, this class may be defined for a topological manifold.

Our problem here is to study the homological properties of the tangential class of mappings that may be expressed by the characteristic cycles of the manifold M^k (see Definition 4).

Let us introduce the following notation.

(C) Let $\chi(i)$ denote a monotone non-increasing integral-valued function of an integral-valued argument $i = 1, \ldots, k$, satisfying the condition $0 \leqslant \chi(i) \leqslant 1$. Then the function $\omega \equiv l - \chi$ defined by the relation

$$\omega(i) = l - \chi(i), \quad i = 1, \ldots, k,$$

satisfies the requirements of Definition 2, and may be used in constructing the set $Z(\omega)$ in $H(k, l)$. Put $Z_\chi = Z(l - \chi)$, then

$$D(Z_\chi) = kl - r(\chi), \quad r(\chi) = \sum_{i=1}^{k} \chi(i). \tag{2}$$

The numbers $\alpha_1, \ldots, \alpha_n, \beta_1, \ldots, \beta_n$ introduced for the function $\omega \equiv l - \chi$ (see (3), Section 2) can be calculated with the help of the function χ. For this purpose it is sufficient to find the jump points of the function χ as we did for the function ω (see (B) Section 2). Then the jump points of both of these functions coincide and the numbers $\alpha_1, \ldots, \alpha_n$ can be expressed using the function χ. The numbers $\beta_1, \ldots, \beta_n$ are determined from the formulas:

$$\beta_h = \chi(i_h) - \chi(i_h + 1), \quad h = 1, \ldots, n-1, \quad \beta_n = \chi(k).$$

The condition $\omega(1) > 0$ is equivalent to the condition $\chi(1) < l$, and the condition $\omega(k) < l$ is equivalent to the condition $\chi(k) > 0$. It is easily verified that the numbers $\alpha_1, \ldots, \alpha_n, \beta_1, \ldots, \beta_n$ determine the function χ. Below we shall show (see (B) Section 5) that when $\omega(1) > 0$, the orientability of $Z(\omega)$ depends only on the numbers $\alpha_1, \ldots, \alpha_n, \beta_1, \ldots, \beta_n$. Thus all the pseudo-manifolds Z_χ, if $l > \chi(1)$, are orientable or not, irrespective of the value of l.

We now rewrite (12), (13), and (14) derived in the last section in terms of our new notation. Notice that if $\omega = l - \chi$, then $\omega'' = l'' - \chi$; therefore we denote the pseudo-manifold $Z(\omega'')$ in $H(k, l'')$ by Z''_χ. Then, in place of (12), (13), and (14) of Section 2, we obtain

$$Z_\chi \cap H(k, l'') \text{ is empty when } \chi(1) > l'', \tag{3}$$

$$Z_\chi \times H(k, l'') = Z''_\chi, \text{ when } \chi(1) \leqslant l'', \quad \chi \neq l'', \text{ and} \tag{4}$$

$$I(Z_\chi, H(k, l'')) = 1 + (-1)^k, \text{ when } \chi(1) \equiv l''. \tag{5}$$

(D) Let θ be a mapping of a p-dimensional complex K^p into the manifold $H(k, l)$, χ a function such that $r(\chi) \leqslant p$, $\chi(1) < l$ (see (C)), and E an arbitrary oriented simplex of dimension $r(\chi)$ of the complex K^p. Set

$$Y^\theta_\chi(E) = I(X_\chi, 0(E)). \tag{6}$$

The right-hand side is the index of intersection of the pseudo-manifold Z_χ with the image of the simplex E in $H(k, l)$. It is considered modulo 2 if the pseudo-manifold Z_χ is non-orientable, and integer valued otherwise. Thus, Y^θ_χ is a function of an oriented simplex of the complex K^p. This function, as is known, is a ∇-cycle of the complex K^p. Below we shall show that the homology class of the ∇-cycle Y^θ_χ depends only on the function χ and on the equivalence class $\{\theta\}$ of θ (see (A)). The ∇-cycle Y^θ_χ is called a *characteristic ∇-cycle of the type χ of the map θ.*

If the pseudo-manifold Z_χ is non-orientable, then we have

$$\Gamma Y^\theta_\chi(E') = I(\Gamma Z_\chi, \theta(E')) \quad \text{(see (F) Section 2).} \tag{7}$$

The ∇-cycle ΓY^θ_χ is called a *secondary characteristic ∇-cycle of the type χ* of the map θ.

If a complex K^p is an oriented manifold M^k, then, as is known, a cycle X^θ_χ of dimension $k-r(\chi)$, defined up to homology, corresponds to the ∇-cycle Y^θ_χ, where X^θ is determined by the condition:

$$Y^\theta_\chi(E) = I(X^\theta_\chi, E), \tag{8}$$

where E is an arbitrary oriented $r(\chi)$-dimensional simplex of M^k. The right-hand side of (8) is the intersection index taken in M^k. The cycle X^θ_χ can also be found directly from the relation:

$$X^\theta_\chi = \theta^{-1}(Z_\chi \times \theta(M^k)). \tag{9}$$

where $Z_\chi \times \theta(M^k)$ is the algebraic intersection calculated in $H(k,l)$ and the inverse operation θ^{-1} sends this intersection back into M^k. In order to calculate this intersection, it is necessary that Z_χ and $\theta(M^k)$ should be in general position in $H(k,l)$. This can be easily achieved by substituting the mapping θ by a mapping sufficiently close to it. If the dimension of $H(k,l)$ is sufficiently large (which is always assumed in the sequel), we can always suppose that θ embeds the manifold M^k into $H(k,l)$ without self-intersections. In this case the inverse operation θ^{-1} acquires a clear meaning. We shall write (9) without making any assumptions about the generality of position, but keeping in mind that in calculations we should always replace θ by an appropriate close mapping.

The cycle X^θ_χ is called a *characteristic cycle of the type* χ of the map θ. If the pseudo-manifold Z_χ is non-orientable, the cycle ΓX^θ_χ (see (F) Section 2) is called a *secondary characteristic cycle of the type* χ of the map θ. Evidently, we have

$$\Gamma X^\theta_\chi = \theta^{-1}(\Gamma Z_\chi \times \theta(M^k)). \tag{10}$$

We now prove that the homology class of the ∇-cycle Y^θ_χ depends only on the function χ and the mapping class $\{\theta\}$.

We first take up the question of orientation. If Z_χ is orientable, then (6) presupposes fixed orientations for Z_χ and $H(k,l)$.

If $2Z_\chi \sim 0$, then $2Y^\theta_\chi \sim 0$, i.e., $-Y^\theta_\chi \sim Y^\theta_\chi$. Thus, the homology class of Y^θ_χ does not depend on the choice of orientation. If, however, $2Z_\chi \not\sim 0$, the orientations of Z_χ and $H(k,l)$ are determined by the orientation of the space R^{k+l}; if the orientation of the space R^{k+l} is changed, the orientations of Z_χ and $H(k,l)$ acquire a factor $(-1)^k$ (see (C) and (D) Section 2). Thus, in this case, too, the choice of the orientations has no effect on the homology class of Y^θ_χ.

Now let θ' and θ'' be two equivalent mappings of the complex K^p into the manifolds $H'(k,l')$ and $H''(k,l'')$, respectively (see (A)). This means that there exist linear nondegenerate mappings a' and a'' of the spaces

$R'^{k+l'}$ and $R''^{k+l''}$ into R^{k+l} such that the mappings $a'\theta'$ and $a''\theta''$ are homotopic in $H(k,l)$. Let θ be some mapping of K^p into $H(k,l)$ homotopic to $a'\theta'$ and $a''\theta''$. Let Y_χ^θ, $Y_\chi^{\theta'}$, and $Y_\chi^{\theta''}$ denote the characteristic ∇-cycles of the maps θ, θ', and θ'', respectively. We show that $Y_\chi^\theta \sim Y_\chi^{\theta'}$, and $Y_\chi^\theta \sim Y_\chi^{\theta''}$. Thus we prove $Y_\chi^{\theta'} \sim Y_\chi^{\theta''}$. Since θ' and θ'' are equivalent, it suffices to show that $Y_\chi^\theta \sim Y_\chi^{\theta''}$.

The ∇-cycle Y_χ^θ calculated for the mapping θ by the formula (6) is determined, up to homology, by the homotopy class of the mapping θ. We can therefore take $a''\theta''$ as the map θ. Put

$$a''(R''^{k+l''}) = R^{k+l''}, \quad a''(H''(k,l'') = H(k,l'').$$

Then

$$Y_\chi^\theta(E) = I(Z_\chi, \theta(E)), \quad Y_\chi^{\theta''}(E) = I(Z''_\chi, \theta(E)).$$

Since the pseudo-manifolds Z_χ and Z''_χ can be chosen arbitrarily in the manifolds $H(k,l)$ and $H(k,l'')$, we can suppose that Z''_χ is expressed using Z_χ by the formula (4). This formula then gives

$$I(Z_\chi, \theta(E)) = I(Z''_\chi, \theta(E)).$$

Thus, the invariance of a characteristic ∇-cycle has been proved.

We now apply these results to tangential mappings.

Definition 4. Let M^k be a closed differentiable oriented manifold, and T a tangential mapping of M^k into $H(k,l)$ (see Definition 3).

The characteristic ∇-cycle Y_χ^T (see (D)) is called a *characteristic ∇-cycle of the type χ* of the manifold M^k and is denoted by $Y_\chi(M^k)$ or simply as Y_χ.

The characteristic cycle X_χ^T is called a *characteristic cycle of the type χ* of the manifold M^k and is denoted by the symbol $X_\chi(M^k)$ or simply as X_χ. Thus, for $\chi(1) < l$,

$$X_\chi(M^k) = X_\chi = T^{-1}(Z_\chi \times T(M^k)), \quad D(X_\chi) = k - r(\chi). \tag{11}$$

The cycle X_χ is assumed to be a cycle modulo 2 if the pseudo-manifold Z_χ is non-orientable, and an integer valued cycle otherwise.

If Z_χ is orientable, and $2Z_\chi \not\sim 0$, the orientations of Z_χ and $H(k,l)$ should be determined from a fixed orientation of the vector space R^{k+l} (see (C) and (D) Section 2).

If Z_χ is non-orientable, the secondary characteristic cycle ΓX_χ of the type χ of the manifold M^k is given by the relation:

$$\Gamma X_\chi = T^{-1}(\Gamma Z_\chi \times T(M^k)), \quad D(\Gamma X_\chi) = k - r(\chi) - 1. \tag{12}$$

Note that in calculating the characteristic cycle X_χ with the help of formula (11), we can disregard the condition $\chi(1)<l$ (see (4)), but

$$X_\chi \sim 0 \text{ for } \chi(1) > l \quad \text{(see (3))}. \tag{13}$$

For $\chi(1)=l$, whether the cycle X_χ is to be regarded as a cycle modulo two or as an integer-valued cycle, depends on whether the pseudo-manifold Z_χ is non-orientable or orientable for $\chi(1)<l$.

(E) The characteristic cycle X_χ for $r(\chi)=k$ is of special interest. Its dimension in this case is zero and its homology class is determined by an integer or residue modulo 2, depending on whether the pseudo-manifold Z_χ is orientable or not. In this case we shall denote by X_χ the integer or the residue respectively. If Z_χ is orientable, but $2Z_\chi \sim 0$, the number X_χ is zero, because $2X_\chi \sim 0$. The characteristic number X_χ in this case is devoid of interest. For the same reason there is also no need to consider the characteristic numbers $\Gamma X_\chi (r(\chi)=k-1)$.

The study of characteristic cycles should lie in finding the relationships between these cycles. These relations may be of two different types. The relations of the first type pertain to characteristic cycles Y_χ^θ and hold true for an arbitrary mapping θ of the complex K^p. Relations of the second type pertain to characteristic cycles Y_χ of the manifold M^k and take into account the specific features of tangential mappings. Undoubtedly, the relations of the second type are deeper than those of the first type. To find the relations of the first type, we first of all have to establish the homological relationships between pseudo-manifolds Z_χ in $H(k,l)$.

In studying the mappings of the complex K^p, we are only interested in the functions χ satisfying the condition $r(\chi)\leqslant p$; but we may also assume that $p\leqslant l-1$ (see (A)), so that $r(\chi)\leqslant l-1$. This condition gives for the function $\omega\equiv l-\chi$

$$r(\omega) \geqslant kl-l+1. \tag{14}$$

Hence, for our purposes it is sufficient to study homology bases of the manifold $H(k,l)$ of dimensions r satisfying the condition

$$r \geqslant kl-l+1 \tag{15}$$

The next two sections are devoted to the study of this question.

4. Cellular Subdivision of the Manifold $H(k, l)$

We shall calculate the homology in $H(k, l)$ with the help of the Ehresmann method by subdividing $H(k, l)$ into cells of a rather general type. Ehresmann applied this method in particular to the Grassmann manifold for which $H(k, l)$ is a two-sheeted universal covering. Thus, the number of cells doubles when we pass from the Grassmann manifold to $H(k, l)$ and the whole construction becomes slightly more complicated. As it is not possible to refer to the final results of Ehresmann, we give here a complete exposition of these results for the convenience of the reader.

(A) In a vector space R^{k+l} choose a fixed basis

$$g_1, g_2, \ldots, g_{k+l} \tag{1}$$

and let Q^m denote a space with the basis

$$g_1, g_2, \ldots, g_m, \quad m \leqslant k + l. \tag{2}$$

A pseudo-manifold $Z(\omega)$ (see Definition 2) consisting of all $R^k \in H(k, l)$ and satisfying the conditions

$$D(R^K \cap Q^{\omega(i)+i}) \geqslant i, \quad i = 1, \ldots, k,$$

is denoted by $Z_0(\omega)$. Furthermore, let

$$e_1 = g_{\omega(1)+1}, \ldots, e_k = g_{\omega(k)+k}. \tag{3}$$

We write the subsequence of the sequence (1) consisting of the vectors not contained in (3) as $f_1, \ldots, f_l$. The vectors

$$e_1, \ldots, e_k, \quad f_1, \ldots, f_l \tag{4}$$

form a basis of the space R^{k+l}. Let R_0^k denote the oriented linear span of the vectors in the system (3). Then R_0^k is an element in general position in $Z_0(\omega)$, and the basis (4) satisfies the conditions imposed in (C) Section 1 on $R_0^k \in Z_0(\omega)$ (see (6) Section 1).

Thus, an oriented neighbourhood $U \cap Z_0(\omega)$ of the element R_0^k in $Z_0(\omega)$ corresponds to the basis (4). Denote this neighbourhood by $U(\omega)$ (see (C) Section 1 and (A) Section 2). The oriented image of the neighbourhood $U(\omega)$ obtained under the mapping $R^k \to \hat{R}^k$ is denoted by $\hat{U}(\omega)$. The pre-image $U(\omega)$ and the image $\hat{U}(\omega)$ are thus $r(\omega)$-dimensional oriented cells of the manifold $H(k, l)$. Below we shall show (see the auxiliary theorem) that these cells form a cellular subdivision of the manifold $H(k, l)$.

(B) Let ω' and ω be two admissible functions (see Definition 2). We write $\omega' \leqslant \omega$ if $\omega'(i) \leqslant \omega(i)$ for $i = 1, \ldots, k$. If, in addition, $\omega'(i) \not\equiv \omega(i)$, we write $\omega' < \omega$.

Let $\omega'' < \omega$ and $r(\omega'') = r(\omega) - 1$. Then it is clear that ω'' will coincide with ω for all values of the argument, except for one value, p say for which ω'' is one less than ω. Since the function ω'' is nondecreasing, the exceptional value p of the argument i cannot be arbitrary, namely $p = i_h + l$, for some $h = 0, 1, \ldots, n-1$ (see (B) Section 1). Take $\omega'' = \gamma^p(\omega) = \omega^h$. Thus,

$$\omega^h(i) = \omega(i), \quad \omega^h(i_h + 1) = \omega(i_h + 1) - 1 \quad \text{for } i \neq i_h + 1.$$

If $\omega(1) = 0$, the function ω^0 takes a negative value ($\omega^0(1) = -l$) and is therefore of no use for our purpose. In order not to repeat mentioning this case, we henceforth assume that $U(\omega^0)$, $\hat{U}(\omega^0)$, $Z_0(\omega^0)$ and $Z(\omega^0)$ are equal to zero when $\omega(1) = 0$. It is easily seen that if $\omega' < \omega$, by applying the operation γ^p to the function ω for different p, we can obtain the function ω'.

(C) For every $R^k \in H(k, l)$, there exists one and only one function ω such that R^k belongs to the sum $U(\omega) \cup \hat{U}(\omega)$. Since $U(\omega)$ and $\hat{U}(\omega)$ do not intersect (see (A) Section 1), the manifold $H(k, l)$ can be expressed as the sum of all cells constructed in (A). Further, the pseudo-manifold $Z_0(\omega)$ can be expressed as the sum of all $U(\omega')$ and $\hat{U}(\omega')$ where $\omega' \leqslant \omega$. Thus, we have

$$Z_0(\omega) = \sum_{\omega' \leqslant \omega} U(\omega') \cup \hat{U}(\omega'). \tag{5}$$

We now prove (C). Consider an integral-valued function $D(m)$ defined as follows:

$$D(m) = D(R^k \cap Q^m), \quad m = 1, \ldots, k + l. \tag{6}$$

It is readily seen that the function $D(m)$ satisfies the following conditions:

$$0 \leqslant D(1) \leqslant 1, \quad D(k + l) = k;$$

$$0 \leqslant D(m + 1) - D(m) \leqslant 1, \quad m = 1, \ldots, k + l - 1. \tag{7}$$

The last of these conditions follows from the fact that the difference of the dimensions of Q^{m+1} and Q^m is equal to one. From (7) it follows that the set of all positive values of $D(m)$ consists of the numbers $1, \ldots, k$. Let m_i denote the least m for which $D(m) \equiv i$; $i \equiv 1, \ldots, k$. Thus, we have

$$D(R^k \cap Q^{m_i}) = i;$$

$$D(R^k \cap Q^m) < i \text{ for } m < m_i; \quad D(R^k \cap Q^m) \geqslant i \text{ for } m \geqslant m_i. \quad (8)$$

This relation puts an element R^k in $H(k, l)$ into correspondence with the sequence

$$m_1, \ldots, m_k, \quad 0 < m_1 < \ldots < m_k \leqslant k+l. \quad (9)$$

Now consider the relationships:

$$R^k \in Z_0(\omega), \quad (10)$$

$$m_i \leqslant \omega(i) + i, \quad i = 1, \ldots, k; \quad (11)$$

we show that they are equivalent.

If (10) holds, then

$$D(R^k \cap Q^{\omega(i)+i}) \geqslant i, \quad i = 1, \ldots, k,$$

and from (8) we obtain (11). If (11) is valid, then from (8) it follows that

$$D(R^k \cap Q^{\omega(i)+i}) \geqslant i, \quad i = 1, \ldots, k,$$

i.e., (10) holds. Thus, (10) and (11) are equivalent.

Consider the relationships:

$$R^k \in U(\omega) \cup \hat{U}(\omega), \quad (12)$$

$$m_i = \omega(i) + i, \quad i = 1, \ldots, k. \quad (13)$$

We show that they are equivalent.

By virtue of (A), the basis (4) of the space R^{k+l} corresponds to the function ω. Let P stand for the linear span of the vectors $f_1, \ldots, f_l$ and define φ as the projection of the space R^{k+l} on R_0^k in the direction of P (see (A) Section 1). Let x_i be a vector in $R^k \cap Q^{m_i}$ that does not belong to $R^k \cap Q^{m_{i-1}}$. Evidently, the system $x_1, \ldots, x_k$ is a basis of the space R^k. Thus, we have

$$x_i = \sum_{m=1}^{m_i} b_i^m g_m, \quad i = 1, \ldots, k. \quad (14)$$

Since $U(\omega) \cup \hat{U}(\omega) \subset Z_0(\omega)$, (10) follows from (12) and, consequently, (11) follows from (12). Thus, if (12) or (13) is true, then (11) is true. Therefore

$$x_i = \sum_{m=1}^{\omega(i)+i} b_i^m g_m, \quad i = 1, \ldots, k. \quad (15)$$

Thus, we obtain

$$\varphi(x_i) = \sum_{s=1}^{i} b_i^{\omega(s)+s} e_s, \quad i = 1, \ldots, k \quad \text{(see (3))}. \tag{16}$$

Put $a_i^s = b_i^{\omega(s)+s}$; the matrix $\|a_i^s\|$ is triangular and (16) is true, if at least one of the relations (12) and (13) is true.

Suppose that (12) is true. Then the determinant of the matrix $\|a_i^s\|$ is non-zero, and, consequently, all the numbers $a_i^i \equiv b_i^{\omega(i)+i}$, $i = 1, \ldots, k$, are also non-zero. This means that the vector x_i is contained in $Q^{\omega(i)+i}$, but is not contained in $Q^{\omega(i)+i-1}$. Hence, by the choice of the vectors x_i, it at once follows that $m_i = \omega(i) + i$, i.e., (13) holds.

Suppose now that (13) is true. Then the vector x_i contained in $Q^{\omega(i)+i}$ cannot be contained in $Q^{\omega(i)+i-1}$. Therefore the number $b^{\omega(i)+i}$ is non-zero (see (15)). Hence the determinant of the matrix $\|a_i^s\|$ is non-zero, and this means that the projection φ of the space R^k on the space R_0^k is non-degenerate. Together with the inclusion $R^k \in Z_0(\omega)$ this yields (12). Thus, the relations (12) and (13) are equivalent.

Because of the equivalence of the relations (12) and (13), the relations (13) uniquely define that function ω for which (12) holds true.

From the foregoing and from the equivalence of the relations (10) and (11), we directly obtain (5).

This completes the proof of (C).

Auxiliary Theorem. The cells $U(\omega)$ and $\hat{U}(\omega)$ defined in (A) form a cellular subdivision of the manifold $\underline{H}(k, l)$. The set-theoretical boundaries $V(\omega) = \bar{U}(\omega) - U(\omega)$ and $V(\omega) = \hat{\bar{U}}(\omega) - \hat{U}(\omega)$ of the cells $U(\omega)$ and $\hat{U}(\omega)$ are given by the relation

$$V(\omega) = \hat{V}(\omega) = \sum_{\omega' < \omega} U(\omega') \cup \hat{U}(\omega'). \tag{17}$$

The algebraic boundaries $\Delta U(\omega)$ and $\Delta\hat{U}(\omega)$ of these cells are given by the relations

$$\Delta U(\omega) = \sum_{h=0}^{n-1} (U(\omega^h) + (-1)^{s(\omega,h)}\hat{U}(\omega^h))(-1)^{t(\omega,h)}, \tag{18}$$

$$\Delta\hat{U}(\omega) = \sum_{h=0}^{n-1} (\hat{U}(\omega^h) + (-1)^{s(\omega,h)}U(\omega^h))(-1)^{t(\omega,h)}, \tag{19}$$

$$\left.\begin{aligned} s(\omega, h) &= \omega(i_h + 1) + i_h + 1 + k = \omega(1) + \alpha_1 + \beta_1 + \\ &\quad \ldots + \alpha_h + \beta_h + k + 1 \qquad \text{(see Section 2, (3))}; \\ t(\omega, h) &\text{ is not calculated} \qquad \text{(see (53))} \end{aligned}\right\} \tag{20}$$

Proof. By virtue of (A), the basis

$$e_1, \ldots, e_k, \quad f_1, \ldots, f_l \tag{21}$$

of the space R^{k+l} corresponds to the function ω. Similarly, the basis

$$e_1^h, \ldots, e_k^h, \quad f_1^h, \ldots, f_l^h \tag{22}$$

of the space R^{k+l} corresponds to the function ω^h (see (B)). Both these bases (21) and (22) are obtained by rearranging the elements of the basis (1). In order to find the relationship between the bases (21) and (22), let us introduce the following notation:

$$p = i_h + 1, \quad q = \omega(i_h + 1), \quad h = 0, 1, \ldots, n-1. \tag{23}$$

Then we have

$$\begin{aligned} e_i^h &= e_i, \quad e_p^h = f_q, \quad \text{if } i \neq p \text{ and} \\ f_j^h &= f_j, \quad f_q^h = e_p, \quad \text{if } j \neq q. \end{aligned} \tag{24}$$

To verify the validity of (24), it suffices to consider how the basis (4) is obtained from the basis (1) with the help of some function ω. This is done as follows: first, the vector $g_{\omega(i)+i}$ in the sequence (1) is denoted by e_i $(i = 1, \ldots, k)$ and the remaining untouched vectors of the sequence are denoted by $f_1, \ldots, f_l$. Since ω and ω^h differ only when $i = p$, where $\omega^h(p) = \omega(p) - 1$, after the first step of our transformation the vector e_p^h precedes the vector e_p, whereas the remaining vectors e_i^h coincide with the vectors e_i. To clarify what happens in the second step of our transformation, let us establish the number of the vector f_j which stands in the sequence (1) just before the vector e_p. Evidently, $j = p + q - p = q$. Thus, in passing from the function ω to the function ω^h, the vector f_q is displaced in the sequence (1) by one step to the right and becomes the vector f_q^h. Thus we deduce the truth of (24).

Now consider *a* linear transformation a_t $(0 \leqslant t \leqslant 2)$ of the space R^{k+l} defined as follows (the vectors e_i are the elements of basis (21)):

$$a_t(e_i) = e_i \text{ if } i \neq p \quad \text{and} \quad a_t(f_j) = f_j, \text{ if } j \neq q. \tag{25}$$

$$\left.\begin{aligned} a_t(e_p) &= \left(\cos \frac{\pi}{2} t\right) e_p + \left(\sin \frac{\pi}{2} t\right) f_q, \\ a_t(f_q) &= -\left(\sin \frac{\pi}{2} t\right) e_p + \left(\cos \frac{\pi}{2} t\right) f_q. \end{aligned}\right\} \tag{26}$$

For $t = 2$ and $t = 1$, we have

$$a_2(e_i) = e_i, \quad a_2(e_p) = -e_p, \quad \text{if } i \neq p, \qquad a_2(f_j) = f_j, \quad a_2(f_q) = -f_q, \quad \text{if } j \neq q, \tag{27}$$

$$a_1(e_i) = e_i^h, \text{if } j \neq q, \quad a_1(f_j) = f_j^h, \quad a_1(f_q) = -f_q^h \quad \text{(see (24))}. \tag{28}$$

It is easy to see that $a_t(Q^{\omega(i)+i}) = Q^{\omega(i)+i}, i = 1, \ldots, k$. Therefore,

$$a_t(Z_0(\omega)) = Z(\omega). \tag{29}$$

As in (A) Section 1, let P denote the linear span of the vectors $f_1, \ldots, f_l$, and φ the operation of projection of the space R^{k+l} on the space R_0^k in the direction of P. Then

$$\varphi(a_t(e_t)) = e_i, \varphi(a_t(e_p)) = \left(\cos\frac{\pi}{2}t\right)e_p \qquad \text{for } i \neq p, \quad \text{(see (25), (26))}; \tag{30}$$

$$\varphi(a_t(j_t)) = 0; \varphi(a_t(f_q)) = -\left(\sin\frac{\pi}{2}t\right)e_p \qquad \text{for } j \neq q, \quad \text{(see (25), (26))}. \tag{31}$$

Let us introduce coordinates in the cell $U(\omega)$ in the same way as we did in (C) Section 1. That is, in the element $R_\xi^k \in U(\omega)$ there exists a basis $e_1', \ldots, e_k'$ which defines its orientation and is given by the following relations

$$e_i' = e_i + \sum_{j=1}^{l} \xi_i^j f_i, \quad i = 1, \ldots, k, \tag{32}$$

$$\xi_j^i = 0 \quad \text{for } j > \omega(i). \tag{33}$$

Here the numbers ξ_i^j satisfying the conditions (33) are the coordinates of $R_\xi^k \in U(\omega)$. These coordinates define the orientation of $U(\omega)$ as stated in (A) Section 1. The oriented image of the cell $U(\omega)$ under the mapping a_t is denoted by $a_t(U(\omega))$.

Let us now find the conditions under which $a_t(R_\xi^k)$ may belong either to $U(\omega)$ or to $\hat{U}(\omega)$. To resolve this question, we have to study the mapping φ of the space $a_t(R_\xi^k)$. We have

$$\left.\begin{aligned} &\text{for } i < p: \varphi(a_t(e_i')) = e_i, \\ &\varphi(a_t(e_p')) = \left(\cos\frac{\pi}{2}t\right)e_p - \left(\sin\frac{\pi}{2}t\right)\xi_p^q e_p, \\ &\text{for } i > p: \varphi(a_t(e_i')) = e_i - \left(\sin\frac{\pi}{2}t\right)\xi_i^q e_p. \end{aligned}\right\} \begin{aligned}&\text{(See (30), (31),}\\ &\text{(32), (33)),}\end{aligned} \tag{34}$$

Thus, the determinant of the mapping φ of the space $a_t(R^k_\xi)$ into the space R^k_0 is

$$\cos\frac{\pi}{2}t - \left(\sin\frac{\pi}{2}t\right)\xi^q_p,$$

and, consequently,

$$\left.\begin{aligned}
&\text{for } \cos\frac{\pi}{2}t - \left(\sin\frac{\pi}{2}t\right)\xi^q_p > 0:\ \varphi(a_t(R^k_\xi)) \in U(\omega),\\
&\text{for } \cos\frac{\pi}{2}t - \left(\sin\frac{\pi}{2}t\right)\xi^q_p < 0:\ \varphi(a_t(R^k)) \in \hat{U}(\omega),\quad \text{(see (29))}\\
&\text{for } \cos\frac{\pi}{2}t - \left(\sin\frac{\pi}{2}t\right)\xi^q_p = 0:\\
&\qquad \varphi(a_t(R^k_\xi)) \in Z_0(\omega) - (U(\omega) \cup \hat{U}(\omega)).
\end{aligned}\right\} \tag{35}$$

From this it follows that

$$a_t(U(\omega)) \cap U(\omega) = \left\{a_t(U(\omega)),\ \cos\frac{\pi}{2}t - \left(\sin\frac{\pi}{2}t\right)\xi_p > 0\right\}, \tag{36}$$

$$a_t(U(\omega)) \cap \hat{U}(\omega) = \left\{a_t(U(\omega)),\ \cos\frac{\pi}{2}t - \left(\sin\frac{\pi}{2}t\right)\xi^q_p < 0,\right\}. \tag{37}$$

The right-hand sides in these relations denote the domains in the cell $a_t(U(\omega))$ given by the inequalities:

$$\cos\frac{\pi}{2}t - \left(\sin\frac{\pi}{2}t\right)\xi^q_p > 0,\quad \cos\frac{\pi}{2}t - \left(\sin\frac{\pi}{2}t\right)\xi^q_p < 0,$$

where the numbers ξ^j_i satisfying the conditions (33) are the coordinates of $a_t(R^k_\xi)$ in the cell $a_t(U(\omega))$. Thus, the intersections (36) and (37) are connected domains in the cell $a_t(U(\omega))$ or are empty sets, depending on the value of the parameter t.

It directly follows from (35) and (37) that the cells $\hat{U}(\omega)$ and $a_2(U(\omega))$ coincide as sets, but they may have different orientations. Let us examine the relationship between their orientations.

Let $\hat{a}_2(R^k_\xi)$ denote the image of the element $a_2(R^k_\xi)$ under the mapping $R^k \to \hat{R}^k$. Since $\hat{a}_2(R^k_\xi) \in U(\omega)$, we have $\hat{a}_2(R^k_\xi) \equiv R^k_{\xi'}$. Let us now examine the relationship between ξ and ξ'. The element $a_2(R^k_\xi)$ has the basis $a_2(e'_1), \ldots, a_2(e'_k)$ (see (32)).

Put

$$\hat{e}_i = a_2(e'_i),\quad \hat{e}_p = -a_2(e'_p) \text{ if } i \neq p. \tag{38}$$

Then the vectors $\hat{e}_1, \ldots, \hat{e}_k$ form a basis of the element $\hat{a}_2(R^k_\xi)$. From (27), (32) and (38), we obtain

$$\hat{e}_i = e_i + \sum_{j=1}^{l} \xi'^j_i f_j, \tag{39}$$

$$\left.\begin{aligned} \xi'^j_i &= \xi^j_i \text{ if } i \neq p, j \neq q, \quad \text{and} \quad \xi'^q_i = -\xi^q_i, \text{ if } i \neq p; \\ \xi'^j_p &= -\xi^j_p, \text{ if } j \neq q; \quad \xi'^q_p = \xi'^q_p. \end{aligned}\right\} \tag{40}$$

We find that the matrix ξ' is obtained from the matrix ξ with the help of (40). The determinant of the transformation $\xi \to \xi'$ is consequently equal to $(-1)^{p+q+k+1}$, and we have

$$a_2(U(\omega)) = (-1)^{p+q+k+1}\, \hat{U}(\omega). \tag{41}$$

We now consider the cell $a_1(U(\omega))$. Its element $a_1(R^k_\xi)$ has the following basis:

$$a_1(e'_1), \ldots, a_1(e'_k).$$

By virtue of (28) and (32), this basis can be rewritten as

$$a_1(e'_i) = e^h_i + \sum_{j=1}^{l} \eta^j_i f^h_j, \tag{42}$$

$$\eta^j_i = \xi^j_i, \quad \text{if } j \neq q; \quad \eta^q_i = -\xi^q_i. \tag{43}$$

Since the numbers ξ^j_i satisfy the condition (33), the numbers η^j_i also satisfy a similar condition:

$$\eta^j_i = 0 \text{ for } j > \omega(i). \tag{44}$$

The equation $\xi^q_p = 0$ is transformed into the equation

$$\eta^q_p = 0. \tag{45}$$

The equations (44) and (45) taken together give

$$\eta^j_i = 0 \text{ for } j > \omega^h(i). \tag{46}$$

Thus, the equation $\xi^q_p = 0$ defines the cell $U(\omega^h)$ in the cell $a_1(U(\omega))$, and so we can write

$$\{a_1(U(\omega)),\ \xi^q_p = 0\} = U(\omega^h). \tag{47}$$

Orientation is not taken into account in the last equation. Hence the cell $U(\omega^h)$ subdivides the cell $a_1(U(\omega))$ into two half-spaces. Taking into account the orientations, we may write

$$\Delta\{a_1(U(\omega)),\ \xi^q_p < 0\} = \epsilon U(\omega^h);$$

$$\Delta\{a_1(U(\omega)),\ \xi_p^q > 0\} = -\epsilon U(\omega^h), \tag{48}$$

where $\epsilon = \pm 1$ and are not calculated.

By virtue of the relations (36) and (37) for $t = 1$, we find that the two half-spaces into which the cell $a_1(U(\omega))$ is subdivided by the cell $U(\omega^h)$ coincide with the intersections $a_1(U(\omega)) \cap U(\omega)$ and $a_1(U(\omega)) \cap \hat{U}(\omega)$, respectively.

Since $a_0(U(\omega)) = U(\omega)$, and for $0 \leqslant t \leqslant 1$ the intersection (36) is connected, the orientations induced in the intersection $a_1(U(\omega)) \cap U(\omega)$ from $U(\omega)$ and from $a_1(U(\omega))$ coincide. Therefore

$$\Delta U(\omega) = \epsilon U(\omega^h) + \ldots \quad \text{(see (48))}. \tag{49}$$

Since $a_2(U(\omega)) = (-1)^{p+q+k+1}\hat{U}(\omega)$ (see (41)), and for $1 \leqslant t \leqslant 2$ the intersection (37) remains connected, the orientations induced in the intersection $a_1(U(\omega)) \cap \hat{U}(\omega)$ from $(-1)^{p+q+k+1}\hat{U}(\omega)$ and from $a_1(U(\omega))$ coincide. Therefore

$$\Delta\hat{U}(\omega) = \epsilon(-1)^{p+q+k} U(\omega^h) + \ldots \quad \text{(see (48))}.$$

Applying the transformation $R^k \to \hat{R}^k$ to the last relation, we obtain

$$\Delta U(\omega) = \epsilon(-1)^{p+q+k}\hat{U}(\omega^h) + \ldots . \tag{50}$$

The relations (49) and (50) taken together give

$$\Delta U(\omega) = \epsilon(U(\omega^h)) + (-1)^{p+q+k}\hat{U}(\omega^h) + \ldots . \tag{51}$$

The relation (51) shows that both the cells $U(\omega^h)$ and $\hat{U}(\omega^h)$ are contained in the set-theoretical boundary $V(\omega)$ of the cell $U(\omega)$. Since $V(\omega)$ is closed, by virtue of (B), we conclude that $V(\omega)$ contains all the cells $U(\omega')$ and $\hat{U}(\omega')$ if $\omega' < \omega$. On the other hand, it is obvious that $V(\omega) \subset Z_0(\omega)$. Hence, by (5), we obtain (17).

By (17), the algebraic boundary of $U(\omega)$ may contain only such cells $U(\omega'')$ and $\hat{U}(\omega'')$, for which $\omega'' < \omega$, $r(\omega'') = r(\omega) - 1$. From (B) we therefore deduce that the algebraic boundary of the cell $U(\omega)$ may contain only terms of the type (51) which are taken into account by (18). Therefore (18) is true (see (23)).

The relation (19) is obtained from (18) by using the transformation $R^k \to \hat{R}^k$. Thus, we have proved the auxiliary theorem.

It should be noted that the auxiliary theorem can also be easily derived from Ehresmann's results. As far as I know, this cannot be said about the results obtained in the next section.

Define a function $\bar{\omega}$ as follows:

$$\left.\begin{aligned}\bar{\omega}(i) &= \omega(i) \quad \text{for } i \leqslant i_h + 1; \\ \bar{\omega}(i) &= \omega(i_h + 1) \quad \text{for } i \geqslant i_h + 1.\end{aligned}\right\} \tag{52}$$

Then

$$t(\omega, h) = r(\bar{\omega}) = \sum_{i=1}^{k} \bar{\omega}(i) \quad \text{(see (20))}, \tag{53}$$

and this relation can be obtained without any difficulty from (1) of Section 1 and (43) of Section 4. But we shall not make use of this relationship in our paper.

5. Homology in $H(k, l)$

In this section we shall construct an r-dimensional canonical basis of homology of the manifold $H(k, l)$ for any r satisfying the inequality:

$$r \geqslant kl - l + 1. \tag{1}$$

The construction of a homology basis for smaller dimensions is difficult and is not required for our purpose (see (15) Section 3).

The cellular subdivision discussed in the previous section underlies our study of the homology of $H(k, l)$, so we shall consider chains which are linear forms with integral coefficients in the oriented cells constructed in (A) Section 4. It is easy to see that (1) implies the condition for the function ω,

$$\omega(1) \geqslant 1 \tag{2}$$

which will henceforth play a key role.

(A) A cellular subdivision of a pseudo-manifold $Z_0(\omega)$ only contains two cells of maximum dimension, namely $U(\omega)$ and $\hat{U}(\omega)$ (see (C) Section 4). We shall regard $Z_0(\omega)$ as a chain, taking

$$Z_0(\omega) = U(\omega) + (-1)^{\omega(1)+k} \hat{U}(\omega). \tag{3}$$

Then

$$\hat{Z}_0(\omega) = (-1)^{\omega(1)+k} Z_0(\omega), \tag{4}$$

$$\Delta U(\omega) = Z_0(\omega^0)(-1)^{t(\omega,0)} + \\ + \sum_{h=1}^{n-1} (U(\omega^h) + (-1)^{s(\omega,h)} \hat{U}(\omega^h))(-1)^{t(\omega,h)}, \tag{5}$$

$$\Delta\hat{U}(\omega) = Z_0(\omega^0)(-1)^{t(\omega,0)+\omega(1)+k-1} + \\ + \sum_{h=1}^{n-1} (\hat{U}(\omega^h) + (-1)^{s(\omega,h)}U(\omega^h))(-1)^{t(\omega,h)}, \quad (6)$$

$$\Delta Z_0(\omega) = \sum_{h=1}^{n-1} (1-(-1)^{\alpha_1+\beta_1+\ldots+\alpha_h+\beta_h}) Z_0(\omega^h)(-1)^{t(\omega,h)}. \quad (7)$$

The relation (4) follows directly from (3). The relations (5) and (6) follow directly from (18), (19), and (20) of Section 4 and from (3) of Section 5. The relation (7) follows from these relations, but we must pay attention to the fact that $\omega^h(1) = \omega(1)$ for $h \geqslant 1$. Thus, the relations (4), (5), (6), and (7) are true.

(B) Let $\omega(1) \geqslant 1$. If the chain $aU(\omega) + b\hat{U}(\omega) = X$ has the property that its boundary ΔX contains the cells $U(\omega^0)$ and $\hat{U}(\omega^0)$ with zero coefficients, then

$$X = aZ_0(\omega). \quad (8)$$

Moreover, the pseudo-manifold $Z(\omega)$ (see Definition 2) is orientable, if and only if

$$\alpha_1 + \beta_1 \equiv \alpha_2 + \beta_2 \equiv \ldots \equiv \alpha_{n-1} + \beta_{n-1} \equiv 0 \pmod 2). \quad (9)$$

If the function ω is a constant identically equal to c, then it has no jumps, i.e., $n = 1$ and thus the condition (9) is satisfied. Hence the pseudo-manifold $Z(c)$ is orientable.

If follows from (5) and (6) that

$$\Delta X = \pm (a + b(-1)^{\omega(1)+k+1})Z_0(\omega^0) + \ldots,$$

so that, if the cells $U(\omega^0)$ and $\hat{U}(\omega^0)$ do not enter into ΔX, we have $b = a(-1)^{\omega^0(1)+k+1}$, and hence (8) is true. Now suppose that the pseudo-manifold $Z_0(\omega)$ is orientable; this means that with a proper choice of signs, the chain $\pm U(\omega) \pm \hat{U}(\omega)$ is a cycle. Since this chain is a cycle, the cells $U(\omega^0)$ and $\hat{U}(\omega^0)$ are not contained in its boundary, and therefore it is equal to $\pm Z_0(\omega)$ (see (8)). By virtue of (7), $Z_0(\omega)$ is a cycle if and only if the condition (9) is satisfied.

Any linear form with integral coefficients $\Sigma_\omega c_\omega Z_0(\omega)$ of chains of the type (3) for $r(\omega) = r$ is called a *Z-chain of dimension r*. The relation (7) shows that the boundary of a Z-chain is also a Z-chain.

The fact that the boundary of a Z-chain is itself a Z-chain suggests that it suffices to consider only the Z-chains in order to calculated the homology in $H(k, l)$. This hypothesis is true for dimensions that satisfy the condition (1), and thus we can narrow the class of chains under consideration.

A chain

$$\sum_{\omega} (a_{\omega} U(\omega) + b_{\omega} \hat{U}(\omega)),$$

where the summation is taken only over such functions ω whose first jump occurs at a point not less than 2 ($i_1 \geqslant 2$) is called a *special chain.* From (7) it follows that the boundary of a special Z-chain is also a special Z-chain.

Lemma. Let X be a chain whose dimension satisfies the condition (1), and whose boundary is a special chain. Then there exists a special Z-chain Y such that $X - Y$ is a cycle homologous to zero.

This lemma shows that it suffices to consider only special Z-chains in order to study the homology for dimensions satisfying the condition (1).

Proof. Let us lexicographically order the set of all functions ω satisfying the conditions of Definition 2, namely, we take $\omega_1 \prec \omega_2$ if for the least i for which $\omega_1(i) \neq \omega_2(i)$ we have $\omega_1(i) < \omega_2(i)$.

If C is an arbitrary chain, it can be written as

$$C = \sum_{p=1}^{q} (a_p U(\omega_p) + b_p \hat{U}(\omega_p)), \tag{10}$$

where for any p, the coefficients a_p and b_p do not simultaneously vanish and $\omega_1 \prec \omega_2 \prec \ldots \prec \omega_q$. Now let

$$C' = \sum_{p=1}^{q'} (a'_p U(\omega'_p) + b'_p \hat{U}(\omega'_p))$$

be some other chain written in a similar manner. If $\omega_1 \prec \omega'_1$, we consider that $C \prec C'$. Further, if $\omega_p = \omega'_p$ for $p < t$, but $\omega_t \prec \omega'_t$, then we set $C \prec C'$. In case $q' < q$ and $\omega_p = \omega'_p$ for $p \leqslant q'$, we also set $C \prec C'$. Thus we partially order the set of all chains. Here there is no order in the usual sense, because it may happen that $q = q'$ and $\omega_p = \omega'_p$ for $p = 1, \ldots, q$, but the chains C and C' may be quite distinct.

Let M be some set of chains. A chain $C \in M$ is said to be *maximal* in M, if there is no chain C' in M such that $C \prec C'$.

Among all the chains C satisfying the condition $X - C \sim 0$, let us choose a maximal chain and denote it by Y. Let

$$Y = \sum_{p=1}^{q} (a_p U(\omega_p) + b_p \hat{U}(\omega_p))\text{+}$$

This expression has been written in the form of (10). We now prove that Y is a special Z-chain.

First, note that, by virtue of the conditions imposed on the dimension of the chain Y (see (1)), we have $\omega_p(1) \geqslant 1$ (see (2)).

Suppose that we have already proved that for $p < t$. the first jump of the function ω_p occurs at a point not less than 2 ($i_1 \geqslant 2$), and

$$a_p U(\omega_p) + b_p \hat{U}(\omega_p) = Y_p = c_p Z_0(\omega_p)$$

(for $t = 1$ this condition makes no sense). We then prove that the first jump of the function ω_t occurs at a point not less than two and that

$$a_t U(\omega_t) + b_t \hat{U}(\omega_t) = Y_t = c_t Z_0(\omega_t).$$

We have

$$\Delta Y_t = dU(\omega_t^0) + e\hat{U}(\omega_t^0) + \ldots .$$

We now show that $d = e = 0$; and this will imply that $Y_t = c_t Z_0(\omega_t)$ (see (B)). Since $\omega_t(1) \geqslant 1$, the first jump of the function ω_t^0 occurs at one, and therefore ΔY_t is a special chain only if $d = e = 0$. Since ΔY is a special chain, the terms $dU(\omega_t^0) + e\hat{U}(\omega_t^0)$ should cancel in calculating ΔY. They may not, however, cancel with terms contained in ΔY_p when $p < t$, because ΔY_p is a special chain for $p < t$ (see (D)). Similarly, the terms $dU(\omega_t^0) + e\hat{U}(\omega_t^0)$ cannot be cancelled by the terms contained in ΔY_p for $p > t$, because $\omega_t^0 \prec \omega_p^h$ for $p > t$. Thus, $d = e = 0$ and $Y_t = c_t Z_0(\omega_t)$.

We now assume that the first jump of the function ω_t occurs at one ($i_1 = 1$); then there exists a function ω such that $\omega_t = \omega^0$, and therefore

$$\Delta U(\omega) = \epsilon Z_0(\omega_t) + \ldots, \quad \epsilon = \pm 1 \quad \text{(see (5))}.$$

It is easy to see that

$$Y - \epsilon c_t \Delta U(w_t) \succ Y,$$

i.e., the chain Y does not satisfy the maximality condition imposed on it. Thus the first jump of the function ω_t occurs at a point not less than two.

Let us now introduce some notation convenient for our further work.

(C) Let Ω' denote the set of all functions ω (see Definition 2) whose first jump occurs at a point not less than two ($i_1 \geqslant 2$) (see (B) Section 1).

Associate the numbers $\alpha_1, \ldots, \alpha_n, \beta_1, \ldots, \beta_n$ (see (3) Section 2) with the function ω. Now consider the sequences:

$$\left.\begin{array}{l}\alpha_1, \beta_1, \alpha_2, \beta_2, \ldots, \alpha_{n-1}, \beta_{n-1}, \alpha_n, \\ \alpha_1, \beta_1, \alpha_2, \beta_2, \ldots, \alpha_{n-1}, \beta_{n-1}.\end{array}\right\} \quad (11)$$

If $\omega(k) < 1$, associate the first of these sequences with the function ω, and if $\omega(k) = 1$ the second.

Let the set Ω consist of the functions $\omega \in \Omega'$ whose corresponding sequence (11) does not contain odd numbers. In particular, the function $\omega \equiv l$ is included in Ω, because for this function we have $n = 1$ and $\omega(k) = l$.

Let the set Ω_α consist of the functions $\omega \in \Omega'$ whose corresponding sequence (11) contains odd numbers of which the first is some α_h.

Let the set Ω_β consist of the functions $\omega \in \Omega'$ whose corresponding sequence (11) contains odd numbers of which the first is some β_h.

Thus, the set Ω' is subdivided into three disjoint sets Ω, Ω_α, and Ω_β.

It follows from (B) that for $\omega \in \Omega$ the pseudo-manifold $Z(\omega)$ is orientable and for $\omega' \in \Omega_\beta$ the pseudo-manifold $Z(\omega')$ is non-orientable. If $\omega \equiv l - \chi$ (see (C) Section 3), then knowing χ, we can determine whether ω belongs to Ω' or not. Similarly, if $\omega \in \Omega'$, then knowing χ, we can say to which of the three sets $\Omega, \Omega_\alpha, \Omega_\beta$ ω belongs. Let X, X_α, and X_β denote the corresponding sets for the functions χ.

Theorem 1. The pseudo-manifold $Z(\omega)$, for $\omega \in \Omega$, is orientable (see (C)), and when taken with some orientation, may be regarded as a cycle of dimension $r(\omega)$. The pseudo-manifold $Z(\omega')$, for $\omega' \in \Omega_\beta$, is not orientable (see (C)) and $\Gamma Z(\omega')$ (see (F) Section 2) is therefore a cycle of dimension $r(\omega') - 1$. A canonical homology basis of dimension $r \geqslant kl - l + 1$ of the manifold $H(k, l)$ consists of the cycles

$$Z(\omega), \ \omega \in \Omega, \ r(\omega) = r; \quad \Gamma Z(\omega'), \ \omega' \in \Omega_\beta, \ r(\omega') - 1 = r, \quad (12)$$

where $Z(\omega)$ are free cycles and $\Gamma Z(\omega')$ are cycles of order 2.

Proof. We prove this in several steps.

(a) Let $\omega \in \Omega_\beta$, then

$$\Delta Z_0(\omega) = \sum_{h=1}^{n-1} a_h Z_0(\omega^h), \quad (13)$$

where the coefficients a_h may take the values ± 2 or 0 (see (7)). It turns out that (13) contains coefficients which are distinct from zero; let a_q be the first such coefficient and put $\omega^q = \Psi(\omega)$. Furthermore, we find

$$\text{if } m > q\colon\ \omega^m \in \Omega_\beta,\quad \Psi(\omega) = \omega^q \in \Omega_\alpha. \tag{14}$$

Moreover, the mapping Ψ of the set Ω_β into the set Ω_α is a one-to-one mapping of Ω_β onto Ω_α.

We now prove (a). Denote the numbers α and β (see (3) Section 2) corresponding to the function ω by $\alpha_1, \ldots, \alpha_n, \beta_1, \ldots, \beta_n$. Denote the numbers that correspond to the function ω^m by $\alpha_1^m, \ldots, \alpha_n^m, \beta_1^m, \ldots, \beta_n^m$. Then, we have

$$\text{if } h < m\colon\quad \alpha_h^m = \alpha_h,\quad \beta_h^m = \beta_h, \tag{15}$$

$$\text{if } \beta_m = 1\colon\quad \alpha_m^m = \alpha_m + 1, \tag{16}$$

$$\text{if } \beta_m > 1\colon\quad \alpha_m^m = \alpha_m,\quad \beta_m^m = \beta_m - 1,\quad \alpha_{m+1}^m = 1. \tag{17}$$

By virtue of (7), the number a_h is non-zero if and only if the number

$$(\alpha_1 + \beta_1) + \ldots + (\alpha_n + \beta_n) \tag{18}$$

is odd. Thus, q is determined as the least value of h for which (18) is odd. Since $\omega \in \Omega_\beta$, β_q is the first odd number in the sequence (11) that corresponds to the function ω.

If $m > q$, it follows from (15) that the first odd number in the sequence (11) that corresponds to the function ω^m is β_q^m, i.e., if $m > q$, we have $\omega^m \in \Omega_\beta$.

Now let $m = q$; then for $\beta_q = 1$, the first odd number in the sequence (11) that corresponds to the function ω^q is α_q^q (see (15) and (16)), i.e., $\omega^q \subset \Omega_\alpha$; if, however, $\beta_q > 1$, the first odd number in the sequence (11) that corresponds to the function ω^q is α_{q+1}^q (see (15) and (17)), i.e., $\omega^q \in \Omega_\alpha$. Thus, we have demonstrated the relation (14).

Let ω^* be an arbitrary function in Ω_α; the numbers α and β that correspond to this function are denoted by $\alpha_1^*, \ldots, \alpha_{n*}^*, \beta_1^*, \ldots, \beta_{n*}^*$. We shall search for a function ω such that for some $h > 0$, we have $\omega^* = \omega^h$. It is readily seen that this function is given by the following expression:

$$\left.\begin{aligned} &\text{if}\quad i \neq \alpha_1^* + \ldots + \alpha_p^*\colon\quad \omega(i) = \omega^*(i),\\ &\text{if}\quad i = \alpha_1^* + \ldots + \alpha_p^*\colon\quad \omega(i) = \omega^*(i) + 1.\end{aligned}\right\} \tag{19}$$

(Here the equality $p = n^*$ is possible only when $\omega^*(k) < 1$). Indeed

$$\omega^* = \omega^{p-1} \text{ if } \alpha_p^* = 1 \text{ and } \omega^* = \omega^p \text{ if } \alpha_p^* > 1. \tag{20}$$

Denoting the numbers α and β corresponding to the function ω by $\alpha_1, \ldots, \alpha_n, \beta_1, \ldots, \beta_n$, we obtain

$$\text{for}\quad h < p-1\colon\ \alpha_h = \alpha_h^*;\ \beta_h = \beta_h^*;\ \alpha_{p-1} = \alpha_{p-1}^*, \tag{21}$$

$$\text{for} \quad \alpha_p^* = 1: \beta_{p-1} = \beta_{p-1}^* + 1, \tag{22}$$

$$\text{for} \quad \alpha_p^* > 1: \beta_{p-1} = \beta_{p-1}^*, \alpha_p = \alpha_p^* - 1, \beta_p = 1. \tag{23}$$

Now let α_t^* be the first odd number in the sequence (11) corresponding to the function ω^*. It is easy to establish that, for $p<t$, the chain $Z_0(\omega^*)$ is contained in $\Delta Z_0(\omega)$ with a zero coefficient. If $p>t$, the function ω belongs to Ω_α, and only when $p=t$ do we have $\omega \in \Omega_\beta$ and $\omega^* = \Psi(\omega)$. Thus, Ψ is a one-to-one mapping of Ω_β onto Ω_α.

(b) If $\omega' \in \Omega_\beta$, we shall determine a cycle $\Gamma Z(\omega')$ of dimension $r(\omega')-1$, taking

$$\Delta Z_0(\omega') = 2\Gamma Z_0(\omega') \quad \text{(see (13))}. \tag{24}$$

Then, the chains

$$Z_0(\omega),\ \omega \in \Omega,\ r(\omega) = r; \quad \Gamma Z_0(\omega'),\ \omega' \in \Omega_\beta,\ r(\omega') - 1 = r;$$
$$Z_0(\omega''),\ \omega'' \in \Omega_\beta,\ r(\omega'') = r, \tag{25}$$

when taken together, form a linearly independent basis of r-dimensional special Z-chains for any r, while the chains

$$Z_0(\omega),\ \omega \in \Omega,\ r(\omega) = r; \quad \Gamma Z_0(\omega'),\ \omega' \in \Omega_\beta,\ r(\omega') - 1 = r, \tag{26}$$

form a linearly independent basis of all r-dimensional special Z-cycles.

We first show that the system (26) is linearly independent. Let

$$\sum_{\omega \in \Omega} a_\omega Z_0(\omega) + \sum_{\omega' \in \Omega_\beta} b_{\omega'} \Gamma Z_0(\omega') = 0. \tag{27}$$

If $b_{\omega'_1} \neq 0$, then (27) contains the term $\pm b_{\omega'_1} Z_0(\Psi(\omega'_1))$ (see (13)) which cannot be cancelled from a similar term $\pm b_{\omega'_2} Z_0(\Psi(\omega'_2))$ in (27) if $\omega'_1 \neq \omega'_2$, because in this case we have $\Psi(\omega'_1) \neq \Psi(\omega'_2)$ (see (a)). Further, the term $+ b_{\omega'_1} Z_0(\Psi(\omega'_1))$ cannot be cancelled by terms in $b_{\omega'_2} \Gamma Z_0(\omega'_2)$ either (see (14) and (24)). Similarly, the term $\pm b_{\omega'_1} Z_0(\Psi(\omega'_1)$ cannot be cancelled by terms of the type $a_\omega Z_0(\omega)$, because $\omega \in \Omega$, $\Psi(\omega'_1) \in \Omega_\alpha$.

Thus, all coefficients $b_{\omega'}$ are equal to zero. Since the system $Z_0(\omega)$, $\omega \in \Omega$ is linearly independent, all the coefficients a_ω are also equal to zero. Thus, we have demonstrated the linear independence of the system (26).

We now prove the linear independence of the system (25). Let

$$\sum_{\omega \in \Omega} a_\omega Z_0(\omega) + \sum_{\omega' \in \Omega_\beta} b_{\omega'} \Gamma Z_0(\omega') + \sum_{\omega'' \in \Omega_\beta} c_{\omega''} Z_0(\omega'') = 0. \tag{28}$$

Taking the boundary of the left-hand side in this expression, we obtain

$$\sum_{\omega'' \in \Omega_\beta} 2c_{\omega''} \Gamma Z_0(\omega'') = 0. \tag{29}$$

Since the system (26) has already been shown to be linearly independent, it follows from (29) that all the coefficients $c_{\omega''}$ are equal to zero. Hence, it is clear that the linear form of chains (25) may be a cycle if it contains only cycles of the system (26).

As we have demonstrated that all $c_{\omega''}$ are zero, the expression (28) is converted into (27); therefore all a_ω and $b_{\omega'}$ are also equal to zero. Thus, the system (25) is linearly independent.

We now prove that any special Z-chain X of dimension r can be expressed as a linear form of the chains (25). Consider a chain $X + Y$, where Y is a linear form of the chains of the system (26). From among all chains of the type $X + Y$, where X is given and Y is a linear form as stated above, let us choose a maximal chain $X' = X + Y'$ (maximal in the sense defined in the proof of the lemma). We shall show that

$$X' = \sum_{\omega'' \in \Omega_\beta} c_{\omega''} Z_0(\omega''). \tag{30}$$

Express the chain X' as follows:

$$X' = \sum_{p=1}^{q} c_p Z_0(\omega_p), \quad \omega_p \in \Omega', \quad p = 1, \ldots, q, \tag{31}$$

where the coefficients c_p are non-zero and $\omega_1 \prec \omega_2 \prec \ldots \prec \omega_q$ (the ordering of the functions $\omega \in \Omega'$ is the same as that used in the proof of the lemma).

We show that $\omega_p \in \Omega_\beta$. Assume that $\omega_p \in \Omega$. Then, obviously, we have

$$X' - c_p Z_0(\omega_p) \succ X'.$$

and, consequently, X' does not satisfy the maximality condition. Thus, ω_p cannot belong to Ω. Now, assume that $\omega_p \in \Omega_\alpha$. Then there exists a function $\omega' \in \Omega_\beta$, such that $\omega_p = \Psi(\omega')$ (see (a)), and if

$$\Gamma Z_0(\omega') = \epsilon Z_0(\omega_p) + \ldots, \quad \epsilon = \pm 1 \quad \text{(see (13), (24))}$$

then clearly,

$$X' - \epsilon c_p \Gamma Z_0(\omega') \succ X',$$

i.e., X' again does not satisfy the maximality condition. Hence the only possibility is that $\omega_p \in \Omega_\beta$ and thus we have demonstrated the truth of (30).

Hence we now have $X = -Y' + X'$, where $-Y' + X'$ is a linear form of the chains of the system (25).

This completes the proof of (b).

(c) Since, by virtue of the lemma, every cycle of dimension $r \geqslant kl - l + 1$ is homologous to a special Z-cycle, and a special Z-cycle of a dimension $r \geqslant kl - l + 1$ homologous to zero is a boundary of a special Z-chain, it follows from (b) that the system (26) is a canonical homology basis for the manifold $H(k, l)$ of dimension $r \geqslant kl - l + 1$ (see (24)), where the cycles $Z_0(\omega)$, $\omega \in \Omega$ are free and the cycles $\Gamma Z_0(\omega')$, $\omega' \in \Omega_p$, are of order two.

(d) In place of the specially chosen cycle $Z_0(\omega)$ we can also take any cycle $Z(\omega)$, $\omega \in \Omega$ (see Definition 2 Section 1). Similarly, in place of the cycle $\Gamma Z_0(\omega')$, we can also take any $\Gamma Z(\omega')$, $\omega' \in \Omega_\beta$.

This completes the proof of our theorem.

It follows from Theorem 1 that all torsion coefficients of dimension $r \geqslant kl - l + 1$ of the manifold $H(k, l)$ are equal to two. Hence, there is hardly any need for considering in $H(k, l)$ modular cycles of these dimensions modulo a number different from two.

(D) If $r \geqslant kl - l + 1$, an r-dimensional homology basis modulo 2 of $H(k, l)$ can be formed from the cycles

$$\left.\begin{aligned} Z(\omega),\ \omega \in \Omega,\ r(\omega) = r;\quad \Gamma Z(\omega'),\ \omega' \in \Omega_\beta,\ r(\omega') - 1 = r; \\ Z(\omega''),\ \omega'' \in \Omega_\beta,\ r(\omega'') = r, \end{aligned}\right\} \tag{32}$$

considered as cycles modulo two (see Theorem 1); a basis can also be formed from the pseudo-manifolds

$$Z(\omega), \quad \omega \in \Omega', \quad r(\omega) = r, \tag{33}$$

regarded as cycles modulo two.

The fact that the system (32) is a homology basis modulo 2 follows from Theorem 1 by virtue of known results.

To verify that the system (33) is also a homology basis modulo two, consider in place of (33) the system

$$Z_0(\omega), \quad \omega \in \Omega', \quad r(\omega) = r, \tag{34}$$

which is equivalent to (33):

$$Z_0(\omega) \sim Z(\omega) \pmod 2.$$

It follows directly from our lemma (considered modulo two) and from the relation (7) that (34) is a homology basis modulo two of the manifold $H(k,l)$.

6. Some Properties of Characteristic Cycles

In this section we shall consider the characteristic cycles of a closed orientable differentiable manifold M^k, i.e., the cycles X_χ, $r(\chi)\leqslant k$ and $\Gamma X_{\chi'}, r(\chi')+1\leqslant k$ (see Definition 4).

The cycle X_χ is a cycle with integral coefficients or a cycle modulo two, depending on whether the pseudo-manifold $Z(l-\chi)=Z$, for $l>\chi(1)$, is orientable or not. By virtue of (B) Section 5, we can distinguish between these two cases on the basis of the properties of the function χ.

(A) Associate numbers $\alpha_1,\ldots,\alpha_n$, $\beta_1,\ldots,\beta_n$ with the function χ (see (C) Section 3). Consider the relation

$$\alpha_1+\beta_1\equiv\alpha_2+\beta_2\equiv\ldots\equiv\alpha_{n-1}+\beta_{n-1}\equiv 0 \pmod 2. \tag{1}$$

Let X_0 denote the set of all functions χ for which relations (1) are satisfied, and X_2 the set of all other functions χ.

It follows directly from (B) Section 5 that for $\chi\in X_0$, X_χ is a cycle with integral coefficients, and for $\chi\in X_2$, X_χ is a cycle modulo 2.

It follows from (C) Section 5 that

$$X\subset X_0,\quad X_\beta\subset X_2. \tag{2}$$

Furthermore, for $\chi\in X_0$, i.e., when Z_χ is orientable, for $l\geqslant\chi(1)$, but $2Z_\chi\not\sim 0$ in $H(k,l)$, we have

$$\chi\in X. \tag{3}$$

This also follows from (B) Section 5 and from (C) Section 2. Thus, if $\chi\in X_0-X$, then

$$2X_\chi\sim 0. \tag{4}$$

Using Theorem 1 we can choose from the set of all characteristic cycles a set of cycles by means of which all other characteristic cycles can be expressed.

(B) The characteristic cycles X_χ, $\chi\in X$ and $X_{\chi'}$, $\chi'\in X_\beta$ (see (C) Section 5) we shall consider as basic cycles. All other characteristic cycles can be expressed in terms of these basic cycle. If $\chi^*\in X_0-X$, then

$$X_{\chi^*} \sim \sum_{\chi \in X_\beta} b_\chi \Gamma X_\chi, \tag{5}$$

where b_χ are residues modulo two determined by the function χ^*.

If $\chi^* \in X_2$, then

$$X_{\chi^*} \sim \sum_{\chi \in X} a_\chi X_\chi + \sum_{\chi \in X_\beta} b_\chi \Gamma X_\chi + \sum_{\chi \in X_\beta} c_\chi X_\chi \quad (\mathrm{mod}\ 2), \tag{6}$$

where a_χ, b_χ, and c_χ are residues modulo two determined by the function χ^*. Finally, if $\chi^* \in X_2$, it follows from (6) that

$$\Gamma X_{\chi^*} \sim \sum_{\chi \in X_\beta} c_\chi \Gamma X_\chi. \tag{7}$$

We now prove (5). Theorem 1 gives that

$$Z_{\chi^*} \sim \sum_{\chi \in X} a_\chi Z_\chi + \sum_{\chi \in X_\beta} b_\chi \Gamma Z_\chi, \tag{8}$$

where a_χ are integers and b_χ are residues modulo two. Since Theorem 1 gives a canonical basis, a_χ and b_χ are determined uniquely by the function $l-\chi^*$. The relation (4) of Section 3 shows that a_χ and b_χ do not depend on the number l and depend only on the function χ^*. Since $\chi^* \in X_0 - X$, we have $2Z_{\chi^*} \sim 0$ (see (A)). Therefore, all a_χ in the relation (8) are equal to zero. Hence, the relation (8) gives (5) (see (11) and (12), Section 3).

We now prove (6). By virtue of (D) Section 5, we have

$$Z_{\chi^*} \sim \sum_{\chi \in X} a_\chi Z_\chi + \sum_{\chi \in X_\beta} b_\chi \Gamma Z_\chi + \sum_{\chi \in X_\beta} c_\chi Z_\chi \quad (\mathrm{mod}\ 2); \tag{9}$$

where a_χ, b_χ, and c_χ are residues modulo two. Since the basis defined in (D) Section 5 is independent modulo two, a_χ, b_χ, and c_χ are uniquely determined by the function χ^*; but, by virtue of (4) Section 3, a_χ, b_χ, and c_χ do not depend on the number l and are determined only by the function χ^*. Thus the relation (9) gives (6) (see (11) and (12), Section 3).

(C) If $\chi(k) > 0$, it follows directly from the condition $r(\chi) \leqslant k$, that $\chi \equiv 1$. Thus, for $\chi(k) > 0$, we always have $\chi \equiv 1$. The dimension of the cycle $X_1 = X_\chi$ $(\chi \equiv 1)$ is zero, and therefore it defines the characteristic number X_1 (see (F) Section 3). In [5] it is proved that the characteristic number $X_1(M^k)$ is the Euler characteristic $\mathscr{E}(M^k)$ of the manifold M^k. The function $\chi \equiv 1$ belongs to X_0; it belongs to X if and only if k is even.

Thus, for odd k, the number X_1 is equal to zero (see (E), Section 3), and therefore obviously coincides with the Euler characteristic.

(D) Among all characteristic cycles X_χ with integral coefficients, of special interest are those cycles for which $\chi \in X$, because the relation $2X_\chi \sim 0$ is violated, in general, only for these cycles (see (4)). We have already considered one such cycle, namely, X_1 for an even k (see (C)). If $\chi \in X$, $\chi \not\equiv 1$, then $\chi(k) = 0$; therefore for the function χ the numbers $\alpha_1, \ldots, \alpha_{n-1}$, $\beta_1, \ldots, \beta_{n-1}$ are all even (see (C) Section 5). Hence, if $\chi \in X, \chi \not\equiv 1$, we have

$$\left.\begin{aligned} \chi(1) &= \chi(2) \geqslant \ldots \geqslant \chi(2p-1) = \chi(2p) \geqslant \chi(2p+1) = \ldots = \\ &= \chi(k) = 0; \quad \chi(i) \equiv 0 \pmod 2; \quad r(\chi) \leqslant k. \end{aligned}\right\} \tag{10}$$

From this it is clear that $r(\chi)$ is divisible by four. All characteristic numbers X_χ are determined by the conditions: $\chi \in X$, $r(\chi) = k$. The characteristic number X_1 may differ from zero only when k is even, and the other characteristic numbers X_χ given by (10) and by the condition $r(\chi) = k$ exist only when k is divisible by four. When $k = 2$, there exists only one characteristic number, namely, X_1. When $k = 4$, there exists, besides X_1, one more characteristic number X_χ, defined by the formulas:

$$\chi(1) = \chi(2) = 2, \quad \chi(3) = \chi(4) = 0. \tag{11}$$

We now investigate the behaviour of characteristic cycles when the orientation of the manifold M^k is reversed, i.e., in passing from M^k to $-M^k$.

Theorem 2. In passing from an oriented manifold M^k to an oriented manifold $-M^k$ of opposite orientation, we have

$$X_\chi(-M^k) \sim X_\chi(M^k) \text{ if } \chi \text{ does not belong to } X, \tag{12}$$

$$X_1(-M^k) \sim X_1(M^k) \text{ for } \chi \equiv 1, \text{ and} \tag{13}$$

$$X_\chi(-M^k) \sim -X_\chi(M^k) \text{ for } \chi \in X, \chi \not\equiv 1. \tag{14}$$

Proof. Let $M^k \subset R^{k+l}$ and let T be a tangential mapping of the manifold M^k into $H(k,l)$. Let $\hat{T}(\chi)$ denote the vector subspace of the space R^{k+l} differing from $T(\chi)$ only in its orientation. Then $\hat{T}$ is a tangential mapping of the manifold $-M^k$ into $H(k,l)$.

Furthermore, set $H = H(k,l)$. The relation (4) Section 5 has been demonstrated for the chain $Z_0(\omega)$, but it holds for a pseudo-manifold

$Z_0(\omega)$ as well. Indeed, if $Z_0(\omega)$ is non-orientable, the relation (4) makes no sense; if, however, $Z_0(\omega)$ is orientable, when taken with the proper orientation and regarded as a cycle, it coincides with the chain $Z_0(\omega)$. Since we can pass from a pseudo-manifold $Z_0(\omega)$ to any pseudo-manifold $Z(\omega)$ by rotating the space R^{k+l}, the relation (4) Section 5 holds for any pseudo-manifold $Z(\omega)$. Thus, from (4) we obtain

$$\hat{Z}_\chi = (-1)^{k+l+\chi(1)} Z_\chi; \quad \hat{H} = (-1)^{k+l} H. \tag{15}$$

In the proof we shall use characteristic ∇-cycles. Let $\hat{Y}_\chi$ denote a characteristic ∇-cycle of the manifold $-M^k$. Then

$$Y_\chi(E) = I_H(Z_\chi, T(E)); \quad \hat{Y}_\chi(E) = I_H(Z_\chi, \hat{T}(E)), \tag{16}$$

where E is an arbitrary oriented simplex of dimension $r(\chi)$ of the manifold M^k. To calculate the right-hand side of the second equation in (16), we apply the homeomorphic transformation $R^k \to \hat{R}^k$ to the manifold $H(k, l)$. Thus, we obtain

$$I_H(Z_\chi, \hat{T}(E)) = I_{\hat{H}}(\hat{Z}_\chi, T(E)) = (-1)^{\chi(1)} I_H(Z_\chi, T(E)) \quad \text{(see (15))}. \tag{17}$$

From (16) and (17) we obtain

$$\hat{Y}_\chi = (-1)^{\chi(1)} Y_\chi. \tag{18}$$

In passing from a ∇-cycle to the corresponding Δ-cycle, the orientation of the manifold M^k is also taken into account (see (8), Section 3). From (18) we obtain

$$X_\chi(-M^k) = (-1)^{\chi(1)+1} X_\chi(M^k). \tag{19}$$

If $\chi \in X_2$, then X_χ is a cycle modulo two, and the relation (19) gives (12). If $\chi \in X_0 - X$, then we have $X_\chi \sim -X_\chi$ (see (4)) and the relation (19) gives (12). If $\chi \equiv 1$, then $\chi(1) = 1$, and the relation (19) implies (13). Finally, if $\chi \in X$, $\chi \not\equiv 1$, then $\chi(k) = 0$ (see (C)), and the numbers $\beta_1, \ldots, \beta_{n-1}$ are all even (see (C) Section 5). Since $\chi(1) = \beta_1 + \ldots + \beta_{n-1} + \chi(k)$, the relation (19) gives (14).

This completes the proof of Theorem 2.

From Theorem 2 we directly obtain the following corollary:

(E) If a manifold M^k admits a homeomorphic differentiable mapping onto itself which inverts the orientation, then every characteristic cycle X_χ, $\chi \not\equiv 1$ of the manifold M^k satisfies the condition $2X_\chi \sim 0$. In particular, every characteristic number X_χ, $\chi \not\equiv 1$ of the manifold M^k is equal to zero.

Theorem 3. If a manifold M^k is the boundary of an oriented differentiable bounded manifold M^{k+1}, the characteristic number X_1 of the manifold M^k is even and all other characteristic numbers and residues of this manifold are equal to zero.

Proof. We assume that $M^{k+1} \subset R^{k+l}$. On the manifold M^{k+1} we can easily define a differentiable numerical function $f(x)$, $x \in M^{k+1}$ satisfying the following conditions:

(a) $f(x) \geqslant 0$,

(b) the equation $f(x) = 0$ gives the boundary M^k of the manifold M^{k+1}, and

(c) the critical points of the function $f(x)$ are isolated and lie outside the manifold M^k.

Let a be an arbitrary noncritical point in M^{k+1}. Let us orient the level surface $f(x) = f(a)$ in such a way that it is the boundary of the set $f(x) \leqslant f(a)$, and draw an oriented tangent T'_a at the point a to this level surface. Let $T'(a)$ denote the oriented vector subspace of the space R^{k+l} that is parallel to the oriented plane T'_a. Then $T'(a) \in H(k, l)$, and we obtain a mapping T' which is well-defined and continuous at any non-critical point $a \in M^{k+1}$, and which is a tangential mapping of the manifold M^k.

Let $a_1, \ldots, a_t$ be the set of all critical points of the function $f(x)$. Let P_j^{k+1} denote the tangent to M^{k+1} at the point a_j, and R_j^{k+1} the vector subspace of the space R^{k+l} parallel to this tangent.

Further, assume that δ is a small positive number. Let E_j^{k+1} denote the spherical neighbourhood of the point a_j in M^{k+1} of radius δ oriented in agreement with M^{k+1}, and S_j^k its boundary taken with a proper orientation.

Furthermore, assume that

$$M_\delta^{k+1} = M^{k+1} - (E_1^{k+1} \cup \ldots \cup E_t^{k+1})$$

Then the mapping T' is well-defined everywhere on the bounded manifold M^{k+1} and, consequently,

$$T'(M^k) \sim T'(S_1^k) + \ldots + T'(S_t^k) \text{ in } H(k, l). \tag{20}$$

Thus, to calculate the characteristic number or the residue X_χ of the manifold M^k, it is sufficient to calculate the intersection index $I(Z_\chi, T'(S_j^k))$ for all $j = 1, \ldots, t$.

If δ is sufficiently small, then for $a \in S_j^k$ the plane $T'(a)$ is close to its projection $T''(a)$ on R_j^{k+1} in the sense of the topology of the manifold $H(k, l)$. Hence the mapping T' of the sphere S_j^k into $H(k, l)$ can be transformed by a small displacement into the mapping T'', such that for $a \in S_j^k$, $T''(a) \subset R_j^{k+1}$.

If $H_j(k, 1)$ denotes all the k-dimensional oriented subspaces of the space R_j^{k+1}, then T'' maps S_j^k into $H_j(k, 1)$. Since the dimension of $H_j(k, 1)$ is k, the mapping T'' of the sphere S_j^k into $H_j(k, 1)$ has a certain degree γ_j, and we obtain

$$I(Z_\chi, T'(S_j^k)) = \gamma_j I(Z_\chi, H_j(k, 1)). \tag{21}$$

We now make use of the relations (3) and (5) of Section 3, assuming that $R_j^{k+1} = R^{k+l''}$, i.e., we take $l'' = 1$. If $\chi \equiv 1$, then from (5) Section 3, we obtain

$$I(Z_1, H_j(k, 1)) = 1 + (-1)^k. \tag{22}$$

If $\chi \not\equiv 1$, then from $r(\chi) = k$ it follows that $\chi(1) > 1$, and the relation (3), Section 3 gives

$$I(Z_\chi, H_j(k, 1)) = 0. \tag{23}$$

The truth of Theorem 3 now follows from (20), (21), (22), and (23).

As the characteristic number X_1 is equal to the Euler characteristic of a manifold, Theorem 3 implies that a manifold M^k which is the boundary of an oriented manifold M^{k+1} has an even Euler characteristic. This assertion can easily be proved for combinatorial manifolds M^k and M^{k+1} by glueing two copies of the manifold M^{k+1} along their boundaries and by calculating the Euler characteristic from the number of simplices.

Thus, there exist four-dimensional oriented manifolds which cannot be the boundaries of five-dimensional oriented manifolds. The simplest example is the complex projective plane whose Euler characteristic is equal to three.

It would be interesting to prove that every three-dimensional oriented manifold is the boundary of a four-dimensional oriented bounded manifold.

The assertions (F) and (G) stated below may be helpful in calculating the characteristic cycles of certain manifolds.

(F) If there exists a regular mapping f of a manifold M^k into the vector space R^{k+1} (see (1) Section 3), then all characteristic cycles of the manifold M^k, except possibly, X_1 (see (D)), are homologous to zero in this manifold, and the characteristic number X_1 is even.

Let $R^{k+1} \subset R^{k+l}$, where l is sufficiently large, then $H(k, 1) \subset H(k, l)$. The manifold $H(k, 1)$ is homeomorphic to a k-dimensional sphere. The tangential mapping T of the manifold M^k constructed from the regular mapping f (see Definition 3) is such that $T(M^k) \subset H(k, 1)$.

If $\chi(1) > 1$, then by (3) Section 3, the pseudo-manifold Z_χ can be so chosen that the intersection $Z_\chi \cap H(k, 1)$ is empty. If $\chi(1) = 1$, but $r(\chi) < k$, then we have $Z_\chi \times H(k, 1) \sim 0$ in $H(k, l)$. Therefore there exists a cycle Z_χ^* homologous to the cycle Z_χ in $H(k, l)$ such that the intersection $Z_\chi^* \cap H(k, 1)$ is empty. In case $\chi(1) = 1$ and $r(\chi) = k$, we have $\chi \equiv 1$. Then, by virtue of (5) Section 3, we have

$$I(Z_\chi, H(k, 1)) = 1 + (-1)^k.$$

From the foregoing and from the fact that $T(M^k) \subset H(k, 1)$, we directly obtain the assertion (F).

(G) Let M^k be an oriented, and possibly, non-connected manifold, and U_1^k and U_2^k two small spherical neighbourhoods of M^k whose closures $\bar{U}_1^k$ and $\bar{U}_2^k$ are disjoint. Let S_1^{k-1} and S_2^{k-1} denote the boundaries of these neighbourhoods; both these boundaries are homeomorphic to a $(k-1)$-dimensional sphere. From the manifold M^k cut out the neighbourhoods U_1^k and U_2^k, and in the bounded manifold thus obtained glue together the boundaries S_1^{k-1} and S_2^{k-1} paying attention to orientation. Let M_1^k denote the closed oriented manifold thus obtained.

It is readily seen that only the Betti groups of dimensions $0, 1, k-1$, and k may change in passing from the manifold M^k to M_1^k. For all other dimensions we may assume that the cycles of the manifolds M^k and M_1^k coincide. As regards the zero-dimensional cycles, they coincide in the sense that their indices coincide. It turns out that all characteristic cycles, except possibly X_1 (see (D)), of the manifolds M^k and M_1^k coincide.

Let R^{k+l} be a vector space of sufficiently large dimension, and $R^{k+1} \subset R^{k+l}$. Then $H(k, 1) \subset H(k, l)$. A regular mapping f of the manifold M^k into R^{k+l} can easily be constructed which satisfies the following conditions:

(a) $f(U_1^k) \subset R^{k+1}, f(U_2^k) \subset R^{k+1}$; and

(b) the mapping f of the bounded manifold $M^k - U_1^k - U_2^k$ realizes a regular mapping of the manifold M_1^k into R^{k+l}.

Choosing the cycles Z_χ and Z_χ^* as in (F), we find that assertion (G) is true.

We now consider certain simple characteristic cycles.

(H) Let χ be a function which has no more than one jump and satisfies the condition $r(\chi) \leqslant k$. Then it is given by the formulas:

$$\chi(1) = \ldots = \chi(p) = q, \quad \chi(p+1) = \ldots = \chi(k) = 0,$$
$$r(\chi) = pq \leqslant k. \tag{24}$$

Since this function χ depends on two numbers p and q, let us denote it by $\chi_{p,q}$. Let Z_{pq} be the pseudo-manifold Z_χ defined by this function, and $X_{p,q}$ the characteristic cycle X_χ. In case χ has no jump at all, i.e., $p = k$, then from $pq \leqslant k$, we find that $q = 1$ or $q = 0$. If $q = 1$, we obtain the characteristic cycle $X_{k,1} = X_1$ which we already know (see (C)). If $q = 0$, we have $Z_{k,0} = H(k, l)$ and $X_{k,0} = M^k$.

If χ has one jump then $p < k$ and $\chi_{p,q}(k) = 0$. Furthermore, $\alpha_1 = p$ and $\beta_1 = q$. Therefore

$$\left.\begin{aligned} &\chi_{p,q} \in X_0 && \text{for } p + q \equiv 0 \pmod 2; \\ &\chi_{p,q} \in X_2 && \text{for } p + q \not\equiv 0 \pmod 2; \\ &\chi_{p,q} \in X && \text{for } p \equiv q \equiv 0 \pmod 2. \end{aligned}\right\} \qquad (25)$$

In all these cases the pseudo-manifold Z_{pq} is determined by only one plane S_1 (see (B), Section 1) of dimension $l + p - q$, namely Z_{pq} consists of all $R^k \in H(k, l)$ satisfying the condition:

$$D(R^k \cap S_1) \geqslant p. \qquad (26)$$

References

1. L. S. Pontryagin, Characteristic cycles of manifolds, *Doklady AN SSSR* **XXXV**, N 2 (1942).
2. H. Hopf, Die Curvatura integra Clifford–Kleinscher Raumformen, *Nachr. Gesell. der Wiss. Gottingen* 131–141 (1925).
3. E. Stiefel, Richtungsfelder in *n*-dimensionalen Mannigfaltigkeiten, *Math. Ann.*, 96 (1927).
4. H. Whitney, *Lectures in Topology,* University of Michigan Press, Ann. Arbor. Mich., 101–141 (1941).
5. L. S. Pontryagin, Vector fields on manifolds, *Matem. sbor., novaya seriya,* vol. 24 (66), N 2, 129–162 (1949).
6. S. Lefschetz, Intersections and transformations of complexes and manifolds, *Trans. Amer. Math. Soc.*, 28 (1926).
7. C. Ehresmann:
   (a) Sur la topologie de certains espaces homogenes, *Ann. of Math.* 35, N 2 (1934);
   (b) Sur la topologie de certaines variétés algebriques réeles, *J. de Math pur et appl.*, 16 (1937).

Received 6 June 1947

On Certain Topological Invariants of Closed Riemannian Manifolds

Izvestiya Akademii Nauk SSSR, Seriya Matematika (1949) 13(2): 125–62

The concept of a characteristic cycle of a differentiable manifold was introduced in my paper [2]. In the present paper the results of [2] are reformulated in terms of differential forms: "characteristic" differential forms, which are topological invariants of a given manifold, are constructed from its Riemannian tensor.

Introduction

In the present paper we give a complete exposition of my results published earlier in [1]. My objective is to reformulate the results obtained in [2], as far as possible, in terms of differential forms. Here I shall make an attempt to proceed as far as possible without using [2], therefore the first three sections can be read without knowing the contents of [2]. Section 1 gives a review of the tensor homology theory written by V. A. Rokhlin on my request. It is included here for the convenience of the reader.

In geometry a key role is played by the spherical mapping technique. In essence it consists of the following.

Let M^k be a k-dimensional differentiable oriented manifold with a continuous tangent in a $(k + 1)$-dimensional Euclidean space R^{k+1}. At a point $x \in M^k$ draw a unit normal N_x to M^k in a direction consistent with the orientation of the manifold M^k. Translate the normal N_x parallel to itself in R^{k+1} in such a way that its origin lies at a fixed point 0 in R^{k+1}. Then its end will lie at a point $N(x)$ on the unit sphere S^k with centre at 0. Thus we obtain a spherical map N of the manifold M^k which sends every point $x \in M^k$ into a point $N(x) \in S^k$.

A study of the spherical map N reveals certain invariants, both differential-geometric and topological, of the manifold M^k.

Let dV denote an infinitesimal volume element at a point x of the manifold M^k, and $N(dV)$ the corresponding infinitesimal volume element in S^k. Then the ratio

$$\frac{N(dV)}{dV} = K(x)$$

represents the curvature of the manifold M^k at x. This curvature depends only on the internal metric of M^k but does not depend on its location in R^{k+1}. The integral of $K(x)$ over the closed manifold M^k, for even k, is a topological invariant of M^k. This integral is equal to $\sigma_k X_1(M^k)/2$, where $X_1(M^k)$ is the Euler characteristic of M^k, and σ_k is the k-dimensional volume of the sphere S^k [3].

In the present paper we construct an analog of the spherical map for a case where the manifold M^k is embedded in a Euclidean space R^{k+1} of arbitrary dimension, and find certain differential-geometric invariants of M^k that are also in some sense topological invariants. This construction is based on tensor homology theory whose essence consists in the following.

Let M be a differentiable closed oriented manifold and $F = \{F_{\alpha_1 \ldots \alpha_r}\}$ a skew-symmetric covariant tensor field of order r on M. The field F is said to be *closed*, if its exterior (alternating) derivative is zero. A closed field F is called *homologous to zero*, if it is equal to the exterior derivative of some other skew-symmetric field on M. Two closed fields are called *homologous*, if their difference is homologous to zero. The homology classes of closed fields of order r form a linear space of dimension p^r, where p^r is the r-dimensional Betti number of the manifold M. This fact was established by de Rham whose paper [4] gives a complete exposition of all the pertaining results.

If M is a homogeneous manifold, i.e., if a compact transitive Lie group Φ of differentiable transformations of M is given, then the concept of a field invariant under the group Φ arises in a natural way. Cartan [5] proved the following fundamental theorem:

For every closed field there exists a homologous invariant closed field. If the invariant closed field is homologous to zero, it is the exterior derivative of some invariant field.

This theorem reduces the study of the homology in a homogeneous manifold M to invariant fields. Since an invariant field is determined by its value at a point $p \in M$, we have to study the constant tensors at this point p.

The study is carried out mainly along the following lines.

Let R^{k+1} be a $(k+1)$-dimensional Euclidean space. Denote by $H(k, l)$ the manifold of all k-dimensional oriented planes passing through a fixed point 0 in R^{k+1} and by Φ the group of all rotations of R^{k+1} about 0. As applied to $H(k, l)$, the group Φ is obviously transitive, and thus $H(k, l)$ becomes a homogeneous manifold.

In Section 2 we determine all closed invariant fields on a homogeneous manifold $H(k, l)$ of an order not greater than k, assuming that $l \geqslant k+1$.

Let M^k be an oriented differentiable manifold in the Euclidean space R^{k+1}. Denote by T_x the oriented k-dimensional tangent plane to M^k at a point x, and by $T(x)$ the oriented k-dimensional plane passing through 0 and parallel to T_x. Then $T(x) \in H(k, l)$, and we obtain a continuous mapping T of M^k into $H(k, l)$. I call it the *tangential mapping* of the manifold M^k. It plays the role of a spherical map, the sphere being replaced by the manifold $H(k, l)$. To the mapping T there corresponds an adjoint mapping T^* which transforms every field F on $H(k, l)$ into a field T^*F on M^k.

In Section 3 we calculate all the fields on M^k that correspond to invariant closed fields on the manifold $H(k, l)$. It turns out that all these fields can be expressed through the Riemannian tensor of the manifold M^k. I call them *characteristic fields* of the Riemannian manifold M^k.

If F is a characteristic field of order r of a Riemannian manifold M^k, and γ^r an arbitrary cycle of dimension r in M^k, then the integral of the field F over the cycle γ^r is an invariant of the homology class of the field F and of the homology class of the cycle γ^r. Moreover, it does not depend on the Riemannian metric of the manifold M^k. For even k, in particular, the integral of one of the characteristic fields of order k taken over the manifold M^k itself is equal to $\sigma_k X_1(M^k)/2$, where $X_1(M^k)$ is the Euler characteristic of M^k, and σ_k is the k-dimensional volume of the k-dimensional unit sphere. This result is established in Section 4. It is, however, not new and has been proved repeatedly (see, for instance, [6]).

In Sections 4 and 5 we establish a relationship between characteristic fields and characteristic cycles of a manifold M^k. Here I shall fully utilize the results of my paper [2].

1. Tensor Homology Theory on a Manifold

This is an introductory section. It gives a survey of the main concepts and the results of the tensor homology theory in a form convenient for subsequent use. Complete exposition of this topic can be found in the works of de Rham [4] and Cartan [5].

(A) *Multilinear forms*. Let R^k be a real vector space of dimension k. A real-valued numerical function $F(u)$ of a vector $u \in R^k$ is called a *linear form*, if for any two vectors $u, v \in R^k$ and for any real number λ,

$$F(u+v) = F(u) + F(v); \quad F(\lambda u) = \lambda F(u). \tag{1}$$

A function $F(u_1, \ldots, u_r)$ of several vectors $u_i \in R^k$ $(i = 1, \ldots, r)$ linear in each vector is called a *multilinear form*.

A multilinear form $F(u_1, \ldots, u_r)$ is called *skew-symmetric*, if its sign changes on permuting any two of its arguments. Such a form is obviously invariant under any even permutation and changes its sign under any odd permutation of the arguments: if σ is an arbitrary permutation of $1, \ldots, r$, and sign σ is the sign of this permutation (i.e., positive for an even and negative for an odd permutation), then

$$F(u_{\sigma(1)}, \ldots, u_{\sigma(r)}) = \text{sign } \sigma F(u_1, \ldots, u_r). \tag{2}$$

A skew-symmetric multilinear form vanishes for any linearly dependent system of its arguments. Any skew-symmetric form of more than k arguments is in particular identically equal to zero.

To every multilinear form F, there naturally corresponds a certain skew-symmetric form $[F]$ of the same arguments. The operation by means of which the form $[F]$ is obtained from the form F is called *alternation*. It consists in taking all possible permutations of the arguments of F and then taking the arithmetic mean of all the forms thus obtained, where the forms obtained by odd permutations are first multiplied by -1:

$$[F](u_1, \ldots, u_r) = \frac{r}{r!} \sum_{\sigma} \text{sign } \sigma F(u_{\sigma(1)}, \ldots, u_{\sigma(r)}). \tag{3}$$

If $F(u_1, \ldots, u_r)$ and $G(v_1, \ldots, v_s)$ are two multilinear forms, their product

$$H(u_1, \ldots, u_r, v_1, \ldots, v_s) = F(u_1, \ldots, u_r) G(v_1, \ldots, v_s) \tag{4}$$

is also a multilinear form. Assuming that the forms F and G are skew-symmetric, we defined their *exterior product* as the form $[H]$ obtained from (4) by alternation. This definition is applicable to any finite number of multipliers. Exterior multiplication is associative and distributive with respect to linear operations on forms.

(B) *Tensors*. If a basis

$$e_1, \ldots, e_k, \tag{5}$$

is chosen in the space R^k, then every vector $u \in R^k$ can be represented in terms of its components:

$$u = \sum_{\alpha=1}^{k} u^\alpha e_\alpha, \tag{6}$$

and for any multilinear form $F(u_1, \ldots, u_r)$, we have

$$F(u_1, \ldots, u_r) = \sum_{\alpha_1, \ldots, \alpha_r = 1}^{k} F_{\alpha_1, \ldots, \alpha_r} u_1^{\alpha_1} \ldots u_r^{\alpha_r}, \tag{7}$$

where

$$F_{\alpha_1, \ldots, \alpha_r} = F(e_{\alpha_1}, \ldots, e_{\alpha_r}). \tag{8}$$

For a given basis (5), the numbers (8) are uniquely determined by the form (7), and these numbers, in turn, uniquely determine the form.

A multilinear form represented by its coefficients is called a *tensor*. The coefficients (8) of the form (7) are called the *components* of the corresponding tensor in the basis (5), and the number r of arguments of the form is called the *order* of the tensor. In changing over from the basis (5) to a new basis

$$e'_1, \ldots, e'_k \tag{5'}$$

the components are transformed according to the tensor rule:

$$F_{\alpha'_1 \ldots \alpha'_r} = \sum_{\alpha_1, \ldots, \alpha_r = 1}^{k} a_{\alpha'_1}^{\alpha_1} \ldots a_{\alpha'_r}^{\alpha_r} F_{\alpha_1 \ldots \alpha'_r},$$

where $\|a_{\alpha'}^{\alpha}\|$ is the transformation matrix from the basis (5) to the basis (5′).

To a skew-symmetric form, there corresponds a skew-symmetric tensor. The condition for a tensor to be skew-symmetric is that for any arbitrary permutation σ

$$F_{\alpha_{\sigma(1)} \ldots \alpha_{\sigma(r)}} = \operatorname{sign} \sigma F_{\alpha_1 \ldots \alpha_r}. \tag{9}$$

A skew-symmetric tensor $F_{\alpha_1 \ldots \alpha_k}$ of order k has, in particular, only one essential component $F_{1 \ldots k}$. Every component of $F_{\alpha_1 \ldots \alpha_k}$ is either zero (if there are repeated indexes among $\alpha_1, \ldots, \alpha_k$) or $+F_{1, \ldots, k}$ (if the sequence $\alpha_1, \ldots, \alpha_k$ is obtained from the sequence $1, \ldots, k$ by an even permutation) or $-F_{1 \ldots k}$ (if the sequence $\alpha_1, \ldots, \alpha_k$ is obtained from the sequence $1, \ldots, k$ by an odd permutation). The form itself is

$$F(u_1, \ldots, u_k) = F_{1, \ldots, k}|u_i^\alpha|.$$

To the alternation of a form corresponds the alternation of a tensor, and to the exterior product of skew-symmetric forms corresponds the exterior product of skew-symmetric tensors. A tensor $F_{[\alpha_1 \ldots \alpha_r]}$ derived from a tensor $F_{\alpha_1 \ldots \alpha_r}$ by alternation is given by the expression:

$$F_{[\alpha_1 \ldots \alpha_r]} = \frac{r}{r!} \sum_\sigma \operatorname{sign} \sigma F_{\alpha_{\sigma(1)} \ldots \alpha_{\sigma(r)}}, \tag{10}$$

and the exterior product $H_{\alpha_1 \ldots \alpha_r \beta_1 \ldots \beta_s}$ of two tensors $F_{\alpha_1 \ldots \alpha_r}$ and $G_{\beta_1 \ldots \beta_s}$ is

$$H_{\alpha_1 \ldots \alpha_r \beta_1 \ldots \beta_s} = F_{[\alpha_1 \ldots \alpha_r} \cdot G_{\beta_1 \ldots \beta_s]}. \tag{11}$$

(C) *Differentiable manifolds. Tangent vector spaces.* Let M^k be a k-dimensional topological manifold, i.e., a topological space each point of which has a neighbourhood homeomorphic to a domain in a k-dimensional Euclidean space E^k. In the sequel we only consider such neighbourhoods. If a definite Cartesian coordinate system is already introduced in E^k, then establishing a definite homeomorphism between a domain in E^k and a neighbourhood $U^k \subset M^k$ is equivalent to introducing a definite coordinate system $\xi^1, \ldots, \xi^k$ in U^k. In this case two distinct coordinate systems $\xi^1, \ldots, \xi^k$ and $\eta^1, \ldots, \eta^k$ in U^k are always related by a one-to-one and bicontinuous transformation:

$$\eta^\beta = \eta^\beta(\xi^1, \ldots, \xi^k) \quad (\beta = 1, \ldots, k). \tag{12}$$

Let us choose an arbitrary but a definite natural number m and assume that the functions (12) are not only continuous, but are also continuously differentiable m times in U^k and have a non-zero Jacobian $|\partial\eta^\beta/\partial\xi^\alpha|$. Under this condition, we put the coordinate systems $\xi^1, \ldots, \xi^k$ and $\eta^1, \ldots, \eta^k$ into the *same class*. Evidently, different classes do not intersect, and each class is completely defined by any coordinate system that belongs to it. If, among all these classes, a particular class is singled out, then the neighbourhood U^k is called *m times continuously differentiable*. By virtue of this definition, two m times continuously differentiable neighbourhoods U^k and V^k always induce two classes of corrdinate systems in their intersection. If these classes coincide, the neighbourhoods U^k and V^k are said to be *consistently* differentiable. If all the neighbourhoods in a manifold M^k are m times consistently differentiable, and any pair of neighbourhoods are consistently differentiable, the manifold M^k is said to be m *times continuously differentiable*.

In the sequel, in place of "m times continuously differentiable manifold" we simply say "differentiable manifold". For this we shall always assume that m is sufficiently large for our purposes.

Let a be a certain point of a differentiable manifold M^k. Any coordinate system defined in some neighbourhood U_a^k of the point a and belonging to the class already specified is called a *local coordinate system at the point a. A vector on the manifold* M^k *at the point a* is a function which sends each local coordinate system at the point a into a system of k real numbers – the components of the vector in this coordinate system – in such a way that the components $u^1, \ldots, u^k$ and $v^1, \ldots, v^k$ of a vector in two different local coordinate systems $\xi^1, \ldots, \xi^k$ and $\eta^1, \ldots, \eta^k$, are connected by the relationship:

$$w^\beta = \sum_{\alpha=1}^{k} a_\alpha^\beta u^\alpha \quad (\beta = 1, \ldots, k), \tag{13}$$

where $\|a_\alpha^\beta\|$ is the Jacobian matrix at the point a of the transformation (12)

$$a_\alpha^\beta = \frac{\partial \eta^\beta(a)}{\partial \xi^\alpha} . \tag{14}$$

Since, in two successive coordinate transformations the corresponding Jacobian matrices are multiplied, the condition (13) is consistent, and the vector is uniquely determined by its components in any local coordinate system. Defining linear operations on vectors as operations on the vector components (which is also obviously consistent), we convert the vectors at the point a into a k-dimensional space R_a^k called the *tangent space* to the differentiable manifold M^k at the point a. To each local coordinate system at a point a, there evidently corresponds a definite basis in the tangent space, in which all vectors have the same components as in the local coordinate system.

(D) *Tensor field on a manifold*. Let M^k be a differentiable menifold, and R_x^k the vector space tangent to M^k at a point $x \in M^k$. A multilinear form F is said to be given on M^k, if some multilinear form $F_x(\underset{x}{u}_1, \ldots, \underset{x}{u}_r)$ $(\underset{x}{u}_i \in R_x^k, i = 1, \ldots, r)$ is defined on each R_x^k. On changing over from the form $F_x(\underset{x}{u}_1, \ldots, \underset{x}{u}_r)$ to its corresponding tensor, we obtain a *tensor field* on M^k. After we have chosen a definite local coordinate system in the neighbourhood of each point $a \in M^k$ (and, consequently, certain bases have been chosen in the spaces R_x^k tangent to M^k at the points of the neighbourhood), this field is represented by a system of its components $F_{\alpha_1 \ldots \alpha_r}$ $(\xi^1, \ldots, \xi^k)$, i.e., by a system of numerical functions of numerical arguments. If these functions are n times (continuously) differentiable, the components of our vector field will also be n times continuously differentiable in any other local coordinate system. Thus, we can speak of the (n *times continuous*) *differentiability of a field*, and so, the differentiability of the corresponding form F. In the sequel we shall assume that all tensor fields studied are continuously differentiable a sufficient number of times.

(E) *Exterior derivative of a skew-symmetric field*. Though we have defined differentiability, it does not yet imply that we have defined the derivative. The ordinary partial derivatives of the components of a field

$$\partial_\alpha F_{\alpha_1 \ldots \alpha_r} = \frac{\partial F_{\alpha_1 \ldots \alpha_r}(\xi^1, \ldots, \xi^k)}{\partial \xi^\alpha}, \tag{15}$$

expressed in some local coordinate system $\xi^1, \ldots, \xi^k$ are not, by themselves, of tensor type. We define a derivative, the so-called *exterior derivative*, only for a skew-symmetric field. The exterior derivative of a skew-symmetric field $F_{\alpha_1 \ldots \alpha_r}$ of order r is a skew-symmetric field $F'_{\alpha\alpha_1 \ldots \alpha_r}$ of order $(r+1)$ given in each local coordinate system by the expression

$$F'_{\alpha\alpha_1 \ldots \alpha_r} = \partial_\alpha F_{\alpha_1\alpha_2 \ldots \alpha_r} - \partial_{\alpha_1} F_{\alpha\alpha_2 \ldots \alpha_r} - \partial_{\alpha_2} F_{\alpha_1 \alpha \ldots \alpha_r} - \ldots$$
$$\ldots - \partial_{\alpha_r} F_{\alpha_1\alpha_2 \ldots \alpha} = (r+1)\partial_{[\alpha} F_{\alpha_1 \ldots \alpha_r]}. \tag{16}$$

By virtue of the alternation on the right-hand side, this definition is consistent, i.e., it gives the same skew-symmetric multilinear form F', the exterior derivative of the skew-symmetric form F, in any local coordinate system.

(F) *Mappings of differentiable manifolds*. Let M^k and N^l be two differentiable manifolds, and ϕ a mapping of the first into the second. At a point $a \in M^k$ choose some local coordinate system $\xi^1, \ldots, \xi^k$ and at the point $b = \phi(a) \in N^l$ choose some local coordinate system $\eta^1, \ldots, \eta^l$. Then in the neighbourhood of the point a, the mapping ϕ is represented as

$$\eta^\beta = \eta^\beta(\xi^1, \ldots, \xi^k) \quad (\beta = 1, \ldots, l). \tag{17}$$

If the functions (17) are (continuously) differentiable a certain number of times, they will also be (continuously) differentiable the same number of times in any other local coordinate system. So we can speak of the (n times continuous) differentiability of the mapping ϕ itself. In the sequel, in dealing with a differentiable mapping we shall always assume that it is continuously differentiable a sufficiently large number of times.

Every differentiable mapping ϕ of a differentiable manifold M^k into a differentiable manifold N^l induces at each point $a \in M^k$ a definite linear *mapping ϕ_a of the vector space R_a^k tangent to the manifold M^k at the point a into the vector space R_a^b tangent to the manifold N^l at the point* $b = \phi(a)$. In other words, if the local coordinate systems chosen at the points a and b are $\xi^1, \ldots, \xi^k$ and $\eta^1, \ldots, \eta^l$, respectively, then to the vector $u \in R_a^k$ with the components u^α ($\alpha = 1, \ldots, k$) in the coordinate system $\xi^1, \ldots, \xi^k$, there corresponds a vector $v \in R_b^l$ with the components

$$v^\beta = \sum_{\alpha=1}^{k} \frac{\partial \eta^\beta(a)}{\partial \xi^\alpha} u^\alpha \quad (\beta = 1, \ldots, l) \tag{18}$$

in the coordinate system $\eta^1, \ldots, \eta^l$. It is readily seen that this definition is consistent, namely, we obtain the same mapping ϕ_a for any choice of the local coordinate systems.

To every differentiable mapping ϕ of a differentiable manifold M^k into a differentiable manifold N^l, there corresponds a definite adjoint mapping ϕ^* which sends any multilinear form (any tensor field) on N^l into a corresponding multilinear form (tensor field) on M^k. Namely, to the form $F_y(\underset{y}{v_1}, \ldots, \underset{y}{v_r})$ on N^l there corresponds a form $G = \phi^* F$ given by the relation:

$$G_x(\underset{x}{u_1}, \ldots, \underset{x}{u_r}) = F_{\phi(x)}(\phi_x(\underset{x}{u_1}), \ldots, \phi_x(\underset{x}{u_r})). \tag{19}$$

It is a simple matter to verify that the mapping ϕ^* transforms a skew-symmetric form into a skew-symmetric form, and a differentiable form into a differentiable form, and that

$$(\phi^* F)' = \phi^* F'. \tag{20}$$

(G) *Differentiable chains*. Before we proceed further, we should specialize here for a differentiable manifold the definitions of the usual concepts of the homology theory, those of oriented singular simplices, chains and cycles constructed from simplices. An ordinary singular r-dimensional simplex in M^k is a class of affine-equivalent continuous mappings of linear r-dimensional simplices into M^k. If, in this statement, differentiability is substituted for continuity, we obtain the definition of a *singular differentiable r-dimensional simplex in M^k*.

We shall only deal with such singular differentiable simplices that lie wholly in one neighbourhood, or those, as we often say, that are covered by one coordinate system. Just like a singular simplex, a singular differentiable simplex always has two orientations. In order to fix one of them, in each linear simplex-preimage we choose one particular class out of the two oriented classes of affine coordinate systems, and only use the systems of this class. Chains are constructed in the usualy way from oriented singular differentiable simplices. These chains are also called *differentiable*. The boundary of a differentiable chain defined in the natural way is also a differentiable chain. Once the boundary is defined, cycles and homology are also defined. Since there always exists for every cycle a homologous differentiable cycle and every differentiable cycle homologous to zero in the usual sense is the boundary of some differentiable cycle, we find that the Betti groups constructed from differentiable cycles are ordinary Betti groups.

(H) *Integral of a skew-symmetric field of order r over an r-dimensional chain*. Let $F_{\alpha_1,\ldots,\alpha_r}$ be a skew-symmetric tensor field of order r on M^k and τ^r an oriented singular differentiable r-dimensional simplex in M^k. After choosing a coordinate system $\xi^1, \ldots, \xi^k$ covering τ^r, any mapping ϕ defining τ^r can be represented as

$$\xi^\alpha = \xi^\alpha(\eta^1, \ldots, \eta^r) \quad (\alpha = 1, \ldots, k), \tag{21}$$

where $\eta^1, \ldots, \eta^r$ are affine coordinates in the oriented linear simplex T^r which is mapped by ϕ into M^k. Consider the only essential component $F^*_{1\ldots r}$ of the skew-symmetric field

$$F^*_{\alpha_1\ldots\alpha_r} = \phi^* F_{\alpha_1\ldots\alpha_r}{}^*.$$

on T^r. Direct verification shows that the integral

$$\int_{\tau^r} F^*_{1\ldots r}\, d\eta^1 \ldots d\eta^r \tag{22}$$

does not depend on the choice of the coordinate systems $\xi^1, \ldots, \xi^k$ and $\eta^1, \ldots, \eta^r$, in other words, it is completely determined by the field $F_{\alpha_1\ldots\alpha_r}$ and the oriented singular differentiable simplex τ^r. It is called *the integral of a skew-symmetric field $F_{\alpha_1\ldots\alpha_r}$ (or of the corresponding multilinear form F) over the simplex τ^r*. This definition is immediately extended to any differentiable r-dimensional chain by linearity and in particular, to any differentiable r-dimensional cycle. To denote the integral of a field $F_{\alpha_1\ldots\alpha_r}$ (or of a form F) over a chain γ^r, we use the symbol

$$\int_{\gamma^r} F. \tag{23}$$

Integration allows us to establish an important relationship between exterior differentiation (denoted by a prime) and Δ, the boundary

* Though in (F) the adjoint mapping was defined only for a differentiable mapping of a manifold into a manifold, it could easily be defined for a mapping of a closed simplex into a manifold as well.

operation. *For every skew-symmetric field* $F_{\alpha_1 \ldots \alpha_r}$ *of order r and any (r + 1)-dimensional chain*

$$\int_{\gamma^{r+1}} F' = \int_{\Delta\gamma^{r+1}} F. \tag{24}$$

(I) *Tensor homology theory*. From (16) it follows that *the second exterior derivative of a skew-symmetric tensor field is identically equal to zero.* This assertion, equivalent to the theorem, that the "*boundary of a boundary is zero*" in the ordinary homology theory, is sufficient for constructing a tensor homology theory.

We shall only deal with skew-symmetric tensor fields on a closed oriented differentiable manifold M^k. A field is called *closed*, if its exterior derivative is zero, and *homologous to zero* if it is the exterior derivative of some other field. Two fields of the same order are said to be *homologous*, if their difference is homologous to zero. The classes of homologous tensor fields of order r on M^k naturally form a real linear space.

The dimension of this space, i.e., the maximal number of homologously (linearly) independent skew-symmetric fields of order r on M^k, *is equal to to the r-dimensional Betti number of the manifold* M^k.

Thus, as we are dealing with weak homology, which is all that can be taken into account by means of the tensor fields considered here, the tensor homology theory gives rise to the same invariants as the usual homology theory.

A deeper connection between the usual homology and tensor homology is obtained as follows. From (24) it follows that the integral of a closed field $F_{\alpha_1 \ldots \alpha_r}$ over a cycle γ^r does not change, if the field $F_{\alpha_1 \ldots \alpha_r}$ is replaced by a homologous field, or if the cycle γ^r is replaced by a homologous cycle. This integral is an invariant of both homology classes. Let $\gamma_1^{k-r}, \ldots, \gamma_p^{k-r}$ be an independent basis of the weak homology of dimension $(k-r)$ in M^k. Then there exists a basis $F^{(1)}, \ldots, F^{(p)}$ of the tensor homology of order r satisfying the condition

$$\int_{\gamma^r} F^{(i)} = J(\gamma_i^{k-r}, \gamma_r) \quad (i = 1, \ldots, p), \tag{25}$$

where γ^r is an arbitrary r-dimensional cycle in M^k and J stands for the intersection index.

From (20) it follows that the mapping ϕ^* adjoint to a differentiable mapping ϕ of a manifold M^k into a manifold N^l (see (F)) transforms every closed field into a closed field and any field homologous to zero into a field homologous to zero. Furthermore, from the definition of the integral (23) it follows that for an arbitrary chain γ^r in M^k and an arbitrary skew-symmetric field $F_{\alpha_1 \ldots \alpha_r}$ on N^l

$$\int_{\gamma^r} \phi^* F = \int_{\phi\gamma^r} F. \tag{26}$$

(J) *The case of a homogeneous manifold.* A closed differentiable manifold M^k is called homogeneous, if there is given a compact transitive Lie group of differentiable transformations of M^k. The multilinear form F on a homogeneous manifold is called *invariant*, if $\phi^* F = F$ for every transformation ϕ of the group mentioned above. Once an invariant form is defined, an invariant tensor field is also defined.

From (20) it follows that the exterior derivative of an invariant field is also an invariant field. Thus, it is possible to construct a homology theory only on the basis of invariant fields. It turns out that this theory is equivalent to the usual homology theory: *For every closed field, it is always possible to find a homologous invariant closed field, and every invariant field homologous to zero is the exterior derivative of some other invariant field.*

Obviously, every invariant field is uniquely determined by its values at a point $a \in M^k$, i.e., by ordinary tensors. Therefore by the theorem formulated above the homology theory on a homogeneous manifold is reduced to the study of the relations between ordinary tensors. These relations should take into account the connection between an invariant form and its exterior derivative. Here the central role is naturally played by that subgroup of the basic Lie group which consists of the transformations for which the point a is fixed.

2. The homogeneous Manifold $H(k, l)$

In this section we construct the homogeneous manifold $H(k, l)$ which plays the main role in the sequel. Moreover, we shall determine all the invariant fields on $H(k, l)$ whose orders are needed later.

Definition 1. Let R^{k+l} be a $(k+l)$-dimensional Euclidean vector space, and G the group of all linear orthogonal transformations of R^{k+l} with a positive determinant. Let $H(k, l)$ denote the manifold of all oriented k-dimensional linear subspaces of the space R^{k+l}. Each transformation

$g \in G$ maps any element R^k of $H(k, l)$ into an element of the same manifold: $g(R^k) \in H(k, l)$. Thus, G may be regarded as a transitive group of transformations of $H(k, l)$, so this manifold becomes homogeneous (see (J), Section 1).

(A) Let R_0^k be an arbitrary but fixed element of $H(k, l)$ and

$$e_1, \ldots, e_k \tag{1}$$

some orthonormal basis in R_0^k. Now, form the orthonormal basis

$$e_1, \ldots, e_k, \quad f_1, \ldots, f_l \tag{2}$$

of the whole space R^{k+l}. Let U denote the set of all $R_\xi^k \in H(k, l)$ such that the orthogonal projection of R_ξ^k on R_0^k is non-degenerate and has the same orientation. Evidently, U is a neighbourhood of the element R_0^k in $H(k, l)$. Let $R_\xi^k \in U$, and denote by e_i' a vector in R_ξ^k such that its orthogonal projection in R_0^k is equal to e_i. Then the vectors

$$e_1', \ldots, e_k' \tag{3}$$

form some basis of the linear space R_ξ^k, and

$$e_i' = e_i + \sum_{j=1}^{l} \xi_{ij} f_j \quad (i = 1, \ldots, k), \tag{4}$$

where the ξ_{ij} are real numbers uniquely determined by the element R_ξ^k of the domain U (the basis (2) is assumed to be fixed). If, on the other hand, the numbers ξ_{ij} are given arbitrarily, then, after determining the vectors (3) from (4) and taking the appropriately oriented linear span of these vectors, we obtain the element R_ξ^k of the domain U. Thus, the elements ξ_{ij} of the matrix $\xi = \|\xi_{ij}\|$ can be taken as the coordinates of the element $R_\xi^k \in U$. From this it is clear that the domain U in ξ_{ij}-coordinates is a linear space of dimension kl. By virtue of linearity, we can now identify the space U with the space tangent to $H(k, l)$ at the point R_0^k.

(B) Let G_0 denote the subgroup of the groups G consisting of those elements of G which transform the oriented space R_0^k into itself. For a $g \in G_0$, we have

$$g(e_i) = \sum_{\alpha=1}^{k} a_{i\alpha} e_\alpha, \quad g(f_j) = \sum_{\beta=1}^{l} b_{j\beta} f_\beta, \tag{5}$$

where $a = \|a_{i\alpha}\|$ and $b = \|b_{j\beta}\|$ are orthogonal matrices with positive determinants of order k and l, respectively. For a fixed basis (2), these matrices are uniquely determined by the transformation g, and these matrices, in turn, uniquely determine the transformation. It turns out that transformation $g \in G_0$ linearly maps the space U onto U (see (A)), and an element $R^k \in U$ is transformed into an element $R^k_\eta = g(R^k_\xi)$ given by the relation

$$\eta = a^{-1}\xi b. \tag{6}$$

Since the elements of the group G_0 linearly transform the space U, this space behaves in relation to G_0 in the same way as the tangent space to $H(k, l)$ at R^k_0, and then therefore takes its place. When the basis (2) is replaced by some other basis, the coordinates ξ_{ij} in U behave as follows. Let

$$\bar{e}_1, \ldots, \bar{e}_k, \quad \bar{f}_1, \ldots, \bar{f}_l \tag{2'}$$

be some orthonormal basis in R^{k+l} satisfying the condition that the vectors $\bar{e}_1, \ldots, \bar{e}_k$ lie in R^k_0; then the bases (2) and (2′) are related by the obvious expressions:

$$\bar{e}_i = \sum \bar{a}_{i\alpha} e_\alpha, \quad \bar{f}_j = \sum \bar{b}_{j\beta} f_\beta. \tag{5'}$$

To the basis (2′) there corresponds a coordinate system in U. Let us denote the coordinates of the element R^k_ξ in this system by $\bar{\xi}_{ij}$. Then the relation between ξ_{ij} and $\bar{\xi}_{ij}$

$$\xi_{ij} = \sum_{\alpha, \beta} \bar{a}_{\alpha i} \bar{\xi}_{\alpha\beta} \bar{b}_{\beta j}. \tag{6'}$$

is quite similar to (6), so we shall restrict ourselves to a proof of the latter.

We now proove (6). Since a basis of the space R^k_ξ consists of the vectors (3) [see (4)], a basis $e''_1, \ldots, e''_k$ of the space R^k is given by the relation:

$$e''_i = \sum_{\alpha=1}^{k} a_{i\alpha} e_\alpha + \sum_{j, \beta=1}^{l} \xi_{ij} b_{j\beta} f_\beta. \tag{7}$$

In place of the basis $e''_1, \ldots, e''_k$, let us introduce the following basis $e^*_1, \ldots, e^*_k$ in R^k:

$$e^*_i = \sum_{\gamma=1}^{k} a_{\gamma i} e''_\gamma. \tag{8}$$

Then, as the matrix a is orthogonal, from (7) and (8) we obtain

$$e^*_i = e_i + \sum_{\gamma, \beta, j} a_{\gamma i} \xi_{\gamma\beta} b_{\beta j} f_j,$$

and this means that (6) holds.

(C) We interpret thc spacc U with the coordinates ξ_{ij} [see (A)] as a vector space, namely, we call each matrix ξ a vector in the space U. The vector space U is acted on by the linear transformation group G_0, each element of which is given by a pair of orthogonal matrices a and b of orders k and l with positive determinants [see (B)]. Here the transformation $g \in G_0$ corresponding to a pair a, b is given by (6). Now in U consider a finite sequence

$$\xi^1 = \|\xi^1_{ij}\|, \ldots, \xi^r = \|\xi^r_{ij}\|, \ldots \tag{9}$$

of arbitrary (variable) vectors. From these vectors we now construct certain multilinear forms invariant under the transformations of the group G_0. For $p \neq q$, put

$$\xi^{pq}_{\alpha\gamma} = \sum_{j=1}^{l} \xi^p_{\alpha j}\xi^q_{\gamma j}, \quad p \neq q. \tag{10}$$

Under the transformation $g \in G_0$ the bilinear form (10) of the vectors ξ^p and ξ^q is transformed by the formula:

$$\eta^{pq}_{hi} = \sum_{\alpha,\gamma=1}^{k} \xi^{pq}_{\alpha\gamma} a_{\alpha h} a_{\gamma i}. \tag{11}$$

In changing over from the basis (2) to the basis (2′), i.e., in changing over from the coordinates ξ_{ij} to the coordinates $\bar{\xi}_{ij}$, the bilinear form ξ^{pq}_{hi} is expressed in terms of the bilinear forms

$$\bar{\xi}^{pq}_{\alpha\gamma} = \sum_{j} \bar{\xi}^p_{\alpha j}\bar{\xi}^q_{\gamma j} \quad (\alpha, \gamma = 1, \ldots, k)$$

as follows:

$$\xi^{pq}_{hi} = \sum_{\alpha,\gamma=1}^{k} \bar{\xi}^{pq}_{\alpha\gamma} \bar{a}_{\alpha h} \bar{a}_{\gamma i}. \tag{11′}$$

Hence, it is clear that the bilinear forms ξ^{pq}_{hi} $(h, i = 1, \ldots, k)$ behave like the components of a double-indexed tensor in the Euclidean space R^k_0 [see (11) and (11′)]. Let $\epsilon_{i_1 \ldots i_k}$ denote a skew-symmetric tensor in the oriented Euclidean space R^k_0 satisfying the condition that $\epsilon_{1 \ldots k} = +1$ in a system of orthogonal coordinates that give the positive orientation to R^k_0. It is easily seen that when the orientation of R^k_0 is changed, this tensor changes its sign. For even k, let us consider the multilinear form

$$\tilde{K}^1 = \sum_{i_1 \ldots i_k} \epsilon_{i_1 \ldots i_k} \xi^{1,2}_{i_1 i_2} \xi^{3,4}_{i_3 i_4} \cdots \xi^{k-1,k}_{i_{k-1} i_k} \tag{12}$$

of vectors $\xi^1, \ldots, \xi^k$. Since $\epsilon_{i_1 \ldots i_k}$ is a tensor, and all ξ^{pq}_{hi} behave like tensors with respect to the lower indices [see (11) and (11′)], the expression (12) is invariant under the transformations of the group G_0 as well as under the substitution of the basis (2) by the basis (2′). When the orientation of the space R_0^k is changed, the sign of (12) changes. Let

$$K^1 = K^1(\xi^1, \ldots, \xi^k) \tag{13}$$

denote the expression derived from (12) by alternation over the vectors $\xi^1, \ldots, \xi^k$. This expression is also invariant under the transformations of the group G_0 and under the substitution of the basis (2) by the basis (2′); and it changes its sign when the orientation of the space is changed. For an integer r divisible by four, put $r = 2s$, and then consider the expression:

$$\widetilde{K}^r = \sum_{\alpha_1, \ldots, \alpha_s = 1}^{k} \xi^{1,2}_{\alpha_1\alpha_2}\xi^{3,4}_{\alpha_2\alpha_3}\xi^{5,6}_{\alpha_3\alpha_4} \cdots \xi^{r-1,r}_{\alpha_s\alpha_1}. \tag{14}$$

It follows directly from (11) and (11′) that the multilinear form $\widetilde{K}^r$ of the vectors $\xi^1, \ldots, \xi^r$ is invariant both under the transformations of the group G_0 and under the substitution of the basis (2) by the basis (2′), and is independent of the orientation of the space R_0^k as well. Let us denote the multilinear form of vectors $\xi^1, \ldots, \xi^r$ derived from (14) by alternation over all these vectors by

$$K^r = K^r(\xi^1, \ldots, \xi^r). \tag{15}$$

Now let

$$\rho = \{r_1, \ldots, r_t\}, \quad r_1 \leqslant r_2 \leqslant \cdots \leqslant r_t, \quad r_1 + \cdots + r_t = r(\rho) = r,$$
$$r_1 \equiv r_2 \equiv \ldots \equiv r_t \equiv 0 \pmod 4 \tag{16}$$

be a sequence of natural numbers. Let us denote the exterior product of the forms

$$K^{r_1}(\xi^1, \ldots, \xi^{r_1}), K^{r_2}(\xi^{r_1+1}, \ldots, \xi^{r_1+r_2}),$$
$$\ldots, K^{r_t}(\xi^{r_1+\cdots+r_{t-1}+1}, \ldots, \xi^{r_1+\cdots+r_t}) \tag{17}$$

by

$$K^\rho = K^\rho(\xi^1, \ldots, \xi^r). \tag{18}$$

The form K^ρ is invariant under the transformations of the group G_0, under the substitution of the basis (2) by the basis (2′) and under a change of orientation of the space R_0^k. In order to unify our notation, let us also denote K^1 by K^ρ, taking $\rho \equiv 1$ and putting $r(\rho) = k$ for $\rho \equiv 1$ K^r is also K^ρ, if we put $\rho \equiv r$.

Theorem 1. Let $k \leqslant l - 2, r \leqslant k + 1$, and let

$$A = A(\xi^1, \ldots, \xi^r) \tag{19}$$

be an arbitrary multilinear skew-symmetric form of the vecotrs $\xi^1, \ldots, \xi^r$ in the space U which is invariant under the group G_0 [see (C)]. Then A is a linear combination of forms of the type (18) with constant coefficients (where the form (13) obviously enters only when $r = k$). Consequently, all non-zero forms (19) are of even order.

The proof of Theorem 1 is based on the following well-known assertion of the theory of invariants.

(D) Let R^n be an n-dimensional oriented Euclidean space, g the group of all orthogonal transformations of R^n preserving its orientation, and

$$x^1, \ldots, x^r \tag{20}$$

a certain sequence of arbitrary (variable) vectors in R^n. Let (x^p, x^q) denote the scalar product of the vectors x^p and x^q, and $D(x^{p_1}, \ldots, x^{p_n})$ the volume of the oriented parallelopiped defined by the vectors $x^{p_1}, \ldots, x^{p_n}$ of the system (20) in the space R^n. Further, let $P = P(x^1, \ldots, x^r)$ be a polynomial in the components of the vectors (20), which is invariant under the group G. Then P can be expressed as a polynomial in expressions of the type (x^p, x^q) and $D(x^{p_1}, \ldots, x^{p_n})$.

We state this assertion (D) without proof; its proof is given in [7].

Proof of Theorem 1. Let u^p be a vector in the k-dimensional Euclidean space R_0^k and v^p a vector in the l-dimensional Euclidean space R_0^l:

$$\begin{aligned} u^p &= (u_1^p, \ldots, u_k^p), \\ v^p &= (v_1^p, \ldots, v_p^l). \end{aligned} \tag{21}$$

Put

$$\xi_{ij}^p = u_i^p v_j^p. \tag{22}$$

The matrix $\xi^p = \|\xi_{ij}^p\|$ is associated with a pair of vectors u^p and v^p by means of the relation (22). With the matrices a and b let us associate transformations of the spaces R_0^k and R_0^l, putting

$$u_i' = \sum_{\alpha=1} u_\alpha a_{\alpha i}, \quad v_j' = \sum_{j=1}^{l} v_\beta b_{\beta j}. \tag{23}$$

Obviously, a transformation of the matrix ξ^p by means of (6) corresponds to the transformation (23) of vectors. Now in the form A, replace each matrix $\xi^1, \ldots, \xi^r$ by its expression in (22). And denote the expression thus obtained in $u^1, \ldots, u^r$ and $v^1, \ldots, v^r$ by

$$A' = A(u^1, \ldots, u^r; v^1, \ldots, v^r).$$

Since, by assumption, the form A is invariant under all transformations of the group G_0, the form A' remains invariant when the spaces R_0^k and R_0^l are subjected to arbitrary othogonal transformations which preserve their orientations. As a result of this relation between the forms A and A', we can reduce our problem to Theorem (D).

In coordinate notation the form A is:

$$A = \sum A_{i_1 j_1 \dots i_r j_r} \xi^1_{i_1 j_1} \dots \xi^r_{i_r j_r}. \tag{24}$$

Here summation is taken over all lower indices, where

$$i_1, \dots, i_r = 1, \dots, k; \quad j_1, \dots, j_l = 1, \dots, l.$$

Furthermore, we have

$$A' = \sum A_{i_1 j_1 \dots i_r j_r} u^1_{i_1} \dots u^r_{i_r} v^1_{j_1} \dots v^r_{j_r}. \tag{25}$$

By virtue of Theorem (D), the form A', as a function of the vectors $v^1, \dots, v^r$, is a polynomial in pairwise scalar products of these vectors, i.e., in expressions of the type (v^p, v^q). Expressions of the type $D(v^{q_1}, \dots, v^{q_l})$ cannot occur in A' since $r < l$. Similarly, the form A', regarded as a function of the vectors $u^1, \dots, u^r$, is a polynomial in expressions of the type (u^p, u^q) and $d(u^{p_1}, \dots, u^{p_k})$, where the latter may occur in A' only when $r \geqslant k$. Since $r \leqslant k+1$, in the case that $D(u^{p_1}, \dots, u^{p_k})$ is contained in A', it cannot occur in a product with an expression of its own type or with an expression of the type (u^p, u^q). Hence, the form A' can be expressed as a linear combination with constant coefficients of forms of the type

$$B' - D(u^{p_1}, \dots, u^{p_k})(v^{q_1}, v^{q_2})(v^{q_3}, v^{q_4}) \dots (v^{q_{k-1}}, v^{q_k}) \tag{26}$$

$$C' = (u^{p_1}, u^{p_2})(u^{p_3}, u^{p_4}) \dots (u^{p_{r-1}}, u^{p_r})(v^{q_1}, v^{q_2})(v^{q_3}, v^{q_4}) \dots (v^{q_{r-1}}, v^{q_r}), \tag{27}$$

where an expression of the type (26) may occur only when $r = k$. Expressions of the types (26) and (27) may obviously be formed only for even r. Furthermore, as the form A' is linear in each u^p and in each v^q, the numbers $p_1, \dots, p_r$, are all distinct as are the numbers $q_1, \dots, q_r$.

We shall say that, by virtue of (24), a composite tensor $A_{i_1 j_1 \dots i_r j_r}$ corresponds to the form A'. Here the indices $i_1, \dots, i_r$ refer to the space R^k_0, and the indices $j_1, \dots, j_r$ to the space R^l_0. Similarly, a composite tensor $B_{i_1 j_1 \dots i_k j_k}$ corresponds to each expression (26) and a composite tensor $C_{i_1 j_1 \dots i_r j_r}$ to the expression (27). Since, as has been shown, A' is a linear combination of expressions of the type (26) and (27), the composite tensor $A_{i_1 j_1 \dots i_r j_r}$ is a linear combination of composite tensors of the type $B_{i_1 j_1 \dots i_k j_k}$ and $C_{i_1 j_1 \dots i_r j_r}$. Now we study these composite tensors $B_{i_1 j_1 \dots i_k j_k}$ and $C_{i_1 j_1 \dots i_k j_k}$.

Let $\delta_{i\alpha}$ denote a tensor in the space R_0^k which is 1 for $i = \alpha$ and 0 for $i \neq \alpha$. And let $\delta'_{j\beta}$ denote a similar tensor in R_0^l. The composite tensors $B_{i_1 j_1 \dots i_k j_k}$ and $C_{i_1 j_1 \dots i_r j_r}$ can be easily expressed in terms of the tensors $\epsilon_{i_1 \dots i_k}$ (see (C)), $\delta_{i\alpha}$ and $\delta_{j\beta}$:

$$B_{i_1 j_1 \dots i_k j_k} = \epsilon_{i_1 \dots i_k} \delta'_{j_{q_1} j_{q_2}} \delta'_{j_{q_3} j_{q_4}} \dots \delta'_{j_{q_{k-1}} j_{q_k}}, \tag{28}$$

$$C_{i_1 j_1 \dots i_r j_r} = \delta_{i_{p_1} i_{p_2}} \delta_{i_{p_3} i_{p_4}} \dots \delta_{i_{p_{r-1}} i_{p_r}} \delta'_{j_{q_1} i_{q_2}} \delta'_{j_{q_3} j_{q_4}} \dots \delta'_{j_{q_{r-1}} j_{q_r}}. \tag{29}$$

Since, as has been proved, the composite tensor $A_{i_1 j_1 \dots i_r j_r}$ is a linear combination of the composite tensors $B_{i_1 j_1 \dots i_k j_k}$ and $C_{i_1 j_1 \dots i_r j_r}$, the form A can be expressed linearly in terms of the forms B and C, which correspond to the tensors $B_{i_1 j_1 \dots i_k j_k}$ and $C_{i_1 j_1 \dots i_r j_r}$, i.e., A can be expressed in terms of forms of the type:

$$B = \sum B_{i_1 j_1 \dots i_k j_k} \xi^1_{i_1 j_1} \dots \xi^k_{i_k j_k}, \tag{30}$$

$$C = \sum C_{i_1 j_1 \dots i_r j_r} \xi^1_{i_1 j_1} \dots \xi^r_{i_r j_r}. \tag{31}$$

As the form A is skew-symmetric, in place of the forms B and C occurring in such an expression, we can substitute the corresponding skew-symmetric forms B_1 and C_1 derived from B and C via alternation. Taking the resulting alternation into account, B can be replaced by any form obtained from B by a permutation of the vectors $\xi^1, \dots, \xi^r$, remembering that the sign of the expression obtained after an odd permutation should be changed. This is also true of the form C.

From the foregoing it is clear that in place of the form B we can take the form

$$\sum \epsilon_{i_1 \dots i_k} \delta'_{j_1 j_2} \delta'_{j_3 j_4} \dots \delta'_{j_{k-1} j_k} \xi'_{i_1 j_1} \dots \xi^k_{i_k j_k}, \tag{32}$$

which, as can be easily verified, coincides with (12). The form derived from (12) by alternation thus coincides with K^1 [see (13)].

We now consider the form C. In order to determine this form it suffices to indicate the two permutations:

$$p_1, p_2; p_3, p_4; \dots ; p_{r-1} p_r \tag{33}$$

and

$$q_1, q_2; q_3, q_4; \dots , q_{r-1} q_r \tag{34}$$

of the integers $1, 2, 3, \dots, r$, or more precisely two ways, (33) and (34), of subdividing the sequence $1, 2, \dots, r$ into pairs. In order to elucidate the combinatorial structure of the form C corresponding to the sequences

(33) and (34), let us assign to each of the numbers $m = 1, 2, \ldots, r$, a line segment a_m with end points b_m and c_m. Now, using the segments a_m, form a one-dimensional complex, joining the vertices b_m and b_n if the pair m, n belongs to the system (33) or joining the end points c_m and c_n if the pair m, n belongs to the system (34). Since in the complex thus obtained, exactly two segments adjoin every vertex, this complex consists of a finite number of simple closed polygons. Number these polygons with the integers 1, 2, ..., t, and let $r_1, r_2, \ldots, r_t$ denote the number of sides in the polygons. Since, as already mentioned, the vectors $\xi^1, \ldots, \xi^r$ can be permuted in any manner, we may assume that the first polygon consists of the segments $a_1, \ldots, a_{r_1}$ arranged in this order, the second polygon, of the segments $a_{r_1+1}, \ldots, a_{r_1+r_2}$ arranged in this order, and so on. We first assume that there is only one polygon made of the segments $a_1, \ldots, a_r$. Under this assumption, the sequences (33) and (34) take the form

$$r, 1; 2, 3; \ldots; r-2, r-1, \text{and} \tag{35}$$

$$1, 2; 3, 4; \ldots, r-1, r, \tag{36}$$

respectively. The corresponding form C is evidently of the type:

$$\sum \delta_{i_r i_1} \delta_{i_2 i_3} \ldots \delta_{r-2, r-1} \delta'_{j_1 j_2} \delta'_{j_3 j_4} \ldots \delta'_{j_{r-1} j_r} \xi^1_{i_1 j_1} \ldots \xi^r_{i_r j_r}. \tag{37}$$

For an r divisible by four, this form coincides with the form K^r [(see (14))]. We now show that for an r not divisible by four, this form vanishes after alternation. Indeed, if the vectors $\xi^1, \ldots, \xi^r$ are replaced by the vectors $\xi^r, \ldots, \xi^1$, respectively, the form (37) does not change; but the number of permutations is equal to $r(r-1)/2$, i.e., is odd for an even r not divisible by four. If there are several polygons, then the form C is equal to the product of forms of the type (37) corresponding to these polygons. Thus, after alternation, a form of the type C [(see (31)] is converted into a form of the type K^ρ [(see (18))].

This completes the proof of Theorem 1.

(E) According to the results obtained in Section 1, to each form A defined in U and invariant under the transformations of the group G_0, there corresponds a unique invariant field $\mathfrak{A}$ on the homogeneous manifold $H(k, l)$ (see Definition 1), whose value $\mathfrak{A}_0$ at the point R^k_0 coincides with A: $\mathfrak{A}_0 = A$. Thus, there also corresponds to the form K^ρ (see (C)) an invariant field $\mathfrak{K}$ on $H(k, l)$ satisfying the condition: $\mathfrak{K}_0 = K^\rho$. It turns out that the value of $\mathfrak{K}^\rho_1$ of the field at any point R^k_1 is calculated in the same way as its value at R^k_0, namely, if

$$\bar{e}_1, \ldots, \bar{e}_k, \quad \bar{f}_1, \ldots, \bar{f}_l \tag{38}$$

is an orthonormal basis of the space R^{k+l}, such that the vectors $\bar{e}_1, \ldots, \bar{e}_k$ lie in R_1^k, and U_1 is the coordinate neighbourhood of the element R_1^k that corresponds to the basis (38) with the coordinates $\bar{\xi}_{ij}$, $\|\bar{\xi}_{ij}\| = \bar{\xi}$, then

$$\mathfrak{R}_1^\rho = \mathfrak{R}_1^\rho(\bar{\xi}^1, \ldots, \bar{\xi}^r) = K^\rho(\bar{\xi}^1, \ldots, \bar{\xi}^r), \tag{39}$$

where $K^\rho(\bar{\xi}^1, \ldots, \bar{\xi}^r)$ is the form made up of the vectors $\bar{\xi}^1, \ldots, \bar{\xi}^r$ by the rules stated in (C).

We now prove (E). Let $g \in G$, where $g(R_1^k) = R_0^k$. Take

$$e_1 = g(\bar{e}_1), \ldots, e_k = g(\bar{e}_k), \quad f_1 = g(\bar{f}_1), \ldots, f_l = g(\bar{f}_l). \tag{40}$$

These vectors (40) may now be taken as the basis (2), which is suitable for constructing the forms K^σ in the linear space U, because the forms K^ρ are, by (C), invariant under the basis (2). Since the mapping g transforms the basis (38) into the basis (40), vectors in the spaces U_1 and U_2, corresponding to each other under the mapping g, have the same coordinates. Therefore

$$K^\rho(g(\bar{\xi}^1), \ldots, g(\bar{\xi}^r)) = K^\rho(\bar{\xi}^1, \ldots, \bar{\xi}^r). \tag{41}$$

On the other hand, by (F) Section 1, we have

$$\mathfrak{R}_1^\rho(\bar{\xi}^1, \ldots, \bar{\xi}^r) = \mathfrak{R}_0^\rho(g(\bar{\xi}^1), \ldots, g(\bar{\xi}^r)) \tag{42}$$

Since $\mathfrak{R}_0 = K$, we obtain (39) on comparing (42) and (41).

(F) To each element $R^k \in H(k, l)$, let us associate an element $\hat{R}^k = \phi(R^k)$ differing from R^k only in orientation. Further, take $\hat{\mathfrak{R}}^\rho = \phi^* \mathfrak{R}^\rho$ (see (F), Section 1). Then

$$\hat{\mathfrak{R}}^1 = -\mathfrak{R}^1; \quad \text{for } \rho \neq 1: \quad \hat{\mathfrak{R}}^\rho = \mathfrak{R}^\rho. \tag{43}$$

Since, by (E), the field $\mathfrak{R}^\rho$ is constructed at all points of the manifold $H(k, l)$ in a similar way, it suffices to demonstrate the relation (43) for any one of the points in $H(k, l)$. We demonstrate it for the point R_0^k. We shall construct the coordinate neighbourhood U of the element R_0^k with the help of the basis (2). In order to apply (E), take $R_1^k = \hat{R}_0^k$. To construct a coordinate system in $U_1 = \hat{U}$, let us again apply the basis (2). Then the vectors ξ and $\hat{\xi}$ corresponding to each other under the mapping ϕ of the space U onto $\hat{U}$ will have the same coordinates. Hence, since the form K^1 changes its sign when the orientation of R_0^k is changed, and the form K^ρ for $\rho \neq 1$ does not depend on the orientation of R_0^k (see (C)), we obtain

$$\begin{aligned} \mathfrak{R}_1^1(\xi^1, \ldots, \xi^r) &= -\mathfrak{R}_0^1(\xi^1, \ldots, \xi^r); \\ \text{for } \rho \neq 1: \quad \mathfrak{R}_0^\rho(\xi^1, \ldots, \xi^r) &= \mathfrak{R}_0^\rho(\xi^1, \ldots, \xi^r). \end{aligned} \tag{44}$$

On the other hand

$$\hat{\mathfrak{K}}_1^\rho(\hat{\xi}^1, \ldots, \hat{\xi}^r) = \mathfrak{K}_0^\rho(\xi^1, \ldots, \xi^r), \tag{45}$$

where the left-hand side of the equation is the value of the field $\hat{\mathfrak{K}}^\rho$ at the point R_1^k. Comparing (44) and (45), we obtain (43) for the point R_0^k.

We now formulate a corollary of Theorem 1 essential for our purposes.

Theorem 2. The invariant fields $\mathfrak{K}^\rho$ of order $r \leqslant k$ (see (E)) defined on the homogeneous manifold $H(k, l)$ are closed and form an independent basis of homology for the closed fields of order r on $H(k, l)$. Moreover every closed homogeneous field of order r on $H(k, l)$ is a linear combination of the fields $\mathfrak{K}^\rho$ with constant coefficients.

Proof. To demonstrate that the fields $\mathfrak{K}^\rho$ are independent we apply a result from my paper [2] that it is not easily proved. But the fact that the fields $\mathfrak{K}^\rho$ are independent is unimportant for understanding Section 2 and 3.

We first show that the fields $\mathfrak{K}^\rho$ are closed. For this purpose it is sufficient to demonstrate that every invariant field $\mathfrak{A}$ of order $r \leqslant k$ on $H(k, l)$ is closed. If $\mathfrak{A}$ is not identically zero, then the corresponding multilinear form A is non-zero, and consequently, by Theorem 1, is of even order. Thus, if the field $\mathfrak{A}$ is not identically zero, it is of even order, and its exterior derivative is of odd order. Therefore this derivative is zero (once again, by Theorem 1). Thus, in either case the exterior derivative of an invariant field $\mathfrak{A}$ is zero, i.e., the field $\mathfrak{A}$ is closed.

That every closed invariant field $\mathfrak{A}$ of order $r \leqslant k$ is a linear combination of the fields $\mathfrak{K}^\rho$ with constant coefficients follows directly from Theorem 1, because, by what has been stated in Section 1, it suffices to consider the multilinear forms A and K^ρ. Moreover, by Section 1, every closed field is homologous to an invariant field, therefore we find that the fields $\mathfrak{K}^\rho$ form a basis of homology of the manifold $H(k, l)$. We now show that this basis is linearly independent.

Let $\mathfrak{A}$ be some linear combination of fields $\mathfrak{K}^\rho$ of order $r \leqslant k$ with constant coefficients. Assume that the field $\mathfrak{A}$ is homologous to zero. Then $\mathfrak{A}$ is the exterior derivative of some invariant field. This derivative is of odd order and is therefore identically zero. Thus, the field $\mathfrak{A}$, being homologous to zero, should also be identically zero. We see, therefore, that to establish the homological independence of the fields $\mathfrak{K}^\rho$ it suffices to demonstrate the linear independence of the forms K^ρ. I did not, however, succeed in proving this fact directly, and therefore I refer the reader to the results of my paper [2].

Each form $\mathfrak{K}^\rho$, for $\rho \not\equiv 1$, is determined by a non-decreasing sequence

$\rho = \{r_1, \ldots, r_t\}$ of natural numbers (see (16)). As $r \leqslant k$ and the numbers $r_1, \ldots, r_t$ are divisible by four, we have $k - 2t \geqslant 0$. We now define a function $\chi(i)$ of an integral-valued argument $i = 1, \ldots, k$, as follows:

$$\chi(1) = \chi(2) = \frac{r_t}{2}, \quad \chi(3) = \chi(4) = \frac{r_{t-1}}{2}, \ldots, \chi(2t-1) = \chi(2t) = \frac{r_1}{2},$$

$$\chi(2t+1) = \ldots = \chi(k) = 0. \tag{46}$$

This function satisfies the conditions (10) Section 6 of my paper [2]. Consequently, there corresponds to this function a cycle Z_χ of dimension $kl - r$ contained in the basis of weak homology of the manifold $H(k, l)$, constructed in Theorem 1 of my paper [2]. It is readily seen that this correspondence between the forms $\mathfrak{K}^\rho$, $\rho \not\equiv 1$ and the cycles Z_χ ($\chi \in X$, $\chi \neq 1$) is one-to-one. When k is even, let us associate the cycle Z_χ, $\chi \equiv 1$ of dimension $kl - k$ with the form K^1. Thus, the number of forms K^ρ of order $r \leqslant k$ is equal to the $(kl - r)$-dimensional Betti number of the manifold $H(k, l)$ or, by virtue of Poincare's theorem, is equal to the r-dimensional Betti number of this manifold. If the forms K of order r were linearly dependent, then the Betti number calculated from the forms would be different from the Betti number calculated from the cycles. And this is impossible. Hence the forms K^ρ are linearly independent. This completes the proof of Theorem 2.

3. Characteristic Fields on a Riemannian Manifold

In this section we construct certain topologically invariant fields on a Riemannian manifold M^k that are expressed in terms of the Riemannian tensor of this manifold. Here we assume that the Riemannian metric of the manifold M^k is obtained as a result of embedding the differentiable manifold M^k into a Euclidean space. Topological invariance is understood in the sense that the closed fields constructed up to homology (see (I) in Section 1) do not depend on the embedding of the differentiable manifold M^k in Euclidean space.

Let U^k be some coordinate neighbourhood of a point a of a k-dimensional differentiable manifold M^k in which the local coordinates $x^1, \ldots, x^k$ have been introduced. Whitney calls a continuous mapping f of the manifold M^k into a Euclidean vactor space R^{k+l} with coordinates $z^1, \ldots, z^{k+l}$ *regular at a point a*, if in the vicinity of the point a, this mapping can be expressed in coordinate form as follows:

$$z^j = z^j(x^1, \ldots, x^k) = z^j(x) \quad (j = 1, \ldots, k+1) \tag{1}$$

where the functional matrix

$$\left\| \frac{\partial z^j(x)}{\partial x^i} \right\| \qquad (i = 1, \ldots, k; j = 1, \ldots, k + l)$$

at the point a is of rank k and has a sufficient number of derivatives. If a mapping f is regular at every point $a \in M^k$, it is called *regular*. Whitney has demonstrated that a differentiable manifold M^k can be mapped regularly into a vector space R^{2k}. The manifold M^k can be mapped regularly and homeomorphically into the vector space R^{2k+1}.

If a mapping f of a manifold M^k into a Euclidean vector space R^{k+l} is regular, then each point $a \in M^k$ has a sufficiently small neighbourhood U^k which is mapped by f homeomorphically onto some subset of the Euclidean space R^{k+l}. As a result, a Riemannian metric is naturally induced in the neighbourhood U^k. Thus a definite Riemannian metric in M^k corresponds to a regular mapping f of the manifold M^k into the Euclidean space R^{k+l}.

If the mapping f in the neighbourhood U^k is expressed by (1), the metric tensor in local coordinates $x^1, \ldots, x^k$ in the neighbourhood U^k is then given by the expression:

$$g_{\alpha\beta} = \sum_{j=1}^{k+1} \frac{\partial z^j}{\partial x^\alpha} \frac{\partial z^j}{\partial x^\beta}, \tag{2}$$

and is therefore differentiable a sufficiently large number of times.

Definition 2. Let f be a regular (possibly not homeomorphic) mapping of a differentiable oriented manifold M^k into a Euclidean vector space R^{k+l}. To a point $a \in M^k$, there corresponds a definite oriented tangent T_a to $f(U^k)$ at the point $f(a)$, where U^k is a small neighbourhood of the point a in M^k. Let us denote the oriented k-dimensional vector subspace of the space R^{k+l} parallel to T_a by $T(a)$. As $T(a) \in H(k, l)$, we obtain a mapping T of the differentiable manifold M^k into $H(k, l)$ which is continuously differentiable a sufficient number of times. We call it the *tangential mapping corresponding to the regular mapping f*.

Definition 3. Let f be a regular mapping of a closed oriented differentiable manifold M^k into a Euclidean vector space R^{k+l}, $l \geqslant k + 2$, $H(k, l)$ the manifold of all k-dimensional oriented vector subspaces of the space R^{k+l}, $\mathfrak{R}^\rho$ a closed invariant field on $H(k, l)$ as defined in (E) Section 2, and T a tangential mapping of M^k into $H(k, l)$ corresponding to the regular mapping f. Put

$$P^\rho = T^* \mathfrak{R}^\rho \tag{3}$$

(see (F) Section 1). The closed fields P^ρ on the manifold M^k are called *characteristic fields*, and the corresponding tensors $P_{i_1 \dots i_r}$ the *characteristic tensors of the manifold* M^k. It turns out that the characteristic fields are, up to homology not dependent on the mapping f; in particular, they do not depend on the number l.

We first show that the forms P^ρ do not depend on the number l. For this, suppose that $R^{k+l} \subset R^{k+l'}$. Let us denote the objects stated in Definition 3 corresponding to the Euclidean vector space $R^{k+l'}$ by $H(k, l')$, $\mathfrak{K}'^\rho$ and T', respectively. To each element $R^k \in H(k, l)$ associate the same element $R^k \in H(k, l')$ and let ϕ denote the embedding of the manifold $H(k, l)$ into $H(k, l')$ thus obtained. From the definition of the form K^ρ, it follows that

$$\mathfrak{K}^\rho = \phi^* \mathfrak{K}'^\rho. \tag{4}$$

Furthermore, obviously we have

$$T' = \phi T. \tag{5}$$

From (4) and (5), it follows that

$$T'^* \mathfrak{K}'^\rho = T^* \mathfrak{K}^\rho. \tag{6}$$

Thus, we have demonstrated that the fields P^ρ do not depend on the number l.

Assume, now, that there are two arbitrary regular mappings f_0 and f_1 of the manifold M^k into the spaces R^{k+l_0} and R^{k+l_1}, respectively. Let us form the direct sum R^{k+l} of the spaces R^{k+l_0} and R^{k+l_1} and then consider the spaces R^{k+l_0} and R^{k+l_1} as two orthogonal subspaces of the space R^{k+l}. In view of what has been proved earlier, we can now regard f_0 and f_1 as two distinct mappings of the manifold M^k into the same vector space R^{k+l}. Denote the tangential maps of the manifold M^k into the manifold $H(k, l)$ corresponding to the mappings f_0 and f_1 by T_0 and T_1, respectively. We prove that the mappings T_0 and T_1 of the manifold M^k into $H(k, l)$ are homotopic.

Consider the mapping f_t $(0 < t < 1)$ of the manifold M^k into $H(k, l)$ given by the relation

$$f_t(x) = (1-t) f_0(x) + t f_1(x), \quad x \in M^k. \tag{7}$$

Obviously, the mapping f_t of the manifold M^k into R^{k+l} is regular. Consequently, the tangential mapping T_t of the manifold M^k into $H(k, l)$ corresponding to the mapping f_t is defined. Since T_t continuously deforms T_0 into T_1, these mappings are homotopic. As T_0 and T_1 are homotopic, we have

$$T_0^* \mathfrak{K}^\rho \sim T_1^* \mathfrak{K}^\rho. \tag{8}$$

Thus, we have demonstrated that the characteristic fields P^ρ do not depend on the initial regular mapping f.

We now express the characteristic fields of the manifold M^k by its Riemannian tensor. Here we mean the Riemannian tensor corresponding to the metric of the manifold M^k which is obtained from any regular mapping of the manifold M^k into a Euclidean space. Since, as a consequence of the foregoing, the characteristic field do not depend, up to homology, on the initial regular mapping, the expressions of the characteristic fields derived below in terms of the Riemannian tensor also do not depend on the Riemannian metric. We shall however assume that the Reimannian metric is obtained from a regular mapping into a Euclidean speace. If the Riemannian metric is given on M^k independent of the mapping of this manifold into a Euclidean space, then the results derived in this paper do not imply that the expressions (14) and (16) yield invariants of the manifolds M^k.

First let us introduce some notation.

(A) Let R_0^k be an oriented Euclidean vector space, the metric in which, under the coordinates already chosen, is given by the metric tensor $g_{\alpha\beta}$. Let h denote the square root of the determinant of the matrix $\|g_{\alpha\beta}\|$:

$$h = +\sqrt{(g_{\alpha\beta})}\,. \tag{9}$$

Now let $x_1, \ldots, x_k$ be a sequence of vectors in R_0^k, where $x_i = (x_i^1, \ldots, x_i^k)$ $(i = 1, \ldots, k)$. Denote the determinant of the matrix $\|x_i^\alpha\|$ by $|x_i^\alpha|$. Then the volume of the oriented parallelopiped on the vectors $x_1, \ldots, x_k$ is

$$V = h|x_i^\alpha|. \tag{10}$$

To express this volume in tensor form, let us, as in (C) Section 2, denote by $\epsilon_{i_1 \ldots i_k}$ the number which depends skew-symmetrically on the indices and satisfies the condition: $\epsilon_{1 \ldots k} = 1$ or -1, depending on whether the coordinates in R_0^k generate its initial orientation or not. Then we have

$$|x_i^\alpha| = \sum \epsilon_{i_1 \ldots i_k} x_1^{i_1} \ldots x_k^{i_k}, \tag{11}$$

and, assuming that

$$h_{i_1 \ldots i_k} = h\epsilon_{i_1 \ldots i_k}, \tag{12}$$

we obtain

$$V = \sum h_{i_1 \ldots i_k} x_1^{i_1} \ldots x_k^{i_k}. \tag{13}$$

From (13) it is clear that $h_{i_1 \ldots i_k}$ is a covariant skew-symmetric tensor. In orthonormal coordinates this tensor is converted into the tensor $\epsilon_{i_1 \ldots i_k}$ (see (C) Section 2).

Theorem 3. If f is a regular mapping of a clsoed oriented differentiable manifold M^k into a Euclidean space R^{k+l}, the characteristic tensors (see Definition 3) corresponding to the mapping f are expressed in terms of the Riemannian tensor $R_{\alpha\beta\gamma\delta}$ corresponding to this mapping as follows:

$$P^1_{i_1 \dots i_k} = \sum (-\tfrac{1}{2})^{k/2} h_{\alpha_1 \dots \alpha_k} R^{\alpha_1 \alpha_2 \cdot \cdot}_{\cdot \cdot [i_1 i_2} R^{\alpha_3 \alpha_4 \cdot \cdot}_{\cdot \cdot i_3 i_4} \dots R^{\alpha_{k-1} \alpha_k \cdot \cdot}_{\cdot \cdot i_{k-1} i_k]}, \tag{14}$$

for

$$\rho \equiv r \neq 1 : P^\rho_{i_1 \dots i_r} = P^r_{i_1 \dots i_r} =$$

$$= \sum (\tfrac{1}{2})^s R^{\cdot \alpha_1}_{\cdot \alpha_2 [i_1 i_2} R^{\cdot \alpha_2}_{\cdot \alpha_3 i_3 i_4} \dots R^{\alpha_s}_{\cdot \alpha_1 i_{r-1} i_r]} \quad \left(s = \frac{r}{2}\right). \tag{15}$$

Here square brackets stand for alternation over all the indices $i_1, \dots, i_k$ in (14) and over all the indices $i_1, \dots, i_r$ in (15) In the general case, where $\rho = \{r_1, \dots, r_t\}$, we obviously have

$$P^\rho_{i_1 \dots i_r} = P^{r_1 \cdot \cdot \cdot}_{[i_1 \dots i_{r_1}} \dots P^{r_t}_{i_{r_1} + \dots + r_{t-1} + 1 \dots i_r]} \tag{16}$$

Proof. Let a be an arbitrary point in M^k and U^k a small neighbourhood of a in M^k. Without loss of generality, we may assume that $f(a)$ is the zero of the vector space R^{k+l}. Let R^k_0 denote the oriented tangent to $f(U^k)$ at the point $f(a) = 0$. In R^{k+l} choose an orthonormal basis $e_1, \dots, e_k, f_1, \dots, f_k$ such that the vectors $e_1, \dots, e_k$ lie in R^k_0. Express the vector $f(x), x \in U^k$ as follows:

$$f(x) = \sum_{i=1}^{k} x^i e_i + \sum_{j=1}^{l} y_j f_j. \tag{17}$$

If U^k is a sufficiently small neighbourhood of the point a, then the orthogonal projection of the set $f(U^k)$ into R^k_0 is a homeomorphic mapping and therefore $f(U^k)$ can be given by the equations:

$$y_j = y_j(x) = y_j(x^1, \dots, x^k) \quad (j = 1, \dots, l). \tag{18}$$

Here $x_1, \dots, x_k$ may be taken as local coordinates in U^k. By virtue of (18) and (2), the metric generated in U^k by the mapping f is given by the metric tensor:

$$g_{\alpha\beta} = g_{\alpha\beta}(x) = \delta_{\alpha\beta} + \sum_{j=1}^{l} \frac{\partial y_j(x)}{\partial x^\alpha} \frac{\partial y_j(x)}{\partial x^\beta}. \tag{19}$$

The vector e'_i tangent ot the coordinate axis x^i in $f(U^k)$ at the point x is obviously expressed as follows:

$$e'_i = e_i + \sum_{j=1}^{l} \frac{\partial y_j(x)}{\partial x^i} f_j. \tag{20}$$

Thus the vector subspace $T(x)$ parallel to the tangent T_x contains the vectors $e'_1, \ldots, e'_k$ and has in the space U coordinates $\xi_{ij}(x)$ defined by the relations:

$$\xi_{ij}(x) = \frac{\partial y_j(x)}{\partial x^i}, \tag{21}$$

this describes the tangent mapping T corresponding to the regular mapping f in the neighbourhood U^k. As the coordinates ξ_{ij} of the element R_0^k in U are zero, we have

$$\frac{\partial y_j(a)}{\partial x^i} = 0 \quad (i = 1, \ldots, k; j = 1, \ldots, l). \tag{22}$$

Hence it immediately follows, by virtue of (19), that

$$g_{\alpha\beta}(a) = \delta_{\alpha\beta}, \quad \frac{\partial g_{\alpha\beta}(a)}{\partial x^i} = 0 \quad (i, \alpha, \beta = 1, \ldots, k). \tag{23}$$

Thus, the Riemannian tensor at the point a is given by the following expression:

$$R_{\alpha\beta\gamma\delta} = \frac{1}{2}\left(\frac{\partial^2 g_{\alpha\gamma}(a)}{\partial x^\beta \partial x^\delta} + \frac{\partial^2 g_{\beta\delta}(a)}{\partial x^\alpha \partial x^\gamma} - \frac{\partial^2 g_{x\delta}(a)}{\partial x^\beta \partial x^\gamma} - \frac{\partial^2 g_{\beta\gamma}(a)}{\partial x^\alpha \partial x^\delta}\right) \tag{24}$$

(see [8]). By virtue of (19), expression (24) gives

$$R_{\alpha\beta\gamma\delta} = -\sum_{j=1}^{l}\left(\frac{\partial^2 y_j(a)}{\partial x^\alpha \partial x^\gamma}\frac{\partial^2 y_j(a)}{\partial x^\beta \partial y^\delta} - \frac{\partial^2 y_j(a)}{\partial x^\alpha \partial x^\beta}\frac{\partial^2 y_j(a)}{\partial x^\beta \partial x^\gamma}\right). \tag{25}$$

Let $x_p = (x_p^1, \ldots, x_p^k)$ be some vector in M^k at the point a. By virtue of the mapping T, vector $\xi^p = \xi_{ij}^p$ given by the relations (see (21))

$$\xi_{ij}^p = \sum \frac{\partial^2 y_j(a)}{\partial x^i \partial x^\alpha} x_p^\alpha \tag{26}$$

corresponds to the vector x_p in U. Hence, by virtue of (25), we have

$$\begin{aligned} T^*\frac{1}{2}(\xi_{\alpha\beta}^{pq} - \xi_{\alpha\beta}^{qp}) &= \frac{1}{2}\sum_{j,\gamma,\delta}\left(\frac{\partial^2 y_j(a)}{\partial x^\alpha \partial x^\gamma}\frac{\partial^2 y_i(a)}{\partial x^\beta \partial x^\delta} - \frac{\partial^2 y_j(a)}{\partial x^\alpha \partial x^\delta}\frac{\partial^2 y_j(a)}{\partial x^\beta \partial x^\gamma}\right) x_p^\gamma x_p^\delta \\ &= -\frac{1}{2}\sum R_{\alpha\beta\gamma\delta} x_p^\gamma x_p^\delta. \end{aligned} \tag{27}$$

As the forms K^ρ are skew-symmetric, we can substitute $\frac{1}{2}(\xi_{\alpha\beta}^{pq} - \xi_{\alpha\beta}^{qp})$ for $\xi_{\alpha\beta}^{pq}$ in the expressions of these forms in terms of the forms $\xi_{\alpha\beta}^{pq}$. Therefore, by (27), we obtain

$$P_{i_1 \ldots i_k}^1 = (-\tfrac{1}{2})^{k/2} \sum \epsilon_{\alpha_1 \ldots \alpha_k} R_{\alpha_1\alpha_2[i_1 i_2} \ldots R_{\alpha_{k-1}\alpha_k i_{k-1} i_k]}, \tag{28}$$

$$P^r_{i_1 \ldots i_r} = (\tfrac{1}{2})^s \sum R_{\alpha_1\alpha_2[i_1i_2} R_{\alpha_2\alpha_3 i_3 i_4} \ldots R_{\alpha_s\alpha_1 i_{r-1}i_r]}, \tag{29}$$

where alternation on the right-hand sides is taken over all the indices $i_1, \ldots, i_k$ and $i_1, \ldots, i_r$, respectively. The validity of (28) and (29) has been demonstrated in special coordinates, in which, incidentally, $g_{\alpha\beta}(ab) = \delta_{\alpha\beta}$. By virtue of this last relation, the equalities (28) and (29) coincide with (14) and (15), respectively, which are of tensor type. Therefore they are true in any coordinates.

This completes the proof of Theorem 3.

We now give some corollaries of Theorem 3.

(B) Let Y be a cycle in M^k with integral coefficients of dimension $r = r(\rho) \leqslant k$. The integral $\int_Y P^\rho$ of the field P^ρ over the cycle Y depends only on ρ, on the weak homology class of the cycle Y, and certainly on the initial manifold M^k. If $r = k$, then the oriented manifold M^k itself can be taken as the cycle Y, so that $\int_{M^k} P^\rho$ is a numerical invariant of the oriented manifold M^k that depends on ρ. In particular, the integral $\int_{M^k} P^1$ is a numerical invariant of the oriented manifold M^k.

Assertion (B) follows directly from what has been stated in (I) Section 1 and Definition 3.

(C) Let $\psi(x)$ be a real-valued numerical function defined on an oriented Riemannian manifold M^k, $g_{\alpha\beta}$ the metric tensor of this manifold, $h_{i_1 \ldots i_k}$ the tensor constructed from $g_{\alpha\beta}$ as stated in (A). Then, a covariant skew-symmetric tensor of order k can be defined as follows:

$$\Psi_{i_1 \ldots i_k} = \psi(x) h_{i_1 \ldots i_k}. \tag{30}$$

Let Ψ denote the corresponding field. The integral $\int_{M^k} \Psi$ of the field Ψ over the oriented manifold M^k is equal to the integral of the function $\psi(x)$ over the Riemannian manifold M^k. Let us denote this integral by $\int_{M^k} \psi(x)\, dV$. The function $\Psi(x)$ can be easily reconstructed from the tensor $\Psi_{i_1 \ldots i_h}$, namely, we have

$$\psi(x) = \frac{\Psi_{1 \ldots k}}{h(x)}, \tag{31}$$

where the right-hand side contains the component of the tensor $\Psi_{i_1 \ldots i_k}$ in coordinates which give a positive orientation to M^k. Using this method, from the characteristic tensors $P_{i_1 \ldots i_k}$ with $r(\rho) = k$ we can extract numerical functions whose integrals over the manifold M^k are topological invariants of M^k. The function $p^1(x)$ corresponding to the tensor $P^1_{i_1 \dots i_k}$ is expressed through the Riemannian tensor of the manifold M^k as follows. Put

$$R^{\alpha_1 \dots \alpha_k}_{i_1 \dots i_k} = R^{[\alpha_1\alpha_2 \cdot}{}_{\cdot \; \cdot [i_1 i_2} R^{\alpha_3\alpha_4}{}_{\cdot \; \cdot \; i_3 i_4} \dots R^{\alpha_{k-1}\alpha_k]}{}_{\cdot \; \cdot \; i_{k-1} i_k]}, \tag{32}$$

where alternation is taken over upper and lower indices separately. Then, we have

$$p^1(x) = (-\tfrac{1}{2})^{k/2} k! \, R^{1 \dots k}_{1 \dots k}. \tag{33}$$

This relation follows directly from (14), (31) and (32).

4. Characteristic Fields and Characteristic Cycles

In this section we make a connection between the characteristic fields (see Definition 3) and the characteristic cycles (see Section 3 in [2]) of a manifold M^k. Therefore, a knowledge of my paper [2] is essential for an understanding this section. In particular, using certain results in my paper [6] we prove an assertion which is already known, (see, for instance [6]) that for even k, the integral of the field P^1 over a manifold M^k is equal to the Euler characteristic of the manifold M^k multiplied by $\sigma_k/2$, where σ_k is the k-dimensional volume of a k-dimensional unit sphere. This result establishes an exact relationship between the characteristic field P^1 and the characteristic cycle X_1 (see Section 6, (C) in [2]).

Since only weak homology can be taken into account when using fields, only characteristic cycles X satisfying the condition

$$\chi \in X, \tag{1}$$

will be considered here, because a cycle Z_χ for a χ that does not satisfy the condition (1) is always weakly homologous to zero in $H(k, l)$. In the sequel we always assume that the condition (1) is satisfied.

In constructing cycles we not only use integral but also real coefficients. The cycles will be obtained as linear forms of integral cycles with real coefficients,

(A) As the fields $\mathfrak{R}^\rho$, $r(\rho) = r$, form a linearly independent basis of homology for closed fields of order r on the manifold $H(k, l)$ (see Theorem 2), and the cycles Z_χ, $r(\chi) = r$, $\chi \in X$, form a linear independent basis of weak homology of dimension $kl - r$ in $H(k, l)$ (see Theorem 1 in [2]), then (see (I) Section 1) to each field $\mathfrak{R}^\rho$, there corresponds a unique linear form $Z(\mathfrak{R}^\rho) = Z^\rho$, of cycles Z_χ, $r(\chi) = r$:

$$Z(\mathfrak{R}^\rho) = \sum_{\chi \in X} a^{\rho\chi} Z_\chi, \tag{2}$$

satisfying the condition:

$$\int_W \mathfrak{R}^\rho = I(Z(\mathfrak{R}^\rho), W), \tag{3}$$

where W is an arbitrary r-dimensional cycle with integral coefficients in $H(k, l)$. The coefficients $a^{\rho\chi}$ are real numbers that do not depend on l, and the determinant of the matrix $\|a^{\rho\chi}\|$ is non-zero.

We now prove that the numbers $a^{\rho\chi}$ do not depend on the number l. Let $R^{k+l''}$ be a linear subspace of the space R^{k+l}, $H(k, l'')$ its corresponding manifold and $\mathfrak{R}''^\rho$ a field defined on $H(k, l'')$. As has already been proved in Section 3 (see (4) Section 3) the form $\mathfrak{R}^\rho$ restricted to the submanifold $H(k, l'')$ of the manifold $H(k, l)$ coincides with $\mathfrak{R}''^\rho$. Thus, for any cycle W in $H(k, l'')$, we have

$$\int_W \mathfrak{R}^\rho = \int_W \mathfrak{R}''^\rho. \tag{4}$$

Furthermore, the relation (4) in Section 3 of my paper [2]

$$Z_\chi \times H(k, l'') = Z''_\chi, \tag{5}$$

shows that for an arbitrary cycle W in $H(k, l'')$, we have

$$I(Z_\chi, W) = I(Z''_\chi, W), \tag{6}$$

where the intersection index on the left-hand side is taken in $H(k, l)$ and the intersection index on the right-hand side in $H(k, l'')$. Thus, putting

$$Z(\mathfrak{R}''^\rho) = \sum_{\chi \in X} a^{\rho\chi} Z''_\chi, \tag{7}$$

we obtain from (4) and (6)

$$\int_W \mathfrak{R}'' = I(Z(\mathfrak{R}''^\rho), W),$$

This means that the coefficients $a^{\rho\chi}$ calculated in $H(k, l)$ and $H(k, l'')$ have the same values.

Theorem 4. Let the $(k - r)$-dimensional cycle (see (A))

$$X(P^\rho) = \sum_{\chi \in X} a^{\rho\chi} X_\chi \tag{8}$$

be associated with the characteristic field P^ρ, $r(\rho) = r$, of the manifold M^k

(see Definition 3). Then, for every r-dimensional cycle Y with integral coefficients in M^k

$$\int_Y P^\rho = I(X(P^\rho), Y). \tag{9}$$

Proof. Let f be a regular mapping of the manifold M^k into R^{k+l}, and T the corresponding tangent map of the manifold M^k into $H(k, l)$ (see Definition 2). Then (see (I) Section 1):

$$\int_{TY} \mathfrak{R}^\rho = \int_Y P^\rho \tag{10}$$

and

$$I(Z_\chi, TY) = I(X_\chi, Y). \tag{11}$$

Taking $W = TY$ in (3) and then comparing the resulting expression with (10) and (11), we obtain (9).

This completes the proof of Theorem 4.

We note that whether the coefficients $a^{\rho\chi}$ are determined by the condition (9) remains obscure, because we do not know to what extent the characteristic cycles are independent. In fact, if some linear combination of characteristic cycles is homologous to zero, then adding it to the right-hand side of (8), we can obtain new values of the coefficients for which condition (9) is still satisfied.

Theorem 5. The coefficients $a^{\rho\chi}$ in (2) satisfy the conditions:

$$a^{\rho 1} = 0 \text{ for } \rho \not\equiv 1; \quad a^{1\chi} = 0 \text{ for } \chi \not\equiv 1. \tag{12}$$

Proof. Transforming the manifold $H(k, l)$ by the mapping $R^k \to \hat{R}^k$ (see (F) Section 2), we find:

$$\hat{\mathfrak{R}}^1 = -\mathfrak{R}^1; \quad \text{for } \rho \not\equiv 1: \hat{\mathfrak{R}}^\rho = \mathfrak{R}^\rho, \tag{13}$$

$$\hat{Z}_\chi = (-1)^{k+l+\chi(1)} Z_\chi; \quad \hat{H}(k, l) = (-1)^{k+l} H(k, l) \tag{14}$$

(see (15), Section 6 in [2]). By virtue of condition (1), we have

$$\chi(1) = 1 \text{ for } \chi \equiv 1; \quad \chi(1) \equiv 0 \ (\text{mod } 2) \text{ for } \chi \not\equiv 1. \tag{15}$$

From (2) and (3), it follows that

$$\int_W \mathfrak{R}^\rho = \sum_{\chi \in X} a^{\rho\chi} I(Z_\chi, W), \tag{16}$$

where W is an arbitrary r-dimensional cycle in $H(k, l)$ with integral coefficients and the intersection indices on the right-hand side are taken in $H(k,l)$. Under the transformation $R^k \to \hat{R}^k$, the relation (16) becomes

$$\int_{\hat{W}} \hat{\mathfrak{R}}^\rho = \sum_{\chi \in X} a^{\rho\chi} I(\hat{Z}_\chi, \hat{W}); \tag{17}$$

where $\hat{W}$ is again an arbitrary r-dimensional cycle in $H(k,l)$ with integral coefficients, so that it can be replaced by W. The intersection indices on the right-hand side are taken in $H(k,l)$. Thus, by virtue of (14), the relation (17) can now be rewritten as

$$\int_W \hat{\mathfrak{R}}^\rho = \sum_{\chi \in X} (-1)^{\chi(1)} a^{\rho\chi} I(Z_\chi, W), \tag{18}$$

where W is an arbitrary r-dimensional cycle in $H(k,l)$ with integral coefficients and the intersection indices are taken in $H(k,l)$. For $\rho \equiv 1$, we obtain from (18), by virtue of (13) and (15),

$$\int_W \mathfrak{R}^1 = a^{11} I(Z_1, W) - \sum_{\chi \not\equiv 1} a^{1\chi} I(Z_\chi, W). \tag{19}$$

Comparing this relation with (16) for $\rho \equiv 1$ and using that the cycles Z_χ are independent and the cycle W is arbitrary, we obtain

$$a^{1\chi} = 0 \quad \text{for } \chi \not\equiv 1. \tag{20}$$

For $\rho \not\equiv 1$, the relation (18) gives by (13) and (15),

$$\int_W \mathfrak{R}^\rho = -a^{\rho 1} I(Z_1, W) + \sum_{\chi \not\equiv 1} a^{\rho\chi} I(Z_\chi, W). \tag{21}$$

Comparing (21) with (16) for $\rho \not\equiv 1$, we obtain

$$a^{\rho 1} = 0 \quad \text{for } \rho \not\equiv 1. \tag{22}$$

This completes the proof of Theorem 5.

An immediate corollary of Theorem 5 is the formula

$$\int_{M^k} P^1 = a^{11} X_1(M^k), \tag{23}$$

where $X_1(M^k)$ is the Euler characteristic of the manifold M^k. Indeed, from Theorems 4 and 5, it follows that

$$\int_{M^k} P^1 = a^{11} I(X_1, M^k) = a^{11} X_1(M^k) \tag{24}$$

(see Theorem 4 in [9]).

Recall that the field P^1 on M^k and, consequently, the coefficient a^{11} are defined only for even k. The numerical value of a^{11} is given in Theorem 6.

Theorem 6. The value of the coefficient a^{11} is given by the formula:

$$a^{11} = \sigma_k/2, \tag{25}$$

where σ_k is the k-dimensional volume of the unit sphere. From (23) and (25) we have

$$\int_{M^k} P^1 = \frac{\sigma_k}{2} X_1(M^k). \tag{26}$$

Proof. By Theorem 5, relations (2) and (3) yield

$$\int_W \mathfrak{K}^1 = a^{11} I(Z_1, W). \tag{27}$$

To calculate the coefficient a^{11}, it is sufficient to take a definite cycle W and carry out the operations in (27) over this cycle.

Let us choose an oriented $(k+1)$-dimensional linear subspace R^{k+1} in the space R^{k+l} and denote by $H(k, 1)$ the oriented manifold (see Section 2 in [2]) consisting of all k-dimensional oriented linear subspaces of the space R^{k+1}. This manifold $H(k, 1)$ is homeomorphic to the k-dimensional sphere and we take it as W: $W = H(k, 1)$.

Let S^k denote the set of all unit vectors $\eta \in R^{k+1}$, which is a k-dimensional sphere of radius 1. We define a homeomorphic mapping ϕ of the manifold $H(k, 1)$ onto the sphere S^k as follows: let $R^k \in H(k, 1)$, and define $\phi(R^k) = \eta \in R^{k+1}$ to be a unit vector orthogonal to R^k and satisfying the condition that if the vectors $e_1, \ldots, e_k$ form a basis in R^k giving a positive orientation to R^k, then the vectors $e_1, \ldots, e_k, \eta$ form a basis in R^{k+1} which defines the positive orientation of R^{k+1}.

Let R_0^k be a fixed element in $H(\mathrm{k}, 1)$ with an orthonormal basis $e_1, \ldots, e_k$ which gives a positive orientation to R_0^k and let $f_1 = \phi(R_0^k)$.

Each vector $\eta \in S^k$ can be expressed as

$$\eta = \eta^0 f_1 + \eta^1 e_1 + \ldots + \eta^k e_k, \tag{28}$$

where $\eta^0, \eta^1, \ldots, \eta^k$ are real numbers the sum of whose squares is equal to one. A neighbourhood U of the element R_0^k in the manifold $H(k, 1)$ (see (A) Section 2) corresponds to the basis $e_1, \ldots, e_k, f_1$. Each element $R^k \in U$ admits a basis $e_1', \ldots, e_k'$ given by the relations

$$e_i' = e_i + x^i f_1 \quad (i = 1, \ldots, k). \tag{29}$$

Here $x^1, \ldots, x^k$ are coordinates of the element R^k in U coinciding with the coordinates introduced in (A) Section 2. It is easily verified that to the element $R^k \in U$ with the coordinates $x^1, \ldots, x^k$, corresponds the element $\eta = \phi(R^k)$ with the coordinates:

$$\eta^0 = \frac{1}{\sqrt{1 + (x^1)^2 + \ldots + (x^k)^2}},$$

$$\eta^i = \frac{-x^i}{\sqrt{1 + (x^1)^2 + \ldots + (x^k)^2}} \quad (i > 0). \tag{30}$$

Thus, the relations (30) define the mapping ϕ in the neighbourhood U. Equations (30) give a parametric representation of the neighbourhood $\phi(U)$ of the point f_1 in S^k, where $x^1, \ldots, x^k$ have been taken as the parameters. From (30), it follows that the metric tensor $g_{\alpha\beta}$ at the point $f_1 \in S^k$ in these parameters satisfies the condition (see (2) Section 3)

$$g_{\alpha\beta} = \delta_{\alpha\beta}. \tag{31}$$

Let $d_p x = (d_p x^1, \ldots, d_p x^k)$ $(p = 1, \ldots, k)$ be an arbitrary sequence of vectors tangent to $H(k, 1)$ at the point R_0^k. It is easily verified that

$$K^1(d_1 x, \ldots, d_k x) = |d_p x^i|. \tag{32}$$

By virtue of condition (31), we have $h(f_1) = 1$, therefore the integral $\int_{H(k,\,1)} \mathfrak{K}^1$ is converted by ϕ into the integral $\int_{S^k} dV$ (see (C), Section 3), i.e.,

$$\int_{H(k,\,1)} \mathfrak{K}^1 = \int_{S^k} dV = \sigma_k. \tag{33}$$

The intersection index $I(Z_1, H(k, 1))$ on the right-hand side of (27) is calculated in [2] (see (5) Section 3 in [2]), and is equal to two. Hence, from (23), we obtain the relation (25).

This completes the proof of Theorem 6.

5. An integral Expression for the Coefficients $a^{\rho\chi}$

In this section we shall express the coefficients $a^{\rho\chi}$ as integrals of the fields $\mathfrak{K}^\rho$ over definite cycles in $H(k, l)$. Here the integration is not actually carried out so that the numerical values of the coefficients remain uncalculated. In order to express the coefficients $a^{\rho\chi}$ in terms of integrals, it is sufficient to choose a maximal linearly independent system of cycles W of dimension $r = r(\chi) = r(\rho)$ and then apply the relations (2) and (3) given in the previous section. A system of cycles convenient for this purpose is given by Theorem 7.

(A) In what follows we give the pseudo-manifold Z_χ in terms of the function ω instead of the function χ: $Z_\chi = Z(\omega)$. The functions χ and ω are connected by the relation (see (C), Section 3 in [2]):

$$\chi(i) + \omega(i) = l \quad (i = 1, \ldots, k). \tag{1}$$

In this new notation, condition (1) of the previous section is expressed as follows (see (C) Section 5 in [2]):

$$\omega \in \Omega. \tag{2}$$

To each function ω let us associate a function $\bar{\omega}$ defined as follows:

$$\omega(i) + \bar{\omega}(k - i + 1) = l \quad (i = 1, \ldots, k) \tag{3}$$

and consider the pseudo-manifolds $Z(\omega)$ along with the pseudo-manifolds $Z(\bar{\omega})$. If $\omega \in \Omega$ and $\bar{\omega}(1) \neq 0$, it turns out that $\bar{\omega}$ is also contained in Ω and the pseudo-manifold $Z(\bar{\omega})$ belongs to the class already investigated; if, however, $\omega \in \Omega$ and $\bar{\omega}(1) = 0$, then the pseudo-manifold $Z(\bar{\omega})$ is orientable and taking the orientation into account we obtain (see (F) Section 2)

$$\hat{Z}(\bar{\omega}) = Z(\bar{\omega}). \tag{4}$$

Let $\alpha_1, \ldots, \alpha_n, \beta_1, \ldots, \beta_n$ denote the numbers corresponding to the function ω, and $\bar{\alpha}_1, \ldots, \bar{\alpha}_n, \bar{\beta}_1, \ldots, \bar{\beta}_n$ the numbers corresponding to the function $\bar{\omega}$ (see (B) Section 2 in [2]). By virtue of (3), we obtain the following relations for these numbers:

$$\bar{\alpha}_1 = \alpha_n, \quad \bar{\alpha}_2 = \alpha_{n-1}, \ldots, \bar{\alpha}_n = \alpha_1, \tag{5}$$

$$\bar{\beta}_1 = \beta_{n-1}, \bar{\beta}_2 = \beta_{n-2}, \ldots, \bar{\beta}_{n-1} = \beta_1. \tag{6}$$

If $\bar{\omega}(1) > 0$, i.e., $\omega(k) \leqslant l$, then for $\omega \in \Omega$

$$\alpha_1 \equiv \beta_1 \equiv \ldots \equiv \alpha_{n-1} \equiv \beta_{n-1} \equiv \alpha_n \equiv 0 \pmod 2 \tag{7}$$

(see (C) Section 5 in [2]). From (5), (6), and (7), we obtain

$$\bar{\alpha}_1 \equiv \bar{\beta}_1 \equiv \ldots \equiv \bar{\alpha}_{n-1} \equiv \bar{\beta}_{n-1} \equiv \bar{\alpha}_n \equiv 0 \pmod 2. \tag{8}$$

This means that $\bar{\omega} \in \Omega$ because $\bar{\omega}(1) > 0$.

If $\bar{\omega}(1) = 0$, i.e., $\omega(k) = l$, then for $\omega \in \Omega$

$$\alpha_1 \equiv \beta_1 \equiv \ldots \equiv \alpha_{n-1} \equiv \beta_{n-1} \equiv 0 \pmod 2 \tag{9}$$

(see (C) Section 5 in [2]). From (5), (6), and (9), we obtain

$$\bar{\beta}_1 \equiv \bar{\alpha}_2 \equiv \bar{\beta}_2 \equiv \ldots \equiv \bar{\alpha}_{n-1} \equiv \bar{\beta}_{n-1} \equiv \bar{\alpha}_n \equiv 0 \pmod 2. \tag{10}$$

Put

$$Z_1(\bar{\omega}) = U(\bar{\omega}) + \hat{U}(\bar{\omega}). \tag{11}$$

From (10), and from (18), (19) and (20) of Section 4 in [2], it follows that

$$\Delta Z_1(\bar{\omega}) = 0. \tag{12}$$

Thus, $Z_1(\bar{\omega})$ is orientable, and along with it, the pseudo-manifold $Z(\omega)$ obtained from $Z_1(\bar{\omega})$ by a rotation of R^{k+l}. The relation (4) now follows from (12).

Theorem 7. Let $\omega \in \Omega$; then the two pseudo-manifolds $Z(\omega)$ and $Z(\bar{\omega})$ are orientable (see (A)). For $\omega \in \Omega$, assign to the pseudo-manifolds $Z(\omega)$ and $Z(\bar{\omega})$ orientations according to the rules stated in [2] (see (B) Section 2 in [2]). As has been demonstrated in [2], the orientation of the pseudo-manifold $Z(\omega)$ only depends on the orientation of the space R^{k+l} (see (C) Section 2 in [2]). It turns out that the orientation of the pseudo-manifold $Z(\bar{\omega})$ does not depend on the arbitrariness in the constructions. Furthermore,

$$I(Z(\omega), Z(\bar{\omega}')) = 2\delta_{\omega\omega'} \text{ if } \omega, \omega' \in \Omega \text{ and } r(\omega) = r(\omega') \tag{13}$$

where

$$\delta_{\omega\omega'} = 0 \text{ for } \omega \not\equiv \omega' \text{ and } \delta_{\omega\omega'} = 1 \text{ for } \omega \equiv \omega'.$$

Proof. We divide the proof into four steps.

(a) Let ω and ω' be two functions satisfying the condition

$$r(\omega) = r(\omega'), \text{i.e. } r(\omega) + r(\bar{\omega}') = kl. \tag{14}$$

Then the pseudo-manifolds $Z(\omega)$ and $Z(\bar{\omega}')$ are of complementary dimensions in $H(k, l)$, and it turns out that if $\omega \not\equiv \omega'$, the pseudo-manifolds $Z(\omega)$ and $Z(\omega')$ can be so chosen that they do not intersect.

Let

$$R_1 \subset \ldots \subset R_k \tag{15}$$

and

$$\bar{R}'_1 \subset \ldots \subset \bar{R}'_k \tag{16}$$

be sequences defining the cycles $Z(\omega)$ and $Z(\bar{\omega}')$, respectively (see, Definition 2 in [2]). Furthermore, let

$$R^k \in Z(\omega) \cap Z(\bar{\omega}'). \tag{17}$$

We have

$$R^k \cap R_i \cap \bar{R}'_{k-i+1} = (R^k \cap R_i) \cap (R^k \cap \bar{R}'_{k-i+1}). \tag{18}$$

Since the spaces $R^k \cap R_i$ and $R^k \cap \bar{R}'_{k-i+1}$ both lie in R^k, and their dimensions are not less than i and $k - i + 1$, respectively (see (17)), we find that the dimension of their intersection is estimated by the inequality:

$$D(R^k \cap R_i \cap \bar{R}'_{k-i+1}) \geqslant 1. \tag{19}$$

Assume that the functions ω and $\bar{\omega}'$, satisfying the condition (14), do not satisfy the condition $\omega \equiv \omega'$. Then there exists an $i = t$ for which

$$\omega(t) + \bar{\omega}'(k-t+1) < 1. \tag{20}$$

The sequences (15) and (16) can be so chosen that R_t and $\bar{R}'_{k-t+1}$ are in general position. Then the dimension of their intersection, being positive (see (19)), is calculated by means of the formula (see (20)):

$$\begin{aligned} D(R_t \cap \bar{R}'_{k-t+1}) &= [t + \omega(t)] + [k - t + 1 + \bar{\omega}'(k-t+1)] - (k+l) \\ &\leqslant 0, \end{aligned} \tag{21}$$

and thus we come to a contradiction. Consequently, there is no element R^k common to $Z(\omega)$ and $Z(\bar{\omega}')$. This completes the proof of assertion (a).

(b) For $\omega \in \Omega$

$$I(Z(\omega), Z(\bar{\omega}')) = \sigma.2, \quad \sigma = \pm 1; \tag{22}$$

thus the cycle $Z(\bar{\omega})$ is not even weakly homologous to zero in $H(k, l)$.

In the space R^{k+l} choose the basis

$$e_1, \ldots, e_k, \quad f_1, \ldots, f_l. \tag{23}$$

Further, form a second basis

$$\bar{e}_1, \ldots, \bar{e}_k, \quad \bar{f}_1, \ldots, \bar{f}_1, \tag{24}$$

by putting

$$\bar{e}_i = e_{k-i+1}, \quad \bar{f}_j = f_{l-j+1} \quad (i = 1, \ldots, k; j = 1, \ldots, l). \tag{25}$$

Using the basis (23), form the pseudo-manifold $Z(\omega)$, and using the basis (24), form the pseudo-manifold $Z(\bar{\omega})$. Accordingly, the space R_i should be determined as the linear span of the vectors

$$e_1, \ldots, e_i, \quad f_1, \ldots, f_{\omega(i)}, \tag{26}$$

and the space $\bar{R}_i$ as the linear span of the vectors

$$\bar{e}_1, \ldots, \bar{e}_i, \quad \bar{f}_1, \ldots, \bar{f}_{\bar{\omega}(i)} \tag{27}$$

(see Definition 2 in [2]). It is readily seen that the intersection $R_i \cap \bar{R}_{k-i+1}$ has a dimension equal to one and contains the vector e_i. Thus, by virtue of (19), the common element R_0^k of the pseudo-manifolds $Z(\omega)$ and $Z(\bar{\omega})$ has the vectors $e_1, \ldots, e_k$ as a basis. Consequently, this element is defined uniquely up to orientation. This means that the intersection $Z(\omega) \cap Z(\bar{\omega})$ consists of two elements: $\hat{R}_0^k$ and R_0^k, where R_0^k is the oriented linear span of the vectors $e_1, \ldots, e_k$.

The basis (23) determines a neighbourhood U of the element R_0^k, where the coordinates in U are ξ_{ij} (see (A) Section 2). In these coordinates the intersection $U \cap Z(\omega) = U(\omega)$ is given by the equations (see (A) and (C) of Section 1 in [2]):

$$\xi_{ij} = 0 \text{ if } j > \omega(i) \quad (i = 1, \ldots, k), \tag{28}$$

and the intersection of $U \cap Z(\bar{\omega}) = U(\bar{\omega})$ by the equations:

$$\xi_{ij} = 0 \text{ for } j \leqslant \omega(i) \quad (i = 1, \ldots, k). \tag{29}$$

Thus, the index σ of intersection of $Z(\bar{\omega})$ and $Z(\omega)$ at the point R_0^k is equal to ± 1.

To find the intersection index of $Z(\omega)$ and $Z(\bar{\omega})$ at R_0^k, let us transform the whole manifold $H(k, l)$, using the transformation $R^k \to \hat{R}^k$. We obtain

$$\hat{Z}(\omega) = (-1)^{\omega(1)+k} Z(\omega), \quad \hat{H}(k, l) = (-1)^{k+l} H(k, l) \tag{30}$$

(see (15) Section 6 in [2]). If $\bar{\omega}(1) > 0$, the pseudo-manifold $Z(\bar{\omega})$ obeys the same rule, namely,

$$\hat{Z}(\bar{\omega}) = (-1)^{\bar{\omega}(1)+k} Z(\bar{\omega}). \tag{31}$$

From (30) and (31), we find that the intersection index of $Z(\omega)$ and $Z(\bar{\omega})$ at the point R_0^k is (see (3) and (7)).

$$\sigma(-1)^{k+l+\omega(1)+k+\bar{\omega}(1)+k} = \sigma(-1)^{\beta_1+\ldots+\beta_{n-1}+\alpha_1+\ldots+\alpha_n} = \sigma.$$

If, however, $\bar{\omega}(1) = 0$, then from (4) and (3) we find that the intersection index of $Z(\omega)$ and $Z(\bar{\omega})$ at R_0^k is (see (9)):

$$\sigma \cdot (-1)^{\omega(1)+k+k+l} = \sigma(-1)^{\beta_1+\ldots+\beta_{n-1}} = \sigma.$$

Thus, the relation (22) is true in both cases. This completes the proof of assertion (b).

(c) From the sequence

$$\bar{R}_1, \ldots, \bar{R}_k \tag{32}$$

defining the pseudo-manifold $Z(\bar{\omega})$, let us choose the subsequence

$$\bar{S}_1, \ldots, \bar{S}_n \tag{33}$$

(see (B) Section 1 in [2]) of spaces on which $Z(\bar{\omega})$ depends. According to [2] (see (B) Section 2 in [2]) the orientation of $Z(\bar{\omega})$ is determined by the orientations of the spaces (33) and of the space R^{k+l}. For $\omega \in \Omega$, the orientation of $Z(\bar{\omega})$ does not depend on the orientations of the spaces (33) and R^{k+l}. Moreover, if in place of (33) we take some other spaces

$$\tilde{S}_1, \ldots, \tilde{S}_n \tag{34}$$

which correspond to the same function $\bar{\omega}$, then the new oriented manifold $\tilde{Z}(\bar{\omega})$ will be homologous to the former manifold.

We first show that the orientation of $Z(\bar{\omega})$ does not depend on the randomly chosen orientation of the space R^{k+l}. Since $\bar{\omega}(k) < l$, on substituting $-\bar{f}_l$ for $\bar{f}_l$, the orientation of R^{k+l} changes but the orientation of $Z(\bar{\omega})$ does not change. Let the spaces of the sequences (33) and (34) be oriented and define the orientations of the pseudo-manifolds $Z(\bar{\omega})$ and $\tilde{Z}(\bar{\omega})$, respectively. Then there exists an orthogonal mapping a with positive determinant of the space R^{k+l} into itself, under which

$$a(\bar{S}_p) = \tilde{S}_p \quad (p = 1, \ldots, n), \tag{35}$$

orientations being taken into account. Since the group of orthogonal transformations with positive determinants is connected, there exists a one-parameter family a_t $(0 \leqslant t \leqslant 1)$ of elements, in which $a_1 = a$, and a_0 is the identity. The spaces

$$a_t(\bar{S}_1), \ldots, a_t(\bar{S}_n) \tag{36}$$

give an oriented pseudo-manifold $Z_t(\bar{\omega})$. Thus, the family $Z_t(\bar{\omega})$, $0 \leqslant t \leqslant 1$, realizes the homology between $Z(\bar{\omega})$ and $\tilde{Z}(\bar{\omega})$.

We now assume that the spaces (34) differ from the spaces (33) at most in orientation. If, in this case, $\tilde{Z}(\bar{\omega}) = -Z(\bar{\omega})$, then by what has been proved, we would have $2Z(\bar{\omega}) \sim 0$, which contradicts (b). This completes the proof of assertion (b).

(d) In (22), $\sigma = +1$.

By numbering the coordinates ξ_{ij} in the coordinate space U according to the columns of the matrix $\|\xi_{ij}\|$, definite orientations arise for the space U itself and of its coordinate spaces $U(\omega)$ and $U(\bar{\omega})$ given by the systems of equations (28) and (29), respectively. The spaces U and $U(\omega)$ thus oriented induce the orientation of the manifold $H(k, l)$ and of the pseudo-manifold $Z(\omega)$. We now prove that the orientation of the space $U(\bar{\omega})$ is also consistent with the orientation of the pseudo-manifold $Z(\bar{\omega})$.

Let $\bar{R}_0^k$ denote an oriented linear span of the vectors $e_1, \ldots, e_k$. Geometrically, R_0^k and $\bar{R}_0^k$ coincide, and

$$\bar{R}_0^k = \epsilon R_0^k, \quad \epsilon = (-1)^{k(k-1)/2}. \tag{37}$$

with regard for orientation. Let $\bar{U}$ denote the coordinate neighbourhood of the point $\bar{R}_0^k$ in $H(k, l)$ given by the basis (24). Its coordinate subspace $\bar{U}(\bar{\omega}) = \bar{U} \cap Z(\bar{\omega})$ is then given by the equations

$$\bar{\xi}_{ij} = 0 \text{ for } j > \bar{\omega}(i) \quad (i = 1, \ldots, k). \tag{38}$$

The orientation generated in $\bar{U}(\bar{\omega})$ by numbering the elements of the matrix $\|\bar{\xi}_{ij}\|$ according to columns induces the orientation of the pseudo-manifold

$Z(\bar{\omega})$ under consideration. Let ϕ denote a mapping of $H(k, l)$ onto itself which is the identity if $\epsilon = +1$ and the mapping $R^k \to \hat{R}^k$ if $\epsilon = -1$. Then

$$\phi(\bar{R}_0^k) = R_0^k,$$

and disregarding orientation, we obtain

$$\phi(\bar{U}) = U, \quad \phi(\bar{U}(\bar{\omega})) = U(\bar{\omega}).$$

If ξ_{ij} and $\bar{\xi}_{ij}$ denote the coordinates of corresponding elements in U and $\bar{U}$ under ϕ, then

$$\bar{\xi}_{ij} = \xi_{k-i+1,\, l-j+1}. \tag{39}$$

This relation can be used to find the sign of the mapping of the oriented space $\bar{U}(\bar{\omega})$ onto the oriented space $U(\bar{\omega})$; namely, we have

$$\phi(\bar{U}(\bar{\omega}) = \epsilon^{\bar{\omega}(1)} U(\bar{\omega}). \tag{40}$$

This relation follows from the fact that in each column of the matrix $\|\xi_{ij}\|$ the number of elements not connected by the relations (38) is even, and in each row the number of such elements is congruent to $\bar{\omega}(1)$ modulo 2 (see (8) and (10)). Furthermore, from (4) for $\bar{\omega}(1) = 0$ and from (31) and (8) for $\bar{\omega}(1) > 0$ it follows that

$$\phi(Z(\bar{\omega})) = \epsilon^{\bar{\omega}(1)} Z(\bar{\omega}). \tag{41}$$

Comparing (40) and (41), we find that the orientation of the space $U(\bar{\omega})$ is consistent with the orientation of the pseudo-manifold $Z(\bar{\omega})$. It only remains to show that

$$I(U(\omega), U(\bar{\omega})) = +1. \tag{42}$$

But this follows from the fact that each column in the matrix $\|\xi_{ij}\|$ contains an even number of elements not connected by the relation (29).

This completes the proof of assertion (d).

From the four assertions (a)–(d) it immediately follows that Theorem 7 is true. For ease in formulating Theorem 8 we introduce the following notation:

$$Z(\bar{\omega}) = \bar{Z}_\chi, \tag{43}$$

where the functions $\bar{\omega}$ and χ are connected by the relation:

$$\bar{\omega}(i) = \chi(k - i + 1) \quad (i = 1, \ldots, k). \tag{44}$$

Theorem 8. The coefficients $a^{\rho\chi}$ are given by the relations:

$$a^{\rho\chi} = \tfrac{1}{2} \int_{\bar{Z}_\chi} \mathfrak{R}^\rho \tag{45}$$

Proof. Putting $W = \bar{Z}_\chi$ in (3), and taking (2) into account, we obtain (see Theorem 7):

$$\int_{\bar{Z}_\chi} \mathfrak{R}^\rho = a^{\rho\chi}\cdot 2,$$

from which (45) immediately follows.

References

1. L. S. Pontryagin, Certain topological invariants of Riemannian manifolds, *Doklady AN SSSR* **XLIII**, N 3, 95–98 (1944).
2. L. S. Pontryagin, Characteristic cycles of differentiable manifolds, *Matem. sbornik,* v. 21(63): 2, 232–284 (1947).
3. H. Hopf, Die Curvatura integra Clifford–Kleinscher Raumformen, *Nachr. Gesellder Wiss. Göttingen*, 131–141 (1925).
4. G. De Rham, Sur l'Anaysis situs des variètés à n dimensions, *J. Math. pures et appl.* **10**, 115–200 (1931).
5. E. Cartan, Sur les invariants intégraux des certains espaces homogènes clos et les propriétés topologiques de ces espaces, *Ann. Soc. Pol. Math.* 8, 181–225 (1929).
6. Chern Shung-shen, A simple proof of the Gauss–Bonnet formula for closed Riemannian manifolds, *Ann. of Math.* **45**, 747–752 (1944).
7. H. Weyl, *Classical Groups, Their Invariants and Representations,* Moscow–Leningrad, 1947 (in Russian).
8. P. K. Rashevskii, *An Introduction to Riemannian Geometry and Tensor Analysis,* Moscow–Leningrad, 1936 (in Russian).
9. L. S. Pontryagin, Vector fields on manifolds, *Matem. sbornik*, vol. 24 (66):2 (1949).

Homotopic Classification of the Mappings of an $(n+2)$-Dimensional Sphere into an n-Dimensional Sphere

Doklady Akademii Nauk SSSR (1950) 70(6): 957–59

I have already published a note [1] on this subject. However a mistake was made, which led to an erroneous result. Using the same method as in [1], but correcting the mistake, I arrive at the following result.

Theorem. For $n \geqslant 2$, the number of classes of mappings of an $(n+2)$-dimensional sphere S^{n+2} into an n-dimensional sphere S^n is two.

Proof. Since there are no more than two classes of mappings of S^{n+2} into S^n, I only prove that the number of these classes is not less than two.

(a) We assume that the spheres S^n and S^{n+k} (starting from c) we take $k = 2$), are ordinary metric spheres in a Euclidean space. Continuous mappings and deformations can be approximated by analytic mappings; in the sequel we shall only deal with such mappings. Under an analytic mapping of some manifold into S^n, a point of the manifold is called *proper*, if the rank of the functional matrix of this mapping at this point is equal to n. We interpret the deformation f_t, $0 \leqslant t \leqslant 1$ as a mapping F of the topological product $T = S^{n+2} \times I$ into S^n, where I is the segment $0 \leqslant t \leqslant 1$. Let $p \in S^n$, and take $f_t^{-1}(p) = P_t$ and $F^{-1}(p) = Q$. The analyticity of the mappings f_0, f_1, and F imply the existence of a point p for which the sets P_0, P_1, and Q consist of only proper points, and are, in particular, analytic manifolds (Q has the boundary $P_0 \times O \cup P_1 \times I$). Let $y = (x, t) \in Q$ ($x \in S^{n+2}$, $t \in I$) and $\theta(y) = t$, where $\theta(y)$ is an analytic function on Q. We can modify the deformation f_t, without changing this deformation at $t = 0$ and 1, so that all the critical points of the function $\theta(y)$ are non-degenerate.

(b) Let R^n denote the tangent to S^n at the point p, and $d_1, \ldots, d_n$ some linearly independent sequence of vectors in R^n. Let t be a fixed *noncritical* value of the function $\theta(y)$, R^{n+k} the tangent to the sphere S^{n+k} at the point $x \in P_t$, and N_x^n the normal in R^{n+k} to P_t at the point x. Since (x, t) is a proper point of the mapping F, and t is a non-critical value of the function $\theta(y)$, then x is a proper point of the mapping f_t and the corresponding linear mapping g of the space N_x^n into R^n is non-degenerate. Put $g^{-1}(d_i) = e_i(x)$ $(i = 1, \ldots, n)$.

(c) Let M denote the manifold of all sequences $x;\, e_0, e_1, \ldots, e_n$, where $x \in S^{n+2}$ and $e_0, e_1, \ldots, e_n$ are linearly independent vectors tangent to S^{n+2} at the point x. The one-dimensional Betti group of the manifold M contains only two elements: 0 and 1.

(d) Let $k = 2$ and t be a fixed non-critical value of the function $\theta(y)$ (see (a)); then P_t is a closed orientable (possibly, disconnected) surface. A finite set of pairwise disjoint simple closed oriented smooth curves is called a *curve on* P_t. The number of components of a curve C is denoted by $\nu(C)$. Let $x \in C$ and $e_0(x)$ be the unit vector tangent to the curve C at the point x. Put $\omega(x) = \{x; e_0(x), e_1(x), \ldots, e_n(x)\}$ (see (b)). This defines a mapping ω of the curve C into the manifold M (see (c)) and with it the homology class $\varphi(C)$ of the curve $\omega(C)$ in M is defined, i.e., 0 or 1 (see (c)). Put

$$\gamma(C) = \nu(C) + \varphi(C) \pmod 2. \tag{1}$$

Clearly, when the curve C is deformed, the functions $\nu(C)$ and $\varphi(C)$, consequently $\gamma(C)$, do not change. It is even true that $\gamma(C)$ is determined by the homology class c modulo 2 of the curve C on P_t: $\gamma(C) = \gamma(c)$. Furthermore, if c_1 and c_2 are two homology classes, then

$$\gamma(c_1 + c_2) = \gamma(c_1) + \gamma(c_2) + I(c_1, c_2), \tag{2}$$

where $I(c_1, c_2)$ is the intersection index (mod 2) of the classes c_1 and c_2.*

To prove this, let us define the operation of addition of curves, the result of which is again a curve. Let C_1 and C_2 be two curves on P_t in general position, i.e., they only have a finite number of intersection points and are not tangent at any of these points. The sum of the curves $C_3 = C_1 + C_2$ differs from the set-theoretic sum $C_1 \cup C_2$ of two oriented curves C_1 and C_2 only in the neighbourhood of their intersection. Figure 1 shows how we can pass from the union $C_1 \cup C_2$ to the sum $C_3 = C_1 + C_2$ in the neighbourhood of an intersection point.

* The mistake committed in [1] is that the term $I(c_1, c_2)$ was omitted in (2).

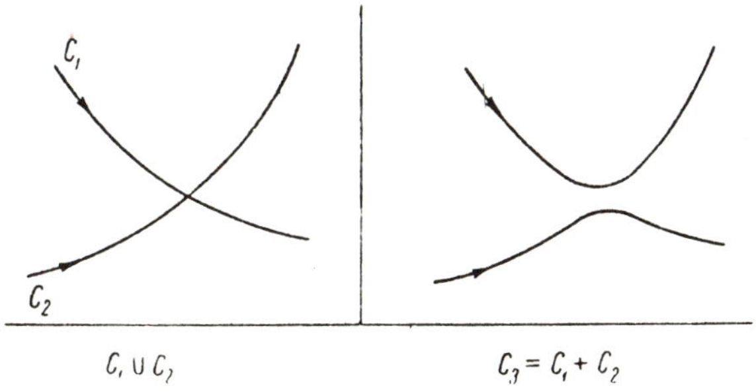

Fig. 1.

Evidently, the homology classes (mod 2) c_1, c_2, c_3 of the curves C_1, C_2, C_3 are connected by the relation

$$c_3 = c_1 + c_2. \tag{3}$$

Furthermore, it is readily seen that

$$\varphi(C_1) + \varphi(C_2) = \varphi(C_1 + C_2),$$
$$\nu(C_1) + \nu(C_2) = \nu(C_1 + C_2) + I(C_1 + C_2) \quad (\text{mod } 2).$$

Hence, from (1), we get

$$\gamma(C_1 + C_2) = \gamma(C_1) + \gamma(C_2) + I(C_1, C_2). \tag{4}$$

It now remains to demonstrate the homological invariance of the function $\gamma(C)$. Suppose that $C_1 \sim C_2$ (mod 2), then the last term in (4) vanishes and so it suffices to prove that $\gamma(C_3) = 0$ if $C_3 \sim 0$ (mod 2) (see (3)). From $C_3 \sim 0$ (mod 2) it readily follows that C_3 is a level curve of some smooth real-valued function $\zeta(x)$ defined on P_t, all of whose critical points are non-degenerate and whose values at different critical points are distinct. Let us denote by $C(s)$ the level curve of this function that corresponds to a non-critical value s. It is a simple matter to prove that $\gamma(C(s))$ does not depend on s. For this, it suffices to examine the passage of the parameter s through a critical value. When the parameter s passes through a critical value we find that the functions $\nu(C(s))$ and $\varphi(C(s))$ both change by one, so that $\gamma(C(s))$ does not at all change. Since, for sufficiently large s, the set $C(s)$ is empty, we have $\gamma(C_3) = \gamma(C(s)) = 0$.

(e) Let $2q$ denote the rank of the one-dimensional Betti group B (mod 2) of the surface P_t. Let us now examine the question: Does there exist in B a linearly independent system of elements $c_1, \ldots, c_q$ satisfying the conditions:

$$I(c_i, c_j) = 0, \quad \gamma(c_i) = 0 \quad (i,j = 1, \ldots, q).$$

Let us define a function $\Gamma(f_t)$ as follows: $\Gamma(f_t) = 0$ if such a system exists and $\Gamma(f_t) = 1$ if such a system does not exist. This function is defined only for non-critical values of t; it is readily verified that this function does not change when passing through critical values. And it is obvious that it remains constant on other intervals of variation of t. Further, $\Gamma(f_t) = 0$, if the surface P_t is homeomorphic to a sphere; therefore for the mapping f_0, which is homotopic to zero, we have $\Gamma(f_0) = 0$. Thus, Γ is an invariant of the homotopy class and it is zero for the zero class.

(f) We now construct a mapping $f = f_0$ of the sphere S^{n+2} into S^n for which $\Gamma(f) = 1$. For this purpose, it is sufficient to give a surface $P = P_0$ in S^{n+2} and to define at each point $x \in P$ a sequence of linearly independent vectors $e_1(x), \ldots, e_n(x)$ orthogonal to P such that the above construction gives $\Gamma = 1$. We shall take the sphere S^{n+2} to be a Euclidean space R^{n+2} with a point at infinity. In R^{n+2} choose a subspace R^3 and assume that P is an ordinary metric torus in R^3 with the usual cyclic coordinates α and β. Let $u_1(x)$ be the unit normal to P at $x = (\alpha, \beta) \in P$ in R^3 and $u_2, \ldots, u_n$ an orthonormal system of vectors in R^{n+2} orthogonal to R^3. Put

$$e_1(x) = u_1(x) \cos(\alpha - \beta) - u_2 \sin(\alpha - \beta),$$

$$e_2(x) = u_1(x) \sin(\alpha - \beta) + u_2 \cos(\alpha - \beta),$$

$$e_i(x) = u_i \quad (i = 3, \ldots, 4).$$

It is readily verified that $\Gamma = 1$.

References

1. L. S. Pontryagin, *Doklady AN USSR* **19**, N 5, 957 (1938).
2. H. Freudenthal, *Comp. Math.* **5**, 299 (1937).

Received 4 January 1950

On the Zeros of Certain Elementary Transcendental Functions (Supplement)

Doklady Akademii Nauk SSSR (1953) 91(6): 1279–80

The results obtained in my paper [1] are fully applicable to functions of a slightly more general type than quasi-polynomials. The proofs in this case do not change much and here I shall confine myself to a detailed formulation of the main result. The results of the present paper and those of [1] are closely connected to control theory.

Let $h(z,t) = \Sigma_{m,n} a_{mn} z^m t^n$ be a polynomial in z and t. The function $H(z) = h(z, e^z)$ is called *quasi-polynomial.* Furthermore, let $l(z) = l_0 z^p + l_1 z^{p-1} + \ldots + l_p$ be a polynomial of degree $p (l_0 \neq 0)$. The function $\hat{H}(z) = H(z)/l(z)$, provided it has no poles, is called *generalized quasi-polynomial.* In this paper we shall formulate necessary and sufficient conditions for all the roots of $H(z)$ to lie to the left of the imaginary axis, i.e., for all the roots to have negative real parts. A function $\hat{H}(z)$ exhibiting this property is called *stable.*

A polynomial $h(z,t)$ is said to have a principal term $a_{rs} z^r t^s$, if $a_{rs} \neq 0$, and for any other non-vanishing term $a_{mn} z^m t^n$, $r \geqslant m$, $s \geqslant n$ where the case of simultaneous equalities $r = m$ and $s = n$ is excluded.

Necessary condition. For the function $\hat{H}(z)$ to be stable, it is necessary that the polynomial $h(z,t)$ have a principal term.

In view of the condition formulated above, we shall assume in the sequel that the polynomial $h(z,t)$ always has a principal term $a_{rs} z^r t^s$. In the polynomial $h(z,t)$ collect all the terms containing z^r, in other words, express the polynomial as

$$h(z,t) = z^r\chi(t) + \sum a_{mn}z^m t^n,$$

where $\chi(t)$ is a polynomial of degree s, and the summation is taken over all $m<r$ and all $n\leqslant s$. Consider the function $\chi(e^{iy})$, where y is a real variable. Separate it into real and imaginary parts

$$\chi(e^{iy}) = \varphi(y) + i\psi(y),$$

where φ and ψ are real-valued functions of y, which are, as can easily be seen, trigonometric polynomials. Consequently, the real zeros of $\varphi(y)$ and $\psi(y)$ are repeated with a period of 2π. So we can choose a real number ϵ (which we use later) such that $2k\pi+\epsilon$ is not a root of either $\varphi(y)$ or of $\psi(y)$ for any k.

Now consider the function $\hat{H}(iy)$, where y is a real variable. Separating it into its real and imaginary parts, we get

$$\hat{H}(iy) = \hat{F}(y) + i\hat{G}(y),$$

where $\hat{F}(y)$ and $\hat{G}(y)$ are real-valued functions of y.

Necessary and sufficient conditions. For the function $\hat{H}(z)$ to be stable, it is necessary and sufficient that (recall that $h(z,t)$ is supposed to have a principal term):

(1) each of the functions $\hat{F}(y)$ and $\hat{G}(y)$ should have, for all sufficiently large integral k, not less than $4ks+r-p$ distinct real zeros in the interval $-2k\pi+\epsilon\leqslant y\leqslant 2k\pi+\epsilon$. (For sufficiently large k, these functions cannot have a greater number of zeros, so that the functions $F(y)$ and $G(y)$ will have, under the condition formulated above, exactly $4ks+r-p$ zeros on the interval $-2k\pi+\epsilon\leqslant y\leqslant 2k\pi+\epsilon$).

(2) The zeros of the functions $\hat{F}(y)$ and $\hat{G}(y)$ should alternate, namely, between any two consecutive zeros of one function there should occur exactly one zero of the other.

(3) It can be proved that under the conditions (1) and (2), the expression

$$\hat{G}'(y)\cdot\hat{F}(y) - \hat{F}'(y)\cdot\hat{G}(y)$$

does not vanish for any real y and therefore has a constant sign. It is necessary that this expression be positive. To verify this, it is sufficient to test its positivity for one value of y.

Note 1. If we know that $\hat{F}(y)$ has the number of zeros stated in Condition (1) and that at least one zero of $\hat{G}(y)$ lies between two zeros of $\hat{F}(y)$, then we can easily show that condition 1 for $\hat{G}(y)$ and condition 2 are satisfied.

Therefore there is no need to verify them. Here the functions $\hat{F}(y)$ and $\hat{G}(y)$ can be interchanged.

Note 2. If we already know that $\hat{F}(y)$ (or $\hat{G}(y)$) has the number of zeros stated in Condition (1), and if we have proved that $\hat{G}'(y)\cdot\hat{F}(y)-\hat{F}'(y)\cdot\hat{G}(y)$ is positive for any y, then it immediately follows that all other conditions are also satisfied, consequently the function $\hat{H}(z)$ is stable.

References

1. L. S. Pontryagin, *Izv. AN SSSR, ser. matem.* N 6, 115 (1942).

Received 10 June 1953

The Asymptotic Behaviour of the Solutions of Systems of Differential Equations With a Small Parameter Attached to Higher Derivatives

Izvestiya Akademii Nauk SSSR, Seriya Matematika (1957)
21(5): 605–26

Asymptotic formulas are derived for the solutions of systems of differential equations with a small parameter attached to higher derivatives in the neighbourhood of a junction point.

Various problems in oscillation theory lead to a study of systems of differential equations containing a small parameter attached to higher derivatives.

Let

$$x = (x^1, \ldots, x^k), \quad y = (y^1, \ldots, y^l)$$

be certain unknown functions of time t, and let

$$\begin{aligned} \epsilon \dot{x}^i &= f^i(x^1, \ldots, x^k, y^1, \ldots, y^l), \quad i = 1, \ldots, k, \\ \dot{y}^j &= g^j(x^1, \ldots, x^k, y^1, \ldots, y^l), \quad j = 1, \ldots, l_j, \end{aligned} \tag{1}$$

be a system of equations which govern their variations, where ϵ is a small positive parameter, and the functions f^i and g^j are differentiable several times with respect to the arguments. The system (1) can be expressed in vector form as follows:

$$\epsilon \dot{x} = f(x, y), \qquad \dot{y} = g(x, y). \tag{1'}$$

When we say that the parameter ϵ is small we have in mind an approximate study of the solutions of system (1), obtained by discarding the quantities of various orders of magnitude of ϵ. Let

$$x = \varphi(t, \epsilon), \quad y = \psi(t, \epsilon) \tag{2}$$

be a solution of the system (1). The question now is, does the solution (2) tend to some limit as $\epsilon \to 0$; in other words, can it be expressed as

$$x = \varphi_1(t) + \Delta_1\varphi(t, \epsilon), \quad y = \psi_1(t) + \Delta_1\psi(t, \epsilon), \tag{3}$$

where the functions $\Delta_1\varphi(t, \epsilon)$ and $\Delta_1\psi(t, \epsilon)$ tend to zero as $\epsilon \to 0$? It might happen that this is possible only on a certain interval of time t, or only for one of the functions: $\varphi(t, \epsilon)$ or $\psi(t, \epsilon)$. If, even one of the functions $\Delta_t\varphi(t, \epsilon)$ or $\Delta_t\psi(t, \epsilon)$, tends to zero as $\epsilon \to 0$, then it would be possible to determine the order of magnitude of this function with respect to ϵ. For instance, it may happen that

$$\Delta_1\varphi(t, \epsilon) = \epsilon^{2/3}\varphi_2(t) + \Delta_2\varphi(t, \epsilon),$$

where $\Delta_2\varphi(t, \epsilon)$ tends to zero more rapidly than $\epsilon^{2/3}$, say, as $\epsilon \ln \epsilon$; then the function $\varphi(t, \epsilon)$ can be written as

$$\varphi(t, \epsilon) = \varphi_1(t) + \epsilon^{2/3}\varphi_2(t) + \epsilon \ln \epsilon \varphi_3(t) + \Delta_3\varphi(t, \epsilon), \tag{4}$$

where $\Delta_3\varphi(t, \epsilon)$ tends to zero faster than $\epsilon \ln \epsilon$. Consequently, it makes sense to raise a question about the asymptotic expansion of the solution (2) as a series or the calculation of a finite number of terms of this series. Here it is quite possible that we have to use different expansions in different time ranges. It might also happen that we would not be able to express $\varphi(t, \epsilon)$ in the form (4), where $\varphi_1(t)$ is not dependent on ϵ, but would only be able to express it in the form:

$$\varphi(t, \epsilon) = \varphi_1(t, \epsilon) + \Delta_1\varphi(t, \epsilon), \tag{5}$$

where $\Delta_1\varphi(t, \epsilon)$ tends to zero with ϵ, and the function $\varphi_1(t, \epsilon)$, although depending on ϵ, could be calculated. In this case it is worth studying the rate at which the function $\Delta_1\varphi(t, \epsilon)$ tends to zero, as well as determining its principal part. This, in general outline, is the formulation of our problem.

The variables x and y in the system of equations (1) are not identical in behaviour: the phase velocity vector v in x, y space decomposes into two vectors:

$$v = \left(\frac{1}{\epsilon}f(x,y), g(x,y)\right) = \left(\frac{1}{\epsilon}f(x,y), 0\right) + (0, g(x,y)), \tag{6}$$

where the second component does not depend on ϵ, while the first tends to infinity as $\epsilon \to 0$ if $f(x,y) \neq 0$. Consequently, the variables in (1) may be called *rapidly-changing variables*, and those in (2), *slowly-changing* variables.

The main approach to system (1) lies in first studying the behaviour of the rapidly-changing variables for constant values of the slowly-changing variables. Thus, initially, the system of equations:

$$\epsilon \dot{x} = f(x,y), \tag{7}$$

is studied where y is a constant vector. We may postulate various assumptions concerning the behaviour of the system (7) for constant y. In this paper we deal with a particular case where the system (7) has only equilibrium states as its stationary solutions and each solution tends to a certain stable equilibrium state as $t \to \infty$. (Another important but natural case, where there are also stable limit cycles among the stationary solutions of the system (7), will not be discussed in this paper).

Let

$$x = \varphi(y) \tag{8}$$

be some stable equilibrium state of the system (7). This state depends on the vector parameter y and remains stable in some domain Γ of variation of y. Substituting (8) into the system

$$\dot{y} = g(x,y), \tag{9}$$

we obtain the system

$$\dot{y} = g(\varphi(y),y). \tag{10}$$

for y. This system no longer contains any unknown function of x. Let

$$y = \psi(t) \tag{11}$$

be its solution. Substituting y into the right-hand side of (8), we obtain

$$x = \varphi(\psi(t)). \tag{12}$$

It may be conjectured that (11) and (12) give an approximate solution of the system (1). This was indeed demonstrated by A. N. Tikhonov in [1]. It can also be shown that this approximate solution differs from the exact solution by an order of magnitude ϵ.

Since the equilibrium state (8) is determined from the equation

$$f(x,y) = 0, \tag{13}$$

the approximate solution (11) and (12) of the system (1) may be taken to be the exact solution of the degenerate system:

$$f(x,y) = 0, \quad \dot{y} = g(x,y), \tag{14}$$

derived from (1) for $\epsilon = 0$.

This operation is feasible as long as the equilibrium state (8) remains stable, i.e., as long as all the eigenvalues of the matrix

$$A = \left\| \frac{\partial f^i}{\partial x^\alpha} \right\| \tag{15}$$

have negative real parts at the respective points ("exponential stability"). We now assume that on the interval $-\alpha < t < 0$ the solution (11) exists and the equilibrium state (8) is exponentially stable, and at $t = 0$ the solution (8) loses its exponential stability and the matrix (15) acquires one zero eigenvalue of multiplicity one, while all other eigenvalues retain their negative real parts. Thus on the interval $-\alpha < t < 0$ the solution

$$x = x_0(t), \quad y = y_0(t) \equiv \psi(t) \tag{16}$$

of the degenerate system (14) is defined.

Let us now consider the phase space R^{k+l} of the system (1). It is naturally decomposed into the direct sum of its subspaces X^k and Y^l, so that each point in the space can be expressed as a pair (x, y). Let us denote the space of all pairs (x, y) with a fixed vector y by X_y^k. It is a phase space of the system (7). When t changes on the interval $-\alpha < t < 0$, the equilibrium state (8) also changes, and at $t = 0$ a degenerate equilibrium state (x_1, y_1) appears in the phase space $X^k_{\psi(0)}$. Since all the eigenvalues of the matrix (15), except one, which is zero, have negative real parts at (x_1, y_1), it follows that at $y = y_1$ the system (10) has (under certain additional assumptions of a general nature) only one trajectory, which approaches the equilibrium state x_1 as $t \to -\infty$. This trajectory lies in the subspace $X^k_{\psi(0)}$ of the space R^{k+l}.

Let us consider the solution

$$x = x(t, \epsilon), \quad y = y(t, \epsilon) \tag{17}$$

of the system (1) with initial values at $t = -\alpha$ which differ from the initial values of the solution (16) of the degenerate system (14) by an order of magnitude ϵ. On the interval $-\alpha < t < -p$, where p is a small positive number not dependent on ϵ, the solution (17), as already noted, differs from the solution (16) by an order of magnitude ϵ. Moreover, it can be easily proved that on this interval the solution (17) can be expanded into an asymptotic series in integral powers of ϵ. But, as $t \to 0$, its behaviour becomes more complicated.

In this paper we shall derive the solution (17) in the form of an asymptotic series for values of t, including $t = 0$, up to terms of the orders $\epsilon^{2/3}$ and $\epsilon \ln \epsilon$, neglecting the terms of order ϵ. We shall also calculate with the same degree of accuracy the deviation of the solution (17) from the k-dimensional plane $X^k_{\psi(0)}$ over a certain finite section of the trajectory.

The main results of this paper were briefly reported earlier in [8].

1. Calculation of Solutions Near Intervals of Slow Motion

Let

$$x^i = x_0^i(t), \quad y^j = y_0^j(t) \tag{1.1}$$

be some solution to the system (14), which, in the interval $\bar{t} \leqslant t \leqslant \bar{\bar{t}}$ ($\bar{\bar{t}} = +\infty$ not excepted) lies on the surface F given by equation (13), at a finite distance (not dependent on ϵ) from the $(l-1)$-dimensional surface Φ whose equation is

$$f(x,y) = 0, \quad \det \left\| \frac{\partial f^i}{\partial x^\alpha} \right\| = 0.$$

Furthermore, let

$$x^i = x^i(t,\epsilon), \quad y^j = y^j(t,\epsilon) \tag{1.2}$$

be some solution of the system (1) with the initial values $x^i(\bar{t},\epsilon) = \bar{x}^i$ and $y^j(\bar{t},\epsilon) = \bar{y}^j$ satisfying the conditions:

$$\bar{x}^i - x_0^i(\bar{t}) = O(\epsilon), \quad \bar{y}^j - y_0^j(\bar{t}) = O(\epsilon).$$

We can easily show that on the interval $\bar{t} \leqslant t \leqslant \bar{\bar{t}}$, the solution (1.2) can be expressed as

$$x^i = x_0^i(t) + M_1^i(t,\epsilon), \quad y^j = y_0^j(t) + N_1^j(t,\epsilon), \tag{1.3}$$

where the functions M_1^i and N_1^j tend to zero as $\epsilon \to 0$. Moreover, (as follows from the results derived by A. N. Tikhonov [1] and A. B. Vasil'eva [2]) these functions are of the order ϵ, and may even be represented in the form:

$$M_1^i = \epsilon x_1^i(t) + M_2^i(t,\epsilon), \quad N_1^j = \epsilon y_1^j(t) + N_2^j(t,\epsilon),$$

where M_2^i and N_2^j are of the order ϵ^2. Proceeding in this way, we obtain the asymptotic formula

$$\begin{aligned} x^i &= x_0^i(t) + \epsilon x_1^i(t) + \ldots + \epsilon^k x_k^i(t) + M_{k+1}^i(t,\epsilon), \\ y^j &= y_0^j(t) + \epsilon y_1^j(t) + \ldots + \epsilon^k y_k^j(t) + N_{k+1}^j(t,\epsilon), \end{aligned} \tag{1.4}$$

for the solution (1.2) where M_{k+1}^i and N_{k+1}^j are of the order $O(\epsilon^{k+1})$. To derive the formulas (1.4), it is, of course, essential that f^i and g^j should be suitably smooth functions.

We shall not prove here the validity of the expansions (1.3) and (1.4), but only note that they can be derived by combining the results of A. N. Tikhonov [1] and A. B. Vasil'eva [2]. Incidentally, in our next communication we shall give a simple but direct proof of the validity of (1.4).

Now assume that the solution (1.1) of the degenerate system (14) hits, for some $t = t_1 \geqslant \bar{\bar{t}}$, a point $s(x_1, y_1)$ where $\|\partial f^i/\partial x^\alpha\| = 0$. Simple calculation shows that, for the values of t sufficiently close to t_1, the sums

$$\sum_{r=0}^{k} \epsilon^r x_r^i(t), \quad \sum_{r=0}^{k} \epsilon^r y_r^j(t)$$

no longer represent the solution (1.2) with the accuracy mentioned above, and for $t = t_1$ they are altogether devoid of meaning.

Our task now is to study the behaviour of the solution (1.2) as $t \to t_1$. To this end, it is more convenient to express the system (1) in a neighbourhood of the point $s(x_1, y_1)$ in a special form.

Assume that the point $s(x_1, y_1)$ (hereafter called a *junction point*) is of "general type", namely, (a) it is not an equilibrium of the system (1), and (b) all eigenvalues (except one, which is zero) of the matrix $\|\partial f^i/\partial x^\alpha\|$ at this point have negative real parts. Then, under certain additional non-degeneracy conditions, in the neighbourhood of the point $s(x_1, y_1)$ the system (1) can be expressed as

$$\left.\begin{aligned} \epsilon\dot{\xi}_1 &= (\xi^1)^2 + \eta^1 + b^1_{\beta'}\eta^{\beta'} + c^1_\beta\xi^1\eta^\beta + d^1_1(\xi^1)^3 + e^1_{\alpha'}\xi^1\xi^{\alpha'} + \dots, \\ \epsilon\dot{\xi}^i &= a^i_{\alpha'}\xi^{\alpha'} + b^i_\beta\eta^\beta + c^i_0(\xi^1)^2 + d^i_1(\xi^1)^3 + e^i_{\alpha'}\xi^1\xi^{\alpha'} + c^i_\beta\xi^1\eta^\beta + \dots, \\ \dot{\eta}^1 &= 1 + \alpha^1_1\xi^1 + \dots, \\ \dot{\eta}^j &= \alpha^j_1\xi^1 + \dots \\ &\qquad (i = 2, \dots, k, \quad j = 2, \dots l), \end{aligned}\right\} \tag{1.5}$$

where $\det \|a^{i'}_{\alpha'}\| \neq 0$. (Here only those terms that are needed in the calculations have been written down. Note that in (1.5), as everywhere in the sequel, summation over a primed index is taken starting from two:

$$a^i_{\alpha'}\xi^{\alpha'} = \sum_{\alpha'=2}^{k} a^i_{\alpha'}\xi^{\alpha'}, \quad b^1_{\beta'}\eta^{\beta'} = \sum_{\beta'=2}^{l} b^1_{\beta'}\eta^{\beta'}$$

etc.). In Section 3 we shall give a suitable coordinate system and find the necessary expressions in terms of the coefficients on the right hand sides of (1.5) in an invariant form. We also state there additional conditions for the junction point $s(x_1, y_1)$ to be non-degenerate.

Rewrite the system (1.5) in compact form as follows:

$$\epsilon\dot{\xi}^i = \Phi^i(\xi^1, \dots, \xi^k, \eta^1, \dots, \eta^l), \quad \dot{\eta}^j = \Psi^j(\xi^1, \dots, \xi^k, \eta^1, \dots, \eta^l). \tag{1.6}$$

Let the solution of the degenerate system

$$\Phi^i(\xi^1, \dots, \xi^k, \eta^1, \dots, \eta^l) = 0, \quad \dot{\eta}^j = \Psi^j(\xi^1, \dots, \xi^k, \eta^1, \dots, \eta^l), \tag{1.7}$$

in the coordinates ξ^i, η^j corresponding to the solution (1.1) be

$$\xi^i = \xi_0^i(t), \quad \eta^j = \eta_0^j(t). \tag{1.8}$$

The origin $\xi = 0, \eta = 0$ is now a junction point of the solution (1.8).

We first calculate a trajectory of the solution (1.8), taking ξ^1 as the independent variable along this trajectory. Clearly,

$$\begin{vmatrix} \frac{\partial \Phi^1}{\partial \eta^1} & \frac{\partial \Phi^1}{\partial \xi^2} & \cdots & \frac{\partial \Phi^1}{\partial \xi^k} \\ \frac{\partial \Phi^2}{\partial \eta^1} & \frac{\partial \Phi^2}{\partial \xi^2} & \cdots & \frac{\partial \Phi^2}{\partial \xi^k} \\ \cdot & \cdot & \cdots & \cdot \\ \cdot & \cdot & \cdots & \cdot \\ \cdot & \cdot & \cdots & \cdot \\ \frac{\partial \Phi^k}{\partial \eta^1} & \frac{\partial \Phi^k}{\partial \xi^2} & \cdots & \frac{\partial \Phi^k}{\partial \xi^k} \end{vmatrix} \neq 0 \tag{1.9}$$

at $\xi = 0, \eta = 0$. Therefore, using the k relations

$$\Phi^i(\xi^1, \ldots, \xi^k, \eta^1, \ldots, \eta^l) = 0 \tag{1.10}$$

we can express $\eta^1, \xi^2, \ldots, \xi^k$ in terms of $\xi^1, \eta^2, \ldots, \eta^l$. It is a simple matter to verify that these expressions take the form:

$$\begin{aligned} \eta^1 &= -(\xi^1)^2 + h^1(\xi^1, \eta^2, \ldots, \eta^l), \\ \xi^i &= K^i(\xi^1)^2 + h^i(\xi^1, \eta^2, \ldots, \eta^l), \quad i = 2, \ldots, k, \end{aligned} \tag{1.11}$$

where the K^i are numerical coefficients, and the functions $h^1, \ldots, h^k$ do not contain terms of the type $p\xi^1$, $q \cdot (\xi^1)^2$. Differentiating the first relation in (1.11) with respect to t, we obtain

$$\dot{\eta}^1 = -2\xi^1 \cdot \dot{\xi}^1 + h^1_{\xi^1} \cdot \dot{\xi}^1 + h^1_{\eta\beta'}\dot{\eta}^{\beta'}. \tag{1.12}$$

Hence

$$\dot{\xi}^1 = \frac{\dot{\eta}^1 - h^1_{\eta\beta'} \cdot \dot{\eta}^{\beta'}}{-2\xi^1 + h^1_{\xi^1}}, \tag{1.13}$$

or, substituting the values of $\dot{\eta}^1, \dot{\eta}^{\beta'}$ from (1.15) into (1.13), we find

$$\dot{\xi}^1 = \frac{1 + \alpha_1^1\xi^1 + \ldots - h^1_{\eta\beta'} \cdot (\alpha_1^{\beta'}\xi^1 + \ldots)}{-2\xi^1 + h^1_{\xi^1}(\xi^1, \eta^2, \ldots, \eta^l)}. \tag{1.14}$$

Further, substituting the expressions (1.11) for $\eta^1, \xi^2, \ldots, \xi^k$ into (1.14) and into the equations

$$\dot{\eta}^j = \Psi^j(\xi^1, \ldots, \xi^k, \eta^1, \ldots, \eta^l), \quad j = 2, \ldots, l, \tag{1.15}$$

we obtain the following system of differential equations for $\xi^1, \eta^2, \ldots, \eta^l$

$$\dot{\xi}^1 = \frac{1 + \alpha_1^1 \xi^1 + \ldots}{-2\xi^1 + \ldots}, \quad \dot{\eta}^j = \alpha_1^j \xi^1 + \ldots, \quad j = 2, \ldots, l \tag{1.16}$$

(the dots here stand for expressions that do not contain terms of the type $p, q\xi^1$). In place of (1.16), let us consider the following system:

$$\frac{d\eta^j}{d\xi^1} = \frac{-2\alpha_1^j (\xi^1)^2 + \ldots}{1 + \alpha_1^1 \xi^1 + \ldots}, \quad j = 2, \ldots, l. \tag{1.17}$$

A solution which passes through the origin is, as can be easily verified,

$$\eta^j = -\tfrac{2}{3}\alpha_1^j (\xi^1)^3 + \ldots, \tag{1.18}$$

where we have not written down higher powers of (ξ^1). Substituting (1.18) into (1.11) and performing some simple calculations, we obtain:

$$\left.\begin{aligned} \eta^1 &= -(\xi^1)^2 + (\tfrac{2}{3}\alpha_1^{\beta'} b_{\beta'}^1 + c_1^1 - d_1^1 - e_{\alpha'}^1 \theta^{\alpha'})(\xi^1)^3 + \ldots, \\ \eta^j &= -\tfrac{2}{3}\alpha_1^j (\xi^1)^3 + \ldots, \\ \xi^i &= \theta^i \cdot (\xi^1)^2 + \pi^i \cdot (\xi^1)^3 + \ldots . \end{aligned}\right\} \tag{1.19}$$

where θ^i and π^i are numerical coefficients whose explicit expressions are not given here.

Thus, we have calculated a trajectory of the solution (1.8), taking ξ^1 as the independent variable along this trajectory.

We now turn our attention to the system (1.5). On the interval $-p \leqslant \xi^1 \leqslant p$, where p is a sufficiently small number not dependent on ϵ, we take ξ^1 as the independent variable and we consider, instead of (1.6), the system

$$\begin{aligned} \frac{d\xi^i}{d\xi^1} &= \frac{\Phi^i(\xi^1, \ldots, \xi^k, \eta^1, \ldots, \eta^l)}{\Phi^1(\xi^1, \ldots, \xi^k, \eta^1, \ldots, \eta^l)}, \quad i = 2, \ldots, k, \\ \frac{d\eta^j}{d\xi^1} &= \frac{\Psi^j(\xi^1, \ldots, \xi^k, \eta^1, \ldots, \eta^l)}{\Phi^1(\xi^1, \ldots, \xi^k, \eta^1, \ldots, \eta^l)}, \quad j = 1, \ldots, l. \end{aligned} \tag{1.20}$$

Subdivide the interval $(-p, p)$ into three subintervals $(-p, -\sigma_1)$, $(-\sigma_1, \sigma_2)$, and (σ_2, p), where $\sigma_1 = \epsilon^{2/7}$ and $\sigma_2 = \epsilon^{2/9}$. The system (1.20) is solved separately on each of these three subintervals. In this section we solve it on the subinterval $-p \leqslant \xi^1 < -\sigma_1$, up to $O(\epsilon)$.

It is natural to attempt to express the solution of the system (1.20),

$$\xi^i = \xi^i(\xi^1, \epsilon), \quad \eta^j = \eta^j(\xi^1, \epsilon), \tag{1.21}$$

whose initial values at the point $\xi^1 = -p$ differ from the initial values of the solution (1.19) calculated for the system (1.7) by a quantity of the order $O(\epsilon)$, approximately in the form of a sum:

$$\begin{aligned} \xi^{i,n} &= \xi_0^i(\xi^1) + \epsilon\xi_1^i(\xi^1) + \ldots + \epsilon^n \xi_n^i(\xi^1), \\ \eta^{j,n} &= \eta_0^j(\xi^1) + \epsilon\eta_1^j(\xi^1) + \ldots + \epsilon^n \eta_n^j(\xi^1). \end{aligned} \tag{1.22}$$

It turns out that such a representation is possible. In other words, for any solution (1.21) satisfying the initial conditions stated above, it is possible on the interval $-p \leqslant \xi \leqslant -\sigma_1$ to construct a sum of the type (1.22) which represents this solution up to a quantity of any order of magnitude in ϵ. The proof of this statement is not simple, but requires tedious calculations and will be communicated separately. Here we restrict ourselves to the construction of the second-order approximations of the solution (1.21),

$$\begin{aligned} \xi^{i,2} &= \xi_0^i(\xi^1) + \epsilon\xi_1^i(\xi^1) + \epsilon^2 \xi_2^i(\xi^1), \\ \eta^{j,2} &= \eta_0^j(\xi^1) + \epsilon\eta_1^j(\xi^1) + \epsilon^2 \eta_2^j(\xi^1) \end{aligned} \tag{1.23}$$

and to the assertion that these approximations represent the solution (1.21) on the interval $(-p, -\sigma_1)$, up to $O(\epsilon)$. Here we define the approximations (1.23) as functions which satisfy the system (1.20) "up to ϵ^2".

The functions $\xi_0^i(\xi^1)$ and $\eta_0^j(\xi^1)$ have already been calculated (see (1.19)). We now calculate the functions $\xi_1^i(\xi^1)$ and $\eta_1^j(\xi^1)$.

Substituting (1.23) into the right- and left-hand sides of the equation

$$\frac{d\eta^1}{d\xi^1} = \epsilon \frac{\Psi^1(\xi^\alpha, \eta^\beta)}{\Phi^1(\xi^\alpha, \eta^\beta)} \tag{1.24}$$

and equating the free terms (in the expansion in powers of ϵ), we find

$$(\eta_0^1)' = \frac{\Psi^1(\xi^1, \xi_0^{\alpha'}(\xi^1), \eta_0^\beta(\xi^1))}{\delta\Phi^1}, \tag{1.25}$$

where

$$\begin{aligned} \delta\Phi^1 &= \frac{\partial}{\partial \xi^{\alpha'}} [\Phi^1(\xi^1, \xi_0^{\alpha'}(\xi^1), \eta_0^\beta(\xi^1))] \cdot \xi_1^{\alpha'} \\ &\quad + \frac{\partial}{\partial \eta^\beta} [\Phi^1(\xi^1, \xi_0^{\alpha'}(\xi^1), \eta_0^\beta(\xi^1))] \cdot \eta_1^\beta. \end{aligned} \tag{1.26}$$

Consequently,

$$\delta\Phi^1 = \frac{1}{(\eta_0^1)'} \Psi^1(\xi^1, \xi_0^{\alpha'}(\xi^1), \eta_0^\beta(\xi^1)). \tag{1.27}$$

Further, substituting (1.23) into the equations

$$\frac{d\eta^j}{d\xi^1} = \epsilon \frac{\Psi^j(\xi^\alpha, \eta^\beta)}{\Phi^1(\xi^\alpha, \eta^\beta)}, \quad j = 2, \ldots, l, \tag{1.28}$$

and then equating the free terms and the coefficients of ϵ (after some simple calculations), we obtain

$$(\eta_0^j)' = \frac{\Psi^j(\xi^1, \xi_0^{\alpha'}(\xi^1), \eta_0^\beta(\xi^1))}{\delta\Phi^1}, \tag{1.29}$$

$$(\eta_1^j)' = \frac{(\eta_0^1)'}{\Psi^1(\xi^1, \xi_0^{\alpha'}, \eta_0^\beta)} \cdot [\delta\Psi^j - (\eta_0^j)'\delta^2\Phi^1]. \tag{1.30}$$

Here $\delta^2\Phi^1$ stands for the coefficient of ϵ^2 in the expansion of the function $\Phi^1(\xi^1, \xi^{\alpha',2}, \eta^{\beta,2})$ as a series in powers of ϵ. This coefficient can be easily calculated. Indeed, as

$$(\eta_0^1)' + \epsilon(\eta_1^1)' = \frac{\Psi^1(\xi^1, \xi_0^{\alpha'}, \eta_0^\beta) + \epsilon\delta\Psi^1 + \dots}{\delta\Phi^1 + \epsilon\delta^2\Phi^1},$$

then

$$\delta\Psi^1 = (\eta_0^1)'\delta^2\Phi^1 + \delta\Phi^1(\eta_1^1)',$$

hence

$$\delta^2\Phi^1 = \frac{\delta\Psi^1 - (\eta_1^1)' \cdot \delta\Phi^1}{(\eta_0^1)'}. \tag{1.31}$$

Substituting this expression into (1.30), we obtain after simple calculations that

$$(\eta_1^j)' = \frac{1}{\Psi^1}[(\eta_0^1)'\delta\Psi^j - (\eta_0^j)'\delta\Psi^1] + \frac{(\eta_0^j)'(\eta_1^1)'}{(\eta_0^1)'}, \quad j = 2, \dots, l. \tag{1.32}$$

Similarly, from

$$\frac{d\xi^i}{d\xi^1} = \frac{\Phi^i(\xi^\alpha, \eta^\beta)}{\Phi^1(\xi^\alpha, \eta^\beta)}, \quad i = 2, \dots, k, \tag{1.33}$$

we find

$$(\xi_0^i)' = \frac{\delta\Phi^i}{\delta\Phi^1}, \tag{1.34}$$

Hence, by virtue of (1.27), we obtain

$$\frac{\partial\Phi^i}{\partial\xi^{\alpha'}} \cdot \xi_1^{\alpha'} + \frac{\partial\Phi^i}{\partial\eta^\beta}\eta_1^\beta = (\xi_0^i)' \frac{\Psi^1(\xi^1, \xi_0^{\alpha'}, \eta_0^\beta)}{(\eta_0^1)'}, \quad i = 2, \dots, k. \tag{1.35}$$

Moreover, on writing out the relation (1.27), it takes the form:

$$\frac{\partial\Phi^1}{\partial\xi^{\alpha'}}\xi_1^{\alpha'} + \frac{\partial\Phi^1}{\partial\eta^1}\eta_1^1 + \frac{\partial\Phi^1}{\partial\eta^{\beta'}}\eta_1^{\beta'} = \frac{\Psi^1(\xi^1, \xi_0^{\alpha'}, \eta_0^{\beta'})}{(\eta_0^1)'}. \tag{1.36}$$

Using the concrete form of the function Φ^1, we can write (1.36) more explicitly as follows:

$$(1+B^1(\xi^1))\eta_1^1 = \frac{\Psi^1}{(\eta_0^1)'} - \frac{\partial\Phi^1}{\partial\eta^{\beta'}}\eta_1^{\beta'} + A_{\alpha'}^1(\xi^1)\xi_1^{\alpha'}, \tag{1.37}$$

where

$$B^1(0) = A_{\alpha'}^1(0) = 0.$$

Combining (1.35), (1.37) and (1.32), we obtain the following system of differential equations for ξ_1^i, $(i = 2, \ldots, k)$ and η_1^j $(j = 1, \ldots, l)$:

$$\left.\begin{aligned} \frac{\partial\Phi^i}{\partial\xi^{\alpha'}}\cdot\xi_1^{\alpha'} + \frac{\partial\Phi^i}{\partial\eta^\beta}\eta_1^\beta &= (\xi_0^i)'\frac{\Psi^1}{(\eta_0^1)'}, \\ (1+B^1(\xi^1))\eta_1^1 &= -\frac{\partial\Phi^1}{\partial\eta^{\beta'}}\eta_1^{\beta'} + A_{\alpha'}^1(\xi^1)\xi_1^{\alpha'} + \frac{\Psi^1}{(\eta_0^1)'}, \\ (\eta_1^j)' &= \frac{1}{\Psi^1}[(\eta_0^1)'\delta\Psi^j - (\eta_0^j)'\delta\Psi^1] + \frac{(\eta_0^j)'\cdot(\eta_1^1)'}{(\eta_0^1)'}. \end{aligned}\right\} \tag{1.38}$$

We now examine this system. In the first $(k-1)$ relations, expressing the functions $\xi_1^{\alpha'}$ in terms of η_1^β and then substituting the resulting expressions

$$\xi_1^{\alpha'} = G_\beta^{\alpha'}(\xi^1)\eta_1^\beta + D^{\alpha'}(\xi^1) \tag{1.39}$$

into the right-hand sides of the last $(l-1)$ equations of (1.38), we see after some simple calculations that

$$(\eta_1^j)' = \widetilde{N}^j(\xi^1) + \widetilde{N}_\beta^j(\xi^1)\eta_1^\beta + \frac{(\eta_0^j)'(\eta_1^1)'}{(\eta_0^1)'}, \tag{1.40}$$

where $\widetilde{N}^j(0) = \widetilde{N}_\beta^j(0) = 0$.

On the other hand, substituting (1.39) into the second relation of (1.38), we readily find that

$$\eta_1^1 = \left\{\left[-\frac{\partial\Phi^1}{\partial\eta^{\beta'}} + \widetilde{A}_{\beta'}(\xi^1)\right]\eta_1^{\beta'} + \widetilde{B}(\xi') + \frac{\Psi^1}{(\eta_0^1)'}\right\}\frac{1}{1+B^1(\xi')}, \tag{1.41}$$

where $\widetilde{A}_{\beta'}(0) = \widetilde{B}(0) = 0$.

The system of equations (1.40) and (1.41) can be solved. Differentiating (1.41) and substituting the resulting expression into (1.40), we find after some calculation that

$$(\eta_1^j)' = P^j(\xi^1) + N_{\beta'}^j(\xi^1)\eta_1^{\beta'} + Q_{\beta'}^j(\xi^1)(\eta_1^{\beta'})', \quad j = 2, \ldots, l, \tag{1.42}$$

where the function $P^j(\xi^1)$ has a first order pole at zero and a principal term $\alpha_2^j/2$, and $N_{\beta'}^j(0) = Q_{\beta'}^j(0) = 0$. We shall seek a solution of the system (1.42) in the form:

$$\eta_1^j = K^j(\xi^1) \ln |\xi^1| + L^j(\xi^1), \tag{1.43}$$

where $K^j(\xi^1)$ are functions which are continuously differentiable on $-p \leqslant \xi^1 \leqslant 0$. Substituting (1.43) into (1.42), we obtain

$$(K^j)' \ln |\xi^1| + \frac{K^j}{\xi^1} + (L^j)' = P^j(\xi^1) + N^j_{\beta'} K^{\beta'} \ln |\xi^1| + N^j_{\beta'} L^{\beta'} + Q^j_{\beta'} \left[(K^{\beta'})' \ln |\xi^1| + \frac{K^{\beta'}}{\xi^1} + (L^{\beta'})' \right]. \tag{1.44}$$

Equating the coefficients of $\ln |\xi^1|$, we obtain a system of differential equations for $K^j(\xi^1)$:

$$(K^j)' = N^j_{\beta'} K^{\beta'} + Q^j_{\beta'} (K^{\beta})'. \tag{1.45}$$

Solving this system for the initial values $K^j(0) = \alpha_1^j/2$, we find unique functions $K^j(\xi^1)$. Then from the following system of ordinary differential equations

$$(L^j)' = P^j(\xi^1) - \frac{K^j(\xi^1)}{\xi^1} + Q^j_{\beta'} \frac{K^{\beta'}}{\xi^1} + N^j_{\beta'} L^{\beta'} + Q^j_{\beta'} L^{\beta'} \tag{1.46}$$

we can determine the functions L^j as well. They do not have singularities.

Thus, we have

$$\eta_1^j = \frac{\alpha_1^j}{2} \ln |\xi^1| + O(1), \quad j = 2, \ldots, l. \tag{1.47}$$

Furthermore, from (1.41), it follows that

$$\eta_1^1 = -\frac{1}{2\xi^1} - \frac{\partial \Phi^1}{\partial \eta^{\beta'}} \frac{\alpha_2^{\beta'}}{2} \ln |\xi^1| + O(1). \tag{1.48}$$

Finally, from the first $(k-1)$ relations in (1.38), we find

$$\xi_1^i = \frac{m^i}{\xi^1} + n^i \ln |\xi^1| + O(1), \quad i = 2, \ldots, k, \tag{1.49}$$

where m^i and n^i are constants (their explicit expressions are not given here).

Thus, we have calculated the following sums

$$\eta^{j,1} = \eta_0^j(\xi^1) + \epsilon \eta_1^j(\xi^1), \quad j = 1, \ldots, l,$$
$$\xi^{i,1}(\xi^1) = \xi_0^i(\xi^1) + \epsilon \xi_1^i(\xi^1), \quad i = 2, \ldots, k.$$

It is rather cumbersome to calculate the functions $\xi_2^i(\xi^1)$ $(i = 2, \ldots, k)$

and $\eta_2^j(\xi^1)$ $(j = 1, \ldots, l)$. But for our purposes it suffices to know their principal terms. After the necessary calculations, we find

$$\eta_2^1(\xi^1) = -\frac{1}{8(\xi^1)^4} + \delta^1(\xi^1), \quad \eta_2^j(\xi^1) = \delta^j(\xi^1), \quad j = 2, \ldots, l,$$

$$\xi_2^i(\xi^1) = \frac{Q^i}{(\xi^1)^4} + \gamma^i(\xi^1), \quad i = 2, \ldots, k, \tag{1.49'}$$

where Q^i are numerical coefficients whose explicit expressions are not given here. The functions $\delta^j(\xi^1)$ $(j = 1, \ldots, l)$ and $\gamma^i(\xi^1)$ $(i = 2, \ldots, k)$ are such that $\epsilon^2\delta^j(\xi^1)$ and $\epsilon^2\gamma^i(\xi^1)$ are quantities of the order $O(\epsilon)$ on the whole interval $-p \leqslant \xi^1 \leqslant -\sigma_1$.

We can prove that on the interval $-p \leqslant \xi^1 \leqslant -\sigma_1$ the solution (1.21) is representable, as already mentioned, in the form:

$$\eta^j = \eta_0^j(\xi^1) + \epsilon\eta_1^j(\xi^1) + \epsilon^2\eta_2^j(\xi^1) + S^{j,2}(\xi^1, \epsilon),$$
$$\xi^i = \xi_0^i(\xi^1) + \epsilon\xi_1^i(\xi^1) + \epsilon^2\xi_2^i(\xi^1) + R^{i,2}(\xi^1, \epsilon),$$

where the functions $S^{j,2}(\xi^1, \epsilon)$ and $R^{i,2}(\xi^1, \epsilon)$ have an order of magnitude on this interval not greater than $O(\epsilon)$. Consequently, on this interval the asymptotic expressions for the solution (1.21) are

$$\left.\begin{aligned}
\eta^1 &= -(\xi^1)^2 + (\tfrac{2}{3}\alpha_1^{\beta'}b_{\beta'}^1 + c_1^1 - d_1^1 - e_{\alpha'}^1\theta^{\alpha'})(\xi^1)^3 + \ldots + \\
&\quad + \epsilon\left(-\frac{1}{2\xi^1} - \tfrac{1}{2}b_{\beta'}^1\alpha_1^{\beta'}\ln|\xi^1|\right) + \epsilon^2\left[-\frac{1}{8(\xi^1)^4}\right] + S^1(\xi^1, \epsilon), \\
\eta^j &= -\tfrac{2}{3}\alpha_1^j(\xi^1)^3 + \ldots + \epsilon\left(\frac{\alpha_1^j}{2}\ln|\xi^1|\right) + S^j(\xi^1, \epsilon), \quad j = 2, \ldots, l, \\
\xi^i &= \theta^i(\xi^1)^2 + \pi^i(\xi^1)^3 + \ldots + \epsilon\left(\frac{m_i}{\xi^1} + n^i\ln|\xi^1|\right) + \epsilon^2\frac{Q^i}{(\xi^1)^4} + R^i(\xi^1, \epsilon).
\end{aligned}\right\} \tag{1.50}$$

In these formulas the dots stand for terms of the order $(\xi^1)^4$ or higher, and the functions $S^j(\xi^1, \epsilon)$, $R^i(\xi^1, \epsilon)$ are of the order $O(\epsilon)$ on the interval $-p \leqslant \xi^1 \leqslant -\sigma_1$.

2. Calculation of Solutions on the Transition Interval

In this section we shall continue the solution (1.21) over the interval $-\sigma_1 \leqslant \xi^1 \leqslant \sigma_2$. As we shall see, on this interval the solution (1.21) deviates much more from the corresponding solution of the degenerate system, namely, by quantities of orders $\epsilon^{2/3}$ and $\epsilon \ln \xi$.

The change of variables

$$\xi^1 = \mu u^1, \quad \xi^i = \mu^2 u^i, \quad \eta^1 = \mu^2 v^1, \quad \eta^j = \mu^3 v^j, \quad t = \mu^3 \tau, \quad \mu^3 = \epsilon \tag{2.1}$$

reduces the system (1.5) to the following form:

$$\left.\begin{aligned} \dot{u}^1 &= (u^1)^2 + v^1 + \mu(b^1_{\beta'}v^{\beta'} + c^1_1 u^1 v^1 + d^1_1 (u^1)^3 + e^1_{\alpha'} u^1 u^{\alpha'}) + \dots, \\ \mu \dot{u}^i &= a^i_{\alpha'} u^{\alpha'} + b^i_1 v_1 + c^i_0 (u^1)^2 + \mu(b^i_{\beta'} v^{\beta'} + d^i_1 (u^1)^3 + c^i_1 u^1 v^1 + \\ &\quad + e^i_{\alpha'} u^1 u^{\alpha'}) + \dots, \\ \dot{v}^1 &= 1 + \mu \alpha^1_1 u^1 + \dots, \\ \dot{v}^j &= \alpha^j_1 u^1 + \dots \end{aligned}\right\} \tag{2.2}$$

(Here a dot over a symbol denotes differentiation with respect to τ). We write the system in compact form as follows:

$$\left.\begin{aligned} \dot{u}^1 &= \varphi^1(u^\alpha, v^\beta, \mu), \\ \mu \dot{u}^i &= \varphi^i(u^\alpha, v^\beta, \mu), \\ \dot{v}^j &= \psi^j(u^\alpha, v^\beta, \mu). \end{aligned}\right\} \tag{2.2'}$$

We take u^1 as the independent variable, and in place of (2.2′) we study the system:

$$\begin{aligned} \mu \frac{du^i}{du^1} &= \frac{\varphi^i(u^\alpha, v^\beta, \mu)}{\varphi^1(u^\alpha, v^\beta, \mu)}, \quad i = 2, \dots, k, \\ \frac{dv^j}{du^1} &= \frac{\psi^j(u^\alpha, v^\beta, \mu)}{\varphi^1(u^\alpha, v^\beta, \mu)}, \quad j = 1, \dots, l. \end{aligned} \tag{2.3}$$

For $\mu = 0$ the system (2.3) degenerates:

$$\left.\begin{aligned} &\text{(a)} & a^i_{\alpha'} u^{\alpha'} + b^i_1 v^1 + c^i_0 (u^1)^2 &= 0, \\ &\text{(b)} & \frac{dv^1}{du^1} &= \frac{1}{(u^1)^2 + v^1}, \\ &\text{(c)} & \frac{dv^j}{du^1} &= \frac{\alpha^j_1 u^1}{(u^1)^2 + v^1}. \end{aligned}\right\} \tag{2.4}$$

This degenerate system has many solutions. For us a special role will be played by one particular solution which we now indicate.

The equation (b) in (2.4) is a Ricatti equation. Its solutions are defined on the interval $-\infty < u^1 < +\infty$. We take a particular solution, $v_0^1(u^1)$, which for negative u^1 is represented in the form $-(u^1)^2 + z_0(u^1)$, where the additional term $z_0(u^1)$ tends to zero as $u^1 \to -\infty$. For this term we have the following differential equation

$$-2u^1 + z_0'(u^1) = \frac{1}{z_0(u_1)} \tag{2.5}$$

and an asymptotic representation for large negative u^1:

$$z_0(u^1)^- = -\frac{1}{2u^1} - \frac{1}{8(u^1)^4} + O\left(\frac{1}{(u^1)^7}\right). \tag{2.6}$$

Thus,

$$v_0^1(u^1)^- = -(u^1)^2 - \frac{1}{2u^1} - \frac{1}{8(u^1)^4} + O\left(\frac{1}{(u^1)^7}\right). \tag{2.7}$$

It is a simple matter to derive an asymptotic representation of the function $v_0^1(u^1)$ for large positive u^1 as well:

$$v_0^1(u^1) = \Omega - \frac{1}{2u^1} + O\left(\frac{1}{(u^1)^3}\right), \tag{2.8}$$

where $\Omega = \text{const.} = v_0^1(\infty)$. (The expansions (2.7) and (2.8) can also be derived with the help of the asymptotic properties of the Bessel functions since the solution of a Ricatti equation is expressed by Bessel functions. But, it is simpler to derive them directly from the Ricatti equation).

Using the solution $v_0^1(u^1)$ already calculated, we can determine the functions $v_0^j(u^1), j = 2, \ldots, l$:

$$v_0^j = \alpha_1^j \int \frac{u^1}{(u^1)^2 + v_0^1(u^1)} du^1. \tag{2.9}$$

We find that v_0^j is determined up to a constant. Let us take a definite v_0^j, namely,*

$$v_0^j(u^1) = \alpha_1^j \int_0^{u^1} \frac{u^1 du^1}{(u^1)^2 + v_0^1(u^1)} + \frac{\alpha_1^i}{2} \ln \mu. \tag{2.10}$$

* If we take

$$v_0^j(u^1) = \alpha_1^j \int_0^{u^1} \frac{du^1}{(u^1)^2 + v_0^1} + \tfrac{1}{2}\alpha_1^j \ln \mu + C^j,$$

where C^j is an arbitrary constant not dependent on μ, then, as we shall show later, the functions $v_1^1(u^1)$ and $u_1^i(u^1)$ calculated below also only change by certain constants not dependent on μ.

It can be directly verified that for large negative v^1, we obtain the following asymptotic expansion

$$v_0^j(u^1)^- = -\tfrac{2}{3}\alpha_1^j(u^1)^3 + \frac{\alpha_1^j}{2}\ln|u^1| + \frac{\alpha_1^j}{2}\ln\mu + \ldots, \tag{2.11}$$

and for large positive u^1

$$v_0^j(u^1)^+ = \alpha_1^j \ln|u^1| + \tfrac{1}{2}\alpha_1^j \ln\mu + \ldots. \tag{2.12}$$

(In (2.11) and (2.12) the dots stand for the terms which remain bounded as $u^1 \to -\infty$ or as $u^1 \to +\infty$, respectively.)

Now the functions $u_0^i(u^1)$ can be determined uniquely from the first $(k-1)$ equations in (2.4). After some simple calculations, we obtain the following asymptotic expansions

$$u_0^i(u^1)^- = \theta^i(u^1)^2 + \frac{m^i}{u^1} + \frac{Q^i}{(u^1)^4} + O\left(\frac{1}{(u^1)^7}\right),$$

$$u_0^i(u^1)^+ = B_0^i(u^1)^2 + O\left(\frac{1}{(u^1)^3}\right) + O(1). \tag{2.13}$$

for large negative and large positive values of u^1, respectively.

Thus, a particular solution of the degenerate system (2.4) has been found on the interval $-\infty < u^1 < +\infty$:

$$\begin{aligned} v^j &= v_0^j(u^1), \quad j = 1, \ldots, l, \\ u^i &= u_0^i(u^1), \quad i = 2, \ldots, k. \end{aligned} \tag{2.14}$$

The system of functions (2.14) is called a "zero-order approximation" of the true solution of the non-degenerate system (2.3) with appropriate initial values.

Using (2.14), we now formally construct "a first-order approximation":*

$$\left.\begin{aligned} v^1 &= v_0^1(u^1) + \mu v_1^1(u^1), \\ v^j &= v_0^j(u^1), \quad j = 2, \ldots, l, \\ u^i &= u_0^i(u^1) + \mu u_1^i(v^1), \quad i = 2, \ldots, k, \end{aligned}\right\} \tag{2.15}$$

where the functions $v_1^1(u^1)$ and $u_1^i(u^1)$ are determined as follows.

Substituting the functions (2.15) into the right- and left-hand sides of the equation

$$\frac{dv^1}{du^1} = \frac{\psi^1(u^\alpha, v^\beta, \mu)}{\varphi^1(u^\alpha, v^\beta, \mu)} \tag{2.16}$$

---

* We are not interested in the first-order approximation of the function v^j.

and then equating the coefficients of μ, we obtain a differential equation

$$\frac{dv^1}{du^1}+\frac{v_1^1}{[(u^1)^2+v_0^1]^2}=H_1(u^1)-\tfrac{1}{2}\ln\mu\,\frac{b^1_{\beta'}\alpha_1^{\beta'}}{[(u^1)^2+v_0^1]^2},\tag{2.17}$$

where

$$H_1(u^1)=-\frac{b^1_{\beta'}\tilde{v}_0^{\beta'}+c_1^1u^1v_0^1+d_1^1(u^1)^3+e^1_{\alpha'}u^1u^{\alpha'}+\alpha_1^1u^1[(u^1)^2+v_0^1]}{[(u^1)^2+v_0^1]^2}.\tag{2.18}$$

Here, we have

$$\tilde{v}_0^{\beta'}=\alpha_1^{\beta'}\int_0^{u^1}\left(\frac{u^1du^1}{(u^1)^2+v_0^1}\right).$$

We determine the function $v_1^1(u^1)$ as a particular solution of this linear equation:

$$v_1^1=\frac{1}{G(u^1)}\int_{-\infty}^{u^1}H_1(u^1)G(u^1)du^1-\tfrac{1}{2}b^1_{\beta'}\alpha_1^{\beta'}\ln\mu,\tag{2.19}$$

where

$$G(u^1)=\exp\left\{\int_0^{u^1}\frac{du^1}{[(u^1)^2+v_0^1]^2}\right\}.\tag{2.20}$$

From (2.19), by virtue of the asymptotic expansions of the functions $v_0^1, v_0^{\beta'}$, and $u_0^{\alpha'}$ (see (2.7), (2.8), (2.11), (2.12), and (2.13)), we obtain the asymptotic expansions of the function $v_1^1(u^1)$ for large negative and for large positive u^1:

$$v_1^1(u^1)^-=(\tfrac{2}{3}\alpha_1^{\beta'}b^1_{\beta'}+c_1^1-d_1^1-e^1_{\alpha'}\theta^{\alpha'})(u^1)^5-$$
$$-\tfrac{1}{2}b^1_{\beta'}\alpha_1^{\beta'}\ln|u^1|-\tfrac{1}{2}b^1_{\beta'}\alpha_1^{\beta'}\ln\mu+\ldots,\tag{2.21}$$

$$v_1^1(u^1)^+=(\alpha_1^1-d_1^1-e^1_{\alpha'}B_0^{\alpha'})\ln u^1-\tfrac{1}{2}b^1_{\beta'}\alpha_1^{\beta'}\ln\mu+\ldots.\tag{2.22}$$

In these formulas the dots stand for the terms which remain bounded when the modulus $|u^1|$ increases without bound.

The functions $u_1^i(u^1)$, $i=2,\ldots,k$, are found in a simpler manner. Namely, substituting (2.15) in the first $(k-1)$ equations of (2.3) and then equating the coefficients of μ, we obtain linear algebraic equations for $u_1^i(u^1)$. From these equations, and using the asymptotic expansions already known for the functions u_0^i, v_0^j and v_1^1, we easily obtain the asymptotic expansions for $u_1^i(u^1)$. Without going into the details of the elementary but rather cumbersome calculations, we give here only the final results:

$$u_1^i(u^1)^-=\pi^i(u^1)^3+n^i\ln|u^1|+n^i\ln\mu+O(1)_{u^1\to-\infty},\tag{2.21$'$}$$

$$u_1^i(u^1)^+=O((u^1)^3)+O(\ln\mu)+O(1)_{u^1\to+\infty}.\tag{2.22$'$}$$

It should be noted that the numerical coefficients π^i and n^i in (2.21′) are the same as the similar coefficients in (1.50).

Thus, we have calculated the "first-order approximation" (2.15). Here the functions (2.15) have been determined as functions that satsify the non-degenerate system (2.3) to a given accuracy. The question of how closely they approximate the true solution of (2.3) has so far not been examined. Actually, it turns out that on the interval $-\omega_1 \leqslant u^1 \leqslant \omega_2$, where $\omega_i = \sigma_i/\mu$ $(i = 1, 2)$ and $\omega_i \to \infty$ as $\epsilon \to 0$, the system of functions (2.15) approximates, with a definite accuracy, any true solution of equation (2.3) whose initial values at the point $u^1 = -\omega_1$ coincide with the initial values of the functions (2.15) to the same degree of accuracy. More precisely, *every solution*

$$v^j = v^j(u^1, \mu), \quad u^i = u^i(u^1, \mu) \tag{2.23}$$

of the equation (2.3) that satisfies the initial conditions

$$\left.\begin{aligned} &v^1(-\omega_1) - [v_0^1(-\omega_1) + \mu v_1^1(-\omega_1)] = O(\mu), \\ &v^j(-\omega_1) - v_0^j(-\omega_1) = 0, \\ &u^i(-\omega_1) - [u_0^i(-\omega_1)] + \mu u_1^i(-\omega_1)] = O(\mu), \end{aligned}\right\} \tag{2.24}$$

can be represented on the interval $-\omega_1 \leqslant u^1 \leqslant \omega_2$ *as*:

$$\left.\begin{aligned} v^1 &= v_0^1(u^1) + \mu v_1^1(u^1) + r^1(u^1, \mu), \\ v^j &= v_0^j(u^1) + r^j(u^1, \mu), \\ u^i &= u_0^i(u^1) + \mu u_1^i(u^1, \mu) + s^i(u^1, \mu), \end{aligned}\right\} \tag{2.25}$$

where, on this interval, the functions $r^1(u^1, \mu)$, $s^i(u^1, \mu)$ *are of the order* $O(\mu)$ *and the functions* $r^j(u^1, \mu), j \geqslant 2$, *are of the order* $O(1)$.

The proof of this assertion requires certain further calculations and will be published separately.

We now return to the old variables ξ^i, η^j. In these variables the interval $-\omega_1 \leqslant u^1 \leqslant \omega_2$ is converted into the interval $-\sigma_1 \leqslant \xi^1 \leqslant \sigma_2$, and the functions (2.15) take the form:

$$\left.\begin{aligned} \eta^1 &= \mu^2 v_0^1\left(\frac{\xi^1}{\mu}\right) + \mu^3 v_1^1\left(\frac{\xi^1}{\mu}\right), \\ \eta^j &= \mu^3 v_0^j\left(\frac{\xi^1}{\mu}\right), \qquad j = 2, \ldots, l, \\ \xi^i &= \mu^2 u_0^i\left(\frac{\xi^1}{\mu}\right) + \mu^3 u_1^i\left(\frac{\xi^1}{\mu}\right), \quad i = 2, \ldots, k. \end{aligned}\right\} \tag{2.26}$$

Using the asymptotic expansions of $v_0^1(u^1)$, $v_1^1(u^1)$, $v_0^j(u^1)$, $u_0^i(u^1)$, $u_1^i(u^1)$ for large negative u^1 (see formulas (2.11), (2.21), (2.13), (2.21′)), we can easily calculate the value of the functions (2.26) at the point $\xi^1 = -\sigma_1$. We now take up these calculations. For this purpose, let us substitute in (2.15) the asymptotic expansions of the functions v_0^1, v_1^1, v_0^j, u_0^i, and u_1^i for large negative u^1. We obtain

$$\begin{aligned} v^1 &= (-u^1)^2 - \frac{1}{2u^1} - \frac{1}{8(u^1)^4} + O\left(\frac{1}{(u^1)^7}\right) + \mu\{(\tfrac{2}{3}\alpha_1^{\beta'} b_{\beta'}^1 + c_1^1 - \\ &\quad - d_1^1 - e_{\alpha'}^1\theta^{\alpha'})(u^1)^3 - \tfrac{1}{2}b_{\beta'}^1\alpha_1^{\beta'} \ln|u^1| - \tfrac{1}{2}b_{\beta'}^1\alpha_1^{\beta'} \ln\mu + \ldots\}, \\ v^j &= [-\tfrac{2}{3}(u^1)^3 + \tfrac{1}{2}\ln(u^1) + \tfrac{1}{2}\ln\mu + \ldots]\alpha_1^j, \quad j = 2, \ldots, l, \\ u^i &= \theta^i(u^1)^2 + \frac{m^i}{u^1} + \frac{Q^i}{(u^1)^4} + O\left(\frac{1}{(u^1)^7}\right) + \\ &\quad + \mu[\pi^i(u^1)^3 + n^i \ln|u^1| + n^i \ln\mu + \ldots]. \end{aligned} \tag{2.27}$$

Changing over to the variables ξ^i, η^j, we obtain the expressions of the functions (2.26) for the values of ξ^1 close to $-\sigma_1$:

$$\left.\begin{aligned} \eta^1 &= (-\xi^1)^2 + (\tfrac{2}{3}\alpha_1^{\beta'} b_{\beta'}^1 + c_1^1 - d_1^1 - e_{\alpha'}^1\theta^{\alpha'})(\xi^1)^3 + \\ &\quad + \epsilon\left[-\frac{1}{2\xi^1} - \tfrac{1}{2}b_{\beta'}^1\alpha_1^{\beta'} \ln|\xi^1|\right] + \epsilon^2 \cdot \frac{-1}{8(\xi^1)^4} + O(\epsilon), \\ \eta^j &= -\tfrac{2}{3}\alpha_1^j(\xi^1)^3 + \epsilon \cdot \tfrac{1}{2}\alpha_1^j \ln|\xi^1| + O(\epsilon), \\ \xi^i &= \theta^i \cdot (\xi^1)^2 + \pi^i \cdot (\xi^1)^3 + \epsilon\left[\frac{m^i}{\xi^1} + n^i \ln|\xi^1|\right] + \epsilon^2 \frac{Q^i}{(\xi^1)^4} + O(\epsilon). \end{aligned}\right\} \tag{2.28}$$

Hence, in particular, we can find the values of the functions (2.26) at the point $\xi^1 = -\sigma_1$.

Let us now compare (2.28) with (1.50). We find that at $\xi^1 = -\sigma_1$

$$\eta^j_{(1.50)} = \eta^j_{(2.28)} + O(\epsilon), \quad \xi^i_{(1.50)} = \xi^i_{(2.28)} + O(\epsilon). \tag{2.29}$$

Consequently, the functions (2.28), which we have calculated, are indeed continuations of the solution (1.50) over the interval $-\sigma_1 \leqslant \xi^1 \leqslant \sigma_2$ with the appropriate accuracy. And since, as we have already mentioned, on this interval they approximate the true solution of the system (1.20), we have solved the problem of continuing the solution (1.50) over the interval $-\sigma_1 \leqslant \xi^1 \leqslant \sigma_2$. This continuation is given by the formulas:

$$
\left.\begin{aligned}
\eta^1 &= \mu^2 v_0^1\left(\frac{\xi^1}{\mu}\right) + \mu^3 v_1^1\left(\frac{\xi^1}{\mu}\right) + H^1(\xi^1,\mu),\\
\eta^j &= \mu^3 v_0^j\left(\frac{\xi^1}{\mu}\right) + H^j(\xi^1,\mu),\\
\xi^i &= \mu^2 u_0^i\left(\frac{\xi^1}{\mu}\right) + \mu^3 u_1^i\left(\frac{\xi^1}{\mu}\right) + G^i(\xi^1,\mu),
\end{aligned}\right\} \tag{2.30}
$$

where the functions $H^j(\xi^1, \mu)$ and $G^i(\xi^1, \mu)$ are of the order $O(\epsilon)$.

Using the asymptotic expansions (2.8), (2.12) and (2.22), we can, in particular, find the values of the functions η^j at the point $\xi^1 = \sigma_2$:

$$
\left.\begin{aligned}
\eta^1(\sigma_2) &= \epsilon^{2/3}\Omega - \tfrac{1}{3}\epsilon\ln\epsilon\cdot(\alpha_1^1 - d_1^1 - e_{\alpha'}^1 B_0^{\alpha'} + \tfrac{1}{2} b_{\beta'}^1 \alpha_1^{\beta'}) +\\
&\quad + \epsilon\ln\sigma_2\cdot(\alpha_1^1 - d_1^1 - e_{\alpha'}^1 B_0^{\alpha'}) - \frac{\epsilon}{\sigma_2} + O(\epsilon),\\
\eta^j(\sigma_2) &= \epsilon\alpha_1^j\ln\sigma_2 - \tfrac{1}{6}\alpha_1^j\epsilon\ln\epsilon + O(\epsilon), \quad j = 2, \ldots, l.
\end{aligned}\right\} \tag{2.31}
$$

We will have a need for these values.

3. The "Displacement Vector" and Its Calculation

Let

$$
\begin{aligned}
\eta^j &= \eta^j(\xi^1, \epsilon), \quad j = 1, \ldots, l,\\
\xi^i &= \xi^i(\xi^1, \epsilon), \quad i = 2, \ldots, k,
\end{aligned} \tag{3.1}
$$

be the exact continuation of the solution (2.30) over the interval $\sigma_2 \leqslant \xi^1 \leqslant p$, where p is no longer dependent on ϵ. We first prove that

$$
\begin{aligned}
\eta^j(\xi^1, \epsilon) &= o_{\epsilon\to 0}(1),\\
\xi^i(\xi^1, \epsilon) &= B_0^i(\xi^1)^2 + \ldots + o_{\epsilon\to 0}(1)
\end{aligned} \tag{3.2}
$$

where dots stand for the terms in higher powers of (ξ^1).

Indeed, the functions (3.1) are solutions of the system (1.20). It can be shown that they depend continuously on the parameter ϵ (this proof will be published later). But, for $\epsilon = 0$, the system (1.20) is transformed into the system:

$$
\begin{aligned}
\frac{d\xi^i}{d\xi^1} &= \frac{\Phi^i(\xi^\alpha, \eta^\beta)}{\Phi^1(\xi^\alpha, \eta^\beta)},\\
\frac{d\eta^j}{d\xi^1} &= 0,
\end{aligned} \tag{3.3}
$$

for which the functions

$$\eta^j = 0, \quad \xi^i = B_0^i(\xi^1)^2 + \dots. \tag{3.4}$$

are the solutions with zero initial values.

Indeed, expressing this solutions as $\xi^{\alpha'} = p^{\alpha'}\xi^1 + q^{\alpha'}(\xi^1)^2 + \dots$ with indeterminate coefficients $p^{\alpha'}$ and $q^{\alpha'}$, we obtain linear algebraic equations

$$a^i_{\alpha'} \cdot p^{\alpha'} = 0, \quad a^i_{\alpha'} q^{\alpha'} + c^i_0 = 0,$$

Hence, we find $p^{\alpha'} = 0$ and $q^{\alpha'} = B_0^{\alpha'}$.

We now calculate the values of the functions $\eta^j(\xi^1, \epsilon), j = 1, \dots, l$ at an arbitrary point ξ^1 in the interval (σ_2, p). By virtue of (3.2), we can write

$$\eta^j(\xi^1, \epsilon) = \eta^j(\sigma_2, \epsilon) + \epsilon \int_{\sigma_2}^{\xi^1} \frac{\Psi^j(\xi^1, B_0^{\alpha'}(\xi^1)^2, 0, \dots, 0)}{\Phi_1^1(\xi^1, B_0^{\alpha'}(\xi^1)^2, 0, \dots, 0)} d\xi^1 + o(\epsilon). \tag{3.5}$$

But, as is easy to calculate,

$$\frac{\Psi^j(\xi^1, B_0^{\alpha'}(\xi^1)^2, 0, \dots, 0)}{\Phi^1(\xi^1, B_0^{\alpha'}(\xi^1)^2, 0, \dots, 0)} = \frac{1}{(\xi^1)^2} + \frac{1}{\xi^1}(\alpha_1^1 - d_1^1 - e^1_{\alpha'}B_0^{\alpha'}) + O(1)_{\xi^1 \to 0}, \tag{3.6}$$

$$\frac{\Psi^j(\xi^1, B_0^{\alpha'}(\xi^1)^2, 0, \dots, 0)}{\Phi^1(\xi^1, B_0^{\alpha'}(\xi^1)^2, 0, \dots, 0)} = \frac{\alpha_1^j}{\xi^1} + O(1)_{\xi^1 \to 0}, \quad j = 2, \dots, l. \tag{3.7}$$

Therefore

$$\eta^1(\xi^1, \epsilon) = \eta^1(\sigma_2, \epsilon) + \epsilon \int_{\sigma_2}^{\xi^1} \frac{1}{(\xi^1)^2} + \frac{1}{\xi^1}(\alpha_1^1 - d_1^1 - e^1_{\alpha'}B_0^{\alpha'}) \; d\xi^1 + O(\epsilon), \tag{3.8}$$

$$\eta^j(\xi^1, \epsilon) = \eta^j(\sigma_2, \epsilon) + \epsilon \int_{\sigma_2}^{\xi^1} \frac{\alpha_1^j}{\xi^1} d\xi^1 + O(\epsilon), \quad j = 2, \dots, l.$$

The values of $\eta^j(\sigma_2, \epsilon)$ have already been calculated in Section 2 (see (2.31)). Substituting these values into (3.8), and then integrating, we obtain expressions for the functions $\eta^j(\xi^1, \epsilon)$ at an arbitrary point in the interval (σ_2, p):

$$\left.\begin{aligned} \eta^1(\xi^1, \epsilon) &= \epsilon^{2/3}\Omega + \epsilon \ln \epsilon [-\tfrac{1}{3}(\alpha_1^1 - d_1^1 - e^1_{\alpha'}B_0^{\alpha'} + \tfrac{1}{2} b^1_{\beta'}\alpha_1^{\beta'})] + \\ &\quad + \epsilon \left[-\frac{1}{\xi^1} + (\alpha_1^1 - d_1^1 - e^1_{\alpha'}B_0^{\alpha'}) \ln \xi^1 \right] + O(\epsilon), \\ \eta^j(\xi^1, \epsilon) &= \epsilon \ln \epsilon [-\tfrac{1}{6}\alpha_1^j] + \epsilon \alpha_1^j \ln \xi^1 + O(\epsilon). \end{aligned}\right\} \tag{3.9}$$

Let $\Delta_1 = (\Delta_1^1, \dots, \Delta_1^l)$ denote a vector whose coordinates are

$$\begin{aligned} \Delta_1^1 &= \epsilon^{2/3}\Omega + \epsilon \ln \epsilon \left[-\tfrac{1}{3}(\alpha_1^1 - d_1^1 - e_{\alpha'}^1 B_0^{\alpha'}) + \tfrac{1}{2} b_{\beta'}^1 \alpha_1^{\beta'}\right], \\ \Delta_1^i &= \epsilon \ln \epsilon \left[-\tfrac{1}{6}\alpha_1^j\right], \quad j = 2, \ldots, l. \end{aligned} \tag{3.10}$$

It is called the *displacement vector at the junction point* $s(x_1, y_1)$. As (3.9) shows, this vector (up to a quantity of the order $O(\epsilon)$), represents the deviation of the representing point of the system of equations (1.5) from the subspace Σ_0 (consisting of the points $(\xi, 0)$).

The system (1.5) is obtained from (1) by a linear coordinate transformation applied in a neighbourhood of the junction point $s(x_1, y_1)$. We now perform this transformation. This transformation does not mix the "rapidly-changing" and "slowly-changing" variables, in other words, it transforms the subspace $X_{y^1}^k$ into the subspace Σ_0, and the subspace $Y_{x_1}^l$ into the subspace H_0^l (consisting of the points $(0, \eta)$). Thus, the displacement vector Δ_1 is a vector which represents the deviation of the calculated solution of the system (1) from the subspace $X_{y^1}^k$ containing the "fast portion" of that trajectory of the degenerate system (3) which passes through the junction point $S(x_1, y_1)$.

The displacement vector $\Delta_1 = (\Delta_1^1, \ldots, \Delta_1^l)$ naturally does not depend on the coordinate system chosen in the neighbourhood of the junction point $s(x_1, y_1)$, and admits an invariant representation. We now proceed to find this invariant expression. For this purpose, in the neighbourhood of $s(x_1, y_1)$ let us reduce the system (1) to the form (1.5).

Expand the functions $f^i(x, y)$ and $g^j(x, y)$ in a neighbourhood of the point $s(x_1, y_1)$ as Taylor series' (only those terms which we need are written below):

$$\begin{aligned} f^i(x,y) &= A_\alpha^i(x^\alpha - x_1^\alpha) + B_\mu^i(y^\mu - y_1^\mu) + A_{\alpha\beta}^i(x^\alpha - x_1^\alpha)(x^\beta - x_1^\beta) + \\ &\quad + A_{\alpha\beta\gamma}^i(x^\alpha - x_1^\alpha)(x^\beta - x_1^\beta)(x^\gamma - x_1^\gamma) + \ldots, \\ g^j(x,y) &= g^j + C_\alpha^j(x^\alpha - x_1^\alpha) \ldots \end{aligned} \tag{3.11}$$

By the assumption made on the nature of the junction point, the matrix

$$\|A_\alpha^i\| \tag{3.12}$$

has one zero eigenvalue of multiplicity one. Let us denote the corresponding eigenvector by

$$m = (m^1, m^2, \ldots, m^k), \tag{3.13}$$

and the eigenvector with zero eigenvalue of the transposed matrix by

$$n = (n_1, n_2, \ldots, n_k). \tag{3.14}$$

Thus, we have

$$A_\alpha^i m^\alpha = 0, \quad A_i^\alpha n_\alpha = 0. \tag{3.15}$$

In addition, introduce the normalization relation:

$$m^{\alpha}n_{\alpha} = 1. \tag{3.16}$$

Now let us take the system of vectors

$$e_1, e_2, \ldots, e_k \quad (e_i = (e_i^1, \ldots, e_i^k)), \tag{3.17}$$

where $e_1 = m$ and the remaining e_i are such that $(e_i \cdot n) = 0$ (but arbitrary in other respects), as the new basis for the space X.

Furthermore, take the system of vectors

$$h_1, h_2, \ldots, h_l \quad (h_j = (h_j^1, \ldots, h_j^l)), \tag{3.18}$$

where $h_1 = g = (g^1, \ldots, g^l)$ and the remaining h_j are arbitrary, as the new basis for the space Y.

In a neighbourhood of the junction point $s(x_1, y_1)$ let us introduce new coordinates $\bar{\xi}^i, \bar{\eta}^j$ defined as follows:

$$\begin{aligned} x - x_1 &= \bar{\xi}^i e_i, \\ y - y_1 &= \bar{\eta}^j h_j. \end{aligned} \tag{3.19}$$

It is readily verified that in the new coordinates the system (1) takes the form

$$\left.\begin{aligned} \epsilon\dot{\bar{\xi}}^1 &= p(\bar{\xi}^1)^2 + q\bar{\eta}^1 + \bar{b}^1_{\beta'}\bar{\eta}^{\beta'} + \bar{d}^1_1(\bar{\xi}^1)^3 + \bar{e}^1_{\alpha'}\bar{\xi}^1 \cdot \bar{\xi}^{\alpha'} + \ldots, \\ \epsilon\dot{\bar{\xi}}^i &= \bar{a}^i_{\alpha'}\bar{\xi}^{\alpha'} + \bar{c}^i_0(\xi^1)^2 + \ldots, \qquad i = 2, \ldots, k, \\ \dot{\bar{\eta}}^1 &= 1 + \bar{\alpha}^1_1\bar{\xi}^1 + \ldots, \\ \dot{\bar{\eta}}^j &= \bar{\alpha}^l_1\bar{\xi}^1 + \ldots, \qquad j = 2, \ldots, l, \end{aligned}\right\} \tag{3.20}$$

where

$$\begin{gathered} p = n_\delta A^\delta_{\alpha\beta}m^\alpha m^\beta, \quad q = n_\delta B^\delta_\mu q^\mu, \\ \bar{b}^1_j = n_\delta B^\delta_\mu h^\mu_j, \quad \bar{d}^1_1 = n_\delta A^\delta_{\alpha\beta\gamma}m^\alpha m^\beta m^\gamma, \quad \bar{e}^1_{\alpha'} = 2n_\delta A^\delta_{\alpha\beta}m^\alpha e^\beta_{\alpha'}. \end{gathered} \tag{3.21}$$

Note that the $\bar{\alpha}^\beta_1$ are the coefficients in the expansion of the vector

$$H(H^1, \ldots, H^l) = \|C^j_\alpha\|m \tag{3.22}$$

in the basis $h_1, h_2, \ldots, h_l$. Obviously $H^j = C^j_\alpha m^\alpha$.

Assume that the junction point $s(x_1, y_1)$ also satisfies the additional nondegeneracy condition: $p \neq 0, q \neq 0$.

Now, using the substitution

$$\left.\begin{aligned} \bar{\xi}^1 &= p^{-2/3}q^{1/3}\xi^1, \quad \bar{\xi}^i = \xi^i, \quad i = 2, \ldots, k, \\ \bar{\eta}^1 &= p^{-1/3}q^{-1/3}\eta^1, \quad \bar{\eta}^j = \eta^j, \quad j = 2, \ldots, l, \\ \bar{t} &= q^{1/3}p^{1/3}t, \end{aligned}\right\} \tag{3.23}$$

we further reduce the system (3.20) to the form (1.5):

$$\left.\begin{aligned} \epsilon\dot{\xi}^1 &= (\xi^1)^2 + \eta^1 + b^1_{\beta'}\eta^{\beta'} + d^1_1(\xi^1)^3 + e^1_{\alpha'}\xi^1\xi^{\alpha'} + \dots, \\ \epsilon\dot{\xi}^i &= a^i_{\alpha'}\xi^{\alpha'} + c^i_0(\xi^1)^2 + \dots, \\ \dot{\eta}^1 &= 1 + \alpha^1_1\xi^1 + \dots, \\ \dot{\eta}^j &= \alpha^j_1\xi^1 + \dots \end{aligned}\right\} \tag{3.24}$$

were a dot over a symbol denotes differentiation with respect to $\bar{t}$, and

$$\left.\begin{aligned} b^1_j &= \bar{b}^1_j p^{1/3}q^{-2/3}, \qquad j = 2, \dots, l, \\ d^1_1 &= \bar{d}^1_1 p^{-5/3}q^{1/3}, \\ \alpha^1_1 &= \bar{\alpha}^1_1 p^{-2/3}q^{1/3}, \\ \alpha^j_1 &= \bar{\alpha}^j_1 p^{-1}, \qquad j = 2, \dots, l. \\ e^1_{\alpha'} &= \bar{e}^1_{\alpha'} p^{-1/3}q^{-1/3}, \\ a^i_{\alpha'} &= \bar{a}^i_{\alpha'} p^{-1/3}q^{-1/3}, \qquad i = 2, \dots, k, \\ c^i_0 &= \bar{c}^i_0 p^{-5/3}q^{1/3}, \qquad i = 2, \dots, k. \end{aligned}\right\} \tag{3.25}$$

Let us return to the displacement vector $\Delta_1 = (\Delta^1_1, \dots, \Delta^l_1)$. Obviously,

$$\Delta_1 = \bar{\Delta}^\beta_1 h_\beta, \tag{3.26}$$

where

$$\bar{\Delta}^1_1 = p^{-1/3}q^{-1/3}\Delta^1_1, \quad \bar{\Delta}^{\beta'}_1 = \Delta^{\beta'}_1, \quad \beta' = 2, \dots, l. \tag{3.27}$$

By virtue of (3.25) and (3.27), we obtain from (3.10) the following expressions for the coordinates $\bar{\Delta}^\beta_1$, $\beta = 1, \dots, l$, of the displacement vector Δ_1:

$$\begin{aligned} \bar{\Delta}^\beta_1 = \epsilon \ln \epsilon \left(-\frac{1}{6p}\alpha^\beta_1\right) + \delta^\beta_1 \Bigg\{ \epsilon^{2/3}\frac{\Omega}{\sqrt[3]{p\cdot q}} + \\ + \epsilon \ln \epsilon \left(-\frac{\bar{\alpha}^{\beta'}_1 \bar{b}^1_{\beta'}}{6pq} + \frac{\bar{d}^1_1}{3p^2} + \frac{\bar{e}^1_{\alpha'}B^{\alpha'}_0}{3\sqrt[3]{p^2q^2}}\right)\Bigg\}, \end{aligned} \tag{3.28}$$

using δ^β_1 for the Kronecker delta.

Substituting these expressions into (3.26), we obtain in invariant representation for the displacement vector:

$$\Delta_1 = \epsilon^{2/3}\frac{\Omega}{\sqrt[3]{pq}}g + \epsilon \ln \epsilon \left\{-\frac{H}{6p} + g\left(-\frac{r}{6pq} + \frac{s}{3p^2} + \frac{k}{3\sqrt[3]{p^2q^2}}\right)\right\}. \tag{3.29}$$

Invariant expressions for the vectors g and H, and for the numbers p, q, and $s = \bar{d}_1^1$ have already been derived in the previous pages. It remains only to find $r = \bar{\alpha}_1^\beta b_\beta^1$ and $k = \bar{e}_{\alpha'}^1 B_0^{\alpha'}$. Obviously,

$$r = n_\delta B_\mu^\delta C_\alpha^\mu m^\alpha. \tag{3.30}$$

Finally, we have to find $k = \bar{e}_{\alpha'}^1 B_0^{\alpha'}$. (Recall that the $B_0^{\alpha'}$ are determined from the equations $a_{\alpha'}^i B_0^{\alpha'} + c_0^i = 0, i = 2, \ldots, k$.) The matrix $\|A_\alpha^i\|$ can be reduced to the form

$$\left\|\begin{matrix} 0 & 0 \ldots . 0 \\ 0 & \\ \vdots & \|\bar{a}_{\alpha'}^{i'}\| \\ 0 & \end{matrix}\right\|. \tag{3.31}$$

Let us denote by A' the matrix

$$\left\|\begin{matrix} 0, & 0 \ldots . 0 \\ 0 & \\ \vdots & \|\bar{a}_{\alpha'}^{i'}\|^{-1} \\ 0 & \end{matrix}\right\|. \tag{3.32}$$

In the initial coordinate system of the space X, let the linear transformation A' be effected by the matrix

$$\|d_\gamma^\beta\|. \tag{3.33}$$

(It is easily seen that this matrix is uniquely determined by the matrix $\|A_\alpha^i\|$.) Then as is simple to calculate,

$$k = \frac{1}{p^{4/3} q^{2/3}} \cdot (-2n_\delta A_{\alpha\beta}^\delta m^\alpha d_\nu^\beta A_{\lambda\mu}^\nu m^\lambda m^\mu). \tag{3.34}$$

References

1. A. N. Tikhonov, A system of differential equations containing small parameters in the derivatives, *Matem. sborn.* **31**(73):3, 574–586 (1952).
2. A. B. Vasil'eva, On differential equations containing small parameters, *Matem. sborn.* **31**(73):3, 587–644 (1952).
3. E. F. Mishchenko and L. S. Pontryagin, Periodic solutions of systems of differential equations close to discontinuous solutions, *Doklady AN SSSR* **102**, N 5, 889–891 (1955).

Received 9 May 1957

Optimal Control Processes*

Uspekhii Matematicheskikh Nauk (1959) 14(1): 3–20

1. Formulation of the Problem

Let Ω be some topological space. We say that a *control process* is given, if there exists a system of ordinary differential equations

$$\frac{dx^i}{dt} = f^i(x^1, \ldots, x^n; u) = f^i(\bar{x}, u) \quad (i = 1, \ldots, n), \tag{1}$$

which in vector notation takes the form

$$\frac{d\bar{x}}{dt} = \bar{f}(\bar{x}; u), \tag{2}$$

where $x^1, \ldots, x^n$ are real-valued functions of time $t, \bar{x} = (x^1, \ldots, x^n)$ is a vector in an n-dimensional vector space $R, u \in \Omega$, and the functions

$$f^i(\bar{x}, u) \quad (i = 1, \ldots, n)$$

are given and continuous for all pairs $(\bar{x}, u) \in R \times \Omega$. It is also assumed that the partial derivatives

$$\frac{\partial f^i}{\partial x^j} \quad (i, j = 1, \ldots, n)$$

* In this communication I present the results obtained in joint collaboration with V. G. Boltyanskiy and R. V. Gamkrelidze [1–3].

are defined and continuous everywhere in the space $R \times \Omega$.

To find a solution of equation (2) in the interval $t_0 \leqslant t \leqslant t_1$, it is sufficient to prescribe a control function $u(t)$ in this interval and the initial value $\bar{x}_0$ of the solution at $t = t_0$. Hence, we say that a *control*

$$U = (u(t), t_0, t_1, \bar{x}_0) \tag{3}$$

of equation (2) is given, if a function $u(t)$, its range of definition $t_0 \leqslant t \leqslant t_1$, and the initial value $\bar{x}_0$ of the solution $\bar{x}(t)$ are given. Henceforth we only deal with piecewise continuous control functions $u(t)$ which admit discontinuities of the first kind, and *continuous* solutions to equation (2). Under these conditions, the control $u(t)$ is assumed to be continuous at the initial point t_0 and continuous on the left, i.e., $u(t-0) = u(t)$, for $t > t_0$. The control (3) is said to transfer the point $\bar{x}_0$ to the point $\bar{x}_1$, if the corresponding solution $\bar{x}(t)$ of equation (2) satisfying the initial condition $\bar{x}(t_0) = \bar{x}_0$ also satisfies the final condition, namely, $\bar{x}(t_1) = \bar{x}_1$).

Now let $f^0(x^1, \ldots, x^n; u) = f^0(\bar{x}, u)$ be a function defined and continuous, together with its partial derivatives

$$\partial f^0/\partial x^j \quad (j = 1, \ldots, n),$$

everywhere on the space $R \times \Omega$. To each control (3), there corresponds a number

$$L(U) = \int_{t_0}^{t_1} f^0(\bar{x}(t), u(t))\, dt.$$

Thus, L is a function of the control (3). A control U is called *optimal*, if, for any control

$$U^* = (u^*(t), t_0^*, t_1^*, \bar{x}_0)$$

which transfers the point $\bar{x}_0$ to the point $\bar{x}_1$, the inequality

$$L(U) \leqslant L(U^*)$$

is valid.

Note 1. If (3) is an optimal control of equation (2), $\bar{x}(t)$ is the solution of equation (2) corresponding to this control, and t_2, t_3 $(t_2 < t_3)$ are two points in the interval $t_0 \leqslant t \leqslant t_1$, then $U' = (u(t), t_2, t_3, \bar{x}(t_2))$ is also an optimal control.

Note 2. If (3) is an optimal control of equation (2) that transfers the point $\bar{x}_0$ to the point $\bar{x}_1$, and τ is an arbitrary number, then

$$U'' = (u(t-\tau), t_0 + \tau, t_1 + \tau, \bar{x}_0)$$

is also an optimal control which transfers the point $\bar{x}_0$ to $\bar{x}_1$.

An important particular case is the function $f^0(\bar{x};u)$ defined by the equation:

$$f^0(\bar{x},u) \equiv 1. \tag{4}$$

In this case we have

$$L(U) = t_1 - t_0$$

and optimality of the control U implies minimality of the time of transfer from the point $\bar{x}_0$ to the point $\bar{x}_1$.

Of great importance for applications is the case where Ω is a closed domain of some r-dimensional Euclidean space E. Then $u = (u^1, \ldots, u^r)$, and one control parameter u is reduced to a system of numerical parameters $u^1, \ldots, u^r$. If Ω is an open set in the space E, the variational problem formulated here is a particular case of the Lagrange problem (see [4], p. 225), and the main result derived below (the maximum principle) coincides with the Weierstrass criterion. The case where the control parameters satisfy inequalities which include equality, for example,

$$|u^i| \leqslant 1 \quad (i = 1, \ldots, r),$$

is important in practical applications. For this case the Weierstrass criterion is evidently erroneous, and the result derived below is new.

2. Necessary Conditions for Optimality (the Maximum Principle)

To formulate a necessary condition for optimality, let us introduce a vector

$$\tilde{x} = (x^0, x^1, \ldots, x^n)$$

in the $(n+1)$-dimensional Euclidean space S and consider the controlled processes:

$$\frac{dx^i}{dt} = f^i(x^1, \ldots, x^n, u) = f^i(\bar{x}, u) = f^i(\tilde{x}, u) \quad (i = 0, 1, \ldots, n) \tag{5}$$

which in vector notion takes the form

$$\frac{d\tilde{x}}{dt} = \tilde{f}(\tilde{x}, u), \tag{6}$$

where $f^0(\bar{x}, u)$ is a function which defines the functional L. If we know the control (3) of equation (2), to obtain a control of equation (6) it is sufficient to give an initial condition $\tilde{x}_0$ for equation (6), starting from the initial value

$$\bar{x}_0 = (x_0^1, \ldots, x_0^n).$$

We shall define the vector $\tilde{x}_0$ by taking

$$\tilde{x}_0 = (0, x_0^1, \ldots, x_0^n).$$

By this method the control (3) of equation (2) uniquely defines a control of equation (6), and we shall simply consider (3) to be a control of equation (6). If, now, the control (3) transfers the initial value $\tilde{x}_0$ of equation (6) into the final value

$$\tilde{x}_1 = (x_0^1, x_1^1, \ldots, x_1^n),$$

we have

$$L(U) = x_1^0.$$

This establishes a relationship between equation (6) and the variational problem already formulated.

Besides the contravariant vector $\tilde{x}$ of the space S, consider the auxiliary covariant vector

$$\tilde{\psi} = (\psi_0, \ldots, \psi_n)$$

of this space and let us construct the function

$$K(\tilde{\psi}, \tilde{x}, u) = (\tilde{\psi}, \tilde{f}(\tilde{x}, u))$$

where the right-hand side is the scalar product of the vectors $\tilde{\psi}$ and $\tilde{f}$.

For a fixed $\tilde{\psi}$ and $\tilde{x}$, K is a function of the parameter u. Let $N(\tilde{\psi}, \tilde{x})$ denote the upper bound of this function. Now, form a Hamiltonian system of equations

$$\frac{dx^i}{dt} = \frac{\partial K}{\partial \psi_i} \quad (i = 0, \ldots, n), \tag{7}$$

$$\frac{d\psi_i}{dt} = -\frac{\partial K}{\partial x^i} \quad (i = 0, \ldots, n). \tag{8}$$

It is immediately clear that the system (7) coincides with the system (5), and the system (8) is

$$\frac{d\psi_0}{dt} = 0, \quad \frac{d\psi_j}{dt} = -\sum_{i=0}^{n} \psi_i \frac{\partial f^i(\tilde{x}, u)}{\partial x^j} \quad (j = 1, \ldots, n) \tag{9}$$

Theorem 1. Let (3) be an optimal control of equation (2) and $\bar{x}(t)$ the corresponding solution of equation (2). Supplement the vector $\bar{x}(t)$ to the vector $x(t)$, putting

$$x^0(t) = \int_{t_0}^{t} f^0(\bar{x}(t), u(t))\, dt.$$

Then there exists a non-vanishing continuous vector function $\tilde{\psi}(t)$, such that

$$K(\tilde{\psi}(t_0), \tilde{x}(t_0), u(t_0)) = 0 \quad (\psi_0(t_0) \leqslant 0), \tag{10}$$

and the functions

$$\tilde{\psi}(t), \quad \tilde{x}(t), \quad u(t)$$

constitute a solution of the Hamiltonian system (7), (8), where

$$K(\tilde{\psi}(t), \tilde{x}(t), u(t)) = N(\tilde{\psi}(t), \tilde{x}(t)); \tag{11}$$

the function $K(\tilde{\psi}(t), \tilde{x}(t), u(t))$ is a constant, so that

$$K(\tilde{\psi}(t), \tilde{x}(t), u(t)) \equiv 0. \tag{12}$$

To formulate the necessary condition for a time-optimal problem (see (4)), let us form the Hamiltonian function

$$H(\bar{\psi}, \bar{x}, u) = (\bar{\psi}, \bar{f}(\bar{x}, u)).$$

For fixed $\bar{\psi}$ and $\bar{x}$, $H(\bar{\psi}, \bar{x}, u)$ is a function of the parameter u. Let $M(\bar{\psi}, \bar{x})$ denote the upper bound of this function. Let us now form the Hamiltonian system

$$\frac{dx^j}{dt} = \frac{\partial H}{\partial \psi^j} \quad (j = 1, \ldots, n), \tag{13}$$

$$\frac{d\psi_j}{dt} = -\frac{\partial H}{\partial x^j} \quad (j = 1, \ldots, n). \tag{14}$$

Evidently, the system (13) coincides with the system (1), and the system (14) is

$$\frac{d\psi_j}{dt} = -\sum_{k=1}^{n} \psi_k \frac{\partial f^k(\bar{x}, u)}{\partial x^j} \quad (j = 1, \ldots, n). \tag{15}$$

Theorem 2. Let (3) be an optimal control of equation (2) for the functional (4) and $\bar{x}(t)$ the corresponding solution of equation (2). Then there exists a non-vanishing continuous vector function $\bar{\psi}(t) = (\psi_1(t), \ldots, \psi_n(t))$, such that

$$H(\bar{\psi}(t_0), \bar{x}(t_0), u(t_0)) \geqslant 0,$$

and the functions $\bar{\psi}(t)$, $\bar{x}(t)$, $u(t)$ satisfy the Hamiltonian system of equations (13), (14), where

$$H(\bar{\psi}(t), \bar{x}(t), u(t)) = M(\bar{\psi}(t), \bar{x}(t)). \tag{16}$$

Moreover, the function $H(\bar{\psi}(t), \bar{x}(t), u(t))$ is a constant, so that

$$H(\bar{\psi}(t), \bar{x}(t), u(t)) \geqslant 0. \tag{17}$$

Theorem 2 follows directly from Theorem 1.

The main substance of Theorems 1 and 2 are the equalities (11) and (16). Therefore, Theorem 2, originally put forward as a hypothesis in [1], was called the *maximum principle*. In this sense Theorem 1 can also be called a maximum principle.

3. Proof of the Maximum Principle (Theorems 1 and 2)

We first prove Theorem 1, using certain of the techniques developed by McShane [5]. Let (3) be some control of equation (6), and $\tilde{x}(t)$ the corresponding solution of equation (6). The system of equations in variations for the system (5) near the solution $\tilde{x}(t)$ is written, as is known, in the form:

$$\frac{dy^i}{dt} = \sum_{j=0}^{n} \frac{\partial f^i(\tilde{x}(t), u(t))}{\partial x^j} y^j \quad (i = 0, 1, \ldots, n). \tag{18}$$

Rewriting the system (18) in vector notation, we obtain the vector:

$$\tilde{y}(t) = (y^0(t), \ldots, y^n(t)).$$

Henceforth we shall only deal with continuous solutions $\tilde{y}(t)$. A system of variational equations, as is known, can be interpreted as follows. Let $\tilde{y}_0$ be an arbitrary vector in the space S. Let us take*

$$\tilde{x}_0 + \epsilon \tilde{y}_0 + \epsilon O(\epsilon)$$

as the initial value for the solution of equation (6). Then the solution of equation (6) under this initial value can be expressed as

$$\tilde{x}(t) + \epsilon \tilde{y}(t) + \epsilon O(\epsilon),$$

where $\tilde{y}(t)$ is the solution of the system (18) for the initial value $\tilde{y}_0$. The solution $\tilde{y}(t)$ of the system (18) is said to be a *transfer* of the vector $\tilde{y}_0$ at the initial point $\tilde{x}_0$ of the trajectory $\tilde{x}(t)$, along the whole trajectory. In this sense the solution $\tilde{y}(t)$ may be regarded as a transfer of the vector $\tilde{y}(\tau)$ at the initial point $\tilde{x}(\tau)$ of the trajectory $\tilde{x}(t)$, along the whole trajectory.

Besides the contravarient vector $\tilde{y}(t)$, which is a solution of the system (18), let us consider a covariant vector $\tilde{\psi}(t)$ which is a solution of the system (8). It is a simple matter verify that

$$\frac{d}{dt}(\tilde{\psi}(t), \tilde{y}(t)) = 0,$$

so that

$$(\tilde{\psi}(t), \tilde{y}(t)) = \text{const}. \tag{19}$$

---

* Henceforth the symbol $O(\epsilon)$ stands for a quantity which tends to zero as $\epsilon \to 0$.

If the covariant vector $\widetilde{\psi}(t)$ is taken to be a plane through the point $\widetilde{x}(t)$, then the plane $\widetilde{\psi}(t)$ may be regarded as a *transfer* of the plane $\dot{\psi}(\tau)$ at the point $\widetilde{x}(\tau)$ of the trajectory, along the whole trajectory. A control $U^* = U^*(\epsilon, \alpha) = (u^*(t), t_0, t_1, \alpha\epsilon, \bar{x}_0)$ is called a *variation* of control (3), if it depends on a parameter ϵ and a real number α defined for all sufficiently small positive ϵ and satisfies the following condition: the solution $\widetilde{x}^*(t)$ of equation (6) corresponding to the control U^* at the point $t = t_1 + \epsilon\alpha$ can be expressed as

$$\widetilde{x}(t_1) + \epsilon\bar{\delta}(U^*) + \epsilon O(\epsilon),$$

where $\bar{\delta}(U^*)$ does not depend on ϵ.

A family Δ of variations of a control (3) is called *admissible*, if, with any two variations $U_1^*(\epsilon, \alpha_1)$ and $U_2^*(\epsilon, \alpha_2)$, it contains, for any non-negative γ_1 and γ_2, a third variation $U^*(\epsilon, \gamma_1\alpha_1 + \gamma_2\alpha_2)$ satisfying the condition:

$$\widetilde{\delta}(U^*) = \gamma_1\widetilde{\delta}(U_1^*) + \gamma_2\widetilde{\delta}(U_2^*). \tag{20}$$

Let us now construct a special variation

$$U^*(\epsilon, \alpha) = V(\epsilon, \alpha, \tau, \sigma, u^*), \tag{21}$$

that depends on a point τ in the semi-open interval $t_0 < t \leqslant t_1$ (for $\alpha < 0$, τ should be less than t_1), on a non-negative number σ, and on a point u^* in the space Ω. We define the variation $V(\epsilon, \alpha, \tau, \sigma, u^*)$ by defining the function $u^*(t)$ with the relations:

$$u^*(t) = \begin{cases} u(t) & t_0 \leqslant t \leqslant \tau - \epsilon\sigma, \\ u^* & \tau - \epsilon\sigma < t \leqslant \tau, \\ u(t) & \tau < t \leqslant t_1, \\ u(t_1) & t_1 < t \leqslant t_1 + \epsilon\alpha \quad (\text{if } \alpha > 0). \end{cases} \tag{22}$$

It is a simple matter to construct an admissible family Δ containing all variations of the type (21). This family Δ lies at the heart of the following constructions.

To each variation U^* of the admissible family Δ, there corresponds a vector $\widetilde{\delta}(U^*)$ starting at the point $\widetilde{x}_1$. The collection of all these vectors fill the convex cone Π (see (20)) with its vertex at the point $\widetilde{x}_1$. Let

$$\widetilde{v} = (-1, 0, \ldots, 0)$$

be a vector starting from the point $\widetilde{x}_1$ and directed along the negative x^0 axis in the space S. If the cone Π contains the end of the vector $\widetilde{v}$ as an internal point, the control U is not optimal. Indeed, let $U^* \in \Delta$ be that variation of the control U for which

$$\tilde{\delta}(U^*) = \tilde{v}.$$

If $\tilde{x}_1^*$ denotes that the point to which the point $\tilde{x}_0$ is transferred by the control U^*, then

$$\tilde{x}_1^* = \tilde{x}_1 + \epsilon\tilde{v} + \epsilon O(\epsilon).$$

Decomposing this equality into a scalar equality for the zero coordinate and a vector equality for the remaining coordinates, we obtain

$$L(U^*) = x_1^{0*} = x_1^0 - \epsilon + \epsilon O(\epsilon) = L(U) - \epsilon + aO(\epsilon),$$

$$\bar{x}_1^* = \bar{x}_1 + \epsilon O(\epsilon).$$

Thus, the functional has been diminished by a quantity of order ϵ, and the end of the trajectory differs from the desired value by $\epsilon O(\epsilon)$. On refining this construction, we obtain a variation $U^{\#} \in \Delta$ for which the end $\tilde{x}_1^{\#}$ of the trajectory $\tilde{x}^{\#}$ satisfies the exact equality $\tilde{x}_1^{\#} = \tilde{x}_1 - \epsilon\tilde{v}$. But this contradicts the assumption that the control U is optimal.

Therefore, whenever we asssume that the control U is optimal, we also assume that the vector $\tilde{v}$ is not an internal point of the cone II. Since this cone is convex, there exists a support plane Γ to Π such that the cone lies in one (closed) half-space determined by this plane, and the vector $\tilde{v}$ in the other half-space. Denoting by $\tilde{\psi}_1$ the covariant vector corresponding to the plane Γ taken with a proper sign, we obtain.

$$(\tilde{\psi}_1, \tilde{\delta}(U^*)) \leqslant 0 \quad (U^* \in \Delta), \tag{23}$$

$$(\tilde{\psi}_1, \tilde{v}) \geqslant 0. \tag{24}$$

It immediately follows from (24) that

$$\psi_{01} \leqslant 0. \tag{25}$$

Let $\tilde{\psi}(t)$ denote the covariant vector obtained by transferring the vector $\tilde{\psi}_1$, given at the point $\tilde{x}_1$, along the whole trajectory $\tilde{x}(t)$. We prove that $\tilde{\psi}(t)$ is precisely that vector function whose existence is asserted in Theorem 1.

Let $V(\epsilon, 0, \tau, \sigma, u^*)$ be an arbitrary special variation (see (22)) of the family Δ and $\tilde{x}^*(t)$ the solution of equation (6) corresponding to this variation. A simple calculation gives

$$\tilde{x}^*(\tau) = \tilde{x}(\tau) + \epsilon[\tilde{f}(\tilde{x}(\tau), u^*) - \tilde{f}(\tilde{x}(\tau), u(\tau))] + \epsilon O(\epsilon).$$

Let $\tilde{y}(t)$ denote the vector obtained by transferring the vector

$$\tilde{y}(\tau) = \tilde{f}(\tilde{x}(\tau), u^*) - \tilde{f}(\tilde{x}(\tau), u(\tau)),$$

at the point $\tilde{x}(\tau)$ along the trajectory $\tilde{x}(t)$. Then we have

$$\tilde{x}^*(t_1) = \tilde{x}_1 + \epsilon\tilde{y}(t_1) + \epsilon O(\epsilon).$$

Since $\tilde{y}(t_1)$ belongs to the cone Π, by (23) we have

$$(\tilde{\psi}_1, \tilde{y}(t_1)) \leqslant 0.$$

Hence, by (19), we obtain

$$(\tilde{\psi}(\tau), \tilde{f}(\tilde{x}(\tau), u^*) - \tilde{f}(\tilde{x}(\tau), u(\tau))) \leqslant 0.$$

Rewriting this inequality in terms of the function K, we obtain

$$K(\tilde{\psi}(\tau), \tilde{x}(\tau), u(\tau)) \geqslant K(\tilde{\psi}(\tau), \tilde{x}(\tau), u^*),$$

which is equivalent to (11).

Now let $U^* = V(\epsilon, \alpha, \tau, 0, u^*)$. Let $\tilde{x}^*(t)$ stand for the solution of equation (6) corresponding to this control U^*. Evidently, we have

$$\tilde{x}^*(t_1 + \alpha\epsilon) = \tilde{x}_1 + \epsilon\tilde{\delta}(U^*) + \epsilon O(\epsilon),$$

where

$$\tilde{\delta}(U^*) = \alpha\tilde{f}(\tilde{x}_1, u(t_1)).$$

Since the vector $\tilde{\delta}(U^*)$ belongs to the cone Π, virtue of (23), we obtain

$$\alpha(\tilde{\psi}_1, \tilde{f}(\tilde{x}_1, u(t_1))) \leqslant 0.$$

As α is an arbitrary real number, the last inequality is possible only when

$$(\tilde{\psi}_1, \tilde{f}(\tilde{x}_1, u(t_1))) = 0,$$

i.e., when

$$K(\tilde{\psi}(t_1), \tilde{x}(t_1), u(t_1)) = 0. \tag{26}$$

Finally, we prove that the function $K(t) = K(\tilde{\Psi}(t), \tilde{x}(t), u(t))$ of the variable t is constant. Let $t_0 \leqslant t_2 < t_3 \leqslant t_1$, where the function $u(t)$ is continuous in $t_2 < t \leqslant t_3$. We now show that the function $K(t)$ is a constant in this subinterval. Take any two points τ_0 and τ_1 in the subinterval $t_2 < t \leqslant t_3$. By virtue of (11), we have

$$K(\tilde{\psi}(\tau_0), \tilde{x}(\tau_0), u(\tau_0)) - K(\tilde{\psi}(\tau_0), \tilde{x}(\tau_0), u(\tau_1)) \geqslant 0,$$

$$-K(\tilde{\psi}(\tau_1), \tilde{x}(\tau_1), u(\tau_1)) + K(\tilde{\psi}(\tau_1), \tilde{x}(\tau_1), u(\tau_0)) \leqslant 0.$$

On adding the difference $K(\tau_1) - K(\tau_0)$ to both sides of these inequalities, we obtain the inequalities

$$\begin{aligned} -K(\tilde{\psi}(\tau_0), \tilde{x}(\tau_0), u(\tau_0)) + K(\tilde{\psi}(\tau_1), \tilde{x}(\tau_1), u(\tau_0)) &\leqslant K(\tau_1) - K(\tau_0) \\ \leqslant K(\tilde{\psi}(\tau_1), \tilde{x}(\tau_1), \tilde{u}(\tau_1)) &- K(\tilde{\psi}(\tau_0), \tilde{x}(\tau_0), u(\tau_1)). \end{aligned} \tag{27}$$

In addition to the system (7), (8), let us consider a similar system of equations. To clarify the difference between these systems, write the system (7), (8) in full as follows:

$$\frac{dx^i}{dt} = \frac{\partial}{\partial \psi_i} K(\psi(t), x(t), u(t)), \tag{28}$$

$$\frac{d\psi_i}{dt} = \frac{\partial}{\partial \psi_i} K(\psi(t), x(t), u(t)). \tag{29}$$

Along with this system, let us consider the system

$$\frac{dx^i_{\tau_0}}{dt} = \frac{\partial}{\partial \psi_i} K(\psi^{\tau_0}(t), x_{\tau_0}(t), u(\tau_0)), \tag{30}$$

$$\frac{d\psi_i^{\tau_0}}{dt} = -\frac{\partial}{\partial x^i} K(\psi^{\tau_0}(t), x_{\tau_0}(t), u(\tau_0)). \tag{31}$$

(The indices τ_0 in (30) and (31) show that the corresponding functions are a solution of the system (7), (8) for a fixed $u = u(\tau_0)$).

In addition to the vectors $\widetilde{X}$, $\widetilde{\psi}$, introduce the vectors

$$\hat{x}(t) = (x^0_{\tau_0}(t), x^1_{\tau_0}(t), \ldots, x^n_{\tau_0}(t)),$$

$$\hat{\psi}(t) = (\psi_0^{\tau_0}(t), \psi_1^{\tau_0}(t), \ldots, \psi_n^{\tau_0}(t)).$$

We now consider the function $K(\hat{\psi}(t), \hat{x}(t), u(\tau_0))$. From (30) and (31) it follows that

$$\frac{d}{dt} K(\hat{\psi}(t), \hat{x}(t), u(\tau_0)) = 0.$$

It can be easily be shown that the functions $K(\psi(t), \widetilde{x}(t), u(\tau_0))$ and $K(\hat{\psi}(t), \hat{x}(t), u(\tau_0))$ differ very little from each other; namely, we have the inequality

$$|K(\widetilde{\psi}(t), \widetilde{x}(t), u(\tau_0)) - K(\hat{\psi}(t), \hat{x}(t), u(\tau_0))' \leqslant |t - \tau_0|\gamma, \tag{32}$$

where

$$\lim_{|t-\tau_0| \to 0} \gamma = 0.$$

Since the function $u(t)$ is continuous on the subinterval under consideration, the solutions of the systems (28), (29) and (30), (31) hardly differ from each other, namely we have the inequalities

$$|\widetilde{x}(t) - \hat{x}(t)| \leqslant |t - \tau_0|\gamma_1, \tag{33}$$

$$|\widetilde{\psi}(t) - \hat{\psi}(t)| \leqslant |t - \tau_0|\gamma_2, \tag{34}$$

where

$$\lim_{|t-\tau_0| \to 0} \gamma_1 = \lim_{|t-\tau_0| \to 0} \gamma_2 = 0.$$

From (33) and (34) we get (32). Now, using (32), we can make the inequality (27) stronger as follows:

$$-(\tau_1-\tau_0)\gamma \leqslant K(\tau_1)-K(\tau_0) \leqslant (\tau_1-\tau_0)\gamma.$$

Dividing this inequality by $\tau_1-\tau_0$, and taking the limit as $(\tau_1-\tau_0)\to 0$, we finally obtain

$$\frac{d}{dt}K(t) = 0.$$

Thus, on the subinterval $t_2 < t \leqslant t_3$ the function $K(t)$ is constant.

We now prove that $N(\tilde{\psi}, \tilde{x})$ is continuous in $\tilde{\psi}, \tilde{x}$. If this is not true, there will exist pairs $(\tilde{\psi}_1, \tilde{x}_1)$ and $(\tilde{\psi}_0, \tilde{x}_0)$, and a constant c, for which

$$N(\tilde{\psi}_1, \tilde{x}_1)-N(\tilde{\psi}_0, \tilde{x}_0) > c > 0$$

for arbitrarily small distances between $(\tilde{\psi}_1, \tilde{x}_1)$ and $(\tilde{\psi}_0, \tilde{x}_0)$. Let u_1 and u_0 be two values of the control u, such that

$$N(\tilde{\psi}_1, \tilde{x}_1) = K(\tilde{\psi}_1, \tilde{x}_1, u_1)$$
$$N(\tilde{\psi}_0, \tilde{x}_0) = K(\tilde{\psi}_0, \tilde{x}_0, u_0).$$

Since u_0 gives the maximum, we have

$$K(\tilde{\psi}_1, \tilde{x}_1, u_1)-K(\tilde{\psi}_0, \tilde{x}_0, u_0) > c$$

which contradicts the continuity of K in the first two arguments. Hence we find that $K(\tilde{\psi}(t), \tilde{x}(t), u(t))$ is also continuous at the point of discontinuity of $u(t)$.

Hence (see (26)), it follows that (12) is valid over the whole interval $t_0 \leqslant t \leqslant t_1$. In particular, this shows the truth of the first relation in (10). The second relation follows from (25) as a consequence of the first equation in (9).

This completes the proof of Theorem 1.

Note. Theorem 1 holds true even when bounded measurable functions are used as the admissible controls $u(t)$. In this case the equality (11) is satisfied almost everywhere.

4. Time-optimal Linear Control

An example which is important in application and illustrates general results well, is the following linear control system

$$\frac{dx^i}{dt} = \sum_{j=1}^{n} a^i_j x^j + \sum_{k=1}^{r} b^i_k u^k \quad (i = 1, \ldots, n),$$

where $\bar{u} = (u^1, \ldots, u^r)$ is a point in a convex closed bounded polyhedron Ω in a linear space E with coordinates $u^1, \ldots, u^r$. In vector notation this system can be expressed as

$$\frac{d\bar{x}}{dt} = A\bar{x} + B\bar{u}, \tag{35}$$

where A is a linear operator in the space R of the variables $x^1, \ldots, x^n$, and B is a linear operator from the space E into the space R. Here we consider only the problem of minimizing the functional $\int_{t_0}^{t_1} dt$, i.e., the transfer time minimization problem.

In order to obtain some uniqueness results, we shall impose the following conditions (A) and (B) on the controlled equation (35), whose purpose will be clear later.

(A) Let $\bar{w}$ be a certain vector directed along some edge of the polyhedron Ω; then the vector $B\bar{w}$ does not belong to any proper subspace of the invariant space R of the operator A. Thus, the vectors

$$B\bar{w}, AB\bar{w}, \ldots, A^{n-1}B\bar{w} \tag{36}$$

are linearly independent in the space R, if $\bar{w}$ is a vector directed along one of the edges of the polyhedron Ω.

(B) The origin of coordinates in the space E is an internal point of the polyhedron Ω.

The function $H(\bar{\psi}, \bar{x}, \bar{u})$ in our case takes the form

$$H = (\bar{\psi}, A\bar{x}) + (\bar{\psi}, B\bar{u}), \tag{37}$$

and the system (15) can be written as follows:

$$\frac{d\psi_j}{dt} = -\sum_{i=1}^{n} a_j^i \psi_i \quad (j = 1, \ldots, n),$$

or in vector notion as

$$\frac{d\bar{\psi}}{dt} = -A^* \bar{\psi}. \tag{38}$$

Obviously, the function H, being a function of $\bar{u} \in \Omega$, attains its maximum simultaneously with the function

$$(\bar{\psi}, B\bar{u}).$$

Hence, let $P(\bar{\psi})$ denote the maximum of the function $(\bar{\psi}, B\bar{u})$, regarded as a function of $\bar{u} \in \Omega$. From Theorem 2, we find that if

$$U = (\bar{u}(t), t_0, t_1, \bar{x}_0)$$

is an optimal control of equation (35), then there exists a solution $\bar{\psi}(t)$ of equation (38), such that

$$(\bar{\psi}(t), B\bar{u}(t)) = P(\bar{\psi}(t)). \tag{39}$$

Since equation (38) does not contain the unknown functions $\bar{x}(t)$ and $\bar{u}(t)$, we can easily find all its solutions. Thus, by the condition (39), we can easily find all optimal controls $\bar{u}(t)$ of equation (35), also the question of the extent to which the condition (39) uniquely determines the control $\bar{u}(t)$ in terms of $\bar{\psi}(t)$, is solved with the help of the following theorem.

Theorem 3. If condition (A) is satisfied, then for a given nontrivial solution $\bar{\psi}(t)$ of equation (38), the relation (39) uniquely determines the control $\bar{u}(t)$, where $\bar{u}(t)$ is a piecewise constant function and its only values are vertices of the polyhedron Ω.

Proof. Since the function

$$(\bar{\psi}(t), B\bar{u}), \tag{40}$$

being a function of the vector $\bar{u}$, is linear, it is either a constant or attains its maximum on a face of the polyhedron Ω. The same consideration is applicable to each face of the polyhedron Ω. Thus, either the function (40) attains its maximum only at one vertex of the polyhedron Ω, or everywhere on a face of the polyhedron Ω. We prove, by virtue of the condition (A), that a maximum can occur everywhere on a face only for a finite number of values of t. Suppose that the function (40) attains its maximum (and, consequently, is a constant) on some face Γ of the polyhedron Ω. Let $\bar{w}$ be a vector directed along some edge of the face Γ. As the function (40) is constant on the face Γ, we have

$$(\bar{\psi}(t), B\bar{w}) = 0.$$

If this relation held true for an infinite number of values of t, it would be an identity in t. Then differentiating step by step with respect to t, we would obtain

$$\left.\begin{aligned} (\bar{\psi}(t), B\bar{w}) &= 0, \\ (A^*\bar{\psi}(t), B\bar{w}) = (\bar{\psi}(t), AB\bar{w}) &= 0, \\ (A^{*2}\bar{\psi}(t), B\bar{w}) = (\bar{\psi}(t), A^2B\bar{w}) &= 0, \\ \cdots\cdots\cdots\cdots\cdots\cdots\cdots\cdots\cdots\cdots\cdots\cdots \\ (A^{*n-1}\bar{\psi}(t), B\bar{w}) = (\bar{\psi}(t), A^{n-1}B\bar{w}) &= 0. \end{aligned}\right\} \tag{41}$$

But, by condition (A), the vectors (36) for a basis of the space R. Therefore, from (41) it would follow that $\bar{\psi}(t)=0$. This contradicts the assumption that the solution $\bar{\psi}(t)$ is non-trivial.

5. Uniqueness Theorems for Linear Controls

Using the method of variations of constants, we shall solve equation (35) as a non-homogeneous equation. For this, let

$$\bar{\phi}_1(t), \ldots, \bar{\phi}_n(t) \tag{42}$$

denote a fundamental system of solutions of the homogeneous equation

$$\frac{d\bar{x}}{dt} = A\bar{x},$$

satisfying the initial conditions $\phi_j^i(t_0)=\delta_j^i$, and let

$$\bar{\psi}^1(t), \ldots, \bar{\psi}^n(t)$$

denote a fundamental system of solutions of the homogeneous system (38) with the initial condition $\psi_i^j(t_0)=\delta_i^j$. We shall search for the general solution of equation (35) in the form:

$$\bar{x}(t) = \sum_{i=1}^{n} \bar{\phi}_i(t)c^i(t).$$

Substituting this solution into (35), we obtain

$$\sum_{i=1}^{n} \bar{\phi}_i(t)\frac{dc^i(t)}{dt} = B\bar{u}(t).$$

Taking the scalar product of this relation with $\bar{\psi}^j$, and using $(\bar{\psi}^j(t), \bar{\phi}_i(t))=\delta_i^j$, we obtain

$$\frac{dc^i(t)}{dt} = (\bar{\psi}^i(t), B\bar{u}(t)). \tag{43}$$

Thus, the solution of equation (35) for an arbitrary control $U=(\bar{u}(t), t_0, t_1, \bar{x}_0)$ is written in the form:

$$\bar{x}(t) = \sum_{i=1}^{n} \bar{\phi}_i(t)\left(x_0^i + \int_{t_0}^{t} (\bar{\psi}^i(t), B\bar{u}(t))\,dt\right). \tag{44}$$

Theorem 4. Assume that the equation (35) satisfies the condition (A) and let

$$U_1 = (\bar{u}_1(t), t_0, t_1, \bar{x}_0), \quad U_2 = (\bar{u}_2(t), t_0, t_2, \bar{x}_0)$$

be two optimal controls of the equation (35) which transfer the point $\bar{x}_0$ to the point $\bar{x}_1$; then

$$t_1 = t_2, \quad \bar{u}_1(t) \equiv \bar{u}_2(t).$$

Proof. Since both controls U_1 and U_2 are optimal, we have $t_1 = t_2$, because if, say, t_1 were less than t_2, then the control U_2 would not be optimal. We thus have

$$\begin{aligned} \bar{x}_1 &= \sum_{i=1}^{n} \bar{\phi}_i(t_1)\left(x_0^i + \int_{t_0}^{t_1} (\bar{\psi}^i(t), B\bar{u}_1(t))\,dt\right) \\ &= \sum_{i=1}^{n} \bar{\phi}_i(t)\left(x_0^i + \int_{t_0}^{t_1} (\bar{\psi}^i(t), B\bar{u}_2(t))\,dt\right). \end{aligned}$$

Since the vectors $\bar{\phi}^1(t_1), \ldots, \bar{\phi}^n(t_1)$ are linearly independent, from the last equality it follows that

$$\int_{t_0}^{t_1} (\bar{\psi}^i(t), B\bar{u}_1(t))\,dt = \int_{t_0}^{t_1} (\bar{\psi}^i(t), B\bar{u}_2(t))\,dt \quad (i = 1, \ldots, n). \tag{45}$$

By virtue of Theorem 3, the vector function $\bar{\psi}(t)$ which is a solution of (38) corresponds to the optimal control U_1.

Let

$$\bar{\psi}_0 = (\psi_{10}, \ldots, \psi_{n0})$$

be the initial value of this function at $t = t_0$; then the solution $\bar{\psi}(t)$ can be expressed as

$$\bar{\psi}(t) = \sum_{i=1}^{n} \psi_{i0}\bar{\psi}^i(t). \tag{46}$$

Multiplying (45) by ψ_{i0}, and then summing over i, obtain

$$\int_{t_0}^{t} (\bar{\psi}(t), B\bar{u}_1(t))\,dt = \int_{t_0}^{t} (\bar{\psi}(t), B\bar{u}_2(t))\,dt. \tag{47}$$

By Theorem 3, the function $\bar{u}_1(t)$ satisfies the condition

$$(\bar{\psi}(t), B\bar{u}_1(t)) = P(\bar{\psi}(t))$$

and is uniquely determined by this condition. Were the function $\bar{u}_2(t)$ not identical with the function $\bar{u}_1(t)$, it would not satisfy the condition

$$(\bar{\psi}(t), B\bar{u}_2(t)) \equiv P(\bar{\psi}(t)),$$

and therefore the function $(\bar{\psi}(t), Bu_2(t))$, being nowhere greater than the function $(\bar{\psi}(t), B\bar{u}_1(t))$, would be less than this function in some interval. Thus, if the identity $\bar{u}_1(t) \equiv \bar{u}_2(t)$ is not true in the interval $t_0 \leqslant t \leqslant t_1$, the equality (47) is impossible.

This completes the proof of Theorem 4.

A control

$$U = (\bar{u}(t), t_0, t_1, \bar{x}_0)$$

is said to be *extremal*, if it satisfies the condition (39), where $\bar{\psi}(t)$ is some nontrivial solution of equation (38).

To find all the optimal controls which transfer the point $\bar{x}_0$ to the point $\bar{x}_1$, we can first find all extremal controls that transfer the point $\bar{x}_0$ to the point $\bar{x}_1$, and then choose from among these extremal controls the only control that effects the transfer in the least time. Then the question naturally arises: can there exist several extremal controls that transfer the point $\bar{x}_0$ to the point $\bar{x}_1$? In general, there may exist several extremal controls. The following theorem singles out an important case of uniqueness.

Theorem 5. Assume that the equation (35) satisfies the conditions (A) and (B), and let

$$U_1 = (\bar{u}_1(t), t_0, t_1, \bar{x}_0), \quad U_2 = (\bar{u}_2(t), t_0, t_2, \bar{x}_0)$$

be two extremal controls that transfer the point $\bar{x}_0$ to the origin $x_1 = 0$ of coordiantes in the space R: then the controls U_1 and U_2 coincide:

$$t_1 = t_2, \quad \bar{u}_1(t) \equiv \bar{u}_2(t).$$

Proof. By assumption, we have

$$\left.\begin{aligned} \sum_{i=1}^{n} \bar{\phi}_i(t_1)\left(x_0^i + \int_{t_0}^{t_1} (\bar{\psi}^i(t), B\bar{u}_1(t))\,dt\right) &= 0, \\ \sum_{i=1}^{n} \bar{\phi}_i(t_2)\left(x_0^i + \int_{t_0}^{t_2} (\bar{\psi}^i(t), B\bar{u}_2(t))\,dt\right) &= 0. \end{aligned}\right\} \tag{48}$$

Since the vectors (42) are linearly independent for any t, we obtain from (48) that

$$-x_0^i = \int_{t_0}^{t_1} (\bar{\psi}^i(t), B\bar{u}_1(t))\,dt = \int_{t_0}^{t_2} (\bar{\psi}^i(t), B\bar{u}_2(t))\,dt. \tag{49}$$

For the sake of definiteness, assume that $t_1 > t_2$ and let $\bar{\psi}(t)$ be that solution of equation (46) for which the identity

$$(\bar{\psi}(t), B\bar{u}_1(t)) = P(\bar{\psi}(t)),$$

defining the function $\bar{u}_1(t)$ is satisfied. Just as in the proof of Theorem 4, let us express the function $\bar{\psi}(t)$ in the form (46). Multiply (49) by ψ_{i0} and then sum over i. We obtain

$$\int_{t_0}^{t_1} (\bar{\psi}(t), B\bar{u}_1(t))\,dt = \int_{t_0}^{t_2} (\bar{\psi}(t), B\bar{u}_2(t))\,dt.$$

From condition (b), it follows that

$$P(\bar{\psi}(t)) \geqslant 0. \tag{50}$$

Indeed, since zero is an internal point of the convex body Ω, the function $(\bar{\psi}(t), B\bar{u})$, as a function of $\bar{u}$, is either identically equal to zero, or may take negative as well as positive values.

By virtue of (50), we have the inequality

$$\int_{t_0}^{t_2} (\bar{\psi}(t), B\bar{u}_1(t))\, dt \leqslant \int_{t_0}^{t_2} (\bar{\psi}(t), B\bar{u}_2(t))\, dt.$$

Hence, just as in the proof of Theorem 4, we obtain

$$\bar{u}_1(t) \equiv \bar{u}_2(t) \quad \text{for } t_0 \leqslant t \leqslant t_2.$$

Furthermore, as $P(\bar{\psi}(t))$ is zero only for certain distinct values of t, we have $t_1 = t_2$.

This completes the proof of Theorem 5.

6. Existence of Optimal Controls for Linear Systems

Theorem 6. If there exists at least one control of the equation (35) that transfers the point $\bar{x}_0$ to the point $\bar{x}_1$, then there also exists an optimal control for the equation (35) that transfers the point $\bar{x}_0$ to the point $\bar{x}_1$.

Proof. Let $\Delta_{\bar{x}_0, \bar{x}_1}$ denote the set of all controls of the type

$$U = (\bar{u}(t), 0, t, \bar{x}_0), \tag{51}$$

that transfer the point $\bar{x}_0$ to the point $\bar{x}_1$. To each control (51) there corresponds a transfer time t. Let t^* denote the lower bound of all such transfer times for $U \in \Delta_{\bar{x}_0, \bar{x}_1}$. We now prove that there exists a control $U^* = (u^*(t), 0, t^*, \bar{x}_0)$ that transfers the point $\bar{x}_0$ to the point $\bar{x}_1$.

From the set $\Delta_{\bar{x}_0, \bar{x}_1}$ choose an infinite sequence of controls

$$U_k = (\bar{u}_k(t), 0, t_k, \bar{x}_0) \quad (k = 1, 2, \ldots),$$

for which

$$\lim_{k \to \infty} t_k = t^*.$$

Evidently, we have

$$\lim_{k \to \infty} \sum_{i=1}^{n} \bar{\phi}_i(t^*) \left(x_0^i + \int_0^{t^*} (\bar{\psi}^i(t), B\bar{u}_k(t))\, dt \right) = \bar{x}_1. \tag{52}$$

Consider the Hilbert space L_2 of all measurable square integrable functions on the interval $0 \leqslant t \leqslant t^*$. The control $\bar{u}_k(t)$ is a vector function. Let $u_k^i(t)$ stand for the i-th coordinate of this function. The function $u_k^i(t)$ belongs to the space L_2 on the interval $0 \leqslant t \leqslant t^*$. The set of all functions $u_k^i(t)$, $k = 1, 2, \ldots$ evidently belongs to some ball in L_2. Therefore we can choose from this set a weakly convergent subsequence. We assume that the sequence

$$u_1^i(t), u_2^i(t), \ldots, u_k^i(t), \ldots \tag{53}$$

itself converges weakly to some function $u^i(t)$, $i = 1, \ldots, r$. We now prove that the vector function

$$\bar{u}^*(t) = (u^1(t), \ldots, u^r(t))$$

satisfies the condition

$$\bar{u}^*(t) \in \Omega$$

for almost all t.

Let

$$b(\bar{u}) = \sum_{i=1}^{r} b_i u^i = b$$

be a hyperplane carrying one of the $(r-1)$-dimensional faces of the polyhedron Ω, where the polyhedron Ω is located in the halfspace

$$b(\bar{u}) \leqslant b.$$

Let m be the set of all t in the interval $[0, t^*]$ for which $b(\bar{u}^*(t)) > b$, and $v(t)$ the characteristic function of the set m. Then we have

$$\lim_{k \to 0} \int_0^{t^*} v(t)[b(\bar{u}^*(t)) - b(\bar{u}_k(t))]\, dt = 0,$$

by virtue of the weak convergence of (53), and since $b(\bar{u}^*(t)) - b(u_k(t)) > 0$ on the set m, we have mes $(m) = 0$.

Thus, by changing the vector function $\bar{u}^*(t)$ on a set of measure zero, we obtain a new function (which we also denote by the symbol $\bar{u}^*(t)$) satisfying the condition $\bar{u}^*(t) \in \Omega, 0 \leqslant t \leqslant t^*$.

By the weak convergence of (52), it follows from (53) that

$$\sum_{i=1}^{n} \bar{\phi}_i(t^*) \left(x_0^i + \int_0^{t^*} \bar{\psi}(^i(t), B\bar{u}^*(t))\, dt \right) = \bar{x}_1.$$

Thus, $U^* = (\bar{u}^*(t), 0, t^*, \bar{x}_0)$ is a measurable optimal control which transfers the point $\bar{x}_0$ to the point $\bar{x}_1$.

By the Note to Theorem 1, if we change the control $\bar{u}^*(t)$ on the set of measure zero, we can transform this control into a control satisfying the maximum principle, i.e., in our case, satisfying the condition:

$$(\bar{\psi}(t), B\bar{u}^*(t)) = P(\bar{\psi}(t)).$$

This condition obviously implies that the function $\bar{u}^*(t)$ is piecewise continuous.

Thus, we have demonstrated Theorem 6.

Theorem 7. If the equation (35) satisfies conditions (A) and (B) and the operator A is stable, i.e., all its eigenvalues have negative real parts, then for each point $\bar{x}_0 \in R$ there exists an optimal control which transfers this point to the origin $O \in R$ of coordinates.

Proof. We first show that there exists a neighbourhood V of the point O in R, each point $\bar{x}_0$ of which can be transferred by some control to the point O. In Ω choose a vector $\bar{v}$ such that $-\bar{v}$ belongs to Ω and such that the vector

$$\bar{b} = B\bar{v}$$

does not belong to any proper subspace of the invariant space R of the operator A. By conditions (A) and (B), such a vector $\bar{v}$ does exist. For sufficiently small positive ϵ, the operators A and $e^{-\epsilon A}$ have coincident invariant subspaces, and therefore the vectors

$$e^{-\epsilon A}\bar{b}, e^{-2\epsilon A}\bar{b}, \ldots, e^{-n\epsilon A}\bar{b}$$

are linearly independent.

Let $\chi(t)$ be an arbitrary real-valued function defined on some interval $0 \leqslant t \leqslant t_1$ and not exceeding one in absolute value; then

$$U = (\bar{v}\chi(t), 0, t_1, \bar{x}_0)$$

is a control of equation (35) and this control transfers the point $\bar{x}_0$ to the following point (see (44))

$$\bar{x}_1 = e^{t_1 A}\left(\bar{x}_0 + \int_0^{t_1} e^{-tA}\bar{b}\chi(t)\,dt\right). \tag{54}$$

Now choose the function $\chi(t)$ depending on the parameters $\xi^1, \ldots, \xi^n$ in such a manner that the point (54), which we denote by $\bar{x}_1(\bar{x}_0; \xi^1, \ldots, \xi^n)$, satisfies the following conditions:

$$\bar{x}_1(0; 0, \ldots, 0) = 0,$$

and the functional determinant

$$\left.\frac{\partial(x_1^1, \ldots, x_1^n)}{\partial(\xi^1, \ldots, \xi^n)}\right|_{x_0=0,\ \xi^1=0, \ldots,\ \xi^n=0}$$

is non-zero. Having constructed such a function $\chi(t)$, we now prove that the equation $\bar{x}_1(x_0, \xi^1, \ldots, \xi^n) = 0$ is solvable in $\xi^1, \ldots, \xi^n$ for all $\bar{x}_0$ belonging to some neighbourhood V of the origin O.

We first define a function $\sigma(t, \tau, \xi)$ of the variable t, $0 \leqslant t \leqslant t_1$, where $0 < \tau \leqslant t_1$ and ξ is a parameter. As a function of t, $\sigma(t, \tau, \xi)$ is everywhere zero outside the interval $[\tau, \tau + \xi]$ and is equal to sign ξ on this interval. Let

$$\chi(t) = \sum_{k=1}^{n} \sigma(t, k\epsilon, \xi^k).$$

Simple calculations show that the point $\bar{x}_1(\bar{x}_0, \xi^1, \ldots, \xi^n)$, for this choice of the function $\chi(t)$, satisfies the conditions formulated above.

Now let $\bar{x}_0$ be an arbitrary point in the space R. First, let it move under the action of the control $\bar{u}(t) = 0$. Since all the eigenvalues of the operator A have negative real parts, the point is transferred into the neighbourhood V after the lapse of some time, and thereafter, it can be transferred, by what has been proved above, to the origin of coordinates. Hence, by virtue of Theorem 6, there exists an optimal control which transfers the point $\bar{x}_0$ to the origin.

This completes the proof of Theorem 7.

7. Synthesis of a Linear Optimal Control

The problem of synthesizing an optimal control can be formulated for an arbitrary controlled system (1). But here we consider this problem only for a controlled linear system (35) satisfying the conditions (A) and (B) with a stable operator A. Existence and uniqueness theorems (Theorems 7 and 5) are valid for such a system, which guarantee that the problem of synthesis is solvable in principle. The considerations described here give a constructive method of solving the problem. But, in each particular case this method calls for several constructions. An optimal control for a linear system (35) was synthesized, using an entirely different method, only for the case of one controlling parameter (i.e., for $r = 1$), by Feldbaum [6] for real roots of the operator A and by Bushaw [7] when $n = 2$ and the eigenvalues of the operator A are complex.

We assume that the equation (35) satisfies the conditions (A) and (B) and has a stable operator A. Then for each point $\bar{x}_0 \in R$, there exists (only one) optimal control

$$U_{\bar{x}_0} = (\bar{u}_{\bar{x}_0}(t), t_0, t_1, \bar{x}_0), \tag{55}$$

which transfers the point $\bar{x}_0$ to the origin $O \in R$. Uniqueness is valid, of course, up to time shift (see Note 2 in 31). Thus, $\bar{u}_{\bar{x}_0}(t_0)$ depends only on the point $\bar{x}_0$ and does not depend on the arbitrarily chosen zero of the time axis t_0. Therefore we can take

$$\bar{v}(\bar{x}_0) = \bar{u}_{\bar{x}_0}(t_0).$$

Let $\bar{x}(t)$ be a solution of equation (55) corresponding to the control (55); then we have $U_{\bar{x}(\tau)} = (\bar{u}_{\bar{x}_0}(t), \tau, t_1, \bar{x}(\tau))$ (See Note 1 in Section 2). Therefore

$$\bar{u}_{\bar{x}_0}(\tau) = \bar{v}(\bar{x}(\tau)).$$

Thus

$$\frac{d}{dt}\bar{x}(t) = A\bar{x}(t) + B\bar{v}(\bar{x}(t)),$$

and we see that the solution of the equation

$$\frac{d\bar{x}}{dt} = A\bar{x} + B\bar{v}(\bar{x}) \tag{56}$$

with an arbitrary initial condition $\bar{x}(t_0) = \bar{x}_0$ gives the rule of optimal transfer of the point $\bar{x}_0$ to the origin. In this sense the function $\bar{v}(\bar{x})$ synthesizes an optimal control which transfers an arbitrary point $\bar{x}_0$ to the origin.

We now give a method of constructing the function $\bar{v}(\bar{x})$. Let $\bar{\psi}(t)$ be a solution of equation (38) which, by virtue of Theorem 2, corresponds to the control (55), so that

$$\frac{d\bar{\psi}(t)}{dt} = -A^*\bar{\psi}(t), \tag{57}$$

and the function $\bar{u}_{\bar{x}_0}(t)$ is determined from the equation

$$(\bar{\psi}(t), B\bar{u}_{\bar{x}_0}(t)) = P(\bar{\psi}(t)). \tag{58}$$

Furthermore, assume that $\bar{x}(t)$ is a solution of equation (35) satisfying the initial condition

$$\bar{x}(t_0) = \bar{x}_0 \tag{59}$$

and the final condition

$$\bar{x}(t_0) = 0, \tag{60}$$

so that

$$\frac{d\bar{x}(t)}{dt} = A\bar{x}(t) + B\bar{u}_{\bar{x}_0}(t). \tag{61}$$

Then the function $\bar{v}(\bar{x})$ satisfies the condition

$$(\bar{\psi}(t_0), B\bar{v}(\bar{x}(t_0))) = P(\bar{\psi}(t_0)). \tag{62}$$

From the existence and uniqueness theorems it follows that there exists only one (up to time shift) pair of functions $\bar{u}_{\bar{x}_0}(t), \bar{x}(t)$ on the interval

$t_0 \leqslant t \leqslant t_1$ satisfying the conditions (57)–(61). Since time can be shifted, the numbers t_0 and t_1 are not uniquely defined by these conditions, but the number $t_1 - t_0$ is defined.

It is not quite clear how to search for functions $\bar{u}_{\bar{x}_0}(t)$ and $\bar{x}(t)$ satisfying all the conditions (57)–(61), but we can easily find all the functions that satisfy only the conditions (57), (58), (60), and (61). For this purpose we proceed as follows: Since time can be shifted arbitrarily, let us fix t_1, taking $t_1 = 0$. Now let $\bar{\chi}$ be an arbitrary covariant vector different from zero and $\bar{\psi}(t, \bar{x})$ a solution of equation (57) satisfying the initial condition

$$\bar{\psi}(0, \bar{\chi}) = \bar{\chi}$$

and defined for $t \leqslant 0$. Further, let us determine the function $\bar{u}(t, \bar{x})$ from the condition

$$(\bar{\psi}(t, \bar{\chi}), B\bar{u}(t, \bar{\chi})) = P(\bar{\psi}(t, \bar{\chi})) \quad (t \leqslant 0),$$

and the function $\bar{x}(t, \bar{\chi})$ from the equation

$$\frac{d\bar{x}(t, \bar{\chi})}{dt} = A\bar{x}(t, \bar{\chi}) + B\bar{u}(t, \bar{\chi}).$$

From the foregoing, we find that the function $\bar{v}(\bar{x})$ is given by the relation

$$(\bar{\psi}(t, \bar{\chi}), B\bar{v}(\bar{x}(t, \bar{\chi}))) = P(\bar{\psi}(t, \bar{\chi})). \tag{63}$$

From the existence theorem (Theorem 7), it follows that the point $\bar{x}(t, \bar{\chi})$ describes the whole space R as t runs through negative values and the vector $\bar{\chi}$ varies arbitrarily. Thus, the relation (63) gives the value of the function $\bar{v}(\bar{x})$ for any point $\bar{x}$ in R.

References

1. V. G. Boltyanskii, R. V. Gamkrelidze, and L. S. Pontryagin, On the theory of optimal processes, *Doklady AN SSSR* **110**, N 1, 7–10 (1956).
2. R. V. Gamkrelidze, On the theory of optimal processes in linear systems, *Doklady AN SSSR* **116**, N 1, 9–11 (1957).
3. V. G. Boltyanskii, The maximum principle in the theory of optimal processes, *Doklady AN SSSR* **119**, N 6 (1958).
4. G. A. Bliss, *Lectures on the Calculus of Variations*, Univ. of Chicago Press, Chicago, 1946.
5. E. J. McShane, On multipliers for Lagrange problems, *Amer. J. Math.* **61**, 809–819 (1939).
6. A. A. Feldbaum, *Avtomatika i Telemekhanika* **16**, N 2, 129 (1955).
7. D. W. Bushaw, *Experimental Towing Tank*, Stevens Institute of Technology, Report 469, Hoboken, New York, 1953.

Linear Differential Games*

Actes du Congrès International des Mathématicians, Paris
(1970) 1: 163–71

Here we deal with linear differential games whose basic model is the pursuit of one controlled object by another controlled object. Of course linear differential games constitute a particular case of general differential games. Even for this particular case the results are nevertheless not trivial. In addition, these results are more effective than the corresponding generalizations to the nonlinear case.

We formulate the problem for a general nonlinear case, but only give the results for a linear problem.

1. Formulation of the Problem

The theory of differential games emerged as a result of the mathematical modelling of technical problems. The models can be effected in several ways. The model should be chosen so that it aims at reflecting the most essential features of the technical problem, but yields to mathematical treatment. Thus, the theory should not be formulated in complete isolation from technical problems.

To illustrate with a concrete example, let us imagine one aircraft pursuing another one. The aim of the former is to catch up to the latter, the aim of the latter is to evade the pursuit. Each pilot controls his plane,

* Invited one-hour plenary address, International Congress Nice, 163–171, 1970.

keeping his end in view and using the information available about the situation. This information consists of two parts: the first is a complete knowledge of the technical potentialities of the aircraft and the second, information about the behaviour of his own plane and about the behaviour of the enemy aircraft. Information about the aircraft behaviour may include various data on the states of the planes for the period preceding a given instant in time. But we cannot however assume that we know something about the future behaviour of the planes, because they are controlled and the pilot may change the position of controls at any instant and thus alter the behaviour of his plane. In reality, each pilot can only receive information about the enemy with some delay. There is no need, however, to include this fact in the mathematical model. On the contrary, we can even suppose that we know the enemy's behaviour in some advance and thus construct a mathematical model on this assumption and then show that the theory thus obtained can be used in solving the real problem approximately.

We now pass to the mathematical description of the pursuit process. Two controlled objects, the pursuer and the evader, take part in this process. The state of each object at any instant is determined by its phase vector. Let x and y denote the phase vectors of the pursuer and of the evader, respectively. Write their equations as usual:

$$\dot{x} = f(x, u), \quad \dot{y} = g(y, v), \tag{1}$$

where dot stands for differentiation with respect to time, and u, v are controls. Since x and y are phase vectors, each decomposes into two parts:

$$x = (x_1, x_2), \quad y = (y_1, y_2),$$

where x_1 and y_1 determine the geometric positions of the objects, x_2 and y_2, their velocities. Pursuit is assumed to terminate as soon as

$$x_1 = y_1 \tag{2}$$

i.e., when the objects coincide geometrically.

The first part of the information mentioned earlier consists of the equations (1). They do not describe the motion of the objects, but only describe their possibilities, because various motions are obtained for various controls $u = u(t)$ and $v = v(t)$. Thus, in our example, equations (1) describe the technical potentialities of the planes.

We can view the pursuit process from two different standpoints.

(1) We are the pursuer. In this case our aim is to terminate the pursuit and the control u is at our disposal to achieve this end. Thus, we have to construct the control $u(t)$ at every instant t, knowing the equations (1),

i.e., the first part of the information, and using the second part in the form of knowledge about the functions $x(s)$, $y(s)$, and $v(s)$ in the interval $t-\theta \leqslant s \leqslant t$, where θ is a properly chosen positive number.

(2) We are the evader. In this case our aim is to evade the termination of pursuit and the control v is at our disposal to achieve this end. Thus, we have to construct the control $v(t)$ at every instant, knowing the equations (1), i.e., the first part of the information, and using the second part in the form of knowledge about the functions $x(s)$, $y(s)$, and $u(s)$ in the interval $t-\theta \leqslant s \leqslant t$.

This is the mathematical pursuit model we consider here and it inevitably splits the problem into two distinct ones, namely, the pursuit problem and the evasion problem. This happens because we use different information in two different approaches. There is also another model, due to Isaacs, in which the same information, namely the knowledge about the functions $x(t)$ and $y(t)$, is utilized both in the pursuit as well as in the evasion problem. In this model it is assumed that there exists an optimal pursuit control $u = u(x, y)$, a function of the states x and y of the objects, and an optimal evasion control $v = v(x, y)$, also a function of the same states. The problem in this model is mathematically very definite: it lies in finding the functions $u(x, y)$ and $v(x, y)$ called the optimal strategies. This definiteness, however, renders the solution tremendously difficult. In particular, in assuming the existence of optimal strategies, we essentially narrow the class of problems under consideration.

2. Differential games

A differential game arises from the pursuit process as a result of the natural tendency to simplify the notation. Namely, in place of two phase vectors x and y, we introduce one vector $z = (x, y)$, and form the phase space R of the game as the direct sum of these two phase spaces. Then the two equations in (1) are written as one equation:

$$\dot{z} = F(z, u, v) \tag{3}$$

and the relation (2) defines some subset M in the vector space R. We are now in a position to define a differential game, irrespective of the initial pursuit process. A differential game is defined, if its vector phase space R and the equation (3) are given, where $z \in R$, F is some function of three variables, u and v being the pursuit and evasion controls, respectively. In addition, a subset M of the space R is given on which the game terminates.

As in the pursuit process, we associate two distinct problems with a differential game.

(1) Our aim is to terminate the game, i.e., to bring the point z into the set M. For this purpose we have a pursuit control u, so that we choose a value $u(t)$ of this control at every instant, using the functions $z(s)$ and $v(s)$ in the interval $t-\theta \leqslant s \leqslant t$. These are the rules of our pursuit game.

(2) Our aim is to prevent the game from being terminated, i.e., to prevent the point z from hitting the set M. For this purpose, we have an evasion control v, so that we choose a value $v(t)$ of this control at every instant using the functions $z(s)$ and $u(s)$ in the interval $t-\theta \leqslant s \leqslant t$. These are the rules of our evasion game.

3. Linear Differential Games

We assume that the phase space R of a linear game is an n-dimensional Euclidean space. The equation of this game is

$$\dot{z} = Cz - u + v, \tag{4}$$

where $z \in R$, C is some linear mapping of the space R into itself, and the controls u and v are vectors in R. These vectors are not arbitrary: they satisfy the following conditions

$$u \in P, \quad v \in Q, \tag{5}$$

where P and Q are compact convex subsets of R (their dimensions are arbitrary). The controls $u = u(t)$ and $v = v(t)$, as functions of time, are measurable. We assume that the set M on which the game terminates is a vector subspace of R. The more general case where M is an arbitrary closed convex subset of R has also been investigated. When this more general case is being considered, we shall make a special mention of it.

(A) Let $\mathscr{L}$ denote the orthogonal complement of M in the space R, and ν its dimension ($\dim \mathscr{L} = \nu$). Let π denote the orthogonal projection from R onto $\mathscr{L}$. Since C is a linear mapping of R into itself, $e^{\tau C}$, where τ is a real number, is a linear mapping of R onto itself, and $\pi e^{\tau C}$ is a linear mapping of R on $\mathscr{L}$. Both these mappings depend analytically on the real parameter τ. Put

$$P_\tau = \pi e^{\tau C} P; \quad Q_\tau = \pi e^{\tau C} Q. \tag{6}$$

The sets P_τ and Q_τ are compact convex subsets of $\mathscr{L}$ continuously dependent on the real parameter τ.

4. Operations on Compact Convex Sets in $\mathscr{L}$

(B) Let A and B be two compact convex sets in $\mathscr{L}$, and α, β real numbers. Let us denote by

$$\alpha A + \beta B \tag{7}$$

the set of all vectors of the type $\alpha x + \beta y$, where $x \in A, y \in B$. The set (7) is evidently compact and convex. If one of the sets A or B is empty, then the set (7) is also empty. It is easily verified that for non-negative α and β, the distributive law holds:

$$(\alpha + \beta)A = \alpha A + \beta A. \tag{8}$$

The set of all compact convex non-empty sets in $\mathscr{L}$ naturally forms a complete metric space Ω. Thus, if $X_\tau = X(\tau)$ is a compact convex set in $\mathscr{L}$ dependent on a real parameter τ, in other words, if $X(\tau)$ is a function of a real variable τ which takes its values in Ω, then we can define the notion of measurability of this function and its Lebesgue integral:

$$\int_{t_1}^{t_s} X(\tau)\,d\tau \quad (t_1 \leqslant t_2), \tag{9}$$

which is also an element of the space Ω. For $t_1 = t_2$ we take the set (9) to consist of the zero in $\mathscr{L}$.

(C) Let A and B be two compact convex sets in $\mathscr{L}$. If there exists a vector $x \in \mathscr{L}$, such that

$$x + B \subset A, \tag{10}$$

then we write

$$B \overset{*}{\subset} A. \tag{11}$$

The set of all vectors x satisfying condition (10) is denoted by

$$A \overset{*}{-} B \tag{12}$$

and is called the geometric difference of the sets A and B. Evidently, the set (12) is compact and convex. It is non-empty if and only if the condition (11) is satisfied.

5. The Pursuit Game

For the game (4) let us form the geometric difference (see (A) and (C)):

$$P_\tau \overset{*}{-} Q_\tau, \tag{13}$$

It turns out that this difference is a measurable function of τ, so that the integral

$$\int_0^t (P_\tau \overset{*}{-} Q_\tau)\, d\tau, \quad 0 \leqslant t. \tag{14}$$

can be defined. For $t=0$, this integral, by assumption, consists of the vector 0. Let I denote the set of all values of t for which (14) is non-empty. Then I consists of either 0 only, or is an interval $0 \leqslant t \leqslant t_0$, or is the half-line $0 \leqslant t$. Let W_t denote the set of all points $z \in R$ for which

$$\pi e^{\tau C} z \in \int_0^t (P_\tau \overset{*}{-} Q_\tau)\, d\tau, \tag{15}$$

and let $T(z)$ denote the minimal t for which the inclusion (15) holds. Evidently,

$$W_0 = M \tag{16}$$

and W_t is non-empty for all $t \in I$.

The following theorem on pursuit holds true [1].

Theorem 1. If $T(z)$ is defined for an initial value z_0 of the game (4), then a pursuit game with the initial value z_0 may be terminated in a time not greater than $T(z_0)$.

This theorem is not quite precise. In fact, in a time period t not greater than $T(z_0)$ the point z_0 can be brought to a position $z(t)$ at a distance not greater than $c\epsilon$ from the set M, where $c>0$ depends on z_0 and $\epsilon>0$ is an arbitrarily small number, depending on how we conduct our pursuit game.

In order to give a hint of the proof and to show where the theorem is not exact, let us study the main property of the function W_t of t.

(D) Let $z_0 \in W$, and $0<\epsilon \leqslant \tau$. Then for any evasion control $v(t)$ defined on the interval $0 \leqslant t \leqslant \epsilon$, it is possible to find a pursuit control $u(t)$ on this interval such that the same (4) involving the controls $u(t)$ and $v(t)$ translates the point z_0 to the point $z_1 = z(\epsilon) \in W_{\tau-\epsilon}$ in a time ϵ. This property of the function W_t is called the *P-property* (pursuit property).

The P-property of the function $W(t)$ permits us to terminate a pursuit game in a time not greater than $T(z_0)$, the required information being the knowledge of the control $v=v(s)$ in advance, namely on the interval $t \leqslant s \leqslant t+\epsilon$, where $\epsilon>0$ is arbitrarily small. Delayed utilization of the information, say, the values of the function $v(s)$ on the interval $t-2\epsilon \leqslant s \leqslant t-\epsilon$, leads to inaccuracies in hitting the set M.

Assertion (D) can be considerably improved [2].

(E) Consider a game (4) whose final set M is an arbitrary closed convex set in R. Then there exists a closed convex set M_t, which can be described structurally, depending on t, $t \geqslant 0$, satisfying the condition $M_0 = M$ and having the P-property (see (D)). The function M_t is maximal with this property.

Assertion (E) makes it possible to prove a similar but much stronger assertion than Theorem 1. If, for a given $z \in R$, there exists a $\tau \geqslant 0$ such that $z \in M$, let $T(z)$ denote the minimal τ for which this inclusion holds. Then if $T(z_0)$ is defined for a given initial value z_0, a pursuit game with this initial value can be terminated in a time not greater than $T(z_0)$.

This result, however, does not give the complete solution to the pursuit problem. Namely, if, for a given z_0, $T(z_0)$ is not defined, a pursuit game with this initial value z_0 may possibly be terminated in a time not greater than a certain number. Further, if $T(z_0)$ is defined, it may not necessarily give the best possible estimate for the pursuit termination time.

The maximal function M_t for a linear game was constructed by me, but its maximality was in fact noticed by N. Krasovskii and A. Subbotin. They also constructed a maximal function M_t possessing the P-property for a nonlinear game of the type (3).

6. The Evasion Game

Let $\hat{\mathscr{L}}$ be some two-dimensional vector subspace of the space $\mathscr{L}$ (see (A)) of the game (4). In analogy with (A), let $\hat{\pi}$ denote the orthogonal projection from R onto $\hat{\mathscr{L}}$, and take

$$\hat{P}_\tau = \hat{\pi} e^{\tau C} P, \quad \hat{Q}_\tau = \hat{\pi} e^{\tau C} Q. \tag{17}$$

Then the following theorem on evasion holds.

Theorem 2.* For the game (4) assume that there exists a two-dimensional vector subspace $\hat{\mathscr{L}}$ of the space $\mathscr{L}$ (see (A)) such that the following two conditions are satisfied:

(a) there exists a real number $\mu > 1$ such that for all sufficiently small positive τ (see (17)):

$$\mu \hat{P}_\tau \overset{*}{\subset} \hat{Q}_\tau; \tag{18}$$

* A complete proof of Theorem 2 is given in [4].

(b) there is no fixed line $\overset{*}{\mathscr{L}}$ in the plane $\hat{\mathscr{L}}$ such that for all sufficiently small positive τ

$$\hat{Q}_\tau \overset{*}{\subset} \overset{*}{\mathscr{L}}. \tag{19}$$

Then, for any initial value z_0 not belonging to M, the evasion game can be so conducted that the point $z(t)$ never hits the space M $(0 \leqslant t < \infty)$, and in addition the distance between the point $z(t)$ and M is given by the estimate (21) (see below).

(F) To write down the estimate (21), let us associate two non-negative numbers ξ, η with each $z \in R$:

$$z \to (\xi, \eta), \tag{20}$$

where ξ and η are the distances of the point z from the sets M and $\mathscr{L}$, respectively. If z_0 is the initial value, the correspondence (20) is $z_0 \to (\xi_0, \eta_0)$, and if $z(t)$ is the current point, then $z(t) \to (\xi(t), \eta(t))$. There exist positive numbers c and ϵ, and a natural number k dependent on the game but not on its progress, such that

$$\xi(t) > \frac{c\xi_0^k}{(1+\eta(t))^k} \quad \text{for } \xi_0 \leqslant \epsilon. \tag{21}$$

Theorem 2 follows from the following assertion.

(G) For the game (4) it is possible to find two positive numbers, θ a time interval and ϵ a distance. Further, to each initial value z_0 for which $\xi_0 \leqslant \epsilon$, and to the control $u(t)$, $0 \leqslant t \leqslant \theta$, a control $v(t)$ is determined (where $v(t)$ is determined by the point z_0 and the function $u(s)$, known in the interval $0 \leqslant s \leqslant t$) in such a way that for the solution $z(t)$ of equation (4) with the controls $u = u(t)$, $v = v(t)$ and the initial value z_0, the following two inequalities are satisfied:

$$\xi(\theta) > \epsilon, \tag{22}$$

$$\xi(t) > \frac{c\xi_0^k}{(1+\eta(t))^k}; \quad 0 \leqslant t \leqslant \theta. \tag{23}$$

The control $v(t)$ is called a *special evasion control.*

The evasion game can be described as follows. Let S denote the set of all points z in R for which $\xi \leqslant \epsilon$, and S' the set of points z for which $\xi = \epsilon$. If the initial state of the game z_0 belongs to the cylinder S, we immediately use the special evasion control (see (G)) on the time interval $0 \leqslant t \leqslant \theta$, at the end of which $z(\theta)$ lies outside the cylinder S (see (22)), and the inequality (23) is satisfied in the interval $0 \leqslant t \leqslant \theta$. If, either at the initial instant $t = 0$ or at some other instant t, $z(t)$ lies outside the cylinder, then we arbitrarily choose an evasion control $v(t)$ and wait until t_0 when the point $z(t_0)$ will be on the surface S'. Now taking $z(t_0)$ as the initial point for the time interval $t_0 \leqslant t \leqslant t_0 + \theta$, we use the special evasion control (see (G)) in this time interval. Then, by virtue of (23), in this time interval the inequality

$$\xi(t) > \frac{c\epsilon^k}{(1+\eta(t))^k}, \quad t_0 \leqslant t \leqslant t_0 + \theta, \tag{24}$$

holds and at its end point, the initial state lies outside the cylinder. Thereafter the entire process is repeated.

Thus, in the course of the game, either (23) or (24) or the inequality $\xi(t) \geqslant \epsilon$ is always satisfied for the point $z(t)$. Taking $\epsilon > c\epsilon^k$ in these inequalities we obtain the estimate (21).

If a control $u(t)$ with delay is used in constructing the special evasion control $v(t)$, namely, if in calculating $v(t)$, we use the function $u(s)$ on the interval $-\delta \leqslant s \leqslant t - \delta$, where

$$0 < \delta < \frac{c_1 \xi_0^l}{(1+\eta_0)^l}, \tag{25}$$

$c_1 > 0$, and l is a natural number, then the estimates (22) and (23) still hold. Thus, delayed information can be used in an evasion game.

Initially, Theorem 2 on evasion was proved in my paper [3] written in collaboration with E. F. Mishchenko, using instead of (a) and (b) the much more stringent conditions (c) and (d).

(c) There exists a number $\mu > 1$ such that for all sufficiently small positive τ (see (A)):

$$\mu P_\tau \overset{*}{\subset} Q_\tau \tag{26}$$

(cf. 18)).

The mapping $\pi e^{\tau C}$, being linear and analytically dependent on τ, can be expanded into a series

$$\pi e^{\tau C} = g_0 + \tau g_1 + \ldots + \tau^m g_m + \ldots \tag{27}$$

(d) There exists an integer $m \geqslant 0$ for which each mapping $g_0, g_1, \ldots, g_{m-1}$ transforms the set Q into some point, and

$$\dim g_m Q = \nu, \quad \nu \geqslant 2 \text{ (see (A))}.$$

Later I improved this result by substituting a weaker condition for the condition (d).

(e) For all sufficiently small positive values of τ,

$$\dim Q_\tau = \nu, \quad \nu \geqslant 2.$$

R. Gamkrelidze, after reading my paper, expressed with confidence that the space $\mathscr{L}$ in the conditions (c) and (d) could be replaced by any two-dimensional subspace $\hat{\mathscr{L}}$ of $\mathscr{L}$. And, in his opinion, Theorem 2 should hold when the conditions (a) and (f) are satisfied.

(f) $\dim \hat{Q}_\tau = 2$ for all small positive τ.

After rechecking the proof. I found that it indeed holds under the conditions (a) and (f). In addition, I found that the condition (f) could be replaced by the condition (b). After this substitution the result acquires the form in which it has been stated here.

Example. In a Euclidean space E of dimension $\nu \geqslant 2$, consider the motion of a "pursuer" x and an "evader" y. Pursuit terminates as soon as $x = y$. Their equations of motions are

$$x^{(k)} + a_1 x^{(k-1)} + \ldots + a_{k-1}\dot{x} + a_k x = u, \tag{28}$$

$$y^{(l)} + b_1 y^{(l-1)} + \ldots + b_{l-1}\dot{y} + b_l y = v \tag{29}$$

Here $x^{(i)}$ and $y^{(i)}$ are the i-th derivatives of the vectors x and y with respect to time t, a_i, for $i = 1, \ldots, k$; and b_j, for $j = 1, \ldots, l$ are linear mappings of the space E into itself, u and v are control vectors belonging to the space E and satisfying the conditions $u \in P$, $v \in Q$, where P and Q are ν-dimensional compact convex subsets of E.

The point y is said to have superior maneuverability to the point x, if one of the following two conditions is satisfied:

(1) $l < k$,

(2) For $l = k$, there exists a $\mu < 1$ such that $\mu P \overset{*}{\subset} Q$.

It turns out that if the evader y has superior maneuverability to the pursuer x, then the pursuit process satisfies the conditions (c) and (d); consequently, if the points x_0 and y_0 do not coincide at the initial instant, the evasion process continues ad infinitum. If, however, the pursuer x has superior maneuverability, then applying Theorem 1, we can find an open set of the initial states in the phase space of the game starting from which the game could always be terminated.

Computations for this example were carried out by A. Mezentsev.

References

1. L. S. Pontryagin, Linear differential games, Part I, *Doklady AN SSSR* **174**, N 6, 1278–1281 (1967).

2. L. S. Pontryagin, Linear differential games, Part II, *Doklady AN SSSR* **175**, N 4, 764–767 (1967).
3. L. S. Pontryagin and E. F. Mishchenko, The problem of evasion of one controlled object from another, *Doklady AN SSSR* **189**, N 4, 721–723 (1969).

Added in proof sheets:

4. L. S. Pontryagin, Linear differential games of evasion, *Trudy Matem. instituta im. V. A. Steklova AN SSSR*, vol. CXII, 30–63 (1971).

Linear Differential Evasion Games

Trudy Matematicheskogo Instituta imeni V. A. Steklova
(1971) 112: 30–63

This paper deals with linear differential games, the basic model of which is the pursuit of one controlled object by another. Even though the linear differential games constitute a particular case of general differential games, the results for this particular case are nevertheless not trivial. Moreover, these results are more concrete than their generalizations to the nonlinear case.

We shall formulate here the problem for the nonlinear case but only give the results for a linear problem.

The theory of differential games emerged in the course of the mathematical modelling of technical problems. Models can be effected in several ways. The model should be chosen so that it aims at reflecting the most essential features of the technical problem, but yields to mathematical treatment. Thus, the theory should not be constructed in complete isolation from technical problems.

To illustrate with a concrete example, let us imagine one aircraft pursuing another one. The aim of the former is to catch up to the latter, the aim of the latter is to evade the pursuit. Each pilot controls his plane, keeping his end in view and using the information available about the situation. The information consists of two parts. The first is the complete knowledge about the technical capabilities of the aircraft; the second part is the information about the behaviour of the aircraft. The information about aircraft behaviour may include various data on the states of the planes for the period preceding a given instant. But we cannot, however, assume that we know the future behaviour of the planes because they are controlled and the pilot may at any instant change the position of the controls and thereby alter the behaviour of the plane. In reality, each pilot can receive information about the enemy only with some delay. There is nevertheless no need to include this circumstance in the mathematical model. On the contrary, we can even suppose that we know the enemy's behaviour in advance and construct a mathematical model on this assumption and then show that the theory thus constructed can be used in solving the real problem approximately.

We now take up the mathematical description of the pursuit process. Two controlled objects, the pursuer and the evader, take part in this process. The state of each object at any instant is determined by its phase vector. Let x and y denote the phase vector of the pursuer and of the evader, respectively. Write their equations as usual:

$$\dot{x} = f(x, u), \quad \dot{y} = g(y, v), \tag{1}$$

where a dot stands for differentiation with respect to time, and u, v are controls. As x and y are phase vectors, each of them decomposes into two parts:

$$x = (x_1, x_2), \quad y = (y_1, y_2),$$

where x_1 and y_1 determine the geometric positions of the objects, x_2 and y_2 their velocities. Pursuit is assumed to terminate as soon as

$$x_1 = y_1, \tag{2}$$

i.e., when the objects coincide geometrically.

The first part of the information already mentioned consists of the equations (1). They do not define the motion of the objects; they only describe their possibilities because various motions are obtained under various controls $u = u(t)$ and $v = v(t)$. Thus, in our example, equations (1) describe the technical capabilities of the planes.

We can view the pursuit process itself from two different standpoints.

(1) We are the pursuer. Our objective in this case is to terminate the pursuit and the control u is at our disposal to achieve this aim. Thus, we have to construct the value $u(t)$ of the control u at every instant t, knowing the equations (1), i.e., the first part of the information, and using the second part in the form of knowledge about the functions $x(s), y(s)$, and $v(s)$ in the interval $t-\theta \leqslant s \leqslant t$, where θ is an appropriately chosen positive number.

(2) We are the evader. Our objective in this case is to evade the termination of the pursuit and the control v is at our disposal to achieve this aim. Thus, we have to construct the value $v(t)$ of the control v at every instant t, knowing the equations (1), i.e., the first part of the infor-

mation, and using the second part in the form of knowledge about the functions $x(s), y(s)$, and $u(s)$ in the interval $t-\theta \leqslant s \leqslant t$.

This is the mathematical model of pursuit that we consider here and it inevitably splits the problem into two distinct ones, namely, the pursuit problem and the evasion problem. This happens because we use different information in two different approaches.

There are also other models, of which one is due to Issacs. In his model the same information, namely the knowledge of the functions $x(t)$ and $y(t)$, is utilized in the pursuit as well as in the evasion problem, under the assumption that there exists an optimal pursuit control $u=u(x, y)$, a function of the states x and y of the objects, and an optimal evasion control $v=v(x, y)$, also a function of the same states. In such a model the problem becomes mathematically very definite: it lies in finding the functions $u(x, y)$ and $v(x, y)$ called the optimal strategies. This definiteness, however, renders the solution very difficult and essentially narrows down the class of problems under consideration.

The differential game arises in considering the pursuit process as a result of the natural tendency to simplify the notation. Namely, in place of two phase vectors x and y, we use one vector $z=(x, y)$, forming the phase space of the game as the direct sum of the phase spaces of both objects. Then the two equations in (1) are written as one equation:

$$\dot{z} = F(z, u, v) \tag{3}$$

and the relation (2) defines some subset M in the vector space R. We are now in a position to define a differential game independently of the initial pursuit process. A differential game is given, if its vector phase space R and equation (3) are given, where $z \in R$, F is some function of three variables, u and v being the pursuit and evasion controls, respectively. In addition, in R a set M is to be given on which the game terminates.

As in the case of pursuit, we associate two distinct problems with a differential game.

(1) Our aim is to terminate the game, i.e., to bring the point z into the set M. For this purpose we have a pursuit control u, so we choose a value $u(t)$ of this control at every instant t, using the functions $z(s)$ and $v(s)$ in the interval $t-\theta \leqslant s \leqslant t$. These are the rules of our pursuit game.

(2) Our aim is to prevent the game from being terminated, i.e., to prevent the point z from hitting the set M. For this purpose we have an evasion control v so we choose a value $v(t)$ of this control at every instant t, using the functions $z(s)$ and $u(s)$ in the interval $t-\theta \leqslant s \leqslant t$. These are the rules of our evasion game.

In this paper we shall study a linear differential evasion game.

1. Linear Differential Games

We assume that the phase space R of our linear game is an n-dimensional Euclidean space. The equation of the game is

$$\dot{z} = Cz - u + v, \tag{1}$$

where $z \in R$, C is some linear mapping of the space R into itself, and the controls u and v are vectors of the space R. These vectors are not arbitrary: they satisfy the conditions

$$u \in P, \quad v \in Q, \tag{2}$$

where P and Q are compact convex subsets of R (their dimensions are arbitrary). The controls $u = u(t)$ and $v = v(t)$, regarded as functions of time t, are measurable. The set M on which the game terminates is assumed to be a vector subspace of R.

(A) Denote by L the orthogonal complement of M in R, and let ν be its dimension ($\dim L = \nu$). For the time being, let W be an arbitrary vector subspace of L. Denote by π the operation of orthogonal projection from R onto W. Since C is a linear mapping of R into itself, $e^{\tau C}$, where τ is a real number, is a linear mapping of R onto itself, and $\pi e^{\tau C}$ is a linear mapping of R onto W. Both these mappings depend analytically on the real parameter τ. Let

$$P_\tau = \pi e^{\tau C} P, \quad Q_\tau = \pi e^{\tau C} Q, \tag{3}$$

where P_τ and Q_τ are compact convex subsets of W continuously dependent on the real parameter τ. To each point $z \in R$ associate two non-negative numbers

$$z \rightarrow (\xi, \eta), \tag{4}$$

where ξ and η are the distances of the point z from M and L, respectively. If z_0 is the initial value, then (4) gives $z_0 \rightarrow (\xi_0, \eta_0)$, and if $z(t)$ is a current point, then (4) gives $z(t) \rightarrow (\xi(t), \eta(t))$.

(B) Let A be a convex set in W and a a point in W, μ a real number. Denote by

$$a + A \tag{5}$$

the set of all points in W which can be represented as $a + x$, where $x \in A$. We say that the set (5) is obtained from the set A by translation. Let

$$\mu A \tag{6}$$

denote the set of all points in W which can be represented in the form μx, where $x \in A$. Evidently, the sets (5) and (6) are convex. If A and B are two convex sets in W, and there exists a point $a \in W$, such that

$$a + A \subset B, \tag{7}$$

then we write

$$A \overset{*}{\subset} B. \tag{8}$$

All that we have said above can obviously be said about any vector space W.

(C) Evasion conditions are said to be satisfied for the game (1), if there exists a two-dimensional vector subspace W of the space L for which

(a) there is no fixed one-dimensional vector subspace W^1 in W such that

$$Q_\tau \overset{*}{\subset} W^1 \tag{9}$$

for all small positive τ, and

(b) there exists a constant $\mu > 1$, such that

$$\mu P_\tau \overset{*}{\subset} Q \tag{10}$$

for all sufficiently small positive τ (see (3), (A) and (B)).

The following theorem then holds.

Evasion Theorem. If the evasion conditions (see (C)) are satisfied for the game (1), then for any initial value z_0 not belonging to M, the evasion game can be so conducted that the point $z(t)$ never hits the space $M(0 \leqslant t < \infty)$, and the distance of the point $z(t)$ from M is given by the following estimate.

(D) If the evasion conditions (C) are satisfied for the game (1), then there exist three positive constants θ, ϵ and c, and a natural number k depending only on the game and not depending on the initial value or the progress of the game, such that the evasion game can be conducted in a manner so that

(1) for $\xi_0 \geqslant \epsilon$,

$$\xi(t) > \frac{c\epsilon^k}{(1+\eta(t))^k}, \quad 0 \leqslant t < \infty; \tag{11}$$

(2) for $\xi_0 \leqslant \epsilon$,

$$\xi(t) > \frac{c\xi_0^k}{(1+\eta(t))^k}, \quad 0 \leqslant t \leqslant 0, \tag{12}$$

$$\xi(t) > \frac{c\epsilon^k}{(1+\eta(t))^k}, \quad \theta \leqslant t < \infty. \tag{13}$$

Our aim in this paper is to prove the evasion theorem. It is proved in the subsequent sections. The assertion (C) in Section 6 coincides with the assertion (D).

I have proved an initial version of the evasion theorem in collaboration with E. F. Mishchenko in [2], where we assumed that conditions (E), more stringent than the evasion conditions (C), are satisfied, and the resulting estimate was worse than the estimate (D).

(E) Assume that $W = L, \nu \geqslant 2$ and take

$$P_m = \pi C^m P, \quad Q_m = \pi C^m Q; \tag{14}$$

then the stronger evasion conditions have the form:

(c) There exists a natural number number k so large that each of the sets P_m and Q_m consists of one point for $m < k$, and

$$\dim Q_k = \dim W; \tag{15}$$

(d) there exists a constant $\mu > 1$ such that

$$\mu P_k \overset{*}{\subset} Q_k. \tag{16}$$

Later I further strengthened this result [3] by substituting weaker conditions for the evasion conditions (E).

(F) On assuming $W = L$ and $\nu \geqslant 2$, we also assume that

(e) for all sufficiently small positive τ,

$$\dim Q_\tau = \dim W; \quad \text{and} \tag{17}$$

(f) there exists a positive constant $\mu > 1$, such that for all sufficiently small positive τ:

$$\mu P_\tau \overset{*}{\subset} Q_\tau. \tag{18}$$

After having read my paper, R. V. Gamkrelidze confidently asserted that any two-dimensional subspace W of the space L can be substituted for the vector space $W = L$ in the evasion conditions (F). After rechecking my proof, I was convinced that it indeed holds under this assumption, and in addition I replaced the condition (e) by the weaker condition (a). Thus, the result finally took the form of the evasion theorem.

To illustrate the evasion theorem, let us give an example due to A. Mezentsev [4], without presenting the details of the calculations.

Example. In a Euclidean space E of a dimension $\nu \geqslant 2$, consider the motion of two points x and y, where x is the "pursuer" and y is the "evader". Pursuit terminates as soon as $x = y$. The motions of the points x and y are given by the equations:

$$x^{(p)} + a_1 x^{(p-1)} + \ldots + a_{p-1}\dot{x} + a_p x = u, \tag{19}$$

$$y^{(q)} + b_1 y^{(q-1)} + \ldots + b_{q-1}\dot{y} + b_q y = v, \tag{20}$$

where $x^{(i)}$ and $y^{(i)}$ are the i-the derivatives with respect to time t, $a_i (i = 1, \ldots, p)$ and $b_j (j = 1, \ldots, q)$ are linear mappings of E into itself, and u and v are control vectors belonging to E and satisfying conditions:

$$u \in P, \quad v \in Q, \tag{21}$$

P and Q being compact convex subsets of E, and $\dim Q = \nu$.

The point y is said to have superior maneuverability to the point x, if

(1) $q < p$, or

(2) if $q = p$, there exists a number $\mu > 1$, such that

$$\mu P \overset{*}{\subset} Q. \tag{22}$$

In case the evader y has superior maneuverability to the pursuer x, the pursuit process then satisfies the conditions (E), and, consequently, the conditions (C); so if the points x_0 and y_0 do not coincide at the initial instant, evasion will continue ad infinitum and the estimate (D) holds for $|x(t) - y(t)|$, where $k = q + 1$.

2. Matrices Analytically Dependent on a Parameter

In Section 1 we have considered a linear mapping $\pi e^{\tau C}$ of a vector space R onto a subspace W. In the sequel we apply this mapping to certain subspaces of R. The mapping $\pi e^{\tau C}$ depends analytically on the parameter τ; so the elements in the corresponding matrix also depend analytically on τ. In this section we shall establish certain properties of these linear mappings and their corresponding matrices that are needed in the sequel.

(A) Let $f(\tau)$ be a function of the variable τ, analytic for small nonzero values of τ, and possibly having a pole at $\tau = 0$. So in the neighbourhood of the zero it can be expressed as

$$f(\tau) = a_k \tau^k + a_{k+1}\tau^{k+1} + \ldots \tag{1}$$

or

$$f(\tau) \equiv 0, \tag{2}$$

where $a_k \neq 0$ and k is an integer. The number k is called the order of $f(\tau)$

and is denoted by $\{f\}$. The order of $f(\tau) \equiv 0$ is taken to be infinity. Let O_τ denote the set of all functions of τ analytic in the neighbourhood of $\tau = 0$, so the order of each $f(\tau) \in O_\tau$ is non-negative. The set O_τ is naturally an algebraic ring, namely, the sum $f(\tau) + g(\tau)$, the difference $f(\tau) - g(\tau)$ and the product $f(\tau) \cdot g(\tau)$ are naturally defined for any two elements $f(\tau)$ and $g(\tau)$ in this set. The quotient $f(\tau)/g(\tau)$ however belongs to O_τ if and only if $\{f\} \geqslant \{g\}$. An element $f(\tau) \in O_\tau$ has an inverse if and only if its order is zero. Thus, a square matrix $A(\tau)$ consisting of elements belonging to O_τ has an inverse $A^{-1}(\tau)$ consisting of elements belonging to O_τ, if and only if the determinant of $A(0)$ is nonzero. Now let g_τ be a linear mapping of some vector space V into some other vector space W, where this mapping is dependent on a parameter τ. By choosing certain coordinate systems in V and W, we naturally associate some matrix with the mapping g_τ. If the elements of this matrix belong to the ring O_τ, the mapping g_τ is said to depend analytically on τ for small values of τ. This property obviously does not depend on the choice of the coordinate systems in V and W.

The ring O_τ is a principal ideal ring, and the following assertion (B) is therefore implied by a well known theorem of lincar algebra. Nonetheless, I give its proof here.

(B) Let g_τ be a matrix with q columns and r rows whose elements belong to O_τ (see (A)). Then, for sufficiently small τ, the matrix g_τ can be expressed as

$$g_\tau = B(\tau)G(\tau)A(\tau), \tag{3}$$

where $A(\tau)$ and $B(\tau)$ are square matrices of orders q and r, respectively, whose elements belong to O_τ, and whose determinants are nonzero for $\tau = 0$. The matrix $G(\tau)$ is canonical, namely it is a diagonal matrix: all non-diagonal elements are zeros and the following functions of τ sit on the diagonal:

$$\tau^{k_1}, \tau^{k_2}, \ldots, \tau^{k_s}, 0, \ldots, 0, \tag{4}$$

where $k_1, k_2, \ldots, k_s$ ($k_1 \leqslant k_2 \leqslant \ldots \leqslant k_s$) are non-negative integers, and s is the rank of the matrix g_τ for small nonzero values of τ.

To prove (B), let us rewrite (3) as follows:

$$G(\tau) = B^{-1}(\tau)g_\tau A^{-1}(\tau). \tag{5}$$

This is always possible because the matrices $A(\tau)$ and $B(\tau)$ have inverses. Putting $A^{-1}(\tau) = C(\tau)$ and $B^{-1}(\tau) = D(\tau)$, we can rewrite (5) as

$$G(\tau) = D(\tau)g_\tau C(\tau), \tag{6}$$

where $C(\tau)$ and $D(\tau)$ are square matrices whose determinants are nonzero for $\tau = 0$. If $C(\tau)$ and $D(\tau)$ satisfy this condition, then the relation (3) immediately follows from (6). Thus, to demonstrate (B), it suffices to find matrices $C(\tau)$ and $D(\tau)$ such that the matrix $G(\tau)$ derived from $\overset{\circ}{g}_\tau$ by (6) with the help of these matrices is of the canonical form stated in assertion (B).

To construct the square matrices $C(\tau)$ and $D(\tau)$, let us consider an arbitrary matrix $x = \|x_i^j\|$ with q columns and r rows consisting of elements belonging to O_τ and introduce certain elementary operations on it. Operation 1 is interchanging any two columns and operation 2 multiplying one column by an arbitrary element $f(\tau)$ of order $k = 0$ (see (A)), operation 3 is adding any column multiplied by an arbitrary element $g(\tau)$ of the ring O_τ to some other column. Obviously, each of the operations 1, 2, and 3 can be performed by multiplying the matrix x on the right by some square matrix $C(\tau)$ of order q, whose determinant is nonzero for $\tau = 0$. Let us denote by 1′, 2′ and 3′ the corresponding operations carried out on the rows of the matrix x. Obviously, each of these operations can be performed by multiplying the matrix x on the left by some square matrix $D(\tau)$ of order r, whose determinant is nonzero for $\tau = 0$.

Thus, to prove (B), it is sufficient to demonstrate that any matrix x can be reduced to canonical form by repeated application of these six operations in succession. We now prove this. Let $k_1 \geqslant 0$ be the least of the orders of the elements of the matrix x. By interchanging the columns and rows, the matrix x can be converted into a matrix y in which the element y_1^1 is of the order k_1. The order of any other element is not less than k_1. The element y_1^1 can be written as $y_1^1 = \tau^{k_1} f(\tau)$, where the order of $f(\tau)$ is zero. Multiplying the first column by $1/f(\tau)$, we obtain a matrix z in which $z_1^1 = \tau^{k_1}$. The order of any other element is not less than k_1, and is therefore divisible by τ^{k_1}. Let z_i^1 $(i > 1)$ be some element in the first row of the matrix z. As it is divisible by τ^{k_1}, we have $z_i^1 = -\tau^{k_1} g(\tau)$. Multiplying the first column of the matrix z by $g(\tau)$ and then adding it to the i-th column, we obtain a matrix u in which zero stands in the i-th column in the first row. Applying this operation to all the elements in the first row, and then to all the elements in the first column, we obtain a matrix v in which $v_1^1 = \tau^{k_1}$, and the elements v_1^j and v_i^1 are zeros for $i \neq 1, j \neq 1$. All other elements of the matrix v are divisible by τ^{k_1}. Deleting the first row and the first column from the matrix v, we obtain a matrix w of $(q-1)$ columns and $(r-1)$ rows. Now applying all these operations to the matrix w, we obtain a new matrix in which the first two diagonal elements are of the form τ^{k_1} and τ^{k_2}, where $k_1 \leqslant k_2$, and all other elements in the first two rows and two columns are zeros. Continuing this process, we finally reduce the matrix x to canonical form. Since the conversion of the matrix g_τ into canonical form $G(\tau)$ with the help of (6) does not change the rank of the matrix for small τ, the last assertion in (b) is also true.

(C) Let V and W be two vector spaces of dimensions q and r, respectively, $\hat{g}_\tau$ a mapping of V into W which is analytic for small τ, the rank of $\hat{g}_\tau$ being s for $\tau \neq 0$. Then there exists an s-dimensional subspace W_1 of W, and V is a direct sum of its subspaces V_1 and V_2 of dimensions s and $(q-s)$, respectively. Furthermore, there exist three mappings ψ_τ, ϕ_τ, and g_τ^* such that ψ_τ maps W_1 into W, and ψ_0 is the identity on W_1; g_τ^* maps V_1 into W_1 (for $\tau=0$ it maps onto W_1); and ϕ_τ maps V onto V_1, ϕ_τ is the identity on V_1 and $\phi_0 V_2 = 0$. Moreover,

$$\hat{g}_\tau = \phi_\tau g_\tau^* \varphi_\tau. \tag{7}$$

As ϕ_τ is the identity on V_1, (7) implies that the mappings $g_\tau = \psi_\tau g_\tau^*$ and $\hat{g}_\tau$ coincide on V_1. Note that the term V_2 in the direct sum is uniquely defined, while the term V_1 can be chosen arbitrarily so that it forms the whole space V as a direct sum with the fixed V_2.

To prove (C), let us introduce certain coordinates in the vector spaces V and W. The mapping $\hat{g}_\tau$ is expressed in these coordinates in the form of a matrix which we also denote by $\hat{g}_\tau$. Its elements belong to the ring O_τ (see (A)). This matrix has q columns and r rows, and its rank is equal to s when $\tau \neq 0$. By virtue of (B), this matrix can be written as

$$\hat{g}_\tau = \hat{B}(\tau)\hat{G}(\tau)\hat{A}(\tau). \tag{8}$$

In the space W we can choose the coordinate system so that $\hat{B}(0)$ is the identity matrix.

Since $\hat{G}(\tau)$ is a diagonal matrix, its last $(q-s)$ columns and $(r-s)$ rows contain only zeros. Deleting these last $(q-s)$ columns and $(r-s)$ rows from $\hat{G}(\tau)$, we obtain a square matrix $G(\tau)$ of order s. Deleting the last $(q-s)$ rows from the matrix $\hat{A}(\tau)$, we obtain a matrix $A(\tau)$ containing s rows and q columns. Deleting the last $(r-s)$ columns from the matrix $\hat{B}(\tau)$, we obtain a matrix $B(\tau)$ containing r rows and s columns. Obviously,

$$\hat{g}_\tau = B(\tau)G(\tau)A(\tau). \tag{9}$$

In the coordinate systems chosen in the spaces W and V, let W_1 denote the set of all vectors of the type

$$w_1 = (w^1, \ldots, w^s, 0, \ldots, 0)$$

and assume that the matrix $B(\tau)$ gives the mapping ψ_τ of the space W_1 into

W. Clearly, ψ_0 maps identically W_1 onto itself, and the dimension of the space $\psi_\tau W_1$ is, for small τ, therefore equal to s. We consider the matrix $G(\tau)A(\tau)$ as giving the mapping of the space V into the space V_1. Let $V(\tau)$ denote the set of all vectors $v \in V$ for which $\hat{g}_\tau(v) = 0$ for small positive τ. Since the mapping ψ_τ of W_1 onto $\psi_\tau W_1$ is isomorphic, the condition $\hat{g}_\tau(v) = 0$ is equivalent to the condition:

$$G(\tau)A(\tau)v = 0.$$

Since, for $\tau \neq 0$, the matrix $G(\tau)$ has a rank equal to s, the condition $\hat{g}_\tau(v) = 0$ is equivalent to the condition:

$$A(\tau)v = 0 \tag{10}$$

for $\tau \neq 0$. For $\tau \neq 0$ this equation defines the space $V(\tau)$. Since, for all small τ, including $\tau = 0$, the matrix $\hat{A}(\tau)$ has a rank equal to q, the matrix $A(\tau)$ has also a rank equal to s for all small τ. Hence, it follows that the dimension of the space $V(\tau)$ of all vectors v given by (10) is equal to $(q - s)$; equation (10) has meaning even for $\tau = 0$, and the space $V(\tau)$, for $\tau \neq 0$, tends to some space $V(0)$ as $\tau \to 0$, where the space $V(0)$ is given by the equation

$$A(0)v = 0. \tag{11}$$

Take $V(0)$ as the space V_2, and any direct complement as V_1. We now assume that the vector basis $e_1, \ldots, e_s, \ldots, e_q$ is so chosen that the first s vectors belong to V_1 and the remaining $(q - s)$ vectors to V_2. Let $A_1(\tau)$ and $A_2(\tau)$ denote the matrices formed from the first s and the last $(q - s)$ columns of $A(\tau)$, respectively.

As the last $(q - s)$ basis vectors of the space V belong to the subspace V_2, by virtue of (11), we have

$$A_2(0) \equiv 0, \tag{12}$$

and as the rank of the matrix $A(0)$ is s, from (12) we find that the rank of the matrix $A_1(0)$ is also s. We have

$$A(\tau)v = A_1(\tau)v_1 + A_2(\tau)v_2, \quad v = v_1 + v_2, \quad v_1 \in V_1, \quad v_2 \in V_2. \tag{13}$$

We assume that the matrix $G(\tau)A_1(\tau)$ defines the mapping g_τ^* of the space V_1 into W_1, and the matrix $A_1^{-1}(\tau)A(\tau)$ the mapping ϕ_τ of the space V onto V_1. Formula (7) hods for these mappings. This completes the proof of (C).

(D) Let U, V, W be three vector spaces of dimensions p, q, and r, respectively, f_τ a linear mapping of the space U into W, and $\hat{g}_\tau$ a linear mapping of V into W. For small τ, both these mappings depend analytically on τ. Let P and Q be convex subspaces of maximal dimensions in U and V, respectively, and let μ be a positive number. Furthermore, it is assumed that

$$\mu f_\tau(P) \subset \hat{g}_\tau(Q) \tag{14}$$

for all sufficiently small τ. By virtue of (C), the space V can be represented as the direct sum of its subspaces V_1 and V_2. It turns out that there exists a linear mapping h_τ of the space U into V_1 depending analytically on τ, such that

$$f_\tau = g_\tau h_\tau. \tag{15}$$

where g_τ is the mapping $\hat{g}_\tau$, considered on V_1, and the condition

$$\mu h_0(P) \subset \phi_0(Q) \quad \text{(see (C))} \tag{16}$$

is satisfied. In the proof we use the notation adopted in (C).

Let

$$W(\tau) = \psi_\tau W_1. \tag{17}$$

For small τ, we have $\dim W(\tau) = s$. Let g_τ denote the mapping $\psi_\tau g_\tau^*$ of the space V_1 into the space $W(\tau)$. For $\tau \neq 0$, it is an isomorphism. Since P and Q are of maximal dimensions, from (14) we get

$$f_\tau(U) \subset g_\tau(V_1) = W(\tau). \tag{18}$$

As the mapping g_τ of V_1 onto $W(\tau)$ is isomorphic for $\tau \neq 0$, the mapping g_τ^{-1} is defined and is analytic for $\tau \neq 0$. But it may have a pole at $\tau = 0$. Thus, the mapping

$$h_\tau = g_\tau^{-1} f_\tau \tag{19}$$

is analytic for $\tau \neq 0$, and may have a pole at $\tau = 0$. From the inclusion (14) we find

$$\mu h_\tau(P) \subset \phi_\tau(Q), \tag{20}$$

from which it follows that there exists a positive number α so large that

$$|h_\tau(u)| < \alpha |u|. \tag{21}$$

Hence, we find that the mapping h_τ has no poles and is therefore analytic for small values of τ. Thus, (16) follows from (20). This completes the proof of (D).

(E) Let g_τ be a linear mapping of the s-dimensional vector space V_1 ($s = 1, 2$) into a two-dimensional vector space W which depends analytically on τ for small τ and has the property that there is no fixed one-dimensional vector subspace W^1 such that

$$g_\tau(V_1) \subset W^1 \tag{22}$$

for all small τ. Then only one of the following two cases can occur:

(a) The number s is equal to two, and under any choice of the coordinates in the vector spaces V_1 and W, the matrix g_τ can be expressed as

$$g_\tau = \tau^k g_0', \tag{23}$$

where g_0' is a matrix of rank two, and k is a non-negative integer.

(b) In the spaces V_1 and W the coordinates can be so chosen that the matrix g_τ is of the form

$$g_\tau = \begin{pmatrix} a(\tau), & a_1(\tau) \\ b(\tau), & b_1(\tau) \end{pmatrix}, \tag{24}$$

where the orders of its elements are given by the relations:

$$\{a\} = k, \quad \{b\} = l, \quad 0 \leqslant k < 1, \quad \{a_1\} > k, \quad \{b_1\} \geqslant l \, (\text{see (A)}). \tag{25}$$

Formally, the case $s = 1$ is also included in these relations, for which we have to take $a_1(\tau) \equiv b_1(\tau) \equiv 0$, i.e., the matrix contains only one column.

We first take up the case of $s = 1$. In arbitrary coordinates write the matrix g_τ as

$$g = \begin{pmatrix} a(\tau) \\ b(\tau) \end{pmatrix}. \tag{26}$$

In case the orders of $a(\tau)$ and $b(\tau)$ are distinct, then interchanging, if necessary, a number of axes in W, we obtain $\{a\} < \{b\}$. If these orders are equal, then after multiplying the element $a(\tau)$ by some number r, we can subtract it from the element $b(\tau)$, so the order of the element $b(\tau) - ra(\tau)$ is greater than the order of $a(\tau)$. The new matrix

$$\begin{pmatrix} a(\tau) \\ b(\tau) \end{pmatrix}$$

corresponds to the mapping g_τ in some other coordinates in the plane W. The case $b(\tau) \equiv 0$ is impossible, because in that case there would exist in W a fixed line W^1 for which (22) holds.

Now let us take up the case $s = 2$. Expand the matrix g_τ as a series in powers of τ:

$$g_\tau = A_0 + A_1\tau + A_2\tau^2 + \ldots \tag{27}$$

where A_0, A_1, A_2 are constant square matrices of order two. Let k be the least integer for which the matrix A_k is not zero. If the rank of A_k is two, then case (a) holds. If the rank of A_k is one, then under a proper choice of the coordinates in V_1 and W, the matrix A_k takes the form

$$A_k = \begin{pmatrix} 1 & 0 \\ 0 & 0 \end{pmatrix}. \tag{28}$$

In these coordinates the matrix

$$g = \begin{pmatrix} g_1^1(\tau) & g_2^1(\tau) \\ g_1^2(\tau) & g_2^2(\tau) \end{pmatrix} \tag{29}$$

has the property that

$$\{g_1^1\} = k, \quad \{g_2^1\} > k, \quad \{g_1^2\} > k, \quad \text{and} \quad \{g_2^2\} > k. \tag{30}$$

If $\{g_1^2\} > \{g_2^2\}$, add the second column of (29) to its first column. For the new matrix thus obtained the relation (30) holds and furthermore $\{g_1^2\} = \{g_2^2\}$. Addition of the second column to the first column is equivalent to a change of coordinates in the plane V_1. In the second row $g_1^2(\tau)$ and $g_2^2(\tau)$ cannot both be identically zero, because then there would exist in the plane W a line W^1 for which (22) holds. Thus the matrix g_τ has been reduced to the form (24).

This completes the proof of (E).

3. Solution of an Auxiliary Integral Equation

In constructing an evasion control in terms of a pursuit control we have to solve a special integral equation which we describe and solve in this section.

Let S be some family of functions given on the interval $I\,(0 \leqslant t \leqslant \theta)$. An operator A defined on this family is called a Volterra operator, if the value of the function Ag $(g \in S)$ at a point $t \in I$ depends only on the values of the function $g(s)$ on the interval $0 \leqslant s \leqslant t$. If t is taken to be time, it means that the function Ag at the instant t depends only on the past and the present of the function g, does not depend on its future. We only use such operators in constructing an evasion control in terms of a pursuit control. An example of a Volterra type operator is the integral $\int_0^t g(\tau)\,d\tau$. Let $A_t g$ denote the value of the function Ag at the point t. We can say that A_t is a functional depending on the parameter t.

Volterra operators, which we study in this section, arise from the operation of convolution of two functions.

Let $f(t)$ and $g(t)$ be two bounded measurable functions given on the interval I. Their convolution is given by the formula

$$h(t) = \int_0^t f(t-\tau)g(\tau)\,d\tau. \tag{1}$$

The convolution $h(t)$ is also a measurable, bounded function and is defined on the interval I. For a fixed f, expression (1) associates a function h with the function g:

$$h = Ag, \quad h(t) = A_t g,$$

so A (or, which is the same, A_t) is an operator. Here A_t is obviously a Volterra operator.

It is known, and it is a simple matter to verify, that convolution is commutative, i.e.,

$$h(t) = \int_0^t f(t-\tau)g(\tau)\,d\tau = \int_0^t f(\tau)g(t-\tau)\,d\tau. \tag{2}$$

To demonstrate (2) if suffices to make the change of variable $\tau = t - \sigma$.

We now establish some simple properties of convolution. Assume that the function $f(t)$ is differentiable, then we have

$$\dot{h}(t) = f(0)g(t) + \int_0^t \dot{f}(t-\tau)g(\tau)\,d\tau = f(0)g(t) + $$
$$+ \int_0^t \dot{f}(\tau)g(t-\tau)\,d\tau \quad \text{(see (2))}. \tag{3}$$

Let the symbol $\int^k$ stand for k successive integrations carried out from zero to t, i.e.,

$$\int^0 f(t) = f(t), \quad \int^k f(t) = \int_0^t \left(\int^{k-1} f(\tau) \right) d\tau, \quad k = 1, 2, \ldots. \tag{4}$$

Assuming that $f(t)$ is differentiable, and integrating by parts the right-hand side of (1), we obtain

$$h(t) = f(0)\int^1 g(t) + \int_0^t \dot{f}(t-\tau)\int^1 g(\tau)\,d\tau = f(0)\int^1 g(t) + $$
$$+ \int_0^t \dot{f}(\tau)\int^1 g(t-\tau)\,d\tau \quad \text{(see (2))}. \tag{5}$$

Now suppose that $f(t)$ is differentiable $(k+1)$ times and that

$$f(0) = 0, \quad f^1(0) = 0, \ldots, f^{(k-1)}(0) = 0. \tag{6}$$

Then applying (5) successively, we obtain

$$h(t) = f^{(k)}(0)\int^{k+1} g(t) + \int_0^t f^{(k+1)}(\tau)\int^{k+1} g(t-\tau)\,d\tau \quad \text{(see (1) and (6))}. \tag{7}$$

This formula gives

$$\int_0^t \tau^k g(t-\tau)\,d\tau = k!\int^{k+1} g(t). \tag{8}$$

(A) Let $x(t)$ and $y(t)$ be two functions of a real variable t of orders k and l, respectively, and analytic on the interval $I(0 \leqslant t \leqslant \theta)$. Here k is an arbitrary integer and $l \geqslant 0$ (see (A), Section 2). Define a Volterra operator A_t^* by the formula

$$A_t^* z = x(t)\int_0^t y(\tau)z(t-\tau)\,d\tau, \tag{9}$$

where $z(t)$ is a bounded measurable function on the interval I. The operator A_t^* is called a one-term AC-operator of order $p^* = k + l + 1$ with order of differentiability $q^* = l + 1$. Here $q^* \geqslant 1$. Let us define a one-term operator of order $p^* = k$ with order of differentiability $q^* = 0$, by putting

$$A_t^* z = x(t)z(t). \tag{10}$$

Let $A_t^1, \ldots, A_t^n$ be a finite set of one-term AC-operators. Define the operator A_t as follows:

$$A_t = A_t^1 + A_t^2 + \ldots + A_t^n. \tag{11}$$

The operator A_t is called an AC-operator. Let P_i be the order of the one-term operator A_t^i and q_i its order of differentiability. Let p be the least of the numbers $p_1, \ldots, p_n$, and q the least of the numbers $q_1, \ldots, q_n$. One would like to say that p is the order of the operator A_t and q its order of differentiability. But this definition is not invariant since the operator A_t can be expressed as the sum of one-term operators in various ways. Therefore we say that the order of the operator A_t is not less than p and the order of its differentiability is not less than q.

(B) Let A_t be an AC-operator of order not less than p (see (A)), and $z(t)$ a function satisfying the estimate

$$|z(t)| \leqslant t^r, \quad r \geqslant 0; \tag{12}$$

then there exists a constant $c > 0$ depending on the operator A_t and on the number r, such that the function $A_t z$ satisfies the estimate

$$|A_t z| \leqslant ct^{p+r}, \tag{13}$$

and this justifies the name "order".

To prove this, it suffices to consider the case where $A_t = A_t^*$ is a one-term AC-operator (see (9) and (10)). As $x(t)$ is a function of order k, the function $x(t)/t^k$ is bounded on the interval I. So it does not exceed a certain constant c_1 and therefore

$$|x(t)| \leqslant c_1 t^k. \tag{14}$$

Similarly, we have

$$|y(t)| \leqslant c_2 t^l. \tag{15}$$

In case A_t^* is given by (10), we obviously have

$$|A_t^* z| \leqslant |x(t)| \cdot |z(t)| \leqslant c_1 t^{k+r} = c_1 t^{p^*+r}.$$

In case A_t^* is given by (9), we obviously have

$$|A_t^* z| \leqslant |x(t)| \int_0^t |y(t)| \cdot |z(t-\tau)| \, d\tau \leqslant c_1 c_2 t^k \int_0^t \tau^l (t-\tau)^r \, d\tau =$$

$$= c_1 c_2 t^k l! \int^{l+1} t^r = ct^{k+l+1+r} = Ct^{p^*+r}.$$

This completes the proof of (B).

(C) Let A_t be an AC-operator whose order is not less than p with order of differentiability not less than q where $q \geqslant 1$. Then there exists an AC-operator $\dot{A}_t$ of order not less than $(p-1)$ and of order of differentiability not less than $(q-1)$, such that

$$\frac{d}{dt}\dot{A}_t a = \dot{A}_t z. \tag{16}$$

It suffices to prove this assertion for a one-term operator $A_t = A_t^*$. Since $q^* \geqslant 1$, the operator A_t^* is of the form (9). Differentiating (9), we obtain, by virtue of (3)

$$\frac{d}{dt}A_t z = \frac{d}{dt}A_t^* z = \dot{x}(t) \int_0^t y(\tau) z(t-\tau) \, d\tau + x(t) \int_0^t \dot{y}(\tau) z(t-\tau) \, d\tau +$$

$$+ x(t) y(0) z(t) = \dot{A}_t z, \tag{17}$$

where A_t is a three-term operator. The order of each term is not less than $(p-1)$ and the order of differentiability of each term is not less than $(q-1)$.

This completes the proof of (C).

(D) Let A_t be an AC-operator whose order is not less than p and whose order of differentiability is not less than q. Here p is an arbitrary integer and q is an arbitrary non-negative integer. Then the operator $B_t = A_t \int^1$ given by the formula

$$B_t z = A_t \int^1 z \tag{18}$$

is of an order not less than $(p+1)$ and its order of differentiability is not less than $(q+1)$.

Proof. This assertion follows directly from (5), (9) and (10).

(E) Assume that the variable t takes values in the interval $0 \leqslant t \leqslant \theta$, and consider the integral equation

$$\int_0^t \tau^k \chi(t-\tau)\, d\tau = \frac{l+1}{(k+1)t^{l-k}} \int_0^t \tau^l \chi(t-\tau)\, d\tau + k!\,(H_t \chi + K_t \xi), \tag{19}$$

where k and l are integers $(0 \leqslant k < l)$, $\chi(t)$ is an unknown function, $\xi(t)$ is a measurable bounded function on the interval I which satisfies the estimate

$$|\xi(t)| \leqslant \alpha, \tag{20}$$

and H_t and K_t are AC-operators whose orders and the orders of differentiability are not less than $(k+2)$. Under these conditions equation (19) has a solution given by

$$\chi(t) = \int^1 S_t \xi, \tag{21}$$

defined on the interval $0 \leqslant t \leqslant \theta'$ $(\theta' > 0)$ and S_t is a linear Volterra operator satisfying the estimate

$$|S_t \xi| < c\alpha. \tag{22}$$

Hence it follows that $\chi(t)$ satisfies the estimate

$$|\chi(t)| \leqslant c\alpha t. \tag{23}$$

Here θ' depends on the numbers k, l, and on the operator H_t, while c depends also on the operator K_t.

To solve (19), let us introduce a new unknown function

$$z(t) = \int^{k+1} \chi(t). \tag{24}$$

Using (7) and (8), rewrite equation (19) as

$$z(t) = \frac{(l+1)!}{(k+1)!t^{l-k}} \int^{l-k} z(t) + L_t z + K_t \xi, \tag{25}$$

where L_t is an AC-operator whose order and order of differentiatiability are not less than unity. We shall solve this equation by successive approximations, putting

$$z_0 \equiv 0, z_{i+1}(t) = \frac{(l+1)!}{(k+1)!t^{l-k}} \int^{l-k} z_i(t) + L_t z_i + K_t \xi, \tag{26}$$

Note that, by virtue of (B), we have

$$|K_t \xi| \leqslant a\alpha t^{k+2}. \tag{27}$$

For $|z| \leqslant t^{k+2}$,

$$|L_t z| \leqslant bt^{k+3}, \tag{28}$$

where a and b are constants that depend on the operators K_t and H_t, respectively. We have

$$|z_1 - z_0| = |z_1| = |K_t \xi| < a\alpha t^{k+2}. \tag{29}$$

Suppose that the estimate

$$|z_i - z_{i-1}| < a_i t^{k+2} \tag{30}$$

is valid. Then we have

$$|z_{i+1} - z_i| \leqslant \frac{(l+1)!}{(k+1)!t^{l-k}} \int^{l-k} |z_i - z_{i-1}| + |L_t(z_i - z_{i-1})| \leqslant$$

$$\leqslant \frac{k+2}{l+2} a_i t^{k+2} + ba_i t^{k+3} \leqslant a_i t^{k+2} \left[\frac{k+2}{l+2} + b\theta' \right].$$

We take $\theta' = 1/(2b(l+2))$, then

$$a_{i+1} = ai \left[\frac{k+2}{l+2} + b\theta' \right] \leqslant a_i \left[1 - \frac{1}{2(l+2)} \right].$$

Thus, the successive approximations for equation (19) converge, and $|z(t)| \leqslant c_0 \alpha t^{k+2}$.

To obtain the function $\chi(t)$ from $z(t)$, let us differentiate (25) successively. Note that the first term

$$\frac{1}{t^{l-k}} \int^{l-k} z(t)$$

is an operator on the function $z(t)$ of order zero and order of differentiability $l - k \geqslant 1$. Thus, (25) can be differentiated and the solution $z(t)$ can be expressed as

$$z(t) = \int^1 u(t). \tag{31}$$

Substituting this expression into (25), we find that the right-hand side has a second derivative. Therefore z can be written as

$$z(t) = \int^2 v(t).$$

Finally we can write $z(t)$ in the form

$$z(t) = \int^{k+2} w(t),$$

where

$$w(t) = S_t \xi.$$

At each step, we obtain an estimate for the derivatives of z:

$$\left| \frac{d^i}{dt^i} z(t) \right| \leqslant c_i \alpha t^{k+2-i}, \tag{32}$$

so $w(t)$ is a bounded measurable function with the estimate

$$|w(t)| \leqslant c_{k+2}\alpha = c\alpha. \tag{33}$$

This completes the proof of (E).

(F) In (A) we used an AC-operator acting on a scalar function. An AC-operator acting on a vector function

$$z(t) = (z^l(t), \ldots, z^m(t)) \tag{34}$$

is defined in the obvious manner. It is a sequence

$$A_t = (A_{1,t}, \ldots, A_{m,t}) \tag{35}$$

of scalar AC-operators. If the order of each operator in (35) is not less than p, and its order of differentiability is not less than 1, the same is also true of the vector AC-operator A_t (see (A)). The operation $A_t z$ is defined as

$$A_t z = A_{1,t} z^1 + \ldots + A_{m,t} z^m \tag{36}$$

and associates a scalar function $A_t z$ with a measurable vector function $z(t)$. Now, instead of (19), we can consider the following equation

$$\int_0^t \tau^k \chi(t-\tau)\, d\tau \; \mathrm{O} \; \frac{l+1}{(k+1)t^{l-k}} \int_0^t \tau^l \chi(t-\tau)\, d\tau + k\,!\,(H_i \chi + K_t \xi), \tag{37}$$

where H_t is, as before, a scalar AC-operator whose order and order of differentiability are not less than $k+2$, K_t is a vector AC-operator whose order and order of differentiability are not less than $k+2$, and $\xi(t)$ is some measurable vector function with the estimate

$$|\xi(t)|<\alpha, \tag{38}$$

where k and l are integers ($0 \leqslant k < l$), and $\chi(t)$ is the unknown function. It immediately follows from (E) that there exists a solution to equation (37) in the form:

$$\chi(t) = \int^1 S_t\xi, \tag{39}$$

defined on some interval $0 \leqslant t \leqslant \theta'$, where θ' depends on the numbers k, l and on the operator H_t. For the operator S_t we have the estimate

$$|S_t\xi| \leqslant c\alpha. \tag{40}$$

where c depends on the numbers k, l and on the operators H_t and K_t.

To prove the existence of the solution $\chi(t)$ to equation (37), it is sufficient to consider a sequence of equations

$$\int_0^t \tau^k\chi^i(t-\tau)\,d\tau = \frac{l+1}{(k+1)t^{l-k}}\int_0^t \tau^l\chi^i(t-\tau)\,d\tau + k\,!(H_t\chi^i + K_{i,\,t}\xi^i),$$

$$i=1,\ldots,m \tag{41}$$

of the type (19) and form a solution of (37) as the sum of the solutions of equations (41).

We can now consider the integral equation which will be used in constructing an evasion control.

(G) Let $a(t)$ and $b(t)$ be two functions analytic on the interval $I(0 \leqslant t \leqslant \theta)$ of orders k and l, respectively, where $0 \leqslant k < l$ (see (A)); let A_t and B_t be two vector AC-operators of dimension m defined on I, whose order and order of differentiability are not less than $k+2$ and $l+2$, respectively; and let $\xi(t)$ be an m-dimensional vector function defined and measurable on I with the estimate

$$|\xi(t)|<\alpha. \tag{42}$$

Then there exist two functions $\chi(t)$ and $\hat{\chi}(t)$ satisfying the equations

$$\int_0^t a(\tau)\chi(t-\tau)\,d\tau + A_t\xi = \hat{\chi}(t)\int^1 a(t), \tag{43}$$

$$\int_0^t b(\tau)\chi(t-\tau)\,d\tau + B_i\xi = \hat{\chi}(t)\int^1 b(t), \tag{44}$$

defined on some interval $0 \leqslant t \leqslant \theta'$, where θ' depends on the functions $a(t)$ and $b(t)$, and satisfying the estimates:

$$|\chi(t)| < \alpha ct, \quad |\hat{\chi}(t)| \leqslant \alpha\hat{c}t, \tag{45}$$

where the constants c and $\hat{c}$ depend on the functions $a(t)$, $b(t)$ and on the operators A_t, B_t.

The proof consists of reducing the equations (43) and (44) to an equation of the type (37) by eliminating the unknown function $\hat{\chi}(t)$.

As each of the equations (43) and (44) can be divided by an arbitrary nonzero number, we can assume that the functions $a(t)$ and $b(t)$ are of the type

$$a(t) = t^k + a_1(t), \quad b(t) > t^l + b_1(t), \tag{46}$$

where the orders of the functions $a_1(t)$ and $b_1(t)$ are not less than $k+1$ and $l+1$, respectively. Furthermore, put

$$\frac{\int^1 a(t)}{\int^1 b(t)} = \frac{l+1}{(k+1)^{l-k}} + f(t),$$

where the order of the function $f(t)$ is not less than $-(l-k)+1$.

Eliminating the unknown function $\hat{\chi}(t)$ from the system (43), (44), we obtain one equation

$$\int_0^t \tau^k\chi(t-\tau)\,d\tau = \frac{l+1}{(k+1)t^{l-k}}\int_0^t \tau^l\chi(t-\tau)\,d\tau +$$

$$+\frac{l+1}{(k+1)t^{l-k}}\int_0^t b_1(\tau)\chi(t-\tau)\,d\tau + \left[\frac{l+1}{(k+1)t^{l-k}} + f(t)\right]B_t\xi -$$

$$-\int_0^t a_1(\tau)\chi(t-\tau)\,d\tau + f(t)\int_0^t b(\tau)\chi(t-\tau)\,d\tau - A_t\xi,$$

which is evidently of the form (37) (see (F)). Therefore it has a solution $\chi(t)$ satisfying the estimate (45). The function $\hat{\chi}(t)$ is obtained from (43), and the estimate (45) obviously holds for it.

This completes the proof of G).

4. An Evasion Maneuver

In this section we present the main construction by which we can obtain the motion of a point $z(t)$, or more exactly, construct an evasion control so that the point $z(t)$ does not hit the set M (see Section 1).

(A) Let Σ' be a finite-dimensional linear family of functions of a variable t analytic on some interval I of the real axis. Then there exists a natural number m depending on Σ' and I, such that any function $\phi(t) \in \Sigma'$, which is not identically zero, has no more than m zeros taking into account their multiplicities.

To prove this assertion introduce an arbitrary Euclidean metric in the vector space Σ', and let S denote the unit sphere in Σ'. Then there exists a positive number r such that $r\phi(t)$ belongs to S. The functions $\phi(t)$ and $r\phi(t)$ have the same number of zeros. Thus, it suffices to consider only the functions belonging to S. Assume that there exists a sequence of functions

$$\phi_1(t), \phi_2(t), \ldots \tag{1}$$

in S, whose numbers of zeros in I increase without bound. Since the set S is compact, we can assume that the sequence (1) converges uniformly to some function $\phi_0(t) \in S$ in some neighbourhood of the interval I in the plane of the complex variable t. Calculating the number of zeros by means of a logarithmic residue, we conclude that $\phi_0(t)$ has infintely many zeros on I, and this is impossible.

(B) Let $\hat{\Sigma}$ be a finite-dimensional linear family of functions analytic on some interval I belonging to the interval $0 \leqslant t \leqslant 1$, W a two-dimensional Euclidean vector space with a fixed orthogonal coordinate system; $w = (w^1, w^2) \in W$, Γ the square defined by the following inequalities:

$$|w^i| \leqslant a, \quad (i = 1, 2, a > 0), \tag{2}$$

and k_1 and k_2 two integers, where $0 \leqslant k_1 \leqslant k_2$. Then there exists a positive number γ such that for any vector $\phi(t) = (\phi^1(t), \phi^2(t))$ whose components belong to $\hat{\Sigma}$, it is possible to find a square $\Gamma' \subset \Gamma$ with sides 2γ such that the point $v(t) \in W$ given by the equations

$$v^i(t) = \phi^i(t) - \alpha^i t k_i, \quad i = 1, 2, \tag{3}$$

for

$$t \in I, \quad \alpha = (\alpha^1, \alpha^2) \in \Gamma', \tag{4}$$

satisfies the condition

$$|v(t)| \geqslant \gamma t^{k_2} \tag{5}$$

To prove this, extend the family $\hat{\Sigma}$ to the family Σ' by adjoining the functions t^{k_1} and t^{k_2}. Let m be a number corresponding to the family Σ' (see (A)). Assume that

$$p > 2m + 1 \tag{6}$$

is an integer. Subdivide the square Γ by equidistant vertical and horizontal lines

$$\begin{aligned} w^i &= a^i_j, \quad i = 1, 2, \quad j = 0, 1, \ldots, p, \\ a^i_0 &= -a, \quad a^i_p = a \end{aligned} \tag{7}$$

into p^2 small squares with sides $2a/p$. In the W plane consider the curve

$$w^i = w^i(t) = \varphi^i(t)/t^{k_i}, \quad i = 1, 2. \tag{8}$$

We now show that the curve (8) cannot pass through all the small subsquares. Indeed, if the curve enters any one of these subsquares, it should intersect one of its sides, i.e., either one of the vertical or horizontal lines of the system (7). Thus, the number of intersections with either the vertical or horizontal lines is not less than $(p^2 - 1)/2$. For the sake of definiteness, assume that the former is true. Since there are $p + 1$ verticals in the system (7), the curve should intersect one of these verticals, say, $w^1 = a^1_j$ at least $(p - 1)/2$ times. This implies that the function $\phi^1(t) - a^1_j t^{k_1}$ has at least $(p - 1)/2 > m$ zeros (see (6)). As this function belongs to the family Σ', it is identically equal to zero. This means that the curve (8) lies completely on the vertical $w^1 = a^1_j$. Hence it cannot pass through all the small squares.

Let Γ'' denote a small square in Γ through which the curve (8) does not pass, γ' one fourth of the sides of this square and Γ' a square with sides 2γ whose center coincides with the center of the square Γ''. As the curve (8) does not pass through the square Γ'', one of the numbers $|w^i(t) - \alpha^i|$, $i = 1, 2$ is, for a fixed t, greater than or equal to γ, if the conditions (4) are satisfied. Hence it follows that

$$|v(t)| \geqslant \gamma t^{k_2}.$$

This completes the proof of (B).

The following assertion (C) is easily derived from (B).

(C) Let Σ be a finite-dimensional linear family of functions analytic on some interval I belonging to the interval $0 \leqslant t \leqslant 1$, where zero is contained in I; let W be a two-dimensional Euclidean vector space with a fixed orthogonal coordinate system $w = (w^1, w^2) \in W$, Δ a square defined by the inequalities

$$|w^i| \leqslant b, \quad i = 1, 2, \quad b > 0, \tag{9}$$

and ω_t a linear mapping of W into itself depending analytically on t for $t \in I$ and non-degenerate for small positive t. Then there exist two positive numbers δ and $\theta \in I$ such that for any vector $\psi(t) = (\psi^1(t), \psi^2(t))$ whose components belong to Σ it is possible to find a square $\Delta' \subset \Delta$ with sides 2δ such that the point

$$w(t) = \psi(t) + \omega_t \beta \tag{10}$$

for

$$0 \leqslant t \leqslant \theta, \quad \beta \in \Delta' \tag{11}$$

satisfies the condition

$$|w(t)| \geqslant \delta t^{k_2}, \tag{12}$$

where k_2 is an integer depending on the matrix ω_t.

The proof lies in reducing assertion (C) to assertion (B). This is based on reducing the matrix $-\omega_t$ to canonical form, i.e., expressing this matrix as follows:

$$-\omega_t = B(t)\Omega(t)A(t) \quad \text{(see (B) Section 2)} \tag{13}$$

where

$$\Omega(t) = \begin{pmatrix} t^{k_1} & 0 \\ 0 & t^{k_2} \end{pmatrix}, \quad 0 \leqslant k_1 \leqslant k_2, \tag{14}$$

The matrices $A(t)$ and $B(t)$ have inverses on some segment $0 \leqslant t \leqslant \theta$, to which we restrict t in what follows. Put

$$v(t) = B^{-1}(t)w(t), \quad \varphi(t) = B^{-1}(t)\phi(t), \quad \alpha = A(t)\beta, \tag{15}$$

then (10) can be rewritten as

$$v(t) = \varphi(t) - \Omega(t)\alpha, \tag{16}$$

which coincides with (3). The components of the vector $\varphi(t)$ belong to the finite-dimensional linear family $\hat{\Sigma}$ which is described as follows. Let $\hat{b}_i^j(t)$ be the elements of the matrix $B^{-1}(t)$. Multiplying all the elements of the family Σ by the function $\hat{b}_i^j(t)$, we obtain some family Σ_i^j, and the family $\hat{\Sigma}$ is formed as the direct sum of the four families Σ_i^j. In order to satisfy all the assumptions of assertion (B), we now have to give a number a defining the square Γ. Then, by virtue of (B), we shall be able to find the number γ. Using the assumptions made in (B), we first determine a, and then using these assumptions and the number γ, we determine the number δ for which the assertion of (C) holds. In so doing, possibly, it may be necessary to diminish the number θ further.

Let us now take up this somewhat laborious procedure. As the matrices $A(t)$ and $B(t)$ have inverses, and the matrix $A(t) - A(0)$ is divisible by t, then there exists a positive number r such that, for any vector $w \in W$, we have

$$|A(t)w| \geqslant r|w|, \tag{17}$$

$$|A^{-1}(t)w| \geqslant r|w|, \tag{18}$$

$$|B(t)w| \geqslant r|w|, \tag{19}$$

$$|(A(t)-A(0))w| \leqslant \frac{\theta'}{r}|w|. \tag{20}$$

Let us now determine the number a, taking

$$a = rb/\sqrt{2}. \tag{21}$$

Then from (17), we find that

$$|A(t)|\Delta \supset \Gamma. \tag{22}$$

By (B), the number γ is constructed from the given number a. Now in the square Γ arbitrarily take a square Γ' with center at α_0 and sides 2γ, and put

$$A^{-1}(0)\alpha_0 = \beta_0. \tag{23}$$

Let K be the set of all points in Δ of the type

$$\beta_0 + w, \quad \text{where } |w| \leqslant \epsilon, \quad \epsilon > 0, \tag{24}$$

i.e., a circle of radius ϵ with center at β_0. We now find the conditions under which

$$A^{-1}(t)\Gamma' \supset K. \tag{25}$$

We have

$$A(t)(\beta_0 + w) = \alpha_0 + v, \tag{26}$$

where

$$v = (A(t)-A(0))\beta_0 + A(t)w. \tag{27}$$

By virtue of the inequalities (18) and (20), we have

$$|v| \leqslant \frac{\theta b\sqrt{2}}{r} + \frac{\epsilon}{r}. \tag{28}$$

Thus, for the inclusion (25) to hold, it is sufficient that

$$\frac{\theta b\sqrt{2}}{r} + \frac{\epsilon}{r} \leqslant \gamma. \tag{29}$$

Hence, after choosing θ and ϵ in such a manner that

$$\theta \leqslant \frac{r\gamma}{2b\sqrt{2}}, \quad \epsilon = \frac{\gamma r}{2}, \tag{30}$$

we guarantee that (25) holds. Take the number θ so that, in addition to conditions already imposed, it also satisfies the condition (30), and assume

$$\delta = \frac{\epsilon}{\sqrt{2}} = \frac{r\gamma}{2\sqrt{2}}. \tag{31}$$

Then inscribe a square Δ' with sides 2δ in the circle K. Thus we find δ, and to each square Γ' in Γ a square Δ' in Δ is assigned

$$\Gamma' \to \Delta', \quad \text{where } A(t)\Delta' \subset \Gamma'. \tag{32}$$

This correspondence is given by the relation (23) between the centers of the squares.

We now show that for the numbers θ and δ so chosen, the assertion (C) holds. Let $\psi(t)$ be an arbitrary vector function whose components belong to the family Σ. By virtue of (15) a function $\phi(t)$ whose components belong to the family Σ corresponds to $\psi(t)$. Therefore, by virtue of (B), there exists a square Γ' in Γ such that for $\alpha \in \Gamma'$,

$$|v(t)| > \gamma t \tag{33}$$

(see (9) and (16)). Hence, by virtue of (19), we find

$$|w(t)| \geqslant r\gamma t \geqslant \delta t, \tag{34}$$

which holds for $\beta \in \Delta'$ (see (32)). This completes the proof of (C).

5. The Evasion Control

For a linear differential game (1) Section 1 satisfying the evasion conditions (C) Section 1, we shall here construct a special evasion control $v(t)$ as a Volterra operator S_t (see Section 3) from the pursuit control $u(t)$, where the operator depends on the initial value z_0 of the game. In place of the linear differential game (1) Section 1, we consider a differential game (A) formally more general but actually coinciding with the game (1). A formally more general expression is used in order that the convex sets P and Q may be arbitrarily translated in the space R.

(A) In an n-dimensional Euclidean vector space R consider the differential game

$$\dot{z} = Cz - u + v + a, \tag{1}$$

where $z \in R$, C is a given linear mapping of R into itself and $a \in R$ is a given vector. The vectors u and v are the pursuit and evasion controls, respectively. They are not arbitrary vectors but satisfy the conditions

$$u \in P, \quad v \in Q, \tag{2}$$

where P and Q are given compact convex subsets of R of dimensions p and q, respectively. The game is assumed to be terminated at the instant when the point z hits a given vector subspace M or the space R. Thus, a game is defined, if the vector space R, a linear mapping of R into itself, a vector a, compact convex subsets P and Q, and a vector subspace M, whose dimension is denoted by $(n-\nu)$, are given. The game does not, however, change, if each of the sets P and Q are subject to a parallel translation in R, and accordingly the vector a is changed. Consequently, we may assume that the sets P and Q contain the origin of R. Then there exist p- and q-dimensional vector subspaces U and V of the space R such that the inclusions

$$P \subset U \quad \text{and} \quad Q \subset V \tag{3}$$

hold.

(B) In the differential game (A), let L denote the orthogonal complement of M in R, and W a two-dimensional vector subspace of the space L, π the operation of orthogonal projection of R onto W. By parallel translations of the sets P and Q we can ensure that the inclusions (3) hold and the sets P and Q contain zero as an interior point. Denote by f_τ and $\hat{g}_\tau$ the linear mapping $\pi e^{\tau C}$ considered on U and V, respectively. If the game satisfies the evasion conditions (a) and (b) Section 1, by additionally translating the sets P and Q, we can insure that the following conditions hold, which are more stringent than the conditions (a) and (b), Section 1:

(a) There does not exist any fixed one-dimensional subspace W^1 of W such that

$$\hat{g}_\tau(V) \subset W^1 \tag{4}$$

for all sufficiently small τ.

(b) There exists a constant $\mu > 1$ such that

$$\mu f_\tau(P) \subset \ddot{g}_\tau(Q). \tag{5}$$

Note that the constant μ in (5) may be slightly less than the constant μ in (10) Section 1.

(C) Let V be a q-dimensional Euclidean vector space, V_1 a s-dimensional vector subspace, ϕ the operation of orthogonal projection of V onto V_1, A a q-dimensional compact convex subset of V, and ρ an arbitrarily small positive number. Then there exists a convex polyhedron B in V_1 which contains the convex set $\phi(A)$ and is contained in the ρ-neighbourhood of this set. Moreover, there exists a piecewise linear mapping γ of the polyhedron B into the ρ-neighbourhood of the convex set A such that

$$\phi\gamma(v_1) = v_1 \tag{6}$$

for every point $v_1 \in B$. An analytic mapping satisfying the condition (6) can be easily substituted for γ.

To prove (C), let us subdivide the space V_1 into mutually congruent cubes of such a size that the main diagonal of each cube is less than a given number ρ. Let B' denote the sum of those cubes which intersect the set $\phi(A)$, and B the convex closure of the polyhedron B'. The polyhedron B thus constructed satisfies the conditions stated in (C). Now take an arbitrary simplicial subdivision of B, and let

$$b_1, b_2, \ldots, b_k \tag{7}$$

be the vertices of this subdivision. Since the distance of any point b_i in the sequence (7) from the set $\phi(A)$ is less than ρ, there exists a preimage of the point b_i under the mapping ϕ, whose distance from the set A is less than ρ. Thus we obtain a sequence of points

$$a_1, a_2, \ldots, a_k$$

in the ρ-neighbourhood of the set A, where $\phi(a_i) = b_i$ $(i = 1, \ldots, k)$. Let T be an arbitrary simplex of the subdivision of the polyhedron B, and $b_1, \ldots, b_r$ be the set of all its vertices. Span a simplex S on the points $a_1, \ldots, a_r$ corresponding to these vertices in the space V and let γ denote the linear mapping of the simplex T onto S under which $\gamma(b_i) = a_i$ $(i = 1, \ldots, r)$. Defining the mapping γ in this way on each of the simplices of the subdivision of the polyhedron B, we obtain a mapping γ of the whole convex polyhedron B.

We now show how to replace the mapping γ by an analytic mapping $\hat{\gamma}$. For this purpose, let V_2 denote the orthogonal complement of V_1 in V. Then V can be represented as the direct sum of V_1 and V_2, so that every element $v \in V$ can be expressed uniquely in the form

$$v = v_1 + v_2, \quad v_1 \in V_1, \quad v_2 \in V_2.$$

Thus, an element $\gamma(v_1)$, where $v_1 \in B$, can also be expressed as

$$\gamma(v_1) = v_1 + \chi(v_1), \text{ where } \chi(v_1) \in V_2.$$

Here χ is a piecewise linear mapping of the polyhedron B into the space V_2. Expressed coordinatewise, the mapping χ is a set of $(q - s)$ piecewise linear scalar functions, each of which depends on scalar variables. Approximating each of these scalar functions by a polynomial of s variables, we obtain an analytic approximation $\hat{\chi}$ of the mapping χ. The mapping $\hat{\gamma}$ given by the formula

$$\hat{\gamma}(v_1) = v_1 + \hat{\chi}(v_1)$$

is an analytic mapping of the polyhedron B into the ρ-neighbourhood of the set A and satisfies the condition

$$\phi\hat{\gamma}(v_1) = v_1,$$

which is the same as the condition (6).

This completes the proof of (C).

(D) In the vector space R consider the differential equation

$$\dot{z} = Cz + b(t), \tag{8}$$

where $z \in R$, C is a linear mapping of R into itself and $b(t)$ is a vector in R that depends on time t. Then, the solution $z(t)$ of equation (8) under the initial condition

$$z(0) = z_0$$

can be written as

$$z(t) = e^{tC}z_0 + \int_0^t e^{\tau C}b(t-\tau)\,d\tau. \tag{9}$$

The proof is obvious.

(E) Assume that the evasion conditions in (B) are satisfied for the game (A). Here we use the notation introduced in (B). By virtue of (C) Section 2, the representation of the space V as the direct sum of its subspaces V_1 and V_2 corresponds to the mapping $\hat{g}_\tau$, where V_2 is determined uniquely, and we take V_1 as its orthogonal complement. Let g_τ denote the mapping $\hat{g}_\tau$ considered on V_1. Then there exist an analytic mapping $\mathbf{v}$ of the set P into the set Q and a positive number ρ such that the following conditions (c) and (d) are satisfied.

(c) For

$$u \in P, v_1 \in V_1, |v_1| < \rho \tag{10}$$

we have

$$v = v_1 + \mathbf{v}(u) \in Q. \tag{11}$$

Thus, if $u = u(t)$ is an arbitrary pursuit control on the interval $0 \leqslant t \leqslant \theta$, and $v_1(t)$ is an arbitrary measurable vector function defined on the same interval where the conditions

$$v_1(t) \in V_1, \quad |v_1(t)| < \rho,$$

are satisfied, then the formula (11), i.e.,

$$v = v(t) = v_1(t) + \mathbf{v}(u(t)), \tag{12}$$

defines the evasion control.

(d) Substituting an arbitrary control $u = u(t)$ and the corresponding control $v = v(t)$ found from (12) into equation (1), we can express the projection $\pi z(t)$ of the solution $z(t)$ of this equation under the initial value z_0 as follows:

$$\pi z(t) = \psi(t) + \zeta_1(t),$$

where

$$\left.\begin{aligned}\psi(t) &= \pi e^{tc} z_0 + \int_0^t \pi e^{\tau C} a d\tau, \\ \zeta_1(t) &= \int_0^t g_\tau(v_1(t-\tau) - h_\tau^*(u(t-\tau)) + \varphi_\tau^* \mathbf{v}(u(t-\tau)))\, d\tau.\end{aligned}\right\} \tag{13}$$

Here h_τ^* is a linear mapping of the space U into the space V_1, and φ_τ^* is a linear mapping of the space V into the space V_1. Both these mappings are defined and analytic on a sufficiently small interval $0 \leqslant \tau \leqslant \theta$ and vanish for $\tau = 0$. The mapping $\mathbf{v}$, the positive numbers ρ, θ and the mappings h_τ^*, φ_τ^* depend only on the game (A) and do not depend on the control $u(t)$, on the function $v_1(t)$ or on the initial value z_0. In the sequel we shall choose the function $v_1(t)$, which satisfies the condition (10) and is arbitrary in other respects, for the purpose of evasion.

To prove (E), let us first make use of the assertion (C), taking

$$A = \frac{1}{\mu} Q.$$

Since the convex set Q contains the origin as an interior point and $\mu > 1$, the set A lies inside the set Q and the distance, denoted by 2ρ, between the set A and the boundary of the set Q is positive. According to assertion (C), an analytic mapping γ is defined on the convex polyhedron B, which contains the set $\varphi(A)$ and is contained in its ρ-neighbourhood, which sends B into the ρ-neighbourhood of the set A. Thus, the distance between $\gamma(B)$ and the boundary of the set Q is not less than ρ. Moreover, we have

$$\varphi\gamma(v_1) = v_1 \quad \text{for } v_1 \in B. \tag{14}$$

We now make use of assertion (D) Section 2. By virtue of this assertion, the mapping f_τ can be represented as

$$f_\tau = g_\tau h_\tau, \tag{15}$$

where

$$h_0(P) \subset \frac{1}{\mu}\varphi(Q) = \varphi(A). \tag{16}$$

Thus, the mapping γ is defined on the set $h_0(P)$ and the distance from the set $\gamma h_0(P)$ to the boundary of the set Q is not less than ρ. Hence, putting

$$\mathbf{v} = \gamma h_0, \tag{17}$$

we obtain the inclusion (11) for $\mathbf{v}$.

Substituting arbitrary controls $u = u(t)$ and $v = v(t)$ in equation (1), let us write the solution of the equation under the initial value z_0 in the form:

$$z(t) = e^{tc}z_0 + \int_0^t e^{\tau C}(v(t-\tau) - u(t-\tau) + a)\,d\tau \quad \text{(see (D))}. \quad (18)$$

Now applying the operation π, we obtain

$$\pi z(t) = \pi e^{tC}z_0 + \int_0^t \tau e^{\tau C} a\,d\tau + \int_0^t (\hat{g}_\tau(v(t-\tau)) - f_\tau(u(t-\tau)))\,d\tau. \quad (19)$$

The linear mapping $\hat{g}_\tau$ can, by virtue of (C) Section 2, be expressed as

$$\hat{g}_\tau = g_\tau \varphi_\tau. \quad (20)$$

where $\varphi_0 = \varphi$, and φ_τ is the identity on V_1,

$$\varphi_\tau(v_1) = v_1. \quad (21)$$

Now put

$$\varphi_\tau = \varphi_0 + \varphi_\tau^*, \quad h_\tau = h_0 + h_\tau^*. \quad (22)$$

From (15), (20), (22), (21), (12), (17) and (14) we obtain (13). This completes the proof of (E).

(F) Using the notation of (E), write

$$\zeta_2 = \int_0^t g_\tau(v_1(t-\tau) - h_\tau^*(u(t-\tau)) + \varphi_\tau^*(v(t-\tau)))\,d\tau, \quad (23)$$

so that after the substitution

$$v(t) = \mathbf{v}(u(t))$$

we obtain

$$\zeta_2(t) = \zeta_1(t) \quad \text{(see (13) (E))}. \quad (24)$$

For the purpose of evasion we choose the vector function $v_1(t)$ on the interval $0 \leqslant t \leqslant \theta$ in the form

$$v_1(t) = L_t(\beta_0, u, v), \quad (25)$$

where β_0 is a constant two-dimensional vector

$$\beta_0 = (\beta_0^1, \beta_0^2), \quad (26)$$

$u = u(t)$ is a vector function in the space U, $v = v(t)$ is a vector function in the space V, L_t is an operator, which depends linearly on these three quantities, and is of Volterra type with respect to the vector function $(u, v) = (u(t), v(t))$. It turns out that the operator L_t can be so chosen that after substituting the expression (25) for $v_1(t)$ into (23), we obtain

$$\zeta_2 = \omega_t(\beta_0 + \beta_1(t)), \quad (27)$$

where ω_t is a square matrix of order two with a nonzero determinant for $t>0$, and

$$\beta_1(t) = (\beta_1^1(t), \beta_1^2(t)) \tag{28}$$

satisfies the estimate

$$|\beta_1^i(t)| \leqslant ct, \quad i=1,2, \tag{29}$$

for $u(t)\in P$, $v(t)\in Q$.

If, however, the vector β_0 satisfies the estimate

$$|\beta_0^i|<b, \quad i=1,2, \tag{30}$$

and the number θ is sufficiently small, then the vector $v_1(t)$ satisfies the estimate

$$|v_1(t)|<\rho. \tag{31}$$

The positive numbers θ, b and c depend only on the game but do not depend on the vector function $(u(t), v(t))$.

We now prove (F). Since condition (a) is satisfied for the game, the linear mapping g_τ satisfies the conditions stated in (E) Section 2. Therefore we assume either case (a) or case (b) holds (see (E) Section 2) so that the matrix g_τ is either of the form (23) or of the form (24) Section 2. The operator L_t is constructed differently in these two different cases. Let

$$\zeta_3(t) = (\zeta_3^1(t), \zeta_3^2(t)) = \int_0^t g_\tau(-h_\tau^*(u(t-\tau)) + \varphi_\tau^*(v(t-\tau)))\,d\tau. \tag{32}$$

Clearly we have

$$\zeta_3^1 = A_t(u,v), \quad \zeta_3^2 = B_t(u,v), \tag{33}$$

where A_t and B_t are AC-operators acting on the vector function (u, v) (see (A) Section 3). For the case (a) (see (E) Section 2), the orders and orders of differentiability of these operators are not less than $k+2$, while for the case (b) those of the operator A_t are not less than $k+2$, and those of the operator B_t not less than $l+2$.

Consider case (a). Here we define the operator L_t as

$$L_t(\beta_0, u, v) \equiv \beta_0 \tag{34}$$

in other words, we choose the vector $v_1(t)$ in the form of a constant vector β_0. From (23) Section 2, we obtain

$$\int^1 g_t = t^{k+1} g_t'', \tag{35}$$

where g_0'' is a matrix of rank two. We now have

$$\zeta_2(t) = t^{k+1} g_t'' \left(\beta_0 + \frac{(g_t'')^{-1}\zeta_3(t)}{t^{k+1}}\right) = \omega_t(\beta_0+\beta_1(t)), \tag{36}$$

where

$$\omega_t = t^{k+1} g_t'', \quad \beta_1(t) = \frac{(g_t'')^{-1}\zeta_3(t)}{t^{k+1}}. \tag{37}$$

Thus, we have demonstrated (F) for the case (a).

We now take up case (b). We shall assume that the matrix g_t is expressed in the form of (24) Section 2. Thus we have functions $a(t)$ and $b(t)$ of order k and l, respectively, as well as the AC-operators A_t and B_t (see (33)). Using this data, let us form and solve equations (43) and (44) (Section 3) for the unknown functions $\chi(t)$ and $\hat{\chi}(t)$. For this we assume that

$$\xi = (u, v). \tag{38}$$

We define the vector $v_1(t) = (v_1^1(t), v_1^2(t))$, putting

$$v_1^1 = \beta_0^1 + t\beta_0^2 + \chi(t), \quad v_1^2 \equiv 0. \tag{39}$$

Then (27) holds, where

$$\omega_t = \begin{pmatrix} \int^1 a(t) & \int^2 a(t) \\ \int^1 b(t) & \int^2 b(t) \end{pmatrix}, \tag{40}$$

$$\beta_1^1(t) = \hat{\chi}(t), \quad \beta_1^2(t) \equiv 0. \tag{41}$$

It is readily verified that the matrix ω_t given by (40) is of rank two for $t > 0$. Thus we have proved (F) for case (b) as well.

(G) If the evasion conditions are satisfied for the game (1), then there exist positive numbers θ, δ, k (k is an integer) and a Volterra operator S_t defined on the interval $0 \leqslant t \leqslant \theta$, depending on $z_0 \in R$, which assigns to the pursuit control $u = u(t)$ an evasion control

$$v = v(t) = S_t(z_0, u). \tag{42}$$

such that the solution $z(t)$ of the equation

$$\dot{z} = Cz - u(t) + S_t(z_0, u) + a \tag{43}$$

with the initial value z_0 satisfies the condition

$$|\pi z(t)| \geqslant \delta t^k \tag{44}$$

for $0 \leqslant t \leqslant \theta$.

The proof is based on assertion (C) Section 4. We define the operator S_t as

$$S_t(z_0, u) = \mathbf{v}(u(t)) + L_t(\beta_0, u, \mathbf{v}(u)), \tag{45}$$

where the vector β_0 is chosen depending on the vector z_0. The projection $\pi z(t)$ of the solution $z(t)$ with the initial value z_0 is written as

$$\pi z(t) = \psi(t) + \zeta_1(t) \quad \text{(see (13))}, \tag{46}$$

where

$$\psi(t) = \pi e^{tC} z_0 + \int_0^t \pi e^{\tau C} a d\tau, \tag{47}$$

$$\zeta_1(t) = \omega_t(\beta_0 + \beta_1(t)) \quad \text{(see (F))}. \tag{48}$$

The set of all functions $\psi(t)$ of the type (47), for fixed C and a, and for arbitrary z_0, obviously belongs to some linear finite-dimensional family Σ of analytic functions. Putting

$$\beta = \beta_0 + \beta_1(t), \tag{49}$$

we may rewrite $\pi z(t)$ as

$$\pi z(t) = w(t) = \psi(t) + \omega_t(\beta). \tag{50}$$

The square Δ is given by the inequalities

$$|\beta^i| \leqslant b \quad \text{(see (30))}. \tag{51}$$

Thus, all the assumptions of assertion (C) Section 4 are satisfied here and therefore there exists a positive number δ as specified in that assertion. Now choose θ so small that

$$|\beta_1^i(t)| \leqslant \delta. \tag{52}$$

Let z_0 be an arbitrary initial value and $\psi(t)$ a function belonging to the family Σ which corresponds to this value z_0 as given by (47). By virtue of (C) Section 4, there exists a square $\Delta' \subset \Delta$ such that, for

$$0 \leqslant t \leqslant \theta, \quad \beta \in \Delta' \tag{53}$$

we have

$$|w(t)| \geqslant \delta t^{k_2}. \tag{54}$$

Taking the center of the square Δ' at β_0, we have

$$\beta_0 + \beta_1(t) \in \Delta'. \tag{55}$$

Therefore

$$|\pi z(t)| \geqslant \delta t^{k_2}$$

which gives the inequality (44) for $k = k_2$. This completes the proof of (G).

6. The Evasion Process

We shall consider here a game (1) Section 5 which satisfies the evasion conditions (B) Section 5. To this game there correspond, by virtue of (G) Section 5, a positive number θ which we fix, and a Volterra operator S_t

defined on the interval $0 \leqslant t \leqslant \theta$ which assigns the control

$$v(t) = S_t(z_0, u) \tag{1}$$

to the initial value z_0 and to the control $u(t)$. We call the control $v(t)$ given by (1) a *special evasion control.*

In this section we use the notation adopted in Section 1 (see (4)).

(A) Applying the special evasion control $v(t)$ (see (1)), whatever the initial value z_0, we bring the game at the end of the period θ to the position $z(\theta)$ satisfying the condition:

$$\xi(\theta) > \epsilon = \delta\theta^k. \tag{2}$$

This follows immediately from (G) Section 5, because

$$\pi z(\theta) \leqslant \xi(\theta).$$

(B) For our game there exists a positive constant c such that, whatever the initial value z_0, with $\xi_0 \leqslant \epsilon$, by applying the special evasion control (1), we move the point $z(t)$ so that

$$\xi(t) > \frac{c\xi_0^k}{(1 + \eta(t))^k}, \qquad 0 \leqslant t \leqslant \theta. \tag{3}$$

on the whole interval $0 \leqslant t \leqslant \theta$.

To prove this assertion, first note that for any controls $u(t)$ and $v(t)$ on the interval $0 \leqslant t \leqslant \theta$, we always have the estimates

$$\frac{1 + \eta(t)}{1 + \eta_0} > \alpha, \tag{4}$$

$$|z(t) - z_0| \leqslant \beta(1 + \eta_0)t, \tag{5}$$

where α and β are two positive constants which depend only on the game and on the number θ.

From the last inequality it follows that for

$$t \leqslant \frac{\xi_0}{2\beta(1 + \eta_0)} \tag{6}$$

we have

$$\xi(t) \geqslant \frac{\xi_0}{2}. \tag{7}$$

Furthermore, if $v(t)$ is the special evasion control (see (1)), then, by virtue of (G) Section 5, for $t \geqslant \xi_0/2\beta(1 + \eta_0)$, we have

$$\xi(t) \geqslant \frac{\delta \xi_0^k}{(2\beta)^k(1+\eta_0)^k}. \tag{8}$$

From (7), (8) and (4), we obtain (3), where c is a suitably chosen constant. This completes the proof of (B).

(C) For an arbitrary initial value an evasion game can be played in such a manner that

$$\xi(t) > \frac{c\epsilon^k}{(1+\eta(t))^k} \tag{9}$$

for

$$0 \leqslant t < \infty. \tag{10}$$

If the initial value z_0 satisfies the condition $\xi_0 \geqslant \epsilon$, then the estimate (9) holds for all t, i.e., for

$$0 \leqslant t < \infty. \tag{11}$$

If the initial value z_0 does not satisfy the condition $\xi_0 \geqslant \epsilon$, then

$$\xi(t) > \frac{c\xi_0^k}{(1+\eta(t))^k} \tag{12}$$

for

$$0 \leqslant t \leqslant \theta. \tag{13}$$

The proof of (C) consists of describing the evasion process. If the initial value z_0 satisfies the condition $\xi_0 \leqslant \epsilon$, then we apply the special evasion control of the time interval $0 \leqslant t \leqslant \theta$ and thus, by virtue of (B), obtain the estimate (12) on this interval. And the estimate (2) holds at the end of the interval θ. Thus the condition

$$\xi_0 > \epsilon, \quad \xi(\theta) > \epsilon \tag{14}$$

is satisfied either for the point z_0 or for the point $z(\theta)$. Beginning from this moment, i.e., from $t=0$ or $t=\theta$, we apply the following control: if $\xi(t) > \epsilon$ for $z(t)$, a control $v(t)$ is chosen arbitrarily until a time t_0 when $\xi(t_0) = \epsilon$. Starting from t_0 we apply the special evasion control to the point $z(t_0)$. This completes the proof of (C).

References

1. R. Issacs, *Differential Games,* Wiley, New York, 1955.
2. L. S. Pontryagin and E. F. Mishehenko, The problem of the evasion of one controlled object from another, *Doklady AN SSSR,* **189**, N 4 (1969).
3. L. S. Pontryagin, A linear differential game of evasion, *Doklady AN SSSR* **191** N 2 (1970).
4. L. S. Pontryagin, Linear differential games, *Proc. Internat. Congress Math.* (Nice, 1970), Paris, 163–172 (1971).

Linear Differential Pursuit Games*

Matematicheskii Sbornik, Novaya Seriya (1980)
112(3): 307–30

1. INTRODUCTION

This paper is concerned with linear differential games whose basic model is the pursuit of one controlled object by another controlled object. Linear differential games indeed constitute a very narrow class of general differential games. Even for these particular cases the results are nevertheless not trivial. Moreover, these results are more effective than the corresponding generalizations to the nonlinear case.

In this paper we shall formulate the problem for a nonlinear case, but give the results for a linear problem.

Formulation of the Problem. The theory of differential games emerged in the course of the mathematical modelling of technical problems. Models can be constructed in several ways. The model should be chosen so that it aims at reflecting the most essential features of the technical problem, but yields to a mathematical treatment. Thus, the theory should not be formulated in complete isolation from technical problems.

To illustrate with a specific example, let us imagine an aircraft pursuing another one. The aim of the former is to catch up the latter, the aim of

* A translation of this article appeared, in a slightly different form, under the title "Linear Differential Games of Pursuit" in *Mathematics of the USSR – Sbornik*, 1981, 40(3):285–303. This translation appears by permission of the American Mathematical Society.

the latter is to evade the pursuit. Each pilot controls his plane, keeping his end in view and using the information available about the situation. This information consists of two parts. The first part is the complete knowledge about the technical capabilities of the aircrafts; the second part is the information about the behaviour of his own plane and about the behaviour of the enemy aircraft. This information about aircraft behaviour may include various data on the states of the planes for the period preceding a given instant of time. But we cannot, however, assume that we know anything about the future behaviour of the planes, because they are controlled and a pilot may change the position of controls at any instant and thereby alter the behaviour of the plane. In reality, each pilot can receive information about the enemy only with some delay. There is no need, however, to include this fact in a mathematical model. On the contrary, we can even suppose that we know the enemy's behaviour in advance, construct a mathematical model on this assumption, and then show that the theory can be used in solving the real problem approximately.

We now take up the mathematical description of the pursuit process. Two controlled objects, the pursuer and the evader, take part in this process. The state of each object at any instant of time is determined by its phase vector. Let x and y denote the phase vectors of the pursuer and of the evader, respectively. Write the equations of the controlled objects as usual:

$$\dot{x} = f(x, u), \quad \dot{y} = g(y, u) \tag{1}$$

where a dot stands for differentiation with respect to time, and u, v are controls, i.e., the parameters on the right-hand sides of the equations. Each parameter belongs to its bounding set

$$u \in P, \quad v \in Q,$$

where P and Q are arbitrary sets. If the control u is a given function of time t, i.e., $u = u(t)$, then the first equation in (1) becomes an ordinary differential equation which can be solved under a given initial value $x(0) = x_0$. The same can also be said of the second equation. Since x and y are phase vectors, each of them decomposes into two parts

$$x = (x_1, x_2), \quad y = (y_1, y_2),$$

where x_1 and y_1 determines the geometric positions of the objects, x_2 and y_2 their velocities. Pursuit is assumed to be terminated as soon as

$$x_1 = y_1,$$

i.e., when the objects coincide geometrically.

The first part of the information which we mentioned earlier consists of the equations in (1). These equations do not define the motion of the objects, but only describe their possible motions, because we obtain various motions for various controls $u = u(t)$ and $v = v(t)$. Thus, in our example, equations (1) describe the technical potentialities of the planes.

We can consider the pursuit process from two different standpoints.

(1) We are the pursuer. In this case our objective is to terminate the pursuit and the control u is at our disposal to achieve this aim. Thus, we have to construct the control $u(t)$ at every instant of time t, knowing the equations (1), i.e., the first part of the information, and using the second part in the form of knowledge about the functions $x(s)$, $y(s)$, and $v(s)$ in the interval $t - \theta \leqslant s \leqslant t$, where θ is a properly chosen positive number.

(2) We are the evader. In this case our objective is to evade the termination of the pursuit and the control v is at our disposal to achieve this aim. Thus, we have to construct the control $v(t)$ at every instant of time, knowing equations (1), i.e., the first part of the information, and using the second part in the form of knowledge about the functions $x(s)$, $y(s)$, and $u(s)$ in the interval $t - \theta \leqslant s \leqslant t$.

This is the mathematical model of pursuit that we consider and it inevitably splits the problem into two distinct ones, namely, the pursuit problem and the evasion problem. This happens because we use different information in the two different approaches. There also exists another model, due to Isaacs, in which the same information, namely a knowledge of the functions $x(t)$ and $y(t)$, is utilized both in the pursuit as well as in the evasion problem. In this model it is assumed that there exist an optimal pursuit control $u = u(x, y)$, which is a function of the states x and y of the objects, and an optimal evasion control $v = v(x, y)$ which is a function of the same states. In this model, the problem is mathematically very definite: it lies in finding the functions $u(x, y)$ and $v(x, y)$ called the optimal strategies. This definiteness, however, renders the solution tremendously difficult. In particular, assuming the existence of optimal strategies, we essentially narrow the class of problems under consideration.

Differential Games. A differential game arises in considering the pursuit process as a result of the natural tendency to simplify the notation. Namely, in place of two phase vectors x and y, we introduce one vector $z = (x, y)$ forming the phase space R of the game as the direct sum of phase spaces of both objects. Then the two equations in (1) are written as one equation

$$\dot{z} = F(z, u, v) \tag{3}$$

and the relation (2) defines some subset M in the vector space R. We are now in a position to define a differential game, irrespective of the initial pursuit process. A differential game is given, if its vector phase space R

and equation (3) are given, where $z \in R$, F is some function of three variables, u and v are the pursuit and evasion controls, and M is the subset of R on which the game terminates.

As in the case of pursuit, we associate two distinct problems with a differential game.

(1) Our aim is to terminate the game, i.e., to bring the point z into the set M. For this purpose we have at our disposal a pursuit control u, so that we choose a value $u(t)$ of this control at every instant of time t, keeping our aim in view and using the functions $z(s)$ and $v(s)$ in the interval $t - \theta \leqslant s \leqslant t$. These are the rules of our pursuit game.

(2) Our aim is to prevent the game from being terminated, i.e., to prevent the point z from entering the set M. For this purpose we have at our disposal an evasion control v, so that we choose a value $v(t)$ of this control at every instant of time t, keeping our aim in view and using the functions $z(s)$ and $u(s)$ in the interval $t - \theta \leqslant s \leqslant t$. These are the rules of the evasion game.

Linear Differential Games. We assume that the phase space R of a linear game is an n-dimensional Euclidean space. The equation of this game is of the form:

$$\dot{z} = Cz - u + v,$$

where $z \in R$, C is some linear mapping of the space R into itself, the controls u and v are vectors of the space R. These vectors are not arbitrary, but satisfy the conditions:

$$u \in P, \quad v \in Q,$$

where P and Q are compact convex subsets of R (the dimensions of P and Q are arbitrary). The controls $u = u(t)$ and $v = v(t)$, being functions of time, are measurable. We assume that the set M on which the game terminates is a closed convex subset of the space R, and in particular cases, a vector subspace of R.

In this paper we only deal with a differential pursuit game. Its solution is based on several concepts related to convex sets, which are dealt with in the next three sections. This communication reports in full the results obtained in [1] and [2].

2. Convex Sets and Their Geometrical Difference

In this section we consider closed convex subsets of a Euclidean vector space R, but will not always say explicitly that a set is closed and convex.

(A) Let A and B be two closed convex subsets of a space R; the set

$$E = \alpha A + \beta B, \tag{1}$$

where α and β are fixed real numbers, is defined to be the totality of all $z = \alpha x + \beta y$, where $x \in A$, $y \in B$. Clearly, the set E is convex. If one of the sets A or B is compact, E is obviously closed. Furthermore, if both A and B are compact, E is also compact. In the particular case where α, $\beta = 1$, the set E is the algebraic sum $A + B$ of two closed subsets of R.

(B) If A is a closed convex subset of R, its *support function*, denoted by $c(A, u)$, is defined to be a function of an arbitrary vector $u \in R$, and is given by the expression

$$c(A, u) = \sup_{x \in A} (x, u).$$

If α and β are two non-negative numbers, then we have

$$c(E, u) = \alpha c(A, u) + \beta c(B, u), \tag{2}$$

which can be easily verified. The following two relations

$$A \subset B, \tag{3}$$

$$c(A, u) \leqslant c(B, u). \tag{4}$$

are equivalent.

(C) **Geometric Difference.** If A and B are two closed convex subsets of the space R, where B is compact, their *geometric difference*

$$D = A \overset{*}{-} B$$

is defined to be the collection of all points $z \in R$ for which $z + B \subset A$. Clearly, the set D is convex and closed. Moreover, $D + B \subset A$, where D is the maximal set satisfying this condition, i.e.,

$$\hat{D} + B \subset A$$

implies

$$\hat{D} \subset D. \tag{5}$$

It turns out that

$$c(D, u) \leqslant c(A, u) - c(B, u). \tag{6}$$

Let X be a closed convex set, and Y a compact set, and put

$$F = (X + Y) \overset{*}{-} Y.$$

Then

$$F = X. \tag{7}$$

We now prove (6). We have $D+B\subset A$. Hence, by virtue of (2), (3), and (4), we have

$$c(D,u)+c(B,u) \leqslant c(A,u)$$

and, consequently,

$$c(D,u) \leqslant c(A,u)-c(B,u),$$

i.e., the formula (6) is true.

We now prove (7). Since $X+Y\subset X+Y$, by virtue of (6), we have

$$X\subset F. \tag{8}$$

Further, by virtue of (6) and (2), we have

$$c(F,u) \leqslant c(X+Y,u)-c(Y,u) = c(X,u).$$

Hence, by the equivalence of (3) and (4), we find that

$$F\subset X. \tag{9}$$

From (8) and (9) we obtain (7).

This completes the proof of assertion (C).

(D) Let A be a closed convex set, U and V compact convex sets. Then

$$(A \overset{*}{-} V)+U\subset (A+U)\overset{*}{-} V. \tag{10}$$

Suppose that a point z belongs to the left-hand side of this relation. Then

$$z = x+y,$$

where

$$x\in A\overset{*}{-} B, \tag{11}$$

$$y\in U. \tag{12}$$

By (C) and (11) it follows that

$$x+V\subset A. \tag{13}$$

Adding (12) and (13), we obtain

$$z+V\subset A+U.$$

Hence, by the definition of geometric difference, we have

$$z\in (A+U)\overset{*}{-} V,$$

and thus we have demonstrated the truth of the inclusion (10).

(E) Let H_r denote a ball of radius r with center at the origin in the space R. The distance between two compact convex subsets X and Y in R is defined to be the minimal number r for which the inclusions

$$X \subset Y + H_r \quad \text{and} \quad Y \subset X + H_r$$

hold. The distance between X and Y is usually denoted by $h(X, Y)$. We can easily prove that all the axioms of a metric hold for this distance. Hence the collection $\Omega(R)$ of all non-empty compact convex subsets in R is a metric space. It is known that if R' is a non-empty compact subset of R, then the set $\Omega(R')$ of all elements of the space $\Omega(R)$ contained in R' is a compact subset of $\Omega(R)$.

(F) Obviously, $H_r \overset{*}{-} H_s = H_{r-s}$, if $r > s$.

(G) If $X(t)$ is a continuous function of a real numerical parameter t which takes its values in the metric space $\Omega(R)$, the integral

$$Y(p, q) = \int_p^q X(t)\, dt \tag{14}$$

can be defined, where it is assumed that $p \leqslant q$. It turns out that $Y(p, q) \in \Omega(R)$ is a continuous function of the integration limits p and q. Further, if r is a number between p and q, i.e., $p \leqslant r \leqslant q$, then

$$\int_p^r X(t)\, dt + \int_r^q X(t)\, dt = \int_p^q X(t)\, dt.$$

The integral (14) is defined as an ordinary Riemann integral. Let

$$S = \{t_0 = p, t_1, \ldots, t_k = q\}$$

be a sequence of points which subdivide the interval pq in such a way that

$$t_0 < t_1 < \ldots < t_k.$$

Then we define the sum

$$\Sigma(S) = \sum_{i=1}^{k} X(\tau_i)(t_i - t_{i-1}),$$

where τ_i lies in the interval $t_{i-1} t_i$, i.e., $t_{i-1} \leqslant \tau_i \leqslant t_i$. This sum $\Sigma(S)$ is, by virtue of (A), a compact convex set. It depends on the subdivision S of the integral pq. Let $\delta(S)$ denote the length of the largest subinterval in the subdivision S of the interval pq, i.e., the largest of the numbers $t_i - t_{i-1}$, $i = 1, \ldots, k$. Then there exists a compact convex set $Y(p, q)$, such that the distance $h(Y(p, q), \Sigma(S))$ tends to zero together with $\delta(S)$. This limit set $Y(p, q)$ is called the integral

$$\int_p^q X(t)\, dt = Y(p, q).$$

(H) It is easy to prove that the convex set $Y(p,q)$ coincides with the set of all points y of the type

$$y = \int_p^q x(t)\,dt,$$

where $x(t)$ is a measurable function of t with its values in R, such that $x(t) \in X(t)$. Further, if $y \in Y'(p,q)$, where $Y'(p,q)$ is the boundary of the set $Y(p,q)$, then the function $x(t)$ corresponding to this value of y satisfies the condition $x(t) \in X'(t)$ for almost all t, where $X'(t)$ is the boundary of the set $X(t)$.

(I) Obviously,

$$\int_p^q H_{r(t)}\,dt = H_{\int_p^q r(t)dt}.$$

3. Estimation of the Geometric Difference

In this section we shall consider compact convex subsets of a Euclidean vector space R. Let H_r denote, as in Section 2, a ball of radius r with center at the origin in R.

(A) Let A be a compact convex subset of the space R. Put

$$A_r = A \overset{*}{-} H_r.$$

Obviously, the set A_r may be non-empty for certain positive r if and only if the dimension of A coincides with the dimension of R. Let α denote the maximal r for which the set A_r is non-empty. Here we assume that α is a positive number and that $r \leqslant \alpha$. Since α depends on the set A, we denote α by $\alpha(A)$. Let δ stand for the diameter of the set A, and let $\delta = \delta(A)$. Furthermore, put

$$k = k(A) = \delta/\alpha.$$

We now evaluate the distance $h(A_r, A)$ between the sets A_r and A, denoting this distance by $\sigma(A)$. We find that

$$\sigma(r) \leqslant kr. \tag{1}$$

We prove (1). Since A_α is a non-empty set, there exists a point c in A, such that $c + H_\alpha \subset A$. And since the set A_r lies in the set A, there exists a point a in the set A such that the distance between this point and the set A_r is equal to the distance between the sets A_r and A. So we have

$$\sigma(r) = \rho(a, A_r).$$

Choose a point b in the interval ac, so that

$$\frac{\rho(a,b)}{\rho(a,c)} = \frac{r}{\alpha}.$$

Since $c + H_\alpha \subset A$, it follows from similar considerations that $b + H_r \subset A$. Thus, $b \in A_r$, and we have

$$\sigma(r) \leqslant \rho(a,b) = \frac{\rho(a,b)}{\rho(a,c)}\,\rho(a,c) = \frac{r}{\alpha}\,\rho(a,c) \leqslant \frac{r}{\alpha}\,\delta = kr.$$

This completes the proof of (1).

(B) Let A and B be two compact convex subsets of the space R, such that their difference $D = A \overset{*}{-} B$ is of the same dimension as the space R. Then there exist two positive numbers γ and l dependent only on $k(D)$, $\alpha(D)$ and $\delta(A)$, such that, for $r \leqslant \gamma$, any two compact convex subsets A' and B' of R having the property that

$$h(A,A') \leqslant r, \quad h(B,B') \leqslant r, \tag{2}$$

have a difference $D' = A' \overset{*}{-} B'$ which satisfies the condition

$$h(D,D') \leqslant lr. \tag{3}$$

We now prove this assertion. From (2) it follows that

$$A' + H_r \supset A, \tag{4}$$

$$B' \subset B + H_r. \tag{5}$$

Subtracting the set H_r from both parts of (4), we obtain by (C) Section 2

$$A' \supset A \overset{*}{-} H_r. \tag{6}$$

Subtracting (5) from (6), we obtain

$$D' = A' \overset{*}{-} B' \supset A \overset{*}{-} B \overset{*}{-} H_{2r} = D \overset{*}{-} H_{2r}. \tag{7}$$

Since D is of the same dimension as R, we have for $2r \leqslant \alpha(D)$, by virtue of (A), that

$$D \overset{*}{-} H_{2r} + H_{2kr} \supset D.$$

Thus,

$$D' + H_{2kr} \supset D,$$

where $k = k(D)$.

Similarly, we can prove that

$$D + H_{2k'r} \supset D',$$

where $k' = k(D')$ can be estimated, using $\delta(A)$ and $\alpha(D)$, for $\gamma = \alpha(D)/3$. Since the set D contains a ball of radius $\alpha(D)$, we obtain from (7) that

$$\alpha' = \alpha(D') \geqslant \alpha(D) - 2r.$$

Further, we have

$$D' + B' \subset A' \subset A + H_r,$$

and hence

$$\delta' = \delta(D') \leqslant \delta(A) + 2r.$$

From these estimates of α' and δ', we find that for $r \leqslant \gamma = \alpha(D)/3$,

$$k' = \frac{\delta'}{\alpha'} \leqslant \frac{\delta(A) + \frac{2}{3}\alpha(D)}{\frac{1}{3}\alpha(D)}.$$

This estimate depends only on $\delta(A)$ and $\alpha(D)$.

Taking the greater of the two numbers $2k$ and $2k'$ as l, we find that (3) is true. From the construction of l we see that it depends only on the sets A and B, provided their geometric difference D is of the same dimension as the space R.

(C) Let A, A' and B, B' be compact convex subsets of the space R. Clearly, if $h(A, A') \leqslant r, h(B, B') \leqslant r$, then

$$h(A + B, A' + B') \leqslant 2r.$$

4. Alternating Integrals

(A) Let A be a closed convex subset of the space R, and let

$$U_1, \ldots, U_n \quad \text{and} \quad V_1, \ldots, V_n \tag{1}$$

be two sequences of compact convex subsets in R. Take

$$A_0 = A, \quad A_i = (A_{i-1} + U_i) \overset{*}{-} V_i, \quad i = 1, \ldots, n. \tag{2}$$

The set A_n is called an *alternating sum* of the sequence (1) with the initial value A. The set A_n is expressed in full form by the formula:

$$A_n = (\ldots((((A + U_1) \overset{*}{-} V_1) + U_2) \overset{*}{-} V_2) + \ldots + U_n) \overset{*}{-} V_n.$$

Put

$$U = U_1 + \ldots + U_n \quad \text{and} \quad V = V_1 + \ldots + V_n.$$

Then, by virtue of (10) Section 2, we have the inclusions

$$(A \overset{*}{-} V) + U \subset A_n \subset (A + U) \overset{*}{-} V. \tag{3}$$

(B) Assume that the set A is compact and that there exists a positive number β such that

$$\alpha(A_i) > \beta, \quad i = 1, \ldots, n. \tag{4}$$

Then the alternating sum A_n is a continuous function of $A, U_1, \ldots, U_n, V_1, \ldots, V_n$ in $\Omega(R)$ for those values of the variables for which (4) holds. This follows from (B) and (C) of the last section.

(C) Let A be a closed convex subset of R, $U(\tau)$ and $V(\tau)$ two compact convex sets in R continuously dependent on the parameter τ in the interval $p \leqslant \tau \leqslant q$; and

$$P = (r_0, r_1, \ldots, r_n)$$

some subdivision of the interval pq, i.e.,

$$r_0 = p < r_1 < \ldots < r_n = q.$$

Put

$$U_i = \int_{r_{i-1}}^{r_i} U(\tau)\, d\tau, \quad V^i = \int_{r_{i-1}}^{r_i} V(\tau)\, d\tau. \tag{5}$$

As in (A), let us construct from the set A and the sequence (5), a sequence of sets $A_0, A_1, \ldots, A_n$. Furthermore, let us use the symbol

$$Y(A, P, r_i) = A_i$$

to signify that A_i depends on the initial set A, the subdivision P and the index i. We could as well write $Y(A,P,r_i) = Y(A,P,r)$, where $r = r_i$. Thus, the set $Y(A,P,r)$ is defined for any r belonging to the sequence P. In case the set A is compact, we assume that the functions $U(\tau)$ and $V(\tau)$ satisfy the following condition. There exists a continuous function $\beta(r)$ defined in the interval pq, positive for all $r > p$, such that for any subdivision P of the interval pq,

$$\alpha(Y(A, P, r)) \geqslant \beta(r). \tag{6}$$

Thus, the possibility is not excluded that $\alpha(A) = 0$. When this condition is satisfied, the set A_n is a continuous function of $r_1, r_2, \ldots, r_{n-1}$. This follows directly from (B). If the closed convex set A is not compact, then take

$$H_\mu \cap A = A^\mu, \quad H_\mu \cap Y = Y^\mu,$$

and in place of (6), let us impose a new condition. Namely, that there exists a large number μ such that

$$\alpha(Y(A^\mu, P, r)) \geqslant \beta(r). \tag{7}$$

Under this assumption, $Y^\mu(A, P, r)$ is a continuous function of $r_1, r_2, \ldots, r_{n-1}$.

We now prove this assertion. It is easily verified that for a given μ, we can choose a large ν such that

$$Y^{\mu}(A,P,r) = Y^{\mu}(A^{\mu},P,r). \tag{8}$$

Since the right-hand side of this inequality is continuous in $r_1, r_2, \ldots, r_{n-1}$, the left-hand side is also continuous in these variables.

(D) Let $\hat{P} = (P_1, \ldots, P_m, \ldots)$ be a sequence of infinitely decreasing subdivisions of the interval pq:

$$P_1 \subset P_2 \subset \ldots \subset P_m \subset \ldots .$$

Then, by virtue of (10) Section 2, we have

$$Y(A,P_1,q) \supset Y(A,P_2,q) \supset \ldots \supset Y(A,P_m,q) \supset \ldots .$$

This is a decreasing sequence of closed convex sets in R. Let $Y(A,\hat{P})$ denote its intersection. The question now is under what conditions does this set not depend on the sequence P of subdivisions of the interval pq? For this purpose, it suffices that the condition (7) be satisfied.

We prove this assertion. By virtue of (8), it is sufficient to demonstrate the truth of this assertion for the case where A is a compact set for which condition (6), instead of (7), is satisfied. Let $\hat{Q} = (Q_1, \ldots, Q_m, \ldots)$ be some other sequence of subdivisions of the interval pq similar to the sequence $\hat{P}$. Suppose that the compact sets $Y(A,\hat{P})$ and $Y(A,\hat{Q})$ do not coincide. Then the distance between these sets is positive. Let 3ϵ be this distance. Then there exists a large positive number m such that for $j \geqslant m$,

$$h(Y(A,P_j,q), Y(A,\hat{P})) \leqslant \epsilon, \quad h(Y(A,Q_j,q), Y(A,\hat{Q})) \leqslant \epsilon.$$

By virtue of (6), the set $Y(A,P_m,q)$ is a continuous function of the terms in the subdivision P_m, except the points p and q which are fixed. Thus there exists a small number δ such that, by shifting the terms of the subdivision P_m by a distance less than δ, we obtain a new subdivision P' satisfying the condition

$$h(Y(A,P_m,q), Y(A,P',q)) \leqslant \epsilon.$$

Now choose a number $j > m$ such that each interval in the subdivision Q_j is less than δ in length. Then, by shifting the elements of the subdivision P_m by a distance less than δ, we obtain a new subdivision P', each element of which belongs to the subdivision Q_j. So we have $P' \subset Q_j$. Hence we find that $Y(A,P',q) \supset Y(A,Q_j,q)$. Since

$$h(Y(A,P_m,q), Y(A,\hat{P})) \leqslant \epsilon,$$

we have
$$Y(A,\hat{P})+H_{2\epsilon}\supset Y(A,P',q)\supset Y(A,Q_j,q)\supset Y(A,\hat{Q}).$$

But as the sequences $\hat{P}$ and $\hat{Q}$ can be interchanged, we have another inclusion relation
$$Y(A,\hat{Q})+H_{2\epsilon}\supset Y(A,\hat{P}).$$
Hence
$$h(Y(A,\hat{P}),\,Y(A,\hat{Q}))\leqslant 2\epsilon.$$

But by assumption this distance is equal to 3ϵ. Thus we have arrived at a contradiction.

(E) If the condition (7) is satisfied, the set $Y(A,\hat{P})$ does not depend on the sequence of subdivisions $\hat{P}$. This set, called the *alternating integral* of the functions $U(\tau)$ and $V(\tau)$ over the interval pq with the initial set A, is denoted as follows

$$Y(A,\hat{P}) = \int_{A,p}^{q}(U(\tau)\,d\tau \overset{*}{-} V(\tau)\,d\tau).$$

From (8) it follows that

$$H_\mu\cap\int_{A,p}^{q}(U(\tau)\,d\tau \overset{*}{-} V(\tau)\,d\tau) = H_\mu\cap\int_{A^\nu,p}^{q}(U(\tau)\,d\tau \overset{*}{-} V(\tau)\,d\tau). \tag{9}$$

(F) Let x be an arbitrary point in the interval pq, i.e., $p\leqslant x\leqslant q$. Put

$$W(x) = W(A,x) = \int_{A,p}^{x}(U(\tau)\,d\tau \overset{*}{-} V(\tau)\,d\tau).$$

The right-hand side is well-defined since the condition (7) is satisfied in the interval pq, and hence is also satisfied in the interval px. Now let x and y ($x\leqslant y$) be two points in the interval pq. Then

$$W(y) = \int_{W(x),x}^{y}(U(\tau)\,d\tau \overset{*}{-} V(\tau)\,d\tau). \tag{10}$$

Furthermore, if $x\geqslant p+\epsilon$, where ϵ is a positive number, and the set A is compact, there exists a positive number λ such that

$$h(W(y),\,W(x))\leqslant\lambda(y-x). \tag{11}$$

This inequality shows that for a compact set A the function $W(x)$ has a uniformly bounded growth in each interval $p+\epsilon\leqslant x\leqslant q$. Hence it readily follows, by virtue of (9), that $H_\mu\cap W(x)$ also has a uniformly bounded growth in the interval $p+\epsilon\leqslant x\leqslant q$.

We now prove assertion (F). Let $P = (r_0, r_1, \ldots, r_n)$ be some subdivision of the interval pq containing x and y as division points: $x = r_k$, $y = r_1$. Put

$$U_i^* = \int_{r_{k+i-1}}^{r_{k+i}} U(\tau)\,d\tau, \quad V_i^* = \int_{r_{k+i-1}}^{r_{k+i}} V(\tau)\,d\tau, \quad i = 1, 2, \ldots, l-k. \tag{12}$$

Further, let B be some closed convex subset of the space R. Let us construct an alternating sum (see (A)) of the sequences (12) with the initial set B. Let $Y(B, P, x, y)$ denote this sequence. Then, obviously

$$Y(A, P, y) = Y(Y(A, P, x), P, x, y). \tag{13}$$

Now let $\hat{P} = (P_1, \ldots, P_j, \ldots)$ be a sequence of decreasing subdivisions of the interval pq such that each subdivision P_j contains x and y as division points. Let $j \geqslant m$ be two natural numbers. Let P be a subdivision of the interval pq that coincides with the subdivision P_j of the interval px and with the subdivision P_m of the interval xy. Then we have the inclusion

$$Y(A, P_j, Y) \subset Y(A, P, y) \subset Y(A, P_m, y).$$

We now pass to the limit in these inclusions as $j \to \infty$. Then, using (13), we obtain

$$W(y) \subset Y(W(x), P_m, x, y) \subset Y(A, P_m, y).$$

The limit process is justified for the middle term of the inclusion, provided the set A is compact, by the assertion (B). If A is not compact, we have to use the (H_μ, H_ν) technique to demonstrate that the limit can be taken. Taking the limit of these inclusions as $m \to \infty$, we obtain

$$W(y) \subset \int_{W(x), x}^{y} (U(\tau)\,d\tau \overset{*}{-} V(\tau)\,d\tau) \subset W(y).$$

Hence we obtain (10). Thus, the first part of the assertion (F) has been proved.

We now prove the second part, i.e., the inequality (11). From (6) it follows that $\alpha(W(x)) \geqslant \beta(x)$, where the function $\beta(x)$ is continuous and positive for all $x > p$. Thus, in the interval $p + \epsilon \leqslant x \leqslant q$, we have $\alpha(W(x)) \geqslant \beta$, where β is a positive number. But we have $\delta(W(x)) \leqslant \delta$, where δ is some positive number. Consequently, $k(W(x)) \leqslant k$, where k is some positive number (see (A) Section 3).

As the functions $U(\tau)$ and $V(\tau)$ are continuous, there exists a number r such that $U(\tau) \subset H_r$ and $V(\tau) \subset H_r$. Hence

$$\int_x^y U(\tau)\,d\tau \subset H_{r(y-x)}, \int_x^y V(\tau)\,d\tau \subset H_{r(y-x)}. \tag{14}$$

Furthermore, from (10) and (3), it follows that

$$\left(W(x) \overset{*}{-} \int_x^y V(\tau)\,d\tau\right) + \int_x^y U(\tau)\,d\tau \subset W(y) \subset$$

$$\subset \left(W(x) + \int_x^y U(\tau)\,d\tau\right) \overset{*}{-} \int_x^y V(\tau)\,d\tau. \tag{15}$$

Making these inclusions stronger, we obtain

$$W(x) \overset{*}{-} \int_x^y V(\tau)\,d\tau \subset W(y) + H_{r(y-x)} \subset$$

$$\subset \left[\left(W(x) + \int_x^y U(\tau)\,d\tau\right) \overset{*}{-} \int_x^y V(\tau)\,d\tau\right] + H_{r(y-x)}.$$

Further strengthening these inclusions, from (14), we obtain

$$W(x) \overset{*}{-} H_{r(y-x)} \subset W(y) + H_{r(y-x)} \subset W(x) + H_{3r(y-x)}.$$

Subtracting $H_{r(y-x)}$ from these inclusions we obtain

$$W(x) \overset{*}{-} H_{2r(y-x)} \subset W(y) \subset W(x) + H_{2r(y-x)}. \tag{16}$$

By virtue of (A) Section 3, for $2r(y-x) \leqslant \beta$, the first of these inclusions gives

$$W(x) \subset W(y) + H_{2kr(y-x)}.$$

As $k > 1$, this inclusion together with the second inclusion in (16) gives the inequality (11), where $\lambda = 2kr$, provided $2r(y-x) \leqslant \beta$. Hence we obtain the inequality (11) for any x, y.

This completes the proof of (F).

Assume that the set A and both the sequences (1) are balls with centers at zero, so that $A = H_1, U_i = H_{r_i}, V_i = H_{s_i}$. Suppose that for every i,

$$\gamma_i = l + (r_1 + r_2 + \ldots + r_i) - (s_1 + s_2 + \ldots + s_i) \geqslant 0.$$

Then, by virtue of (2), we have $A_i = H_{\gamma_i}$, so that the alternating sum A_n is given by the expression $A_n = H_{\gamma_n}$.

If $A = H_l$, $U(t) = H_{r(t)}$, $V(t) = H_{s(t)}$, then, assuming

$$\gamma(t) = l + \int_p^t r(\tau)\,d\tau - \int_p^t s(\tau)\,d\tau \geqslant 0$$

for $p < t \leqslant q$, we obtain

$$\int_{H_l, p}^q (H_{r(\tau)}\,d\tau \overset{*}{-} H_{s(\tau)}\,d\tau) = A_{\gamma(q)}.$$

5. Linear Differential Pursuit Games

Consider a linear differential game defined by the equation

$$\dot{z} = Cz - u + v, \tag{1}$$

where $z \in R$ is the phase vector of the game, R is the vector phase space of our game, u is the pursuit control and v the evasion control. Here we have the condition

$$U \in P \quad \text{and} \quad v \in Q,$$

where P and Q are compact convex subsets of R. The game is considered terminated, when the point z enters the set M, where M is a closed convex subset of R.

Note that if $u(t)$ and $v(t)$ are two given functions, then the solution of equation (1) for $u = u(t)$ and $v = v(t)$ under the initial condition $z(0) = z_0$ can be expressed as

$$z(t) = e^{tC} z_0 + \int_0^t e^{sC} [v(t-s) - u(t-s)]\, ds. \tag{2}$$

Let us construct the alternating integral

$$W(t) = \int_{M,0}^{t} (e^{\tau C} P\, d\tau \overset{*}{-} e^{\tau C} Q\, d\tau).$$

Let z_0 be an arbitrary point in R. Consider the hypothetical inclusion

$$e^{tC} z_0 \in W(t). \tag{3}$$

If there exists a non-negative t such that this relation holds, then let us denote the minimal t for which (3) holds by $t_0 = T(z_0)$. Then the following result is valid.

(A) Let ϵ be a small positive number and $v(t)$ an evasion control defined in the interval $0 \leqslant t \leqslant \epsilon$. Take an arbitrary pursuit control $u(t)$ in the same interval, and consider the solution of equation (1) in this interval, taking the functions $v(t)$ and $u(t)$ for v and u, respectively. By virtue of (2), the solution of equation (1) under the initial condition $z(0) = z_0$ at $t = \epsilon$, is written as follows:

$$z_1 = z(\epsilon) = c^{\epsilon C} z_0 + \int_0^{\epsilon} e^{sC} (v(\epsilon - s) - u(\epsilon - s))\, ds. \tag{4}$$

Put

$$t_1 = T(z_1).$$

We assume that the control $v(t)$ is given on the interval $O\epsilon$, and the control $u(t)$ is, for the time being, chosen arbitrarily in the same interval. Then t_1 is a functional of the function $u(t)$ given on the interval $O\epsilon$. Now choose a control $u(t)$ on the interval $O\epsilon$ such that t_1 attains its minimum and denote this minimum also by t_1. Then

$$t_1 \leqslant t_0 - \epsilon.$$

We now prove this assertion. For the time being, let t_1 be an arbitrary number. Consider the following hypothetical inclusion

$$e^{(t_1+\epsilon)C} z_0 \in W(t_1 + \epsilon). \tag{5}$$

Clearly, it holds when $t_1 = t_0 - \epsilon$. In place of (5), let us consider a weaker inclusion (see (15) Section 4)

$$e^{(t_1+\epsilon)C} z_0 \in \left(W(t_1) + \int_{t_1}^{t_1+\epsilon} e^{\tau C} P \, d\tau \right) \overset{*}{-} \int_{t_1}^{t_1+\epsilon} e^{\tau C} Q \, d\tau.$$

This inclusion is also true for $t_1 = t_0 - \epsilon$. It is equivalent to the following inclusion:

$$e^{(t_1+\epsilon)C} z_0 + \int_{t_1}^{t_1+\epsilon} e^{\tau C} Q \, d\tau \subset W(t_1) + \int_{t_1}^{t_1+\epsilon} e^{\tau C} P \, d\tau. \tag{6}$$

For the set which is the second term on the left-hand side of this inclusion, let us substitute a definite element from this set, which we write as

$$e^{t_1 C} \int_0^{\epsilon} e^{sC} v(\epsilon - s) \, ds,$$

where the $v(t)$ in this integral is the evasion control $v(t)$ already given. From (6) we obtain

$$e^{(t_1+\epsilon)C} z_0 + e^{t_1 C} \int_0^{\epsilon} e^{sC} v(\epsilon - s) \, ds \in W(t_1) + \int_{t_1}^{t_1+\epsilon} e^{\tau C} P \, d\tau.$$

This inclusion holds for $t_1 = t_0 - \epsilon$. Now take for t_1 the minimal t_1 for which this inclusion holds. Then there exists a definite element of the second term on the right-hand side of this inclusion for which the inclusion holds. Let us write this element as

$$e^{t_1 C} \int_0^{\epsilon} e^{sC} u(\epsilon - s) \, ds,$$

where the $u(t)$ under the integral sign is a certain control $u(t)$ given in the interval $O\epsilon$. Making this substitution and then transferring the second term to the left-hand side, we obtain the inclusion:

$$e^{t_1C}\left[e^{\epsilon C}z_0+\int_0^{\epsilon} e^{sC}(v(\epsilon-s)-u(\epsilon-s))\,ds\right]\in W(t_1).$$

The left-hand side of this inclusion can, by virtue of (4), be expressed as $e^{t_1C}z_1 \in W(t_1)$. This completes the proof of our assertion.

Assertion (A) gives a means by which, using a control $v(t)$ defined in the interval $O\epsilon$, we can construct $u(t)$ that minimizes the equality $T(z_1)=t_1$, where $t_1 \leqslant t_0-\epsilon$. If, however, we know the control $v(t)$ in the interval $\epsilon 2\epsilon$, then from the already given initial value z_1, we can, using the above method, obtain a control $u(t)$ on the interval $\epsilon 2\epsilon$ that minimizes $T(z_2)=t_2$, and $t_2 \leqslant t_1-\epsilon$. Repeating this process, we can construct a control $u(t)$ from the known $v(t)$ step by step, in such a way that at a certain moment of time $t^* \leqslant T(z_0)$ the point $z(t)$ hits the set M. But the control $u(t)$ at the instant t is constructed under the assumption that the control $v(s)$ is given on the interval $Ot+\epsilon$, i.e., we use knowledge of the control $v(t)$ in advance. We correct this defect as follows. Let $\hat{v}(t)$ be some control which becomes known with the passage of time. Take

$$v(t) = \hat{v}(t-\epsilon).$$

Thus, in order to know the control $v(s)$ on the interval $0 \leqslant s \leqslant t+\epsilon$, it is sufficient to know the control $\hat{v}(s)$ on the interval $-\epsilon \leqslant s \leqslant t$, i.e., without knowing it in advance. But the control $\hat{v}(s)$ is not given on the interval $(-\epsilon, 0)$. Therefore we shall specify it arbitrarily. Let $z(t)$ denote the solution of equation (1) for the initial value z_0 and for $u=u(t)$ and $v=v(t)$. And let $\hat{z}(t)$ denote the same solution for $u=u(t)$ and $v=\ddot{v}(t)$. We know that $z(t^*)\in M$. We now examine $\hat{z}(t^*)$. For this purpose, let us estimate the difference $\hat{z}(t)-z(t)$. We have

$$\hat{z}(t)-z(t) = \int_0^t e^{sC}\hat{v}(t-s)\,ds - \int_0^t e^{sC}v(t-s)\,ds =$$

$$= \int_0^t e^{sC}\hat{v}(t-s)\,ds - \int_0^t e^{sC}\hat{v}(t-s-\epsilon)\,ds. \qquad (7)$$

Substituting τ for $s+\epsilon$, write the second integral in (7) as:

$$\int_0^t e^{sC}\hat{v}(t-s-\epsilon)\,ds = \int_{\epsilon}^{t+\epsilon} e^{(\tau-\epsilon)C}\hat{v}(t-\tau)\,d\tau.$$

Furthermore, this expression can be rewritten as

$$\int_{\epsilon}^{t+\epsilon} e^{(\tau-\epsilon)C}\hat{v}(t-\tau)\,d\tau = \int_0^t e^{(\tau-\epsilon)C}\hat{v}(t-\tau)\,d\tau -$$

$$-\int_0^{\epsilon} e^{(\tau-\epsilon)C}\hat{v}(t-\tau)\,d\tau+\int_t^{t+\epsilon} e^{(\tau-\epsilon)C}\hat{v}(t-\tau)\,d\tau.$$

Substituting τ for the integration variable s in the first integral of (7), we obtain

$$\hat{z}(t)-z(t) = \int_0^t (e^{\tau C}-e^{(\tau-\epsilon)C})\,\hat{v}(t-\tau)\,d\tau+$$

$$+\int_0^{\epsilon} e^{(\tau-\epsilon)C}\hat{v}(t-\tau)\,d\tau-\int_t^{t+\epsilon} e^{(\tau-\epsilon)C}\hat{v}(t-\tau)\,d\tau.$$

From this expression it is clear that

$$|\hat{z}(t)-z(t)| \leqslant \epsilon_c,$$

where c depends on the interval Ot and the game. Since the value of $t = t^* \leqslant T(z_0)$ is important for us, the constant c for $t = t^*$ can be estimated from $T(z_0)$ and other quantities that depend on the game. Thus, choosing ϵ *a priori* sufficiently small, we can effect our pursuit in such a way that the point $\hat{z}(t^*)$ occurs at a distance not exceeding ϵc from the set M, where c depends on z_0 and on the game, but does not depend on ϵ.

Summing up this result, we conclude that a differentiable pursuit game starting at the point z_0 may be terminated in a time not greater than $T(z_0)$. Though we do not in reality hit the set M, nonetheless, we approach it to within a distance not greater than ϵ_c. But, as ϵ can be chosen in advance to be arbitrarily small, the above assertion says that the pursuit terminates in a time not greater than $T(z_0)$.

So far it has not been possible to get rid of the inaccuracy ϵc, though it is natural to expect an exact result as ϵc is arbitrarily small. Considerable difficulties apparently have to be overcome before we could obtain an exact result.

The fact that a pursuit game can be terminated in a time exactly equal to $T(z_0)$ can easily be demonstrated without taking recourse to the ϵ-step constructions made above. But this construction gives us a means of improving the result, namely, a means of reducing the pursuit time, when the evasion control $v(t)$ is not chosen in the best possible way.

6. A Simplified Linear Differential Game

In this section we consider a differential game defined by equation (1) Section 5 with a set M which is a vector subspace of the space R. Let L

denote the orthogonal complement of M in R, and π the orthogonal projection of R on the subspace L. Put

$$P(t) = \pi e^{tC} P, \quad Q(t) = \pi e^{tC} Q,$$

and assume that the set $S(t) = P(t) \overset{*}{-} Q(t)$ is of dimension equal to the dimension of the subspace L for all positive $t < T$, where T can, in particular, be equal to infinity. We shall only study the set $S(t)$ in the interval $0 \leqslant t < T$. Furthermore, put

$$\bar{W}(t) = \int_0^t S(\tau)\, d\tau. \tag{1}$$

Consider now the hypothetical inclusion $\pi e^{tc} z_0 \in \bar{W}(t)$. If, for some t, this relation holds, let $t_0 = T(z_0)$ denote the minimal t for which this relation holds.

(A) Let $v(t)$ be a control defined in the interval $0 \leqslant t \leqslant \epsilon$. Now take an arbitrary control $u(t)$ in this interval and let $z(t)$ be a solution of the differential equation of our game for these controls with the initial value z_0. Let $z_1 = z(\epsilon)$ and consider $T(z_1)$. It is a functional of the control $u(t)$ in the interval $O\epsilon$ which we have not chosen yet. Now choose a control $u(t)$ in this interval such that $t_1 = T(z_1)$ attains its minimum. Then we find that $t_1 \leqslant t_0 - \epsilon$.

We prove this assertion. For the time being let t_1 be an arbitrary number. Consider the following hypothetical inclusion

$$\pi e^{(t_1+\epsilon)C} z_0 \in \bar{W}(t_1 + \epsilon). \tag{2}$$

It holds when $t_1 = t_0 - \epsilon$. In place of (2), let us consider another equivalent inclusion

$$\pi e^{(t_1+\epsilon)C} z_0 \in \bar{W}(t_1) + \int_{t_1}^{t_1+\epsilon} S(\tau)\, d\tau. \tag{3}$$

We have

$$S(\tau) + Q(\tau) \subset P(\tau).$$

This inclusion can be integrated. Integrate it from t_1 to $t_1 + \epsilon$. Thus we obtain

$$\int_{t_1}^{t_1+\epsilon} S(\tau)\, d\tau + \int_{t_1}^{t_1+\epsilon} Q(\tau)\, d\tau \subset \int_{t_1}^{t_1+\epsilon} P(\tau)\, d\tau.$$

From this expression and from (3) we get

$$\pi e^{(t_1+\epsilon)C} z_0 + \int_{t_1}^{t_1+\epsilon} Q(\tau)\, d\tau \subset \bar{W}(t_1) + \int_{t_1}^{t_1+\epsilon} P(\tau)\, d\tau.$$

In place of the second term on the left-hand side of this inclusion substitute one of its elements, i.e.,

$$\pi e^{t_1 C} \int_0^{\epsilon} e^{sC} v(\epsilon - s)\, ds,$$

where $v(t)$ is the control that is already given in the interval $O\epsilon$. We obtain

$$\pi e^{(t_1+\epsilon)C} z_0 + \pi e^{t_1 C} \int_0^{\epsilon} e^{sC} v(\epsilon - s)\, ds \in \bar{W}(t_1) + \int_{t_1}^{t_1+\epsilon} P(\tau)\, d\tau. \quad (4)$$

Choose now the minimal $t_1 \leqslant t_0 - \epsilon$ for which (4) holds. Then there exists a definite element of the second term on the right-hand side of (4) for which this inclusion holds. Write this element as follows

$$\pi e^{t_1 C} \int_0^{\epsilon} e^{sC} u(\epsilon - s)\, ds,$$

where $u(t)$ is an element in P. Then (4) can be rewritten as

$$\pi e^{t_1 C} \left[e^{sC} z_0 + \int_0^{\epsilon} e^{sC} (v(\epsilon - s) - u(\epsilon - s))\, ds \right] \in \bar{W}(t_1).$$

The left-hand side can be rewritten in a different way. We get

$$\pi e^{t_1 C} z_1 \in \bar{W}(t_1),$$

and this completes the proof of our assertion.

The results can be summed up in the same way as in the previous section.

7. Application to Concrete Pursuit Cases

(1) Let E be a Euclidean space of dimension $\nu \geqslant 2$, x and y the geometric positions of the pursuer and of the evader. Their motions are given by the equations

$$\ddot{x} + \alpha \dot{x} = a, \quad |a| \leqslant \rho, \quad (1)$$

$$\ddot{y} + \beta \dot{y} = b, \quad |b| \leqslant \sigma. \quad (2)$$

Here α, β, ρ, and σ are positive numbers, **a** and **b** are control vectors belonging to the space E and satisfying the above inequalities, but in other respects they are arbitrary. We assume that pursuit terminates when $x = y$.

We construct a differential game corresponding to the pursuit process described above as follows. Take

$$z^1 = x - y, \quad z^2 = \dot{x}, \quad z^3 = \dot{y}.$$

Here z^1, z^2, and z^3 are vectors of the space E. The phase vector $\mathbf{z}$ is given by the expression $z = (z^1, z^2, z^3)$, so that $\mathbf{z}$ is defined by three vector components.

We now rewrite the set of two equations (1) and (2) as follows:

$$\dot{z}^1 = z^2 - z^3, \quad \dot{z}^2 = -\alpha z^2 + a, \quad \dot{z}^3 = -\beta z^3 + b. \tag{3}$$

The controls u and v are given by the expressions:

$$u = (0, -a, 0), \quad v = (0, 0, b).$$

Consequently,

$$P = \{(0, -a, 0): |a| \leqslant \rho\}, \tag{4}$$

$$Q = \{(0, 0, b): |b| \leqslant \sigma\}, \tag{5}$$

$$M = \{z: z^1 = 0\}, \quad L = \{(z^1, 0, 0): z^1 \in E\}.$$

Thus, we can identify the space L with the space E, taking $(z^1, 0, 0) \triangleq z^1$.

The operator π has the form $\pi(z^1, z^2, z^3) = (z^1, 0, 0) = z^1$. The matrix C is

$$C = \begin{pmatrix} 0 & 1 & -1 \\ 0 & -\alpha & 0 \\ 0 & 0 & -\beta \end{pmatrix}.$$

Solving the homogeneous equation corresponding to the system (3) we obtain the matrix e^{tC}. Thus,

$$e^{tC} = \begin{pmatrix} 1 & \dfrac{1 - e^{-\alpha t}}{\alpha} & -\dfrac{1 - e^{-\beta t}}{\beta} \\ \cdots & \cdots & \cdots \\ \cdots & \cdots & \cdots \\ \cdots & \cdots & \cdots \end{pmatrix}.$$

We have not written the second and the third rows in this matrix because we are only interested in the operator πe^{tC} which, by what has been said above, is expressed as follows:

$$\pi e^{tC}(z_0^1, z_0^2, z_0^3) = z_0^1 + \frac{1 - e^{-\alpha t}}{\alpha} z_0^2 - \frac{1 - e^{-\beta t}}{\beta} z_0^3. \tag{6}$$

We can use this expression to calculate the sets $P(t) = \pi e^{tC}P$, and $Q(t) = \pi e^{tC}Q$. By virtue of (4), (5), and (6), we have

$$P(t) = \left\{\frac{1-e^{-\alpha t}}{\alpha}(-a)\colon |a| \leqslant \rho\right\}, \quad Q(t) = \left\{-\frac{1-e^{-\beta t}}{\beta}b\colon |b| \leqslant \sigma\right\}.$$

Thus the set $P(t)$ is a ball of radius $r(t) = (1-e^{-\alpha t})\,\rho/\alpha$, and $Q(t)$ is a ball of radius $s(t) = (1-e^{-\beta t})\,\sigma/\beta$. And the geometric difference of these sets $S(t)$ is a ball of radius

$$\omega(t) = \frac{1-e^{-\alpha t}}{\alpha}\rho - \frac{1-e^{-\beta t}}{\beta}\sigma.$$

The set $\bar{W}(t)$ defined by (1) Section 6 is a ball of radius

$$\gamma(t) = \int_0^t \left(\frac{1-e^{-\alpha \tau}}{\alpha}\rho - \frac{1-e^{-\beta t}}{\beta}\sigma\right) d\tau.$$

It can be easily proved that for $\omega(t)$ and $\gamma(t)$ to be positive for all $t > 0$, it is sufficient that

$$p > \sigma, \quad \frac{\rho}{\alpha} > \frac{\sigma}{\beta}. \tag{7}$$

Under these conditions, $\gamma(t)$ increases without bound as t increases without bound. If one of the inequalities in (7) is replaced by an opposite inequality, then $\gamma(t)$ will definitely take negative values.

In order to find $T(z_0)$ $(z_0 = (z_0^1, z_0^2, z_0^3))$, we have to find the minimal t for which the point

$$\pi e^{tC}z_0 = z_0^1 + \frac{1-e^{-\alpha t}}{\alpha}z_0^2 - \frac{1-e^{-\beta t}}{\beta}z_0^3$$

belongs to the ball of radius $\gamma(t)$. Thus, $T(z_0)$ is the least positive root of the transcendental equation

$$\gamma(t)^2 = (\pi e^{tC}z_0, \pi e^{tC}z_0).$$

(2) Let x be the pursuit point and y the evasion point in a Euclidean space E of dimension $\nu \geqslant 2$. Their equations of motion are

$$\ddot{x} = a, \quad |a| \leqslant \rho,$$

$$\dot{y} = b, \quad |b| \leqslant \sigma.$$

where **a** and **b** are control vectors belonging to the space E and are subjected to the inequalities given above. The pursuit game terminates when $|x-y| \leqslant l$, where $l > 0$ (l is capture).

The differential game of this pursuit process is defined as follows:

$$z^1 = x - y, \quad z^2 = \dot{x}.$$

Thus, the phase vector of the game z is $z = (z^1, z^2)$, so that the vector z is determined by two vector components. The differential equation of the game is

$$\dot{z}^1 = z^2 - b, \quad \dot{z}^2 = a.$$

The controls u and v are:

$$u = (0, -a), \quad v = (-b, 0).$$

Consequently,

$$P = \{(0, -a)\colon |a| \leqslant \rho\}, \quad Q = \{(-b, 0)\colon |b| \leqslant \sigma\},$$

$$M = \{z = (z^1, z^2)\colon |z^1| \leqslant l\}.$$

Let L denote the set $L = \{(z^1, 0)\colon z^1 \in E\}$. Thus we can identify the space L with the space E, taking

$$(z^1, 0) \triangleq z^1.$$

The matrix C is of the form

$$C = \begin{pmatrix} 0 & 1 \\ 0 & 0 \end{pmatrix},$$

and the matrix e^{tC} is given by the expression

$$e^{tC} = \begin{pmatrix} 1 & t \\ 0 & 1 \end{pmatrix}.$$

Since the set M is not a vector subspace of the space R, we have to solve the problem using the technique described in Section 5. But here we can project the whole construction of the alternating integral orthogonally on the subspace L of the space R. The projection operator is

$$\pi(z^1, z^2) = (z^1, 0) \triangleq z^1.$$

Under this operator the set M is transformed into a ball H_l in $E = L$, and the projections π of the sets $e^{tC}P$ and $e^{tC}Q$ are balls of radius ρt and σ, respectively, i.e., balls $H_{\rho t}$ and H_σ. We now have to calculate the alternating integral

$$\bar{W}(t) = \int_{H_l, 0}^{t} (H_{\rho\tau}\, d\tau \stackrel{*}{-} H_\sigma\, d\tau) \triangleq H_{\gamma(t)},$$

where $\gamma(t) = 1 + \frac{1}{2}\rho t^2 - \sigma t$. For $t > 0$, the condition $\gamma(t) > 0$ is satisfied, provided $l > \frac{1}{2}\sigma^2/\rho$. If this condition is satisfied, then the radius of the ball $\bar{W}(t)$ is always positive for $t > 0$ and increases without bound as t increases without bound.

The quantity $T(z_0)$ is now determined as the minimal t for which the ball $H_{\gamma(t)}$ contains the point

$$\pi e^{tC}(z_0^1, z_0^2) = z_0^1 + tz_0^2 .$$

Thus, to find $T(z_0)$ we have to find the least positive root of the equation

$$(l + \tfrac{1}{2}\rho t^2 - \sigma t)^2 = |z_0^1|^2 + 2(z_0^1, z_0^2)t + |z_0^2|^2 t^2 .$$

References

1. L. S. Pontryagin, Linear differential games, Part I, *Doklady AN SSSR* **174,** N 6, 1278–1280 (1967).
2. L. S. Pontryagin, Linear differential games, Part II, *Doklady AN SSSR* **175,** N 4, 764–766 (1967).

Received 10 April 1980

Bibliography of Basic Publications of L. S. Pontryagin

1. Zum Alexanderschen Dualitätssatz. Nachr. Gesel. Wiss. Göttingen. Math. Phys., Kl., 1927, H. 4, S. 315–322.
2. Zum Alexanderschen Dualitätssatz. Zweite Mitteilung. Ibid., S. 446–456.
3. Ein Knotensatz mit Anwendung auf die Dimensionstheorie. Math. Ann., 1930, Bd. 102, H. 55, S. 785–789 (in Gemeinschaft mit F. Frankl).
4. Sur une hypothèse fondamentale de la théorie de la dimension. C.R. Acad. Sci. Paris, 1930, t. 190, N 19, p. 1105–1107.
5. Einfacher Beweis eines dimensionstheoretischen Überdeckungssatzes. Ann. Math., 1931, v. 32, N 4, p. 761–762.
6. Über den algebraischen Inhalt topologischer Dualitätssätze. Math. Ann., 1931, Bd. 105, Nr. 2, S. 165–205.
7. Beweis des Mengerschen Einbettungssatzes. Math. Ann., 1931, Bd. 105, H. 5, S. 734–745 (in Gemeinschaft mit G. Tolstowa).
8. Der allgemeine Dualitätssatz für abgeschlossene Mengen. In: Verhandlungen des Internationalen Mathematiker-Kongresses, Zürich, 1932, Bd. 2, Sektions-Vorträge. Zürich, Leipzig etc., 1932, S. 195–197.
9. Sur une propriété métrique de la dimension. Ann. Math., 1932, v. 33, N 1, p. 156–162 (en collaboration avec L. Schnirelmann).
10. Über stetige algebraiche Körper. Ibid., p. 163–174.
11. A statistical approach to dynamical systems. Zhur. eks teor. fiz.,

1933, v. 3, N 3, p. 165–180 (in collaboration with A. Andronov and A. Vitt) (in Russian).

12. Les fonctions presque périodiques et L'Analysis situs. C.R. Acad. Sci. Paris, 1933, t. 196, N 17, p. 1201–1203.
13. On dynamical systems close to Hamiltonian systems. Zhur eks. teor. fiz., 1934, v. 4, N 9, p. 883–885 (in Russian).
14. Statistische Auffassung dynamischer Systeme. Phys. Z. Sowj. Un., 1934, Bd. 6, H. 1–2, Sonderdruck, S. 1–24 (in Gemeinschaft mit A. Andronoff und A. Witt).
15. Über Autoschwingungssysteme, die den Hamiltonschen nahe liegen. Phys. Z. Sowj. Un., 1934, Bd. 6, H. 1–2, Sonderdruck, S. 25–28.
16. The theory of topological commutative groups. Ann. Math., 1934, v. 35, N 2, p. 361–388.
17. Sur les groupes abéliens continus. C.R. Acad. Sci. Paris, 1934, t. 198, N 4, p. 328–330.
18. The general topological theorem of duality for closed sets. Ann. Math., 1934, v. 35, N 4, p. 904–914.
19. Sur les groupes topologiques compacts et le cinquième problème de M. Hilbert. C.R. Acad. Sci. Paris, 1934, t. 198, N 3, p. 238–240.
20. The Betti numbers of compact Lie groups. Doklady AN SSSR, 1935, v. 1, N 7–8, p. 433–437 (in Russian and English).
21. Sur les nombres de Betti des groupes de Lie. C.R. Acad. Sci. Paris, 1935, t. 200, N 15, p. 1277–1280.
22. The structure of compact topological groups. In: Trudy Vtorogo Vsesoyuznogo matematischeskogo s''ezda, Leningrad, 24–30 iyunya 1934 g., v. 2, Sektsionnye doklady, Leningrad–Moscow, 1936, p. 135 (in Russian).
23. The structure of locally compact commutative groups. Ibid., p. 136.
24. Linear representations of topological groups, Us. mat. nauk, 1936, N 2, p. 121–143 (in Russian).
25. The theory of topological commutative groups. Ibid., p. 177–195 (in Russian).
26. Linear representations of compact topological groups. Matem. sbornik. Novaya seriya, 1936, v. 1, N 3, p. 267–272 (Abstract in Russian).
27. Les variétés à n dimensions généralisées. C.R. Acad. Sci. Paris, 1936, t. 202. N 15, p. 1327–1329 (en collaboration avec P. Alexandroff).
28. Sur les transformations des sphères. In: C.R. du Congrès International des Mathématiciens Oslo, 1936, T. 2, Conférences de Sections. Oslo, 1937, p. 140.

29. Rough systems. Doklady AN SSSR, 1937, v. 14, N 5, p. 247–250 (in collaboration with A. Andronov) (in Russian).
30. Syst`mes grossiers. C.R. Acad. Sci. URSS, 1937, v. 14, N 5, p. 247–250 (en collaboration avec A. Andronov).
31. Über den Brouwerschen Dimensions-begriff. Comp. Math., 1937, v. 4, fas. 2, S. 239–255 (in Gemeinschaft mit P. Alexandroff und H. Hopf).
32. Continuous groups. Moscow–Leningrad, 1938, 315 p. (Matematika v monografiyakh. Osnovn. ser., Kn. 3), Bibliogr.: p. 314–315 (in Russian).
33. Lie groups. Usp. mat. nauk. 1938, N 4, p. 165–200 (in Russian).
34. The classification of continuous mappings of a complex into a sphere. Communication I. Doklady AN SSSR, 1938, v. 19, N 3, p. 147–149 (in Russian).
35. The classification of continuous mappings of a complex into a sphere. Communication II. Doklady AN SSSR, 1938, v. 19, N 5, p. 361–363 (in Russian).
36. Classification des transformations d'un complexe n + 1 dimensionnel dans une sphère n-dimensionnelle. C.R. Acad. Sci. Paris, 1938, t. 206, N 20, p. 1436–1438.
37. Homologies in compact Lie groups. Matem. sbornik, Novaya ser., 1939, v. 6, N 3, p. 389–422 (abstract in Russian).
38. Topological groups. Princeton: Princeton Univ. Press, 1939, 299 p. (Princeton Mathematical Series, N 2). Bibliogr.: p. 295–296.
39. Über die topologische Struktur der Lieschen Gruppen. Comment. Math. Helv., 1940/1941, v. 13, fas. 4, S. 277–283.
40. A classification of mappings of the three-dimensional complex into the two-dimensional sphere. Matem. sbornik, Novaya ser., 1941, v. 9, N 2, p. 331–363 (Abstract in Russian).
41. Products in complexes. Matem. sbornik, Novaya ser., 1941, v. 9, N 2, p. 321–330 (Abstract in Russian).
42. On the zeros of certain elementary transcendental functions. Izv. AN SSSR, Ser. matem., 1942, v. 6, N 3, p. 115–134 (Abstract in English) (in Russian).
43. Mappings of a three-dimensional sphere into an n-dimensional complex. Doklady AN SSSR, 1942, v. 34, N 2, p. 39–41, Bibliogr.: p. 41 (in Russian).
44. Mappings of a three-dimensional sphere into an n-dimensional complex. C.R. Acad. Sci. URSS, 1942, v. 34, N 2, p. 35–37.
45. Characteristic cycles of manifolds. Doklady AN SSSR, 1942, v. 35, N 2, p. 35–39 (in Russian).

46. A method of calculation of homology groups. Matem. sbornik, Novaya ser., 1942, v. 11, N 1–2, p. 3–14 (Abstract in Russian).
47. Hermitian operators in a space with an indefinite metric. Izv. AN SSSR, Ser. mat., 1944, v. 8, N 6, p. 243–280 (Abstract in English) (in Russian).
48. Certain topological invariants of Riemannian manifolds. Doklady AN SSSR, 1944, v. 43, N 3, p. 95–98 (in Russian).
49. The classification of certain skew products. Doklady AN SSSR, 1945, v. 47, N 5, p. 327–330 (in Russian).
50. Characteristic cycles. Doklady AN SSSR, 1945, v. 47, N 4, p. 246–249 (in Russian).
51. Topological groups. Princeton: Princeton Univ. Press, 1946, 299 p. (Princeton Mathematical Series, V. 2). Bibliogr.: p. 295–296.
52. Foundations of combinatorial topology. Moscow–Leningrad, Gostekhizdat, 1947, 143 p. (in Russian).
53. Intersections in manifolds. Usp. mat. nauk, 1947, v. 2, N 1, p. 58–155 (in collaboration with M. Glezerman) (in Russian).
54. Topological duality theorems. Usp. mat. nauk, 1947, v. 2, N 2, p. 21–44: Bibliogr.: p. 44 (in Russian).
55. A general topological duality theorem for closed sets. Ibid., p. 44–55, Bibliogr.: p. 55 (in Russian).
56. Characteristic cycles of differentiable manifolds. Matem. sbornik, Novaya ser., 1947, v. 21, N 2, p. 233–284, Bibliogr.: p. 284 (in Russian).
57. On certain topological invariants of closed Riemannian manifolds. Izv. AN SSSR, Ser. mat., 1949, v. 13, N 2, p. 125–162, Bibliogr.: 162 (in Russian).
58. On a relation between homology and homotopy. Izv. AN SSSR, Ser. mat., 1949, v. 13, N 3, p. 193–200, Bibliogr.: 200 (in Russian).
59. The homotopy group $\pi^{n+1}(K_n)$ $(n \geqslant 2)$ of dimension $(n+1)$ of a connected finite polyhedron K_n of an arbitrary dimension, whose fundamental group and Betti groups of dimensions $2, \ldots, (n-1)$ are trivial. Doklady AN SSSR, 1949, v. 65, N 6, p. 797–800 (in Russian).
60. On the classification of four-dimensional manifolds. Usp. mat. nauk, 1949, v. 4, N 4, p. 157–158 (in Russian).
61. Vector fields on manifolds. Matem. sbornik, Novaya ser., 1949, v. 24, N 2, p. 129–162, Bibliogr.: p. 162 (in Russian).
62. Classification of the mappings of an $(n+1)$-dimensional sphere into a polyhedron K_n whose fundamental group and Betti groups of

dimensions 2, . . . , $(n-1)$ are trivial. Izv. AN SSSR, Ser. mat., 1950, v. 14, N 1, p. 7–44, Bibliogr.: p. 44 (in Russian).

63. Homotopic classification of the mappings of an $(n+2)$-dimensional sphere into an n-dimensional sphere. Doklady AN SSSR, 1950, v. 70, N 6, p. 957–959 (in Russian).
64. A local method of investigating the mappings of a sphere S^{n+k} into S^n. Usp. mat. nauk, 1951, v. 6, N 4, p. 214–216 (jn Russian).
65. Foundations of combinatorial topology. Rochester; New York: Graylock Press, 1952, 99 p.
66. On the zeros of certain elementary transcendental functions (supplement). Doklady AN SSSR, 1953, v. 91, N 6, p. 1279–1280 (in Russian).
67. Continuous groups. 2nd ed. revised and enlarged. Moscow: Gostekhizdat, 1954. 515 p. Bibliogr.: 509–511 (in Russian).
68. Continuous groups, Bol'shaya sovetskaya entsiklopediya, 2nd ed., 1954, v. 29, p. 454–456 (in Russian).
69. Smooth manifolds and their applications in homotopy theory. Trudy Mat. instituta, 1955, v. 45, 139 p., Bibliogr.: p. 137 (in Russian).
70. Periodic solutions of systems of differential equations close to discontinuous solutions. Doklady AN SSSR, 1955, v. 102, N 5, p. 889–891 (in collaboration with E. F. Mishchenko) (in Russian).
71. Grundzüge der kombinatorischen topologie. Berlin: Dt. Verl. Wiss., 1956, 133 S., Bibliogr.: S. 128.
72. On the stability of equilibrium states of a "bang–bang" system of ordinary differential equations. In: Trudy Tret'ego Vsesoyuznogo matematicheskogo s"ezda. Moskva, iyun'-iyul' 1956. V. I. Sektsionnye doklady. Moskva, 1956, p. 217–218 (in collaboration with V. G. Boltyanskii) (in Russian).
73. On the theory of optimal processes. I. Ibid., p. 218 (in collaboration with V. G. Boltyanskii and R. V. Gamkrelidze) (in Russian).
74. Periodic solutions of systems of differential equations with a small parameter attached to derivatives. Ibid., p. 224 (in collaboration with E. F. Mishchenko) (in Russian).
75. Systems of ordinary differential equations with a small parameter attached to higher derivatives. In: Trudy Tret'ego Vsesoyuznogo matematicheskogo s"ezda. Moskva, iyun'-iyul' 1956. T. 2, Kratkoe soderzhanie obzornykh i sektsionnykh dokladov. Moscow, 1956, p. 93–95 (in Russian).
76. On the theory of optimal processes. Doklady AN SSSR, 1956, v. 110, N 1, p. 7–10 (in collaboration with V. G. Boltyanskii and R. V. Gamkrelidze) (in Russian).

77. Pavel Sergeevich Aleksandrov (on the 60-th Birth Anniversary and 40 Years of Scientific Activity). Usp. mat. nauk, 1956, v. 11, N 4, p. 183–187 (in collaboration with E. F. Mishchenko) (in Russian).
78. Topologische Gruppen. Teil I. Leipzig: B. G. Teubner Verlagsgesellschaft, 1957. 263 S., Bibliogr.: S. 258–259.
79. Certain mathematical problems in the theory of optimal automatic control systems. In: Sesiya AN SSSR po nauchnnim problemam avtomatizatsii proizvodstva. 15–20 oktyabrya 1956. T. 2, Osnovnye problemy avtomaticheskogo regulirovaniya i upravleniya. Moscow, 1957. p. 107–117 (in Russian).
80. The asymptotic behaviour of the solutions of systems of differential equations with a small parameter attached to higher derivatives. Izv. AN SSSR, Ser. mat., 1957, v. 21, N 5, p. 605–626 (in Russian).
81. Topologische Gruppen. Teil 2, Leipzig: B. G. Teubner Verlagsgesellschaft, 1958, 308 S., Bibliogr.: S. 303–305.
82. Topological groups. Princeton University Press, 1958. XI + 229 p.
83. Systems of ordinary differential equations with a small parameter attached to higher derivatives. In: Trudy Tret'ego Vsesoyuznogo matematicheskogo s''ezda. Moskva, iyun'-iyul' 1956 g. T. 3. Obzornye doklady. Moscow, 1958, p. 570–577. Bibliogr.: p. 577 (in Russian).
84. The proof of certain asymptotic formulas for solving differential equations with a small parameter. Doklady AN SSSR, 1958, v. 120, N 5, p. 967–969 (in collaboration with E. F. Mishchenko) (in Russian).
85. Derivation of certain asymptotic estimates for solutions of differential equations with a small parameter attached to higher derivatives. Izv. AN SSSR. Ser. mat., 1959, v. 23, N 5, p. 643–660 (in collaboration with E. F. Mishchenko) (in Russian).
86. A statistical problem of optimal control. Doklady AN SSSR, 1959, v. 128, N 5, p. 890–892 (in collaboration with E. F. Mishchenko) (in Russian).
87. Optimal control processes. Usp. mat. nauk, 1959, v. 14, N 1, p. 3–20. Bibliogr.: p. 20 (in Russian).
88. Smooth manifolds and their applications in homotopy theory. Amer. Math. Soc., Translations Ser. 2, 1959, v. 11, p. 1–114.
89. Optimal control processes. In: Proceedings of the International Congress of Mathematicians, 14–21 August 1958. Cambridge Univ. Press, 1960, p. 182–202. Bibliogr.: p. 202.
90. The theory of optimal processes: I. Maximum principle. Izv. AN SSSR, Ser. mat., 1960, v. 24, N 1, p. 3–42. Bibliogr.: p. 42 (in

collaboration with V. G. Boltyanskii, and R. V. Gamkrelidze) (in Russian).

91. An approximate solution of a system of ordinary differential equations with a small parameter attached to derivatives. Doklady AN SSSR, 1960, v. 131, N 2, p. 255–258 (in collaboration L. V. Rodygin) (in Russian).

92. Periodic solution of a system of ordinary differential equations with a small parameter attached to derivatives. Doklady AN SSSR, 1960, v. 132, N 3, p. 537–540 (in collaboration with L. V. Rodygin) (in Russian).

93. Differential equations with a small parameter attached to higher derivatives and some problems in the theory of oscillations. IRE Trans. Circuit Theory. CT-7, 1960, N 4, p. 527–535. Bibliogr.: p. 535 (in collaboration with E. F. Mishchenko).

94. The mathematical theory of optimal processes. Moscow: Fizmatgiz. 1961. 391 p. Bibliogr.: p. 389–391 (in collaboration with V. G. Boltyanskii, R. V. Gamkrelidze, and E. F. Mishchenko) (in Russian).

95. Ordinary differential equations. A college textbook. Moscow: Fizmatgiz., 1961, 311 p. (in Russian).

96. The maximum principle in optimal control theory. In: Trudy I Mezhdunarodnogo kongressa mezhdunarodnoi federatsii po avtomaticheskomu upravleniyu. Moskva 27 iyunya–7 iyulya 1960 g. T. 2. Teoriya diskretnykh, optimal'nykh i samonastraivayushchikhsya sistem. Moscow, 1961, p. 457–467. Bibliogr.: p. 466–467 (in collaboration with V. G. Boltyanskii, R. V. Gamkrelidze, and E. F. Mishchenko) (in Russian).

97. On a statistical optimal control problem. Izv. AN SSSR, Ser. mat., 1961, v. 25, N 4, p. 477–498 (in collaboration with E. F. Mishchenko) (in Russian).

98. Optimal regulation processes. Amer. Math. Soc. Trans. Ser. 2, 1961, v. 18, p. 321–339.

99. The mathematical theory of optimal processes. New York; London: Intersci. Publ., 1962, 360 p. Bibliogr.: p. 354–356 (in collaboration with V. G. Boltyasnkii, R. V. Gamkrelidze, and E. F. Mishchenko).

100. Ordinary differential equations. Reading (Mass.): Addison-Wesley Publ. Co., 1962, 298 p.

101. On a probability optimal control problem. Doklady AN SSSR, 1962, v. 145, N 5, p. 993–995 (in collaboration with A. N. Kolmogorov and E. F. Mishchenko) (in Russian).

102. The mathematical theory of optimal processes. In: Trudy chetvertogo vsesoyuznogo matematicheskogo s"ezda. Leningrad, 3–12

iyulya 1961 g. T. I. Plenarnye doklady. Leningrad, 1963. p. 214–218 (in collaboration with V. G. Boltyanskii, R. V. Gamkrelidze, and E. F. Mishchenko) (in Russian).

103. Relaxation oscillations and differential equations containing a small parameter in the higher derivatives. In: Calcutta mathematical society: Golden Jubilee Commemoration Volume (1958–1959). Pt. I. India, Calcutta, 1963. p. 141–150 (in collaboration with E. F. Mishchenko), Bibliogr.: p. 150.

104. The mathematical theory of optimal processes. New York; Paris: Pergamon Press, 1964. VII + 340 p. (in collaboration with W. G. Boltjanskii, R. V. Gamkrelidze, and E. F. Mishchenko).

105. Mathematische Theorie optimaler Prozesse. Berlin etc.; VEB Dt. Verl. Wiss., 1964. 340 S. Bibliogr.: p. 339–340 (in Gemeinschaft mit V. G. Boltyanskij, R. V. Gamkrelidze und E. F. Misčenko).

106. On certain differential games. Doklady AN SSSR, v. 156, N 4, p. 738–741 (in Russian).

107. Ordinary differential equations. A college textbook. 2nd Ed. Revised. Moscow: Nauka, 1965. 331 p. (in Russian).

108. Gewönliche Differentialgleischungen. Berlin: VEB Dt. Verl. Wiss., 1965. 263 S. (Math. für Naturwissenschaft und Technik. Bd. II). Bibliogr.: S. 255–258.

109. On some differential games. J. SIAM Control Ser., A, 1965, v. 3, N 1, p. 49–52.

110. Topological groups. 2nd ed. New York etc.: Gordon & Breach, 1966. XV + 543 p.

111. On the theory of differential games. Ups. mat. nauk, 1966, v. 21, N 4, p. 219–274 (in Russian).

112. A course in ordinary differential equations. Delhi: Hindustan Publ., 1967. 10 + 333 p.

113. Mathematische Theorie optimaler Prozesse. 2. verb. Aufl. Wien; München. Oldenburg, 1967. 340 S. (in Gemeinschaft mit W. G. Boltjanski, R. W. Gamkrelidze und E. F. Misčenko).

114. Linear differential games. Mathematical theory of control. In: Proceedings of the Conference, University, Southern California. Los Angeles, 1967, p. 330–334.

115. Linear differential games. Doklady AN SSSR, 1967, v. 174, N 1, p. 27–29 (in collaboration with E. F. Mishchenko) (in Russian).

116. On linear differential games, Part 1. Doklady AN SSSR, 1967, v. 174, N 6, p. 1278–1280 (in Russian).

117. On linear differential games, Part 2. Doklady AN SSSR, 1967, v. 175, N 4, p. 764–766 (in Russian).

118. Homology in compact Lie groups (Trans. from: Matem. sbornik, 1938, N 6), Usp. mat. nauk, 1968, v. 23, N 6, p. 151–185. Bibliogr.: p. 185 (in Russian).
119. The mathematical theory of optimal processes. 2nd ed., Moscow, 1969. 384 p. Bibliogr.: p. 383–384 (in collaboration with V. G. Boltyanskii, R. V. Gamkrelidze, and E. F. Mishchenko) (in Russian).
120. Equations différentielles ordinaires. Moscow: Mir, 1969. 352 p.
121. The problem of evasion of one controlled object from another. Doklady AN SSSR, 1969, v. 189, N 4, p. 721–723. Bibliogr.: p. 723 (in collaboration with E. F. Mishchenko) (in Russian).
122. Ordinary differential equations. A college textbook. 3rd ed. stereotype. Moscow: Nauk, 1970. 331 p. (in Russian).
123. Le problème d'un mobile controlable poursuivi par un autre. In: Symposium on optimization held in Nice, June 29–July 5, 1969. Berlin, etc., 1970, p. 307–310 (Lecture Notes in Mathematics; vol. 132). Bibliogr.: p. 310 (in collaboration with E. F. Mishchenko).
124. A linear differential game of evasion. Doklady AN SSSR, 1970, v. 191, N 2, p. 283–285 (in Russian).
125. The problem of evasion in linear differential games. In: Problemy prikladnoi matematiki i mekhaniki. Moskva, 1971, p. 17–26. Bibliogr.: p. 26 (in collaboration with E. F. Mishchenko) (in Russian).
126. Les jeux différentiels linéaires. In: Actes du Congrès international des mathématiciens. Nice, 1970, T. I. Paris, 1971, p. 163–171.
127. Linear differential evasion games. Trudy mat. in-ta im. V. A. Steklova, 1971, v. 112, p. 30–63 (in Russian).
128. The problem of evasion in linear differential games. Differents. uravneniya, 1971, v. 7, N 3, p. 436–445. Bibliogr.: p. 445 (in collaboration with E. F. Mishchenko) (in Russian).
129. Continuous groups. 3rd ed. Revised. Moscow: Nauka, 1973. 519 p. Bibliogr.: p. 515–516 (in Russian).
130. Ecuaciones diferenciales ordinarias. Madrid: Aguilar, Tolie Lege, 1973. 399 p.
131. Ordinary differential equations. A college textbook. 4th ed. Moscow: Nauka, 1974. 331 p. (in Russian).
132. Theorie mathematique des processus optimaux. Moscow: Mir, 1974. 317 p. (en collaboration avec V. Boltianci, R. Gamkrelidze et E. Michtchenko).
133. Continuous groups. Bol'shaya sovetskaya entsiklopediya, 3rd ed. 1974, v. 17, p. 493–494 (in Russian).
134. On the evasion process in differential games. In: Applied mathematics and optimization. New York, 1974, v. 1, N 1, p. 5–19.

135. Equations différentielles ordinaires. Moscow: Mir, 1975. 347 p.
136. The mathematical theory of optimal processes. 3rd ed., Moscow: Nauka, 1976. 392 p. Bibliogr.: p. 391–392 (in collaboration with V. G. Boltyasnkii, R. V. Gamkrelidze and E. F. Mishchenko) (in Russian).
137. Smooth manifolds and their applications in homotopy theory. 2nd ed., Moscow: Nauka, 1976. 175 p. Bibliogr.: p. 174 (in Russian).
138. Foundations of combinatorial topology. 2nd ed., Moscow, Nauka, 1976. 136 p. (in Russian).
139. Acquaintance with higher mathematics. The coordinate method. Moscow: Nauka, 1977. 135 p. (in Russian).
140. Optimization and differential games. Vestn. AN SSSR, 1978, N 7, p. 10–17 (in Russian).
141. Acquaintance with higher mathematics. Analysis of infinitesimals. Moscow: Nauka, 1980. 256 p. (in Russian).
142. Mathematical analysis for school students. Moscow: Nauka, 1980. 86 p. (in Russian).
143. Linear differential pursuit games. Matem. sbornik, 1980, v. 112, N 3, p. 307–330 (in Russian).